Verification, Validation, and Uncertainty Quantification in Scientific Computing

Can you trust results from modeling and simulation? This text provides a framework for assessing the reliability of and uncertainty included in the results used by decision makers and policy makers in industry and government.

The emphasis is on models described by partial differential equations and their numerical solution. Procedures and results from all aspects of verification and validation are integrated with modern methods in uncertainty quantification and stochastic simulation. Methods for combining numerical approximation errors, uncertainty in model input parameters, and model form uncertainty are presented in order to estimate the uncertain response of a system in the presence of stochastic inputs and lack of knowledge uncertainty.

This new edition has been extensively updated, including a fresh look at model accuracy assessment and the responsibilities of management for modeling and simulation activities. Extra homework problems and worked examples have been added to each chapter, suitable for course use or self-study.

William L. Oberkampf has more than 50 years of experience in research and development in fluid dynamics, heat transfer, and solid mechanics. Over the past 25 years he has focused on research and teaching of verification, validation, and uncertainty quantification in modeling and simulation. He is a fellow of the American Institute of Aeronautics and Astronautics and a fellow of NAFEMS.

Christopher J. Roy is Professor in the Kevin T. Crofton Department of Aerospace and Ocean Engineering at Virginia Tech. He has worked primarily in the area of computational fluid dynamics but has participated in or organized multiple validation experiments. He has taught more than 50 short courses in the field of verification, validation, and uncertainty quantification and has more than 200 publications in the field.

Verification, Validation, and Uncertainty Quantification in Scientific Computing

SECOND EDITION

William L. Oberkampf
W. L. Oberkampf Consulting

Christopher J. Roy
Virginia Tech

CAMBRIDGE
UNIVERSITY PRESS

CAMBRIDGE
UNIVERSITY PRESS

Shaftesbury Road, Cambridge CB2 8EA, United Kingdom

One Liberty Plaza, 20th Floor, New York, NY 10006, USA

477 Williamstown Road, Port Melbourne, VIC 3207, Australia

314–321, 3rd Floor, Plot 3, Splendor Forum, Jasola District Centre, New Delhi – 110025, India

103 Penang Road, #05-06/07, Visioncrest Commercial, Singapore 238467

Cambridge University Press is part of Cambridge University Press & Assessment,
a department of the University of Cambridge.

We share the University's mission to contribute to society through the pursuit of
education, learning and research at the highest international levels of excellence.

www.cambridge.org
Information on this title: www.cambridge.org/9781316516133

DOI: 10.1017/9781009031004

When citing this work, please include a reference to the DOI 10.1017/9781009031004

First published 2010
Second edition 2025

A catalogue record for this publication is available from the British Library.

Library of Congress Cataloging-in-Publication Data
Names: Oberkampf, William L., 1944- author. | Roy, Christopher J., author.
Title: Verification, validation, and uncertainty quantification in scientific computing / William L. Oberkampf,
W. L. Oberkampf Consulting, Christopher J. Roy, Virginia Tech.
Other titles: Verification and validation in scientific computing
Description: Second edition. | Cambridge, United Kingdom ; New York, NY, USA : Cambridge University Press, 2024. |
Originally published: Verification and validation in scientific computing. New York : Cambridge University Press,
2010. | Includes bibliographical references and index. | Summary: "This updated edition provides a framework for
assessing the reliability and uncertainty of modeling and simulation results used in industry and government. With extra
worked examples and homework problems for graduate students, and new material on topics including model accuracy
assessment and management responsibilities for modeling activities"– Provided by publisher.
Identifiers: LCCN 2024023119 (print) | LCCN 2024023120 (ebook) | ISBN 9781316516133 (hardback) |
ISBN 9781009031004 (epub)
Subjects: LCSH: Science–Data processing. | Numerical calculations–Verification. | Computer programs–Validation. |
Decision making–Mathematical models.
Classification: LCC Q183.9 .O24 2024 (print) | LCC Q183.9 (ebook) | DDC 502.85–dc23/eng/20240617
LC record available at https://lccn.loc.gov/2024023119
LC ebook record available at https://lccn.loc.gov/2024023120

ISBN 978-1-316-51613-3 Hardback

To my children Marni and Brandon, and my grandchild Benjamin – WLO
To my children Reagan and Cameron – CJR

Contents

Preface

Modeling and simulation (M&S) are used in a myriad of ways in research, business, and government. M&S is applied in science, engineering and technology, industry, governmental regulations, product and plant safety, design of military systems, financial investing, and governmental planning and policy making. In all of these applications, a mental construct is formulated that describes how we believe the activity functions and how it is influenced by internal and external factors. A mathematical model of the activity of interest is formulated based on many different types of abstractions, approximations, and assumptions. The model is then programmed for numerical solution on a digital computer, and the computer produces a simulation result. The simulation result may have high fidelity to the reality of interest, or it may be nonsense. The question is: How can we tell which is which? This book deals with various technical and procedural methods and tools that can be used to assess the fidelity of the modeling, the accuracy of the numerical techniques, the reliability of the software, and the uncertainty quantification of the simulation result. Our focus is on modeling physical processes occurring in a broad range of systems and processes in the natural sciences and engineering, but the principles and procedures discussed can be applied to any field of application requiring a numerical solution.

The tools and procedures discussed here are centered on mathematical models that are represented by differential and/or integral equations. Many of these mathematical models occur in physics, chemistry, biology, astronomy, earth sciences, and all fields of engineering, but they also occur in other diverse applications. The topics addressed in this book are all related to the principles and practices involved in assessing the credibility of the models and estimating the uncertainty of the simulation results. We do not deal with the specific application details of modeling the physical process or system of interest, but with assessment procedures related to the fidelity of the models and the accuracy of the simulation results. These procedures are typically described by the terms *model verification* and *validation*. Heavy emphasis is also placed on *uncertainty quantification of simulation results*, which is complementary to the activities of verification and validation (V&V). The perspective of uncertainty quantification (UQ) in this text is the consideration of all possible sources of uncertainty that could impact the simulation results, e.g., uncertainties related to assumptions concerning system environments and scenarios, numerical uncertainties, and model input and model form uncertainties. Some uncertainties are due to randomness, some due to lack of knowledge, and some are a mixture of the two.

Because of very active research and applications in VVUQ over the last decade, this second edition is updated with the latest techniques and technology in each of these fields. *Verification* can simply be described as "solving the equations right" and *validation* as "solving the right equations." V&V are both built on the concept of quantitative accuracy assessment relative to some type of reference quantity. V&V activities address restricted questions of simulation credibility and reliability, but they are also the foundation on which M&S results are built. V&V could be described as the processes that provide evidence of the correctness and/or accuracy of computational results for a limited number of test cases. In verification, reference quantities may be available as exact solutions, but for general situations, the numerical errors must be estimated. In model validation, the reference quantities are high-quality experimental measurements of simplified or related systems of interest with highly characterized conditions so that model accuracy can be critically assessed.

During the last two decades, the field of UQ has grown even more rapidly than the fields of V&V. As a result, this second edition has increased the coverage, explanations, and examples in UQ. In our view, the reason for this remarkable growth in UQ is that researchers and computational analysts are increasingly challenged by the customers who use their simulation results. For example, "How comprehensive are the simulation results regarding possible system environments and scenarios that may impact my decision making?" And, "What are the largest contributors to the uncertainty in the simulation results and how might these be reduced?" The approach to UQ in this text is to recognize and respect the two different types of uncertainty: uncertainty due to randomness, as opposed to uncertainty due to lack of knowledge.

Because M&S is prevalent in essentially every field of science and engineering, researchers, computational analysts, and customers of simulation results need to become more familiar with the concepts and practices of VVUQ. The philosophical foundation of VVUQ is candid *skepticism*. Hence, VVUQ is the evidence for simulation credibility, reliability, and completeness. If M&S results do not include appropriate discussion and explanation of VVUQ activities, then the reliability of the results should be seriously questioned. If consequential design, certification, or policy decisions are informed by simulation results, then the requirement for VVUQ activities and documentation should increase significantly. In the last three chapters of the book, methods are discussed for maturity assessment and adequacy assessment of M&S activities, particularly VVUQ activities.

As humankind steadily moves from decision making primarily based on system testing and operational experience to decision making based heavily on M&S, increased understanding of VVUQ, as well as prudence and caution, are in order.

Acknowledgments for the Second Edition

Assistance, encouragement, and support for this second edition was generously provided by a number of colleagues and friends. William Rider, Joshua Kaizer, Timothy Trucano, Martin Pilch, and Ben Thacker have provided technical insights in every aspect of verification, validation, and uncertainty quantification (VVUQ).

Barna Szabo, Jean-Francois Imbert, Luís Eça, Barton Smith, Jeffery Bodner, Christopher Rumsey, Joseph Powers, Patrick Roache, Charles Hirsch, Casimir Suchyta, and Tina Morrison have shared their knowledge and experience in the fields of verification and validation.

Scott Ferson, Jon Helton, Michael Balch, and Thomas Paez have a remarkable depth of understanding of uncertainty quantification, which advanced our understanding of the topic.

Weicheng Xue skillfully contributed computational and programming support for several of the examples in the text.

Reviewers of the manuscript have provided invaluable constructive criticism, corrections, and suggestions for improvements. Timothy Trucano, Joshua Kaizer, Martin Pilch, John (Jack) Benek, Weicheng Xue, Aniruddha Choudhary, Joris Brouwer, K. Todd Lowe, David Higdon, and William Tyson reviewed one or more chapters and helped immeasurably in improving the quality, correctness, and clarity of the material. Excellent technical editing of most of the chapters was provided by Riki Graves. We take full responsibility for any errors or misconceptions still remaining.

Our wives, Sandra and Rachel, supported us not only during the years of writing this second edition but also over the years since the first edition was completed.

Acknowledgments for the First Edition

Although only two names appear on the cover of this book, we recognize that if other people had not been there for us, and many others had not helped, this book would have never been written. These people provided training and guidance, created opportunities, gave advice and encouragement, corrected us when we were wrong, and showed the way to improved understanding of the subject. Although there were many pivotal individuals early in our lives, here we only mention those who have contributed during the last decade when the idea for this book first came to mind.

Timothy Trucano, Frederic Blottner, Patrick Roache, Dominique Pelletier, Daniel Aeschliman, and Luís Eça have been critical in generously providing technical insights for many years. We have benefited from their deep knowledge of verification and validation, as well as a number of other fields. Jon Helton and Scott Ferson have guided our way to an understanding of uncertainty quantification and how it is used in risk-informed decision making. They have also provided key ideas concerning how to connect quantitative validation results with uncertainty estimates in model predictions. Without these people entering our technical lives, we would not be where we are today in our understanding of the field.

Martin Pilch created opportunities and provided long-term funding support at Sandia National Laboratories, without which we would not have been able to help advance the state of the art in V&V. He, along with Paul Hommert, Walter Rutledge, and Basil Hassan at Sandia, understood that V&V and uncertainty quantification were critical to building credibility and confidence in modeling and simulation. They all recognized that both technical advancements and changes in the culture of people and organizations were needed so that more reliable and understandable information could be provided to project managers and decision makers.

Many colleagues provided technical and conceptual ideas, as well as help in working through analyses. Although we cannot list them all, we must mention Mathew Barone, Robert Croll, Sharon DeLand, Kathleen Diegert, Ravi Duggirala, John Henfling, Harold Iuzzolino, Jay Johnson, Cliff Joslyn, David Larson, Mary McWherter-Payne, Brian Rutherford, Gary Don Seidel, Kari Sentz, James Stewart, Laura Swiler, and Roger Tate. We have benefited from outstanding technical editing support through the years from Rhonda Reinert and Cynthia Gruver. Help from students Dylan Wood, S. Pavan Veluri, and John Janeski in computations for examples and/or presentation of graphical results was vital.

Reviewers of the manuscript have provided invaluable constructive criticism, corrections, and suggestions for improvements. Edward Allen, Ryan Bond, James Carpenter, Anthony Giunta, Matthew Hopkins, Edward Luke, Chris Nelson, Martin Pilch, William Rider, and

William Wood reviewed one or more chapters and helped immeasurably in improving the quality and correctness of the material. Special recognition must be given to Tim Trucano, Rob Easterling, Luís Eça, Patrick Knupp, and Frederick Blottner for commenting on and correcting several draft chapters, or in some cases, the entire manuscript. We take full responsibility for any errors or misconceptions still remaining.

We were blessed with encouragement and patience from our wives, Sandra and Rachel. They tolerated our long hours of work on this book for longer than we deserved.

1

Introduction

Numbers tell a story, but they do not make a decision. —James O'Rourke

Contents of Chapter 1

Although identifying the beginning of anything is a matter of perspective, Chapter 1 starts with the stunning invention of calculus in order to briefly sketch the historical beginnings of modeling and simulation (M&S). This chapter discusses how the steadily increasing performance and decreasing costs of computing have been the critical drivers in advancing M&S in modern times. The discussion follows with a few key contributors to the credibility of M&S and the preliminary concepts of verification and validation. Chapter 1 closes with an outline of the book and suggestions on how the book might be used by undergraduate and graduate students, as well as professionals in any field of M&S.

1.1 Historical and Modern Role of Modeling and Simulation

1.1.1 Historical Role of Modeling and Simulation

For millennia, the primary method for designing a physical system has been to incrementally improve the design of an existing system. During and after the system was built, it would be gradually tested in a number of ways. In order to begin to understand the characteristics and responses of the new system, the first tests would usually be done during the building process. This new system was commonly a modification in the old system's geometrical character, materials, fastening or assembly techniques, or a combination of all of these changes. If the system was intended to be used in some new environment such as a longer bridge span, a taller structure, or a move at higher speed, the system was always tested first in environments where the experience base already existed. These tests would hopefully reveal design or assembly weaknesses and flaws, thus modifications to the system could often be made during the building and testing process. Sometimes a catastrophic failure of a monumental project would occur and the entire process would begin again; occasionally after attending the funeral of the previous chief designer and his apprentices (de Camp, 1995). In ancient times, chief designers understood the consequences of a major design failure because they were usually personally invested. This understanding is bluntly captured in the Code of Hammurabi, a seventeenth-century BC Babylonian King (Driver and Miles, 1955): "If a builder has built a house for a man and has not made his work sound, and the house which he has built has fallen down and so caused the death of the householder, that builder shall be put to death."

After the invention of calculus by Newton and Leibniz around 1700, the mathematical modeling of physics slowly began to have an impact on concepts for the understanding of nature and the design of physical systems. The second key ingredient that had an impact on mathematical physics was the invention of logarithms by John Napier about 1594 (Kirby et al., 1956). A mathematical model is of little practical use until it is exercised, currently referred to as obtaining a *simulation result*. Until the existence and use of logarithms, it was not practical to conduct routine simulations. Not long after the invention of logarithms, the slide rule was invented by William Oughtred. This device provided a simple mechanical method for adding and subtracting logarithms and enabling rapid multiplication and division of numbers. The slide rule and mechanical calculators not only revolutionized simulation, but also fields such as surveying, navigation, and astronomy. The previous combination of mathematical theory and computing machines provided the opportunity for the beginning of massive changes in science, engineering, and technology.

During the 1800s in England, the Industrial Revolution's impact on modeling and simulation in engineering and design began to grow. However, during the Industrial Revolution, M&S was always an adjunct to experimentation and testing of engineered systems, if it was done at all. The primary reason for this was that computations were typically done by hand on a slide rule or mechanical calculator. By the early 1960s, programmable digital computers began to appear in a wide number of industrial, academic, and governmental organizations. During this time period, the number of arithmetic calculations commonly done for a simulation grew from hundreds or thousands to millions of calculations. Because of the digital computer, the 1960s should be considered as the beginning of widespread scientific computing. The term *scientific computing* is restricted in this book to mean the numerical solution of models given by differential equations or integrodifferential equations. During the 1960s, computer power reached the level at which scientific computing began to have a significant effect on the design and decision making of engineered systems, particularly aerospace and military systems. Scientific computing initially grew as a field within the topic of M&S, which now includes economic and investment modeling, as well as modeling of individual and social behavior.

Nuclear weapon design in the United States during the 1940s and 1950s was one important exception where scientific computing began to significantly influence designs. The initial impetus for building much faster computers was the Cold War between the US and the Soviet Union. (See Edwards, 1997, for a perspective of the early history of electronic computing and its influence.) M&S activities were primarily modeling activities in which models were simplified until it was practical to obtain simulation results in an acceptable time period while still having an impact on the design of a system or research activity. These models were extremely simplified relative to today's models because computing power was minimal. However, this simplicity does not denigrate the M&S conducted during the 1940s or the century before. Indeed, M&S conducted before the 1960s was more creative and insightful than present-day simulation because the modeler had to deeply understand the physical coupling and mathematical character to decide what should be kept and what could be ignored.

One of the most remarkable scientific computing articles to appear during the 1960s was "Computer Experiments in Fluid Dynamics" by Harlow and Fromm (1965). This article suggested that scientific computing should be thought of as the third pillar of science, along with theory and experiment. During the 1970s and 1980s, many science traditionalists strongly resisted this suggestion, but that resistance faded as the power of scientific computing became dominant in essentially every field of science and engineering. The scientific community now widely accepts that simulation does indeed provide the third pillar of science; with its own unique strengths and weaknesses.

From a historical perspective, it is important to recognize that we are *only beginning* to build this third pillar. Some could argue that the pillar of experiment, and the related field of metrology, have been built and continually refined since the beginning of the Italian Renaissance in the 1400s. Some could also argue that this pillar has much earlier historical roots: millennia BC with the Mesopotamian, Egyptian, Babylonian, and Indus Valley civilizations. The pillar of theory, i.e., mathematical physics in its general sense, has been built, tested, and refined since the late 1700s. Understanding the strengths and weaknesses of each of these pillars has not come without major controversies. For example, the importance of uncertainty estimation in experimental measurements, particularly the importance of using different measurement techniques, is well understood and documented. Even in modern times, history has shown the bitter and sometimes destructive debates that occur when there is a paradigm shift, e.g., the shift from Newtonian mechanics to relativistic mechanics. In a century or so, when present day human egos and organizational and national agendas have faded, science and engineering will admit that the pillar of simulation is *just now* beginning to be constructed. Thus, the weaknesses and failings that impact the reliability of computational simulation are beginning to be better understood. At this point in history, the current youthful enthusiasm of computational simulation must be tempered with the centuries of experience and errors encountered during the building of the pillars of experiment and theory.

1.1.2 Changing Role of Simulation in Engineering

1.1.2.1 Changing Role of Simulation in Design, Performance, and Safety of Engineering Systems

The capability and impact of simulation has increased at an astounding pace. For example, scientific simulations that were published in research journals in the 1990s are now given as homework problems in undergraduate and graduate courses. Similarly, what was at the competitive leading edge in simulation applied to engineering system design in the 1990s is now common design practice in the industry. The impact of simulation has also increased with regard to helping designers and project managers improve their decision making, as well as in the assessment of the safety and reliability of manufactured products and public works projects. During most of this simulation revolution, system design and development were based primarily on testing and experience in the operating environment of the system, while simulation was commonly a secondary contributor in both preliminary and final design. For example, if there was some type of unexpected behavior, system failure, or safety issue that could not be solved quickly by testing, simulation was frequently used for assistance and insight. Another common mode for the use of simulation was to reduce the number of design-then-test-then-redesign iterations that were needed for a product to perform better than competing products or to meet reliability or safety requirements. Specialized mathematical models for components or features of components were commonly constructed to better understand specific performance issues, flaws, or sensitivities of the components. For example, models were made to study the effect of joint stiffness and damping on structural response modes. Similarly, specialized mathematical models were built so that certain impractical, expensive, or restricted tests could be eliminated, such as tests of high-speed entry of a space probe into the atmosphere of another planet or the structural failure of a full-scale containment vessel of a nuclear power plant.

As simulation steadily moves from a supporting role to a leading role in physical system design and evaluation, new terminology has been introduced. Terminology such as *virtual prototyping* and *virtual testing* are now being used in engineering development to describe simulation used in the evaluation and testing of new components and subsystems, and even entire systems. For relatively simple components, manufacturing processes, or low-consequence systems, such as simple consumer products, virtual prototyping can greatly reduce the time to market of new products. However, for complex, high-performance systems, such as gas turbine engines, commercial and military aircraft, and rocket engines, these systems continue to go through a long and careful development process based on testing, modification, and retesting. For these complex systems, it would be fair to say that simulation plays a supporting role.

The substantial trend toward using simulation in engineering systems is driven by increased competition in many markets, particularly aircraft, automobiles, propulsion systems, military systems, and systems for the exploration for oil and gas deposits. The competitive drive to decrease the time and cost of bringing products to market is intense. For example, simulation is relied on to reduce the high cost and time required to test components, subsystems, and complete systems. In addition, simulation is used in the highly industrialized nations of the world to improve automated manufacturing processes. These industrialized nations increasingly rely on simulation to improve their competitiveness against nations that have much lower labor costs.

The safety aspects of products or systems also represent an important, sometimes dominant, element of both simulation and experimental testing. The potential legal and liability costs of hardware failures can be staggering to a company, the environment, or the public. This is especially true in the litigious culture of the US. The engineering systems of interest are both existing or proposed systems that operate at design conditions, off-design conditions, conditions of misuse, and failure-mode conditions in accident scenarios. In addition, after the terrorist attacks on September 11, 2001, simulation is being used to analyze and improve the safety of a wide range of civil systems that may need to function in hostile environments.

Simulation is used in assessing the reliability, robustness, and safety systems in two rather different situations. The first situation, which is by far the most common, is to *supplement* test-based engineering; for example, to supplement crash-worthiness testing of automobiles to meet federal safety regulations. In fact, crash worthiness has become so important to some customers that automobile manufacturers now use computational simulation in marketing their vehicles. The second situation includes the use of simulation for reliability, robustness, and safety assessment of high-consequence systems that cannot be tested in fully representative environments and scenarios. Some examples are: failure of a large-scale dam due to an earthquake; explosive failure of the containment building of a nuclear power plant; underground storage of nuclear waste; and a nuclear weapon in a transportation accident. These types of high-consequence system analyses attempt to predict events that very rarely, if ever, occur in reality. Thus, simulation is used to assess the reliability, robustness, and safety of systems where little or no direct experimental data exists.

For both of these situations, the burden of credibility and confidence that is required of simulation is dramatically higher than when confidence is built based on extensive testing and operational experience with similar systems. The methodologies and techniques for attaining a higher level of credibility and confidence in simulation have been developed over the last four to five decades and these are discussed in detail in this book. However, it is fair to say that these practices are not implemented in most engineering organizations. Major improvements need to be made in the transparency, understandability, and maturity of all of the elements of simulation credibility building so that simulation-informed decision making can be improved. These improvements are especially required in order to progress toward Certification by Analysis (CbA) of engineering systems. Decision makers, regulatory authorities, and stakeholders need to be informed of the limitations, weaknesses, and uncertainties of M&S, as well as the strengths. The needed improvements are not just technical in terms of the burden of evidence, but also major changes in simulation culture and procedures.

1.1.2.2 Interaction of Simulation and Experimental Investigations

Interactions of simulation and experimental investigations have traditionally been very much one-way: from experiments to simulation. For example, experimental measurements were made, and then mathematical models were formulated, or experimental measurements were used to assess the accuracy of a simulation result. Until recently, given the limited capabilities of simulation, this was an appropriate relationship. With the dramatic improvements in computing capabilities, however, the relationship between simulation and experiment is in the midst of change. However, the changes have been slow and sometimes painful. When viewed from a historical as well as a human perspective, the slow rate of change is perhaps understandable.

The fields of computational fluid dynamics (CFD) and computational solid mechanics (CSM) have pioneered many of the theoretical, practical, and methodological developments in simulation. The relationship between experiment and simulation in each of these fields, however, has been quite different. In CSM, there has been a long-term and consistent tradition of a constructive and symbiotic relationship. Because of the nature of the physics modeled, CSM is fundamentally and critically dependent on experimental results for the construction of the mathematical models being used. A simple example includes predicting the linear elastic vibrational modes of a built-up structure, e.g., a structure that is constructed from a number of individual beams that are fastened together by bolts. A mathematical model is formulated for the elastic beams in the structure and the joints between the beams are simply modeled as torsional springs and dampers. The stiffness and damping of the joints are treated as calibrated model parameters, along with the internal and external damping of the structure. The physical structure is built and then tested by excitation of many of the modes of the structure. Using the results of the experimental measurements, the stiffness and damping parameters in the mathematical model are optimized, i.e., calibrated, so that the results of the model best match the measurements of the experiment. This example demonstrates that the computational model could not be reliable without the information obtained from experimental testing.

The relationship between experiment and CFD has not always been as collegial. Very early in the development of CFD, an article was published entitled "Computers vs. Wind Tunnels" (Chapman et al., 1975). This article by influential CFD leaders set a very negative and competitive tone early on in the relationship between simulation and experiment. The authors of this article simply gave voice to the brash claims of some CFD practitioners in the 1970s and 1980s, such as, "Wind tunnels will be used to store the output from CFD simulations." These attitudes often set a competitive and frequently adversarial relationship between experimentalists and CFD practitioners, which led to a lack of cooperation between the two groups. Where close cooperation has occurred, it is usually in small research teams forming voluntarily or in industrial settings where engineering project needs were paramount.

The most efficient and rapid progress in simulation and experiment is obtained through a synergistic and cooperative environment. While obvious to proponents of this viewpoint, human and organizational attitudes that will work against this type of environment still remain. Practical issues can also hinder progress in the synergistic coupling of simulation and experiment. The following are two examples of practical difficulties: one related to simulation and one related to experiment.

Simulation practitioners commonly view that comparison of computational results and experimental results, referred to as the *validation* step, can be accomplished through comparison to existing data. These data are normally documented in corporate or institute reports, conference papers, and archival journal articles. However, experience has shown that this approach is commonly less quantitative and complete than desired. Almost invariably, critical details are missing from published data, particularly in journal articles where the length of an article is highly constrained. When important details such as precise boundary conditions and initial conditions are missing, the simulation practitioner commonly uses this lack of knowledge as freedom to adjust unknown quantities in the model to obtain the best agreement with experimental data. That is, the comparison of computational results with experimental data begins to take on the character of a *calibration* of a computational model, as opposed to the evaluation of the predictive accuracy of the model. Many simulation practitioners will argue that this is unavoidable. Although this calibration mentality is prevalent, an alternative methodology can be used which directly addresses the uncertainties in the simulation.

Experimentalists now face the difficulty of finding funding sources due to the rapid increase in the visibility and importance of simulation. In addition, both governmental and industrial funding sources consider simulation to provide all of the important breakthroughs in research and technology, *not* experimental activities. This attitude over the last few decades has produced a decrease in the number of experimental research projects, including funding for graduate students, and a dramatic decrease in the number of experimental facilities. With restricted funding for experimental activities, there is also less research into the development of new experimental diagnostic techniques. We believe this has had an unintended detrimental effect on the growth of simulation. That is, with less high-quality experimental data available for validation activities, the ability to critically assess computational results will decrease, or worse, there will be a false sense of confidence in those simulations. For example, major efforts are being initiated in multi-scale and multi-physics computational modeling. This type of modeling commonly bridges at two or more spatial scales. Spatial scales are usually divided into the macro-scale (e.g., meter scale), the meso-scale (e.g., millimeter scale), the micro-scale (e.g., the micrometer scale), and the nano-scale (e.g., nanometer scale). Without new experimental diagnostic techniques and highly detailed and complete experimental data for multi-scale models, the confidence in simulation should be seriously questioned.

1.1.3 Changing Role of Simulation in Various Fields of Science

Beginning around the 1960s, simulation has had an ever-increasing impact on several scientific fields. Computational physics should be primarily noted. Although there is significant overlap between computational physics and computational engineering, there are certain areas of physics that are now dominated by simulation. Some examples are nuclear physics, solid state physics, quantum mechanics, high energy/particle physics, condensed matter physics, and astrophysics.

A second major area where simulation has become a factor is environmental science. Some of the environmental areas where simulation is having a dominant impact are atmospheric science, ecology, oceanography, hydrology, and environmental assessment. Atmospheric science has received worldwide attention with the debate over global warming. Environmental assessment, particularly when it deals with long-term, underground storage of nuclear waste, has also achieved very high visibility. The predictions in fields such as global warming and underground storage of nuclear waste are extremely challenging because large uncertainties exist and because the prediction time scales are on the order of centuries. The accuracy of these predictions cannot be confirmed or falsified for many human generations. Because of the widespread, potentially catastrophic effects studied in environmental science, the credibility of computational results is being scrutinized far more closely than in the past. Computational results can affect public policy, the well-being of entire industries, and the determination of legal liability in the event of loss of life or environmental harm. With this major level of impact of computational results, the credibility and uncertainty quantification in these areas must be greatly improved and standardized. If this is not done, hubris and the political and personal agendas of the participants will take precedence.

1.2 Credibility of Simulation

1.2.1 Computer Speed and Maturity of Simulation

The speed of computers over the last 50 years has consistently increased at a rate that can only be described as *stunning*. Figure 1.1 shows the computing speed of the fastest computer in the world, the 500th fastest computer, and the sum of computing speed of the 500 fastest computers in the world as of February 2022 (Wikipedia, 2022). The speed of the fastest computer has consistently increased by roughly a factor of 10 every four years. Over the last few decades, this remarkable increase in speed has been sustained by dramatically increasing the number of parallel processors. For example, one of the computers in the top 10 has roughly 10 million computer cores! There is now wide agreement that the massively parallel approach cannot maintain the past growth rate of computing speed. Completely new paradigms, such as DNA-based computing and quantum computing, are currently areas of intense research.

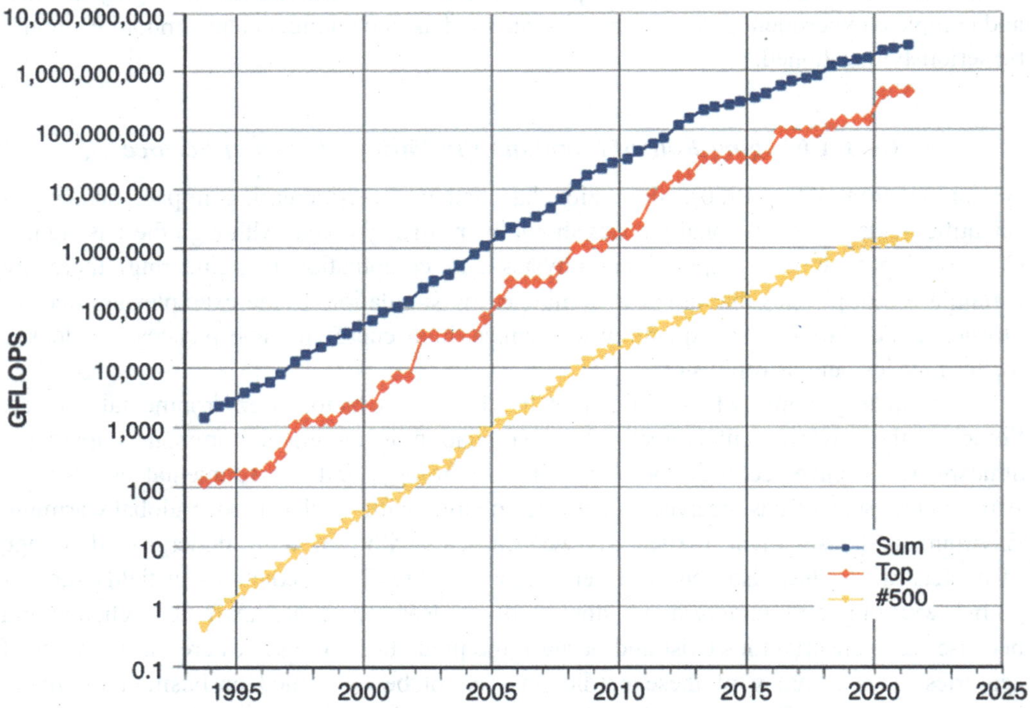

Figure 1.1 Computing speed of the 500 fastest computers in the world.
(Wikipedia, 2022)

Many people, particularly non-technically trained individuals, feel that computer speed is directly related to the impact and maturity of simulation. There is a relationship between computer speed and the broad impact of simulation, but it is far from direct. Simulation maturity relates to issues such as credibility, trust, and reliability of the computational results. Impact of simulation relies directly on its trustworthiness as perceived by the customers of the simulation, in addition to many other issues considered by the customer. For example, in industry, some of the key measures of simulation are: Can it reduce the cost of product development? Can it significantly reduce the time to market of new products? Can it improve management decision making and reduce risks in terms of both product development and corporate liability? For a government regulator, impact of simulation is commonly assessed in these terms: Can it reduce the risk of poor decision making for assessing the safety of a product? Can it help the identification and understanding of possible policy approaches to an environmental issue? In academia, simulation impact is measured in terms of new understanding and knowledge created.

In 1986, the US National Aeronautics and Space Administration (NASA) funded a study conducted by the National Research Council to study the maturity and potential impact of CFD (NaRC, 1986). This study, chaired by Richard Bradley, was one of the first to broadly examine the field of CFD from a business and economic competitiveness perspective. The study sent questionnaires to a wide range of individuals in industry and government to evaluate the maturity of CFD. Although they focused on CFD, their high-level analysis is equally applicable to *any* field product development. In this study, the committee identified five stages of maturity of computational simulation. These stages, along with their descriptive characteristics, are:

- Stage 1: Developing enabling technology – scientific papers published, know-how developed.
- Stage 2: Demonstration and confidence building – limited pioneering applications, subject to surprise.
- Stage 3: Putting it all together – major commitment made, gradual evolution of capability.
- Stage 4: Learning to use effectively – changes the engineering process, value exceeds expectations, skilled user groups exist.
- Stage 5: Mature capability – dependable capability, cost effective for design applications, most analyses done without supporting experimental comparisons.

The individuals surveyed ranked the maturity according to a matrix of elements. The matrix was formed on one side by increasing levels of complexity of the modeling approach to CFD and on the other by increasing levels of complexity of engineering systems that would be of interest. A maturity score of 0 meant that scientific papers have *not been* published and know-how has *not been* developed for that particular element in the matrix. A score of 5 meant that a mature capability existed, indicated by Stage 5 described above. Rather surprisingly, they found that depending on the model complexity and on the system complexity, the scores ranged from 0 to 5 over the matrix.

It is possible that if the survey was conducted today, 30 years after the original survey, the maturity levels would be higher for essentially all of the elements in the matrix. However, there would still be a very wide range of scores in the matrix. Even within a well-developed field of simulation, such as CFD and CSM, the range of maturity varies greatly, depending on the modeling approach and the application of interest. Claims of high maturity for complex systems, whether from commercial software companies, government, or academia, are unfounded. Companies that sell software and organizations that promote self-serving policies based on colorful graphics and flashy video animations, as opposed to critical assessment of simulation credibility, have no skin in the game.

1.2.2 Perspectives on Credibility of Simulation

People tend to think that their own perspective on requirements for credibility or believability of a prediction is similar to most other individuals. However, broader experience shows that this view is fundamentally mistaken. With regard to the present topic of simulation, a wide range of perspectives exist regarding what is required for various individuals to say, "I believe this simulation is credible and I am comfortable making the needed decision." In human nature, a key factor in decision making is the heavier weighting on downside risk as opposed to upside gain (Tversky and Kahneman, 1992; Kahneman and Tversky, 2000). Hence, a person's loss, pain, or embarrassment from a decision is weighed much more heavily than the expected gain. For example, when a decision must be made based on the results from a simulation, the individual's perspective is weighted more heavily toward the personal consequences of a poor decision because of a deficient or misleading simulation instead of the personal gains that may result from a trustworthy simulation. However, if there is little downside risk, individuals and organizations will more easily convince themselves of the strengths of a simulation than its weaknesses. An analyst conducting a simulation will normally work toward attaining the personal required level of confidence in the simulation, given the time and resources available. If an analyst is working in a team environment, each member of the team will have to make a personal value judgment concerning the risk versus gain in how the simulation result is used.

When simulation results are submitted to an archival journal for publication, the individual(s) authoring the work may ask themselves a slightly more demanding question: "Will other people (editor, reviewers, readers) believe the results?" This is usually a more demanding test of credibility of the work than the judgment of most individuals. Within the group of editor, reviewers, and readers, there will certainly be a wide range of perspectives toward the credibility of the work presented. However, the final decision maker concerning credibility and whether the article is published is the journal editor. In this regard, several well-known journals, such as the ASME *Journal of Fluids Engineering* and *Journal of Verification, Validation, and Uncertainty Quantification*, as well as all of the AIAA journals, have been increasing the requirements for credibility of simulation results.

If the computational results are to be used as an important element in the decision making regarding some type of engineering project, then the issue of credibility of the computational results becomes more important. For this situation, presume that the individual computational analysts have satisfied themselves with the credibility of their results. The engineering project manager must assess the credibility of the computational results based on the manager's own personal requirements for credibility and own experience. This judgment will involve not only the computational results themselves, but also any knowledge the manager might have concerning the individuals involved in the analysis. Personal knowledge of the individuals involved is a great advantage to the project manager. However, for large-scale projects or projects that have contributions from individuals or teams from around the country or around the world, this type of information is very rare. To judge the credibility of the computational results, one question the project manager might consider is whether or not the results are worth betting the success of the project (the manager's career or company) on these results. These kinds of perspectives of the project manager are rarely appreciated by all of the contributors to the project.

Certain projects are of such a magnitude that the effects of the success or failure of the project, or decisions made on the system, have major consequences beyond the project/system itself. This type of situation is referred to as a *high-consequence system*. An example of this situation is the US Nuclear Regulatory Commission's use of simulation to analyze the safety of nuclear power reactors. Similarly, the US Environmental Protection Agency uses extremely long-term simulation of underground storage of nuclear waste. These simulations commonly deal with the immediate effects of a possible accident, as well as the effects on the environment for hundreds and thousands of years. For these situations, analysts and managers commonly consider if they are willing to bet the public's safety and catastrophic damage to the environment for possibly centuries to come based on the results of the simulation.

High-consequence systems are not mentioned to dramatize the importance of simulation in decision making, but to point out that there is an exceedingly wide range of impact of simulation. Typically, scientists and engineers work in one or two technical fields, e.g., research or system design. Individuals rarely consider the wide range of impact that simulation has on decision making for real systems, especially those involved in academic research.

1.2.3 How Is Credibility Built in Simulation?

The *credibility* of computational results means that the results of a simulation are worthy of belief or confidence. The fundamental elements that build credibility in computational results are (a) quality and experience of the analysts conducting the work, (b) fidelity and completeness of the physics modeling, (c) verification and validation activities, (d) uncertainty quantification and sensitivity analyses, and (e) adequacy of time and resources to obtain results. While *all* of these elements are necessary for simulation credibility, none is sufficient in itself. *Simulation is a tool for generating information about some physical situation, process, or system.* The reliability and completeness of the information depends on how it was developed. However, the wisdom or prudence of the decision made based on the information depends on many other factors. Two key factors are: (a) the customer's depth of understanding the reliability and credibility of the information produced and (b) the appropriateness and completeness of the information for its intended use. Although it is beyond the scope of this book to discuss how the information might be used, methods for assessing the maturity and completeness of the information are discussed in later chapters of this book.

1.2.3.1 Quality and Experience of the Analysts

This book refers to *analysts* as the group of individuals that: (a) construct the conceptual model for the problem of interest, (b) formulate the mathematical model, (c) choose the discretization and numerical solution algorithms, (d) program or choose the software to compute the numerical solution, (e) compute the simulation on a digital computer, and (f) analyze and prepare the results from the simulation. On small-scale simulations of subsystems or components, or on research projects, a single individual may conduct all of these tasks. On any significantly-sized effort, a team of individuals conducts all of these tasks.

The quality of the analysts encompasses their training, experience, sound technical judgment, and understanding the customers' needs from simulation results. Some customers believe the quality of a computational effort is entirely centered on the quality of the analysts involved. For example, a customer may say, "I have such confidence in this analyst that whatever simulation he/she produces, I'll believe it." No one would argue against the extraordinary value added by the quality and experience of the analysts involved. However, many large projects and organizations have painfully learned that they cannot completely depend on a few extraordinarily talented individuals for long-term success. Large organizations must develop and put into place business, technology, training, and management processes for all the elements that contribute to the quality and on-time delivery of a product, whether it is information from a simulation or a commercial product. In addition, many modern large-scale projects will typically involve groups that are physically separated and culturally dissimilar, such as groups from around the nation or around the world. For these situations, users of the computational information will have little or no personal knowledge of the individual contributors, their backgrounds, or value systems.

1.2.3.2 Fidelity and Completeness of the Physics Modeling

Fidelity and completeness of the physics modeling refers to the appropriate level and breadth of physics embodied in the mathematical model representing the relevant physics taking place in the system of interest. These modeling decisions are made in the formulation of the conceptual and mathematical models of the system of interest and its interactions with its surroundings. Two contrasting levels of physics model fidelity are: (a) at the low end, fully-empirical models that are entirely built on statistical fits of experimental data with no fundamental relationship to physics-based principles and (b) at the high end, physics-based models that are reliant on PDEs or integrodifferential equations that represent conservation of mass, momentum, and energy in the system. Either extreme of model fidelity may be adequate for a given application, depending on the information needs specified by the simulation customer. Modeling completeness refers to the number of different types of relevant physics modeled in the system, the level of coupling of the various physical processes, and the extent of possible environments and scenarios that are considered for the system. For example, in CSM, modeling completeness would consider a variety of failure modes such as buckling, plastic deformation, fracture, fatigue, and creep.

This book does *not* say that the highest possible quality level of physics modeling and completeness should be used for every application. Efficiency, cost effectiveness, and schedule should dictate the appropriate level of physics modeling to meet the information needs of the simulation customer. Stated differently, the analysts should understand the needs of the simulation customer, decide what physical features should be included in the simulation, and then formulate the *simplest* level of physics modeling fidelity that is needed to meet those needs. Accomplishing this requires significant experience from the analysts for the specific problem at hand, very clear communication on the customer's needs, and the intended use of the simulation results. The fidelity and completeness of the physics modeling needed is a *problem-specific judgment*. There is no simple standard; it requires judgement and understanding of the needs of the customer.

1.2.3.3 Verification and Validation Activities

Verification is the process of assessing software correctness and numerical accuracy of the solution to a given mathematical model. *Validation* is the process of assessing the physical accuracy of a mathematical model based on comparisons between computational results and experimental measurements. While verification and validation (V&V) are the primary processes for assessing and quantifying the accuracy of computational results, it must be recognized that the accuracy assessment is quite limited compared to how a mathematical model is commonly used. The perspective of V&V is strongly on the side of skepticism, i.e., critical assessment and evidence of the procedures and results obtained. In verification, the relationship of the simulation to the real world is *not* an issue. In validation, the relationship between the mathematical model and the real world (experimental data) *is* the central issue. Frederic Blottner captured the essence of each in the compact expressions (Blottner, 1990): "Verification is solving the equations right. Validation is solving the right equations." These expressions follow similar definitions by Boehm (1981).

The pragmatic philosophy of V&V is fundamentally built on the concept of *accuracy assessment*. While this may sound obvious, in Chapter 2 "Fundamental Concepts and Terminology," it will become clear that there are wide variations on the fundamental concepts of V&V. In this present context, accuracy assessment is clearly a necessary building block of how credibility is built in computational simulation. V&V could be described as processes that develop and present evidence of the accuracy of computational results, when accuracy can be quantitatively assessed. To measure accuracy, however, accurate benchmarks or reference values must exist. In verification, the primary benchmarks are highly accurate solutions to specific mathematical models. In validation, the benchmarks are high-quality experimental measurements with detailed characterization of the experiments. Given this restrictive perspective of V&V, it stressed that a critical additional element is needed in simulation credibility: estimation of accuracy when *no reliable reference* is available.

1.2.3.4 Uncertainty Quantification and Sensitivity Analyses

Uncertainty quantification (UQ) is the process of identifying, characterizing, and quantifying those factors in a simulation that could affect the computational results. Uncertainties can arise from many sources, but they are commonly identified and addressed in three major steps of the M&S process: (a) construction of the conceptual model, (b) formulation of the mathematical model, and (c) computation of the simulation results. Most sources of uncertainty are spawned in the assumptions and approximations of either the conceptual or mathematical model. Examples include uncertainties in the initial conditions or boundary conditions for the PDEs, uncertainties in the model input parameters, and the uncertainty of mathematical model itself. Model input uncertainties are propagated through the computational model to obtain uncertainties in the simulation results. This text uses the term *propagated* to mean that the sources of uncertainty, wherever they originate, are mathematically mapped to the uncertainties in the simulation results. The issue of quantifying the uncertainty due to the model itself will be one of the most difficult sources to estimate and will be discussed in later chapters of this book. *Predictive capability* of a model is a much broader and more complex topic than V&V. Predictive capability deals with the estimation of total uncertainty in a simulation result for situations where no experimental data or physical observations are available. This situation is clearly the primary reason for the development and use of M&S.

Sensitivity analysis is typically viewed as the process of determining how the simulation results, i.e., the outputs, depend on the input uncertainties that make up the computational model. Both global and local sensitivity analyses are briefly discussed in Chapter 13. This chapter will stress that sensitivity analyses can be used more broadly than their traditional uses. For example, a sensitivity analysis can be used to understand the impact of various assumptions or approximation made in the construction of the conceptual and mathematical models. Similarly, a sensitivity analysis could be used to estimate the change in system safety due to examining various failure modes of a system of interest. Uncertainty quantification and sensitivity analysis critically contribute to informing the customer of the uncertainty in the simulation results and what factors are the most important in the results, respectively.

1.2.3.5 Adequacy of Time and Resources

Adequacy of time and resources for a computational simulation is self-evident, but this book stresses two aspects to accentuate what this means in practice. First, *adequacy* is a crucial concept in engineering, technology, and business. Adequacy refers to the activity, i.e., the information generation, meeting the specifications dictated by the customer. This concept is divorced from the ambiguous value system of a simulation being good, bad, or accurate. The focus is entirely on whether or not the requirements are met. Once this perspective is accepted, the project manager of the simulation can concentrate on the issues of time and resources needed to satisfy the specified requirements.

Second, representing the reliability and completeness of the computational simulation result to the customer must also be emphasized. The computational analysts must estimate the total uncertainty in each of the simulation results provided and determine the limitations of the results. During the last few decades, as simulation has taken a more important role in providing information to decision makers, the topic of uncertainty quantification has increased in importance. However, expressing the completeness of a simulation has received very little attention. As discussed by Ayyub (2001), there are two broad categories of incompleteness or ignorance: *conscious ignorance* and *blind ignorance*. Whether the customer of the simulation requires completeness information or not, the simulation analyst should provide some discussion and possible impact of the assumptions, approximations, and neglected factors concerning the simulation. All of these issues are in the category of conscious ignorance. The possible impact of blind ignorance cannot be quantified in practice but must be addressed by procedures such as questioning accepted assumptions and cultural taboos. Unfortunately, time, resources, and documentation are rarely spent on the topic of incompleteness of the results.

1.3 Outline and Use of the Book

1.3.1 Structure of the Book

This book is structured into five parts. Part I: Fundamental Concepts (Chapters 2–3) deals with the development of the foundational principles of V&V&UQ (VVUQ), the fundamentals of modeling and simulation, error vs. uncertainty, UQ vs. risk, and the six phases of computational simulation. Part II: Code Verification and Software Engineering (Chapters 4–6) deals with the relationship between software quality assurance and code verification, different methodological approaches to code verification, software testing to determine the observed order of solution convergence, and the method of manufactured solutions. Part III: Solution Verification (Chapters 7–9) covers fundamental concepts of solution verification, discretization error, iterative convergence error, finite-element-based error estimation procedures, extrapolation-based error estimation procedures, and practical aspects of mesh refinement and adaption. Part IV: Model Validation and Predictive Capability (Chapters 10–13) addresses the fundamental principles of model validation, the design and execution of model validation experiments, quantitative assessment of model accuracy, and the eight steps of nondeterministic simulation. Part V: Planning, Management, and Implementation Issues (Chapters 14–16) discusses planning and prioritization of modeling activities and VVUQ, maturity assessment of modeling and simulation, and finally, responsibilities for, management of, and implementation of VVUQ activities.

This book covers both the fundamental and practical issues of VVUQ, with the emphasis on how the concepts are applied. Theoretical issues are discussed as they are needed to understand and implement practical procedures. VVUQ focuses on simulation quality control, procedures for assessing simulation credibility, and best practices for estimating uncertainty in simulation results. This book emphasizes how VVUQ can improve the information content of the simulation, including its limitations; both of which are of interest to customers of the simulation. Since VVUQ is still a relatively new field of formal technology and practice, there are various methods and divergent opinions on many of the topics discussed. This book covers the most common approaches and references other approaches in the final section of each chapter. The goal is to provide readers with enough detail on a few methods so that they can be used in practical applications of simulation. The methods' strengths and weaknesses are noted, and cautions are given where methods should not be used. Most chapters give detailed examples of the principles and procedures discussed.

The concepts of VVUQ are *not* associated with specific application areas, such as physics, astronomy, chemistry, life science, earth science, social science, and engineering. VVUQ is a fascinating mixture of computer science, numerical analysis, probability, statistics, and information theory. As a result, the concepts and practices of VVUQ are rarely addressed in these fields, nor in the application areas just mentioned. Consequently, there is a strong need to compile and explain these concepts and practices in a separate university course or as part of continuing education for those involved in M&S. The application area clearly influences what VVUQ procedures are used on a particular problem, but the application area is not part of the concepts of VVUQ. The practitioners of the application area should bring the needed technical knowledge with them.

This book is written so that it can be used either as a textbook in a university semester course or by professionals working in computational simulation. The emphasis of the book is directed toward computational models that are represented by differential equations, particularly PDEs. Most chapters require basic knowledge of probability and statistics. The book does not require that any particular computer software or programming language be used. To complete most of the examples, the reader should have experience in some programming

Table 1.1 *Book structure and suggested chapters for a university course*

Part I: Fundamental Concepts *Ch. 2: Fundamental Concepts and Terminology* *Ch. 3: Modeling and Computational Simulation*	
Part II: Code Verification and Software Engineering *Ch. 4: Software Engineering for Scientific Computing* Ch. 5: Code Order-of-Accuracy Verification Ch. 6: Exact and Manufactured Solutions	**Part III: Solution Verification** *Ch. 7: Numerical Uncertainty Estimation* Ch. 8: Iterative Error Ch. 9: Discretization Error
Part IV: Model Validation and Predictive Capability *Ch. 10: Model Validation Fundamentals* *Ch. 11: Design and Execution of Model Validation Experiments* Ch. 12: Model Accuracy Assessment *Ch. 13: Predictive Capability*	**Part V: Planning, Management, and Implementation Issues** Ch. 14: Planning and Prioritization in Modeling and Simulation Ch. 15: Maturity Assessment of Modeling and Simulation Ch. 16: Verification, Validation, and Uncertainty Quantification Responsibilities and Management
Undergraduate course chapters in *italics* Graduate course: Parts I through V	

language and should have access to a general-purpose software package, such as MATLAB or Mathematica. For a graduate course or for professionals, software can be written or commercial software packages can be used that solve PDEs.

1.3.2 Use of the Book in Undergraduate and Graduate Courses

For senior-level undergraduates to get the most out of the book, they should be familiar with some type of computer programing language, as well as basic numerical methods, ordinary differential equations (ODEs), as well as probability and statistics. Chapters that could be studied at the undergraduate level are Chapters 1–4, 7, 10, 11, and 13. Table 1.1 shows these chapters in *italics*. Depending on the background of the students, these chapters could be supplemented with any needed background material. Homework problems are given at the end of most chapters with the emphasis on the practice of VVUQ concepts in application areas of interest to the reader. Those problems dealing with ODEs are well suited to undergraduate students, either working individually or in small teams.

For a graduate course, students should be experienced with the numerical solution of PDEs and probability and statistics. If the students have not had a course in probability and statistics, they may need supplementary material provided or cited by the instructor. Chapters 1–13 could be covered in a one-semester graduate course. Assigned homework problems or semester projects are ideally suited to teams of individuals. For a semester project, each team could pick an application area of interest to the team, with the approval of the instructor. The semester projects could be defined such that each element of the project builds on the previous elements completed. Each element of the project could deal with specific topics in various chapters of the book, and each could be graded separately. This approach would be similar in structure to that commonly used in engineering fields for the senior design project. Teams of individuals are very beneficial because team members experienced in one area can assist other team members lacking in knowledge in that area. Also, learning to work in a team environment is exceptionally important in any science or engineering field.

1.3.3 Use of the Book by Professionals

Professionals working in their particular application area would use the book quite differently than students in a classroom environment. Professionals typically scan through a book, and then concentrate on particular topics of interest at the moment. In the following list, five groups of professionals are identified, and chapters are suggested that may be of particular interest:

- Code builders and software developers: Chapters 1–9
- Builders of mathematical models of physical processes: Chapters 1–3, 10–12
- Computational analysts working in all aspects and application areas of M&S: Chapters 1–13
- Experimentalists: Chapters 1–3, 10–12
- Project managers, line managers of computational simulation organizations, and decision makers: Chapters 1–3, 10, 13–16.

1.4 References

Ayyub, B. M. (2001). *Elicitation of Expert Opinions for Uncertainty and Risks*. Boca Raton, FL: CRC Press.

Blottner, F. G. (1990). Accurate Navier–Stokes Results for the Hypersonic Flow over a Spherical Nosetip. *Journal of Spacecraft and Rockets*, **27**(2), 113–122.

Boehm, B. W. (1981). *Software Engineering Economics*. Saddle River, NJ: Prentice Hall.

de Camp, L. S. (1995). *The Ancient Engineers*. New York, NY: Ballantine Books.

Chapman, D. R., H. Mark, and M. W. Pirtle (1975). Computer vs. Wind Tunnels. *Astronautics & Aeronautics*, **13**(4), 22–30.

Driver, G. P., and J. C. Miles (1955). *The Babylonian Laws*, Vol. II. London: Oxford University Press.

Edwards, P. N. (1997). *The Closed World: Computers and the Politics of Discourse in Cold War America*. Cambridge, MA: The MIT Press.

Harlow, F. H., and J. E. Fromm (1965). Computer Experiments in Fluid Dynamics. *Scientific American*, **212** (3), 104–110.

Kahneman, D., and A. Tversky (eds.). (2000). *Choices, Values, and Frames*. Cambridge: Cambridge University Press.

Kirby, R. S., S. Withington, A. B. Darling, and F. G. Kilgour (1956). *Engineering in History*. New York, NY: McGraw-Hill.

NaRC (1986). *Current Capabilities and Future Directions in Computational Fluid Dynamics*. Washington, DC: National Research Council.

Tversky, A., and D. Kahneman (1992). Advances in Prospect Theory: Cumulative Representation of Uncertainty. *Journal of Risk and Uncertainty*, **5**(4), 297–323.

Wikipedia (2022). Top 500 Supercomputers. https://en.wikipedia.org/wiki/TOP500.

Part I

Fundamental Concepts

Chapter 2: *Fundamental Concepts and Terminology* and Chapter 3: *Modeling and Computational Simulation* form the foundation of the book. These chapters are recommended reading for all individuals interested in any aspect of verification, validation, and uncertainty quantification (VVUQ) of mathematical models and computational simulations. Chapter 2 discusses the basic concepts of modeling and simulation (M&S) and defines all of the key terms related to VVUQ. The discussion presents clear arguments on why the concepts and terminology are logical and useable in all applications of M&S. Chapter 2 closes with a discussion of how all of the aspects of VVUQ are sequentially completed in an M&S application.

Chapter 3 discusses the concepts in M&S in more depth, with the emphasis on the physical sciences and engineering. The terms *system*, *surroundings*, *environments*, and *scenarios* are formally defined. The concept of nondeterministic simulation is introduced, and an explanation is given as to why the concept is important in the analysis of real systems and processes. Two fundamentally different types of uncertainty in the system, surrounding, environments, and scenarios are discussed: uncertainty due to inherent randomness and uncertainty due to our lack of knowledge. Chapter 3 closes with a discussion of the conceptual framework of the six phases of computational simulation.

2

Fundamental Concepts and Terminology

Things always become obvious after the fact. —Nassim Taleb

Contents of Chapter 2

This chapter discusses the fundamental concepts and terminology associated with verification, validation, and uncertainty quantification (VVUQ) in modeling and simulation (M&S). The beginning includes a brief history of the philosophical foundations of verification and validation (V&V) so that the reader can better understand why there are a wide variety of interpretations and definitions of V&V. Although the terminology is moving toward common understanding within some communities, there are still significant differences in various application areas of M&S. Awareness of these differences in terminology will help minimize confusion and unnecessary technical disagreements, as well as help anticipate possible difficulties in contractual and regulatory obligations in business and government. This chapter also discusses a number of important and closely related terms in M&S. Examples are predictive capability, model calibration, certification, error, and uncertainty. The chapter ends with a discussion of a framework for integrating the steps associated with simulation of an application of interest with the activities of verification, validation, and predictive capability. Although there are different frameworks for integrating these concepts, the framework discussed here has proven very helpful in understanding how the various activities are typically ordered and how they interact.

2.1 Development of Concepts and Terminology

Philosophers of science have been struggling with the fundamental concepts underlying verification and validation (V&V) for at least two millennia. Some researchers in engineering and the applied sciences have used several of the modern concepts in the philosophy of science to develop the fundamental principles and terminology of V&V; see Kleindorfer et al. (1998) for an excellent historical review of the philosophy of science viewpoint of validation. When the viewpoint of "absolute truth" is carried to the extreme (Oreskes et al., 1994), the following position remains: One can only *disprove* or *fail to disprove* theories and laws of nature. Hence, theories and laws *cannot be proved*; they can only be falsified (Popper, 1969). This view is epistemologically satisfying, but it is nearly useless for assessing the credibility of computational results in engineering and technology. Engineering and technology must deal with practical decision making that is usually focused on satisfying requirements, system adequacy, schedule, and cost. During the last half-century, a workable and constructive approach to the concepts, terminology, and methodology of V&V has been developed, but it was based on practical realities in technology and business, *not* the issue of absolute truth in the philosophy of nature.

2.1.1 Early Efforts of the Operations Research Community

The first applied technical discipline that struggled with the methodology and terminology of V&V was the operations research (OR) community, also referred to as systems analysis or simulation community; see Balci and Sargent (1984) for an extensive list of early work in V&V and Sheng et al. (1993) for a historical review of V&V concepts from the OR viewpoint. In the OR activities, the systems analyzed could be extraordinarily complex, e.g., industrial production models, models of business or governmental processes, marketing models, national and world economic models, and military conflict models. These complex models commonly involve a strong coupling of complex physical processes, human

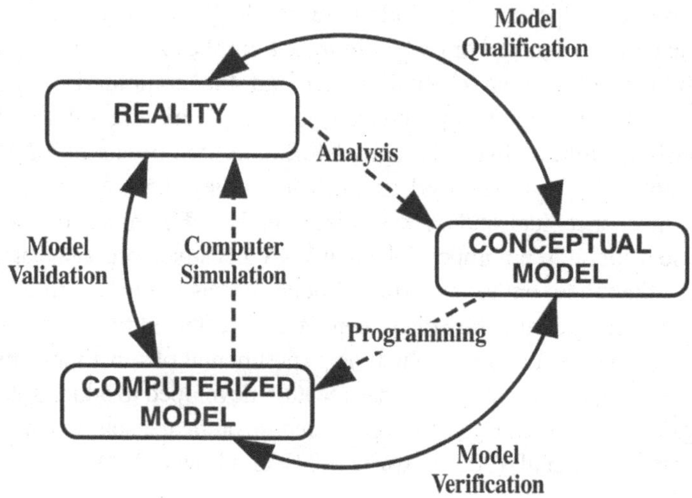

Figure 2.1 Phases of M&S and the role of V&V (Schlesinger, 1979).
(© 1979 by Simulation Councils, Inc.)

behavior in a wide variety of conditions, and computer-controlled systems that adapt to changing system characteristics and environments. For such complex systems and processes, fundamental conceptual issues immediately arise with regard to (a) defining the system as opposed to its external influences, (b) issues of causality, (c) human behavior, (d) measuring system responses, and (e) assessing the accuracy of the model.

A key milestone in the early work by the OR community was the publication of the first definitions of V&V by the Society for Computer Simulation (SCS) in 1979 (Schlesinger, 1979).

Model Verification: substantiation that a computerized model represents a conceptual model within specified limits of accuracy.
Model Validation: substantiation that a computerized model within its domain of applicability possesses a satisfactory range of accuracy consistent with the intended application of the model.

The SCS definition of verification, although brief, is quite informative. The main implication is that the computerized model, i.e., the computer code, must accurately mimic the model that was originally conceptualized. The SCS definition of validation, although instructive, appears somewhat vague. Both definitions, however, contain a critical concept: *substantiation* or *evidence of correctness*.

Along with these definitions, the SCS published the first useful diagram depicting the role of V&V within M&S (Figure 2.1).

Figure 2.1 identifies two types of models: a conceptual model and a computerized model. The *conceptual model* comprises all relevant information, modeling assumptions, and mathematical equations that describe the physical system or process of interest. The conceptual model is produced by analyzing and observing the physical system of interest. The SCS defined *Qualification* as follows: "Determination of adequacy of the conceptual model to provide an acceptable level of agreement for the domain of intended application." The *computerized model* is an operational computer program that implements a conceptual model using computer programming. Modern terminology typically refers to the computerized model as the computer model or code. Figure 2.1 emphasizes that *verification* deals with the relationship between the conceptual model and the computerized model, and that *validation* deals with the relationship between the computerized model and reality. These two key relationships are not always recognized in other definitions of V&V.

The OR community clearly recognized, as it still does today, that V&V are tools for assessing the accuracy of the conceptual and computerized models. For much of the OR work, the assessment was so difficult, if not impossible, that V&V became more associated with the issue of credibility, i.e., the model either is worthy of belief or has the power to elicit belief. In the applied sciences and engineering, however, there is a critical need to better quantify estimated accuracy for the common situation in which no experimental observations are available. The concept of predictive capability of a model has recently developed to address this need and will be discussed later in this chapter. For a conceptual and theoretical discussion of V&V in modern texts on M&S, see Zeigler et al. (2000), Roza (2004), and Law (2014).

2.1.2 IEEE and Related Communities

During the 1970s, computer-controlled systems started to become crucial and widespread in commercial, defense, and public systems. Some examples are manufacturing control systems, automatic flight-control systems for aircraft, and control of high-consequence systems, such as nuclear power reactors. In response to this interest, the Institute of Electrical and Electronics Engineers (IEEE) defined verification as follows (IEEE, 1984):

Verification (software): (A) the process of determining whether the products of a given phase of the software development cycle fulfill the requirements established during the previous phase; (B) formal proof of program correctness.

The first alternative definition is quite general, and it is also strongly referential in the sense that the descriptive value of the definition directly depends on the specification of "requirements established during the previous phase." Because those requirements are not stated in the definition, the definition does not contribute to an intuitive understanding of verification or specify the primary issue addressed in verification. The second alternative definition is much more informative from the perspective of scientific computing. It is a broad and demanding statement of correctness of the software in the sense of a mathematical proof or formal evidence of correctness.

Also in 1984, IEEE defined validation as follows (IEEE, 1984):

Validation (software): the process of evaluating the software at the end of the software development process to ensure compliance with software requirements.

The definition of validation is also referential because of the phrase "compliance with software requirements." Because specific requirements are not defined as part of the definition, the definition of validation is not very informative, particularly from a scientific computing perspective. The substance of the meaning must be provided in the specification of additional or subsidiary information concerning requirements. For example, requirements for the output of a feedback control system may specify that, given a range of possible inputs, the output should be controlled within a given range. Furthermore, note that the IEEE definitions emphasize that V&V are processes, i.e., ongoing activities, that are relevant to the application under consideration.

An additional important organization addressing terminology is the International Organization for Standardization (ISO). Their definitions for V&V are given as follows (ISO, 2005):

Verification: confirmation, through the provision of objective evidence, that specified requirements have been fulfilled.
Validation: confirmation, through the provision of objective evidence, that the requirements for a specific intended use or application have been fulfilled.

It is clear that these definitions are a generalization of the definitions formulated by IEEE. This generalization is important not only from a world trade perspective that must address an extremely wide range of products and services but also from a systems engineering perspective. The systems engineering community has recently begun to address the dichotomy between the concepts of V&V that are needed in systems engineering, as opposed to what is needed in scientific computing (ASSESS, 2021; INCOSE, 2021). For example, as pointed out by Shaw (2021), the V&V definitions of ISO take the perspective of usefulness, fit for purpose, or utility of models or processes. Whereas the SCS, and more importantly the DoD, AIAA, and ASME definitions of V&V to be discussed next, take the perspective of correctness and accuracy of simulation models. Both perspectives are appropriate and useful, but they serve different purposes.

2.1.3 US Department of Defense Community

In the early 1990s, the US Department of Defense (DoD) began to recognize the importance of establishing terminology and procedures for V&V that would serve the DoD's very broad range of needs in M&S (Davis, 1992; Hodges and Dewar, 1992). The DoD tasked the Defense Modeling and Simulation Office (DMSO) with studying the terminology created by the IEEE and to determine if the IEEE definitions would serve the DoD's needs. The DMSO obtained the expertise of researchers in the fields of OR, operational testing of combined hardware and software systems, physical sciences-based simulation, and warfare simulation. They concluded that the IEEE definitions were too ambiguous as to what technical issue was addressed by verification, as opposed to validation. In 1994, the DoD/DMSO published their definitions of V&V (DoD, 1994):

Verification: the process of determining that a model implementation accurately represents the developer's conceptual description of the model.
Validation: the process of determining the degree to which a model is an accurate representation of the real world from the perspective of the intended uses of the model.

From a comparison of these definitions with those codified by the IEEE, it is clear there was a major break in conceptualization of V&V by the DoD. The DoD definitions could be referred to as *modeling and simulation V&V*, as opposed to the IEEE definitions of *software V&V*. The DoD definitions are actually similar to those formulated by the SCS in 1979.

There are two key features of the DoD definitions that make them such an important step forward in the practical conceptualization of V&V for simulation models. First is the emphasis on *accuracy*, which is not mentioned in the IEEE definitions. This feature assumes that a measure of accuracy or correctness can be determined. Accuracy, from the DoD perspective, can be measured relative to any accepted referent. In verification, the referent could be either well-accepted solutions to specified problems or expert opinions as to the reasonableness of the solution. In validation, the referent could be either experimentally measured data or expert opinions defining a reasonable or credible result of the model. Second, the concept of verification is distinctly independent from that of validation. Verification deals with the issue of the correctness of the model implementation relative to "the developer's conceptual description of the model." That is, verification addresses the issue of consistency between the model implementation and the conceptual model. However, the concept of validation deals with the issue of fidelity of the model results relative to "the real world." That is, validation addresses the issue of correspondence between a model result and physical reality in nature.

As noted in the discussion of the IEEE definitions, the DoD definitions also stress that both V&V are "process[es] of determining." V&V are ongoing activities that do not have a clearly defined completion point unless additional specifications are given in terms of intended uses of the model and adequacy. The definitions include the ongoing nature of each process because of an unavoidable, but distressing, fact: The veracity, correctness, and accuracy of a computational model cannot be demonstrated for all possible conditions and applications, except for trivial models. For example, verification *cannot* prove that even a moderately complex computer code has no errors.

2.1.4 AIAA and ASME Engineering Societies

Most applied science and engineering communities focus on their particular types of applications, as opposed to the very broad range of DoD systems. Specifically, scientific computing concentrates on modeling physical systems and processes, as opposed to modeling human interactions, such as in the social sciences. Typically, the mathematical model of the system of interest is dominated by physical processes that are described by ordinary and partial differential equations (PDEs) or integrodifferential equations. Furthermore, the mathematical model of the physical processes are strongly rooted in quantified causes and quantified effects, as well as conservation principles, such as conservation of mass, momentum, and energy. Human interactions with a physical system, as well as the effect of computer control systems, is explicitly given by way of the boundary conditions, initial conditions, system excitation, or other auxiliary submodels. The effect of this narrow focus of the natural science and engineering communities will be apparent in the further development of V&V terminology, concepts, and methods.

The computational fluid dynamics (CFD) community, primarily through the American Institute of Aeronautics and Astronautics (AIAA), was the first engineering community to seriously begin developing concepts and procedures for modeling and simulation V&V methodology. For a more complete history of the development of V&V concepts in CFD, see Oberkampf and Trucano (2002).

2.1.4.1 AIAA

In 1992, the AIAA Committee on Standards for Computational Fluid Dynamics (AIAA COS) began a project to formulate and standardize the basic terminology and methodology in V&V for CFD simulations. The committee was composed of representatives from academia, industry, and government, with representation from the US, Canada, Japan, Belgium, Australia, and Italy. After six years of discussion and debate, the committee's project culminated in the publication of the *Guide for the Verification and Validation of Computational Fluid Dynamics Simulations* (AIAA, 1998), referred to herein as the *AIAA Guide*. This guide defines a number of key terms, discusses fundamental concepts, and specifies general procedures for conducting V&V in CFD.

The *AIAA Guide* (AIAA, 1998) slightly modified the DoD definition for verification, giving the following definition:

Verification: the process of determining that a model implementation accurately represents the developer's conceptual description of the model and the solution to the model.

The DoD definition of verification did not make it clear that the accuracy of the numerical solution to the computational model should be included in the definition. However, science and engineering communities are keenly interested in the accuracy of the numerical solution relative to the exact solution to the mathematical model. As a result, the AIAA and the engineering societies are more demanding in the assessment of accuracy of the implemented model with regard to the referent.

The fundamental strategy of verification is the identification, quantification, and reduction of errors in the computer code that solves the mathematical model. Verification provides evidence or substantiation that the conceptual model, e.g., the differential equation model, is accurately solved by the discrete mathematics model embodied in the computer code. To quantify computer coding errors, a highly accurate, reliable benchmark solution must be available. Highly accurate solutions, unfortunately, are only available for simplified models. Verification does *not* deal with the relationship of the conceptual model to the real world. Roache notes that "verification is a mathematics issue; not a physics issue" (Roache, 1998a). *Validation*, to be discussed next, addresses the physics fidelity of the model. Figure 2.2, from AIAA (1998), depicts the verification process of comparing the numerical solution from the code in question with various types of highly accurate solutions.

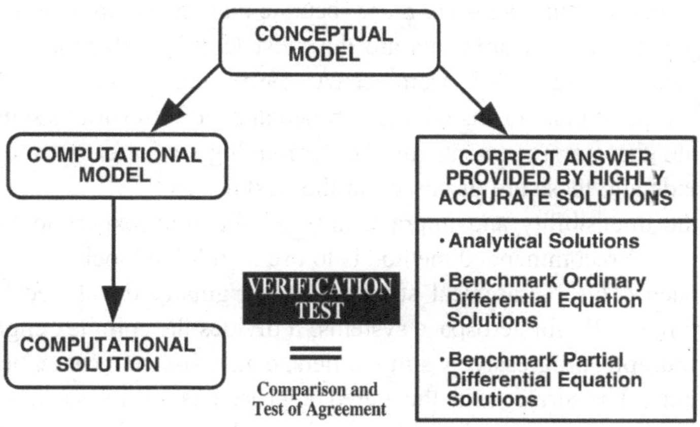

Figure 2.2 Verification process.
(AIAA, 1998)

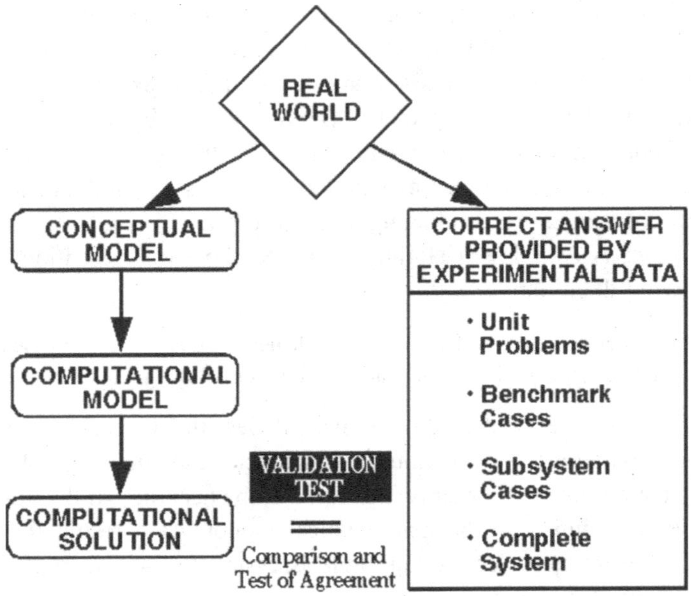

Figure 2.3 Validation process.
(AIAA, 1998)

Although the *AIAA Guide* adopted verbatim the DoD definition of *validation*, a significant break was made from the DoD perspective in terms of the types of comparisons or referent allowed for accuracy assessment with respect to "the real world." The *AIAA Guide* specifically required that assessment of the accuracy of computational results be *only* allowed using experimental measurements. The fundamental strategy of validation involves identification and quantification of the error and uncertainty in the conceptual and computational models. The procedure involves the comparison of simulation results and the experimental data, taking into account experimental uncertainty. Hence, physics modeling accuracy is measured in relation to experimental data, the best measure of reality. This strategy *does not* assume that the experimental measurements are more accurate than the computational results; it only asserts that experimental measurements are the most faithful reflections of reality for the purposes of validation. Figure 2.3, from AIAA (1998), depicts the validation process of comparing the computational results with experimental data from various sources of physical complexity. Note that improvements in the terminology of the types of models were subsequently made and these are discussed in the next section.

Because of the infeasibility and impracticality of obtaining experimental data on most complex systems, the recommended method is to use a building block or system complexity hierarchy approach. This hierarchical strategy was originally developed by Sindir et al. (1996) and Marvin (1995) for aerospace systems. It divides the complex engineering system of interest into multiple, progressively simpler tiers, e.g., subsystem cases, benchmark cases, and unit problems. The strategy in the tiered approach is to assess how accurately the computational results compare with the experimental data (with quantified uncertainty estimates) at multiple degrees of physics coupling and geometric complexity (Figure 2.4). The strategy is clearly constructive in that it: (a) recognizes a hierarchy of complexity in

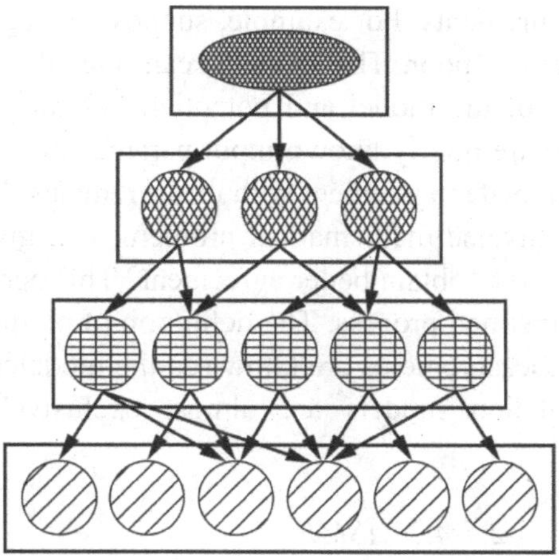

Figure 2.4 Validation tiers of a system hierarchy.
(AIAA, 1998)

systems and simulations, (b) recognizes that the quantity and accuracy of information obtained from experiments varies radically over the range of tiers, and (c) guides the accumulation of validation evidence with the focus always being on the complete system. While additional building-block tiers beyond the four discussed here could be defined, additional tiers would not fundamentally alter the hierarchical strategy. The hierarchical strategy is similar to the methodology developed to analyze severe accidents in nuclear power plants (Zuber et al., 1998). However, the nuclear power methodology focused on how non-dimensional scaling in both physical size and time could be used to relate integral effects and separate effects physics to subscale experiments.

The *AIAA Guide*'s fundamental concept of validation (depicted in Figure 2.3) is the comparison of computational results with experimental measurements for the purpose of model accuracy assessment. Defining validation in this restrictive sense requires addressing the practical situation where *no* experimental data are available, which is the crucial reason why M&S is even conducted. The *AIAA Guide* addresses this issue with the formal definition of *prediction*:

Prediction: use of a computational model to foretell the state of a physical system under conditions for which the computational model has not been validated.

A prediction refers to the computational simulation of a specific case of interest that is *different* in some way from cases where experimental data are available. The term *foretell* commonly refers to time-dependent systems, but it can also refer to steady-state systems which have not been observed before. This definition is highly specific since it refers only to *pre*diction, not *retro*diction or *post*diction (replication of previously obtained results). An accurate prediction is a much stronger test of the model's explanatory value because the experimental data are unknown beforehand to the modeler. From this perspective, it becomes clear that validation (as depicted in Figure 2.3) may, or may not, be a prediction as just defined.

The distinction between prediction and retrodiction (or postdiction) is important because the computational analyst has great flexibility in tuning or adjusting the model to achieve better agreement with known measurements. For example, suppose the agreement between simulation and measurements is rather poor. The modeler can alter the assumptions and approximations in the formulation of the model and potentially obtain better agreement. Another example would be if there are poorly known input parameters in the model, these parameters can be adjusted to obtain better agreement with measurements. Retrodiction is, to various degrees, a solution of the inverse mathematical problem, i.e., given the measurements, how can the model be adjusted to obtain better agreement? This perspective does *not* say that retrodiction is a bad or unjustified process. The definition of prediction accentuates the issue of whether experimental measurements are known to the modeler, or not. Clearly, real world applications of simulation models are almost exclusively in the realm of predictions.

2.1.4.2 ASME

In the late 1990s, members of the solid mechanics community became interested in the concepts and methodology of V&V. The first V&V committee within the Codes and Standards branch of the American Society of Mechanical Engineers (ASME) was formed in 2001 and designated the Committee on Verification and Validation in Computational Solid Mechanics. After five years of discussion and debate, the *ASME Guide for Verification and Validation in Computational Solid Mechanics*, herein referred to as the *ASME Guide*, was completed (ASME, 2006). During the next decade, the concepts of V&V were more refined, debated, and clarified and the ASME Guide was revised. The revision was published as the *ASME Standard for Verification and Validation in Computational Solid Mechanics* (ASME, 2019), herein referred to as the *ASME Standard*.

The *ASME Standard* slightly modified the definition of verification as formulated by the *AIAA Guide*:

Verification: the process of determining that a computational model accurately represents the underlying mathematical model and its solution.

The *ASME Standard* clarified the definition of validation as formulated by the DoD, given in Section 2.1.3, and used by the *AIAA Guide*. In its definition of validation, the *ASME Standard* made it very specific that validation is a comparison of simulation results with commensurate experimental measurements:

Validation: the process of determining the degree to which a model is an accurate representation of corresponding physical experiments from the perspective of the intended uses of the model.

Instead of graphically showing V&V as separate entities as in the *AIAA Guide*, the *ASME Standard* constructed a comprehensive diagram showing both activities, along with other complementary activities (Figure 2.5). It must be stressed that the ASME diagram and all of the activities shown can be applied to *any tier* of a validation hierarchy, or to *any type* of physical system or process, as long as experimental measurements are available. Analogous activities in both the M&S and the physical experimentation branches of the diagram are clearly shown, along with their interactions. The conceptual model, the mathematical model, and the computational model are all shown in the figure, as well as being defined in the *ASME Standard*. The separation of the concepts and activities in each of these three types of models significantly improved the understanding of not only the V&V process, but also the

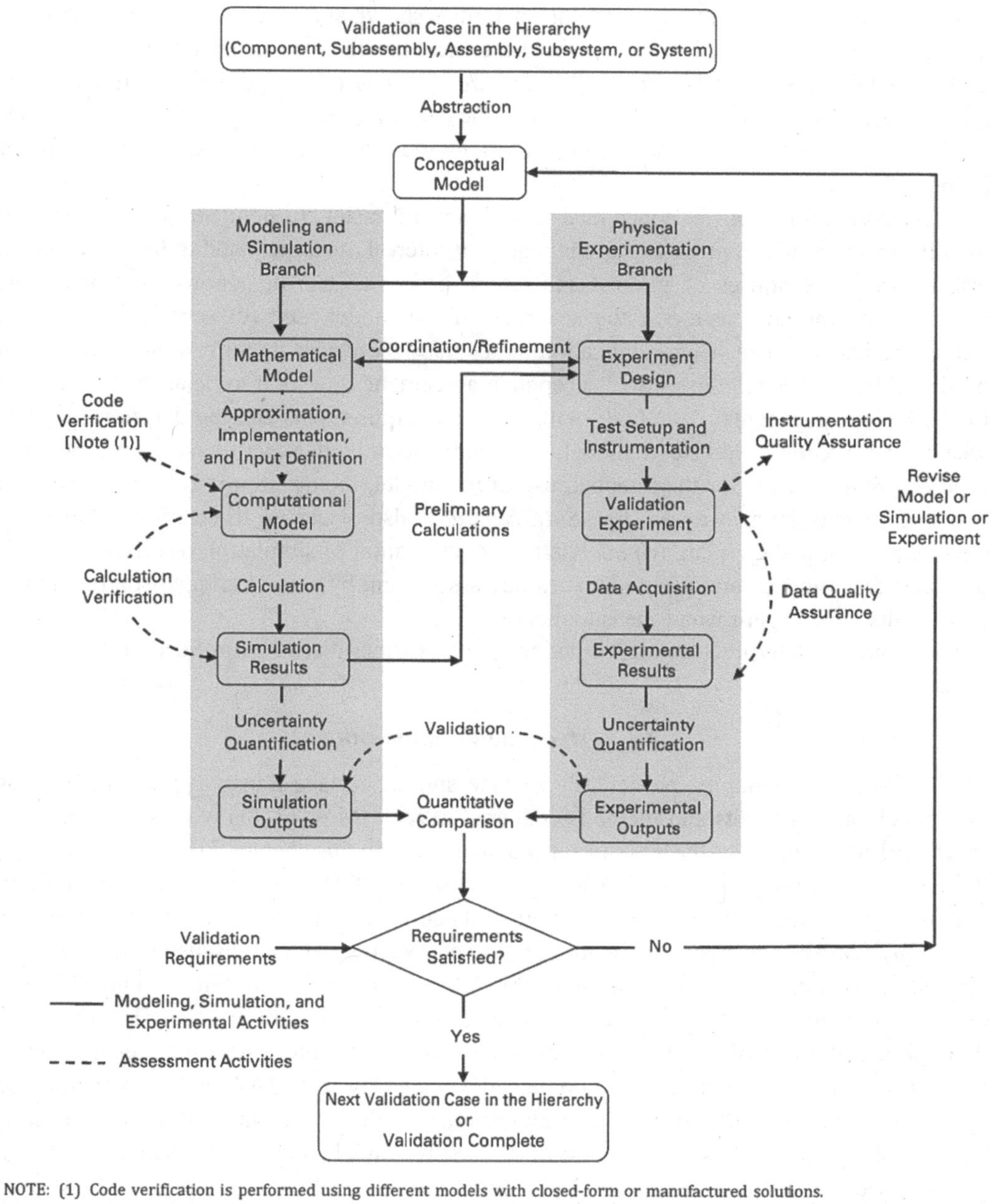

Figure 2.5 ASME diagram for V&V activities and producs.
(ASME, 2019)

M&S process. Two elements of verification are specifically called out in the diagram: code verification and calculation verification. This separation of verification activities followed the pioneering work of Blottner (1990) and Roache (1998b). With this separation of verification activities, a much clearer set of techniques could be discussed to improve coding reliability and numerical accuracy assessment of the simulation results.

An important decision point in the V&V activities shown at the bottom of Figure 2.5 is the answer to the question: Is there acceptable agreement between the simulation results and the experimental measurements? The *ASME Standard* discusses how this decision should be made, specifically regarding the key phrase in the definition of validation: "the intended uses of the model." Some M&S application communities use the equivalent phrase "fit for purpose."

In the formulation of the conceptual model, several tasks are defined, among them: (a) identify which physical processes in the reality of interest are anticipated to have significant effects on the responses of interest and which processes can be ignored, (b) determine requirements for demonstrating the accuracy of the model, and (c) specify the model's domain of intended use. The specification of accuracy requirements for responses of interest produced by the model allows the "acceptable agreement" question to be answered relative to available experimental data. Only with accuracy requirements can the decision be made whether to accept or revise a model, or obtain additional experimental data. Without accuracy requirements for the intended use of the model, the question: "How good is good enough?" cannot be answered. The *ASME Standard* also discussed three critical topics: (a) development of a V&V plan, (b) uncertainty quantification of simulation results and experimental measurements, and (c) model accuracy assessment by way of comparison of simulation results with experimental measurements.

Key issues of interpretation of validation will be further discussed in Section 2.2.3.

2.1.5 Hydrology Community

The hydrology community, particularly surface and subsurface transport phenomena, was also developing concepts and methods concerning V&V. Most of this work was developed independently of many of the activities discussed earlier in this chapter. For a good review of the latest in hydrology V&V, see Anderson and Bates (2001). The work of the hydrology community is significant for two reasons. First, it addresses validation for complex processes in the physical sciences where validation, in the restricted sense just discussed, is extremely difficult, if not impossible. One reason for this difficulty is remarkably limited knowledge of the specific underground transport characteristics and material properties associated with the underground domain of interest. For such situations, calibration or parameter estimation in models instead of the concept of validation developed by the AIAA and ASME must be dealt with more explicitly. Hence, the only practical path forward in hydrology, as well as with most complex physical processes, is to solve the inverse mathematical problem. Second, because of the limited knowledge about the physical characteristics of the system under consideration, the hydrology field has strongly adopted statistical methods of model parameter calibration/estimation. In hydrology, it is not just calibration of scalar parameters, but also scalar and tensor fields. Parameter calibration will be discussed in Section 2.2.5.

The hydrology community in Europe (Rykiel, 1996; Beven, 2002; Refsgaard and Henriksen, 2004) has independently developed ideas about V&V that are very similar to those developed in the United States. Rykiel (1996) makes an important practical point, especially to analysts and decision makers, about the difference between the philosophy-of-science viewpoint and the practitioner's view of validation: "Validation is not a procedure for testing scientific theory or for certifying the 'truth' of current scientific understanding. . . . Validation means that a model is acceptable for its intended use because it meets specified performance requirements." Refsgaard and Henriksen (2004) recommended terminology and fundamental procedures for V&V that are similar to the *AIAA Guide* and *ASME Standard*. They define model validation as: "Substantiation that a model within its domain of applicability possesses a satisfactory range of accuracy consistent with the intended application of the model." Refsgaard and Henriksen (2004) also stressed another crucial issue that is corroborated by the *AIAA Guide* and the *ASME Standard*: "Validation tests against independent data that have not also been used for calibration are necessary in order to be able to document the predictive capability of a model." In other words, the major challenge in validation is to perform an assessment of the model in a blind test with experimental data, whereas the key issue in calibration is to adjust the physical modeling parameters to improve agreement with experimental data. It is difficult, and sometimes impossible, to make blind test comparisons, e.g., when well-known validation data are available for comparison. As a result, caution is encouraged when making conclusions about the predictive capability of models when the computational analyst is aware of the experimental data. Knowing the correct answer beforehand is extremely seductive, even to a saint.

2.2 Primary Terms and Concepts

This section discusses the concepts and underlying principles behind the formal terms just introduced in more detail. This book will use the definitions of V&V as given by the *ASME Standard*. The terms code verification, solution verification, predictive capability, calibration, certification, and accreditation are also discussed in this section. Definitions for these terms will be primarily drawn from ASME, AIAA, IEEE, and DoD.

The modern scientific method is very much aligned with the philosophy of science approach referred to as deductivism. Deductivism is the method of drawing conclusions by logically combining new ideas with facts that have been accepted as true. Deductivism argues from the general to the particular, or reasons from known general principles to deduce previously unobserved or unknown phenomena. This perspective can be most clearly seen in the manner in which scientists and engineers are trained, for example, learning general principles such as conservation of mass, momentum, and energy in physical processes. V&V, however, is aligned with inductive reasoning processes, i.e., processes that present the correctness of individual pieces of evidence to support the conclusion of correctness of a generalization. Stated differently, the philosophical perspective of V&V must be one of *fundamental skepticism*: if any claim cannot be demonstrated or proven, then it is not accepted as true. The dichotomy of perspectives between the training of scientists and engineers, as opposed to the philosophy of V&V, is sometimes at the root of the lack of interest, or even open resistance, to many V&V activities by some scientists and engineers.

2.2.1 Code Verification

The *ASME Standard* (ASME, 2019) defines *code verification* as:

Code verification: the process of determining that the numerical algorithms are correctly implemented in the computer code and of identifying errors in the software.

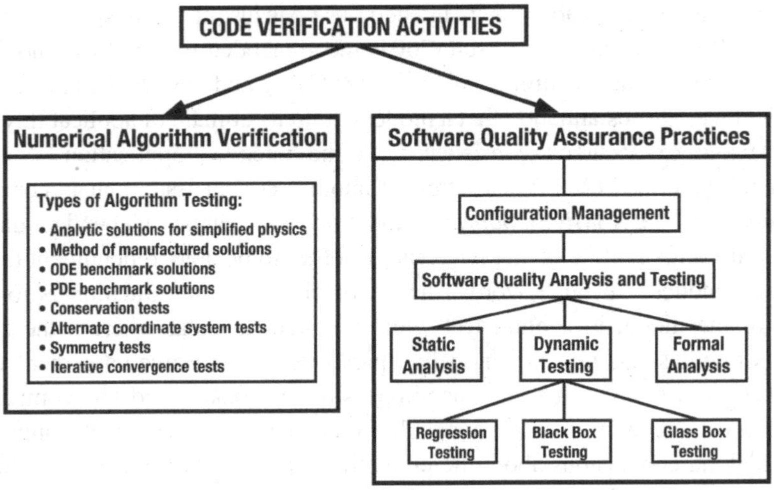

Figure 2.6 Integrated view of code verification in M&S.
(Oberkampf et al., 2004)

An equivalent definition of code verification is given in (AIAA, 2021; ASME, 2022). Code verification can be segregated into two activities: numerical algorithm verification and software quality assurance (SQA), as shown in Figure 2.6. Numerical algorithm verification addresses the mathematical correctness in the software implementation of all the numerical algorithms that affect the numerical accuracy of the computational results. The major goal of numerical algorithm verification is to accumulate evidence that demonstrates that the numerical algorithms in the code are implemented correctly and that they are functioning as intended. As an example, numerical algorithm verification would demonstrate that a spatial discretization method would produce the expected formal convergence rate, as the mesh is refined for the specific PDE being tested. The emphasis in SQA is on determining whether or not the code, as part of a software system, is implemented correctly and whether it produces repeatable results on specified computer hardware and in a specified software environment. Such environments include computer operating systems, compilers, function libraries, data communication software, etc.

Numerical algorithm verification is fundamentally based on software solution testing, consistency testing, and analyses of code results for individual executions of the code. It focuses on detailed investigations of numerical aspects, such as spatial and temporal convergence rates, spatial convergence in the presence of discontinuities in solution variables, independence of solutions to coordinate transformations, and symmetry tests related to various types of boundary conditions (BCs). Analytical or formal error analysis is inadequate in numerical algorithm verification because the code itself must *demonstrate* the analytical and formal results of the numerical analysis. Numerical algorithm verification is usually conducted by comparing computational solutions with highly accurate solutions, which are commonly referred to as verification benchmarks. Oberkampf and Trucano (2007) divided the types of highly accurate solutions into four categories (listed from highest to lowest in accuracy): manufactured solutions, analytical solutions given by an infinite series, specialized solutions of PDEs given by numerical solutions to ordinary differential equations, and numerical solutions to PDEs. Methods for numerical algorithm verification will be discussed in detail in Chapters 5–6.

SQA activities consist of practices, procedures, and processes that are primarily developed by researchers and practitioners in the computer science and software engineering communities. Conventional SQA emphasizes processes (i.e., management, planning, design, acquisition, supply, development, operation, and maintenance), as well as reporting, administrative, and documentation requirements. A key element of SQA is software configuration management, which is composed of configuration identification, configuration and change control, and configuration status accounting. As shown in Figure 2.6, software quality analysis and testing can be divided into static analysis, dynamic testing, and formal analysis. Dynamic testing can be further divided into such elements as regression testing, black box testing, and glass box testing. From a SQA perspective, Figure 2.6 could be reorganized so that all types of algorithm testing categorized under numerical algorithm verification could be moved to dynamic testing. Although this perspective is common among software specialists, it fails to stress the importance of numerical algorithm verification that is critical in the numerical solution of differential equations. SQA is a necessary element of code verification, but it is not the only element. SQA methods will be discussed in Chapter 4 "Software Engineering for Scientific Computing."

2.2.2 Solution Verification

Solution verification, also called calculation verification, is defined as:

Solution verification: the process of determining the correctness of the input data, the numerical accuracy of the solution obtained, and the correctness of the output data for a particular simulation.

Solution verification attempts to identify and quantify three sources of errors that can occur in the exercise of a computer simulation code (Figure 2.7). The first source includes errors, blunders, or mistakes made by the computational analysts in preparation of the input data for the simulation. The second source involves numerical errors resulting from computing the discretized solution of the mathematical model on a digital computer. Finally, the third source includes errors, blunders or mistakes made by the computational analysts in processing or representing the output data that is produced by the simulation. An example of an error in representing the output data is the construction and use of response surface methods. *None* of the error sources shown in Figure 2.7 refers to errors or approximations made in the formulation or construction of either the conceptual or mathematical model. Note that sometimes the definition of solution verification only refers to the second source of error shown in Figure 2.7 (ASME, 2022).

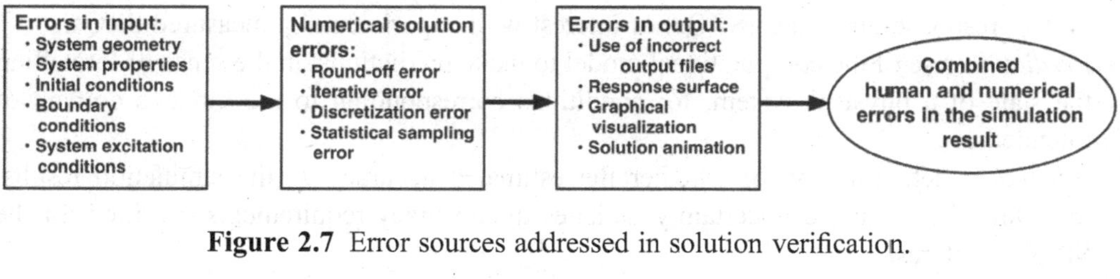

Figure 2.7 Error sources addressed in solution verification.

Most of the error sources in the first and third sources refer to human errors, in the sense of not doing what was intended, exclusive of any other sources. Human errors can be very difficult to detect in large-scale computational analyses of a complex system. Even in relatively small-scale simulations, human errors can go undetected if procedural or data-checking methods are not employed to detect possible errors. For example, if an analysis requires tens of CAD/CAM files, perhaps hundreds of different material models, and thousands of Monte Carlo simulation samples for a nondeterministic simulation, human errors, even by the most experienced and careful practitioners, commonly occur.

The second source, numerical solution errors, is concerned with (a) computer round-off errors, (b) iterative solution errors usually resulting from a repetitive approximate solution to a set of nonlinear equations, (c) spatial and temporal discretization errors in the numerical solution of PDEs, and (d) statistical sampling error. The importance and difficulty of numerical error estimation has increased as the complexity of the physics and mathematical models has increased, e.g., mathematical models given by nonlinear PDEs with singularities and discontinuities, and solutions with features of mathematical chaos.

The two basic approaches for estimating the discretization error in the numerical solution of a PDE are *a priori* and *a posteriori* error estimation techniques. An *a priori* technique only uses information about the numerical algorithm that approximates the given differential equation and the given initial conditions (ICs) and BCs. *A priori* error estimation is a significant element of classical numerical analysis for linear PDEs, especially in analyzing finite element methods. An *a posteriori* approach can use all the *a priori* information as well as the computational results from related numerical solutions, e.g., solutions using different mesh resolutions or solutions using different order-of-accuracy methods. During the last decade, it has become clear that the only way to achieve a useful quantitative estimate of numerical error in practical cases for nonlinear DEs is by using *a posteriori* error estimates. Estimation of numerical solution errors are discussed in detail in Chapters 7–9.

2.2.3 Model Validation and Prediction

Even though the DoD, the *AIAA Guide*, and the *ASME Standard* use the same formal definition for validation, the discussion in Section 2.1 hinted at differences in the interpretation and implications of the term. For example, Section 2.1 noted that the *AIAA Guide* and the *ASME Standard* require *experimental data* when comparisons are made with simulations, whereas in the DoD interpretation this is not required. To help clarify critical aspects of simulation terminology, Figure 2.8 depicts three central activities as follows (Oberkampf and Trucano, 2008):

- *Validation*: quantification of the accuracy of simulation results by comparing the computed system response quantities (SRQs) of interest with experimentally measured SRQs.
- *Prediction*: use of the computational model to make predictions, in the sense of foretelling the state of a physical system, for conditions corresponding to the model's domain of intended use.
- *Adequacy*: determination of whether the estimated accuracy of the simulation results, including the estimated uncertainty, satisfies the accuracy requirements specified for the SRQs of interest.

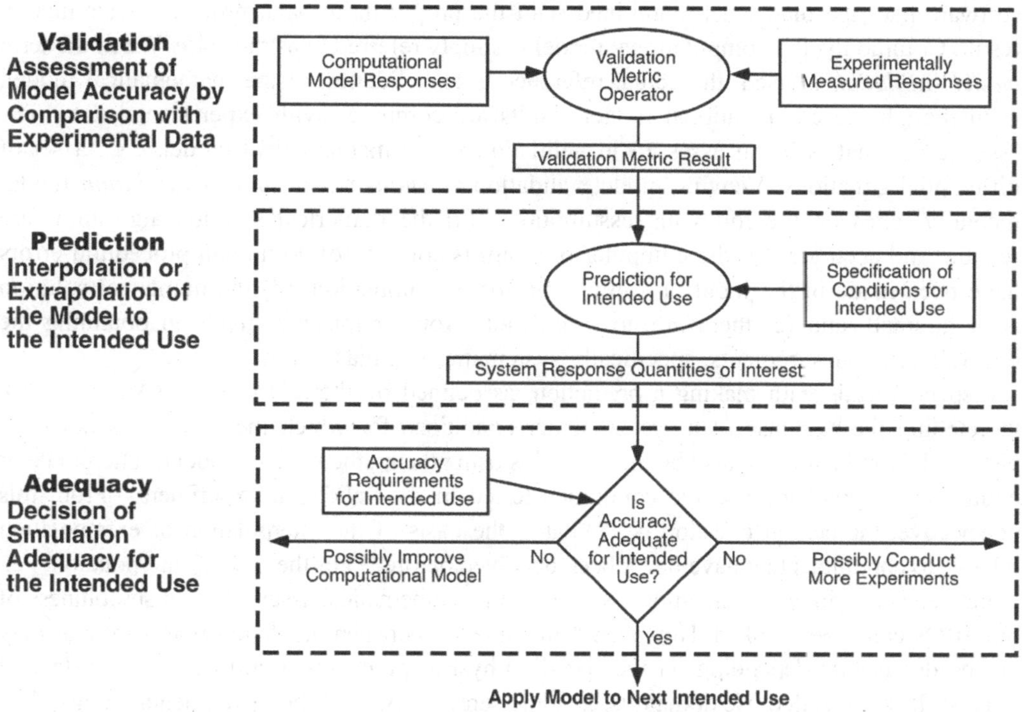

Figure 2.8 Three aspects of model validation.
(Oberkampf and Trucano, 2008)

As depicted in Figure 2.8, Aspect 1 deals with assessing the accuracy of results from the model by comparisons with available experimental data. The assessment could be conducted for the actual system of interest or *any* related system of interest, such as any tier of the system hierarchy. Some examples are: (a) the actual system at normal operational conditions of the system, (b) the actual system operating in some type of accident or failure environment, or (c) subsystems or components of the actual system that have been identified in the system hierarchy. Model accuracy is quantitatively estimated using a *validation metric operator*. This operator computes a difference between the simulation results and the experimental results for individual SRQs as a function of the input or control parameters occurring for the validation conditions. The validation metric operator can also be referred to as a *mismatch* operator between the simulation results and the experimental results over the multidimensional space of all model input parameters that are tested. In general, the operator must be able to input simulation results and the experimental results that are not just single numbers, but distributions of numbers (e.g., probability distributions) and even interval-valued quantities. Validation metrics are discussed in detail in Chapter 12 "Model Accuracy Assessment."

A *computational model* is derived from the numerical implementation of the mathematical model, a process that results in a set of discretized equations and solution algorithms, and then these equations and algorithms are programmed into a computer. Another way to describe the computational model is that it is a mapping of the mathematical model into a software package that, when combined with the proper input data, produces a simulation result. Commonly, the computational model is simply referred to as the *code*. When the term *model validation* is used, the actual reference is to validation of the mathematical model, even though the computational model results are compared with experimental data. The essence of what is being assessed in validation is the mathematical model, e.g., a set of differential equations. Viewing model validation as *mathematical model validation* fundamentally relies on the following assumptions: (a) the numerical solution algorithms are reliable and accurate, (b) the computer program is correct, (c) no human procedural errors have been made in the input or output data for the simulation, (d) the numerical solution error is small, and (e) there are no significant errors or mistakes made in obtaining the experimental measurements, specifically systematic, i.e., bias, errors.

Aspect 2 deals with making a prediction, as defined by the *AIAA Guide* (AIAA, 1998), "foretelling the response of a system under conditions for which the model has not been validated." Prediction can also be thought of as using the mathematical model to interpolate or extrapolate beyond the specific conditions tested in the validation experiments. From this perspective, the central question is: "What is the basis of the interpolation or extrapolation relative to conditions that have never been observed before?" For the case of interpolation, i.e., in between experimental validation conditions, the mathematical assumption of smoothness of the SRQs could be invoked. However, "smoothness" between validation points may or may not be defensible; depending on the specific physical processes occurring in the system of interest. If the intended use domain is, in some sense, *outside* of the experimental points, then the issue of extrapolation must be addressed. The only rational basis for extrapolation is the fidelity of the physics captured within the mathematical model. Stated differently, an explanatory model of a physical process has the potential to foretell states of nature that have never before been observed. Use of the model to interpolate or extrapolate to these previously unobserved conditions involves an additional type of uncertainty discussed in the next section.

Aspect 3 deals with the issue of model adequacy for the intended use of the model. To make the decision of adequacy or inadequacy, a comparison of the estimated accuracy of the model relative to the accuracy requirements of the model must be made over the domain of the model's intended use. The more general perspective of model adequacy typically depends on many factors, such as computer resource requirements, speed with which the model can be recomputed for new input data, and ease of use of the software for the given experience level of the analysts involved. The adequacy decision in Aspect 3 *only* refers to whether the model satisfies or does not satisfy the accuracy requirements specified. An accuracy requirement may be stated as: the maximum of the total estimated uncertainty for specified SRQs cannot exceed a fixed absolute value over the domain of the intended use domain. To estimate the total uncertainty of a model prediction over the entire intended use domain, due to all contributing sources, is a daunting task both computationally and conceptually. The maximum allowable uncertainty over the intended use of the model could be an absolute-value quantity or a relative-uncertainty quantity (i.e., the uncertainty is scaled by the magnitude of the quantity). If the adequacy decision is positive, then the model could be employed over the intended use domain. If the decision is negative, then a few options are available: (a) conduct more experiments to expand the validation domain for better coverage of the intended use domain, (b) improve the predictive capability of the model, or (c) eliminate the portion of the intended use domain where the model accuracy is inadequate.

An important conceptual point should be mentioned regarding Aspect 3. The decision model adequacy assessment is only concerned with the accuracy of the simulation, *not* the performance of the engineering system being analyzed. Whether the system of interest, e.g., a gas turbine engine or a flight vehicle, meets its performance, safety, or reliability requirements is a completely separate topic from the aspects discussed relative to Figure 2.8. Simply put, a computational model of a system could be accurate, but the system itself could be lacking in performance, safety, or reliability because of inadequate design.

As a final point, the terminology regarding the meaning of *model validation* in technical discourse could generally be described as pandemonium. This has caused great confusion, misunderstanding, and conflict; much of it unnecessary because of the wide range of meaning of the concept. Many individuals use the concept of validation to mean all three aspects depicted in Figure 2.8. This text refers to this concept as the *encompassing view of validation*. The US DoD community commonly takes this encompassing view of validation, as well as most in common language use of the term *validation*. Thus, Figure 2.8 was constructed roughly a decade ago to understand and focus the discussion on a specific issue represented in the graphic. This book will consistently use the term validation to *only mean* Aspect 1 shown in Figure 2.8. Sometimes in the literature this concept is referred to as the *restricted view of validation*.

2.2.4 Predictive Capability

The term *predictive capability* is a rather new term that became popular during the last two decades. The term grew primarily from the Accelerated Strategic Computing Initiative (ASCI), which was started in the mid-1990s; a large-scale research and development program of the US Department of Energy. The term captured the need to move the purpose of simulation from insight for the system response to quantitative prediction of the response. This term is defined similar to the *AIAA Guide* definition of prediction, as well as the definition given by (ASME, 2022):

Predictive capability: anticipated accuracy of a computational model to foretell the state of a physical system, including relevant uncertainties, under conditions for which the computational model has not been validated.

To better explain the relationship of predictive capability to model validation, consider Figure 2.9. The axes of the figure are two model input parameters representing some characteristic of the system or the surroundings of the system. Examples of these parameters are (a) initial speed and angle of impact of an automobile in a crash environment, (b) Mach number and Reynolds number of a flight vehicle, (c) amplitude and frequency of vibrational excitation of a structure, and (d) degree of initial damage to a system that is designed to fail safely. The Xs represent the values of the two model input parameters for the conditions of the validation experiments. At each of these conditions, the SRQs of interest have been measured, along with an experimental uncertainty estimate. Examples of SRQs are crush-up distance in an automobile crash, maximum range or speed of a flight vehicle, and maximum deflection or stress in the deformation of a solid. The validation domain is the interior region of all the experimental data obtained that are relevant to the system of interest. Specifically, the validation domain is defined as the *convex hull* of the space of all input parameters of the

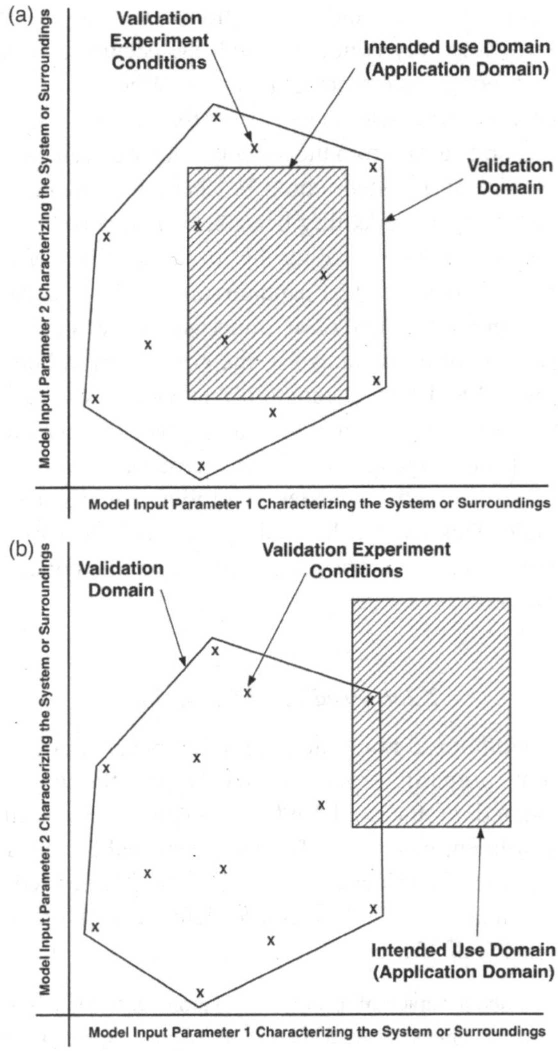

Figure 2.9 Two examples of the relationships of the validation domain to the application domain. (a) Complete overlap of the application domain and the validation domain. (b) Partial overlap of the application domain and the validation domain.

model that represent the conditions of the validation experiments. (The convex hull of a set of points, also known as the minimum convex polygon or as the intersection of all convex sets of the points, is the envelope of all points such that the enclosure is convex everywhere (Leonard and Lewis, 2016).) The intended use domain, also referred to as the application domain or the context of use domain, of the system is shown as the cross-hatched area. The concept of the validation domain and its relationship to the application domain was originated by Trucano et al. (2002).

For the purpose of our discussion, presume the following two features concerning the experimental and computational results shown in Figure 2.9. First, the validation domain includes some level of confidence that the relevant processes are modeled at a level that is commensurate with the needs of the application. For example, at one extreme, the model may be a physics-based model, in the sense that it conserves mass, momentum, and energy, and it may contain model parameters that have been previously calibrated. At the other extreme, the model may be an entirely empirical model calibrated using previous observations of a similar process, as well as the experimental data within the validation domain. Second, the accuracy of the model has been quantitatively estimated by comparing the experimental and simulation results for the SRQs at the indicated points in the validation domain. Thus, a validation metric is used to calculate the difference between the experimental and computational SRQ at each of the experimental conditions, incorporating the uncertainty in both the experiment and the simulation. The comparison quantity is referred to as the *model form uncertainty*. It is the uncertainty due to the assumptions and approximations in the formulation of the model and it is separate from the uncertainty due to the model input parameters. Depending on the fidelity of the model, the model form uncertainty may be rather small, meaning close agreement between experiment and simulation, or it may be rather large.

Figure 2.9a depicts the desirable situation for predictive capability of a model that is applied to some system of interest; that is, the complete overlap of the validation domain with the application domain. From a conceptual perspective, Figure 2.9a suggests that, regardless of the type of model, the model has some degree of representation of the SRQs over the application domain. For all points inside the validation domain, the notion of function interpolation can be argued. It must be recognized, however, that *any* interpolation point away from the existing experimental points is actually a prediction. Figure 2.9a characterizes the vast majority of engineering systems in history. Engineering systems over the centuries have been designed, their performance and reliability determined, and then the system is commonly redesigned or modified in some way to better meet requirements.

Figure 2.9b depicts the situation where there is very little overlap between the validation domain and the application domain. Outside of the validation domain, the notion of extrapolation or model-based projection is required. As is well known from function approximation theory, for example, regression analysis, extrapolation is not recommended because there is *no explicit information* concerning the region outside of the existing data. As a result, the issue focuses on the basis of a prediction outside of the present experience. If the model is physics-based, then there is at least some reason to believe its predictive capability outside of the validation domain. The basis for the prediction is *not only* in our present observational experience, but it is in the fidelity of the physics that is captured within the model. A physics-based model is an explanatory model that has, at least to some degree, the capability to describe underlying processes and interactions that cause the resulting outcome. An empirical model, such as a statistical fit of the experimental data, has little or no claim of predictive capability outside of the validation domain. For example, a neural net model or a data mining model is an empirical model that does not need to obey any type of general principles, such as conservation laws or coupling of physical phenomena. For the case of extrapolation, it reinforces our need for models to be critically judged on the basis of achieving the right answers for the right reasons.

Two final points should be emphasized regarding the concept of predictive capability. First, the concept of distance outside of the validation domain, or even inside the validation domain must be considered. Distance is well characterized with regard to the change in magnitude of input parameters and the change in the magnitude of an SRQ from one experimental condition to another. The concept of distance with regard to the change in the interaction of physical processes, however, is not well defined. For example, suppose at one experimental condition inside the validation domain chemical diffusion is dominant, whereas outside the validation domain convective transport is dominant. How could the distance between these two physical phenomena be measured? It could possibly be measured in terms of the distance between two points in the multi-dimensional space of input parameters. However, this type of measure of distance would not make sense because, in general, the dimensional units of the parameters would be different. Perhaps the question could be viewed as a topological question in some type of multi-dimensional space composed of magnitude of physical transport processes.

Second, the number of system or environmental input parameters in a simple physical system numbers 5 to 10. In even a moderately complex physical system, the number of system and environmental parameters number into the hundreds. For this high dimensional space, it is very common for large portions of the application domain to be *outside* the validation domain. The volume of the validation domain as compared to the volume of the application domain is typically miniscule in real systems. As a result, physics-based models are required to extrapolate outside of the validation domain.

2.2.5 Parameter Calibration

The *ASME Standard* (ASME, 2019) gives the definition of model parameter *calibration* as:

Calibration: the process of adjusting physical modeling parameters in the computational model to improve agreement with experimental results.

Calibration is primarily directed toward improving the agreement of simulation results with existing experimental data. Model calibration is also referred to as model updating or model tuning. This goal is *not directly* related to model validation or predictive capability. The goal of calibration is more focused on addressing local or condition-specific issues. Because of technical issues (such as limited experimental data or poor understanding of the physics) or practical issues (such as constraints in programmatic schedules and fiscal budgets), calibration may be a more appropriate process than increasing the fidelity of the model to achieve better agreement with measurements. The distinction between calibration and validation would usually be clear if the concentration was on experiments and simulations for only simple processes, one component, or unit problem. However, when examining a complete system, the upper tiers of the system hierarchy are found to focus more on model calibration, while the lower tiers are focused on model validation. As a result, both calibration and validation commonly occur during different phases of the computational simulation of a complete system. When calibration is done simply for expediency, as will be discussed in more detail, it negatively impacts the predictive capability of the model. Calibration of model parameters typically confounds a variety of M&S weaknesses in a computational model, thereby resulting in decreased predictive capability. *How* model calibration impacts predictive capability, particularly model extrapolation, is difficult to determine and is presently an active research topic.

Parameter calibration can be considered as part of the broader field of parameter estimation. Parameter estimation refers to procedures for estimating any type of parameter in a model using available data, e.g., experimentally measured data or computationally generated data. The estimated parameter can be either a deterministic value, such as a single value determined by some optimization process, or it can be a nondeterministic quantity, such as a random variable.

Calibration is regularly required in the modeling of complex physical processes, when dealing with incomplete or imprecise measurements in experiments, or when physical input parameters cannot be directly measured in an experiment. Examples of technical fields that routinely use model calibration are multi-phase fluid flow, solid mechanics, structural dynamics, fracture mechanics, meteorology, hydrology, and reservoir engineering. These fields demonstrate that there is a symbiotic relationship between parameter calibration and model validation. Sometimes the parameters requiring calibration result from a phenomenological model or approximations made in mathematical modeling, e.g., effective quantities or subscale physics effects. Phenomenological models are those that express mathematically the results or impact of observed phenomena without paying detailed attention to the physical causes. These types of models are commonly used as submodels in describing complex physical processes. The parameters that need calibration are often not independently or physically measurable at all, but only exist as adjustable parameters in a mathematical model. Although the definition of calibration refers to a parameter, note that the parameter can be a scalar, a scalar field, a vector, or a tensor field.

Because of the wide range in which parameter calibration and estimation can enter a simulation, these procedures should be considered as a spectrum of activities. Figure 2.10 shows a three-level spectrum that can constructively segregate these different activities. The left, or more confident end of the spectrum, has *parameter measurement*. This means that the determination of physically meaningful parameters can be measured using simple, independent models. There are many physical parameters in this category, including mechanical properties, electrical properties, thermal properties, and chemical properties. Mechanical properties include Young's modulus, tensile strength, hardness, mass density, viscosity, and porosity. Electrical properties, consist of electrical conductivity, dielectric constant, and piezoelectric constants. Thermal properties include thermal conductivity, specific heat, vapor

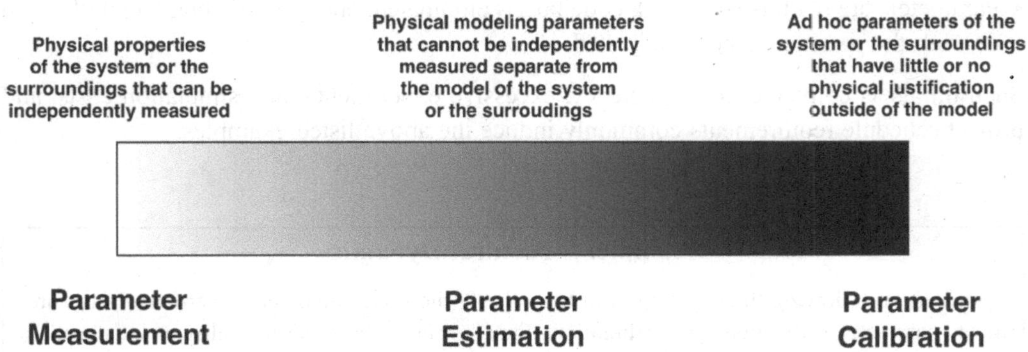

Figure 2.10 Spectrum of parameter measurement, estimation, and calibration.

pressure, and melting point, while chemical properties consist of pH, surface energy, and reactivity.

Parameter estimation is in the middle of the spectrum. This means the determination of physically meaningful parameters can only be determined by solving an inverse model problem. Some examples are: (a) internal dynamic damping in a material, (b) aerodynamic damping of a structure, (c) stiffness and damping of assembled joints in a multi-element structure, (d) effective reaction rate in turbulent reacting flow, and (e) effective surface area of droplets in multiphase flow.

Parameter calibration is at the right end of the spectrum. This is the adjustment of a parameter that has little or no physical meaning outside of the model in which it is used. Some examples are: (a) most parameters in fluid dynamic turbulence models, (b) parameters obtained by regression fits of experimental data, and (c) ad hoc parameters that are added to a model to simply obtain agreement with experimental data.

The spectrum shown in Figure 2.10 can aid in judging the physical soundness of how parameters are determined. When moving to the right in this spectrum, the confidence in using the model for extrapolation decreases significantly. Predictive capability decreases rapidly when extrapolating using a model that has heavily relied on parameter estimation and especially parameter calibration. Concerning parameter adjustment in models versus blind prediction using models, Lipton made a graphic comparison: "Accommodation [parameter calibration] is like drawing the bull's-eye afterwards, whereas in prediction the target is there in advance" (Lipton, 2005).

There are situations where the spectrum shown in Figure 2.10 is distorted because of the procedure that is used to determine a parameter. As a result, when certain parameter adjustment procedures are used that are normally considered as parameter measurement or parameter estimation, this can cause a fundamental shift toward the parameter calibration category. Some examples are:

- Well-known, physically meaningful parameters are changed simply to obtain agreement with newly obtained system-level experimental measurements.
- Parameters are readjusted when physically unrelated submodels are changed.
- Parameters are readjusted when spatial mesh refinement or discretization time step are changed.
- Parameters are readjusted when numerical algorithms are changed.
- Parameters are readjusted when a code bug is eliminated, and the code bug had nothing to do with the parameters being adjusted.

Such things as convenience, expediency, excessive experimental and simulation costs, and project schedule requirements commonly induce the above listed examples.

Example: Structural Dynamics

Consider the following three examples to help clarify the issues involved in parameter measurement, parameter estimation, and calibration. First, suppose an experimentalist is interested in determining Young's modulus, also known as the modulus of elasticity of a material, in solid mechanics. Young's modulus, E, is defined as

$$E = \frac{tensile\,stress}{tensile\,strain}. \tag{2.1}$$

An experiment is conducted in which the tensile stress and tensile strain are measured over the linear elastic range of the material and then a value for E is computed. Although a mathematical model is used to define E, it would be inappropriate to say that the value of E is calibrated because the physics of the process is very well understood. The appropriate term for this activity would be *measurement* of E. If a large number of material samples were drawn from some production batch of material, then parameter estimation methods could be used to characterize the sample-to-sample variability of the production batch. The result would then be given as a probability distribution to describe the variability in E from sample-to-sample in the production batch.

Second, suppose that a structural dynamics simulation is conducted on a structure that is constructed from several structural members and all of these members are bolted together. All of the structural members are made from the same batch of material as the previously discussed experiment to measure E. A finite element model is made for the vibration of the structure and an experiment is conducted in which the structure is excited over the linear elastic range. Various vibration modes of the structure are measured and these are used to estimate the joint stiffness and damping in the mathematical model. Hence, the inverse solution to the mathematical model is used to optimize the input parameters for stiffness and damping to best match the experimental data. Assume that all of the attachment joints are of the same design and the pre-load torque on all of the bolts is the same. This procedure to determine the stiffness and damping in the bolted joints is referred to as *parameter estimation*. It is obvious that these two parameters, joint stiffness and damping, cannot be measured independently from the model for the vibration of the structure, i.e., the structural members must be bolted together before the structure exists. As a result, the term *parameter estimation* would properly describe this procedure in the spectrum shown in Figure 2.10.

Third, consider a similar structural dynamics simulation to before, but now the geometry of the structure is more complex with many structural members of varying thicknesses and cross-sections, all bolted together. All of the structural members, however, are made from the same batch of material as the above experimental measurement of E. Also, the deflection of the structure is small so that it remains in the linear elastic range. If the value of E in the simulation of the vibration of this structure were allowed to be an adjustable parameter, then E would be considered as a *calibrated* parameter. Therefore, the parameter is allowed to change simply due to expediency in the simulation. For this simple example, there is no physical reason to claim that E has changed. Confidence in the predictive capability of the calibrated model is degraded and the uncertainty in the predictions for similar structures would be difficult to estimate.

A more detailed discussion of model parameter calibration procedures is given in Chapter 13 "Predictive Capability."

2.2.6 Certification and Accreditation

The IEEE (IEEE, 1991) defines *certification* as:

Certification: A written guarantee that a system or component complies with its specified requirements and is acceptable for operational use.

For the purpose of this book, the system or component will be considered either a model, a code, or a simulation. For simplicity, all of these will be referred to as an *entity*. In certification of an entity, the written guarantee of acceptable performance can be generated by anyone who is willing to accept the responsibility or legal liability associated with the guarantee. Model developers, code developers, code assessors, or organizations could choose to provide the written guarantee required for certification. For example, a national laboratory, a governmental organization, or a commercial code company could certify their own codes. The documentation for certification is normally done in a more formal and comprehensive manner than is typical for documentation for a code, a model, or a simulation.

The DoD (1997) defines *accreditation* as:

Accreditation: the official certification that a model or simulation is acceptable for use for a specific purpose.

The definition of accreditation uses the phrase "model or simulation," whereas the definition of certification uses the phrase "system or component." However, this is not the crux of the difference between these two terms. The fundamental difference between the terms certification and accreditation is the phrase "written guarantee" versus "official certification" in certification and accreditation, respectively. These terms suggest that the focus is changing from technical credibility issues to legal, control authority, and liability issues when moving from certification to accreditation. Note that the DoD does not formally use the term *certification*, and the IEEE does not formally use the term *accreditation*.

In accreditation, *only* officially designated individuals or organizations can provide the guarantee of "acceptable for use for a specific purpose." Typically, the customer (or potential customer) or a separate legal representative has the authority to select the individual or organization that can accredit the entity. The accrediting authority is *never* the entity developer, anyone from the developers' organization, or anyone else who might have a vested interest in the performance, accuracy, or sale of the entity. Considering high-consequence public safety risks and environmental impact, there is an explicit need for the formality of accreditation of entities. The fundamental difference between accreditation and certification is the level of authority, independence, and responsibility to guarantee the performance or accuracy of the entity. In addition, when compared to certification, the accreditation of an entity is generally more formal, involves more in-depth entity testing, and requires much more extensive documentation. Note that commercial software companies *never* make any statement of certification or accreditation. In fact, the *Conditions of Use* statement that one must agree to specifically states, "No warranty is expressed or implied with this product." However, it is doubtful that this would absolve the software company of legal liability if a high-consequence system failed due to a software bug.

Certification and accreditation of an entity can also be viewed as increasing levels of independence of assessment in VVUQ activities. A number of practitioners over a wide range of fields of simulation have pointed out the importance and value of independent VVUQ (see, e.g., Lewis, 1992; Arthur and Nance, 1996). The levels of independence in the VVUQ assessment of simulation entities can be viewed as a continuum (Figure 2.11). The least independent evaluation, i.e., no independence, occurs when the entity developer conducts assessment activities. Essentially all research activities published in archival journals are conducted at this first level of assessment. The proper perspective for critical appraisal only occurs with some level of independence between the entity developer and the assessor. For example, it is common that an individual's professional esteem or reputation, or the public image or future business opportunities of the sponsoring organization, are intertwined with the entity.

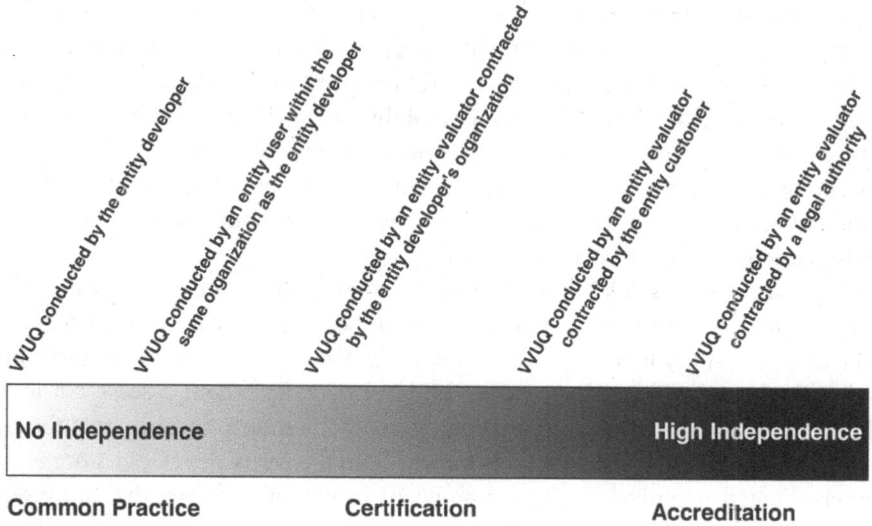

Figure 2.11 Spectrum of assessment independence of VVUQ levels applied to a model, a code, or a simulation.

Higher levels of independence should be required for any production or commercial activity or any simulation results that can have a significant organizational, safety, or security impact.

At the second level, the VVUQ evaluation is conducted by a user of the entity who is in the same or a closely related organization as the developer of the entity. Thus, a user can also be an entity evaluator, but the user *cannot* be one of the developers. This level of independence in evaluation is a major step not commonly appreciated by research organizations that develop scientific computing entities. The entity evaluator at this second level can have various degrees of independence from the entity developer. If the entity evaluator is in the same group or team, e.g., the lowest-level organizational unit, then the independence of the entity evaluator is marginal. If the entity evaluator is in a group separated laterally by two or three lines of management from the entity developer's management, then the evaluation has a much-improved independence. For example, the entity evaluator could be a potential user of the entity in a design group that conducts simulations for product design or manufacturing processes. The suggested minimum level of evaluation independence that should be considered for certification is separation of the entity developer and the entity evaluator by two to three lines of management. Some organizations refer to this type of internal evaluation as a *non-advocate evaluator.*

At the third level, the VVUQ evaluation is conducted by an entity evaluator who is contracted by the entity developer's organization. This level of evaluation typically provides considerable independence for the entity evaluator because an external contractor is commonly hired for the task. At this level and at the higher levels discussed later, information concerning the credentials of the contractor should be obtained to ensure that the contractor is objective in the evaluation and has the proper expertise for the evaluation. Occasionally, a monetary bonus is paid to the contractor if the contractor demonstrates exceptional thoroughness and energy in evaluating the entity. For example, the contractor may be paid a bonus for each coding error, data input error, or failure to meet a specification claimed by the entity. If national security classification or extraordinary proprietary issues are a concern, the evaluator could be employed from a subsidiary or sister organization. For this case, the evaluator would not have any organizational connection to the entity developer's organization or to the anticipated users of the entity. This level of independent VVUQ provides fresh perspectives on the entity's performance, robustness, applicability, and reliability. In addition, this level of independent VVUQ commonly provides helpful and constructive ideas for improvements in the entity's performance or documentation. This level of independence could be viewed as strong certification, but not accreditation because the entity developer's organization is still in control of all of the information obtained in the assessment.

At the fourth level, the VVUQ evaluation is conducted by an entity evaluator who is contracted by the customer or potential customer of the entity. The term *customer* in this text means a user of the entity that is an independent organization from the developer's organization. This shift is a significant increase in the level of independent assessment and would normally be considered part of accreditation. Here, the authority to guarantee the performance, accuracy, or quality of the entity has moved from the entity developer's organization to a customer-oriented organization. This amount of insulation between the entity developer and the entity evaluator is appropriate for certain situations, such as systems where major liability is at stake, but can also cause technical and practical problems. These problems are discussed as part of the next level of independence. The interpretation of accreditation commonly assumes that the assessment authority moves from the developer to the customer of the entity. If the developer and customer of the entity are essentially the same, then our assumption of independence does not apply. For example, in many DoD simulation activities the developer and the customer are essentially the same, or very closely related. As a result, this arrangement would not adhere to the interpretation of accreditation independence.

At the fifth level of independence, the VVUQ evaluation is conducted by an entity evaluator who is contracted by an *independent legal authority* or *governmental organization*. The evaluation authority has now moved not only further from the entity developer, but also moved from the entity customer, i.e., the user. The amount of insulation between the entity developer and the entity evaluator at the fifth level can be quite beneficial to the independent legal authority or governmental organization responsible for performance and safety assessment of high-consequence systems. However, this insulation can have a detrimental effect on the quality of the entity produced, e.g., it can weaken the scientific quality of the entity as a by-product. For example, any changes to the entity, even those intended to improve accuracy, efficiency, or robustness, cannot be made unless the entity is re-accredited. As a result, modifying an entity becomes a very time-consuming and expensive process. The degree to which a slow, bureaucratic, and expensive re-accreditation procedure can obstruct the advancement and innovation of simulations is illustrated by the history of the safety assessment of nuclear power reactors in the United States. Currently, it is not clear how to achieve a better balance between the need for improving the quality of entities and the need for adequate assurance of public and environmental safety.

2.3 Error in a Quantity

There are certain situations in scientific computing and in experimental measurements where the concept of *error* proves to be quite useful. This text uses the common dictionary definition of error.

Error in a quantity: deviation from the true value of the quantity.

This definition is also used in a number of modern metrology texts (Rabinovich, 2005; Grabe, 2014). To be more specific, let y_T be the true value of the quantity y, and let $y_{obtained}$ be the obtained value of the quantity y. *Obtained value* means that it can be derived from any source, e.g., numerical solution, computational simulation, experimental measurement, or expert opinion. For this discussion, it is assumed that y_T and $y_{obtained}$ are fixed numbers, as opposed to random quantities. Then the error in $y_{obtained}$ is defined as

$$\varepsilon_{obtained} = y_{obtained} - y_T. \tag{2.2}$$

Many texts and technical articles use the terms *error* and *uncertainty* interchangeably. However, this is not only epistemologically incorrect, it also produces a great deal of confusion and misinterpretation in practical settings. Interchangeable use of the terms error and uncertainty can lead to a misrepresentation of results, causing misguided efforts directed at reduction or elimination of the source of the error or uncertainty.

The concept of error does not address the nature of the source of the error, but solely concentrates on the identification of the true value. The true value can be defined in a number of different ways. For example, the true value of a physical constant, such as the gravitational constant or the speed of light in a vacuum, can be defined in multiple ways depending on the accuracy needed for the situation. The true value can also be defined as a reference standard. For example, the reference standards for length, mass, and time are set by the International System of Units. In scientific computing, it is sometimes convenient to define a true value as a floating-point number with single or double precision in a computer, e.g., an analytical solution to a PDE that is *not* given by an infinite series. In most situations, however, the true value in not known and in experimental measurements of engineering and scientific quantities the true value is *never* known.

As a result, the usefulness of the concept of error in practical applications depends on the definition and accuracy of the true value. If the true value is given by an appropriate definition, or the accuracy of the true value is known to be much higher than $y_{obtained}$, then the concept of error is quite useful, both conceptually and practically. For example, consider the case in code verification where an analytical solution to the differential equations can be computed with high accuracy, e.g., known to the double precision accuracy of the computer. The error could be defined as the difference between the particular solution obtained by a code and the highly accurate analytical solution. However, if the accuracy of the analytical solution is not computed very accurately, e.g., using an insufficient number of terms in an infinite series expansion, then the concept and terminology associated with uncertainty are more useful in practice.

For example, consider the case of two computer codes solving the same mathematical models, but using different numerical solution procedures. Suppose that one of the codes has been traditionally accepted as producing the correct result and the other code is relatively new. The error could be defined as the difference between the new code result and the traditional code result. Unless the traditional code result has been thoroughly investigated and the accuracy carefully documented, it would be foolish to assume that it produces a solution more accurate than the new code.

2.4 Types and Sources of Uncertainties

The history of science has tended to strongly emphasize what we know, or think we know, instead of what is uncertain. Our perspective of computational simulation is to identify and properly characterize the different types of uncertainties that can occur in a model and in the simulation results. A large number of researchers and practitioners in risk assessment (Cullen and Frey, 1999; Haimes, 2015; Hubbard, 2020), system reliability (Melchers, 1999; Ayyub and Klir, 2006; Bernardini and Tonon, 2010), and information theory (Klir et al., 1997; Cox, 1999; Augustin et al., 2014) have dealt with defining and categorizing types of uncertainty. Consistent with these references, a general definition of uncertainty is given as:

Uncertainty: imperfectly known information concerning an object, process or issue.

Many of the categorizations that have been constructed tend to mix the nature or essence of a type of uncertainty with *how* it might occur in computational simulations. For example, some taxonomies would have randomness as one type and model form uncertainty as another type. However, this categorization does not address the root cause. A sound taxonomy would *only* categorize uncertainty types according to their fundamental essence, as opposed to how they occur or how they are mathematically represented. For types of uncertainty that can be identified and characterized in some way, the computational simulation that incorporates these uncertainties will result in nondeterministic outcomes. *Nondeterministic outcomes* mean those outcomes that are possible realizations of the computational simulation. Nondeterministic simulation is sometimes referred to as *stochastic simulation*. Although these outcomes are more difficult to interpret than deterministic outcomes, the goal of nondeterministic simulation is to improve the understanding of the system of interest and the environments to which it might be exposed. With this improved description and understanding, the design, safety, performance, and decision making related to the system can be improved.

The information theorists and the nuclear reactor safety community have developed the most workable and effective categorization of uncertainties: aleatory and epistemic uncertainties. The benefits of distinguishing between aleatory and epistemic uncertainty include improved interpretation of simulation results by analysts and decision makers, as well as improved strategies on how to decrease system response uncertainty when both are present. As will be discussed, the fundamental nature of each is different. Hence, different approaches are required to characterize and reduce each type of uncertainty.

2.4.1 Aleatory Uncertainty

Consistent with the references just given, *aleatory uncertainty* is defined as:

Aleatory uncertainty: uncertainty due to inherent randomness.

Aleatory uncertainty is also referred to as stochastic uncertainty, variability, inherent uncertainty, uncertainty due to chance, and Type A uncertainty. The fundamental nature of aleatory uncertainty is randomness, e.g., resulting from a stochastic process. In principle, randomness can be reduced, e.g., by improved control of a random process, but it can also be removed by assumption or definition. Aleatory uncertainty can exist due to inter-individual differences, such as random heterogeneity in a population, and it can exist spatially or temporally. Sources of aleatory uncertainty can commonly be singled out from other contributors to uncertainty by their representation as randomly distributed quantities that may take on values in a known range, but for which the exact value will vary by chance from unit to unit, point to point in space, or time to time. The mathematical representation or characterization most commonly used for aleatory uncertainty is a probability distribution.

Aleatory uncertainty can be embodied in two ways in computational simulation: in the model formulation itself and in parameters of the model. If the mathematical model is given by a differential operator, then aleatory uncertainty in the model formulation can be expressed as a stochastic differential operator. Although there have been some applications of stochastic differential operators to engineering systems, this type of modeling is in its very early stages (Oksendal, 2013; Han and Kloeden, 2017; Sarkka and Solin, 2019). However, stochastic differential equations are used in fields such as financial investing, game theory, and socioeconomic modeling. Aleatory uncertainty in parameters is a much more common situation in computational simulation. Aleatory uncertainty in parameters can occur in the mathematical description of system characteristics, initial conditions, boundary conditions, or excitation of the system. Typically, aleatory uncertainty occurs in a scalar quantity appearing in the differential equation, but it can also appear as a field quantity, such as a stochastic variable over the domain of the differential equation. Some examples of scalar parameters having random variability are: variability in geometric dimensions of manufactured parts; variability in material properties; and spatial and temporal variability of the atmospheric temperature on a given day.

Example: Thermal Conductivity

Consider a simple example of a scalar variability in a heat conduction analysis. Suppose the analyst was interested in heat conduction through a homogenous, isotropic, material whose thermal conductivity varied from unit to unit due to a manufacturing process. Assume that a large number of samples have been randomly drawn from the material population produced by the manufacturing process and the thermal conductivity has been measured on each of these samples. Assume that the number of samples is so large that a continuous random variable can be constructed, as shown in Figure 2.12. The probability density function (PDF) is shown in Figure 2.12a and the cumulative distribution function (CDF) is shown in Figure 2.12b representing the variability of the thermal conductivity. The PDF shows the local variation of probability density over the range of possible values of the quantity of interest. It is *not* the probability of a given outcome. The CFD is simply the integral of the PDF from negative infinity up to a give value of the quantity of interest. Since the CDF is the integral of a PDF up to a given value, the CDF is the probability of obtaining the given value, or less, for a large number of random draws from the sample population. The PDF and the CDF both represent the variability of the thermal conductivity of the population, but each show it in a different way. The variability of the population could also be shown as a histogram, which is analogous to a PDF, but this is a more intrusive and less precise method to represent variability. The PDF (Figure 2.12a) shows the probability density of any chosen value of conductivity x. Thus, it shows the *probability per unit variation* in conductivity for any value x. The CDF (Figure 2.12b) shows the fraction of the population that would have a conductivity *less than or equal to* the particular value of conductivity chosen x. For example, the probability is 0.37 that all possible thermal conductivity values will be 0.7 or lower for a large number of samples chosen.

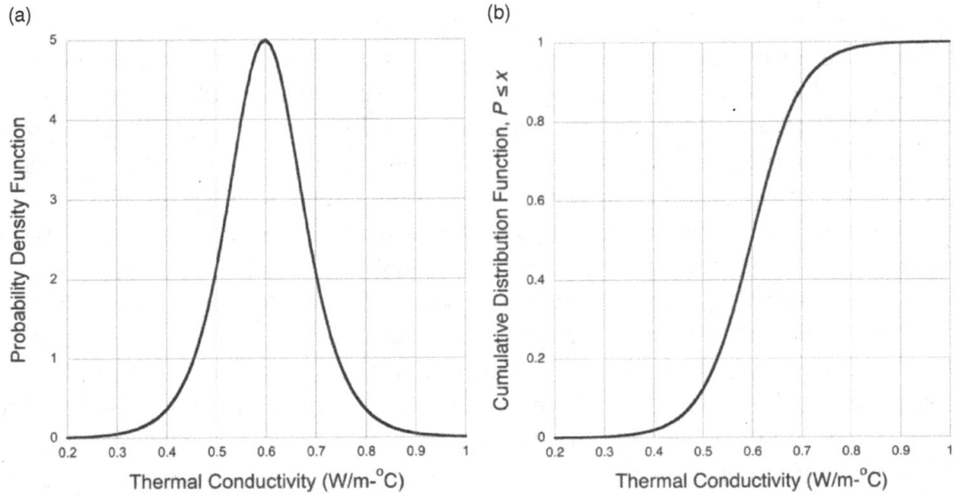

Figure 2.12 Examples of PDF and CDF for variability of thermal conductivity. (a) Probability density function. (b) Cumulative distribution function

2.4.2 Epistemic Uncertainty

Consistent with the references cited above, *epistemic uncertainty* is defined as:

Epistemic uncertainty: uncertainty due to lack of knowledge.

Epistemic uncertainty is also referred to as reducible uncertainty, knowledge uncertainty, and subjective uncertainty. In the risk assessment community, it is common to refer to epistemic uncertainty simply as *uncertainty* and aleatory uncertainty as *variability.* The fundamental source of epistemic uncertainty is incomplete information or incomplete knowledge of any type that is related to a quantity or process of interest. Incomplete information or knowledge includes the situation of inconsistent or contradictory information or knowledge. Epistemic uncertainty is a *characteristic of the modeler* or observer, whereas aleatory uncertainty is a characteristic of the quantity of interest. The lack of knowledge can be related to modeling issues for the system, model input parameters, computational issues in the solution of the model, or experimental data needed for a simulation. Modeling issues include lack of knowledge of characteristics or processes in the system, the initial state of the system, the surroundings of the system, or the environment to which the system is exposed. If very little information is known about model input parameters, they will have large epistemic uncertainty. Computational issues include programming mistakes, estimation of numerical solution errors, and numerical approximations in algorithms. Experimental data issues include incomplete knowledge of experimental information that is needed for simulation of the experiment and unexplained changes or shifts in the measurement of a given quantity. An increase in knowledge or information, which always requires time and/or resources, can lead to a reduction in epistemic uncertainty.

From a complementary perspective, it is seen that the fundamental characteristic of epistemic uncertainty is ignorance; specifically, ignorance by an individual, or group of individuals. Smithson (1989) points out that ignorance is a social construction, analogous to the creation of knowledge. Ignorance can only be discussed by referring to the viewpoint of one individual (or group) with respect to another. Smithson gives the following working definition of ignorance:

> A is *ignorant* from B's viewpoint if A fails to agree with or show awareness of ideas which B defines as actually or potentially valid.

This definition avoids the absolutist problem by placing the onus on either A or B to define what he/she means by ignorance. It also permits self-attributed ignorance, since A and B can be the same person.

Ayyub (2001), following Smithson, divides ignorance into two types: conscious ignorance and blind ignorance. Conscious ignorance is defined as a self-ignorance recognized through reflection. Conscious ignorance would include any assumptions or approximations made in modeling, the use of expert opinion, recognition that programming errors could exist in software, numerical solution errors, and bias errors in experiments. For conscious ignorance, the term *recognized uncertainty* is used to mean any epistemic uncertainty that has been recognized that it could exist in some way. Blind ignorance is defined as ignorance of self-ignorance; more commonly referred to as *unknown unknowns.* For blind ignorance, the term *blind uncertainty* is used to mean any epistemic uncertainty that has *not* been recognized in some way. Figure 2.13 shows the categorization of uncertainty just described and that is used in this book. Recognized uncertainty and blind uncertainty are discussed in detail, along with where and how they occur in computational simulation.

2.4.2.1 Recognized Uncertainty

This text defines *recognized uncertainty* more formally as:

Recognized uncertainty: an epistemic uncertainty for which a conscious decision has been made to either characterize or deal with it in some way, or to ignore it for practical reasons.

For example, when making decisions concerning the modeling of a system, the assumptions and approximation concerning what physics will be included in the model and what will be ignored must be made. Whether a certain type of physical phenomenon is included or

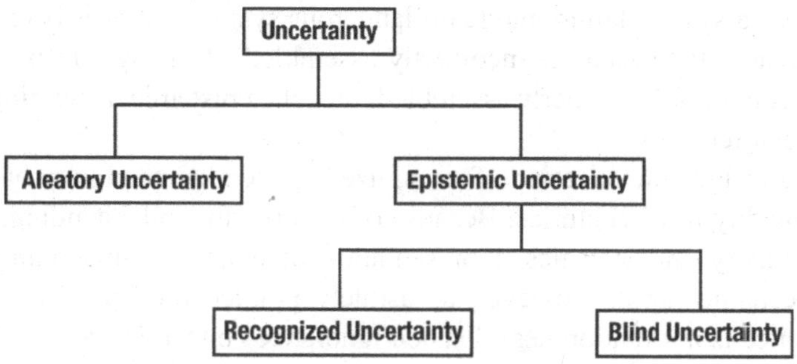

Figure 2.13 Classification of uncertainties.

ignored, or a specific type of conceptual model is chosen, these are recognized uncertainties. The impact of these assumptions and approximations on the simulation results are referred to as model form uncertainties, i.e., uncertainties due to the assumptions made in the modeling of a system. Depending on the complexity of the system being modeled, a modeler could conceptually change the assumptions or approximations and possibly estimate the magnitude of the effect on system response quantities of interest. Regardless of what level of modeling fidelity is chosen, there are *always* spatial and temporal scales of physics, as well as coupled physics, that are ignored. A balance must be decided between what physics should be included, and what should be ignored, in the modeling of a system in order to meet the goals of the simulation. Some modelers take the view that "the more physics the better." This is generally a mindless approach because the higher the model fidelity, the more detailed input data is needed for the model and the higher the computational costs.

Obtaining opinions from experts when experimental data are not available is another example of a recognized uncertainty. For example, suppose an expert is asked to provide an opinion on his/her belief of a scalar input parameter in the system that is a fixed quantity, but the value of the quantity is not known. The expert may provide an opinion in the form of a single number, but more likely the opinion would be given as an interval in which the true value is believed to be. Similarly, suppose an expert is asked to provide an opinion on an input parameter that is characterized by a random variable. The expert would probably provide a named probability distribution family for the characterization, along with estimated fixed values for the parameters of the family. Alternately, the expert could also provide interval values for the parameters of the family. In either case, the scalar parameter in the system would be a mixture of aleatory and epistemic uncertainty, because it represents expert opinion combined with the characterization of a random variable.

Since the root cause of a recognized uncertainty is incomplete knowledge, increasing the knowledge base can reduce the epistemic uncertainty. Epistemic uncertainty can be reduced by an action that generates relevant information such as allowing for a stronger level of physics coupling in a model, accounting for a newly recognized failure mode of a system, changing a calculation from single precision to double precision arithmetic, and performing an experiment to obtain knowledge of system parameters or boundary conditions imposed on the system. Epistemic uncertainty can also be reduced by eliminating the possibility of the existence of certain states, conditions, or values of a quantity. Reducing the collection (or sample space) of possible events reduces the magnitude of uncertainty due to ignorance. For example, suppose a system failure mode or dangerous system state has been identified such that it could occur if the system is incorrectly assembled. If the system is redesigned such that the system cannot be improperly assembled, then the epistemic uncertainty in the system response has been reduced.

The mathematical characterization of recognized epistemic uncertainty has been an issue of debate for roughly three centuries. Because of the growth, understanding, and popularity of probability theory and statistics, representation of epistemic uncertainty as a random variable has been the norm. However, as astutely pointed out by Ferson and Ginzburg (1996); "Ignorance is not randomness." When ignorance concerning a quantity or a simulation result is represented as a random variable, the representation implies more knowledge than is actually known.

Example: Poorly Known Quantity

As a simple example of why traditional probability theory can give an inappropriate representation of a purely epistemic uncertainty, consider the following. Suppose that the only information known about a quantity is that the true value is known to exist within an interval. Suppose the interval is given as [3.3, 6.1]. The least informative mathematical characterization that can be (easily) represented in traditional probability theory is to define a uniform PDF over the range of the interval. The interpretation of this characterization is that all possible values within the interval are equally likely. It is clear that this statement of knowledge contains *much more knowledge* than simply stating that the true value occurs within the interval. In other words, the range of possible values given by traditional probability theory is correct, but it has improperly added the likelihood of outcomes. The original interval information contained *no information* about any likelihood of any value within the interval.

During the last four decades, great strides have been made toward improved characterization of epistemic uncertainty. The field that deals with more general characterizations of both aleatory and epistemic uncertainty is referred to as imprecise probability theory. Although it is beyond the scope of this book, several references are given in Section 2.6 for more detailed information on imprecise probability theory.

2.4.2.2 Blind Uncertainty

This book's formal definition of *blind uncertainty* is:

Blind uncertainty: an epistemic uncertainty for which it is not recognized that the knowledge is incomplete and that the knowledge is relevant to modeling the system of interest.

Adding knowledge can reduce blind uncertainty, just as with recognized uncertainty. However, the approach and the procedures are quite different because one is attempting to identify unknown unknowns. The most common causes of blind uncertainty are human errors, blunders, mistakes in judgment, preconceived ideas, and the inability to think outside the box. Some examples are: programming errors made in software, mistakes made in the preparation of input data or post-processing of output data, blunders made in recording or processing experimental data, and not recognizing how a system could be misused or damaged so that the system could be very dangerous to operate. Blind uncertainty can also be caused by inadequate communication between individuals contributing to M&S activities, for example: (a) between those providing expert opinion and those interpreting and characterizing the information for input to the simulation, (b) between computational analysts and experimentalists working together on model validation, and (c) a simulation customer inadequately describing how the simulation results will be used. In experimental activities, some additional examples of blind uncertainty are unrecognized bias errors in diagnostic techniques or experimental facilities and improper procedures in using a reference standard in the calibration of experimental equipment.

There are *no* reliable methods for estimating or bounding the magnitude of blind uncertainties, their impact on a model, its simulation, or on the SRQs produced. As a result, the primary approach for dealing with blind uncertainties is to try to identify them through techniques such as: (a) independent procedures and protocols for operations or analyses, (b) various software and hardware testing procedures, (c) use of different facilities to conduct experiments, (d) use of diverse range of experts for obtaining input data, and (e) use of independent teams to identify unusual or surprising failure modes of a system. If independent analyses or teams are not practical, a helpful approach is to carefully list all of the assumptions, approximations, and procedures that are used. An open-minded examination of this list can help identify potential weaknesses and blind uncertainties. Once blind uncertainties are identified or a hint of their existence is recognized, they can be pursued or dealt with in some way or the impact of their effect could possibly be estimated or removed. As discussed earlier in code verification, testing of numerical algorithms and SQA practices has proven effective in finding algorithm deficiencies and code bugs. Methods have been developed to estimate the frequency of coding errors, e.g., average number of static or dynamic faults per hundred lines of code. However, these measures do not address the possible *impact* of undetected coding errors. Human mistakes made in input preparation for simulations and mistakes in processing of output data are most commonly detected by having separate individuals check or prepare the data. For centuries, experimental science has been built on the crucial importance of independent reproducibility of experimental results and measurements. Scientific computing has a great deal to learn from this venerable tradition.

To stress the personal or social aspect of blind uncertainty, Ayyub (2001) gives several thought-provoking examples of root causes of blind uncertainty: knowledge that is dismissed as irrelevant (yet it is relevant); knowledge or experience that is intentionally ignored (yet it should not be ignored); relevant knowledge that could be pursued, but is not because of time, schedule, or indolence; and knowledge or questioning that is avoided or shunned because it is socially, culturally, or politically considered taboo. This personal aspect of blind uncertainty and some of those mentioned earlier can be countered, to some extent, by independent and/or external peer reviews of a simulation effort. The effectiveness of an external review depends on the independence, creativeness, expertise, and allowed authority of the external reviewers. If external reviews are encouraged to focus on finding weaknesses, errors, or deficiencies in a system, assumptions, or a simulation, they are commonly referred to as *Red Team* reviews. Sometimes Red Teams have such high enthusiasm and zeal, it is not known if they are friends or enemies.

2.5 Integration of Verification, Validation, and Prediction

As discussed earlier in this chapter, VVUQ activities contribute evidence to the credibility and estimated uncertainty in computational simulation results. The philosophy of VVUQ is one of fundamental skepticism. This is a philosophy where the degree of credibility in simulation results is *entirely* built on the evidence and detail documented VVUQ activities. This philosophy is shared by government regulators that use simulation results in high-consequence decision-making, e.g., the US Nuclear Regulatory Commission, the US Environmental Protection Agency, and the US Food and Drug Administration. For high-consequence decision-making heavily based on simulation, the focus is on documented evidence of VVUQ activities, in addition to the experience and expertise of the computational analysts.

Figure 2.14 depicts an integrated view of the nine elements of M&S, with the emphasis on the activities of VVUQ that contribute credibility to simulation results. Each of these elements is discussed in detail in various chapters of this book; however, a brief description

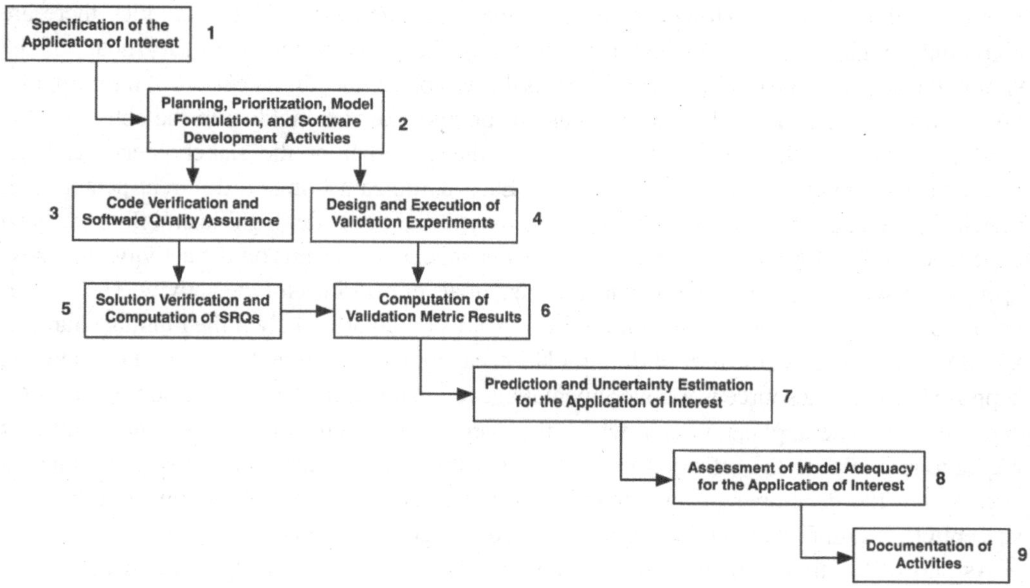

Figure 2.14 Integrated view of the elements of M&S, with emphasis on VVUQ. (adapted from Trucano et al., 2002)

of each element will be given here. Many of the concepts described in each element are based on Trucano et al. (2002). The level of formality and documentation associated with each element should depend on: (a) the level of consequence if the system of interest does not achieve its intended performance, safety, and reliability, and (b) the level of responsibility that simulation assumes with regard to decision making in the design and predicted performance, safety, and reliability of the system of interest.

Note that although the activities in Figure 2.14 are shown sequentially, the process is commonly an iterative one in practice. For example, when the computational results are compared with the experimental measurements in Element 6, computational results may be found to not be as accurate as expected. Several options for iterative adjustment of previously completed elements may be present: (a) alteration of the application domain specified in Element 1; (b) alteration of the modeling assumptions made in Element 2; (c) reprioritization of certain VVUQ activities in Element 2 so as to better address the cause of the problem; (d) performance of additional code verification activities in Element 3 if the software has not been thoroughly tested; (e) conducting additional experimental measurements in Element 4 if measurements uncertainty is large; and (f) computing solutions on finer mesh resolutions in Element 5 if the solution error is poorly estimated.

2.5.1 Specification of the Application of Interest

The first element of the M&S process describes what physical process or engineering system is of interest. The specific purpose for which the M&S process is being undertaken should be defined in some detail. Commonly, the customer for the computational results specifies how the results will be used, such as system design or optimization, or investigating the safety of a system in some failure scenario. However, in both small-scale and large M&S activities there are commonly inadequate communications concerning the primary and secondary goals of the simulation activity. The key participants in this discussion are the ultimate users of the computational information generated (referred to as customers), the stakeholders in the activities, the funding source for the work (which could be the customer or the stakeholders), and the computational analysts conducting the work. Each participant brings a different perspective, priorities, and agenda to the discussion. It is common that only after significant effort has been expended on the M&S activity, or there are cost or schedule overruns on the activity, that these groups discover that each has somewhat different goals or priorities for the activity. These types of miscommunication, or lack of communication, are particularly likely if the funding source of the M&S activity is *not* the user of the simulation capability, but some third party. For example, suppose the funding source is a governmental agency that is interested in developing an M&S capability in some application area where the agency has control of governmental statues or regulations. Suppose the intended user of the capability is a contractor to the governmental agency, but the developer of the capability is a different contractor. This triad is especially susceptible to failure due to miscommunication of the goals of the activity.

As part of the first element, a description should be given for the application domain or conditions of use for the model. This would include, for example, specification of the environments that the system could be exposed to and that which the M&S is intended to address. System environments are typically categorized as normal operating environments, abnormal (or accident) environments, and hostile environments, i.e., the system is under some type of attack. The system environment would include a general specification of all of the initial conditions, boundary conditions, and excitation conditions to which the system might be exposed. If different environments are of interest, then it is appropriate to construct completely separate diagrams such as Figure 2.14; one for each environment.

An important part of the first element is the specification of all of the SRQs that are needed from the simulation. Some examples of SRQs that may be of interest are: (a) temperature distribution inside or on the surface of a component, (b) maximum stress level within a component or group of components, (c) maximum acceleration as a function of time at any point in or on a structural system, and (d) concentration level of a contaminant or toxic waste along a specified outflow boundary. The specification of the predictive accuracy requirements that are needed from the M&S effort, as well as which SRQs are most important, are closely associated with specification of the SRQs of interest. It is the customer and/or the funding source of the simulation effort that should define the predictive accuracy requirements. However, sometimes the customer either (a) fails to do this or (b) is overly demanding of the accuracy requirements. This situation should result in a constructive dialogue and negotiation between the customer and the analysts concerning the cost and schedule required for certain levels of predictive accuracy. Estimates for costs and schedule, as well as estimated predictive accuracy, are often very poorly known early in an effort. Thus, it is critical to have these discussions early in the planning so that all parties, including stakeholders and experimentalists that may need to provide experimental validation data, have some feel for the trade-offs involved. Too often, understanding of trade-offs between cost, schedule, and achieved predictive accuracy occur very late in the effort after significant resources, time, and modeling efforts have been expended or wasted.

The activities in this element are discussed in detail in Chapters 3, 10, and 12–13.

2.5.2 Planning, Prioritization, Model Formulation, and Software Development Activities

Planning, prioritization, model formulation, and software development listed in Element 2 address a wide range of the activities which impact the remaining seven elements shown in Figure 2.14. On a large M&S project, this requires significant resources, time, and effort from a wide range of individuals and, commonly, both internal and external organizations. The M&S-VVUQ Plan should be appropriate to the magnitude of the project and to the consequences of the decisions made based on the simulation results. Preparation of the VVUQ portion of the Plan is discussed in the *ASME Standard* (ASME, 2019).

The focus of the activities listed in Element 2 should always be:

> *Given the resources available (people, time, money, computational facilities, experimental facilities, etc.), what is the appropriate level of effort for each listed activity that is needed to achieve the goals of the M&S effort identified in Element 1?*

Some examples of the types of questions addressed in Element 2 are the following:

- What physical phenomena are important and what level of coupling of the various phenomena are appropriate for the goals of the simulation?
- Are the capabilities needed for the simulation available internally to the organization or does software need to be developed or licenses purchased?
- Given the specified application domain of the model, what is an appropriate validation domain; one that is sufficiently relevant and affordable?
- What is the estimated predictive accuracy of the SRQs of interest specified by the customer?
- What code verification and SQA activities are appropriate for the application of interest?
- Are existing numerical error estimation techniques adequate?
- Are new mesh generation capabilities needed for the simulation?
- Are new experimental diagnostics or facilities needed for model validation activities?
- Do new model validation metric operators need to be developed?
- If model accuracy or experimental measurements are found to be lacking, what alternatives or contingency plans should be considered?

The most commonly used approach for planning and prioritization is the Phenomena Identification and Ranking Table (PIRT). PIRT was originally developed to identify physical phenomena, and the coupling of physical phenomena, that could affect nuclear power plant safety in accident environments. PIRT should be viewed as a process as well as a collection of information. As stressed by Boyack et al. (1990), the PIRT is most certainly *not* set in stone once it is formulated and documented. While a given formulation of a PIRT guides M&S activities, it must also adapt to reflect the information gathered during the conduct of those activities.

An additional planning and prioritization process is a simulation capability gap analysis. The gap analysis begins with the results of the PIRT process and attempts to answer the question:

> *Where does the M&S effort presently stand relative to the phenomena and SRQs that have been identified as important, as well as the expected accuracy requirements of the customer?*

In the gap analysis portion of the process, the emphasis shifts from improving the understanding of the environments, scenarios, system, and physical phenomena, to understanding the possible gap between the *present* simulation capabilities available and the *required* capabilities to achieve the goals of the simulation. Answers to the gap analysis can directly aid in the planning and prioritization regarding modeling, computer codes, verification, validation, and uncertainty quantification.

M&S activities are discussed in detail in Chapters 3 and 13. Planning and prioritization of M&S, specifically PIRT & gap analyses, are discussed in detail in Chapter 14.

2.5.3 Code Verification and Software Quality Assurance

Code verification and software quality assurance (SQA) activities are conducted in the third element. Both of these activities can be viewed as the accumulation of evidence to support the belief that: (a) the numerical algorithms are functioning as intended; (b) the source code is implemented correctly and it is functioning as intended; and (c) the computer system hardware and software environment is functioning as intended. It is well known, although rarely stated, that these three essentials (algorithms, source code, and system hardware and software) *cannot* be proven to be correct and functioning as intended. Most computational analysts assume that credibility exists in results that are produced by software that has been in production use for a number of years. For example, it is commonly argued that legacy software or commercial software is reliable because "It has been used by hundreds or thousands of users over decades." As shown in a landmark study of production software by Hatton (1997), this claim is *not* justified.

Figure 2.14 does call attention to the critical foundation that code verification and SQA activities play in assessing the credibility of a simulation result. For example, if a code bug is found later on in the M&S process, say in Element 7, *all of the effort* devoted to Elements 5 through 7 must be reevaluated to determine if the bug affected the previous work. If the bug did have an effect on the earlier results, then much of the work is wasted and most has to be redone. The far more dangerous situation is if a relevant code bug was *not found* and trust was placed in the simulation. For example, if parameter calibration is able to yield good agreement between computational and experimental results in Element 6, little interest, energy, and resources are found to conduct code verification and SQA activities. By using misleading simulation results, decision makers may then have unknowingly made erroneous decisions on the safety, performance, or reliability of a system.

Individual or organizational tolerance levels for potential errors in software primarily depend on two very different factors. First, how sensitive or averse is the individual to the severity and frequency of errors and unreliability in the software? Second, what are the individual's options for using other software packages to accomplish their job? For example, if a user has a low tolerance for software bugs or computational crashes, *and* they have the option to change to other similarly capable software providers, they will change software providers. On the other hand, a user: (a) may tolerate buggy, unreliable, software if they have no options, e.g., if there is a near monopoly in the software market, or (b) may be forced to use certain software packages because of corporate or organizational mandates (Platt, 2007).

Given the perspective of the balance between individual tolerance of errors and unreliability on one hand, and the limited availability of software options on the other, our observation is that computer users show a high tolerance for errors and lack of robustness, as long as the modeling features and capabilities they need for their simulations are (apparently) met. Accordingly, computational analysts place little value on the accumulation of evidence for code verification and SQA, as opposed to the value they place on the software having the features and capabilities they need to do their job. This commonly held value-system is at the root of why Element 3 primarily receives weak testing when it competes for resources in code development organizations. Code development groups, whether they are in-house groups or commercial software companies, understand the value system of the computational analyst and they respond accordingly. This is unfortunate from a simulation credibility perspective, but not unexpected.

Code verification and SQA are discussed in detail in Chapters 4–6.

2.5.4 Design and Execution of Validation Experiments

Element 4 deals with the design and execution of validation experiments, as well as the more common situation of using existing experimental data in validation activities. Before briefly discussing the design and execution of validation experiments, a few comments should be made concerning how a validation experiment is different from traditional types of experiments. Traditional experiments can generally be grouped into three broad categories (Aeschliman and Oberkampf, 1998; Oberkampf and Blottner, 1998; Oberkampf and Trucano, 2002). The first category comprises experiments that are conducted primarily for the purpose of improving the fundamental understanding of some physical process. Sometimes these are referred to as physical-discovery or phenomena-discovery experiments. Examples are experiments that investigate (a) turbulent reacting flow, (b) decomposition of materials in high temperature environments, (c) micromechanics processes underlying crack growth in solids, and (d) multiphase flow.

The second category of traditional experiments consists of those conducted primarily for constructing or improving mathematical models of fairly well understood physical processes. Sometimes these are called model development or model calibration experiments. Examples are experiments to (a) measure the chemical reaction-rate parameters in a model for reacting flows, (b) determine the joint-attachment damping and the aerodynamic damping parameters in the vibration of a built-up structure, (c) determine the parameters in a model for crack propagation in a certain class of composite materials, and (d) calibrate the constitutive parameters in a material model for reinforced concrete.

The third category of traditional experiments includes those that determine the reliability, performance, or safety of components, subsystems, or complete systems. Sometimes these are called reliability tests, performance tests, safety tests, certification tests, or qualification tests. Examples are (a) tests of a new compressor or combustor design in a gas turbine engine, (b) tests of a new propellant formulation for a solid rocket motor, (c) tests of the crash worthiness of a new automobile design, and (d) qualification tests of a new submarine design submerging to maximum operational depth.

In contrast, a validation experiment is conducted for the primary purpose of assessing the accuracy of a mathematical model. In other words, a validation experiment is designed, executed, and analyzed for the purpose of quantitatively determining the ability of a mathematical model to simulate a well-characterized physical process. Thus, in a validation experiment, *the computational analyst is the customer* or *the code is the customer* for the experiment as opposed to, for example, a physical phenomenon researcher, a model builder, or a system project manager. M&S has matured over the last few decades to the point where it could even be considered as a viable customer for an experiment. As simulation capability continues to expand, there will be increased competitive pressure for engineering systems to be certified, and possibly even fielded, primarily based on M&S. As a result, M&S itself will increasingly become more influential and have the resources to become a customer of experimental activities, specifically, validation experiments.

Since a validation experiment, as defined here, is a relatively new concept, most experimental data generated in Element 4 will be from different types of traditional experiments. Therefore, use of experimental data from traditional experiments when used for validation assessment must deal with a number of technical and practical difficulties. These are discussed in Chapters 10 and 11. A new validation experiment can be used to assess the predictive capability of the mathematical model if certain procedures are used in the design and execution of the experiment. Specifically, the measured SRQs should *not* be made available to the computational analysts for new validation experiments. If possible, the goal should be for a *prediction* as opposed to a *retrodiction* in the use of the model.

Validation experiments can be designed and executed without a specific application of interest in mind. However, the focus here is on validation experiments directed toward a specific application driver, as discussed in Element 1. Referring back to Figure 2.9, the issue should be raised concerning the time and resources required for conducting a new validation experiment within the application domain, versus the value of using experimental data known to the computational analyst. The approach to these trade-off studies should be: For a given quantity of time and resources expended, which option most improves the predictive capability of the SRQs of interest? Even though it is constructive to frame the question as an optimization problem, it must be realized that it is still a very difficult question to answer. Some of these reasons are: (a) the time and resources needed to achieve a specified estimated uncertainty goal are poorly known; (b) it is only vaguely known how the model form uncertainty may vary over the application domain; and (c) the dependencies between the model input parameters may be poorly known in trade-off space, i.e., all of the coordinates in the model input space are not orthogonal.

The primary guidelines for the design and execution of validation experiments have been formulated by Aeschliman and Oberkampf (1998), Oberkampf and Blottner (1998), and Oberkampf and Trucano (2002). These six guidelines are:

1. A validation experiment should be jointly designed by experimentalists, model developers, code developers, and code users working closely together throughout the program, from inception to documentation, with complete candor about the strengths and weaknesses of each approach.
2. A validation experiment should be designed to capture the essential physics of interest, measure all relevant physical modeling data, including initial and boundary conditions, and system excitation information required by the model.
3. A validation experiment should strive to emphasize the inherent synergism attainable between computational and experimental approaches.

4. Although the experimental design should be developed cooperatively, independence must be maintained in obtaining the computational and experimental system response results, e.g., withholding measured SRQs until after simulation results are presented.
5. Experimental measurements should be made of a hierarchy of system response quantities, for example, from globally integrated quantities to local quantities.
6. The experimental design should be constructed to analyze and estimate the components of random (precision) and systematic (bias) experimental uncertainties.

A more detailed discussion of both traditional experiments and validation experiments are given in Chapters 10 and 11.

2.5.5 Solution Verification and Computation of the System Response Quantities

Solution verification activities should be conducted on all computational solutions obtained, but in Element 5 the emphasis is on those that are used to compare with experimental data. As discussed regarding Figure 2.7, two different types of solution verification are conducted: verification of the input and output data processing and verification of the numerical solution accuracy. In most simulations, considerable computational effort is required toward estimating numerical convergence errors (space, time, and iterative). *A posteriori* methods are the most general and effective approach for estimating numerical solution error in the SRQs of interest. If the SRQs of interest are global quantities over the solution domain, such as lift or drag on a flight vehicle, or total strain energy in a solid, these quantities converge relatively quickly compared to local SRQs. If the SRQs of interest are field quantities, such as local pressure and temperature in the solution domain of the differential equation, these are known to be much more demanding in terms of the discretization size needed to obtain convergence.

If a nondeterministic simulation is computed, it must be recognized that the numerical solution error for each SRQ of interest changes as a function of the input data conditions chosen for each simulation. For example, suppose a Monte Carlo procedure is used to sample the uncertain input data for the model to obtain a given solution in the ensemble of solutions for the nondeterministic simulation. For each set of input data sampled, the numerical solution error for each of the SRQs will change from sample to sample because each solution to the differential equation is different.

In order to compare simulation results with experimental measurements, the experimentalist must measure the input data needed to compute a solution to the mathematical model. Examples of the input data needed for the simulation are the boundary conditions, initial conditions, geometric details, material properties, and system excitation due to the surroundings. The information provided by the experimentalist should be accompanied by uncertainty estimates for each quantity provided. The estimates of experimental uncertainties for both the SRQs, as well as the input quantities needed for the simulation must be stressed here. This is one of the important characteristics of high-quality validation experiments. The uncertainty estimates provided by the experimentalist could be characterized in several different ways, e.g., either probability density functions (PDFs), cumulative distribution functions (CDFs), or simply interval-valued quantities with no likelihood information provided. However the uncertainty is characterized, a nondeterministic simulation of the validation experiment must be computed. Section 2.5.7 discusses the concept of nondeterministic simulation and Chapter 3 discusses the basic computational issues involved.

As pointed out earlier, the SRQs measured in the experiment should *not* be provided to the computational analysts before the simulations are completed. The optimum situation is for the analysts to be provided only with the input quantities needed for the simulation, so that a blind prediction of the validation experiment can be made. However, for existing experiments, particularly well-known experiments in the validation database, this is not possible. It is our view that predictive capability is *not* assessed when the experimental measurements of the SRQs are known to the analyst beforehand. Generally, when the analysts know the measured responses, they are influenced in many ways, some obvious and some not so obvious. These tendencies are closely related to the human trait of *confirmation bias* (Pohl, 2004). Some examples of influence are (a) modification in modeling assumptions and approximations, (b) choice of numerical algorithm parameters, (c) mesh or temporal resolution, and (d) adjustment of free parameters or poorly known (i.e., poorly measured) parameters in the model.

Solution verification is discussed in detail in Chapters 7–9, and nondeterministic simulation and model validation are discussed in Chapters 3, and 10–13.

2.5.6 Computation of Validation Metric Results

Element 6 deals with the quantitative comparison of computational and experimental results by using a validation metric operator. It is common practice in all fields of engineering and science to compare computational results and experimental data using graphs. Graphical comparisons are usually made by plotting a computational SRQ along with the experimentally measured SRQ over a range of some parameter. Common practice has been that if the computational results generally agree with the experimental data over the range of measurements, the computational model is commonly declared "validated." However, comparing simulation results and experimental data on a graph is only incrementally better than making a subjective comparison. In a graphical comparison, the quantification of numerical solution error or quantification of uncertainties in the experiment or the simulation is rarely seen. Uncertainties arise from experimental measurement uncertainty, uncertainty due to variability in experimental conditions, and uncertainty in the simulation due to needed input quantities, but the quantities are not reported by the experimentalist. The experimental conditions that are unreported or unmeasured in the experiment are commonly considered as free parameters in the simulation and, as a result, they are adjusted to obtain better agreement with the experimental measurements. As discussed earlier, this is a type of model parameter calibration, *not* model validation.

The topic of validation metrics has received a great deal of attention during the last two decades, primarily initiated by researchers associated with Sandia National Laboratories. A validation metric operator can be viewed as a difference operator between computational and experimental results for the same SRQ. The validation metric operator could also be described as an operator that measures the mismatch between simulation and experiment. The output from the difference operator is called the *validation metric result* and it is a measure of the model-form bias error for the specific conditions of the validation experiment. The validation metric result is a *quantitative* statement of the difference between the model result and the experimental measurements. The validation metric result is of significant practical value because it is an objective measure, as opposed to subjective personal opinions as to "good" or "bad" or "acceptable" agreement. At the conditions of the validation experiment, a validation metric result can be directly compared with the model accuracy requirements specified in Element 1. If the simulation comparisons with experimental data are blind predictions, then the comparisons are directly a measure of the predictive capability of the model. In addition, the aggregation of validation metric results over the entire validation domain can be used to form the basis for characterization of model form uncertainty for extrapolations outside of the validation domain. Validation metric results are based on *observed* performance of the model, but they can also be used to estimate model form uncertainty for extrapolations to other conditions of interest, either inside or outside of the validation domain.

The construction of validation metric operators is relatively new and there are different opinions as to what they should include and exclude, and how they should be constructed. The following recommendations give one perspective on a constructive approach to formulation of validation metrics (Oberkampf and Trucano, 2002; Oberkampf and Barone, 2006):

1. A metric should either: (a) explicitly include an estimate of the numerical error in the SRQ of interest resulting from the computational simulation or (b) exclude the numerical error in the SRQ of interest, but only if the numerical error was previously estimated, by some reasonable means, to be small.
2. A metric should be a quantitative evaluation of simulation accuracy of the SRQ of interest, including all of the combined modeling assumptions, physics approximations, and previously obtained physical parameters embodied in the model.
3. A metric should include, either implicitly or explicitly, an estimate of the error resulting from postprocessing of the experimental data to obtain the same SRQ that results from the model.
4. A metric should incorporate, or include in some explicit way, an estimate of the measurement errors in the experimental data for the SRQ that is compared with the model.
5. A metric should generalize, in a mathematically rigorous way, the concept of a difference between two scalar quantities that have no uncertainty to quantities that can have both aleatory and epistemic uncertainty.
6. A metric should exclude any indications, either explicit or implicit, of the level of adequacy in agreement, or satisfaction of accuracy requirements, between computational and experimental results.
7. A validation metric should be a true metric in the mathematical sense, i.e., retaining essential features of a true distance measure.

A detailed discussion of the construction and use of validation metrics is given in Chapter 12.

2.5.7 Prediction and Uncertainty Estimation for the Application of Interest

In the simulation of the performance, safety, and reliability of many systems, predictions are viewed strictly as deterministic. Hence, uncertainties in the input data to the model, as well as uncertainty in the model itself, are considered small or they are simply ignored. To account for uncertainties that exist, a safety factor is commonly added to various design features of the system or a margin is included to better insure the required performance of the system is achieved (Elishakoff, 2004). A second approach that is sometimes used is to try and identify the worst operating condition, or the most demanding combination of operational conditions, under which the system might be required to operate. The system is then designed to operate successfully and/or safely under those conditions. Depending on the needs of the simulation and the systems involved, either approach can be appropriate and cost effective.

During the last three decades, the fields of nuclear reactor safety and underground storage of toxic and radioactive materials have pioneered modern approaches to nondeterministic simulation and risk assessment. The performance and risk analysis of high-consequence systems such as these required the development of new and more credible nondeterministic methods. The mathematical model of the system, which includes the influence of the surroundings on the system, is considered nondeterministic in the sense that: (a) the model can produce multiple system responses because of the existence of uncertainty in the input data for the model; (b) the analysis may consider multiple possible environments and scenarios that the system may experience; and (c) there is uncertainty in the mathematical model itself, separate from the uncertainty of the input data to the model. This book uses the term *nondeterministic simulation* instead of *stochastic simulation* because it is the more general term. For example, if only epistemic uncertainties exist in a simulation, i.e., there are no random variables used to represent uncertainty, then there is only epistemic uncertainty in the simulation result, i.e., there are only intervals. The mathematical models are assumed to be deterministic such that when all necessary input data for a designated model are specified, the model produces *only one value* for each SRQ. That is, there is a one-to-one correspondence from input data to output data of the model. This book excludes mathematical models that are given by stochastic differential equations because they address the mapping of entire stochastic functions for the system response. To predict the nondeterministic response of the system, it is necessary to evaluate the mathematical model, alternative, or competing mathematical models, of the system multiple times using different input data for the model. Note that different input data for the model are needed to represent different environments and scenarios of the system and its surroundings.

Element 7 deals with the nondeterministic prediction of the SRQs for the application of interest by incorporating into the mathematical model any uncertainties that have been identified. The most common strategy for incorporating uncertainty directly into the computational simulation involves three basic steps (Figure 2.15). The first step is called *characterizing the sources of uncertainty*. The uncertainties can be characterized as either an aleatory uncertainty or as a recognized epistemic uncertainty (Figure 2.13). If an uncertainty is purely aleatory, then it is characterized as a PDF or a CDF (Figure 2.12). If it is characterized as purely epistemic, then it is characterized as an interval-valued quantity, which has no likelihood information. An uncertainty can also be characterized as mixture of aleatory and epistemic uncertainty, i.e., some portions of the characterization are probability distributions and some are given as intervals.

In the second step, called *propagation of the uncertainty*, the uncertain input quantities specified in the previous step are propagated through the model to obtain uncertain output

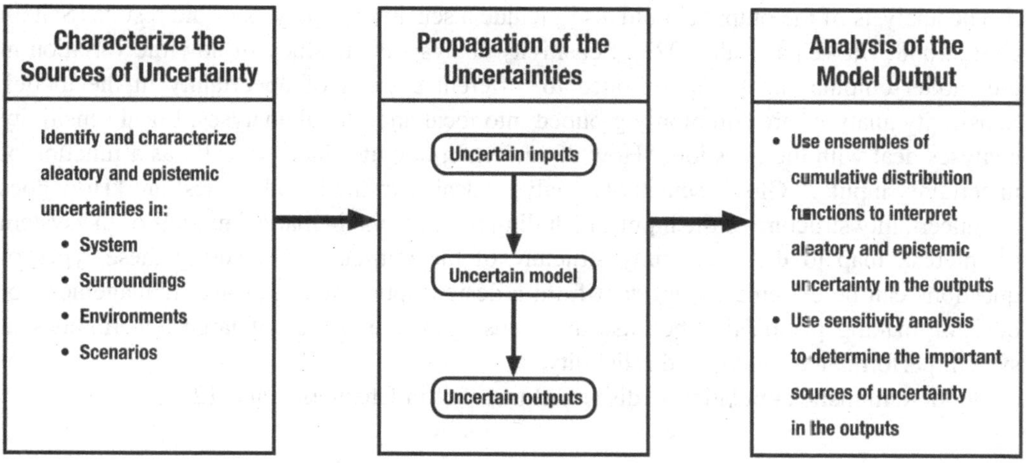

Figure 2.15 Basic steps in a nondeterministic simulation.

quantities. There are a number of propagation methods available to compute the mapping of input quantities to output quantities. For two primary reasons, this text, concentrates on using sampling procedures, such as Monte Carlo or Latin Hypercube Sampling. First, sampling methods are conceptually straightforward to understand and apply in practice. They can be used as a pre-processor to any type of mathematical model, in the sense that they can be used as an *outer loop* or *wrapper* to an existing deterministic simulation code. Second, sampling methods can easily accommodate both aleatory and epistemic uncertainties. Samples are taken from both the aleatory and epistemic uncertainties, but each type of uncertainty is treated separately and kept segregated in the simulation and in the interpretation of the results. Samples from aleatory uncertainties represent stochastic uncertainty or variability and, as a result, these samples represent aleatory uncertainty in the SRQs. Samples from epistemic uncertainties only represent *possible* realizations in the SRQs. Hence, no probability or likelihood is associated with any samples taken from epistemically uncertain input quantities. Note that if alternative mathematical models are used to estimate the model form uncertainty, then the results from each model are also epistemic uncertainties. The complete set of all samples for the SRQs is sometimes called an *ensemble of calculations*. In deterministic simulation, a single simulation is performed to obtain a result. For nondeterministic simulation, a potentially large number of simulations must be performed to obtain a result that characterizes the uncertainty in the SRQs of interest.

After the ensemble of calculations has been generated, the third step, *analysis of the model output*, is performed. This step involves analysis and interpretation of the ensemble of calculations for the SRQs of interest. The typical form of the ensemble of calculations is a family of CDFs. Multiple CDFs are produced because of the existence of epistemic uncertainty. Each CDF represents the results from sampling all of the aleatory uncertainties, but each CDF is only associated with *one* sample from all of the epistemically uncertainty quantities. Stated differently, each one of the probability distributions represents a *possible* probability distribution of the SRQs because each distribution is associated with a specific sample of the epistemic uncertainties. If the simulation *only* contains aleatory uncertainties, then only *one* CFD is required to capture the nondeterministic result of the simulation.

The analysis of the output should also include a sensitivity analysis of the results (Saltelli et al., 2008; Da Veiga et al., 2021). Sensitivity analysis is the study of how the variation in the model outputs can be apportioned to different sources of uncertainty in the model. Sensitivity analyses are commonly grouped into local and global analyses. Local sensitivity analyses deal with the question: "How do uncertain outputs change locally as a function of uncertainty inputs?" Global sensitivity analyses deal with the broader question: "How does the uncertainty structure of the inputs, including multiple mathematical models of the system of interest, map to the uncertainty structure of the outputs?" Answering these types of questions can be extremely important from a design optimization, project management, or decision-making perspective because the focus is on the *causes* of large uncertainties in system performance, safety, and reliability.

Nondeterministic simulation is discussed in detail in Chapters 3 and 12–13.

2.5.8 Assessment of Model Adequacy for the Application of Interest

The assessment of model adequacy conducted in Element 8 primarily deals with the assessment of estimated model accuracy as compared to required model accuracy specified in Element 1. As mentioned earlier, many other practical and programmatic issues enter into the decision of model adequacy for an intended application, such as computer resource requirements and software availability to conduct a nondeterministic simulation. Element 8 relies directly on the results of Element 7, which provides predictions of the SRQs of interest over the application domain, along with the estimated uncertainty for each SRQ. Accuracy requirements should be given over the entire application domain for all of the SRQs of interest. Since there is commonly a broad range in SRQs of interest in a simulation and their importance to the system performance, reliability, and safety varies widely, the accuracy requirements usually vary from one SRQ to another. In addition, the accuracy requirements for a given SRQ can also vary over the application domain of the system. For example, in regions of the application domain that are unimportant from a system performance or reliability perspective, the accuracy requirements may be relatively low. In regions where system performance is critical, such as fuel consumption of an aircraft at cruise conditions or failure modes of a structure, then accuracy requirements on the SRQs are quite demanding.

As discussed regarding Figure 2.9, the application domain can be enclosed within the validation domain or there can be portions of the application domain that are outside of where experimental data for the system are available. The most common situation in simulation is that the application domain is *far from* the validation domain. Thus, the model must be used to make predictions based on the explanatory ability of the model to describe the processes occurring within the system. This process is referred to as use of the model to extrapolate to conditions of interest that previously have not been directly observed. For physical systems, this capability must be based on the *fidelity of the physics* that is captured in the mathematical model describing the system.

If the application domain is far from the validation domain, then the extrapolation method can introduce large uncertainty because of the limitations of the fidelity of the physics in the model. This source of uncertainty is referred to here as model form uncertainty because it is due to the assumptions and approximations in the formulation of the model. As a result, it is an epistemic uncertainty because it is due to lack of knowledge, both recognized and blind epistemic uncertainties, reflected in the model. To stress this critical point more emphatically, model form uncertainty is *not random*, it is a special type of bias, or systematic, error due to the model itself. Additional uncertainty sources that contribute to total estimated uncertainty, such as numerical solution error, are discussed in Chapters 3 and 13.

Conceptually, the model adequacy assessment approach outlined is logically well founded and most importantly, it directly ties model accuracy assessment to the application-specific requirements. However, technical difficulties may arise when using this conceptualization when large extrapolation is required using the model. One is mentioned here, but a more complete discussion is given in Chapters 10, 12–13. Consider dealing with a situation where there is a system or surroundings parameter that cannot be considered as a continuous model input variable. For example, suppose that the system of interest, nor anything quite like it, has ever been built before. Furthermore, suppose that experiments can only produce relevant physical conditions on components or subsystems that have been fabricated during the design phase of the system. As a result, all of the validation domain data exists at lower tiers in the system hierarchy (Figure 2.4). That is, the validation domain of the complete system *does not exist*; thus, the experimentalist is only dealing with validation domains of components and subsystems.

The concept of model-based extrapolation could still be used, but the concept must be thought of as extrapolation upward in the hierarchy to the complete system. The essence of this type of extrapolation would be the coupling of physics embodied in each of the models representing each subsystem. The model input variables for an element at one tier of the hierarchy would *only be a subset* of those of an element at a higher tier. The dimensionality of the input space increases rapidly as one moves upward through the system hierarchy because of the increase in the types of physics occurring, as well as the new coupled physics that occurs. For example, in predictions for global climate change and long-time predictions of the safety of underground storage of nuclear waste, a similar increase in complexity occurs.

2.5.9 Documentation of M&S Activities

Although the topic of documentation in M&S usually generates little enthusiasm (at best) among analysts, some level of documentation is always needed. The magnitude of the documentation effort usually depends on the size and goals of the M&S project, as well as the cost and consequences of the simulation-informed decision making. For example, at the minimal end of the documentation spectrum would be quick response in-house studies or informal questions asked by a design group. In the middle part of the spectrum would be documentation requirements for a corporate product that had either performance guarantees associated with the product, or some aspect of legal liability. At the extreme end of the spectrum would be documentation requirements for high-consequence systems that could affect large portions of the population or the environment if a failure occurred. The comments here are directed at the middle to high end of the documentation spectrum.

The goals of documentation are usually discussed in terms of the need for reproducibility, traceability, and transparency of the M&S-VVUQ activities. *Reproducibility* means that sufficient detailed information is available concerning the assumptions and approximations of the model, system and surroundings specification, environments considered, and numerical procedures used so that the simulation could be reproduced by an independent team of experts. *Traceability* means that all significant aspects of M&S-VVUQ are recorded and dated so that each of the steps and processes can be tracked. *Transparency* means that all important aspects of the M&S activity can be examined, probed, and understood by technically qualified individuals. However, when proprietary models or software are used, transparency suffers greatly. Documentation goals mentioned are important in any serious simulation effort, particularly those that support certification or regulatory approval of the safety and reliability of high-consequence systems. Some examples are the performance and risk assessment for nuclear reactor safety, large-scale public structures such skyscrapers and dams, and large-scale environmental projects.

Since the documentation goals of reproducibility, traceability, and transparency seem rather aloof and uninteresting to most personnel involved in M&S activities, we give the following examples that may be more motivational to various individuals are given. These examples are relevant to the elements shown in Figure 2.14

- Element 1: definitive documentation of the application of interest (including system environments and scenarios considered) and the accuracy requirements for SRQs over the application domain.
- Element 2: documented justification for the planning and prioritization of the M&S-VVUQ activities, assumptions and approximations made in the model formulation, system failure modes considered, as well as those ignored.
- Element 3: documentation of code verification testing that has been completed with regard to options in the code that are used for the application of interest, and SQA activities that have (and have not) been conducted.
- Element 4: documentation of the design and execution of validation experiments, particularly all input data and uncertainty estimation needed for simulation.
- Element 5: detailed documentation of numerical error estimation procedures used for each SRQ of interest and a description of how uncertainties in the simulation of the validation experiments were characterized.
- Element 6: documentation of the validation domain, particularly its relationship to the application domain, as well as model accuracy assessment for various SRQs over the validation domain.
- Element 7: documentation of the predicted SRQs and the sources of uncertainty that contribute to total uncertainty for the conditions of the application of interest, and description of how model parameter calibration was conducted.
- Element 8: documentation of model adequacy (and any identified inadequacies) relative to the prediction accuracy requirements for the application of interest.

Whatever the level or type of documentation generated, an electronic records management system (RMS) or simulation data management (SDM) system should be used. A set of commercial RMS/SDM software are available, but on large-scale projects it is common to construct a tailor-made system. The RMS/SDM could be organized in a tree or folder/subfolder structure so that at the base level would be the particular M&S project, then it would divide into the eight elements shown in Figure 2.14, followed by any further appropriate subdivisions. The RMS/SDM system should be searchable by key words within any portion of any information element. The search engine could operate much like that found in Google or Wikipedia. Functionality could be expanded to include a relevancy-ranking feature that would further improve the search-and-retrieval capability. The overall system design could include searchable elements such as design configurations, computer codes, experimental facilities, system failure modes, and personnel involved. After the results were retrieved, they could be sorted according to their relevance to the words input to the search. The high-level results retrieved should be embedded with hyperlinks. The hyperlinks could be selected to pursue more detailed information of interest, including photographic images and audio/video records of experimental activities, and data input files for simulations.

2.6 Problems

1. Discuss the advantages and disadvantages of deductivism versus inductivism in the development of scientific knowledge.
2. Formulate a conceptual model of some simple system or process in your field of study. Based on this conceptual model, give the mathematical model and list all of the input parameters needed for the model. Determine the dimensionality of in the model input parameter space.
3. The following measurements for the modulus of elasticity were obtained from a given production lot of steel: 31.2E6, 29.8E6, 28.5E6, 30.7.E6, 27.9E6, 31.4E6, 30.9E6, 29.7E6, 30.3E6, 30.5E6, 29.5E6, 30.8.E6, 26.9E6, 31.3E6, and 32.1E6 psi. Show three graphical methods for displaying these measurements.
4. Given the data shown in Problem 3, determine a good approximation to these data using an appropriate continuous PDF and CDF. Characterize the uncertainty of the numerical features of the approximate PDF and CDF.
5. The following measurements for the thermal contact resistance were obtained from parts assembled by one supplier: 0.6, 0.4, 0.3, 0.7, 0.5, 0.4, 0.8, 0.3, 0.9, and 0.2 m^2 °C/kW. The following measurements were obtained from parts assembled by another supplier: 0.1, 0.4, 0.3, 0.5, 0.4, 0.1, 0.3, 0.2, 0.1, and 0.2 m^2 °C/kW. Show three graphical methods for displaying the measurements from both suppliers combined. What can be inferred from these data from the two suppliers and what could be the causes for the difference?
6. Given the data shown in Problem 4, determine a good approximation to these data using an appropriate continuous PDF and CDF. Characterize the uncertainty of the numerical features of the approximate PDF and CDF.
7. The only information known about a quantity of interest is that the true value is believed to be within the interval [0.3, 1.5]. Plot the cumulative distribution function for this information.
8. Consider making a wager on rolling a pair of dice. Consider two states of knowledge concerning the dice; (a) you have not seen the dice and you know nothing about the dice, and (b) you have seen each face of each die and you observe that each die has six unique number of dots, but you have not observed any rolls of the dice. For each state of knowledge, discuss what are the odds of winning or losing a wager placed on rolling the dice.

9. Give two examples of aleatory and epistemic uncertainties in measurements from your field of study.
10. Give two examples of a blind epistemic uncertainty in the M&S of a system from your field of study.
11. Write a discussion of the types of uncertainties that fundamentally contributed to the loss of the Space Shuttle Challenger in 1986.
12. Write a discussion of the types of uncertainties that fundamentally contributed to the loss of the Space Shuttle Columbia in 2003.

2.7 References

Aeschliman, D. P., and W. L. Oberkampf (1998). Experimental Methodology for Computational Fluid Dynamics Code Validation. *AIAA Journal*, **36**(5), 733–741.

AIAA (1998). Guide for the Verification and Validation of Computational Fluid Dynamics Simulations. *American Institute of Aeronautics and Astronautics*, AIAA-G-077-1998, Reston, VA.

AIAA (2021). Recommended Practice: Code Verification in Computational Fluid Dynamics. *American Institute of Aeronautics and Astronautics*, AIAA R-141-2021, Reston, VA.

Anderson, M. G., and P. D. Bates (eds.). (2001). *Model Validation: Perspectives in Hydrological Science*. New York, NY: John Wiley & Sons.

Arthur, J. D., and R. E. Nance. (1996). Independent Verification and Validation: A Missing Link in Simulation Methodology? *1996 Proceedings of the Winter Simulation Conference*, 229–236, Coronado, CA.

ASME (2006). Guide for Verification and Validation in Computational Solid Mechanics. American Society of Mechanical Engineers, ASME Standard V&V 10-2006, New York, NY.

ASME (2019). Standard for Verification and Validation in Computational Solid Mechanics. *American Society of Mechanical Engineers*, ASME Standard V&V 10-2019, New York, NY.

ASME (2022). Verification, Validation, and Uncertainty Quantification Terminology in Computational Modeling and Simulation. *American Society of Mechanical Engineers*, ASME VVUQ 1-2022, New York, NY.

ASSESS (2021). Analysis, Simulation & Systems Engineering Software Strategies Initiative. www.assessinitiative.com/.

Augustin, T., F. P. A. Coolen, G. de Cooman, and M. C. M. Troffaes (eds.) (2014). *Introduction to Imprecise Probabilities*. Chichester: Wiley & Sons.

Ayyub, B. M. (2001). *Elicitation of Expert Opinions for Uncertainty and Risks*. Boca Raton, FL: CRC Press.

Ayyub, B. M., and G. J. Klir (2006). *Uncertainty Modeling and Analysis in Engineering and the Sciences*. Boca Raton, FL: Chapman & Hall.

Balci, O., and R. G. Sargent (1984). A Bibliography on the Credibility Assessment and Validation of Simulation and Mathematical Models. *Simuletter*, **15**(3), 15–27.

Bernardini, A., and F. Tonon (2010). *Bounding Uncertainty in Civil Engineering*. New York, NY: Springer.

Beven, K. (2002). Towards a Coherent Philosophy of Modelling the Environment. *Proceedings of the Royal Society of London, Series A*, **458**(2026), 2465–2484.

Blottner, F. G. (1990). Accurate Navier–Stokes Results for the Hypersonic Flow Over a Spherical Nosetip. *Journal of Spacecraft and Rockets*, **27**(2), 113–122.

Boyack, B. E., I. Catton, R. B. Duffey, P. Griffith, K. R. Katsma, G. S. Lellouche, et al. (1990). Quantifying Reactor Safety Margins Part 1: An Overview of the Code Scaling, Applicability, and Uncertainty Evaluation Methodology. *Nuclear Engineering and Design*, **119**, 1–15.

Cox, E. (1999). *The Fuzzy Systems Handbook: A Practitioner's Guide to Building, Using, and Maintaining Fuzzy Systems*, 2nd ed. San Diego, CA: AP Professional.

Cullen, A. C., and H. C. Frey (1999). *Probabilistic Techniques in Exposure Assessment: A Handbook for Dealing with Variability and Uncertainty in Models and Inputs*. New York, NY: Plenum Press.

Da Veiga, S., F. Gamboa, B. Iooss, and C. Prieur (2021). *Basics and Trends in Sensitivity Analysis: Theory and Practice in R*. Philadelphia, PA: Society for Industrial and Applied Mathematics.

Davis, P. K. (1992). Generalizing Concepts and Methods of Verification, Validation, and Accreditation (VV&A) for Military Simulations. RAND, R-4249-ACQ, R-4249-ACQ, Santa Monica, CA.

DoD (1994). DoD Directive No. 5000.59: Modeling and Simulation (M&S) Management. www.msco.mil.

DoD (1997). DoD Modeling and Simulation Glossary. www.msco.mil.

Elishakoff, I. (2004). *Safety Factors and Reliability: Friends or Foes?* Norwell, MA: Kluwer Academic Publishers.

Ferson, S., and L. R. Ginzburg (1996). Different Methods Are Needed to Propagate Ignorance and Variability. *Reliability Engineering and System Safety*, **54**(2–3), 133–144.

Grabe, M. (2014). *Measurement Uncertainties in Science and Technology*, 2nd ed. New York, NY: Springer.

Haimes, Y. Y. (2015). *Risk Modeling, Assessment, and Management*, 4th ed. New York, NY: John Wiley.

Han, X., and P. E. Kloeden (2017). *Random Ordinary Differential Equations and Their Numerical Solution*. Singapore: Springer Nature.

Hatton, L. (1997). The T Experiments: Errors in Scientific Software. *IEEE Computational Science & Engineering*, **4**(2), 27–38.

Hodges, J. S., and J. A. Dewar (1992). Is It You or Your Model Talking? A Framework for Model Validation. RAND, R-4114-AF/A/OSD, Santa Monica, CA.

Hubbard, D. W. (2020). *The Failure of Risk Management: Why It's Broken and How to Fix It*, 2nd ed. New York, NY: John Wiley.

IEEE (1984). IEEE Standard Dictionary of Electrical and Electronics Terms. ANSI/IEEE Std 100-1984, New York.

IEEE (1991). IEEE Standard Glossary of Software Engineering Terminology. IEEE Std 610.12-1990, New York.

INCOSE (2021). International Council on Systems Engineering. www.incose.org/.

ISO (2005). ISO 9000:2005 Quality Management Systems – Fundamentals and Vocabulary. International Organization for Standardization, ISO 9000:2005(E), Switzerland.

Kleindorfer, G. B., L. O'Neill, and R. Ganeshan (1998). Validation in Simulation: Various Positions in the Philosophy of Science. *Management Science*, **44**(8), 1087–1099.

Klir, G. J., U. St. Clair, and B. Yuan (1997). *Fuzzy Set Theory: Foundations and Applications*. Upper Saddle River, NJ: Prentice Hall PTR.

Law, A. M. (2014). *Simulation Modeling and Analysis*, 5th ed. New York, NY: McGraw-Hill.

Leonard, I. E., and J. E. Lewis (2016). *Geometry of Convex Sets*. New York, NY: John Wiley.

Lewis, R. O. (1992). *Independent Verification and Validation*, 1st ed. New York, NY: John Wiley.

Lipton, P. (2005). Testing Hypotheses: Prediction and Prejudice. *Science*, **307**, 219–221.

Marvin, J. G. (1995). Perspective on Computational Fluid Dynamics Validation. *AIAA Journal*, **33**(10), 1778–1787.

Melchers, R. E. (1999). *Structural Reliability Analysis and Prediction*, 2nd ed. New York, NY: John Wiley.

Oberkampf, W. L., and M. F. Barone (2006). Measures of Agreement between Computation and Experiment: Validation Metrics. *Journal of Computational Physics*, **217**(1), 5–36.

Oberkampf, W. L., and F. G. Blottner (1998). Issues in Computational Fluid Dynamics Code Verification and Validation. *AIAA Journal*, **36**(5), 687–695.

Oberkampf, W. L., and T. G. Trucano (2002). Verification and Validation in Computational Fluid Dynamics. *Progress in Aerospace Sciences*, **38**(3), 209–272.

(2007). Verification and Validation Benchmarks. *Sandia National Laboratories*, SAND2007-0853, Albuquerque, NM.

(2008). Verification and Validation Benchmarks. *Nuclear Engineering and Design*, **238**(3), 716–743.

Oberkampf, W. L., T. G. Trucano, and C. Hirsch (2004). Verification, Validation, and Predictive Capability in Computational Engineering and Physics. *Applied Mechanics Reviews*, **57**(5), 345–384.

Oksendal, B. (2013). *Stochastic Differential Equations: An Introduction with Applications*, 6th ed. New York, NY: Springer.

Oreskes, N., K. Shrader-Frechette, and K. Belitz (1994). Verification, Validation, and Confirmation of Numerical Models in the Earth Sciences. *Science*, **263**, 641–646.

Platt, D. S. (2007). *Why Software Sucks ... and What You Can Do About It.* Upper Saddle River, NJ: Addison-Wesley.

Pohl, R. (ed.) (2004). *Cognitive Illusion: A Handbook on Fallacies and Biases in Thinking, Judgement and Memory.* New York, NY: Psychology Press.

Popper, K. R. (1969). *Conjectures and Refutations: The Growth of Scientific Knowledge.* London: Routledge and Kegan.

Rabinovich, S. G. (2005). *Measurement Errors and Uncertainties: Theory and Practice*, 3rd ed. New York, NY: Springer.

Refsgaard, J. C., and H. J. Henriksen (2004). Modelling Guidelines–Terminology and Guiding Principles. *Advances in Water Resources*, **27**(1), 71–82.

Roache, P. J. (1998a). *Verification and Validation in Computational Science and Engineering.* Albuquerque, NM: Hermosa Publishers.

(1998b). Verification of Codes and Calculations. *AIAA Journal*, **36**(5), 696–702.

Roza, Z. C. (2004). *Simulation Fidelity, Theory and Practice: A Unified Approach to Defining, Specifying and Measuring the Realism of Simulations*. Delft, The Netherlands: Delft University Press.

Rykiel, E. J. (1996). Testing Ecological Models: The Meaning of Validation. *Ecological Modelling*, **90**(3), 229–244.

Saltelli, A., M. Ratto, T. Andres, F. Campolongo, J. Cariboni, D. Gatelli, et al. (2008). *Global Sensitivity Analysis: The Primer.* New York, NY: John Wiley.

Sarkka, S., and A. Solin (2019). *Applied Stochastic Differential Equations*. New York, NY: Cambridge University Press.

Schlesinger, S. (1979). Terminology for Model Credibility. *Simulation*, **32**(3), 103–104.

Shaw, S. (2021). M&S Validation Hierarchies: A Tool for Creating Shared Meaning. From ASSESS Insight Webinar, Oct. 14, 2021.

Sheng, G., M. S. Elzas, T. I. Oren, and B. T. Cronhjort (1993). Model Validation: A Systemic and Systematic Approach. *Reliability Engineering and System Safety*, **42**, 247–259.

Sindir, M. M., S. L. Barson, D. C. Chan, and W. H. Lin (1996). On the Development and Demonstration of a Code Validation Process for Industrial Applications. *27th AIAA Fluid Dynamics Conference*, AIAA Paper 96-2032, New Orleans, LA.

Smithson, M. (1989). *Ignorance and Uncertainty: Emerging Paradigms*. New York, NY: Springer.

Trucano, T. G., M. Pilch, and W. L. Oberkampf (2002). General Concepts for Experimental Validation of ASCI Code Applications. *Sandia National Laboratories*, SAND2002-0341, Albuquerque, NM.

Zeigler, B. P., H. Praehofer, and T. G. Kim (2000). *Theory of Modeling and Simulation: Integrating Discrete Event and Continuous Complex Dynamic Systems*, 2nd ed. San Diego, CA: Academic Press.

Zuber, N., G. E. Wilson, M. Ishii, W. Wulff, B. E. Boyack, A. E. Dukler, et al. (1998). An Integrated Structure and Scaling Methodology for Severe Accident Technical Issue Resolution: Development of Methodology. *Nuclear Engineering and Design*, **186**(1–2), 1–21.

3

Modeling and Computational Simulation

Mathematicians are all about generalisation, aren't they? So why do the probabilists refuse to talk about sets of probability distributions or generalising probabilities in any way? It's like a mathematician who only uses equal signs and never allows greater-than or less-than. —Dominik Hose

Contents of Chapter 3

Chapter 3 discusses the fundamental concepts of modeling and computational simulation as they relate to the physical sciences and engineering. The specification of the system being modeled, as well as the surroundings of the system, are central to the concept of modeling. The mathematical model describes the interactions occurring within the system and how the surroundings impact the system. Models of systems are usually categorized considering the system in a normal, abnormal, or hostile environment. Further specification of the environment may be given in terms of a particular scenario that is being addressed, e.g., the scenario of a gas turbine engine losing a fan blade in an abnormal environment. Many of the difficult issues that must be dealt with in verification, validation, and uncertainty quantification (VVUQ) can be traced back to ambiguities or inconsistencies in the model formulation, the mapping of the mathematical model to a discrete mathematics model, and vague, incomplete or improper characterization of uncertainties. The structure of nondeterministic simulation is formally introduced in this chapter and the basic concept of propagation of uncertainties through a model is discussed, along with two detailed examples. Distinctions are made between system risk and failure, including the difficulty of estimating risk in terms of potential negative consequences on an organization, a business sector, or the environment. The six phases of computational simulation are discussed in detail, with emphasis on how aleatory, epistemic, and blind uncertainties enter during each of the six phases of simulation. To demonstrate the activities that are conducted in each of the phases, a missile flight example is discussed.

3.1 Fundamentals of System Specifications

3.1.1 Systems and Surroundings

The concept of a system is probably the most important element in modeling. The definition of a system that is most useful for modeling physical systems is:

System: a set of physical entities that interact and are observable, where the entities can be a specified quantity of matter or a volume in space.

For those with a background in thermodynamics, this definition is similar to that given for a system in that field. As in thermodynamics, this definition allows for the system to be closed or open. A closed system means there is no exchange of mass with the surroundings of the system. An open system can have mass flow into and out of the system. A system can have forces acting on it, produce work or have work done on it, and have energy exchanged with the surroundings. Also, a system can be time-invariant (static) or time-variant (dynamic).

This text's definition of a system, although very broad, is actually more restrictive than that used in many fields, for example, Operations Research (Neelamkavil, 1987). This definition excludes any type of human organization, social science systems, or human mental processes. All of these types of systems can be considered as living or sentient systems and are, by almost any measure, much more complex systems than physical systems of interest in this text. In using the present definition of a system, however, the physical body of a person, or any part or organ of the body, could be considered as a system. For example, the physiological, mechanical, and chemical changes of an organ or cell exposed to electromagnetic or acoustic excitation could be considered within this text's definition of a system.

The state of a system can be influenced as a result of (a) processes internal to the system, i.e., endogenous processes and (b) processes or activities external to the system, i.e., exogenous effects. Influences or activities not considered as part of the system are considered as part of the surroundings of the system. Since the complement to the system is the surroundings, it is important that a precise definition be given:

Surroundings: all entities and influences that are physically or conceptually separate from the system.

A system is influenced by the surroundings, but the surroundings are *not* modeled as part of the system itself. In most models, the system *responds* to the surroundings, but the surroundings are *not usually* influenced by the system. When the surroundings are influenced by the system, one of two possible modeling changes must occur:

- A *separate* mathematical model is constructed for the surroundings. Then the surroundings become another system that interacts with the first system.
- A weak coupling is constructed between the system and the surroundings. The surroundings are not considered another system, but they can respond in very simple, specific ways. In weak coupling, the interaction of the surroundings with the system can change in a well-defined manner, typically as a result of a change in the mathematical function that represents how the surroundings react along the boundary of the system. An example is the change in the shear stress function used in free surface flow simulations when waves, especially breaking waves, form on the surface.

The distinction between the system and the surroundings should not be simply thought of as a physical boundary or location external to the physical system. The distinction between the system and the surroundings can be entirely conceptual. The decision of what physical entities and features should be considered as part of the system, as well as what should be considered part of the surroundings, depends on the purpose of the simulation. System-surroundings specifications are not always well thought out and, as a result, they can cause subtle modeling errors or conceptual inconsistencies in the formulation of a mathematical model.

Finally, humans can be a conceptual part of a system. When humans are part of the system, they are referred to as *actors* (Bossel, 1994). The term *actor* means an entity of a system that can influence the system in some physical way, or can respond to events or activities occurring in the system in a conscious manner. *Conscious manner* usually means with a goal or purpose in mind, but it does not necessarily mean what would normally be considered rational or logical. The actor can also be unpredictable, unreliable, or acting with an unknown agenda or with some malicious intent. Actors can become important elements in many modern physical systems. For example, actors are present in human-controlled systems or systems that are controlled by a combination of a human and a computer, such as a safety control system. Another example includes accidental or unanticipated human involvement in a system that is normally thought of as isolated from humans or under computer control. When an actor is considered as part of a system, then the only aspect considered is the action taken that impacts the system, e.g., the actor disabled the safety control system.

Example 1: Orbiting Spacecraft

Consider a spacecraft in orbit around the Earth as a system. Assume the system behavior of interest is the three-degree-of-freedom orbital dynamics of the spacecraft. That is, the spacecraft would be considered as a point-mass system and the primary characteristics of the system would be the position, velocity, and acceleration of the spacecraft. The surroundings would be represented by the forces acting on the craft: (a) the gravitational force of the Earth, Moon, Sun, and other planets; (b) the aerodynamic or molecular drag on the craft; and (c) the solar wind and electrostatic forces on the craft. If the spacecraft has a thrust control system onboard to change its orbital parameters, then the force exerted by the thrusters on the spacecraft would be part of the *surroundings*. However, the mass of the thrusters and their propellants are *part of the system*. As the thrusters are fired, the mass of the system would change due to consumption of propellants. As a result, the system is considered an open system since mass is leaving the system.

Example 2: Static Beam Deflection

Consider the deflection of a beam clamped on one end and free on the other. Assume the system behavior of interest is the static deflection of the beam under a specified loading. The mass, material, and geometric properties of the beam would be considered as characteristics of the system. The surroundings would include the static load distribution, how the clamped end affects the deflection of the beam, and the action of gravity. For example, the clamped end may be assumed to be perfectly rigid, i.e., no deflection or rotation occurs at the clamped end. Alternatively, the clamped end could be considered to have no translational deflection, but rotational deflection occurs around three orthogonal axes. For example, three rotational spring stiffnesses could be used to represent the clamped end as part of the surroundings. The system is influenced by the surroundings, but the surroundings are not influenced by the system.

Example 3: Dynamic Beam Deflection with Change in the Surroundings

Consider the dynamic deflection of a beam, that is, the decay of the vibrational modes of the beam. Beyond the impact of the surroundings just mentioned, there would also be aerodynamic damping on the beam. The internal damping acting on the beam is part of the system. Also consider that the beam can have rotational stiffness and damping at the clamped end, which is part of the surroundings. Suppose further that the spring stiffness depends on the vibrational history of the beam. This can occur due to strain softening of a material or the growth of fatigue cracks in the clamping material holding the beam. To a first-order approximation, this effect could be modeled as an integral of the amplitude and number of cycles of the beam deflection history. This model of the surroundings would be based on a previously observed correlation of data from historical beam deflection experiments. The clamp is still part of the surroundings, but the surroundings could change during the vibration in a very specific way due to processes occurring in the system. The mathematical model of the vibration of the beam would be given by an integrodifferential equation.

Example 4: Electronic Circuit

Consider the electronic circuitry of a common television set as a system. Assume the physical behavior of interest for the system is the current flow through all of the electrical components of the television circuitry when the TV is switched on. The initial condition of the system is considered as that before it is switched on. In addition, assume that the TV is plugged into an electrical power outlet before it is switched on. That is, electrical power is applied to certain parts of the circuitry, but not all, in what is commonly called a *stand-by* or *ready mode* of the circuit. The final state of the system is considered to be the current flow in the circuitry after the TV is switched on for some time. One type of simulation would be to consider the functionality of all of the electrical components as elements of the system and everything else as part of the surroundings. The form of the mathematical model would be an initial value problem, given by a system of ordinary differential equations (ODEs) describing the currents and/or voltages in the circuitry.

Example 5: Electronic Components

Consider the model of the electronic components in the TV, where the interest is the electrical, magnetic, and thermal heating of the components and the circuit boards as a function of time. This would be a much more complex simulation that would model the current flow, electrical and magnetic fields, and heating and deformation of wires, electronic components, and circuit boards. The simulation would consider the thermo-physical properties and the geometric features of each of the components and circuit boards as part of the system. For this type of simulation, the surroundings would be (a) the air around each of the components and circuit boards, which would allow for natural convection and/or forced convection heat transfer if a fan cooling system was part of the system, and (b) other physical features near the components and circuit boards, such as holding brackets and other wood, plastic, and metal structures. Radiation heat transfer between the surroundings, components, and circuit boards would also need to be considered. The form of the mathematical model of the system would be an initial-boundary value problem given by a system of partial differential equations (PDEs) which coupled electrical, magnetic, solid mechanics, and heat transfer phenomena between elements of the system, and how they are impacted by the surroundings.

3.1.2 Environments and Scenarios

Simulation is often used to address the performance, safety, or reliability of a system that is exposed to a wide variety of environments. This text uses the following definition.

Environment: the external condition or situation which the system can be exposed to, specifically, normal, abnormal, or hostile conditions.

Although these three environments have been recognized in the past by some system-design communities, they were first formally defined as part of the simulation of the safety, performance, and reliability of nuclear weapons in the US (AEC, 1966). Sometimes all three environments are implied in the concept of “foreseeable conditions of use.” However, it is prudent from a risk assessment perspective to explicitly address each of these three environments.

The *normal* system environment refers to either one of the following two conditions: (a) the operating environment in which the system is typically expected to operate or function and achieve its performance or reliability goals or (b) storage, shipping, or at-the-ready condition for potential use of the system. A normal operating environment for a system depends *entirely* on what the expected operating conditions of the specific system of interest are. For example, what may be considered a normal operating environment for one system may be considered an abnormal environment for another system. Examples of what could be considered a normal operating condition of some engineering systems are high temperature, pressure, or humidity; chemical or corrosive environments; vacuum conditions; a system covered or infiltrated with ice, snow, or sand; and a high intensity electromagnetic or radiation environment. However, other systems of interest could consider any of these environments as abnormal. Examples of typical storage, shipping, or at-the-ready conditions include a spacecraft either in ground storage before launch or in orbit in a non-operational storage mode; a gas turbine engine being shipped from the manufacture or refurbishment facility to the user of the system; and safety or emergency systems for a nuclear power reactor that are at-the-ready. When a system is analyzed in a normal operating environment, the most common characteristics of interest are its performance and reliability. For systems in storage or shipping, the most common characteristics of interest are possible degradation or aging of the system due to the environment. For systems in at-the-ready environments, the most common characteristics of interest are such things as the response time of the system to full capability and the degradation of the system performance or reliability as a function of time at-the-ready.

An *abnormal* environment of a system refers to either: (a) some type of accident or damaged-state environment or (b) a very unusual condition that could put the system in jeopardy to perform its mission or cause it to be unsafe, even if the system is not operational at the time. Examples of abnormal environments are: exposure of a nuclear power plant to an earthquake; loss of electrical or hydraulic power during flight of an aircraft; loss of a fan, compressor, or turbine blade on a gas turbine engine; operation of the system at temperatures or pressures outside of its normal operating conditions; lightning strike on a system during operation, shipping, or storage; contaminants or foreign material in the lubricant of mechanical equipment; and flight worthiness of a hypersonic vehicle with damage to the thermal protection system. Whether an environment is considered an unusual normal environment or an abnormal environment, usually depends on the potential negative consequences or legal liability of failure or under performance of the system. Two examples are: over-riding the safety control systems during a safety check or proof-testing of a system; and the operator of the system is disoriented or distracted. When a system is analyzed in an abnormal environment, the most common characteristic of interest is the safety of the system, as opposed to performance issues.

A *hostile* environment is one in which the system is under any type of attack, in the sense that the intent of the attacker is to do harm to the system, defeat or disable the system, or render it unsafe. The hostile environment can expose the system to attack from either inside the system or by way of the surroundings of the system. Types of attack include the physical or mechanical attack of the system, modifying or taking over computer control of the system, altering the physical or electronic security or safety of the system, or electromagnetic or acoustic attack of the system. Military systems have always been evaluated with respect to

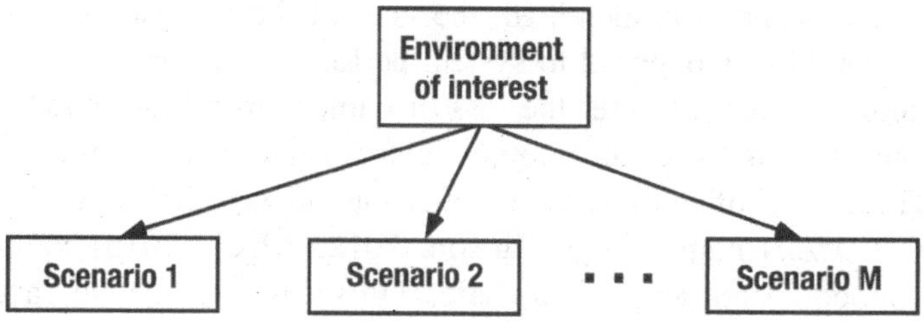

Figure 3.1 Environment-scenario tree.

performance in hostile environments. When military systems are analyzed for a hostile environment, the most common characteristic of interest is potential degradation of system performance. When civilian or public facilities are analyzed for a hostile environment, system safety is the most common concern, but continued operation can also be important. Before the terrorist attacks on the US in September 2001, very few privately owned or public facilities were analyzed with respect to the effects of hostile environments. The phenomenal increase in systems connected to the internet has focused a great deal of attention on software attack from active intruders, as well as degradation or defeat of electronic systems due to software viruses, malware, Trojan horses, and ransomware. Although it is outside the scope of this book to deal with computer security and software issues, modern analysis of hostile environments must consider the safety and the potential degradation of performance of systems that are initiated through the internet.

Given any environment that the system could be exposed to, various scenarios that could occur for an environment should be considered. This text uses the following definition.

Scenario: a possible event, or sequence of events, to which a system in a given environment could be exposed.

Figure 3.1 depicts a given environment of interest and M scenarios that might be considered for simulation of a system. Scenarios should be identified at the specification of the application of interest, as discussed in Chapter 2, or in the conceptual modeling phase discussed later in this chapter. Identification of scenarios is a creative process that focuses on conceiving plausible scenarios that are expected to occur, or could possibly occur, given a specific environment. For a normal operating environment, scenarios would specify expected operating conditions of the system, as well as the extreme conditions that could occur, within the normal operating environment of the system. Some examples include the operation of a transport aircraft at takeoff, climb to altitude, cruise, descent, and landing conditions; operation of instruments and equipment after a dust storm on Mars; and operation of a gas turbine engine on a commercial aircraft during takeoff at high altitude, high temperature, and maximum weight conditions. For an abnormal environment, scenarios are usually identified regarding how the system could be unsafe in various accident conditions. These examples include the safety of passengers and emergency rescue personnel after a serious crash of a battery-powered automobile or the safety of equipment after a loss of electrical power. For a hostile environment, scenarios are commonly identified regarding the type of attack to which the system is exposed. For example: Is the system under physical or electronic attack, is the attack initiated by an outsider or insider, and can the attack be quickly detected and thwarted?

When a particular scenario is identified, the goal of the analysis may focus on system safety and controllability, as opposed to system performance. Some examples are controllability of a commercial aircraft after the loss or damage to a flight condition sensor; and detection and controllability when an automobile driver falls asleep while driving in a cruise control mode. These types of analyses are practiced in the well-developed *Failure Mode and Effects Analysis (FMEA)* community (Carlson, 2012; AIAA, 2015). Some examples of desired failure mode responses are fail-safe, fail-secure, fail-active, and fail-graceful. Identifying a number of scenarios for a given environment does not necessarily mean that each scenario will be simulated. The environment-scenario tree only tries to identify possible conditions of the system that could be analyzed, for example, depending on the likelihood of the failure and the possible consequences associated with each scenario.

3.2 Fundamentals of Models and Simulations

3.2.1 Models and Simulations

Diverse types of models are used in the natural sciences and engineering. Neelamkavil (1987) gives a general and well-founded definition that covers many different types of models for physical systems.

Model: a representation of a physical system or process intended to enhance our ability to understand, predict, or control its behavior.

This text is focused on representations that are given by a mathematical model of the system or process.

There are several variants of the definition of simulation in the literature, but this text uses the following concise definition.

Simulation: the exercise or use of a mathematical model to produce a result.

Simulation of the behavior of a system can be achieved by two types of mathematical models (Bossel, 1994). The first type is referred to as an empirical, phenomenological, or descriptive model of the system. This type of mathematical model is based on observations of how the system responds under different input or stimulation conditions. Empirical models do not attempt to describe any of the detailed processes involved inside the system or determine why the system responds in the way it does. They are devoid of the concept of causality based on the internal workings of the system. The system is considered to be a *black box* and the only issue is the global relationship between observed inputs and observed outputs of the system. The observed behavior of the system to the perceived influences on the system is formulated using some type of mathematical representation, such as statistical correlation, regression fit, or data mining method. An example of this type of model is the dynamic response of a structure to a bolted or riveted joint. If the system is only considered to be the joint or mechanical attachment, then an empirical model can be constructed of how the structure responds dynamically to the joint. The model would represent, say, the structural stiffness and mechanical damping of the joint. The information to construct this type of model is usually obtained from experimental measurements of the dynamic response of the structure. Parameter identification or parameter calibration methods are applied to the observed structural response to determine the input–output relationship over a range of conditions.

The second type of mathematical model is the physical law or explanatory model. For this type of model, a great deal of information must be known about the actual processes occurring inside the system. In the physical sciences and engineering, this type of model is of principal interest. Past observations of the behavior of the system are primarily of value in determining what physical processes and laws should be included and what can be ignored in the system, and secondarily, how physical modeling parameters can be calibrated to best represent the response of the system. Examples of physical law models are: Hamiltonian mechanics, Fourier's law of heat conduction, electric charge conservation, the Navier–Stokes equations, Maxwell's equations, and Boltzman's equation. Many physical law models were devised one or two centuries ago, and they form the foundations of the modern simulation of physical systems. With the creation of extremely powerful computers, the technological impact of these fundamental laws is bearing fruit that could not be imagined centuries ago.

The general strategy of model building should include *only* the elements and processes that are important to achieve the goals of the simulation. Often, preliminary simulations show that changes and improvements in the modeling approach are needed to achieve the goals of the simulation. Albert Einstein's classic advice in this matter was: "Make the model as simple as possible, but no simpler." This strategy of using the simplest possible theory to explain reality is also referred to as Occam's Razor. The predictive power of a model depends on its ability to correctly identify the dominant controlling factors and their influences, *not* upon its completeness. A model of limited, but known, applicability is generally more useful from a system design or decision-making perspective than a more complete model that requires more detailed information and computing resources. As wisely stated by Coombs (1964), "We buy information with assumptions." Sometimes, in hindsight, we recognize that the cost is more than we could afford.

Modeling and simulation activities are always conducted by individuals, or teams of individuals. For these types of people, the following general term is introduced.

Computational analysts: individuals who formulate, exercise, produce, and/or interpret mathematical models and computational simulation results.

Sometimes these individuals are referred to as *simulation analysts*, or simply as *analysts*.

Some computational analysts ignore the argument for using appropriately complex models and prefer to use unnecessarily complex models. They make the argument that with continually increasing computing power, analysts should develop increasingly complex models, so as to include all the possibly relevant processes, effects, and interactions. This argument appears reasonable, but it fails to recognize the central axiom in modeling and simulation: *model building is always a trade-off between model complexity and usefulness of the model to achieve the intended goals of the simulation*. Increasing physics modeling complexity is *inevitably* at the expense of simulation result timeliness, nondeterministic simulation, uncertainty quantification, investigation of possible environments and scenarios, and sensitivity analyses. Most fields of physical sciences and engineering are still entrenched in deterministic thinking, so they do not appreciate the need for many simulations in order to conduct uncertainty quantification and sensitivity analysis; identify system failure modes;

and find potential weaknesses in the simulation. As a result of this entrenched view, most fields in physical sciences and engineering have used the vast increases in computer power to increase modeling complexity, often leading to only limited improvement in simulation-informed decision making.

Without exception, the construction of mathematical models of systems involves simplifications of physical reality. Modeling simplifications and assumptions can usually be thought of as one of three types: omission, aggregation, and substitution (Pegden et al., 1995). Omission simply means that certain physical characteristics, processes, features, or events of a system are ignored. For example, suppose the analyst is interested in modeling the heat flux from the surface of a heated solid. If convective heat transfer is the dominant heat transfer mechanism, then the radiation heat transfer might be neglected in the model of the system. Simplification of a model by aggregation means that a characteristic or effect is not ignored but is combined or lumped together into a roughly equivalent effect. In fluid dynamics, if the mean free path of the atoms is much less than the characteristic length of the geometric features in the flow field, then a continuum fluid is normally assumed. Simplification of a model by substitution means that some complex characteristics are replaced by a simpler characteristic. For example, in modeling of hydrocarbon fuel combustion, the number of intermediate gas species that exist is typically in the hundreds. Depending on the needs of the simulation, a simplified combustion model for the number of gas species may only number 10 to 20. The benefit of model simplification is the decreased need for computer resources; thus, allowing time and resources for nondeterministic simulation and improved information quality to the customer.

The mathematical models of interest in this book are primarily given by differential equations or integrodifferential equations. These equations result in either initial value, boundary value, or initial-boundary value problems. If the differential equations are PDEs, they can be elliptic, parabolic, or hyperbolic in character, have two or more independent variables, and have one or more dependent variables. Given the effect of the surroundings on the system, the PDEs describe the relationship between the dependent variables within the system. Information provided about the surroundings, by way of boundary conditions and system excitation conditions, must be *independent* information needed for the solution of the differential equations. The differential equations can be solved by a wide variety of numerical methods, such as, Runge–Kutta, finite element, finite difference, or finite volume methods.

In addition to the primary differential equations of interest, there are commonly submodels or auxiliary models that are required. These submodels can be stated in a variety of mathematical forms: algebraic, transcendental, table-lookup, matrix, differential, or integral equations. Examples of submodels are PDEs for modeling fluid dynamic turbulence, integrodifferential equations for material constitutive properties in shock physics, and integral equations for linear viscoelasticity models in solid mechanics. The submodels can also be stated, all or in part, in tabular form so that numerical interpolation functions are used to construct the required functional relationships.

Figure 3.2 depicts a system of interest and its surroundings. For most systems of interest, the key types of information describing the system are geometry, initial conditions, and physical modeling parameters. For simple systems, the geometry can be specified by engineering drawings and information concerning how the system is assembled. However, for most engineering systems, all the minute details in the geometry are specified in a

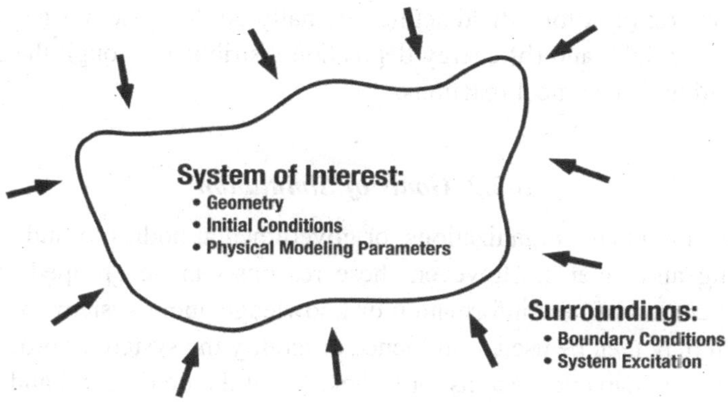

Figure 3.2 Types of information in the system and the surroundings.

computer-aided design (CAD) software package. In addition, computer-aided manufacturing (CAM) software is used for more detail on the actual manufacturing and assembly process, such as deburring and finishing procedures, riveting and bolting specifications, and electrical cable bending and pulling procedures. For systems that are modeled as an initial value problem, initial conditions (ICs) provide required information concerning (a) the initial state of all the dependent variables in the differential equations and (b) the initial state of all other physical modeling parameters, including geometric parameters, that could be dependent on time. If the model is given by PDEs, the IC data could be a function of the remaining independent variables in the PDEs. For example, in a time-dependent heat transfer problem, the initial temperature and thermal conductivity properties could depend on the spatial coordinates in the solution domain. The physical modeling parameters encompass the final element of information characterizing the model of the system. Examples of physical modeling parameters are Young's modulus, mass density, electrical conductivity, thermal conductivity, parameters in constitutive equations, damping and stiffness of assembled joints in a structure, chemical reaction rates, and thermal contact resistance in heat transfer. Some parameters can describe global characteristics of the system, and some can vary as a function of both the independent and dependent variables in the differential equations. Chapter 13 discusses *predictive capability* in detail and how some parameters can be measured independent of the system being modeled and how some must be inferred based on the particular model of the system being used, combined with measurements of the response of the system.

Two types of information must be provided concerning the surroundings: boundary conditions (BCs) and system excitation. BCs provide independent information concerning the dependent variables of the differential equations along the boundary of the domain. If the differential equations are PDEs, then the BCs can be dependent on one or more of the independent variables of the differential equations. These independent variables are typically spatial dimensions, angular coordinates, and time. For example, in a structural dynamics problem, the loading on the structure by way of the BCs can be time dependent. Examples of different types of BCs are: Dirichlet, Neumann, Robin, mixed, periodic, and Cauchy. System excitation refers to how the surroundings affect the system, *other than* through the boundaries. System excitation effects are commonly placed on the right hand side of the differential equations to represent the effect of the surroundings on the system. Common examples of system excitation are: (a) a force field acting internally on the system, such as gravity or an electric or magnetic field; and (b) energy deposition distributed through the system, such as by electrical heating or chemical reactions.

3.2.2 Goals of Simulation

The reasons for individuals, organizations, or governmental bodies to undertake simulation are wide-ranging and diverse. However, these reasons can be grouped into the general objective of generation of new information or knowledge about systems or processes. This new information may then be used to influence or modify the system or process analyzed, to design new and more capable systems, or to help avoid the detrimental and liability impact of a system failure. Or, the information may be used to influence an organization's or society's view of a system or process. The following is a general categorization of the motivations for simulation in engineering and the physical sciences (Bossel, 1994):

(1) *Scientific knowledge*: scientific knowledge means knowledge generated solely for the improved understanding of the universe and humankind's place in it. A good example is the generation of scientific knowledge in astrophysics. The knowledge generated in astrophysics improves human understanding of the universe, without any aspect of influence on the system or use of the knowledge toward other practical or earthly applications.
(2) *Technological knowledge*: technological knowledge means the generation of knowledge used in some way for the creation of applied knowledge or the creation of new physical systems or processes. Technological knowledge is the most common type of knowledge generated from simulation in the applied physical sciences and engineering.

The majority of technological knowledge is used to improve the efficiency or effectiveness of existing systems, design new systems, or assess the impact of existing or proposed systems. For systems not yet in existence, key drivers in simulation are issues such as: (a) creation and design of more capable systems than are presently on the market; (b) creation of new materials and manufacturing processes to reduce costs and time to market of new systems; and (c) prediction of the potential environmental impact of new systems or manufacturing processes, not only at the present time, but also in the future. Examples of generation of new technological knowledge for systems already in existence are: (a) improvement of the performance, safety, or reliability of existing systems; (b) optimization of chemical processes for improved production output; (c) reduction in fuel consumption of transportation vehicles; and (d) safety improvements to existing and future nuclear power plants, particularly improvements to address new risks to the plant or the environment.

Simulation is taking on increased public scrutiny, particularly for risk assessment of high-consequence systems. The catastrophic failure of nuclear reactor number four at the Chernobyl nuclear power plant in Ukraine in 1986 riveted worldwide attention on the dangers of nuclear power. Although simulation primarily deals with technological issues, these issues commonly become convolved with public perceptions of risk, national responsibilities for the impact of technologies on the global climate, and organizational responsibilities for product and environmental liability. As is clear from these examples, the importance of simulation for the generation of reliable technological knowledge is greatly expanding. Correspondingly, there is a compelling need to construct models that are technically sound, where the assumptions and uncertainties in the models are clearly revealed, and the modeling results are comprehensible to a wide range of audiences. The technical difficulties in achieving these goals in simulation are daunting in themselves; however, within the inevitable human, social, cultural, and political context, achieving them becomes nearly impossible.

The obvious alternative, and historically the traditional method to the generation of new technological knowledge, is the execution of a physical experiment. There are advantages and disadvantages to the physical experiment route, just as there are to the simulation approach. The cost and/or time schedule required to conduct a physical experiment may be considerably more, or even impossible, compared to that of simulation. The uncompetitive nature or impossibility of physical experiments may include testing to failure of large-scale public works projects, such as skyscrapers, bridges, and dams. However, because of the time scales and costs involved in some physical processes, it may be unrealistic to consider a physical experiment. For example, a physical experiment is unrealistic when considering the long-term disposal of radioactive nuclear wastes and long-term environmental impact of CO_2 production.

3.3 Nondeterministic Simulation

3.3.1 Importance of Nondeterministic Simulation

In many science and engineering communities, particularly within research activities, simulation is viewed strictly as deterministic. Commonly, the purpose of these investigations is the discovery or understanding of new physical phenomena or the characteristics of a system. Nondeterministic characteristics of the phenomena or system are commonly of secondary importance. (Note that the term *stochastic* is commonly used in the same sense as nondeterministic.) Sometimes in these analyses, it is explicitly stated or implied that since the investigator is only interested in *nominal* responses of the system, the nondeterministic features can be ignored. The nominal values of the uncertain inputs may be specified as the mean or most commonly observed value of each of the simulation inputs. However, it is *rarely true* that the statistical mean of the output quantities can be determined by performing a calculation for a single set of inputs chosen to be the statistical mean of each of the uncertain inputs. Thus, the mean value of the output *cannot* be computed by performing a simulation using the mean value of all input parameters, except when the solution is *linear* in the parameters of the model. Linearity of the model parameters in the solution essentially never occurs when the mapping of inputs to outputs is given by a differential equation, *even a linear* differential equation. The magnitude of error in this approximation depends on the nonlinearity of the parameters in the solution to the differential equation, as well as the nonlinearity of the differential equations themselves.

In most engineering applications, as well as applications in the physical sciences, deterministic simulations are unacceptable approximations. The effort devoted to estimating nondeterministic effects can vary widely depending on the goals of the simulation; the expected performance, safety, and reliability of the system; and the possible consequences of the system not achieving its goals and the consequences of system failure. An example of a system that might not require nondeterministic simulation is the design and operation of a simple consumer product. The simulation of the operation or use of the product may only consider a few important design conditions, and the analyst may use experience-based conservative design margins. This type of design is commonly referred to as *rule-based design*. This is an effective and efficient design procedure of simple parts or products where a deterministic simulation is satisfactory, given that adequate design margins are incorporated in the design and there are no significant risks related to injury of misuse of the item.

Simulation of complex engineering systems with demanding performance and safety requirements and high-consequence systems must include the nondeterministic features of the system and the surroundings. Sources of uncertainty can vary widely depending on whether the simulation is for normal, abnormal, or hostile environments. Several fields that regularly employ nondeterministic simulations are nuclear reactor safety, mechanical and electrical power supply systems, underground transport of toxic and radioactive waste materials, structural dynamics, weather forecasting, and environmental impact assessment. For example, the term *cone of uncertainty* in hurricane track forecasting is now well recognized even by the general public. The emphasis in most of these fields has been directed toward representing and propagating input parameter uncertainties through the mathematical model to obtain uncertain system responses. The majority of this work has used traditional probabilistic methods, where no distinction is made in the mathematical characterization of aleatory and epistemic uncertainty.

3.3.2 Concept of Nondeterministic Simulation

The central concept in nondeterministic simulations is that a single solution to the mathematical model is no longer sufficient to represent the response of the system. A set, or ensemble, of calculations must be performed to map the uncertain inputs to the uncertain outputs. This is also referred to as *stochastic simulation* or *ensemble simulation*, instead of nondeterministic simulations. Figure 3.3 depicts the propagation of input uncertainties through the model to obtain output uncertainties. The number of individual calculations needed to accurately accomplish the mapping depends on three key factors: (a) the nature, e.g., whether they are aleatory or epistemic uncertainties, and magnitude of the uncertainties; (b) the nonlinearity of the differential equations; and (c) the numerical methods used to compute the mapping of input to output. The number of mapping evaluations (i.e., individual, deterministic, numerical solutions of the mathematical model) can range from tens to

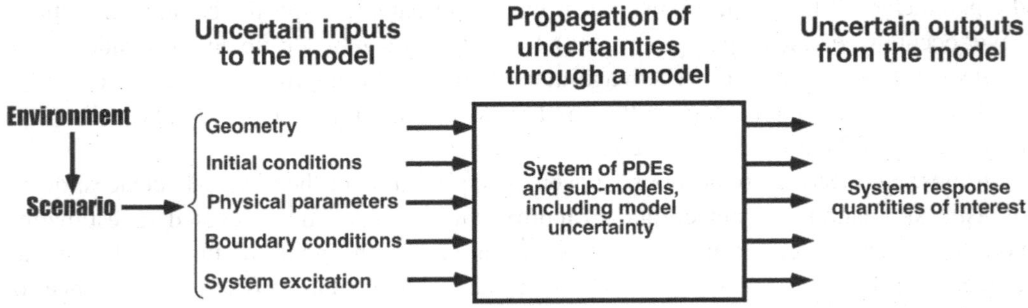

Figure 3.3 Propagation of input uncertainties to obtain output uncertainties.

hundreds to thousands. Obviously, thousands of sampled solutions are shocking to those accustomed to a single solution to a set of differential equations.

The formal mathematical structure that maps model inputs to model outputs, i.e., SRQs, can be written as

$$\mathbb{M}(\mathcal{E}, S; \mathcal{G}, \mathcal{IC}, \mathcal{MP}, \mathcal{BC}, \mathcal{SE}) \to SRQ. \quad (3.1)$$

$\mathbb{M}$ is the specification of the mathematical model given by the differential or integrodifferential equations, $\mathcal{E}$ is the environment of the system, S is the scenario of the system, $\mathcal{G}$ is the geometry of the system, $\mathcal{IC}$ are the initial conditions of the system, $\mathcal{MP}$ are the model parameters of the system, $\mathcal{BC}$ are the boundary conditions imposed by the surroundings, and $\mathcal{SE}$ is the system excitation imposed by the surroundings. $\mathcal{G}$, $\mathcal{IC}$, $\mathcal{MP}$, $\mathcal{BC}$, and $\mathcal{SE}$ are all conditional on the specified environment $\mathcal{E}$ and the scenario of interest S. If $\mathcal{G}$, $\mathcal{IC}$, $\mathcal{MP}$, $\mathcal{BC}$, and $\mathcal{SE}$ are all completely specified, either deterministically or nondeterministically, then a solution to the mathematical model $\mathbb{M}$ can be computed. The form of the model as given by Equation (3.1) is referred to as the *strong definition of a model* (Leijnse and Hassanizadeh, 1994).

If the model is an explanatory type model, e.g., it conserves mass, momentum, and energy, the predictive capability of the model can be quite powerful. This is because the specification of $\mathcal{G}$, $\mathcal{IC}$, $\mathcal{MP}$, $\mathcal{BC}$, and $\mathcal{SE}$ for the application domain of the input data are rarely within the validation domain, as discussed in Chapter 2. The application domain is sometimes referred to as the intended use domain or the context of use domain. The predictive capability of the model to extrapolate to conditions that have never been observed before, i.e., outside of the validation domain, relies on the explanatory capability of the model. This does *not* mean that the explanatory capability of the model is fully reliable. As discussed in Chapter 2, the model form uncertainty that can be estimated at the experimental conditions defining the validation domain is a measure of the observed reliability of the model. To be more precise, as the model form uncertainty approaches zero at the available experimental conditions, the evidence of model predictive capability increases. This is especially true if the model predictions for the experimental conditions are blind predictions.

SRQs can simply be the dependent variables in the differential equations in the mathematical model, for example, local temperature, flow velocity, strain, or chemical concentration. They can also be more complex quantities, such as, derivatives of dependent variables, e.g., local heat transfer rate or mass diffusion rate. Furthermore, they can be functionals of dependent variables, such as, aircraft lift and drag, and total energy absorbed by deformation, tearing, and recontact friction of a structure during collapse. Generally, if *any* input quantity to the model is uncertain, then the SRQs are uncertain.

The weak definition of a model is one where only the differential or integrodifferential equations are specified, given $\mathcal{E}$ and S (Leijnse and Hassanizadeh, 1994). The weak definition of a model could then be written

$$\mathbb{M}(\mathcal{E}, S) \to SRQ. \quad (3.2)$$

For a model given by Equation (3.2), the SRQs cannot be determined because of the lack of specificity in the input data needed for a solution. Hence, without the specification of $\mathcal{G}$, $\mathcal{IC}$, $\mathcal{MP}$, $\mathcal{BC}$, and $\mathcal{SE}$, there are an infinite number of solutions that can satisfy the differential or integrodifferential equations. Additionally, the weak definition of a model result *cannot* be assessed for accuracy.

Many techniques exist for propagating input uncertainties through the mathematical model to obtain uncertainties in the SRQs. The major categories are: sampling techniques, such as Monte Carlo sampling; perturbation methods, such as first-order and second-moment methods; and stochastic spectral methods, such as stochastic Galerkin methods (Smith, 2014; McClarren, 2018). Sampling techniques are the most common approach because of a number of advantages: (a) they can be applied to essentially any type of mathematical model, regardless of the model's complexity or nonlinearity; (b) they can be applied to both aleatory and epistemic uncertainties, regardless of the magnitude of the uncertainties; and (c) they are not intrusive to the numerical solution of the mathematical model, i.e., sampling is done outside the numerical solution to the differential equations. Their key disadvantage is that they are computationally expensive because the number of mapping solutions can be very large in order to obtain the needed sampling statistics for the SRQs.

3.3.3 Propagation of Input to Output Uncertainties

Although more computationally efficient methods are available for propagating input to output uncertainties, this book focuses on the use of sampling methods. This is done primarily because sampling methods are easier to understand for those new to nondeterministic simulation, and they are relatively easy to incorporate into any type of existing deterministic modeling software. This last advantage is a result of the relationship between deterministic and nondeterministic simulation. Specifically, the sampling method can be an external software driver, or software wrapper, outside of the existing deterministic model. Referring to Figure 3.4, a sequence of samples can be drawn from the uncertain input data ($\mathcal{G}$, $\mathcal{IC}$, $\mathcal{MP}$, $\mathcal{BC}$, and $\mathcal{SE}$), where each sample is used to obtain a solution to the model, resulting in a sequence of output samples for the SRQs.

In deterministic simulation, a set of *input scalars* ($\mathcal{G}$, $\mathcal{IC}$, $\mathcal{MP}$, $\mathcal{BC}$, and $\mathcal{SE}$) is mapped using the model to obtain a set of *output scalars*, the SRQs. In nondeterministic simulation, a set of *functions* ($\mathcal{G}$, $\mathcal{IC}$, $\mathcal{MP}$, $\mathcal{BC}$, and $\mathcal{SE}$) is mapped using the model to obtain a set of *output functions*, the SRQs. The inputs and outputs are both functions in a nondeterministic simulation because they characterize uncertainties, for example, probability distributions and intervals. Since efficient sampling techniques are different for propagating aleatory uncertainties versus epistemic uncertainties, each propagation technique is introduced separately in this chapter.

3.3.3.1 Propagation of Aleatory Uncertainty

To introduce the concept of propagation of uncertain inputs to obtain uncertain outputs, Equation (3.1) is simplified as

$$y = f(x), \tag{3.3}$$

where y is the single SRQ, $f(\bullet)$ is the mathematical model, and x is the only uncertain input quantity. All other input quantities are considered deterministic. Assume that x is a random variable that is given by some probability density function (PDF) or cumulative distribution function (CDF). Simple or basic Monte Carlo sampling (MCS) will be used to generate a sequence of samples of x which will be propagated through the model. The sequence of propagated samples will then be used to generate a probability distribution for y.

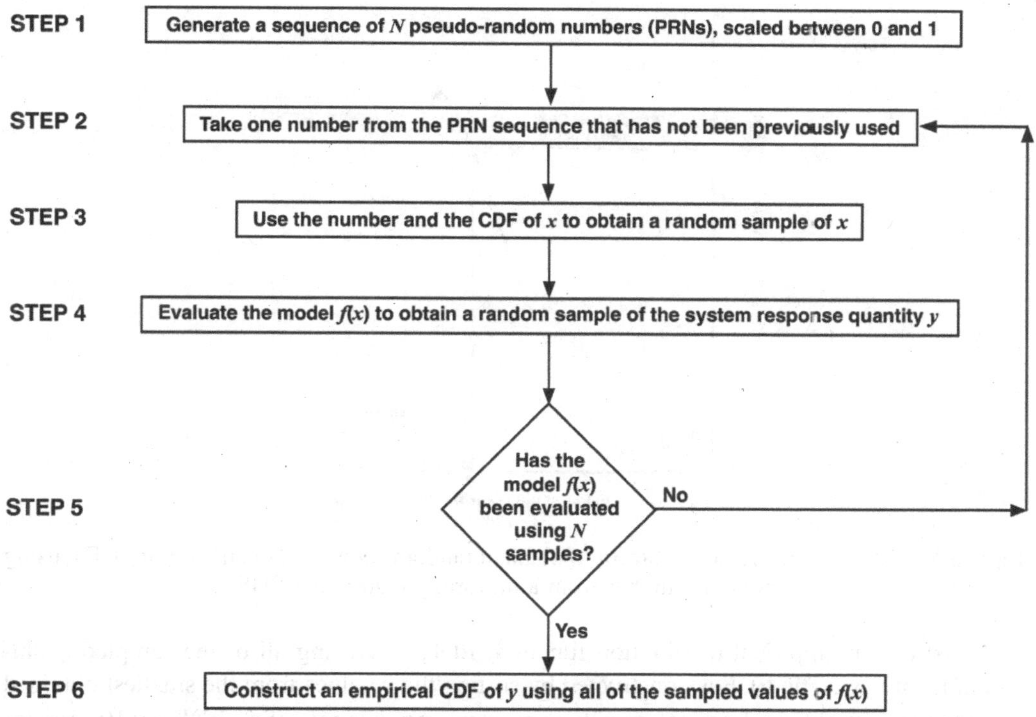

Figure 3.4 Flowchart for simple Monte Carlo sampling of one aleatory uncertainty.

Figure 3.4 shows a flowchart of the steps to compute a probability distribution for y, given the input random variable x, and the mathematical model $f(\bullet)$. The following explains each of the steps in more detail:

1. Generate a sequence of N pseudo-random numbers (PRN) that are scaled between zero and unity. A PRN generator produces a sequence of random numbers that can be repeated, given that the same seed is provided to the generator. Most modern software packages have PRN generators.
2. Pick the first number from the PRN sequence generated in Step 1.
3. Use the number picked in Step 2, combined with the given CDF of x, to generate a random sample of x. The procedure to generate a random sample of x, given a PRN, is based on the probability integral transform theorem (Angus, 1994). The sequence of PRN can be viewed as random draws from a uniform PDF over the interval (0,1). The probability integral transform theorem proves that the inverse transform can be used to convert a random variable from a uniform distribution to a random variable from an arbitrary distribution, if the CDF of the arbitrary distribution is known. The inverse transform process is depicted in Figure 3.5.
4. Evaluate the mathematical model $f(\bullet)$ for the value of x obtained in Step 3. The evaluation of $f(\bullet)$ would usually require the solution of a differential or integrodifferential equation.
5. Test if all N PRN from Step 1 have been used. If no, return to Step 2 and use the next PRN in the sequence. If yes, go to Step 6.

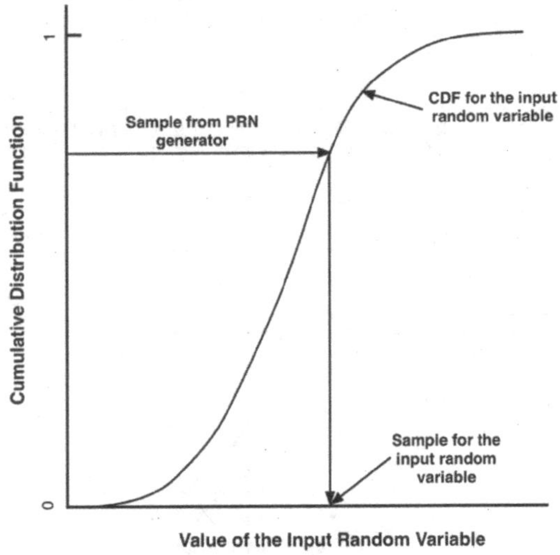

Figure 3.5 Use of the inverse transform to obtain a random sample from an arbitrary CDF using a random number from a uniformly distributed PDF.

6. Construct an empirical distribution function, EDF, of y using all of the sampled evaluations of $f(x)$. The EDF is constructed by sorting the y values from the smallest obtained to the largest obtained, monotonically increasing. An example of an EDF for 10 samples of y is shown in Figure 3.6.

Since 10 samples of y have been obtained in Figure 3.6, the increment in probability must be 1/10. Given the axioms of traditional probability theory, the total probability of the sample set must be unity. The computed EDF of y approximates the exact CDF of y. The numerical solution error, indicated by the difference between the two, is referred to as *sampling uncertainty.* As the number of samples increases, the accuracy of the EDF improves. The statistical features of the EDF converge at a rate of $1/\sqrt{N}$. However, for a given number of samples, the statistical mean is more accurate than the standard deviation of the SRQ, i.e., the second moment of the SRQ. In the statement of the simulation requirements, there may be a requirement for the numerical accuracy, or statistical confidence, in various statistical features of the SRQs. Chapter 13 discusses various types of sampling techniques in more detail.

3.3.3.2 Example of Propagation of Aleatory Uncertainty

Consider the simulation of the oscillation of a mass-spring-damper system that is acted upon by a time-dependent excitation force (Figure 3.7). For a linear spring and damper, the ODE describing the oscillation of the system is given by

$$m\frac{d^2x}{dt^2} + c\frac{dx}{dt} + kx = F(t) \quad \text{Initial conditions: } x(0) = x_0 \text{ and } \left(\frac{dx}{dt}\right)_{t=0} = \dot{x}_0. \tag{3.4}$$

$x(t)$ is the displacement of the mass as a function of time, m is the mass of the system, c is the damping coefficient, k is the spring constant, and $F(t)$ is the external forcing function.

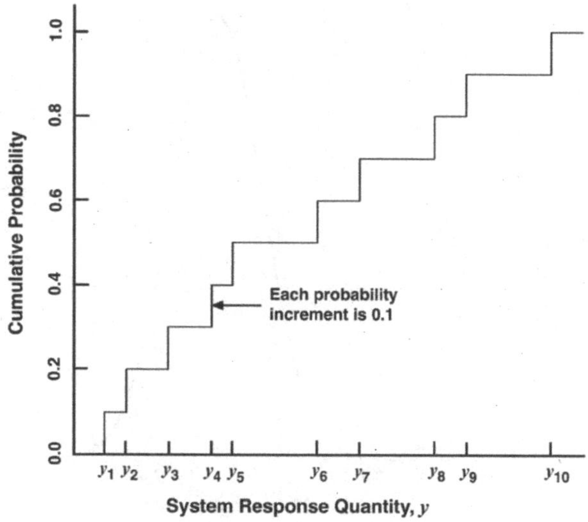

Figure 3.6 Example of an empirical distribution function, EDF, constructed from an ordered sequence of 10 random samples of y.

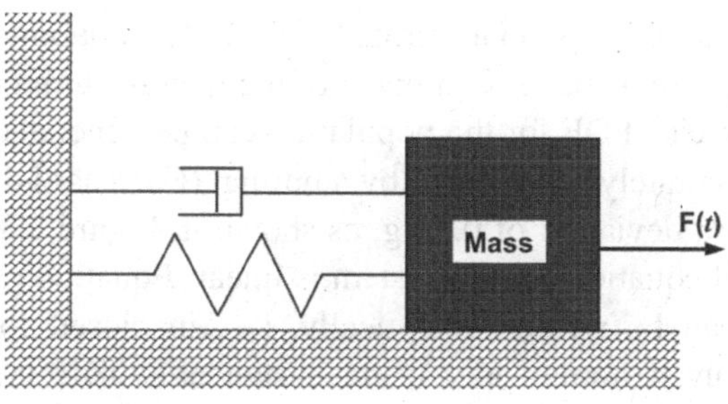

Figure 3.7 Example of a mass-spring-damper system.

Assume that the damping coefficient, c, the spring constant, k, the initial state of the system, x_o *and* $\dot{x}_o$, and the forcing function, $F(t)$, are all precisely known, i.e., they are deterministic. These values are given as

$$c = 1\ \text{N/m/s}, \quad k = 100\ \text{N/m}, \quad x_o = 10\ \text{m}, \quad \dot{x}_o = 0\ \text{m/s}, \tag{3.5}$$

and

$$F(t) = \begin{cases} 0 & \text{for } 0 \leq t < < 2\ \text{s}, \\ 1{,}000\ \text{N} & \text{for } 2 \leq t \leq 2.5\ \text{s}, \\ 0 & \text{for } 2.5\ \text{s} < < t. \end{cases} \tag{3.6}$$

As can be seen from Equations (3.5) and (3.6), the initial conditions are a displacement of 10 m and a velocity of zero. In addition, it is seen that the excitation function, $F(t)$, only comes into play during the time period of 2 to 2.5 s.

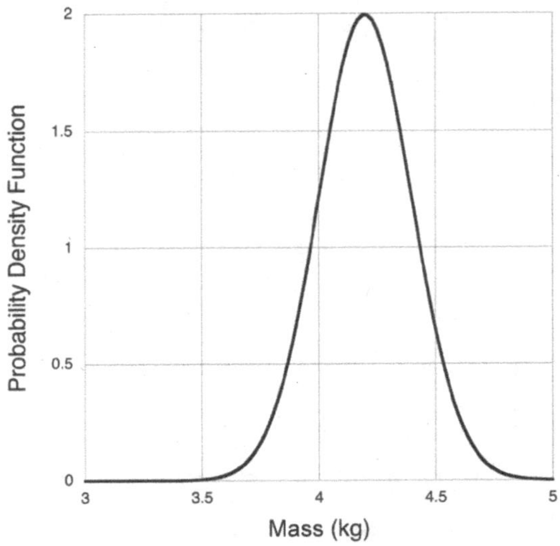

Figure 3.8 Probability density function of the system mass population.

The mass of the system, m, is nondeterministic due to variability in its manufacturing. A large number of inspections have been made of the manufactured masses that are used in the system so that a precise PDF for the population can be generated. It was found that the population could be accurately represented by a normal (Gaussian) distribution with a mean of 4.2 kg and a standard deviation of 0.2 kg, as shown in Figure 3.8.

Since the differential equation for the system is linear, Equation (3.4), the solution to the mathematical model can be written analytically, i.e., in closed form. However, for our purposes it will be solved numerically using a standard ODE solver. Equation (3.4) is converted into two first-order ODEs, one for the time-dependent displacement of the mass and one for the time dependent velocity of the mass. The equations are then solved numerically using MATLAB's Runge–Kutta 4(5) method, *ode45*. The numerical solution error criterion for each time-step advancement is set to be less than 1E-3 for the relative error and less than 1E-6 absolute error for each of the dependent variables.

Since the nondeterministic nature of the system is purely aleatory, traditional sampling methods can be used to propagate the mass uncertainty into uncertainty of the SRQs of interest. Basic Monte Carlo sampling is used and incorporated in MATLAB's normal distribution sampler *randn* in order to obtain samples for the mass. The mean, μ, is set to 4.2 kg, and the standard deviation, σ, is set to 0.2 kg. In nondeterministic simulations using sampling, a random-number seed is required so that the analyst can reproduce precisely the same sequence of random numbers using the PRN generator. In the MATLAB program *randn*, the default seed of 0 is used for 10 samples and a seed of 1 is used for 100 samples. By changing the seed, a different sequence of random numbers is generated for each nondeterministic simulation. This technique is referred to as *replicated Monte Carlo sampling*, which means that an independent sequence of samples is generated, but each sequence has the same statistical properties.

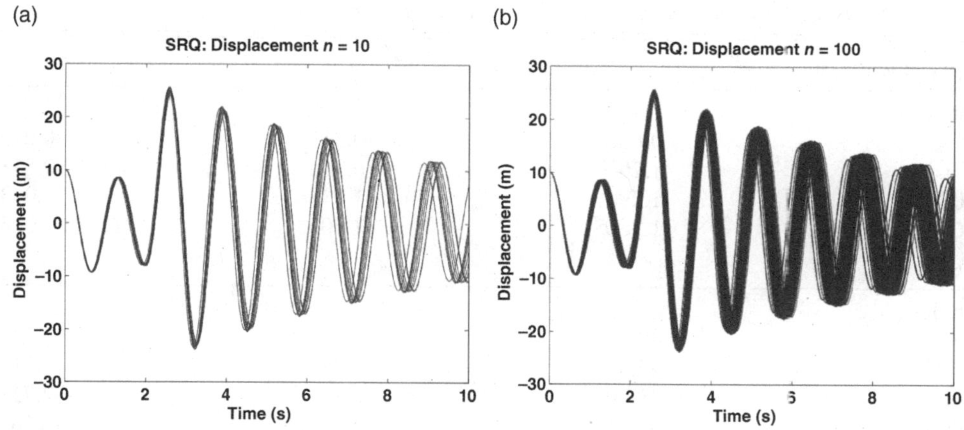

Figure 3.9 Mass displacement as a function of time for aleatory uncertainty. (a) 10 Monte Carlo samples. (b) 100 Monte Carlo samples.

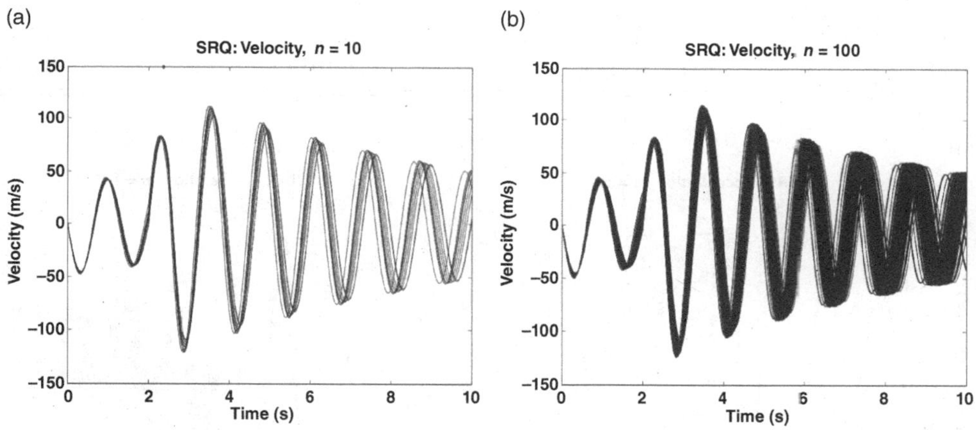

Figure 3.10 Mass velocity as a function of time for aleatory uncertainty. (a) 10 Monte Carlo samples. (b) 100 Monte Carlo samples.

Three common SRQs computed for this type of system are position, velocity, and acceleration, $x(t)$, $\dot{x}(t)$, and $\ddot{x}(t)$, respectively. Figure 3.9, Figure 3.10, and Figure 3.11 show $x(t)$, $\dot{x}(t)$, and $\ddot{x}(t)$, respectively, for time up to 10 s, for $N = 10$ and 100. The expected oscillatory motion is seen in each of the SRQs. In each plot, every solution to the ODE is shown from the Monte Carlo samples. As a result, it is difficult to see any of the individual numerical solutions in the plot for $N = 100$. It is clear that after about 10 s, essentially nothing can be predicted about the displacement, velocity, and acceleration of the system because of the cumulative impact of the variability in the input mass. As a final point, the effect of the excitation function during the time period of 2 to 2.5 s cannot be seen in the displacement plot, it is just barely noticeable in the velocity plot, and is clearly visible in the acceleration plot.

Since the nondeterministic simulation of each of the SRQs is an ensemble of results, it is commonly appropriate to summarize the results in terms of various statistical measures.

Table 3.1 *Statistics for displacement, velocity, and acceleration for* t = *1 s*

Displacement x, m			
Samples (N)	$\hat{\mu}$	$\hat{\sigma}$	$\hat{\gamma}$
10	0.65	0.76	−0.022
100	1.20	0.962	0.548
1,000	1.290	1.033	0.02363
10,000	1.2868	1.0063	0.08675
Velocity $\dot{x}$, m/s			
Samples (N)	$\hat{\mu}$	$\hat{\sigma}$	$\hat{\gamma}$
10	42	0.29	−1.1
100	42.4	0.462	−3.21
1,000	42.40	0.4580	−2.758
10,000	42.411	0.47759	−3.1633
Acceleration $\ddot{x}$, m/s^2			
Samples (N)	$\hat{\mu}$	$\hat{\sigma}$	$\hat{\gamma}$
10	−25	19	−0.098
100	−39.7	25.3	−0.782
1,000	−42.15	26.78	−0.2513
10,000	−42.004	26.1612	−0.34747

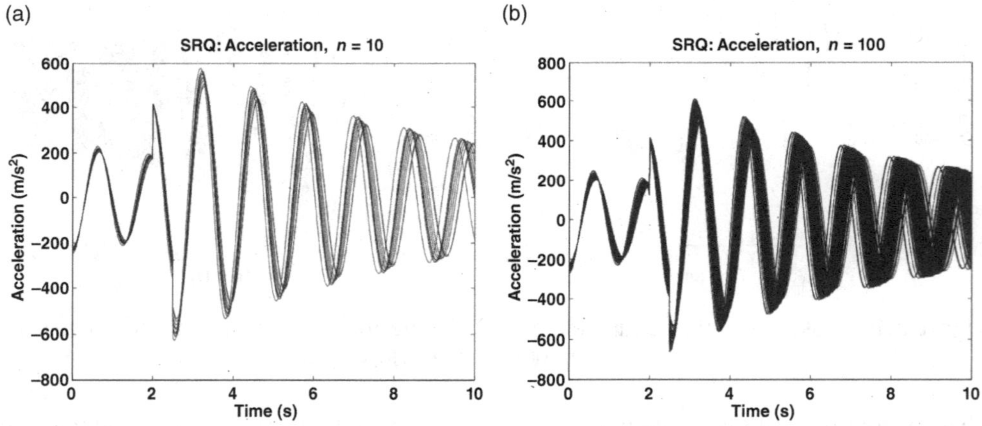

Figure 3.11 Mass acceleration as a function of time for aleatory uncertainty. (a) 10 Monte Carlo samples. (b) 100 Monte Carlo samples.

Tables 3.1 and 3.2 show the estimated mean, standard deviation, and skewness of $x(t)$, $\dot{x}(t)$, and $\ddot{x}(t)$ at $t = 1$ and $t = 5$ s, respectively. These three statistical measures are the first three moments of the data samples. Recall that the mean (μ) is a measure of the central tendency of the data, the standard deviation (σ) indicates the variation or dispersion of the data, and the skewness (γ) is a measure of the statistical asymmetry of the data. Each table shows the first three moments as a function of the number of samples computed. For $n = 10$; 100; 1,000; and 10,000, using seed values of 0, 1, 2, and 3, respectively, in MATLAB. Consequently, each ensemble of output results will be a *different set of realizations* for the system. The (^) symbol indicates that the values for μ, σ, and γ are estimates of

Table 3.2 *Statistics for displacement, velocity, and acceleration for* t = *5 s*

Displacement x, m			
Samples (N)	$\hat{\mu}$	$\hat{\sigma}$	$\hat{\gamma}$
10	10	4.3	−0.36
100	12.7	4.10	−0.556
1,000	13.05	4.690	−0.9870
10,000	13.091	4.5261	−1.0697
Velocity $\dot{x}$, m/s			
Samples (N)	$\hat{\mu}$	$\hat{\sigma}$	$\hat{\gamma}$
10	71	14	−1.2
100	57.3	26.2	−1.39
1,000	53.95	26.89	−0.9136
10,000	54.398	26.492	−1.0608
Acceleration $\ddot{x}$, m/s^2			
Samples (N)	$\hat{\mu}$	$\hat{\sigma}$	$\hat{\gamma}$
10	−260	105	0.27
100	−320	105	0.403
1,000	−329.4	118.1	0.8020
10,000	−330.04	113.90	0.86000

the quantities, as opposed to exact values of the population. Since Monte Carlo sampling is an unbiased estimator of the statistics, the bias error of each set of samples approaches zero as the number of samples becomes large. In the limit as the number of samples increases, $\hat{\mu} \to \mu$, $\hat{\sigma} \to \sigma$, and $\hat{\gamma} \to \gamma$.

Two observations are made regarding the data shown in Tables 3.1 and 3.2. First, a rule of thumb concerning MCS is that one digit of accuracy can generally be obtained with 10 samples for each of the first two moments. This rule is seen to generally hold for the two moments shown in the tables. Furthermore, for each factor of 10 increase in the number of samples, one additional digit of accuracy is generally added. For example, as seen in the tables, 10,000 samples are required to obtain four digits of accuracy for the two moments. The convergence of skewness as a function of number of samples is very erratic because of the sensitivity of this statistical parameter to changes in the tails of the distribution. Second, the mean and the standard deviation increase substantially from 1 s to 5 s. As mentioned earlier, this reflects the inability to predict the system response as time increases. Since this is a common feature of nondeterministic vibrational systems, other more robust SRQs are commonly of interest; specifically, the frequency and amplitude of the vibration.

A more detailed method of presenting the results of a nondeterministic simulation is to show a plot of the computed EDF of each of the SRQs. The EDF shows the fraction of the *sampled* population that would have a value equal to, or less than, a particular value of the SRQ. Figures 3.12 and 3.13 show the EDF of $x(t)$, $\dot{x}(t)$, and $\ddot{x}(t)$ at $t = 1$ and 5 s, respectively, for 10, 100, and 1,000 samples computed. The most notable feature in each graph is the stair-step nature of the curve for $n = 10$ samples. With only 10 samples, each sample *must* represent a probability jump of 0.1. Similarly for 100 samples, the jump in probability is 0.01. For a low number of samples, it is clear that the tails of the distribution are the most inaccurate

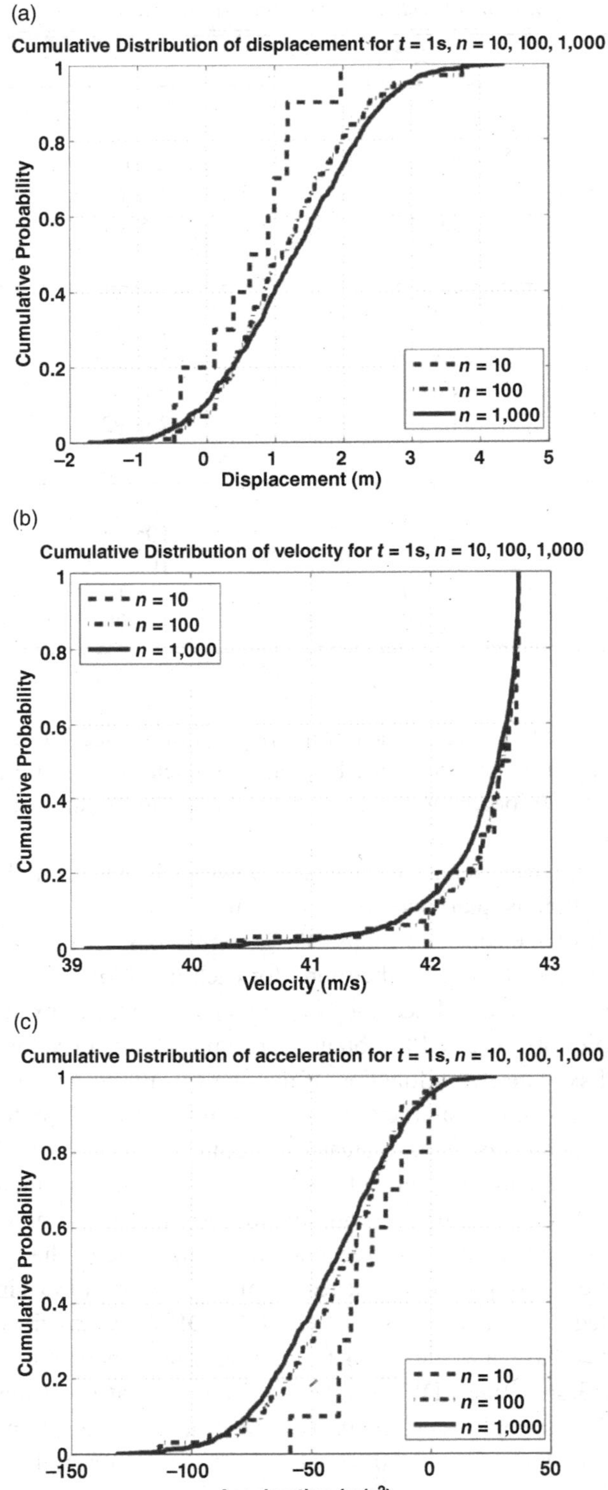

Figure 3.12 Empirical distribution functions for $t = 1$ s for aleatory uncertainty. (a) Empirical distribution function for displacement for $t = 1$ s. (b) Empirical distribution function for velocity for $t = 1$ s. (c) Empirical distribution function for acceleration for $t = 1$ s.

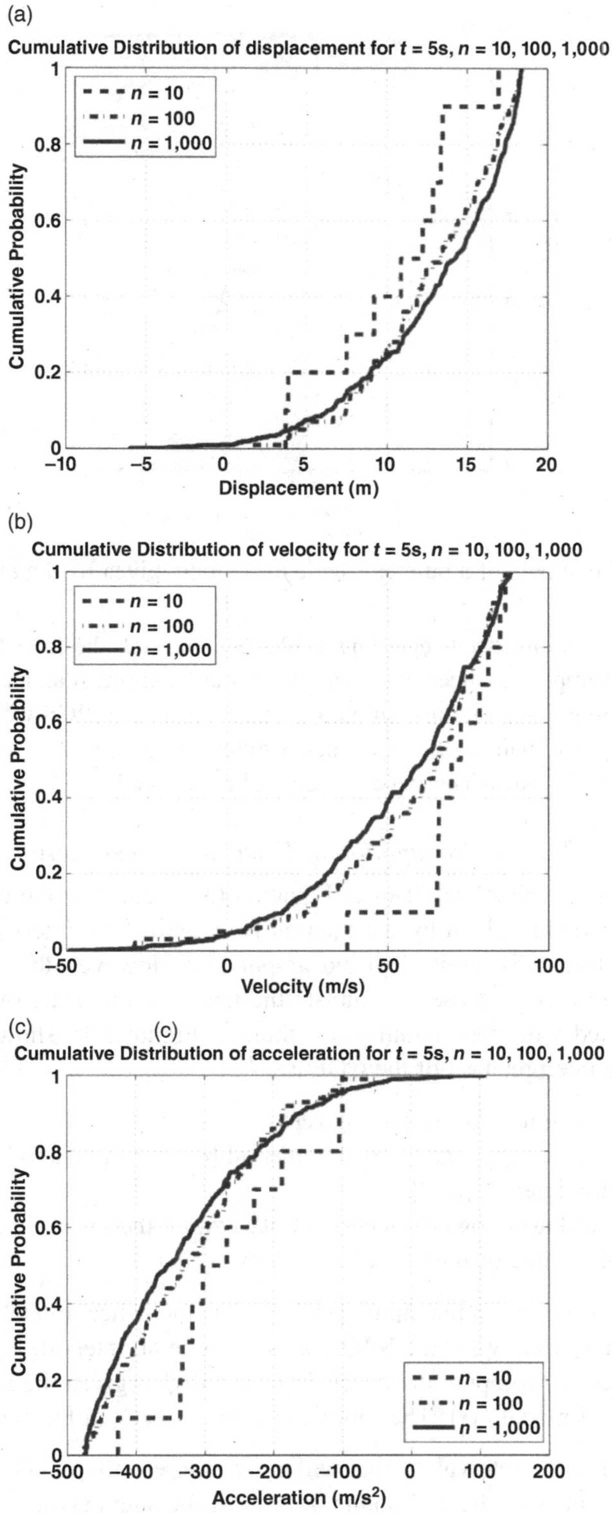

Figure 3.13 Empirical distribution functions for $t = 5$ s for aleatory uncertainty. (a) Empirical distribution function for displacement for $t = 5$ s. (b) Empirical distribution function for velocity for $t = 5$ s. (c) Empirical distribution function for acceleration for $t = 5$ s.

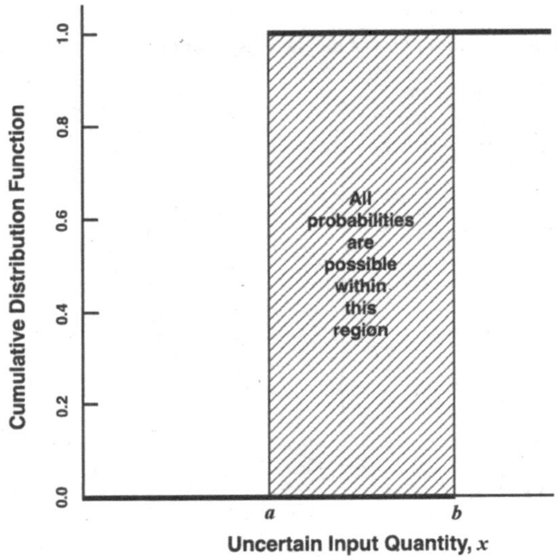

Figure 3.14 Example of a pure epistemic uncertainty given by the interval $[a, b]$.

feature of the EDF. As discussed regarding Tables 3.1 and 3.2, the finer the statistical detail required, the more samples are needed to estimate it. For example, if an analyst was interested in estimating the chance of a one-in-a-million event, then simple MCS would require about a million samples. For estimation of rare events, a much more efficient approach is the use of variance reduction techniques. These are discussed in Chapter 13.

3.3.3.3 Propagation of Epistemic Uncertainty

Consider the simplified model as given in Equation (3.3), but now the uncertain input is a pure epistemic uncertainty given by the interval $[a, b]$. Since an interval is such a simple mathematical structure, it is rarely depicted graphically. However, for the purpose of this text, it is useful to do so because it contrasts the near-vacuous state of knowledge of an interval as contrasted with a probability distribution. Figure 3.14 shows the CDF of the interval $[a, b]$. The interpretation of the CDF is:

- the probability of a value of a, or less, is zero,
- the probability of a value x, where x is the interval $[a, b]$, can be anything, as long as the probability is in the interval $[0, 1]$,
- there is no claim of knowledge of likelihood of x within the interval $[a, b]$, and
- the probability of a value of b, or greater, is unity.

Since the only uncertainty in the input data for the mathematical model $f(\bullet)$ is an interval, it is clear that the uncertainty for the SRQ y *must also* be an interval. Figure 3.15 shows a flowchart of the steps to compute the interval-valued SRQ y, given the input interval $[a, b]$ and the mathematical model $f(\bullet)$. The following explains each of the steps in more detail:

1. Pick the number of subintervals N that will be used to equally divide the interval $[a, b]$.
2. Calculate the x values of the end points of each of the subintervals. This sequence of x values would begin with a and end with b.

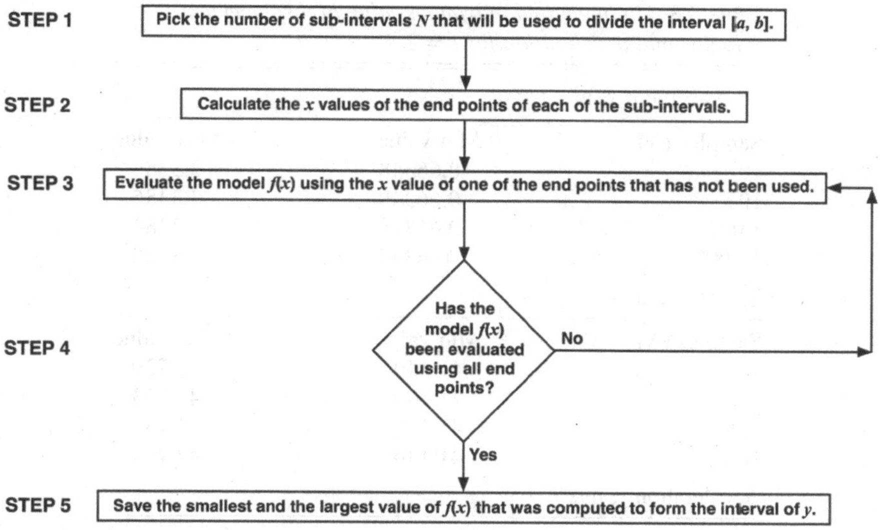

Figure 3.15 Flowchart for equal-partition sampling of one epistemically uncertain input.

3. Evaluate the mathematical model $f(\bullet)$ using one of the x values calculated in Step 2 that have not been used. The evaluation of $f(\bullet)$ would usually require the solution of a differential or integrodifferential equation.
4. Test if all of the $N + 1$ values of subinterval end points have been used. If No, return to Step 3 and use the next x value in the sequence. If Yes, go to Step 6.
5. Find the smallest and largest value of $f(x)$ that has been evaluated. The smallest and largest values will form the SRQ interval for y.

Sampling an interval uncertainty can be done in a number of ways. Equal-partition sampling is recommended, as opposed to MCS (Ferson and Ginzburg, 1996; Beer et al., 2013). The equal-partition sampling stresses that the sampling procedure is *completely unrelated* to any likelihood structure of the uncertainty. As mentioned earlier, there is no informational structure within the interval. The equal-partition sampling procedure ensures sampling of all end points of the subintervals, as well as the end points of the full interval. A general statement of sufficiently refined coverage over the interval cannot be made without making assumptions concerning the smoothness of the mathematical model with respect to the uncertain input. Hence, the convergence rate of sampling an interval-valued uncertainty cannot be estimated in general because it depends entirely on how $f(x)$ depends on the interval. However, it can be proven that

$$\lim_{N \to \infty} f(x_N) \to [y_{\min}, y_{\max}], \tag{3.7}$$

where x_N are the sampled points over the input interval $[a, b]$, and $y_{\min}$ and $y_{\max}$ are the exact limits on the SRQ interval, given the uncertain input interval.

3.3.3.4 Example of Propagation of Epistemic Uncertainty

Consider the mass-spring-damper system discussed earlier, except now the uncertainty of the mass is characterized as an interval. As an example of how such little information would be

Table 3.3 *Minimum and maximum value of displacement, velocity, and acceleration for* $t = 1$ s

Displacement x, m		
Samples (N)	Min value	Max value
10	−0.66348	3.3386
101	−0.66348	3.3386
1,002	−0.66348	3.3386
10,003	−0.66348	3.3386
Velocity $\dot{x}$, m/s		
Samples (N)	Min value	Max value
10	41.140	42.729
101	41.140	42.733
1,002	41.140	42.733
10,003	41.140	42.733
Acceleration $\ddot{x}$, m/s^2		
Samples (N)	Min value	Max value
10	−98.685	5.3404
101	−98.685	5.3404
1,002	−98.685	5.3404
10,003	−98.685	5.3404

known in practice, consider the following acceptance process for the masses produced. A manufactured mass passes inspection if it was found to have a mass greater than a minimum value and a mass less than a maximum value. A probability distribution for the accepted masses exists, but it was decided that knowledge of the distribution was not worth the time and expense of making all of the measurements. As a result, the only statement that can be confidently made is that an accepted mass is in an interval. For the mass-spring-damper system of interest, suppose that the interval for the mass is [3.8, 4.6] kg, which is $\pm 2\sigma$ for the previous aleatory uncertainty.

A similar number of samples, N, are used as the previous example with the aleatory uncertainty: 10; 101; 1,002; and 10,003. Using this sequence of number of sample points, which includes the end points of the interval, forces the sampled points to be different for each ensemble, except for the end points of the interval. This is analogous to replicated MCS because every ensemble result is different. The same Runge–Kutta 4(5) method in MATLAB, *ode45*, is used to integrate the ODE describing the oscillation of the mass. Furthermore, the same numerical solution error criterion as given earlier is used, i.e., the relative error for displacement and velocity for each time step is less than 1E-3, and the absolute error for displacement and velocity for each time step is less than 1E-6.

The results for the interval-valued uncertainty of $x(t)$, $\dot{x}(t)$, and $\ddot{x}(t)$ at $t = 1$ and 5 s are shown in Tables 3.3 and 3.4 for the sequence of samples computed. Two observations are made concerning the computed results. First, the convergence characteristics for the SRQs as a function of the number of samples is quite different compared to the sampling of an aleatory uncertainty. Five significant figures are shown in each table to make it clear that in most cases there is no change in the interval-valued result, regardless of the number of samples. This occurs because one of the end points of the input interval produces the end

Table 3.4 *Minimum and maximum value of displacement, velocity, and acceleration for* t = *5 s*

Displacement x, m		
Samples (N)	Min value	Max value
10	1.8311	18.330
101	1.8071	18.377
1,002	1.6675	18.390
10,003	1.6485	18.392
Velocity $\dot{x}$, m/s		
Samples (N)	Min value	Max value
10	−12.170	87.098
101	−12.170	88.043
1,002	−12.284	88.064
10,003	−12.456	88.081
Acceleration $\ddot{x}$, m/s^2		
Samples (N)	Min value	Max value
10	−472.85	−58.740
101	−475.27	−58.527
1,002	−475.39	−55.409
10,003	−475.40	−55.993

points of the output interval. Of course, this does not occur in general, but it occurred for this example. Second, the magnitude of the interval for each SRQ is similar to the corresponding aleatory uncertainty when compare with the $\mu \pm 2\sigma$ result. However, it should not be concluded that results for the SRQ for the aleatory and epistemic uncertainty are similar in an informational sense. For the epistemic uncertainty, there is *no likelihood or probability structure* to the uncertainty of the SRQ result. Stated differently, the statement of knowledge of the SRQ is much weaker than that for an aleatory uncertainty. Plots of the CDFs of these quantities are not shown because there is no internal structure to the intervals.

As a final comment on the two mass-spring-damper examples, if readers attempt to reproduce the results given in either the aleatory or epistemic uncertainty examples, they may not be able to reproduce exactly the same numerical values shown here. If the reader uses MATLAB, all of the same numerical input values given here, and the same ODE integrator with the same error controls, then the reader should be able to reproduce the results shown. However, if a different software package is used, particularly a different random number generator, then the results could vary noticeably. The differences in the present results and a reader's results would be most noticeable for a low number of samples.

3.3.3.5 Propagation of Both Aleatory and Epistemic Uncertainty

For a general nondeterministic model, the propagation of both aleatory and epistemic input uncertainties through the model must be considered. In traditional nondeterministic simulation, aleatory and epistemic uncertainties are both characterized as a random variable. As a result, there is no difference in the sampling procedure for the two types of uncertainties, i.e., they are both sampled as discussed above for aleatory uncertainty. When a genuine distinction is made between the two, as presented in this book, it is clear that each one must be sampled *separately* because they each represent different types of knowledge. Aleatory uncertainty is a statement about likelihood of occurrence, whereas epistemic uncertainty is a much weaker statement of knowledge; only what is possible, given the present state of knowledge. Propagating aleatory uncertainty through a model yields information about the *likelihood of occurrence* of system responses. Propagating epistemic uncertainty through a model yields *bounds on what could occur* regarding system responses. When both aleatory and epistemic uncertainties exist and are propagated through a model, intuition says that the result should be bounds on probabilities, i.e., probability is no longer a single-valued function but an interval-valued function. This is precisely what happens.

A major breakthrough in the theoretical development of multivalued probabilities came in 1967 with the publication by Dempster (1967). The ideas in this article were developed further by Shafer (1976) and Nguyen (1978) into what is now called Dempster–Shafer theory, evidence theory, or the theory of random sets. Since that time, mathematicians and information theorists have greatly expanded these concepts into what is now called *imprecise probability theory*. (See Section 3.8 for a more detailed list of references.) These theories have the ability to express more subtle features of uncertainty and knowledge than a precisely known random variable and an interval. They are able to express such concepts as: (a) "Is the available information contradictory, duplicative, or ambiguous in some way?" (b) "My belief is stronger that a certain feature is true, but alternative features may also be true." and (c) "A fair price for a gamble cannot be set because there is asymmetric information concerning buying and selling the gamble."

The particular approach used in this text to characterize uncertainty is widely considered to be the simplest version of imprecise probability theory (Bernardini and Tonon, 2010). It is referred to as *probability bounds analysis* (PBA) and was created by Ferson (Ferson, 1996; Ferson and Ginzburg, 1996). PBA is an approach that can be concisely described as a combination of traditional probability theory and interval analysis. PBA stresses the following perspectives: (a) aleatory uncertainty is characterized as a random variable and epistemic uncertainty is characterized as an interval; (b) an uncertainty can be characterized as a mixture of both aleatory and epistemic uncertainty; (c) input uncertainties can be propagated through the model, typically using sampling techniques, but each type of uncertainty must be kept separate; and (d) portray the uncertainty in the SRQs as a bounded-set of probabilities; a probability box, (p-box). A p-box is a special type of cumulative distribution function that represents the bounds on *all possible CDFs* that can exist within the constraints of the p-box. As a result, probabilities can be multivalued quantities, specifically, interval-valued quantities. A p-box expresses both epistemic and aleatory uncertainty in a way that does not confound the two. If a p-box is used to characterize an input uncertainty that is a mixture of aleatory and epistemic uncertainty, then the p-box represents the bounds on all possible CDFs that satisfy the constraints of the interval-valued parameters.

Figure 3.16 shows an example of an SRQ p-box obtained by propagating both aleatory and epistemic input uncertainties through a model, but keeping the sampling of each separate. For each sample of a set of epistemic uncertainties, a CDF is constructed to characterize the variability due to the set of input random variables. The p-box can be interpreted in two complementary ways. First, it can be interpreted as an interval-valued cumulative probability for the occurrence of a given value of the SRQ, or less. Second, it can be interpreted as an

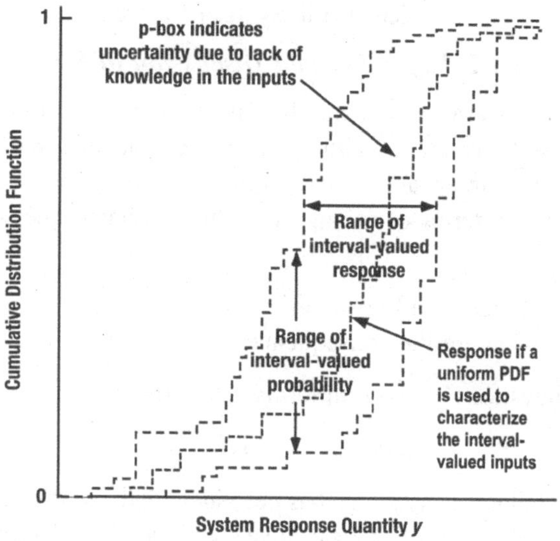

Figure 3.16 Example of a p-box obtained from sampling both aleatory and epistemic input uncertainties.

interval-valued SRQ for a given cumulative probability. Whichever interpretation is used for a given situation or customer, the p-box is the *most precise statement* that can be given for the response of the system, given the present state of knowledge. When a decision maker is presented with information that segregates aleatory and epistemic uncertainty as given in a p-box, more enlightened and better decisions or actions can be taken. For example, if the system response is dominated by epistemic uncertainty, then the decision maker has two choices to reduce the uncertainty of the response. First, more knowledge could be added to the important epistemic inputs so that their range can be decreased. Second, constraints could be imposed on the important epistemic inputs so that the epistemic uncertainty in the response is reduced. For example, poorly known system design input data or poorly known operating conditions due to the surroundings could be restricted so as to achieve the desired performance, safety, or reliability of the system.

If the nondeterministic analysis just discussed was computed, but each of the epistemic uncertainties were characterized as a uniform PDF, a very different picture of the system response is obtained. As shown by the short-dashed line near the center of the p-box in Figure 3.16, the SRQ would be *one CDF.* The reason for this dramatic change in the simulation result is that the uniform PDF characterization of epistemic uncertainty has fundamentally changed the state of knowledge from *what is possible* for the system response, to *what is likely* for the response. Thus, the uniform PDF is a statement that all possibilities over the interval are *equally likely,* as opposed to no knowledge of likelihood given by the interval. As a result, this would produce a *much reduced* picture of uncertainty to a decision maker; one that could lead to misinformed decisions concerning performance, safety, and reliability of the system.

Sections 3.6.3 and 3.6.6 discuss the computation of p-boxes using an example problem and a more detailed discussion of PBA is given in Chapter 13.

3.4 Risk and System Failure

3.4.1 Definition and Discussion of Risk

This text refers to *risk* in various contexts, but the concept is so important that it must be discussed further. Essentially all modern quantitative risk analyses use the foundational concepts formulated by Kaplan and Garrick (1981). Kaplan and Garrick formulated the issue of quantitative risk in terms of asking three failure-related questions:

- What can go wrong?
- What is the likelihood that it will go wrong?
- If it does go wrong, what are the consequences?

The answer to these questions is given in terms of the risk triplet

$$\langle s_i, p_i, x_i \rangle \quad i = 1, 2, \ldots, n, \tag{3.8}$$

where s_i is the specific failure scenario that has been identified and is being considered, p_i is the probability that the failure occurs, x_i is the negative consequence or detrimental impact from that failure occurring, and n is the number of different failure scenarios that have been identified and are being considered. x_i is a scalar that attempts to quantify the magnitude or impact of the negative consequences in terms of a comprehensive and inclusive measure. The most commonly used measure of negative consequence is monetary value, e.g., dollars or euros. Todinov (2018) groups the categories of negative impact as: (a) loss of life, physical injury, or damage to health; (b) losses associated with damage to a community or the environment; (c) direct financial losses such as warranty costs, loss of capital equipment, and cost of repair; and (d) loss of reputation or loss of public/political support, such as loss of market share, loss of customers, and loss of confidence in the company or a technology. The probability of failure, p_i, could be stated in different ways, for example, for each use of the system, over a fixed period of time, or over the lifetime of the system.

The conceptual framework of the risk triplet is very informative, but it is mathematically clumsy to deal with in probabilistic risk assessment (PRA) (also referred to as quantitative risk assessment). Many alternatives could be defined so that the risk could be computed by combining the three terms in Equation (3.8). The most common method of defining risk is simply to take the product of the probability of failure occurring and the negative consequence of failure (Vose, 2008; Haimes, 2015; Modarres et al., 2016),

$$Risk = p_i \cdot x_i \quad i = 1, 2, \ldots, n. \tag{3.9}$$

The value of risk computed in Equation (3.9) is a dimensional quantity, e.g., dollars or euros. For example, Equation (3.9) could be used to estimate the total negative impact of a system failure or multiple failures of the same system when used by multiple customers. PRA has been used successfully in the analysis of essentially every major engineered system. In complex systems, one of the major aspects of the analysis is the construction of event and fault trees in order to identify how subsystems, components, computer controllers, and human control interact. The history of PRA, however, has also shown the difficulty in accurately identifying and quantifying the total negative impact of system failure, particularly for high-visibility or high-consequence systems (Jasanoff, 1999). For example, the risk computed from Equation (3.9) for a high-consequence, very low probability of failure can be similar to a very low-consequence, high probability of failure.

Todinov (2018) points out that a distinction should be made between *expected loss from a failure* versus *potential loss from a failure*. If Equation (3.9) is applied to the situation of multiple failures of the same scenario by multiple customers, then the risk equation is an estimate of the long-term average value of the potential loss from multiple failures. One example is the potential cost to a company to provide free replacement or repair for failures of its system. However, suppose that one of these failures causes a much more serious consequence, such as a fatality, multiple fatalities, or loss in confidence in the company. Then the negative consequence for that single failure could dwarf the cost of replacement and repair of the system, and potentially put the company at risk of insolvency. This type of risk is referred to as potential loss from failure because of the unexpected major loss from one failure; potentially a blind consequence.

Although it is beyond the scope of this book to address the weaknesses of PRA, two well-recognized weaknesses are mentioned here. First, when little or no experimental data are available for input data to a PRA, expert opinion must be used. Although experts in a field are held in high regard, their opinions have been shown to be strongly biased toward overconfidence (Plous, 1993; Lin and Bier, 2008). Plous (1993) even states that "no problem in judgment and decision making is more prevalent and more potentially catastrophic than overconfidence." To make matters worse, the degree of overconfidence tends to be worse the more unskilled the individual in the relevant topic. This cognitive bias is referred to as the Dunning-Kruger effect (Wikipedia, 2022).

During the investigation of the loss of the US Space Shuttle Challenger, Presidential Commissioner Richard Feynman observed: "It appears that there are enormous differences of opinion as to the probability of failure with loss of vehicle and human life. The estimates range from 1 in 100 to 1 in 100,000. The higher [likelihood of failure] figures come from working engineers, and the very low figures from management" (p. 274 of Vaughan, 1996). After the loss of Space Shuttle Columbia, a new PRA was conducted and, in hindsight, it was found that the estimated probability of loss of crew and vehicle at the beginning of the Space Shuttle flights was 1:12 (Hamlin et al., 2011). With improved input data, primarily from operational experience, and better understanding of how this complex system could fail, a very different picture of risk emerged. After the loss of Columbia and her crew, NASA embarked on a major reevaluation of the entire system, redesign of subsystems, and improvements in reporting safety concerns so that near the end of the program the risk was estimated to be 1:90 (Hamlin et al., 2011).

Second, the magnitude of the negative consequence of failure of a system can be unrecognized, grossly misunderstood, and underestimated in certain situations. This is particularly true in high-consequence systems or systems that have highly emotional or political aspects. A good example of this is the accident at the Three Mile Island nuclear power plant in 1979. PRAs of nuclear power plants have focused, and continue to focus, on loss of life and long-term health effects on individuals. Although there was no loss of life and almost no environmental damage from Three Mile Island, the accident utterly changed the entire course of nuclear power development in the US and had a major impact on nuclear power around the world. Some refer to this type of consequence as unforeseen consequences or unexpected consequences. However, a negative consequence that cripples an entire business sector for decades is an unrecognized or blind consequence in the estimation of risk. Tenner (1996) and Mansfield (2010) discuss the inability of some individuals, organizations, and governments to identify short-term and long-term detrimental unintended consequences. They point out that closed groups of individuals and/or hierarchical organizations tend to focus on the benefits of successful systems or their agendas, as opposed to potential long-term risks to the broader community and society.

3.4.2 System Failures and Root Causes

Haimes (2015) and Lawson (2005) give an in depth discussion of the four fundamental sources of failure in systems. They identify these as hardware failure, software failure, individual human failure, and organizational failure. Hardware failures are commonly a result of inadequate system inspection or maintenance, or lack of needed repairs or improvements. The public, for example, does not recognize that the extraordinary safety record of commercial aviation is primarily due to the excellent inspection, maintenance, and repair culture within the airlines and the regulatory authorities. Software failures are frequently due to inadequate software testing, particularly for unusual situations related to failure of input data sensors used for automatic control or system operators doing something really stupid. Human failure can be of many types: e.g., system operator using the equipment outside of the recommended conditions; lack of proper safety training; the system operator is impaired or distracted by talking on a cell phone; or the system operator ignoring warning alarms because they think it is a false alarm.

Organizational failure can occur due to carelessness or malfeasance, but it is commonly caused by neglect, omission, or drift into failure (Dekker, 2011). As pointed out by Haimes (2015) and Lawson (2005), organizational failure is caused by humans in the organization, but the dominant feature of the failure is due to the culture and traditions within an organization. Some examples of organizational failure are: (a) overlooking system weaknesses or delaying maintenance because of competitive or schedule pressures; (b) poor communication from staff to management concerning potential risks; (c) filtered communication from lower levels of management to upper levels; (d) lack of incentives for a project group manager to identify weaknesses in the system's design; or (e) external political pressures on an organization causing impaired or biased judgment of management. Each of these four fundamental sources of failure is interconnected to the others in direct and indirect ways. These connections can occur during the proposal preparation phase for a new system, design phase, testing, and operation and maintenance of the system.

A number of researchers and recent investigations of root causes of high-visibility system failures convincingly argue that organizational failures are the dominant cause of most complex system failures (Duffey and Saull, 2008; Reason, 2008; Dekker, 2019). As stated in the Space Shuttle Columbia accident investigation report (Gehman et al., 2003) "the Board presents its view that NASA's organizational culture had as much to do with this accident as foam did." Organizational failures are typically difficult to identify and isolate, particularly for large organizations or within government. Organizational failure is usually connected to unintentional neglect of an issue, commonly brought on by competitive, schedule, budgetary, cultural, political, or face-saving pressures. However, there are also examples of failures of high-visibility organizations due to *willful,* i.e., *deliberate,* ignorance and malfeasance (Weisberg, 2014; Leggett, 2018; Hertwig and Engel, 2020). News media coverage of failures tends to focus on the technical or individual human failure that initiated the event, such as: "The captain of the ship was intoxicated while on duty," or "The train engineer was distracted by his cell phone." However, most complex systems require multiple contributing failures to occur before some type of disaster is precipitated. Many of these multiple contributors can be traced back to organizational failures.

3.5 Phases of Computational Simulation

The operations research (OR) and systems engineering (SE) communities have developed many of the general principles and procedures for modeling and simulation (M&S). Researchers in this field have made significant progress in defining and categorizing the various activities and phases of M&S. For recent texts in this field, see Raczynski (2006), Law (2014), and Zeigler et al. (2018). The areas of emphasis in OR and SE include specification of the problem entity, a clear description of the goals of the simulation, definition of the conceptual model, assessment of data and information quality, and methods for using simulation results as an aid in decision making. From a computational simulation perspective, however, this work is not focused on models that require complex numerical solutions.

Oberkampf et al. (2000, 2002) developed a comprehensive, new framework of the general phases of computational simulation. This structure is composed of six phases that represent a synthesis of the tasks recognized in the operations research, risk assessment, and numerical methods communities. Figure 3.17 depicts the phases of computational simulation appropriate to systems analyzed by the numerical solution of differential equations. The physical system can be an existing system or process, or it can be a system or process that is being proposed. The phases represent collections of activities required in a generic large-scale simulation. The ordering of the phases implies an information and data flow indicating which tasks are likely to impact analyses, decisions, and methodologies occurring in later phases. Each succeeding phase could be described as a *mapping of the preceding phase into a new phase* of activities. Any assumptions, approximations, and uncertainties introduced in one phase are then propagated to *all* succeeding phases. Suppose, for example, it is

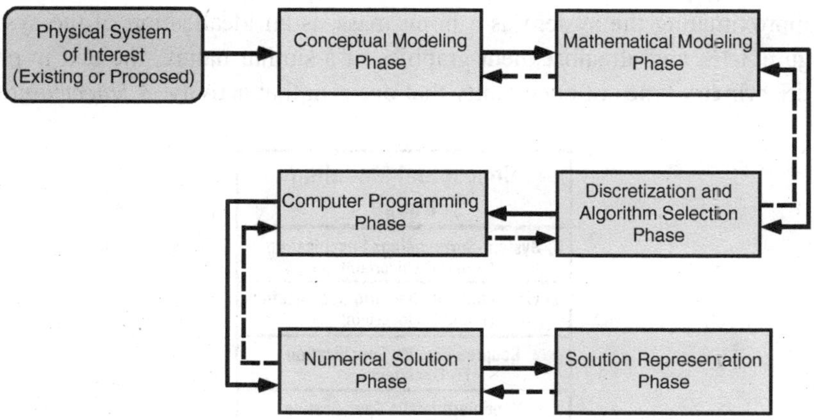

Figure 3.17 Phases of computational simulation.
(Oberkampf et al., 2000, 2002)

discovered in some latter phase that an assumption or approximation was inappropriate, or a blind uncertainty (e.g., oversight or mistake) was introduced at some earlier phase. The analyst must then return to that phase and re-evaluate all subsequent phases to determine what changes must be made. This type of feedback interaction between the phases is shown by the dashed lines in Figure 3.17.

In the following sections, characteristics and activities of each of the phases are discussed. The emphasis in the discussion is on the identification and propagation of different types of uncertainty through the computational simulation process. In Section 2.4, taxonomy of uncertainties was constructed with the critical separation made between aleatory uncertainties and epistemic uncertainties. Epistemic uncertainties were further divided into (a) recognized uncertainty, i.e., uncertainty due to incomplete knowledge in which a conscious decision has been made to either characterize it in some way, or to ignore it with justification; and (b) blind uncertainty, i.e., uncertainty due to incomplete knowledge, but where it is *not* recognized that the knowledge is incomplete, relevant, or in error. The distinction between aleatory and epistemic uncertainty is important not only in assessing how each contributes to an estimate of total predictive uncertainty, but also how each should be mathematically represented and propagated through each phase of the simulation.

3.5.1 Conceptual Modeling Phase

Activities conducted in the conceptual modeling phase are shown in Figure 3.18, along with the primary sources of uncertainty introduced in each activity. The formulation of the conceptual model involves the formal statement of the assumptions and approximations made concerning the physical system of interest. Assumptions made concerning the system of interest typically deal with *abstraction* because irrelevant and non-essential physics and features are eliminated from consideration. Approximations concerning the system of interest deal with *idealizations* of the system. As stated by Godfrey-Smith (2009), abstraction addresses "leaving things out," whereas idealization addresses "treating things as having features they clearly do not have" and "fictionalizes in the service of simplification." For example, ignoring gravity in the formulation of a conceptual model is an abstraction, whereas approximating the system as a point mass, is an idealization of the system. Note that in Figure 3.18, and all subsequent graphics of a similar nature, the text in parentheses indicates the primary types of uncertainty that occur in that activity. *A Uncertainty* indicates

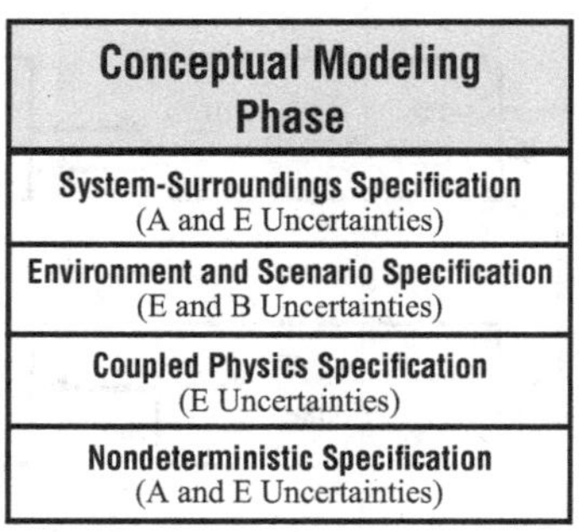

Figure 3.18 Conceptual modeling phase and primary sources of uncertainty.

an aleatory uncertainty. *E Uncertainty* indicates a recognized epistemic uncertainty. *B Uncertainty* indicates a blind epistemic uncertainty, also known as unknown-unknowns.

The first activity is the specification of the physical system of interest and its surroundings. The primary conceptualization that must be specified in this activity is the demarcation between the system and the surroundings. As discussed in Section 3.1.1, the surroundings are not modeled as part of the system. The system responds to the surroundings. The uncertainties associated with the system and surroundings specification consist of both aleatory and epistemic uncertainties that arise in defining the scope of the problem. For a complex engineered system, epistemic uncertainties commonly occur because of factors relating to the particular environment (normal, abnormal, or hostile) that is being examined, such as: (a) Was the system incorrectly manufactured or assembled? (b) How well was the system maintained or has it been damaged previously? (c) Is the simulation focused on system performance in a normal operating environment, or on system safety in some type of failure mode? The system and surroundings specification would also contain aleatory uncertainties, such as variability due to manufacturing, systems exposed to weather conditions, and random excitation of a system by its surroundings. Unless a large amount of experimental data is available to characterize the aleatory uncertainties, it is likely that the uncertainties would be a mixture of aleatory and epistemic uncertainty.

The second activity is the determination of the environments and the scenarios that will be considered in the computational simulation. These must be consistent with the goals of the simulation and the specification of the SRQs and their required accuracy. Different conceptual models are almost always required if more than one environment is considered. Scenario specification identifies physical events or sequences of events that could *possibly* be considered for a given environment, as in Figure 3.1. Identifying possible scenarios or event sequences is similar to developing an event-tree or fault-tree structure in the probabilistic risk assessment. Event and fault trees include not only hardware and software failures, but also human actions that could be taken either within or outside the system, i.e., as part of the surroundings. Even if a certain sequence of events is considered extremely remote, it should still be included as a possible event sequence in the fault tree. Whether or not the event sequence will eventually be analyzed is not a factor that impacts its inclusion in the conceptual modeling phase. In this activity, both epistemic and blind (epistemic) uncertainties are the most likely to occur. This is particularly true in identifying possible scenarios within abnormal and hostile environments. The inability to identify possible scenarios in abnormal and hostile environments is the most significant weakness in virtual testing or virtual prototyping of systems. Creativity and imagination are especially useful qualities of individuals involved in identifying scenarios in abnormal and hostile environments. Qualities that are particularly damaging to the objectivity of an analysis of abnormal and hostile environments are hope, trust, and optimism.

The third activity is the specification of the possible types of coupling of different physical processes that will be incorporated in the modeling. During the coupled physics specification, no mathematical equations are written. After the system and surroundings are specified, options for various levels of possible physics couplings should be identified, even if it is considered unlikely that all such couplings will be considered subsequently in the simulation. If a physics coupling is not considered as relevant in this phase, it cannot be addressed later in the process.

The fourth activity is the specification of all aspects in the modeling that will be considered as nondeterministic versus those that are considered deterministic. The nondeterministic specification applies to every aspect of the first three activities considered in conceptual modeling, assuming that the activity can be characterized as an aleatory uncertainty or a recognized epistemic uncertainty. These determinations must be based on the goal and requirements for the simulation. The question of what mathematical characterization will be used for the uncertainty is deferred until later phases. Blind uncertainties are not considered here because they are not even recognized as an uncertainty to be characterized.

3.5.2 Mathematical Modeling Phase

The primary task in this phase is to develop a precisely stated mathematical model, i.e., mathematical statements based on the conceptual model, or models, formulated in the previous phase. The four activities in mathematical modeling are shown in Figure 3.19. The number of mathematical models and associated analyses to be conducted depends on how many combinations of environments and scenarios were identified in the conceptual modeling phase. For large-scale simulations, the number could be quite large and, as a result, prioritization of the more important analyses needs to be conducted. Typically, this prioritization is based on the risk (probability times consequence) that the environment-scenario pair represents to achieving the goals of the simulation.

The complexity of the ODE or PDE models depends on the physical complexity of each phenomenon being considered, the number of physical phenomena being considered, and the level of coupling of different types of physics. The system-surroundings specification and the physics-coupling specification should have been completed in the conceptual modeling phase. Some examples of epistemic uncertainties that occur in assumptions and approximations of the physics modeling are: (a) fracture dynamics in any type of material; (b) coupling of the liquid, solid, and fluid phases in multiphase flow; (c) phase change of materials that are not in equilibrium; and (d) choosing to model a problem in 2-D instead of 3-D.

A complex mathematical model given by a set of differential equations is usually complemented by a number of mathematical submodels. Examples of submodels are: (a) analytical equations or tabular data for mechanical, thermodynamic, electrical, and optical properties of

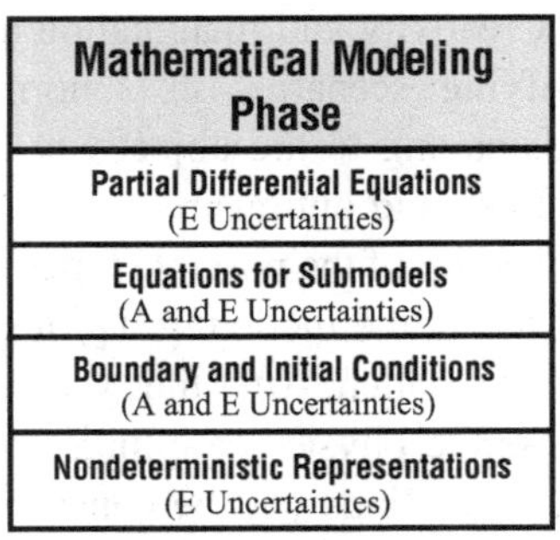

Figure 3.19 Mathematical modeling phase and primary sources of uncertainty.

materials; (b) algebraic, ODEs, or PDEs for constitutive properties of materials; and (c) turbulence modeling in fluid dynamics. The submodels, along with the boundary conditions, initial conditions, and any system excitation equations, complete the equation set for the system. BCs, ICs, and system excitation commonly exhibit aleatory and epistemic uncertainties. For abnormal and hostile environments, BCs, ICs, and system excitation are typically dominated by epistemic uncertainties.

Any mathematical model, regardless of its physical level of detail, is *by definition* a simplification of reality. Stated differently, any physical process or engineering system contains phenomena that are *not* represented in the model. As a result, specification of the mathematical models involves approximations and assumptions. These both result in epistemic uncertainties being introduced into the modeling process. Sometimes in large-scale simulation projects, statements are heard, such as: "The project will use such large-scale, massively parallel computers, that full physics simulations will be computed." These kinds of statements can only be considered as advertising hyperbole. The enduring truth of modeling was succinctly stated many years ago by George Box (1980): "All models are wrong, some are useful."

Another activity addressed during this phase of simulation is selecting appropriate representations for the nondeterministic elements of the problem. Several considerations might drive these selections. Restrictions set forth in the conceptual modeling phase of the analyses may put constraints on the range of values or types of representations that might be used in the simulation. Within these constraints, the quantity and/or limitations of available or obtainable data will play a dominant role. If sufficient sampling data is available for aleatory uncertainties, then a PDF or CDF can be constructed. In the absence of data, expert opinion or an independent simulation may provide useful information. For this type of information, the analyst could either represent the information as a pure epistemic uncertainty, i.e., an interval, or as a mixture of aleatory and epistemic uncertainty, i.e., a p-box.

3.5.3 Discretization and Algorithm Selection Phase

The discretization and algorithm selection phase maps the mathematical model developed in the previous phase into a fully discrete mathematical model. Figure 3.20 shows the four activities that are completed in this phase. These activities are grouped into two general tasks

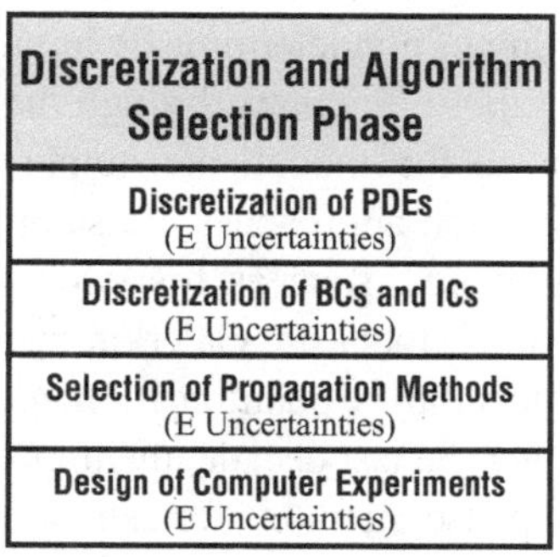

Figure 3.20 Discretization and algorithm selection phase and primary sources of uncertainty.

related to converting the mathematical model into a form that can be addressed through computational simulation. The first task involves conversion of the ODEs and PDEs from the mathematical model into a discrete model. Simply stated, the mathematics is translated from a calculus problem into an arithmetic problem. In this phase, all of the spatial and temporal discretization methods are specified for discretization of the domain of the differential equations, including the geometric features, mathematical submodels, BCs, ICs, and system excitation. In computational solid mechanics, it is common to focus entirely on the discrete model, for example in finite element modeling, as opposed to the mathematical models. This can cause inconsistencies and inability to obtain a converged numerical solution (Szabo and Actis, 2012; Szabo, 2015). The discrete form of the PDEs is typically given by finite element, finite volume, or finite difference equations. The second task is analogous to the first, but here the discretization is applied to the BCs and ICs. In these two tasks, the discretization algorithms and methods are prescribed, but the spatial mesh and temporal step size are simply given as quantities that have not yet been specified. The discretization phase focuses on the conversion of the mathematical model from calculus, i.e., derivatives and integrals, to discrete mathematics.

Although the analyst may not consciously specify all of the discretization methods in some analysis, such as when using commercial software packages, the analyst should be cognizant of the methods because it can greatly assist in detecting certain types of numerical errors. This conversion process is the root cause of certain difficulties in the numerical solution of differential equations. If the mathematical model contains singularities, is given by a stiff differential equation, or the solution is in a chaotic state, then much more care is required when choosing the numerical algorithms. Singularities commonly exist in mathematical models, but their existence in discrete models is even more troubling because of the lack of the critical concept of continuity. One of the key issues in the conversion process is how the impact of a singularity shows itself in the discrete model. Yee and Sweby (1995), Yee et al. (1997), Yee and Sweby (1998), and others have investigated the numerical solution of nonlinear ODEs and PDEs that exhibit chaotic behavior. They have clearly shown that the numerical solution of these equations can be quite different from exact analytical solutions of the mathematical model, even when using established methods that are well within their numerical stability limits. Although it is beyond the scope of this book, much more research is needed in numerical methods for chaotic solutions.

The third task of this phase of the simulation is the specification of uncertainty propagation methods for nondeterministic simulation. As discussed earlier in this chapter, the basic approach is to convert the mapping of nondeterministic input functions through the model to mapping deterministic values of these functions through the model. When combining these multiple deterministic solutions to the model, the output map of the input functions is constructed. Examples of propagation approaches are sampling techniques, such as Monte Carlo sampling; perturbation methods, such as first-order, second-moment methods; and stochastic spectral methods, such as stochastic Galerkin methods (Smith, 2014; McClarren, 2018). Traditional emphasis in uncertainty quantification and risk assessment is on propagation of input uncertainties, but over the last decade, the importance of model-form uncertainties is better appreciated. In this phase, methods for estimating model-form uncertainties are also specified. If any methods are used that approximate the propagation of input to output uncertainties, for example, surrogate models, then they should also be specified in this phase.

The design of computer experiments, i.e., statistical experiments, is driven by the need for, and the availability of, computer resources, as well as the requirements of the simulation. Establishing an experimental design often involves more than just implementation of the propagation method specified above (Box et al., 2005; Box and Draper, 2007; Hinkelmann and Kempthorne, 2008). The computer requirements and statistical complexities associated with large analyses can often be significantly reduced by using a global sensitivity analysis. A global sensitivity analysis informs the analyst how the uncertainty structure of the SRQs depends on the input data to the model, as well as any other features of the model. With regard to features other than the input data, consider the following example concerning the selection of discretization methods. Suppose the mathematical model is given by a stiff ODE, e.g., there are two extremely different time scales exhibited in the mathematical model. For these types of problems, the analyst must have especially flexible ODE integrators to deal with the stiff nature of the ODE. A global sensitivity analysis could be performed by using various ODE integrators to determine the sensitivity of the SRQs to the different integrators. Section 3.5.6 will discuss sensitivity analyses in more detail.

As noted in Figure 3.20, the primary type of uncertainty introduced in this phase is the introduction of epistemic uncertainty. These uncertainties are a specific type of epistemic uncertainty, i.e., they are due to the reliability of the numerical methods chosen to map the calculus problem to a discrete mathematics problem. As a result, they are analogous to choices for constructing mathematical models of physical processes. These numerical approximations are *not* analogous to numerical solution errors, such as mesh resolution errors, because numerical solution errors can usually be ordered in terms of accuracy. Choices of numerical methods and algorithms cannot be ordered with respect to anticipated accuracy, reliability, and robustness.

3.5.4 Computer Programming Phase

The computer programming phase maps the discrete mathematical model formulated in the previous phase into software instructions executable on a digital computer. Figure 3.21 lists the activities conducted in this phase, as well as the primary sources of uncertainty introduced in this phase. These activities are divided into two groups: preparation of input data for the computer code and computer programming activities. Preparation of input data is part of this phase because it sets all of the numerical values of the input quantities that will be used in the actual computation. The dominant uncertainty that occurs in the preparation of input activity is the

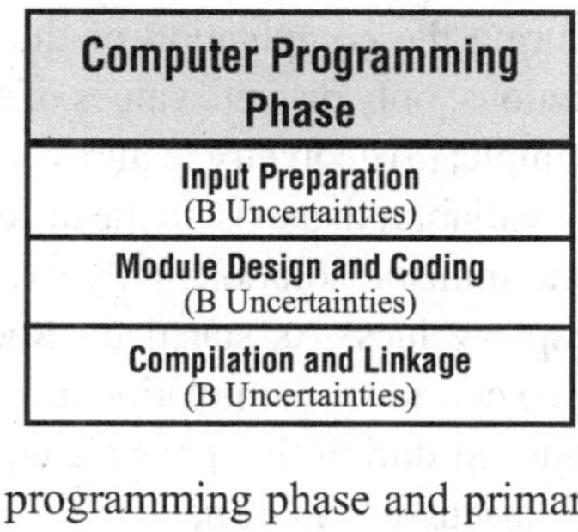

Figure 3.21 Computer programming phase and primary sources of uncertainty.

introduction of blind uncertainties, e.g., mistakes or blunders in the preparation of the input. Some researchers and analysts experienced only with relatively simple model problems do not appreciate the concern with human-caused input errors. This is, however, an important source of blind uncertainties when dealing with a complex and wide range of physical, modeling, and numerical details in a large code, multiple codes that are sequentially coupled, simulations that heavily rely on geometries specified by sophisticated solid modeling software, and tens or hundreds of material models needed for input (Reason, 1997). The complexity of the input data and the resulting opportunity for error with such codes is quite high. The importance of input preparation has been recognized for some time in high-consequence systems, e.g., safety analyses for nuclear power plants. Formal, structured, and rigorous procedures have been developed in this field to ensure the input data accurately reflects the intended input.

The second and third activities relate to all of the software elements used in the simulation, but here we concentrate on the application code itself. In the application code, the computer program modules are designed and implemented through a high-level programming language, such as FORTRAN or C++. This high-level source code is then compiled into object code and linked to the operating system and libraries of additional object code to produce the final executable code. These activities are becoming more prone to blind uncertainties due to massively parallel computers, including elements such as (a) optimizing compilers, (b) message passing and memory sharing, and (c) the effect of individual processors or memory units failing during a computer simulation. The correctness of the computer-programming activities is most influenced by blind uncertainties. In addition to programming errors, there is the subtle problem of undefined variables. This occurs when particular code syntax is undefined within the programming language, leading to executable code whose behavior is compiler dependent. Compilation and linkage introduce the potential for further errors unbeknownst to the simulation analyst. Primary among these are bugs and errors in the numerous libraries of object code linked to the application. Such libraries allow the analyst to reuse previously developed data handling and numerical analysis algorithms. Unfortunately, the analyst also inherits the undiscovered or undocumented errors in these libraries. There is also the possibility that the analyst misunderstands how the library routines should be used or makes an error in the values that are needed by the library routines.

3.5.5 Numerical Solution Phase

This phase maps the complete software bundle programmed in the previous phase into a set of numerical solutions using a digital computer. Figure 3.22 shows the various activities that are conducted during this phase. During the computation of the solution, *no* quantities are left arithmetically undefined or continuous; only discrete values of all quantities exist, all with finite precision in the computer. For example: (a) geometry features only exist as a collection of points; (b) all independent and dependent variables that exist in the differential equations now only exist at discrete points; and (c) nondeterministic solutions only exist as an ensemble of individual deterministic solutions, i.e., sample values. As stated by Raczynski (2006), "In the digital computer nothing is continuous, so continuous simulation using this hardware is an illusion."

The primary uncertainty introduced during this phase is epistemic uncertainty, specifically numerical solution errors. As discussed in Chapter 2, numerical solution errors are

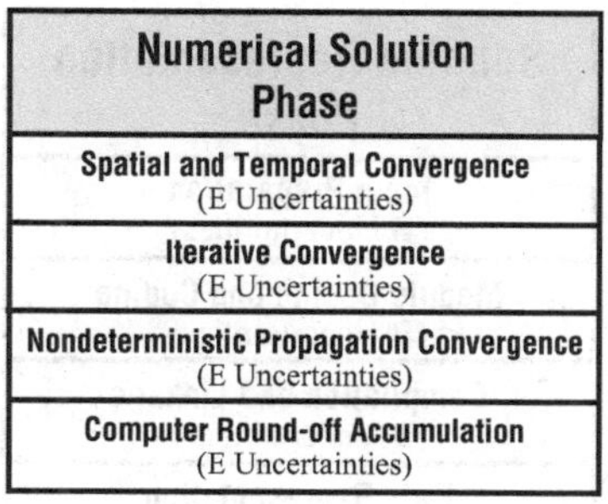

Figure 3.22 Numerical solution phase and primary sources of uncertainty.

characterized as an epistemic uncertainty as opposed to an error because the analysts *do not know the true value* of the solution. As a result, the solution error must be estimated, which necessarily involves lack of knowledge. However, there is a critical feature of numerical solution errors that can be used effectively; they are usually *ordered errors* (Roache, 1998). Most of the contributing errors in the four activities shown in Figure 3.22 can be ordered in terms of magnitude of their effect on the simulation outcomes; e.g., discretization size in terms of space or time, number of iterations in an implicit numerical procedure, number of computed samples in the propagation of input-to-out uncertainties, and word length of the computer system. This ordered nature allows the use of asymptotic analysis to estimate the magnitude of the error.

The numerical solution errors introduced in the four activities shown in Figure 3.22 can be divided into three categories. The first category contains those that are due to the spatial and temporal discretization of the differential equations. The second category contains those that are due to the approximate solution of the discrete equations. Iterative convergence methods and computer round-off errors are of this type, and they account for the difference between the exact solution to the discrete equations and the computed solution. For example, iterative solution errors can be due to iterative solution of a nonlinear matrix equation or the iterative solution each time step of a nonlinear time-dependent solution. The third category contains those that are due to the finite number of individual deterministic solutions obtained, i.e., a statistical sampling error, when computing a nondeterministic solution. The difference between using a finite number of solutions computed and the exact nondeterministic solution is the nondeterministic solution error. The most common example is the error due to a finite number of Monte Carlo samples used to approximate a nondeterministic solution. If stochastic spectral methods are used for uncertainty propagation, e.g., stochastic Galerkin methods or discrete projection methods, then the nondeterministic solution error depends on the number of terms that are included in the solution to the discrete equations (Ghanem and Spanos, 2003; Smith, 2014). Multiple solutions can also be required if the mathematical modeling phase includes the nondeterministic effect of alternative mathematical model forms in order to estimate model-form uncertainty.

3.5.6 Solution Representation Phase

This final phase maps the raw numerical solutions, i.e., the numbers themselves, which are computed in the previous phase into results usable by humans. Figure 3.23 shows the

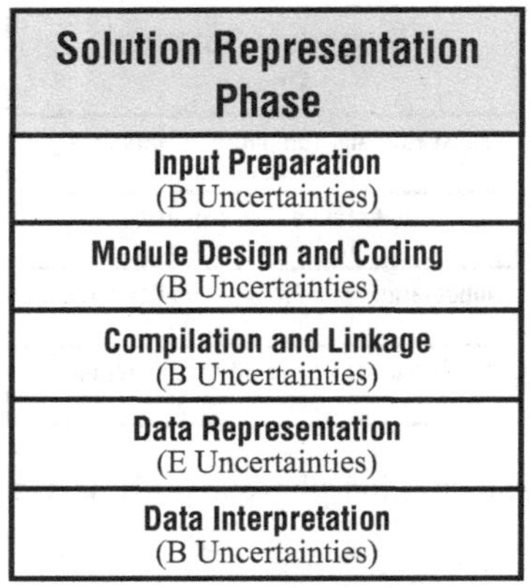

Figure 3.23 Solution representation phase and primary sources of uncertainty.

activities and the primary source of uncertainty introduced in each activity. The solution representation phase is included in the six phases of computational simulation because of (a) the sophisticated post-processing that is increasingly done to comprehend complex simulations and (b) the recognition that this phase can introduce unique types of uncertainties that can affect interpretation of results. Input preparation, module design and coding, and compilation and linkage refer to the same types of activities listed in the computer-programming phase, but here they refer to all of the post-processing software that is used. As discussed earlier in Section 3.5.4, all three of these activities have uncertainty contributions primarily from blind uncertainties.

Data representation is concerned with the construction of the functions or visualizations that represent the simulation results. For example, these can be the dependent variables from the differential equations or any post-processing of the dependent variables to obtain other SRQs of interest. Post-processing includes three-dimensional graphical visualization of solutions, spectral analysis software, animation of solutions, use of surrogate or response surface models, and use of virtual reality tools, such as *going into the solution space*. Epistemic uncertainties are introduced in data representation, primarily ordered numerical errors, which can result in the inaccurate or inappropriate representation of either the dependent variables or other SRQs of interest. Some examples of these numerical errors are: (a) oscillations of the representation function in-between discrete solution points due to the use of a high-order polynomial function in the post-processor, (b) inappropriate interpolation of the discrete solution between multi-block grids, (c) inappropriate interpolation of the discrete solution when the solution to the PDEs is a discontinuous function, and (d) too few samples in the construction of surrogate or response surface models.

Concern for errors in data representation can be better understood by posing the question: What is the mathematically correct or accurate reconstruction of the functions using the discrete solution points to the differential equation, as well as the number of statistical samples available for a nondeterministic simulation? When viewed from this perspective, the analyst better recognizes the potential reconstruction errors that could occur. Note that this perspective is *not* shared in modern data representation or visualization packages. The view of these general-purpose packages is that the reconstruction is based on ease of use, computation speed, convenience, and robustness of the package. Two examples of this perspective are the following. First, we are not aware of any software package that provides an estimate of the numerical error due to the reconstruction of either a deterministic or nondeterministic response surface. Second, in data visualization packages, there is no concern with respect to ensuring that the interpolation function obeys any physical principles such as conserves mass, momentum, or energy, or that the physical quantity must be positive.

Data interpretation errors are made by the user or customer of the simulation results, based on observations of the representation of the solution. For example, the interpreter of the results could be the computational analysts using the code, a system designer, or decision maker relying on the results. Data interpretation errors are blind uncertainties introduced by individuals or an analyst team using the simulation results. Two examples of interpretation error are: (a) concluding that a computed solution is chaotic when it is not (and vice versa) and (b) not correctly recognizing the important frequency and amplitude content in a complex SRQ, such as the vibrational modes of a structure. Importantly, this text's definition of data interpretation errors does *not* include poor decisions made by the user based on the simulation, such as incorrect design choices for a system or inept policy decisions by a government bureaucrat.

If a nondeterministic simulation is produced, then the likelihood of data interpretation errors increases significantly. This occurs not only because the quantity of simulation results can increase by a factor of a hundred or more, but also because of the subtleties that can exist and the questions that can be asked concerning nondeterministic simulations. As discussed earlier in this chapter, two of the primary goals of nondeterministic simulation are to address uncertainty quantification and sensitivity analyses. Uncertainty quantification deals with the question: "How uncertain are the SRQs of interest with regard to all of the sources of uncertainty that have been identified?" This is a difficult question to answer because of all of the uncertainty sources that have been discussed here. In order to estimate the uncertainty in the SRQs, samples must be computed over the range or sources of uncertainty. If the uncertainty is an aleatory input uncertainty, then the samples are taken from the probability distribution of the uncertainty, resulting in probability distributions of the SRQs. If the uncertainty is an epistemic input uncertainty, then the samples are taken from the intervals characterizing the uncertainty, resulting in intervals of SRQs. If there is a mixture of aleatory and epistemic uncertainties, then the result is a set of probability distributions whose bounds are characterized as a p-box.

When an uncertainty analysis is complete, it is commonly followed by a sensitivity analysis. A sensitivity analysis uses the nondeterministic results computed for the uncertainty analysis, but it attempts to answer different questions related to the system of interest. Sometimes sensitivity analyses are referred to as *what-ifs* or perturbation analyses of the system. The computational expense added by a sensitivity analysis is typically minimal compared to the uncertainty analysis because additional sampling is usually not needed. Two of the most common questions raised in a sensitivity analysis follow.

First, what is the rate of change of the SRQs of interest with respect to the input data quantities? Here the focus is on evaluating the partial derivatives of the SRQs with respect to input quantities, all other input quantities remaining either being fixed or considered as random variables. This type of analysis is referred to as a *local sensitivity analysis*, because it addresses the *local rate of change of the SRQ* with respect to a specific quantity. When these derivatives are computed for a variety of input quantities, one can then rank the magnitude of the sensitivity of the output quantity with regard to the various input quantities. Note that these derivatives, and the resulting ranking of input quantities, can strongly depend on the values chosen for the fixed input quantities. That is, the sensitivity derivatives typically vary widely over the range of the input quantity, which can be either deterministic or nondeterministic. For a detailed discussion of local sensitivity analyses, see Saltelli et al. (2004) and McClarren (2018).

The second type of sensitivity analysis asks the question: What uncertain inputs have the largest effect on the SRQs of interest? Here the focus is *not* on the uncertainty of the SRQs, but on which input uncertainties have the largest global effect on the SRQs of interest. This type of analysis is usually referred to as a *global sensitivity analysis*. Here *global* refers to a specific environmental condition and a specific scenario, but it includes all of the uncertainties that have been sampled. For example, suppose there are 10 uncertain input quantities in a design study of the performance of a system in a normal operating environment and a given scenario. A global sensitivity analysis could rank in order the largest sources of uncertainty in the system and the surroundings for a given SRQ. With this knowledge, the designer could possibly restrict some of these uncertainties to obtain better performance, safety, and reliability of the system. For a detailed discussion of global sensitivity analyses, see Saltelli et al. (2008), Smith (2014) and Da Veiga et al. (2021).

3.6 Example Problem: Missile Flight Dynamics

A system-level example is given here of the flight dynamics of a rocket-boosted, aircraft-launched missile to demonstrate each of the phases of computational simulation. This example is adapted from Oberkampf et al. (2000, 2002). For a detailed discussion of this example, as well as an example of a system in an abnormal (i.e., an accident) environment, see these references. The missile is a short range, unguided, air-to-ground rocket. The missile is powered by a solid fuel rocket motor during the initial portion of its flight and is unpowered during the remainder of its flight. The simulation considers the missile flight to be in the unspecified future. Thus, we are attempting to simulate future anticipated flights, not analyze an event in the past, such as a system failure or accident investigation. To keep this example relatively simple, the only SRQ of interest is the range of the missile. Figure 3.24 shows all six phases of computational simulation and the activities just discussed.

3.6.1 Conceptual Modeling Phase

Figure 3.25 shows three possible system-surroundings specifications in the top tier of the graphic for the missile flight example. Other specifications could be identified, but these give examples of options that could be used for various types of simulations. The specifications are listed from the most physically inclusive, with regard to the system specification and the physics that could be coupled, to the least inclusive. For each row or tier of blocks shown in Figure 3.25, the most physically inclusive are given on the left, with decreasing physical complexity moving toward the right. System-surroundings Specification 1 considers the missile and the atmosphere near the missile to be part of the system, whereas the launching

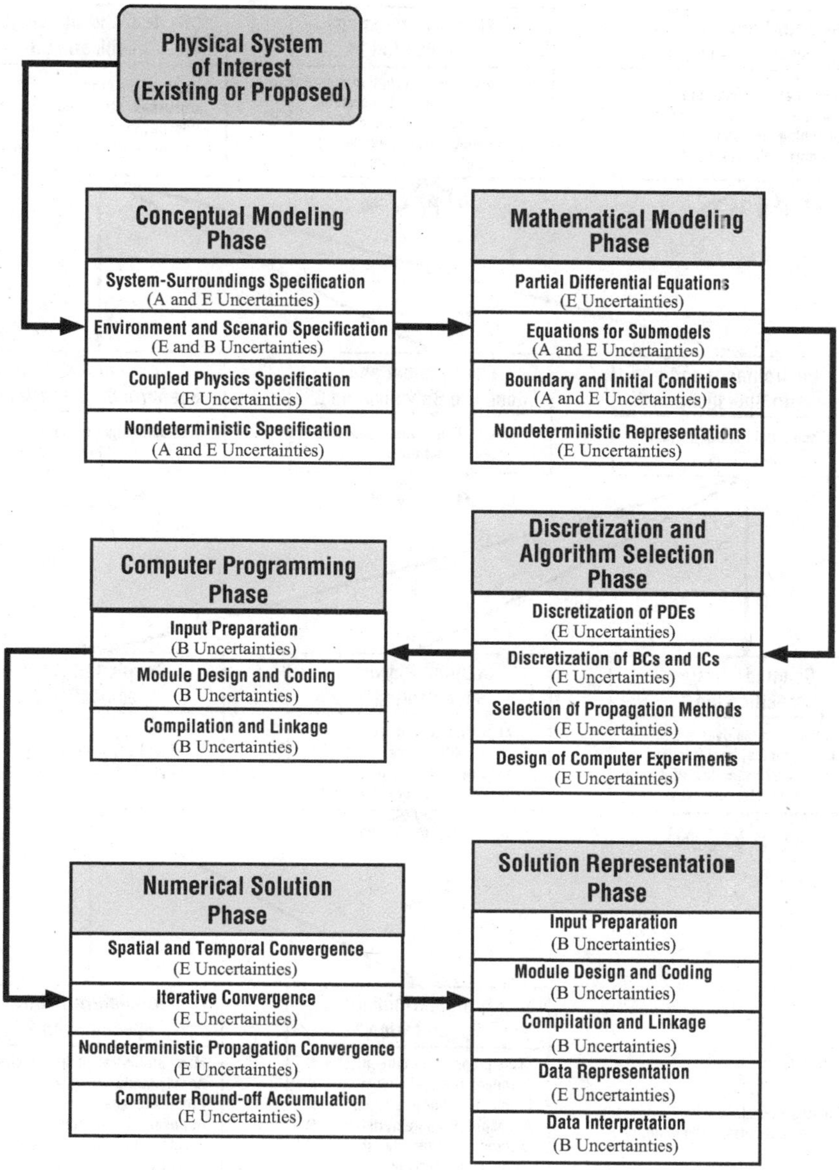

Figure 3.24 Phases and activities in computational simulation.

aircraft and target are considered to be part of the surroundings. An example of a simulation that would be allowed with this specification is where the missile, the flow field of the missile, and the rocket exhaust are coupled to the flow field of the launching aircraft. Thus, the missile and the rocket exhaust could be influenced by the presence of the aircraft and its flow field, but the aircraft structure, for example, could not deform due to the rocket exhaust. For example, the aircraft geometry would serve as BCs for the flow field computed around the missile. Another example allowed by this specification would be the simulation of the missile flight inside a structure, e.g., a launch from inside of a building or tunnel.

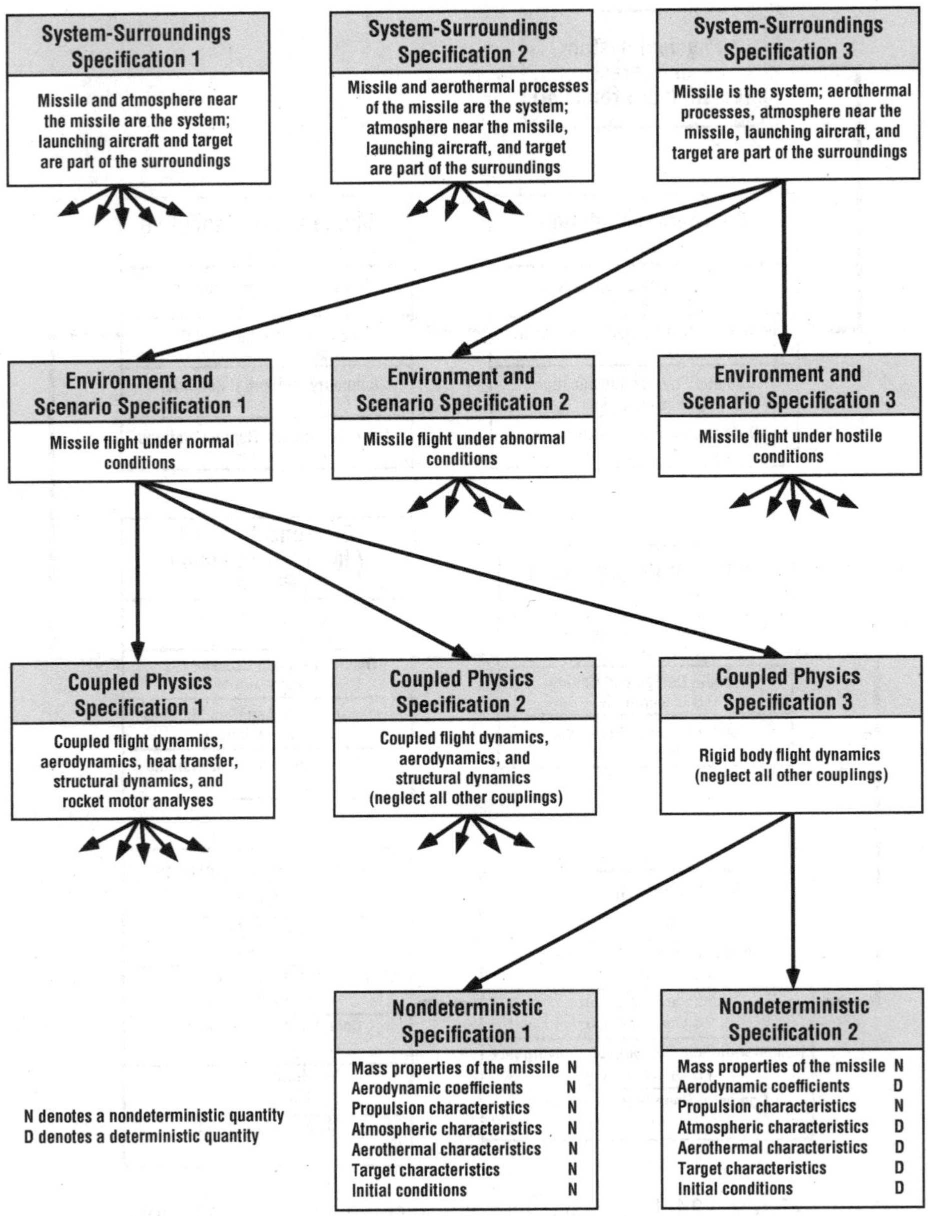

Figure 3.25 Conceptual modeling activities for the missile flight example.

System-surroundings Specification 2 considers the missile and the aerothermal processes occurring on the surface of the missile to be part of the system, whereas the atmosphere near the missile, the launching aircraft, and the target are considered part of the surroundings. This specification allows analyses that couple the missile and the aerothermal effects on the missile. For example, one could consider the structural deformation of the missile due to aerodynamic loading and thermal heating of the structure. Then one could couple the missile deformation and the flow field so that the aerodynamic loading and thermal heating could be simulated on the deformed structure.

System-surroundings Specification 3 considers the missile to be the system, whereas the aerothermal processes on the surface of the missile, the atmosphere near the missile, the launching aircraft, and the target are considered part of the surroundings. Even though this is the simplest specification considered here, it still allows for significant complexities in the simulation. The missile flight example presented here will only pursue system-surroundings Specification 3.

The environment specification (shown in the second tier of Figure 3.25) identifies three general environments: normal, abnormal, and hostile. For each of these environments, the analyst then identifies all possible scenarios, physical events, or sequences of events that may affect the goals of the simulation. For relatively simple systems, isolated systems, or systems with very controlled surroundings or operational conditions, this activity can be straightforward. However, complex engineered systems are commonly exposed to a myriad of scenarios within each of the normal, abnormal, and hostile environments. Constructing environment and scenario specifications for these complex systems is a major undertaking. A many-branched event-tree and/or fault-tree can be constructed with each scenario having a wide range of likelihoods and consequences. Even though the risk (probability times consequence) for many scenarios may be low, these scenarios should be identified at this phase. Often, when various scenarios are identified, other scenarios are discovered that would not have been discovered otherwise. The decision of which scenarios to pursue should be made *after* a very wide range of plausible scenarios has been identified.

Normal environment conditions are those that can be reasonably expected during nominal operations of the aircraft and missile system. Some examples are: (a) typical launch conditions from various types of aircraft that are expected to carry the missile, (b) nominal operation of the propulsion and electrical system, and (c) expected weather conditions while the missile is attached to the aircraft. Examples of flight under abnormal conditions are: (a) improperly assembled missile components or subsystems; (b) explosive failure of the propulsion system during operation, particularly while still attached or very near the aircraft; and (c) flight through adverse weather conditions, like hail or lightning. Examples of flight under hostile conditions are: (a) detonation of nearby defensive weapons; (b) damage to missile components or subsystems resulting from small-arms fire; and (c) damage, either structural or electrical, from laser or microwave defensive systems. Note that the missile flight example will only pursue Environment Specification 1, normal environment. Furthermore, no unusual conditions will be considered within the realm of normal conditions.

In the third tier of Figure 3.25, three levels of physics coupling are identified, although more alternatives could be identified. Coupled Physics Specification 1 couples essentially all relevant physics that could exist in this decision thread of the simulation, i.e., System-surroundings Specification 3 and Environment and Scenario Specification 1. For example, this specification could couple the structural deformation and dynamics with the aerodynamic loading and thermal loading due to atmospheric heating. It could also couple the deformation of the solid-fuel rocket motor case due to combustion pressurization, heat transfer from the motor case into the missile airframe, and the effect of nonrigid-body flight dynamics on the missile.

Coupled Physics Specification 2 couples the missile flight dynamics, aerodynamics, and structural dynamics, neglecting all other couplings. This coupling permits the computation of the deformation of the missile structure due to inertial loading and aerodynamic loading. For example, the analyst could then recompute the aerodynamic loading and aerodynamic damping due to the deformed structure. Coupled Physics Specification 2 would result in a time-dependent, coupled fluid/structure interaction simulation.

Coupled Physics Specification 3 assumes every subsystem of the missile is rigid and non-heat conducting. Not only is physics coupling disallowed within the missile, but every aspect of the missile structure is assumed rigid and thermal effects are neglected, i.e., only rigid-body dynamics are considered. The missile is allowed to respond only to forces and moments from the atmosphere, i.e., the surroundings. Note that the missile flight example will only pursue Coupled Physics Specification 3.

Before addressing the last activity of conceptual modeling, a few comments should be made concerning the possible sources of epistemic and blind uncertainty that could occur in the three activities discussed to this point. Epistemic uncertainties arise primarily because of (a) situations, conditions, or physics that are poorly known or understood; (b) situations or physics that are consciously excluded from the simulation; and (c) approximations made in situations or physics considered. Blind uncertainties arise primarily because of situations or physics that are *not* imagined or recognized but are possible. The more complex the system, the more possibilities exist for blind uncertainties to occur.

A common weakness of modern simulation, sometimes referred to as *virtual testing* or *virtual prototyping*, is either intentionally or unintentionally overlooking unusual events, physical effects, or possibilities that could occur. Two examples are included. First, in an assembled structure using bolts, screws, or rivets, these fasteners can loosen with age. This situation can be assumed not to happen or easily overlooked in virtual testing, resulting in failure to identify a possibly very dangerous condition during the operation of the system. Second, many modern systems have automatic feedback control systems designed to ensure safe operation of the system. Virtual testing can fail to investigate the safety of the system if sensors providing feedback to the control system either fail or provide erroneous data to the controller. For systems in abnormal or hostile environments, the likelihood of blind uncertainties increases greatly compared to normal environments.

In the bottom tier of Figure 3.25, two alternative nondeterministic specifications are listed. Nondeterministic Specification 1 includes the following (indicated by an N in the figure): mass properties of the missile, aerodynamic force and moment coefficients, propulsion characteristics, atmospheric characteristics, aerothermal heating characteristics, target characteristics, and initial conditions at missile launch. Nondeterministic Specification 2 considers only two parameters as uncertain, the mass properties of the missile and the propulsion characteristics of the motor. All other parameters are considered as deterministic (indicated by a D). The missile flight example will only pursue Nondeterministic Specification 2.

3.6.2 Mathematical Modeling Phase

For the missile flight example, two alternative mathematical models are chosen: a six-degree-of-freedom (6-DOF) model and a three-degree-of-freedom (3-DOF) model (Figure 3.26). Both models are consistent with the conceptual model being analyzed: System-surroundings Specification 3, Environment Specification 1, Coupled Physics Specification 3, and Nondeterministic Specification 2 (Figure 3.25). Although it is not

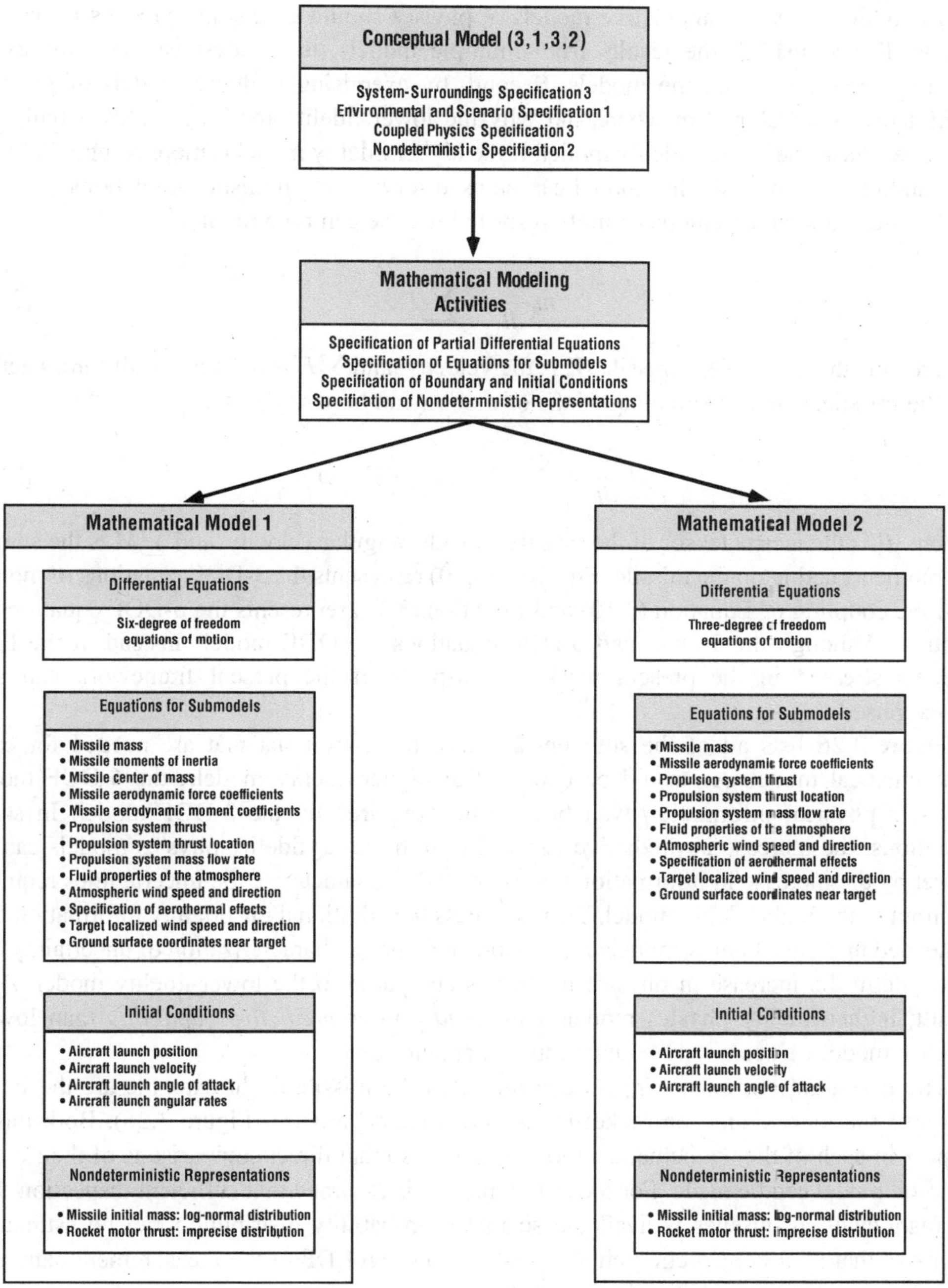

Figure 3.26 Mathematical modeling activities for the missile flight example.

always possible to order the accuracy of alternative models according to the physics fidelity, for this case of 3-DOF and 6-DOF mathematical models of flight dynamics, the analyst can unequivocally order these two models. The ability to clearly order the physics fidelity of multiple models can be used to great advantage in the following situations. First, there are often conditions where alternative models of physics should give similar results for certain SRQs. By comparing the results from multiple models, the analyst can use this as an informal check between the models. Second, by exercising multiple models of physics, confidence is developed in where and why the lower fidelity model gives essentially the same results as the higher fidelity model. If the higher fidelity model is more computationally demanding, the lower fidelity model can be used for nondeterministic simulations.

The translational equations of motion for the missile can be written as

$$m\frac{d\vec{V}}{dt} = \sum \vec{F}, \tag{3.10}$$

where m is the mass of the missile, $\vec{V}$ is the velocity, and $\sum \vec{F}$ is the sum of all forces acting on the missile. The rotational equations of motion can be written as

$$[I]\frac{d\vec{\omega}}{dt} = \sum \vec{M} + \vec{\omega} \times \left\{[I] \cdot \vec{\omega}\right\}, \tag{3.11}$$

where $[I]$ is the inertia tensor of the missile, $\vec{\omega}$ is the angular velocity, and $\sum \vec{M}$ is the sum of all moments acting on the missile. Equation (3.10) represents the 3-DOF equations of motion and the coupling of Equation (3.10) and Equation (3.11) represents the 6-DOF equations of motion. Although the 3-DOF and 6-DOF equations are ODE models instead of the PDE models stressed in the present work, key aspects of the present framework can still be exercised.

Figure 3.26 lists all of the submodels and initial conditions that are needed for each mathematical model. As would be expected of higher-fidelity models, the 6-DOF model requires physical information well beyond that required by the 3-DOF model. In some situations, the increase in predictive capability from higher fidelity physics models can be offset by the increase in information that is needed to characterize the uncertainties required as input to the high-fidelity model. That is, unless the additional uncertainty information that is needed in higher-fidelity models is available, the poorer characterization of uncertainty can overwhelm the increase in physics fidelity as compared to the lower-fidelity model. As a result, higher-fidelity physics models *may yield poorer predictive capability* than lower-fidelity models, a seemingly counterintuitive conclusion.

The two nondeterministic inputs considered in the missile flight example are the initial mass of the missile and the rocket motor thrust characteristics (Figure 3.26). Both inputs appear in each of the mathematical models chosen so that direct comparisons of their effect on each model can be made. For the initial mass, it is assumed that sufficient inspection data of manufactured missiles is available so that a probability distribution can be estimated. Suppose that after constructing either a histogram or an EDF of the measurement data, it is found that a log-normal distribution with precisely known mean and standard deviation could be used (Bury, 1999; Krishnamoorthy, 2006).

For the thrust of the solid rocket motor, suppose that a number of newly manufactured motors have been fired so that variability in thrust can be well represented by a two-parameter Gamma distribution. One parameter of the Gamma distribution is referred to as the shape parameter, λ, and the second is referred to as the scale parameter, σ, which characterizes the variability of the distribution. It is well known that propulsion characteristics can vary substantially with the age of the solid propellant. Suppose that a number of motors with various ages have also been fired. It is found that a Gamma distribution with a shape parameter of $\lambda = 3$ can be used for motors of all ages, but the age significantly affects the variability of the thrust. Because of the small number of motors fired, it is decided to use an interval for the scale parameter of the Gamma distribution, σ_G. As a result, the uncertainty in thrust is represented as a mixture of aleatory and epistemic uncertainty. The aleatory portion of the uncertainty is due to manufacturing variability of the motor and the epistemic uncertainty is due to the unknown age of a specific motor that could be fired. In the flight dynamics simulation, it is clear that the epistemic uncertainty in thrust could be reduced if information is added concerning the age of the specified motors. For example, if all of the missiles that may be launched are known to be from a single production lot, then the epistemic uncertainty could be eliminated because it would be known when the production lot is manufactured. The two parameters of the Gamma distribution would then become precisely known values.

3.6.3 Discretization and Algorithm Selection Phase

The discretization method chosen to solve the ODEs of both mathematical models is a Runge–Kutta 4(5) method (Press et al., 2007). The RK4(5) method is fifth-order accurate at each time step and the integrator coefficients of Cash and Karp (1990) were used. The local discretization error is estimated by comparing the fourth-order accurate solution with the fifth-order accurate solution. With an estimate of the local discretization error, and a per-step error criterion, the integration time step can be controlled to satisfy the error criterion.

The method chosen for propagation of the uncertainties through the model is PBA. There are two aleatory uncertainties, as characterized by the log-normal and Gamma distributions, and there is one epistemic uncertainty as characterized by the σ parameter of the Gamma distribution. Simple Monte Carlo sampling is used for the two aleatory uncertainties and equal partition sampling is used for σ_G. Since there are both types of input uncertainty, the sampling structure must properly account for these differences. That is, samples from an aleatory uncertainty are random draws from a probability distribution (in the present case two distributions), whereas samples from an epistemic uncertainty are samples that are all possible with no likelihood associated with the sample. Stated differently, the samples from the epistemic uncertainty all have a likelihood of unity, i.e., *every possible sample could occur*. As a result, the sampling structure of the aleatory and epistemic uncertainties *must be kept separate*. The most common structure used is double-loop sampling, where the outer loop samples the epistemic uncertainties and the inner loop samples the aleatory uncertainties. A sample from the outer loop epistemic uncertainty is used for the inner loop samples that are used to construct a probability distribution for the SRQ from the inner loop. Then the process is repeated with another sample from the outer loop and a new set of inner loop samples is computed. This double-loop sampling could be described as each probability distribution computed in the inner loop is conditional on the sample from the outer loop epistemic uncertainty. This process yields a set of probability distributions, one for each sample of the outer loop. The p-box of the SRQ is computed to be the outer bound of all of the probability distributions computed.

3.6.4 Computer Programming Phase

A computer code (TAOS) developed at Sandia National Laboratories was used to compute the trajectories of the missile flight example (Salguero, 1999). This general-purpose flight dynamics code has been used for a wide variety of guidance, control, navigation, and optimization problems for flight vehicles. Only the ballistic flight options were used to solve the 3-DOF and 6-DOF equations of motion.

3.6.5 Numerical Solution Phase

The numerical solution method used a variable time step so that the local discretization error could be directly controlled at each step. The local discretization error is estimated at each step for each state variable for each system of differential equations. For the 6-DOF model, there are 12 state variables and for the 3-DOF model, there are six state variables. Before a new time step can be accepted in the numerical solution, a relative error criterion must be met for each state variable. If the relative error criterion for each of the state variables is not satisfied, then the time step is halved, and the solution is attempted again. If the largest local discretization error of all the state variables is significantly less than the error criterion, then the step size is increased for the next time step.

For propagation of input uncertainties, 10 samples were used for the outer loop epistemic uncertainty and 50 samples were used for the inner loop aleatory uncertainties. Five hundred samples were computed for each of the 3-DOF and 6-DOF models.

3.6.6 Solution Representation Phase

For this relatively simple example, the representation of solution results is rather straightforward. The SRQ of interest is the range of the missile. The most common method of showing nondeterministic results is to plot the CDF for the SRQ of interest. If only aleatory uncertainties exist in a nondeterministic simulation, then only one CDF exists for any given SRQ. If epistemic uncertainty also exists, as it does in this example, then an ensemble of CDFs must be computed. One CDF results from *each sample* of the epistemic uncertainty. To compute the p-box of the SRQ, one determines the minimum and maximum probability from *all of the CDFs* that were computed at each value of the SRQ. If alternative mathematical models are used, as in the present case, then a p-box is computed for each model.

Figure 3.27 shows a representative result for the p-box resulting from one of the mathematical models. The p-box characterizes both the aleatory uncertainty due to manufacturing variability and the epistemic uncertainty due to the unknown age of the rocket motor. The p-box is computed to be the envelope of *all 10 CDFs* that were computed as part of the inner-loop sampling of the aleatory uncertainty. Since each CDF from the inner loop is possible, there is no likelihood information within the p-box. The inner-loop sampling of the aleatory uncertainty contains 50 samples, so that each increment in probability is 0.02. It is seen that the age of the motor increases the uncertainty of the range beyond that due to the manufacturing variability of the motor. For example, the lack of knowledge concerning the age of the motor contributes about 0.1 km uncertainty in range, separate from the manufacturing variability of the motor. A different way of interpreting the p-box is to pick a value of

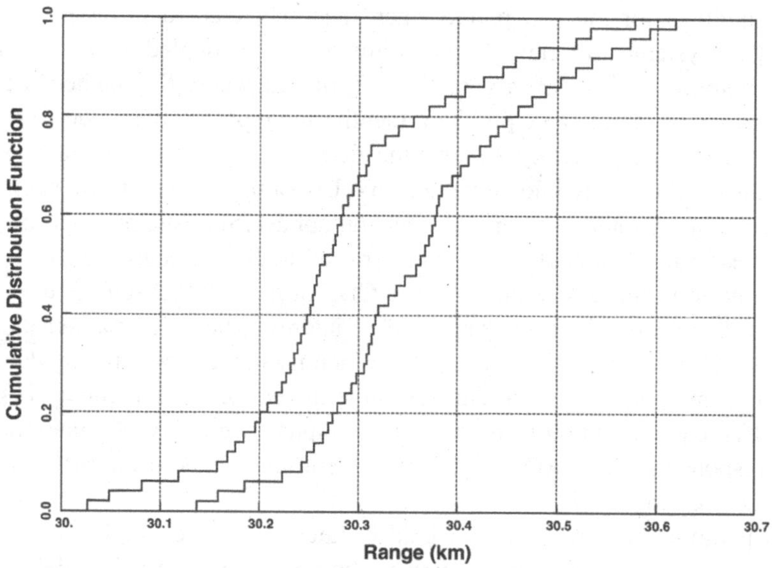

Figure 3.27 Representative p-box for range of the missile.

range and then read the interval-valued probability. For example, the probability of attaining a range of 30.4 km or less, can vary from 0.68 to 0.85, depending on the age of the motor.

3.7 Problems

1. Consider a high-speed automobile crashing into a concrete bridge support. The goal of the simulation is to evaluate the effectiveness of a new design for an airbag for the driver. Construct and discuss two specifications for the system and surroundings for the driver in this type of crash. One system/surroundings should involve more coupled physics than the other.
2. Consider a gas turbine engine with a turbofan on a commercial aircraft flying at cruise conditions. The goal of the simulation is to evaluate the effectiveness of a nacelle covering the fan blades when a fan blade fractures. Construct and discuss two specifications for the system and surroundings for the turbofan portion of the engine and the nacelle. One system/surroundings should involve more coupled physics than the other.
3. Consider a small drone to be used for package delivery to residences. The drone is completely autonomous with an on-board automatic control system with a GPS navigation system. The goal of the simulation is to estimate the maximum wind conditions for which it is safe to deliver packages. Construct and discuss two specifications for the system and surroundings for the drone while it is near the delivery location. One system/surroundings should involve more coupled physics than the other.
4. Consider the structural integrity of a modern cell phone that is subjected to the following loading condition. Suppose the phone is in the back pocket of an individual who sits down in a sturdy wooden chair. Construct and discuss two specifications for the system

and surroundings for the cell phone when it is subjected to this type of mechanical loading. One system/surroundings should involve more coupled physics than the other.

5. Consider a simple system from your field of interest where the mathematical model is given by an ODE, initial value problem. Identify the inputs to the model and specify one input as an aleatory uncertainty. Assume the aleatory uncertainty is given by a log-normal distribution with precisely known parameters. Use simple Monte Carlo sampling for the input uncertainty. Simulate the time-dependent, nondeterministic response of one SRQ of interest. Present the results of the SRQ in terms of the first three moments of the response at a given point in time. Also compute the CDF of the SRQ. Discuss the results.
6. Consider a simple system from your field of interest where the mathematical model is given by an ODE, boundary value problem. Identify the inputs to the model and specify one input as an epistemic uncertainty. Assume the epistemic uncertainty is given by an interval. Use equal-partition sampling for the input uncertainty. Simulate the nondeterministic response of one SRQ of interest. Present the results of the SRQ in terms of an interval. Discuss the results.
7. Consider a simple system from your field of interest. Give examples of model abstractions and model idealizations in the construction of the conceptual model.

3.8 References

AEC (1966). Atomic Energy Commission – Atomic Weapon Safety Program. *U.S. Atomic Energy Commission, Memorandum No. 0560*. Washington, DC.

AIAA (2015). Standard: Capability-Based Product Failure Mode, Effects and Criticality Analysis (FMECA) Requirements. *American Institute of Aeronautics and Astronautics*, AIAA S-102.2.4-2015, Reston, VA.

Angus, J. E. (1994). The Probability Integral Transform and Related Results. *SIAM Review*, **36**(4), 652–654.

Beer, M., S. Ferson, and V. Kreinovich (2013). Imprecise Probabilities in Engineering Analyses. *Mechanical Systems and Signal Processing*, **37**, 4–29.

Bernardini, A., and F. Tonon (2010). *Bounding Uncertainty in Civil Engineering*. New York, NY: Springer.

Bossel, H. (1994). *Modeling and Simulation*. Wellesley, MA: A. K. Peters.

Box, G. E. P. (1980). Sampling and Bayes' Inference in Scientific Modeling and Robustness. *Journal of the Royal Statistical Society: Series A*, **143**(A), 383–430.

Box, G. E. P., and N. R. Draper (2007). *Response Surfaces, Mixtures, and Ridge Analyses*. New York, NY: John Wiley.

Box, G. E. P., J. S. Hunter, and W. G. Hunter (2005). *Statistics for Experimenters: Design, Innovation, and Discovery*, 2nd ed. New York, NY: John Wiley.

Bury, K. (1999). *Statistical Distributions in Engineering*. Cambridge: Cambridge University Press.

Carlson, C. S. (2012). *Effective FMEAs: Achieving Safe, Reliable, and Economical Products and Processes Using Failure Mode and Effects Analysis*. New York, NY: John Wiley.

Cash, J. R., and A. H. Karp (1990). A Variable Order Runge–Kutta Method for Initial-Value Problems with Rapidly Varying Right-Hand Sides. *ACM Transactions on Mathematical Software*, **16**(3), 201–222.

Coombs, C. H. (1964). *A Theory of Data*. New York, NY: John Wiley.

Da Veiga, S., F. Gamboa, B. Iooss, and C. Prieur (2021). *Basics and Trends in Sensitivity Analysis: Theory and Practice in R*. Philadelphia, PA: Society for Industrial and Applied Mathematics.

Dekker, S. (2011). *Drift into Failure: From Hunting Broken Components to Understanding Complex Systems*. Boca Raton, FL: CRC Press.

Dekker, S. (2019). *Foundations of Safety Science: A Century of Understanding Accidents and Disasters*. Boca Raton, FL: CRC Press.

Dempster, A. P. (1967). Upper and Lower Probabilities Induced by a Multivalued Mapping. *Annals of Mathematical Statistics*, **38**(2), 325–339.

Duffey, R. B., and J. W. Saull (2008). *Managing Risk: The Human Element.* New York, NY: John Wiley.

Ferson, S. (1996). What Monte Carlo Methods Cannot Do. *Human and Ecological Risk Assessment*, **2**(4), 990–1007.

Ferson, S., and L. R. Ginzburg (1996). Different Methods Are Needed to Propagate Ignorance and Variability. *Reliability Engineering and System Safety*, **54**(2–3), 133–144.

Gehman, H. W., J. L. Barry, D. W. Deal, J. N. Hallock, K. W. Hess, G. S. Hubbard, et al. (2003). Columbia Accident Investigation Board Report Volume I. *National Aeronautics and Space Administration*, Washington, DC.

Ghanem, R. G., and P. D. Spanos (2003). *Stochastic Finite Elements: A Spectral Approach*, Rev. ed. Mineola, NY: Dover Publications.

Godfrey-Smith, P. (2009). Abstractions, Idealization, and Evolutionary Biology. In A. Barberousse, M. Morange, and T. Pradeu (eds.), *Mapping the Future of Biology*: Springer.

Haimes, Y. Y. (2015). *Risk Modeling, Assessment, and Management*, 4th ed. New York, NY: John Wiley.

Hamlin, T., E. Thigpen, J. Kahn, and Y. Lo (2011). Shuttle Risk Progression: Use of the Shuttle Probabilistic Risk Assessment (PRA) to Show Reliability Growth. *AIAA Space 2011 Conference & Exposition*, AIAA 2011-7353, Long Beach, CA.

Hertwig, R., and C. Engel (eds.). (2020). *Deliberate Ignorance: Choosing Not to Know.* Cambridge, MA: The MIT Press.

Hinkelmann, K., and O. Kempthorne (2008). *Design and Analysis of Experiments: Volume 1 – Introduction to Experimental Design*, 2nd ed. New York, NY: John Wiley.

Jasanoff, S. (1999). The Songlines of Risk. *Environmental Values*, **8**(2), 135–152.

Kaplan, S., and B. J. Garrick (1981). On the Quantitative Definition of Risk. *Risk Analysis*, **1**(1), 11–27.

Krishnamoorthy, K. (2006). *Handbook of Statistical Distribution with Applications.* Boca Raton, FL: Chapman & Hall.

Law, A. M. (2014). *Simulation Modeling and Analysis*, 5th ed. New York, NY: McGraw-Hill.

Lawson, D. (2005). *Engineering Disasters – Lessons to Be Learned.* New York, NY: ASME Press.

Leggett, T. (2018). How VW Tried to Cover Up the Emissions Scandal. *BBC News: Business.* www.bbc.com/news/business-44005844.

Leijnse, A., and S. M. Hassanizadeh (1994). Model Definition and Model Validation. *Advances in Water Resources*, **17**, 197–200.

Lin, S.-W., and V. M. Bier (2008). A Study of Expert Overconfidence. *Reliability Engineering and System Safety*, **93**, 711–721.

Mansfield, J. (2010). *The Nature of Change or the Law of Unintended Consequences.* London: Imperial College Press.

McClarren, R. G. (2018). *Uncertainty Quantification and Predictive Computational Science.* New York, NY: Springer.

Modarres, M., M. Kaminskiy, and V. Krivtsov (2016). *Reliability Engineering and Risk Analysis; A Practical Guide*, 3rd ed. Boca Raton, FL: CRC Press.

Neelamkavil, F. (1987). *Computer Simulation and Modelling.* New York, NY: John Wiley.

Nguyen, H. T. (1978). On Random Sets and Belief Functions. *Journal of Mathematical Analysis and Applications*, **65**(3), 531–542.

Oberkampf, W. L., S. M. DeLand, B. M. Rutherford, K. V. Diegert, and K. F. Alvin (2000). Estimation of Total Uncertainty in Computational Simulation. *Sandia National Laboratories*, SAND2000-0824, Albuquerque, NM.

Oberkampf, W. L., S. M. DeLand, B. M. Rutherford, K. V. Diegert, and K. F. Alvin (2002). Error and Uncertainty in Modeling and Simulation. *Reliability Engineering and System Safety*, **75**(3), 333–357.

Pegden, C. D., R. E. Shannon, and R. P. Sadowski (1995). *Introduction to Simulation Using SIMAN*, 2nd ed. New York, NY: McGraw-Hill.

Plous, S. (1993). *The Psychology of Judgment and Decision Making.* New York, NY: McGraw-Hill.

Press, W. H., S. A. Teukolsky, W. T. Vetterling, and B. P. Flannery (2007). *Numerical Recipes: The Art of Scientific Computing*, 3rd ed. New York, NY: Cambridge University Press.

Raczynski, S. (2006). *Modeling and Simulation: The Computer Science of Illusion.* New York, NY: John Wiley.

Reason, J. (1997). *Managing the Risks of Organizational Accidents*. Burlington, VT: Ashgate Publishing.
Reason, J. (2008). *The Human Contribution: Unsafe Acts, Accidents and Heroic Recoveries*. Burlington, VT: Ashgate Publishing.
Roache, P. J. (1998). *Verification and Validation in Computational Science and Engineering*. Albuquerque, NM: Hermosa Publishers.
Salguero, D. E. (1999). Trajectory Analysis and Optimization Software (TAOS). *Sandia National Laboratories*, SAND99-0811, Albuquerque, NM.
Saltelli, A., M. Ratto, T. Andres, F. Campolongo, J. Cariboni, D. Gatelli, et al. (2008). *Global Sensitivity Analysis: The Primer*. New York, NY: John Wiley.
Saltelli, A., S. Tarantola, F. Campolongo, and M. Ratto (2004). *Sensitivity Analysis in Practice: A Guide to Assessing Scientific Models*. New York, NY: John Wiley.
Shafer, G. (1976). *A Mathematical Theory of Evidence*. Princeton, NJ: Princeton University Press.
Smith, R. C. (2014). *Uncertainty Quantification: Theory, Implementation, and Applications*. Philadelphia, PA: Society for Industrial and Applied Mathematics.
Szabo, B. (2015). Simulation Governance: An Idea Whose Time Has Come. *IACM Expressions Magazine*, **37**, 6–8.
Szabo, B., and R. Actis (2012). Simulation Governance: Technical Requirements for Mechanical Design. *Computer Methods in Applied Mechanics and Engineering*, **249–252**, 158–168.
Tenner, E. (1996). *Why Things Bite Back: Technology and the Revenge Effect*. New York, NY: Alfred A. Knopf.
Todinov, M. (2018). *Methods for Reliability Improvement and Risk Reduction*. New York, NY: John Wiley.
Vaughan, D. (1996). *The Challenger Launch Decision: Risky Technology, Culture, and Deviance at NASA*. Chicago, IL: The University of Chicago Press.
Vose, D. (2008). *Risk Analysis: A Quantitative Guide*, 3rd ed. New York, NY: John Wiley.
Weisberg, H. I. (2014). *Willful Ignorance: The Mismeasure of Uncertainty*. New York, NY: John Wiley.
Wikipedia (2022). Dunning-Kruger Effect. https://en.wikipedia.org/wiki/Dunning%E2%80%93Kruger_effect.
Yee, H. C., and P. K. Sweby (1995). Dynamical Approach Study of Spurious Steady-State Numerical Solutions of Nonlinear Differential Equations II. Global Asymptotic Behavior of Time Discretizations. *Computational Fluid Dynamics*, **4**, 219–283.
Yee, H. C., and P. K. Sweby (1998). Aspects of Numerical Uncertainties in Time Marching to Steady-State Numerical Solutions. *AIAA Journal*, **36**(5), 712–724.
Yee, H. C., J. R. Torczynski, S. A. Morton, M. R. Visbal, and P. K. Sweby (1997). On Spurious Behavior of CFD Simulations. *13th AIAA Computational Fluid Dynamics Conference*, AIAA Paper 97-1869, Snowmass, CO.
Zeigler, B. P., A. Muzy, and E. Kofman (2018). *Theory of Modeling and Simulation: Discrete Event & Iterative System Computational Foundations*, 3rd ed. London: Academic Press.

Part II

Code Verification and Software Engineering

Code verification ensures that the computer program (alternatively referred to as the computer code) is a faithful representation of the original mathematical model and its numerical solution. Chapter 4 discusses the basic concepts of software engineering practices, including software development models, programming languages, version control, static analysis, dynamic software testing, and reproducibility. Chapter 5 explains the powerful code verification technique referred to as order-of-accuracy, or order-of-convergence, testing. Order-of-accuracy testing empirically determines the convergence rate of the numerical procedures used to solve the differential or integral equations of the mathematical model. Order-of-accuracy testing can only be used if exact or highlyaccurate solutions to the mathematical model are available. Chapter 6 discusses various approaches to obtaining such highly-accurate solutions, with primary emphasis on the Method of Manufactured Solutions. Since code verification deals purely with issues of mathematical correctness, there is no mention of the actual behavior of real-world systems or experimental data.

4

Software Engineering for Scientific Computing

The first principle is that you must not fool yourself – and you are the easiest person to fool. —Richard Feynman

Contents of Chapter 4

This chapter addresses software engineering applied to scientific computing at a high level. It addresses the topics of software development models, programming language, version control, static analysis, software testing, software development practices, verification and validation, and reproducibility. The interested reader is encouraged to see the cited references for additional details.

4.1 Introduction

Over the last half-century, scientific computing has emerged as the third pillar of science, joining theory and experiment. In fact, scientific computing simulations are sometimes referred to as "virtual" or "in silico" experiments, especially when used to support "system testing" and scientific discovery. However, in many cases, such simulations do not meet the rigorous reproducibility standards that exist for experimental work, a fundamental tenet of the scientific method. One of the reasons is the lack of rigorous software engineering practices during the development of scientific computing software.

Software engineering encompasses the tools and methods for defining requirements for designing, programming, testing, and managing software. It consists of monitoring and controlling both the software processes and the software products to ensure reliability. Software engineering was developed by the computer science community, primarily for business and information technology applications. Its use is essential for large software development projects and for high-assurance software systems such as those for aircraft control systems, nuclear power plants, and medical devices (e.g., pacemakers).

One may wonder why the software engineering techniques applied successfully to traditional software applications have not found their way into scientific software development. Kelly (2007) writes of a "chasm" that has formed between the scientific computing and software engineering communities. The discipline of computer science focuses on general, domain-independent solutions; whereas scientific computing software has unique needs that are not addressed by traditional software engineering approaches. The solution appears to be to educate software engineers on the unique needs of scientific computing software, while at the same time providing training for computational scientists in appropriate software engineering techniques (Kelly, 2007). We agree wholeheartedly with this observation and recommend additional training both at universities and in industry/government labs. This book is an effort to address the former on the scientific computing side.

Before addressing specific software engineering strategies, it is important to introduce some terminology. Three definitions that will be used throughout this chapter are those for software defects, faults, and failures (Hatton, 1997b). A *software defect* is a coding mistake (bug) or the misuse of a coding construct that could potentially lead to a software failure. A *software fault* is a defect which can be detected without running the code, i.e., through static analysis (Section 4.5). Examples of defects that can lead to software faults include dependence on uninitialized variables, mismatches in parameter arguments, and unassigned pointers. A *software failure* occurs when the software returns an incorrect result or when it terminates prematurely due to a run-time error (overflow, underflow, division by zero, etc.). Some examples of catastrophic software failures are given by Hatton (1997a).

The purpose of this chapter is to lightly touch on the 20% of software engineering aspects that will provide 80% of the increase in reliability and performance for scientific computing software. While we focus strictly on scientific computing software, larger software projects, and those involving high-consequence systems, will no doubt benefit from more traditional software engineering methods (e.g., see the books by McConnell, 2004; Pressman and Maxim, 2015; Sommerville, 2016). For scientific computing, the reader is directed to the series of workshops on software engineering for science and high-performance computing (SE4Science, 2021) and the book edited by Carver et al. (2017).

4.2 Software Development Models

The traditional approach to software development is the linear *waterfall model*, in which the various aspects of the software development process (requirements specification, architectural design, programming, testing, etc.) are decomposed into separate phases; with each phase beginning only after the previous phase is completed (Sommerville, 2016). However, the waterfall model is poorly suited for scientific software development. Scientific computing software tends to have life spans on the order of decades rather than years (Heaton and Carver, 2015). In addition, the software may be used for investigating the underlying physics of new, complex phenomena using rapidly-evolving models. For these reasons, the requirements for scientific software are very difficult to specify up front and often change and evolve over the software's life span.

In response to criticisms of the waterfall software development model, a competing approach called *iterative and incremental development* (also called the spiral model) was proposed (Larman and Basili, 2003). This iterative, or evolutionary, development model is based on the idea of interweaving each of the steps in the software development process, thus allowing customer feedback early in the development process through software prototypes which may initially have only limited capabilities. These software prototypes are then refined based on the customer input, resulting in software with increasing capability. The iterative nature of this development model makes it more amenable to the long lifespans of scientific computing software and frequently changing requirements. The humorous (although not untrue) view of the software development process for (small) scientific computing projects is shown in Figure 4.1. This figure depicts the highly iterative nature of scientific software development, the often poorly defined requirements, and general loose notion of software testing.

The *agile software development* approach is based on the iterative and incremental development model and appears to be well suited for scientific computing software (Carver et al., 2007; Derlaga et al., 2013; Heaton and Carver, 2015). Agile methods (also referred to as rapid software development) allow requirements specification, design, implementation, and testing to occur simultaneously (Sommerville, 2016). They are iterative in nature, and the main goal is to develop useful software quickly. Key features of agile programming methods include:

- concurrency of development activities,
- minimal or automatic design documentation,
- only the high-priority user requirements are specified up front,
- significant user involvement in the development process, and
- incremental software development.

One of the intended advantages of agile programming is to provide rapid software delivery (although initially with reduced capability) that allows for user involvement and feedback

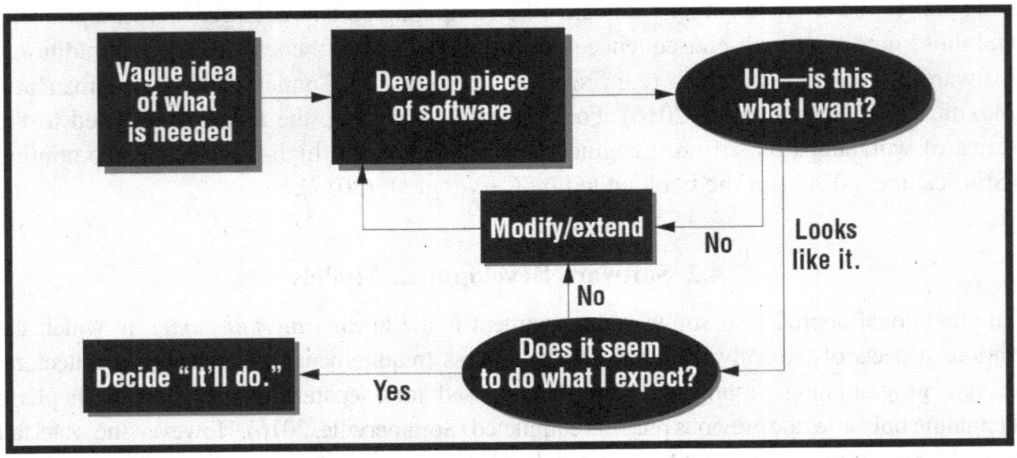

Figure 4.1 Model of stereotypical scientific software development.
(reproduced from Segal and Morris, 2008)

during the software design process; not just after the final software product has been delivered. Agile programming methods are recommended for small and medium-size scientific software development efforts. However, their efficiency for larger software development efforts, which generally require more coordination and planning, is questionable (Sommerville, 2016).

A popular form of the agile programming approach is extreme programming (Beck, 2004), and is so-called because it takes software engineering best practices to their extreme. In extreme programming, requirements are expressed as potential scenarios that lead to software development tasks, which are then programmed by a pair of developers working as a team. Evidence suggests that pair programming productivity is similar to that of solo programmers (Williams and Kessler, 2003), but results in fewer errors since any code produced has necessarily undergone an informal software inspection process (Pressman and Maxim, 2015). In addition, pair programming facilitates software engineering knowledge transfer, which is particularly useful for scientists and engineers who typically have little or no training in software engineering methods. Unit tests (see Section 4.6.2.1) must be developed for each component task before the code is written (an approach called test driven development). All unit tests must be successfully executed before integration into the software system, a process known as continuous integration testing (Duvall et al., 2007). This type of test-first procedure provides an implicit definition of the interface as well as proper usage examples of the component being developed. Software development projects employing extreme programming usually have frequent releases and undergo frequent refactoring (see Section 4.7.4) to improve quality and maintain simplicity. For an example of extreme programming applied to scientific computing, see Wood and Kleb (2003).

4.3 Programming Languages

There are a variety of factors to consider when choosing a programming language. The two main programming paradigms used in scientific computing are procedural programming and

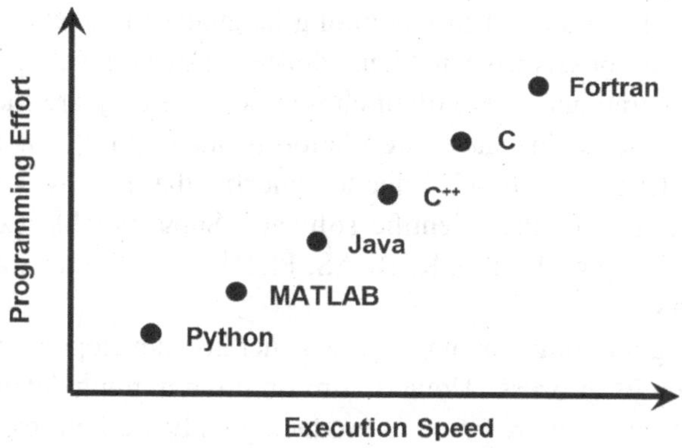

Figure 4.2 *Qualitative* example of programming effort versus computing execution speed.
(adapted from Software Carpentry, 2023)

object-oriented programming. *Procedural programming* relies on calls to different procedures (routines, subroutines, methods, or functions) to execute a series of sequential steps in a given programming task. A significant advantage of procedural programming is that it is modular, i.e., it allows for reuse of procedures when tasks must be performed multiple times. In *object-oriented programming*, the program is decomposed into objects which interact with each other through the sending and receiving of messages. Objects typically make use of private data which can only be accessed through that object, thus providing a level of independence to the objects. This independence makes it easier to modify a given object without impacting other parts of the code. Object-oriented programming also allows for the reusability of components across the software system.

Most modern, higher-level programming languages used for scientific computing support both procedural and object-oriented programming. Programming languages that are primarily procedural in nature include BASIC, C, Fortran, MATLAB, Pascal, and Perl. Languages that are primarily (but not wholly) object-oriented include C^{++}, Java, and Python. Procedural programming is often used when mathematical computations drive the design, whereas object-oriented programming is preferred when the problem is driven by complex data relationships.

Low level computing languages such as machine language and assembly language (often found in simple electronic devices) execute extremely fast but require additional time and effort during the programming and debugging phases. One factor to consider is that high-level languages make use of more natural language syntax and varying levels of programming abstraction. This has the advantage of making programming complex software projects easier, but the higher-level language will not execute as fast as a lower-level programming language. A qualitative comparison of selected programming languages is shown in Figure 4.2 which compares programming effort to floating point execution speed for some typical languages used in scientific computing. The higher-level programming languages such as Python, MATLAB, and Java are ideal for small projects and prototyping, while production-level codes are usually programmed in C, C^{++}, or (modern) Fortran due to the faster execution speeds.

When it comes to the choice of programming language for scientific computing codes, developers tend to be conservative in their choices. Lower level (primarily) procedural languages such as Fortran and C are often chosen because they are likely to be supported well into the future on evolving hardware platforms, although C^{++} is also quite common. Third-party libraries tend to be avoided due to concerns that they will not be supported for the entire (long) life span of the scientific software. Some notable exceptions are linear/iterative solver libraries (e.g., LAPACK, BLAS, PETSc, and Trilinos) and parallel libraries (e.g., MPI and OpenMP).

Standards for programming languages are generally developed by a time intensive, prolonged, and complex process. However, most programming language standards still contain coding constructs that are prone to producing software failures. These failure-prone constructs can arise in a number of different ways (Hatton, 1997a). For example, they can arise due to oversight by the standards process, due to a lack of agreement on the content of the standards, because the decision was explicitly made to retain the functionality provided by the construct, or even due to errors in the programming language standards documentation. In some cases, less fault-prone subsets of a programming language exist which reduce or eliminate the presence of the dangerous coding constructs. One example of a safe subset for the C programming language is Safer C (Hatton, 1995).

4.4 Version Control

Version control tracks changes to source code or other text-based software products. A good version control system can tell you what was changed, who made the change, and when the change was made. It allows a software developer to undo changes to the code, going back to any prior version. This capability can be particularly helpful when one seeks to reproduce results from an earlier analysis, report, or project, and merely requires documentation of the version number or the date the results were generated. Version control also provides a mechanism for incorporating changes from multiple developers, an essential feature for large software projects or projects with geographically remote developers. All source code should be maintained in a version control system, regardless of how large or small the software project (Eddins, 2006).

There are two main types of version control systems. *Centralized version control* systems (e.g., SVN, CVS) provide for a single, centralized repository to which developers commit code updates. Parts of the code that are checked out for modification are typically locked (i.e., other developers cannot access them), and thus centralized version control is recommended only for very small development teams or even single developers. *Distributed version control* systems (e.g., Git, SVK) allow a number of distributed repositories that are all tied into a master repository. Thus, distributed version control makes it easier for a number of developers to perform their work on their own projects then integrate these changes back into the master repository once it appears to be working correctly. This feature helps prevent developers from "breaking" the main version of the code during development. In a distributed repository system, a developer does not need network access to the master repository in order to do development. When developers are not in the same geographical location, web-based version control platforms such as GitHub (https://github.com/) and GitLab (https://gitlab.com/gitlab-com) can be used.

Some key concepts pertaining to version control are discussed below (Chacon and Straub, 2019). Note that the generic descriptor "file" is used which could represent not only source code and other software products, but also any other text-based file stored on a computer.

Repository – a location where the current and all prior versions of the files are stored; in distributed version control systems, there is a master repository which can be copied (or "cloned") locally for development.

Working copy – the local copy of a file from the repository which can be modified and then checked in or "committed" to the repository.

Check-out – the process of creating a working copy from a repository (either the current version or an earlier version).

Check-in – (or commit) occurs when changes made to a working copy are merged into a repository.

Push – the merging of changes from a local repository to the master repository (for distributed version control systems).

Diff – a summary of the differences between two versions of a file, often taking the form of the two files side-by-side with differences highlighted.

Conflict – occurs when two or more developers attempt to make changes to the same file and the system is unable to reconcile the changes (note: conflicts generally must be resolved by either choosing one version over the other or by integrating the changes from both into the repository by hand).

Update – merges recent changes to a repository into a working copy (for centralized version control systems).

Pull – merges recent changes to the master repository into a local repository (for distributed version control systems).

Version – a unique identifier assigned to each version of the file held in the repository which is generated by the check-in process.

The basic steps that are used to initiate a distributed version control system (such as Git) are as follows.

1. Create an original "master" repository.
2. Clone the repository to allow development.
3. Add a directory structure and/or files into the local repository.
4. Check-in or "commit" the directory structure and/or files to the local repository.
5. Push these modifications of the local repository into the master repository.
6. Edit/modify the files in the working copy and examine the differences between the modified files and those in the repository (i.e., diff).
7. Check-in or "commit" the file modifications to the local repository.
8. Pull any modifications from the original repository into the local repository (in case someone has modified some files and your files are out of date).
9. Push these modifications of the local repository into the master repository.

Version control is one of the most important aspects of software engineering for scientific computing software. Extensive use of version control, along with frequent check-ins, can allow for better reproducibility of scientific computing results, which is important for scientific publications and simulation-informed decision making. When used with frequent testing (see Section 4.6), it can improve software quality and aid in debugging.

4.5 Static Analysis

Static analysis is any type of assessment of software correctness that does not require program execution. Examples of static analysis methods include software inspection, peer review, compiling of the code, and the use of automatic static analyzers. Hatton (1997a) estimates that approximately 40% of software failures are due to static faults. Some examples of static faults are:

- dependency on uninitialized or undeclared variables,
- interface faults: too few, too many, or wrong type of arguments passed to a function/ subroutine,
- casting a pointer to a narrower integer type (e.g., C), and
- use of non-local variables in functions/subroutines (e.g., Fortran).

Many of these static faults, as well as others that have their origins in ambiguities in the programming language standards, can be mitigated by using static analysis.

4.5.1 Software Inspection

Software inspection (or review) refers to the act of reading through the source code and other software products to find defects. Although software inspections are time intensive, they are surprisingly effective at finding software defects (Sommerville, 2016). Other advantages of software inspections are that (a) they are not subject to interactions between different software defects (i.e., one defect will not hide the presence of another one), (b) incomplete and non-functional source code can be inspected, and (c) they can find other issues besides defects such as coding inefficiencies or lack of compliance with coding standards. The rigor of the software inspection depends on the technical qualifications of the reviewer, their diligence during the inspection, and their level of independence from the software developers. Pair programming (discussed earlier) provides a form of software inspection during the development process.

4.5.2 Compiling the Code

Any time the code is compiled it goes through some level of static analysis. The level of rigor of the static analysis often depends on the options used during compilation. However, there is a trade-off between the level of static analysis performed by the compiler and the execution speed. Many modern compilers provide different modes of compilation (e.g., debug mode, release mode, user-specified compiler optimization) that perform different levels of static analysis. Due to differences in compilers and operating systems, many software developers make it standard practice to compile the source code with a range of compiler options, different compilers, and on different hardware platforms.

4.5.3 Automatic Static Analyzers

Automatic static analyzers are external software tools that are meant to complement the checking of the code by the compiler. They are designed to find inconsistent or undefined use of a programming language that the compiler will likely overlook, as well as coding constructs that are generally considered as unsafe. Some static analyzers available for C/ C^{++} include the Safer C Toolset, Cleanscape, and PC-lint/FlexeLint. Static analyzers for Fortran include Cleanscape, floppy/fflow, and ftnchek. There is also a recently-developed static analyzer for MATLAB called checkcode (previously called M-Lint) (MATLAB, 2020).

4.6 Software Testing

Software (or dynamic) testing can be defined as the "dynamic verification of the behavior of a program on a finite set of test cases . . . against the expected behavior" (SWEBOK, 2004). It includes any type of testing activity which involves running the code, thus run-time compiler checks (e.g., array bounds checking, pointer checking) fall under the heading of software testing. The types of testing discussed in this section include defect testing and regression testing, both of which can be done at the unit, component, and complete system level. Also discussed in this section are test harnesses and code coverage.

4.6.1 Is Software Testing Really Necessary?

In the early 1990s, Hatton undertook a broad study of commercially available, scientific software reliability known collectively as the "T Experiments" (Hatton, 1997b). This study was broken into two parts: the first (T1) examined codes from a wide range of scientific disciplines using static analysis, while the second (T2) examined codes in a single discipline using dynamic testing.

The T1 study used static deep-flow analyzers to examine more than 100 different codes in 40 different application areas. All codes were written in C, FORTRAN 66, or FORTRAN 77, and the static analyzers used were QA C (for the C codes) and QA Fortran (for the FORTRAN codes). The main conclusion of the T1 study was that the C codes contained approximately 8 serious static faults per 1,000 lines of executable code, while the FORTRAN codes contained approximately 12 faults per 1,000 lines. A serious static fault is defined as a statically-detectable defect that is likely to cause the software to fail. For more details on the T1 study, see Hatton (1995).

The T2 study examined a subset of the codes from the T1 study in the area of seismic data processing which is used in the field of oil and gas exploration. This study examined nine independent, mature, commercial codes which employed the same algorithms, the same programming language (FORTRAN), the same user-defined parameters, and the same input data. Hatton refers to such a study as N-version programming since each code was developed independently by a different company. Each of the codes consisted of approximately 30 sequential steps, 14 of which used unambiguously defined algorithms, referred to in the study as primary calibration points. Agreement between the codes after the first primary calibration point was within 0.001% (i.e., approximately machine precision for single-precision computations); however, agreement after primary calibration point 14 was only within a factor of two. It is interesting to note that distribution of results from the various codes was found to be non-Gaussian with distinct groups and outliers, suggesting that the output from an N-version programming test should not be analyzed with Bayesian statistics. Hatton concluded that the disagreements between the different codes are due primarily to software errors. Such dismal results from the T2 study prompted Hatton to conclude that *"the results of scientific calculations carried out by many software packages should be treated with the same measure of disbelief researchers have traditionally attached to the results of unconfirmed physical experiments."* For more details on the T2 study, see Hatton and Roberts (1994).

These alarming results from Hatton's "T Experiments" highlight the need for employing good software engineering practices in scientific computing. At a minimum, the simple techniques presented in this chapter such as version control, static analysis, dynamic testing, and the recommended programming practices should be employed for all scientific computing software projects to improve quality and reliability.

4.6.2 Defect Testing

Defect testing is performed to uncover the presence of a software defect; however, defect testing *cannot* be used to prove that no errors are present. Once a defect is discovered, the process of finding and fixing the defect is usually referred to as debugging. In scientific computing, it is convenient to decompose defect testing into three levels: unit testing which occurs at the smallest level in the code, component testing which occurs at the sub-model or algorithm level, and system testing during which the desired output from the software is evaluated. While unit testing is generally performed by the code developer, component and system-level testing are more reliable when performed by someone outside of the software development team. System level testing is discussed in detail in Section 4.8.

The benefits of code testing are that it increases reliability/reproducibility, provides examples of correct usage, encourages smaller procedures with short interfaces, and moves the tedious process of debugging to higher levels in the code. However, numerous studies have shown that current testing procedures for scientific computing software are generally inadequate (e.g., see Heaton and Carver, 2015).

4.6.2.1 Unit Testing

Unit testing is used to verify the execution of a single routine (e.g., function, subroutine, object class) of the code (Eddins, 2006). Unit tests are designed to check for the correctness of routine output based on a given input. They should be easy to write and run and should execute quickly. Properly designed unit tests also provide examples of proper routine use such as how the routine should be called, what type of inputs should be provided, and what type of outputs can be expected.

While it does take additional time to develop unit tests, this extra time in code development generally pays off later in reduced time debugging. The authors' experience with even small scientific computing code development in university settings suggests that the typical ratio of debugging time to programming time for students who do not employ unit tests and version control is *at least* five to one. The wider the unit testing coverage (i.e., percentage of routines that

Table 4.1 *Example of a component-level test fixture for Sutherland's viscosity law (adapted from Kleb and Wood, 2006)*

Input: T (K)	Output: μ (kg/s-m)
$200 \le T \le 3{,}000$	$B^* \frac{T^{1.5}}{T+110.4}$
199	error
200	1.3285589×10^{-5}
2000	6.1792781×10^{-5}
3000	7.7023485×10^{-5}
3001	error

*where $B = 1.458 \times 10^{-6}$

have unit tests), the more reliable the code is likely to be. In fact, some software development strategies such as extreme programming (Section 4.2) require tests to be written before the actual routine to be tested is created. Such strategies require the programmer to clearly define the interfaces (inputs and outputs) of the routine before programming even begins.

4.6.2.2 Component Testing

Kleb and Wood (2006) make an appeal to the scientific computing community to implement the scientific method in the development of scientific software. Recall that in the scientific method, a theory must be supported with a corresponding experiment that tests the theory and must be described in enough detail that the experiment can be reproduced by independent sources. For application to scientific computing, they recommend testing at the component level, where a component is considered to be a submodel or algorithm. Furthermore, they strongly suggest that model and algorithm developers publish *test fixtures* with any newly proposed model or algorithm. These test fixtures are designed to clearly define the proper usage of the component, give examples of proper usage, and give sample inputs along with correct outputs that can be used for testing the implementation in a scientific computing code. An example of such a test fixture for Sutherland's viscosity law is presented in Table 4.1.

Component-level testing can be performed when the sub-model or algorithm are algebraic since the expected (i.e., correct) solution can be computed directly. However, for cases where the sub-model involves numerical approximations (e.g., most models for fluid turbulence involve differential equations), then the expected solution will necessarily be a function of the chosen discretization parameters. As a result, more sophisticated testing (i.e., code verification) methods discussed in Chapters 5 and 6 should be used. For models that are difficult to test at the system level (e.g., the `min` and `max` functions significantly complicate the code verification process discussed in Chapter 5), component-level testing of the models (or different parts of the model) can be used. Finally, even when all components have been successfully tested individually, one should not get a false sense of security about how the software will behave at the system level. Complex interactions between components can only be tested at higher levels.

4.6.2.3 System-Level Testing

System-level testing addresses the code as a whole. For a given set of inputs to the code, what is the correct code output? In software engineering, system level testing is the primary means by which one determines if the software requirements have been met (software verification) as well as the customer's actual needs (software validation). For nonscientific software, it is often possible to *a priori* determine what the correct output of the code should be. However, for scientific computing software, where partial differential or integral equations are solved numerically, the "correct" output is generally *not known* ahead of time. This issue is discussed in more detail in Section 4.8.

4.6.3 Regression Testing

Regression testing involves the comparison of code or software routine output to the output from earlier versions of the code. This is in stark contrast to defect testing, where output is compared to the "correct" value, since the earlier version may itself be incorrect. Regression tests are designed to prevent the introduction of coding mistakes by detecting unintended consequences of changes in the code. Regression tests are *not* defect tests; they are consistency tests. They can be implemented at the unit, component, and system level. In fact, all of the defect tests previously described can also be implemented as regression tests. Careful regression testing combined with defect testing and version control can minimize the chances of introducing new software defects during code development and maintenance.

Consider the following example of successful regression testing in structural mechanics. A code developer is implementing a new model for plastic deformation in a structural mechanics code. After completing the programming of the new model, a regression test for linear elasticity is run and fails, i.e., the code no longer produces the same result for a linear elasticity problem that was found before the new plasticity model was implemented. Since the implementation of a plasticity model should have no effect on the behavior of the code when running a linear elasticity problem, the developer knows that there have been unintended consequences of their code modifications. By examining the changes made using a "diff" inside the version control system, the changes which unintentionally affected the linear elasticity behavior can be identified.

Many commercial vendors of scientific computing software tout their extremely large suite of regression tests. However, regardless of how many tens of thousands of regression tests are run, the user must be cognizant of the fact that regression tests do *not* provide evidence of code correctness. Defect testing is the means by which software correctness should be assessed, and the challenges of defect testing scientific computing software at the system level is discussed in detail in Section 4.8.

4.6.4 Test Harness and Test Suites

Many different types of software tests have been discussed in this section. For larger software development projects, it would be extremely tedious if the developer had to run each of the tests separately and then examine the results. In such cases, automation of software testing is a must.

A *test harness* is the combination of software and test data used to test the program or component by automatically running it under various conditions (Eddins, 2006). A test harness is usually composed of a test manager, test input data, test output data, a file comparator, and an automatic report generator. While it is certainly possible to create your own test harness, there are a variety of test harnesses that have been developed for a wide range of programming languages. Many of these test harnesses derive their structure from Smalltalk's SUnit, and collectively go by the name xUnit. Open source xUnit frameworks include FUnit (Fortran), pFUnit (parallel Fortran), CUnit (C), CppUnit (C^{++}), and JUnit (Java).

Once a suite of tests has been set up to run using a test harness, it is recommended that these tests be run automatically at specified intervals. Shorter tests can be run in a nightly test suite, while larger tests which require more computer time or memory may be set up in weekly or even monthly test suites. In addition, an approach called *continuous integration testing* (Duvall et al., 2007) requires that specified test suites be run before any new code modifications are checked into the master repository.

Hovy and Kunkel (2016) recently developed an approach for automatically generating unit tests called *capture* and *replay*. Specifically, they developed the tool *FortranTestGenerator* (FTG) which runs the Fortran application, automatically creates the test driver, and extracts test data from the application during run time. FTG captures input variables for a routine and then is able to call that routine, with the captured input, from the automatically-generated test driver. This approach is extremely valuable for complicated scientific computing codes because it automatically creates application-relevant unit tests from actual code runs. While the utility of FTG for automatically creating regression tests is self-evident, additional work must be done by the user/developer to extend the test to "correct" results required for defect testing.

4.6.5 Code Coverage

Regardless of how software testing is done, one important aspect is the coverage of the tests. *Code coverage* can be defined as the percentage of code components (and possibly their interactions) for which tests exist. While testing at the unit and component levels is relatively straightforward, system-level testing must also address interactions between different components. Large, complex scientific computing codes generally have a very large number of options for models, submodels, numerical algorithms, numerical parameters, boundary conditions, loading conditions, etc. Assume for the moment that there are 100 different options in the code to be tested, which is a conservative estimate for most production-level scientific computing codes. Testing each option independently (although generally not possible) would require 100 different system-level tests. Testing pair-wise combinations for interactions between these different options would require 4,950 system level tests. Testing the interactions between groups of three would require 161,700 tests. While this is clearly an upper bound since many options may be mutually exclusive, it does provide a sense of the magnitude of the task of achieving complete code coverage of model/submodel/algorithm interactions.

Table 4.2 provides a comparison of the number of system-level tests required to ensure code coverage with different numbers of option interactions for codes with 10, 100, and 1,000 different code options. Clearly testing the three-way interactions for our example of 100 coding options is impractical, as would be testing all pair-wise interactions when 1,000 coding options are available, a number not uncommon for large scientific computing

Table 4.2 *Number of system-level tests required for complete code coverage for codes with different numbers of options and option combinations to be tested*

Number of options	Option combinations to be tested	System-level tests required
10	1	10
10	2	45
10	3	720
100	1	100
100	2	4,950
100	3	161,700
1,000	1	1,000
1,000	2	499,500
1,000	3	$\sim 1.7 \times 10^8$

codes. One possible approach for addressing this combinatorial explosion of tests for component interactions is application-centric testing (Knupp and Ober, 2008), where only those components and component interactions that impact a specific code application are tested. This approach could be used in conjunction with the capture and replay technique discussed above for automatically generating a test driver and application-specific data.

4.7 Software Development Practices

4.7.1 Performance versus Portability and Maintainability

As discussed earlier, the long life spans of scientific computing codes are at odds with rapidly changing hardware architectures, especially in the case of accelerators such as graphics processing units (GPUs). Achieving good performance (i.e., computing speed) generally requires lower level languages with careful control of memory usage, cache usage, etc. However, achieving portability and maintainability generally requires a higher level of abstraction between the code programmed by the scientist/engineer and hardware-specific optimizations. For scientific computing, correctness of the code output and performance are the top two code qualities, followed then by portability and maintainability (Carver et al., 2007; Johanson and Hasselbring, 2018).

4.7.2 Modularity and Reuse

Designing software for modularity and reuse can be expensive and takes time (Heaton and Carver, 2015). While modularity/reuse in scientific computing is not as common as for traditional business and information technology software, there are a number of areas in scientific computing where reuse is commonly found. Some examples of software reuse via libraries are: mathematical function and subroutine libraries (e.g., Press et al., 2007), parallel message passing libraries (e.g., MPI, OpenMP), pre-packaged linear solvers such as the Linear Algebra Package (LAPACK) or the Portable Extensible Toolkit for Scientific Computation (PETSc), and graphics libraries. Scientific software developers are generally reluctant to incorporate libraries due to concerns for long-term library support (Heaton and Carver, 2015).

For scientific software, programming in object-oriented programming languages can provide modularity with the added benefit of variable/routine encapsulation and easier unit testing (Jackson et al., 2019). Object-oriented programming can be attractive, especially in an academic setting, since it allows for code improvements without requiring knowledge of the whole code base since high level routines can be agnostic to the underlying physical models (Tyson et al., 2019).

4.7.3 Programming Practices

Some recommended software programming practices for scientific computing are found in Derlaga et al. (2013) and Tyson et al. (2019). These practices are summarized briefly here.

- Write self-commenting code with descriptive variable and routine names.
- Use explicit (rather than implicit) typing, in other words, explicitly define the type (floating point, integer, character, logical, etc.) of every variable and function used.
- Use private data whenever possible to minimize the chance of accidentally overwriting important data.
- Even when using free-form programming languages, enforce line length restrictions for readability, avoid trailing white space (which can cause trivial code diffs), and use indentation (but avoid using tab spacing).
- Use tools for performance profiling (e.g., TAU, 2023, Vampir, 2023) and memory checking (e.g., Valgrind, 2020).

4.7.4 Refactoring

Often, at the end of a software development effort, the developer realizes that choices made early in the software design phase have led to computationally inefficient or cumbersome programming. Refactoring is the act modifying software such that the internal software structure changes, but the outward behavior does not. For example, if a given function or subroutine is supplied a certain set of inputs, it should produce the same outputs both before and after refactoring. Refactoring can reduce the complexity, computational time, and/or memory requirements for scientific software. However, refactoring should not be undertaken until a comprehensive regression test suite (see Section 4.6.4) is in place to ensure that the external behavior is not modified and that new programming errors are not introduced.

4.8 Software V&V versus Scientific Computing V&V

4.8.1 Verification and Validation for Software

The software engineering community has developed definitions for verification and validation that are appropriate for general software (e.g., ISO, 1991; IEEE, 1991). In software engineering, verification is defined as ensuring that software meets its stated requirements. Software verification generally involves assessing the software using tests created by the software developers, and these tests are driven by the software requirements. Software validation goes a step further and has the goal of ensuring that software actually meets the customer's needs. It usually involves customer-supplied data and can be considered both an assessment of the correctness of the software and the software requirements.

4.8.2 Verification and Validation for Scientific Computing

The definitions accepted by AIAA (1998) and ASME (2006) for verification and validation (V&V) as applied to scientific computing address the mathematical accuracy of a numerical solution (verification) and the physical accuracy of a given model through comparisons with experimental data (validation). Oberkampf (2012) makes the point that the software engineering definitions for V&V focus on software requirements for input and output data and hardware. However, scientific computing definitions of V&V instead focus on the consistency of the software to the mathematical model (for verification) and the correspondence of the mathematical model to physical reality (for validation). Stated differently, scientific computing definitions of V&V address the correctness of the software and the correctness of the mathematical model, but the correctness of either are not known in general. While it may be possible to try to map the software engineering definition for V&V to their scientific computing counterparts, attempting to do so simply covers up the fact that rigorous V&V is much more challenging for scientific software than it is for software designed for general applications such as business and information technology.

4.8.3 Why Software V&V Fails for Scientific Computing

To understand why V&V testing is so difficult in scientific software, consider Figure 4.3, which is an adaption of Hinsen's (2015) "approximation tower." At the bottom, there is some real-world phenomenon that the scientist/engineer would like to simulate. The first steps relate to model validation and include making experimental observations of the phenomenon along

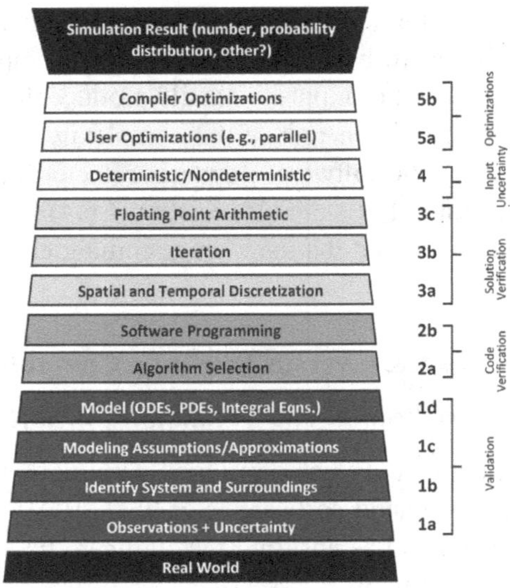

Figure 4.3 Approximation tower in scientific computer software testing.
(adapted from Hinsen, 2015)

with its associated uncertainty (1a), choosing what will make up the system and its surroundings (1b), deciding on some appropriate level of model assumptions and approximations (1c), and coming up with a mathematical model (typically differential or integral equations) of the phenomenon (1d). Level two includes choosing an appropriate numerical algorithm (2a) and programming that algorithm into software (2b). The third level includes discretization of the spatial and temporal domain, thus giving rise to discretization errors (3a), performing some number of iterations if an iterative method is used (3b), and choosing an appropriate level of floating point precision (3c). If there are any uncertainties in the system, these are propagated through the model/simulation (Level 4). Finally, there may be user optimizations such as a choice of parallelism via MPI or OpenMP (5a) and compiler optimizations that may occur unbeknownst to the scientist/engineer (5b). All of these approximations can contribute to errors and uncertainties in the final simulation result relative to the true value in nature, which might take the form of a single number, a probability distribution, or even some looser uncertainty structure, such as a probability box (see Chapter 12).

Performing the validation assessment on the scientific computing software can result in large differences between the output of the code and the experimental test data. These differences could come from any stage in the approximation tower. For example, there may be bias errors in the experimental data itself (1a), the modeling assumptions may have been grossly in error (1c), there may be software programming errors (2b), the spatial grid may have been too coarse (3a), or aggressive compiler optimizations may have caused a loss of accuracy (5b). Furthermore, it is often the case that the scientific software is being used for conditions where no experimental data are available, thus undermining the approximation tower from the beginning.

Given the challenges of comparing simulation results directly with real-world observations as previously discussed, a new approach is needed. In this book, we argue that the approximation tower shown in Figure 4.3 should be decomposed into smaller, more manageable parts that should be tested/evaluated separately. For example, to test Level 5, the sensitivity of the code results to different compiler optimization levels and user optimizations should be assessed. For Level 4, methods are presented in Chapter 7 for estimating errors incurred due to finite sampling sizes. For Level 3, techniques are presented in Chapters 7–9 for estimating the various sources of numerical error. For Level 2, we recommend the use of code order of accuracy testing (Chapter 5) using relatively simple exact and manufactured solutions (Chapter 6) that exercise the options used in the code. The most challenging level is the first one, where model validation (Chapter 10), validation experiments (Chapter 11), and model accuracy assessment (Chapter 12) are needed.

4.8.4 Challenges in Scientific Computing V&V

Computational scientists tend to focus too much on model validation (Heaton and Carver, 2015); if the simulation results agree with the data, then it is often assumed that the software, algorithm, numerics, etc. are correct. Conversely, when there is disagreement, the blame is often placed on the modeling assumptions. In traditional software engineering applications, V&V is performed using test "oracles," which are external means of obtaining the "correct" answer from the software. However, in scientific computing, the generation of test oracles is impossible. The closest thing we have is an exact solution to the underlying PDEs (i.e., the model); however, the output of the code will depend on the numerical discretization method, grid resolution and quality, the time step, the iterative convergence tolerances, the digits of precision, etc. Each of these factors causes the software to produce numerical errors of unknown magnitude relative to the exact solution to the PDEs. In the context of model validation, experimental data always contain errors, both random and bias, and thus also cannot be used as test oracles.

4.9 Reproducibility

Reproducibility in experimental science is a fundamental tenet of the scientific method. Problems with experimental reproducibility have been tied to differences in experimental setups, statistical analysis errors, poor reporting of methodologies and data quality in journals, and even scientific misconduct (Hutton et al., 2016). However, experimental methods for reproducibility have been around for many centuries and have reached a reasonable level of maturity.

For the relatively young field of scientific computing, reproducibility has generally been quite poor for a number of reasons (Merali, 2010). Scientists often perform limited or no testing of their codes, documentation is generally sparse, scientific codes are rarely released when publishing results, and scientists generally focus on their domain science rather than the software. In fact, poor software engineering practices have led to researchers retracting articles, as discussed in Merali (2010) and Kanewala and Bieman (2014).

In order to fix this reproducibility crisis in scientific computing, Hutton et al. (2016) recommend that the following steps be taken.

1. Scientific software codes should be made readable and reusable.
2. Well-documented workflows should be developed which meld scientific codes with their required input data.
3. Codes and workflows should be made available and easy to use.
4. Persistent identifiers such as DOIs should be used to reference the scientific code (including version number) and workflows.

Implementing these recommendations requires a culture change in the use of good software engineering techniques in scientific computing. Hutton et al. (2016) argue that both scientific journals and funding agencies have a role to play in effecting this culture change.

Mufano et al. (2017) offer a broader discussion of reproducibility which extends beyond software. They identify a number of threats to reproducibility in science and offer several recommendations to improve it. Specifically, they cite the need for transparency in methods and data, as well as the processes for "methodology implementation, data analysis and outcome interpretation." They argue for open access not only of research publications, but the data, materials, and software used to obtain the research results. See Mufano et al. (2017) for details.

4.10 Problems

1. Discuss the advantages and disadvantages of various programming languages that are used in scientific computing.
2. Explain how regression testing differs from defect testing.

3. Pick a commercial software package in your field and determine the number of possible combinations of input choices that would be required to achieve complete testing coverage of the software.
4. Contrast software verification and validation with scientific computing verification and validation.
5. Discuss why system-level testing is so challenging for scientific computing.
6. Give two reasons why the traditional waterfall software development model fails for scientific computing.

4.11 References

AIAA (1998). Guide for the Verification and Validation of Computational Fluid Dynamics Simulations. American Institute of Aeronautics and Astronautics, AIAA-G-077-1998, Reston, VA.

ASME (2006). Guide for Verification and Validation in Computational Solid Mechanics. *American Society of Mechanical Engineers*, ASME V&V 10-2006, New York, NY.

Beck, K. (2004). *Extreme Programming Explained: Embrace Change*, 2nd ed. Boston, MA: Addison-Wesley.

Carver, J. C., N. P. Chue Hong, and G. K. Thiruvathukal (eds.) (2017). *Software Engineering for Science*. New York, NY: Chapman & Hall/CRC.

Carver, J. C., R. P. Kendall, S. E. Squires, and D. Post (2007). Software Development Environments for Scientific and Engineering Software: A Series of Case Studies. *Proceedings – International Conference on Software Engineering*, 550–559.

Chacon, S., and B. Straub (2019). *Pro Git*, 2nd ed. Apress, Version 2.1.108, 2019-01-31. https://git-scm.com/book/en/v2.

Derlaga, J. M., T. S. Phillips, and C. J. Roy (2013). SENSEI Computational Fluid Dynamics Code: A Case Study in Modern Fortran Software Development. *21st AIAA Computational Fluid Dynamics Conference*, AIAA Paper 2013-2450, San Diego, CA, June 24–27.

Duvall, P. F., S. M. Matyas, and A. Glover (2007). *Continuous Integration: Improving Software Quality and Reducing Risk*. Upper Saddle River, NJ: Addison-Wesley.

Eddins, S. (2006). Taking Control of Your Code: Essential Software Development Tools for Engineers. *International Conference on Image Processing*, Atlanta, GA, October 9. http://blogs.mathworks.com/images/steve/92/handout_final_icip2006.pdf.

Hatton, L. (1995). *Safer C: Developing Software for High-Integrity and Safety-Critical Systems*. Maidenhead, Berkshire: McGraw-Hill International.

Hatton, L. (1997a). Software Failures: Follies and Fallacies. *IEEE Review*, March, pp. 49–52.

Hatton, L. (1997b). The T Experiments: Errors in Scientific Software. *IEEE Computational Science and Engineering*, **4**(2), 27–38.

Hatton, L., and A. Roberts (1994). How Accurate Is Scientific Software? *IEEE Transactions on Software Engineering*, **20**(10), 785–797.

Heaton, D., and J. C. Carver (2015). Claims about the Use of Software Engineering Practices in Science: A Systematic Literature Review. *Information and Software Technology*, **67**, 207–219.

Hinsen, K. (2015). The Approximation Tower in Computational Science: Why Testing Scientific Software Is Difficult. *Computing in Science & Engineering*, **17**(4), 72–77.

Hovy, C., and J. Kunkel (2016). Towards Automatic and Flexible Unit Test Generation for Legacy HPC Code. *4th International Workshop on Software Engineering for High-Performance Computing in Computational Science and Engineering (SE-HPCCSE)*, Salt Lake City, UT, pp. 1–8.

Hutton, C., T. Wagener, J. Freer, H. Dawei, C. Duffy, and B. Arheimer (2016). Most Computational Hydrology Is Not Reproducible, so Is It Really Science?" *Water Resources Research*, **52**(10), 7548–7555.

IEEE (1991). IEEE Standard Glossary of Software Engineering Terminology. IEEE Std 610.12-1990, New York, NY.

ISO (1991). ISO 9000-3: Quality Management and Quality Assurance Standards – Part 3: Guidelines for the Application of ISO 9001 to the Development, Supply and Maintenance of Software. International Organization for Standardization, Geneva, Switzerland.

Jackson, C. W., W. C. Tyson, and C. J. Roy (2019). Turbulence Model Implementation and Verification in the SENSEI CFD Code. *AIAA SciTech*, AIAA Paper 2019-2331, January 2019.

Johanson, A. N., and W. Hasselbring (2018). Software Engineering for Computational Science: Past, Present, Future. *Computing in Science & Engineering*, **20**(2), 90–109.

Kanewala, U., and J. M. Bieman (2014). Testing Scientific Software: A Systematic Literature Review. *Information and Software Technology*, **56**(10), 1219–1232.

Kelly, D. F. (2007). A Software Chasm: Software Engineering and Scientific Computing. *IEEE Software*, **24**, 118–120.

Kleb, B., and B. Wood (2006). Computational Simulations and the Scientific Method. *Journal of Aerospace Computing, Information, and Communication*, **3**(6), 244–250.

Knupp, P. M., and C. C. Ober (2008). A Code-Verification Evidence-Generation Process Model and Checklist. *Sandia National Laboratories Report*, SAND2008-4832.

Larman, C., and V. R. Basili (2003). Iterative and Incremental Development: A Brief History. *Computer*, **36** (6), 47–56.

MATLAB (2020). *MATLAB® Programming Fundamentals*, R2020b, The Mathworks, Inc., Natick, MA. www.mathworks.com/help/pdf_doc/matlab/matlab_prog.pdf.

McConnell, S. (2004). *Code Complete: A Practical Handbook of Software Construction*, 2nd ed. Redmond, WA: Microsoft Press.

Merali, Z. (2010). Computational Science: Error, Why Scientific Programming Does Not Compute. *Nature*, **467**(7317), 775–777.

Mufano, M. R., B. A. Nosek, D. V. M. Bishop, K. S. Button, C. D. Chambers, N. P. du Sert, U. Simonsohn, E.-J. Wagenmakers, J. J. Ware, and J. P. A. Ioannidis (2017). A Manifesto for Reproducible Science. *Nature: Human Behaviour*, **1**(21), 1–9.

Oberkampf, W. L. (2012). Practical and Technical Challenges in Verification and Validation. *ASME Verification and Validation Symposium*, Plenary Presentation, Las Vegas, NV, May 2012.

Press, W. H., S. A. Teukolsky, W. T. Vetterling, and B. P. Flannery (2007). *Numerical Recipes: The Art of Scientific Computing*, 3rd ed. Cambridge: Cambridge University Press.

Pressman, R. S., and B. R. Maxim (2015). *Software Engineering: A Practitioner's Approach*, 8th ed. New York, NY: McGraw-Hill.

SE4Science (2021). Web page. https://se4science.org.

Segal, J., and C. Morris (2008). Developing Scientific Software. *IEEE Software*, **25**(4), 18–20.

Software Carpentry (2023). Software Carpentry. https://software-carpentry.org.

Sommerville, I. (2016). *Software Engineering*, 10th ed. London: Pearson Education Ltd.

SWEBOK (2004). *Guide to the Software Engineering Body of Knowledge: 2004 Edition*. In Borque, P. and R. Dupuis (eds.). Los Alamitos, CA: IEEE Computer Society. www.swebok.org.

TAU (2023). Tuning and Analysis Utilities (TAU). www.cs.uoregon.edu/research/tau/home.php.

Tyson, W. C., C. W. Jackson, and C. J. Roy (2019). BlueRidge: A Higher-Order Finite-Volume Solver. *AIAA SciTech*, AIAA Paper 2019-1405, January 2019.

Valgrind (2020). Valgrind User Manual, Release 3.16.0, May 2020. http://valgrind.org/docs/manual/manual .html.

Vampir (2023). Vampir Performance Optimization. https://vampir.eu/.

Williams, L., and R. Kessler (2003). *Pair Programming Illuminated*. Boston, MA: Addison-Wesley.

Wood, W. A., and W. L. Kleb (2003). Exploring XP for Scientific Research. *IEEE Software*, **20**(3), 30–36.

5

Code Order-of-Accuracy Verification

Computers are incredibly fast, accurate, and stupid. Human beings are incredibly slow, inaccurate, and brilliant. Together they are powerful beyond imagination. —Albert Einstein

Contents for Chapter 5

This chapter addresses rigorous approaches to code verification which ensure that there are no mistakes in the software programming that affect the solution and that appropriate numerical algorithms are used. It begins with a discussion of both the criteria to be used, with a focus on the order of accuracy test, as well as the recommended quantities to be assessed. Important definitions are given such as those for convergence, discretization error, and truncation error. The order of accuracy is addressed in both a theoretical sense (the formal order) and an empirical sense (the observed order). Requirements are laid out for the systematic refinement of the mesh and the general order of accuracy verification procedure is described. The chapter concludes with a discussion of who bears the responsibility for ensuring that rigorous code verification has been conducted.

5.1 Introduction

In scientific computing, the goal of code verification is to ensure that the code is a faithful representation of the underlying mathematical model. This mathematical model generally takes the form of partial differential or integral equations along with associated initial conditions, boundary conditions, and auxiliary relationships. Code verification thus addresses both the correctness of the chosen numerical algorithm and the correctness of the instantiation of that algorithm into written source code, i.e., ensuring there are no coding mistakes or "bugs."

A computer program, referred to here simply as a code, is a collection of instructions for a computer written in a programming language. As discussed in Chapter 4, in the software engineering community code verification is called software verification and is comprised of software tests which ensure that the software meets the stated requirements. When conducting system-level testing of *nonscientific* software, in many cases it is possible to exactly determine the correct code output for a set of given code inputs using test "oracles." However, in scientific computing, the code output depends on the numerical algorithm, the spatial mesh, the time step, the iterative tolerance, and the number of digits of precision used in the computations (see parts 2 and 3 of Figure 4.3). Due to these factors, it is not possible, in general, to know the correct code output (i.e., numerical solution) *a priori.* The developer of scientific computing software is thus faced with the difficult challenge of determining appropriate system-level software tests.

This chapter discusses procedures for verifying scientific computing codes. Although a formal proof of the "correctness" of a complex scientific computing code is probably not possible (Roache, 1998), code testing using the order verification procedures discussed in this chapter can provide a high degree of confidence that the code will produce the correct solution. An integral part of these procedures is the use of systematic mesh and time step refinement. For rigorous code verification, an exact solution to the underlying governing equations (i.e., the mathematical model) is required. We defer the issue of how to obtain these exact solutions until Chapter 6, and for now simply assume that an exact solution to the mathematical model is available. Finally, unless otherwise noted, the code verification procedures discussed in this chapter do not depend on the discretization approach, thus we will defer our discussion of the different discretization methods (finite difference, finite volume, finite element, etc.) until Chapter 9. The reader is also referred to the recent recommended practices document on code verification by AIAA (2021).

5.2 Code Verification Criteria

There are a number of different criteria that can be used for verifying a scientific computing code. In order of increasing rigor, these criteria are:

1. simple tests,
2. code-to-code comparisons,
3. discretization error quantification,
4. convergence tests, and
5. order-of-accuracy tests.

The first two, simple tests and code-to-code comparisons, are the least rigorous but have the advantage that they can be performed for cases where an exact solution to the mathematical model is not available. The remaining criteria require that an exact solution to the mathematical model be available, or at the very least a demonstrably accurate reference solution. These five criteria are discussed in more detail below.

5.2.1 Simple Tests

Simple tests for symmetry, conservation, Galilean invariance, etc., while not a replacement for rigorous code verification studies, are still often helpful as part of the code development process. They have the advantage that they can be applied even when an exact solution to the mathematical model is not available since they are applied directly to the numerical solution. Further discussion of these simple tests can be found in Oberkampf and Roy (2010).

5.2.2 Code-to-Code Comparisons

Code-to-code comparisons are one of the most common approaches used to assess code correctness. A *code-to-code comparison* occurs when the output (numerical solution or system response quantity) from one code is compared to the output from another code. Following Trucano et al. (2003), code-to-code comparisons are only useful when (1) the two codes employ the exact same mathematical models and (2) the "reference" code has undergone rigorous code verification assessment or some other comparable type of code verification. Even when these two conditions are met, code-to-code comparisons should be used with caution. For additional discussion of the appropriate use of code-to-code comparisons, see the discussion of numerical benchmark solutions in Chapter 6, as well as the book chapter by Rider (2019).

If the same models are not used in the two codes, then differences in the code output could be due to model differences and not coding mistakes. Likewise, agreement between the two codes could occur due to the serendipitous cancellation of errors due to coding mistakes and differences due to the model. A common mistake made while performing code-to-code comparisons with codes that employ different numerical schemes (i.e., discrete equations) is to assume that the codes should produce the same (or very similar) output for the same problem with the same spatial mesh and/or time step. On the contrary, the code outputs will only be the same if exactly the same algorithm is employed, and even subtle algorithm differences can produce different outputs for the same mesh and time step.

For the case where the reference code has not itself been rigorously verified, agreement between the two codes *does not* imply correctness for either code. The fortuitous agreement (i.e., a false positive for the test) can occur due to the same algorithm deficiency being present in both codes. Even when the above two requirement have been met, "code comparisons do not provide substantive evidence that software is functioning correctly" (Trucano et al., 2003). Thus, code-to-code comparisons should not be used as a substitute for rigorous code verification assessments.

5.2.3 Discretization Error Evaluation

Discretization error evaluation is the traditional method for code verification that can be used when an exact solution to the mathematical model is available. This test involves the quantitative assessment of the error between the numerical solution (i.e., the code output) and an exact solution to the mathematical model (or a demonstrably accurate numerical benchmark solution) using a single mesh and/or time step. The main drawback to this test is that once the discretization error has been evaluated, it requires a subjective judgment of whether or not the error is sufficiently small.

For example, it is commonly stated in testing of commercial software that obtaining agreement between the numerical solution and the exact solution within eight significant digits for a system response quantity is satisfactory evidence for code verification. This subjective statement seems reasonable, given the general perspective that eight significant digits is all that can be expected from computational simulation. However, this claim of sufficient accuracy is actually inadequate and misleading from the perspective of numerical correctness of a software package. The perspective of code verification is one of assessing the strength of the evidence that the code is without error. For example, researchers have found that a code can appear to be converging to the correct answer, but then with further mesh refinement the numerical solution *diverges* from the correct answer.

5.2.4 Convergence Tests

A *convergence test* is performed to assess whether the error in the discrete solution relative to the exact solution to the mathematical model (i.e., the discretization error) reduces as the mesh and time step are refined. (The formal definition of convergence will be given in Section 5.4.5.) As was the case for discretization error evaluation, the convergence test also requires an exact solution to the mathematical model. However, in this case, it is not just the magnitude of the discretization error that is assessed, but whether or not that error reduces with increasing mesh and time step refinement. The convergence test is the minimum criterion that should be used for rigorous code verification.

5.2.5 Order-of-Accuracy Tests

The most rigorous code verification criterion is the *order-of-accuracy test*, which examines not only the convergence of the numerical solution, but also whether or not the discretization error reduces at the theoretical rate when the mesh and/or time step are refined. This theoretical rate is called the *formal order of accuracy* of the entire numerical solution procedure and is usually found by performing a truncation error analysis of the numerical scheme (see Section 5.5.1). The actual rate at which the discretization error is reduced by the software package is called the *observed order of accuracy*. The calculation of the observed order of accuracy requires: (a) the exact solution to the mathematical model and (b) numerical solutions on two systematically refined meshes and/or time steps. The procedure for computing the observed order of accuracy is presented in Section 5.5.2.

The order-of-accuracy test is the most difficult test to satisfy; therefore, it is the most rigorous of the code verification criteria. It is extremely sensitive to even small mistakes in the code and deficiencies in the numerical algorithm. The order-of-accuracy test is the most reliable code verification criterion for finding coding mistakes and algorithm deficiencies which affect the computed solutions.

5.3 Code Verification Quantities

With the code verification criteria established, we now address the code outputs to be tested. The first code outputs to be considered are the dependent variables in the mathematical model. For all but the simplest code verification criteria, we will compare the solution to a reference solution, ideally an exact solution to the mathematical model. While one could examine the solution at any point in the domain, it is better to examine the solution over the entire domain. In this case, we can convert the difference between the code output and the reference solution over the entire domain into a single, scalar error measure using a norm.

If a continuous representation of the numerical solution is available (e.g., from a finite element method), then a continuous norm can be used. For time independent (i.e., steady) problems, the L_1 norm of the solution error over the domain is given by

$$\left\|u - u_{ref}\right\|_1 = \frac{1}{\Omega}\int_{\Omega}\left|u - u_{ref}\right|d\omega, \tag{5.1}$$

where u is the numerical solution, u_{ref} the reference solution, and Ω is the domain of interest. The L_1 norm is the most appropriate norm to use when discontinuities or singularities exist in the solution. For time dependent problems, the norm should be computed over both the spatial domain and the temporal domain.

When a discrete representation of the numerical solution is available (e.g., from a finite difference or finite volume method), then a discrete norm of the error can be used. The discrete L_1 norm provides a measure of the average absolute error over the domain and can be defined as

$$\left\|u - u_{ref}\right\|_1 = \frac{1}{\Omega}\sum_{n=1}^{N}\omega_n\left|u_n - u_{ref,n}\right|, \tag{5.2}$$

where the subscript n refers to a summation over all N cells of size ω_n in both space and time. Note that for uniform meshes (i.e., those with constant cell spacing in all directions), the local cell sizes cancel resulting in simply:

$$\left\|u - u_{ref}\right\|_1 = \frac{1}{N}\sum_{n=1}^{N}\left|u_n - u_{ref,n}\right|. \tag{5.3}$$

Another commonly used norm for evaluating the discretization error is the L_2 (or Euclidean) norm, which effectively provides the root mean square of the error. For a general mesh, the discrete L_2 norm is given by

$$\left\|u - u_{ref}\right\|_2 = \left(\frac{1}{\Omega}\sum_{n=1}^{N}\omega_n\left|u_n - u_{ref,n}\right|^2\right)^{1/2}. \tag{5.4}$$

The max (or infinity) norm returns the maximum absolute error over the entire domain, and is generally the most sensitive measure of error:

$$\left\|u - u_{ref}\right\|_\infty = \max\left|u_n - u_{ref,n}\right|, \quad n = 1 \text{ to } N. \tag{5.5}$$

For the different discretization schemes, the approach for computing the local errors is different. For finite difference methods, the numerical solution only exists at the nodes; thus, the reference solution must also be computed at the nodes. For finite volume methods, the solution represents an average over the cell, not the value at a cell centroid (a common misconception). The reference solution should thus be averaged over the cell using an integration method that is of higher-order accuracy than the numerical scheme. For example, for a second-order scheme, the average of the reference solution could be found with Gaussian quadrature that is third-order accurate or higher. For finite element methods, the representation of the reference solution and the integration over an element should both be performed to a higher order of accuracy than the underlying finite element solution.

In addition to the dependent variables, one should also examine any system response quantities that may be of interest to the code user. These quantities usually take the form of integrals of the dependent variables (e.g., drag on an object, net heat flux through a surface), but can also take other forms such as natural frequency of vibration, maximum deflection or maximum stress for a structure, and so on. All system response quantities that may potentially be of interest to the user should be included as part of the code verification process to ensure that both the dependent variables and the procedures for obtaining the system response quantities are verified. For example, a numerical solution for the dependent variables may be verified, but if the subsequent numerical integration used for a system response quantity contains a mistake, then incorrect values of that quantity will be produced.

5.4 Definitions

The definitions presented in this section follow the standard definitions from the numerical analysis of partial differential and integral equations (e.g., Richtmyer and Morton, 1967; Morton and Mayers, 2005). A firm grasp of these definitions is needed before moving on to the concepts behind the formal and observed orders of accuracy used in the order-verification procedure.

5.4.1 Truncation Error

The *truncation error* is the difference between the discretized equations and the original partial differential (or integral) equations. It is *not* the difference between a real number and its finite representation for storage in computer memory; this "digit truncation" is called *round-off error* and is discussed in Chapter 7. Truncation error occurs whenever a mathematical model is approximated by a discretization method. The form of the truncation error can usually be found by performing Taylor series expansions of the dependent variables then inserting these expansions into the discrete equations. Recall the general Taylor series representation for the smooth function $u(x)$ expanded about the point x_0:

$$u(x) = u(x_0) + \left.\frac{\partial u}{\partial x}\right|_{x_0}(x - x_0) + \left.\frac{\partial^2 u}{\partial x^2}\right|_{x_0}\frac{(x - x_0)^2}{2} + \left.\frac{\partial^3 u}{\partial x^3}\right|_{x_0}\frac{(x - x_0)^3}{6} + O\left[(x - x_0)^4\right],$$

where the $O\left[(x - x_0)^4\right]$ term denotes that the leading term that is omitted is proportional to $(x - x_0)$ to the fourth power. This expansion can be represented compactly as:

$$u(x) = \sum_{k=0}^{\infty} \left.\frac{\partial^k u}{\partial x^k}\right|_{x_0}\frac{(x - x_0)^k}{k!}.$$

Example: Derivation of the Truncation Error

For a simple example of truncation error analysis, consider the following mathematical model for 1D unsteady heat conduction

$$\frac{\partial T}{\partial t} - \alpha\frac{\partial^2 T}{\partial x^2} = 0, \tag{5.6}$$

where the first term is the unsteady contribution, and the second term represents thermal diffusion with a constant diffusivity α. Let $L(T)$ represent this partial differential operator and let $\tilde{T}$ be the exact solution to this mathematical model assuming appropriate initial and boundary conditions. Thus, we have

$$L(\tilde{T}) = 0. \tag{5.7}$$

For simplicity, we will omit the initial and boundary conditions from the following discussion.

Equation (5.6) can be discretized with a finite difference method using a forward difference in time and a centered second difference in space. The result is the simple explicit numerical scheme

$$\frac{T_i^{n+1} - T_i^n}{\Delta t} - \alpha\frac{T_{i+1}^n - 2T_i^n + T_{i-1}^n}{(\Delta x)^2} = 0, \tag{5.8}$$

where the i subscripts denote spatial location, the n superscripts denote the temporal step, Δx is the constant spatial distance between nodes, and Δt is the time step. We can represent this discrete equation compactly using the discrete operator $L_h(T)$. Using this discrete operator, Equation (5.8) can be written as

$$L_h(T_h) = 0, \tag{5.9}$$

where T_h is the discrete solution to the equation. The variable h is a single parameter that is used to denote systematic mesh refinement, i.e., refinement over the entire spatial domain, in all spatial coordinate directions, and in time (for unsteady problems). For the current example, this parameter is given by

$$h = \frac{\Delta x}{\Delta x_{ref}} = \frac{\Delta t}{\Delta t_{ref}}, \tag{5.10}$$

where Δx_{ref} and Δt_{ref} refer to some arbitrary reference spatial node spacing and time step, respectively. Later, the h parameter will be extended to consider different refinement ratios in time and even in different spatial coordinate directions. For the current purposes, the important point is that when h goes to zero, it implies that Δx and Δt also go to zero at the same rate. Note that for this finite difference discretization, T_h represents a *vector* of temperature values defined at each spatial node and time step.

In order to find the truncation error for the numerical scheme given in Equation (5.8), we can expand the above temperature values in a Taylor series about the temperature at spatial location i and time step n (assuming sufficient differentiability of T):

$$T_i^{n+1} = T_i^n + \left.\frac{\partial T}{\partial t}\right|_i^n \frac{\Delta t}{1!} + \left.\frac{\partial^2 T}{\partial t^2}\right|_i^n \frac{(\Delta t)^2}{2!} + \left.\frac{\partial^3 T}{\partial t^3}\right|_i^n \frac{(\Delta t)^3}{3!} + O(\Delta t^4)$$

$$T_{i+1}^n = T_i^n + \left.\frac{\partial T}{\partial x}\right|_i^n \frac{\Delta x}{1!} + \left.\frac{\partial^2 T}{\partial x^2}\right|_i^n \frac{(\Delta x)^2}{2!} + \left.\frac{\partial^3 T}{\partial x^3}\right|_i^n \frac{(\Delta x)^3}{3!} + O(\Delta x^4)$$

$$T_{i-1}^n = T_i^n + \left.\frac{\partial T}{\partial x}\right|_i^n \frac{(-\Delta x)}{1!} + \left.\frac{\partial^2 T}{\partial x^2}\right|_i^n \frac{(-\Delta x)^2}{2!} + \left.\frac{\partial^3 T}{\partial x^3}\right|_i^n \frac{(-\Delta x)^3}{3!} + O(\Delta x^4).$$

Substituting these expressions into the discrete equation and rearranging yields

$$\underbrace{\frac{T_i^{n+1} - T_i^n}{\Delta t} - \alpha \frac{T_{i+1}^n - 2T_i^n + T_{i-1}^n}{(\Delta x)^2}}_{L_h(T)} = \underbrace{\frac{\partial T}{\partial t} - \alpha \frac{\partial^2 T}{\partial x^2}}_{L(T)} + \underbrace{\left[\frac{1}{2}\frac{\partial^2 T}{\partial t^2}\right]\Delta t + \left[-\frac{\alpha}{12}\frac{\partial^4 T}{\partial x^4}\right](\Delta x)^2 + O(\Delta t^2, \Delta x^4)}_{\text{Truncation Error: } TE_h(T)}. \tag{5.11}$$

Thus, we have the general relationship that the discrete equation equals the mathematical model plus the truncation error. In order for this equality to make sense, it is implied that either (1) the continuous derivatives in $L(T)$ and $TE_h(T)$ are restricted to the nodal points, or (2) the discrete operator $L_h(T)$ and the truncation error $TE_h(T)$ are mapped onto a continuous function space. As an example of the latter, the truncation error over a local set of nodes could be used to create a polynomial interpolation over that region, thus converting the quantity that exists only at the finite difference nodes to a continuous function.

Generalized Truncation Error Expression (GTEE)

Using the operator notation discussed above and substituting in a generic (but sufficiently smooth) dependent variable u into Equation (5.11) yields

$$L_h(u) = L(u) + TE_h(u). \tag{5.12}$$

We again assume an appropriate mapping of the operators onto either a continuous or discrete space. We refer to Equation (5.12) as the *Generalized Truncation Error Expression*, or *GTEE* (Roy, 2009) and it is used extensively in this chapter as well as in Chapter 9. It relates the discrete equations to the mathematical model in a very general manner and is one of the most important equations in the evaluation, estimation, and reduction of discretization errors in scientific computing. When the numerical solution is inserted for u, the right-hand side of the *GTEE* can be thought of as the actual discrete mathematical model that is solved by the discretization scheme $L_h(u_h) = 0$. The *GTEE* is the starting point for determining the *consistency* and the *formal order of accuracy* of the numerical method. It can be derived even for nonlinear mathematical models and shows the relationship between the truncation error and the discretization error.

As will be shown in Chapter 9, the *GTEE* can also be used to provide estimates of the truncation error. Finally, this equation provides a general relationship between the discrete equation and the (possibly nonlinear) mathematical model since we have not specified what the function u is, only that it satisfies certain differentiability constraints. It is relatively straightforward (although somewhat tedious) to show that general polynomial functions such as

$$u(x,t) = \sum_{i=0}^{N_x} a_i x^i + \sum_{j=0}^{N_t} b_j t^j$$

will satisfy Equation (5.12) exactly (see Oberkampf and Roy, 2010, for more details).

Most authors formally define the truncation error only when the exact solution to the mathematical model is inserted into Equation (5.12), thus resulting in

$$L_h(\tilde{u}) = TE_h(\tilde{u}) \tag{5.13}$$

since $L(\tilde{u}) = 0$. However, for the purposes of this book, we will employ the more general idea of truncation error, the *GTEE* found from Equation (5.12), since it will provide more flexibility in how the truncation error is used.

5.4.2 Discretization Error

Discretization error is formally defined as the difference between the exact solution to the discrete equations and the exact solution to the mathematical model. Using our earlier notation, we can thus write the discretization error for the general dependent variable u as

$$\varepsilon_h = u_h - \tilde{u}. \tag{5.14}$$

The h subscript denotes the exact solution to the discrete equations and the overtilde denotes the exact solution to the mathematical model. While some authors refer to this quantity as the global truncation error, we find this terminology confusing and will thus refer to the error in the discrete solution as the discretization error.

5.4.3 Consistency

For a numerical scheme to be *consistent*, the discretized equations $L_h(\bullet)$ must approach the mathematical model equations $L(\bullet)$ in the limit as the discretization parameters $(\Delta x, \Delta y, \Delta z, \Delta t$, denoted collectively by the parameter $h)$ approach zero. In terms of the truncation error discussion above, a consistent numerical scheme can be defined as one in which the truncation error vanishes in the limit as $h \rightarrow 0$. Not all numerical schemes are consistent, and one notable example is the DuFort-Frankel finite difference method applied to the unsteady heat conduction equation, which has a leading truncation error term proportional to $(\Delta t/\Delta x)^2$ (Tannehill et al., 2021). This scheme is only consistent under the restriction that Δt approach zero at a faster rate than Δx.

5.4.4 Stability

For initial value (i.e., hyperbolic and parabolic) problems, a discretization scheme is said to be *stable* if numerical errors do not grow unbounded in the marching direction. The numerical errors are typically considered to come from computer round-off (Ferziger and Peric, 2002), but can arise from any source. The idea of numerical stability originally derives from initial value problems for hyperbolic and parabolic partial differential equations (Crank and Nicolson, 1947), but the concepts can also be applied to iterative methods for elliptic problems (e.g., see Hirsch, 2007). It is important to note that the concept of numerical stability applies only to the discrete equations (Hirsch, 2007), and should not be confused with natural instabilities that can arise in the mathematical model itself.

Most approaches for analyzing numerical stability apply only to linear partial differential equations with constant coefficients. The most popular approach for determining stability is von Neumann stability analysis (Hirsch, 2007). Also referred to as Fourier stability analysis, von Neumann's method employs a Fourier decomposition of the numerical error and neglects the boundary conditions by assuming these error components are periodic. The fact that von Neumann's method neglects the boundary conditions is not overly-restrictive in practice and it results in a fairly straightforward stability analysis (e.g., see Hirsch, 2007). However, the restriction to linear differential equations with constant coefficients is significant. Time and time again, we will find that many of our tools for analyzing the behavior of numerical schemes are only applicable to linear equations. We are thus left in the uncomfortable situation of hoping that we can simply extend the results of these methods to the complicated, nonlinear mathematical models of interest, and this fact should not be forgotten. As a practical matter when dealing with nonlinear problems, a stability analysis should be performed for the linearized problem to provide initial guidance on the stability limits; then numerical tests should be performed to investigate the stability restrictions for the nonlinear problem.

5.4.5 Convergence

Convergence addresses whether or not the exact solution to the discrete equations approaches the exact solution to the mathematical model in the limit of decreasing mesh spacing and time step size. Whereas convergence and consistency both address the limiting behavior of the discrete method relative to the mathematical model, convergence deals with the *solution* while consistency deals with the *equations*. This definition of convergence should not be confused with convergence of an iterative method (see Chapter 8), which we will refer to as iterative convergence.

For marching problems, convergence is determined by Lax's equivalence theorem, which is again valid only for linear equations. Lax's theorem states that given a well-posed initial value problem and a consistent numerical scheme, stability is the necessary and sufficient condition for convergence (Richtmyer and Morton, 1967). When used for code verification purposes, convergence is demonstrated by examining the actual behavior of the discretization error ε_h as $h \to 0$. Since there is no general convergence theorem for non-linear mathematical models, a discrete solution may converge as $h \to 0$, but there is no guarantee that the discrete solution converges to the exact solution to the mathematical model. (See Yee and Sweby, 1998; Yee, 2004; and Nakao et al., 2019, for a detailed discussion of this topic.)

Work with finite volume methods (Despres, 2004) suggests that some modifications to (or perhaps clarifications of) Lax's equivalence theorem may be needed. In his work, Despres claims that the finite volume method is formally inconsistent for 2D triangular meshes, but he found it to be convergent assuming certain solution regularity constraints. This finding brings into question the "necessary" requirement of Lax's theorem as applied to unstructured meshes. It is interesting to note that while Despres does provide theoretical developments, numerical examples are not included. Given the work of Despres (2004) and the references cited therein, it is possible that Lax's theorem should be augmented with the additional assumption of systematic mesh refinement (discussed in Section 5.6) along with mesh topology restrictions. Additional research is needed to understand these mesh quality and topology issues as they relate to the consistency and convergence of discretization schemes.

5.5 Order of Accuracy

The term *order of accuracy* refers to the rate at which the discrete solution approaches the exact solution to the mathematical model in the limit as the discretization parameters go to zero. The order of accuracy can be addressed in either a theoretical sense (i.e., the order of accuracy of a given numerical scheme assuming it has been implemented correctly) or in an empirical manner (i.e., the actual order of accuracy of discrete solutions produced by the software). The former is called the *formal order of accuracy*, while the latter is the *observed order of accuracy*. These two terms are discussed in detail in the next sections.

5.5.1 Formal Order of Accuracy

The *formal order of accuracy* is the theoretical rate of convergence of the discrete solution u_h to the exact solution to the mathematical model $\tilde{u}$. This theoretical rate is defined only in an asymptotic sense as the discretization parameters ($\Delta x, \Delta y, \Delta t$, etc., currently represented by the single parameter h) go to zero in a systematic manner. It will be shown next that the formal order of accuracy can be related back to the truncation error; however, we are limited to linear (or linearized) equations to show this relationship.

The key relationship between the discrete equation and the mathematical model is the *GTEE* given by Equation (5.12), which is repeated here for convenience:

$$L_h(u) = L(u) + TE_h(u). \tag{5.12}$$

Inserting the exact solution to the discrete equation u_h into Equation (5.12), then subtracting the original mathematical model equation $L(\tilde{u}) = 0$, yields

$$L(u_h) - L(\tilde{u}) + TE_h(u_h) = 0.$$

If the mathematical operator $L(\bullet)$ is linear, then $L(u_h) - L(\tilde{u}) = L(u_h - \tilde{u})$. The difference between the discrete solution u_h and the exact solution to the mathematical model $\tilde{u}$ is simply the discretization error ε_h defined by Equation (5.14). As a result, for linear models we find that the discretization error and the truncation error are related by

$$L(\varepsilon_h) = -TE_h(u_h). \tag{5.15}$$

For nonlinear mathematical models, one could either employ a linearization step or develop a nonlinear form of the error transport equations (see Chapter 9).

Equation (5.15) governs the transport of the discretization error and is called the *continuous discretization error transport equation* since it employs the continuous mathematical operator (Roy, 2009). According to this equation, the discretization error is propagated in the same manner as the original solution u. For example, if the original mathematical model contains terms governing the convection and diffusion of the solution u, then the discretization error ε_h will also be convected and diffused through the domain. More importantly in the context of our present discussion, Equation (5.15) also shows that the truncation error serves as the local source for the discretization error (Ferziger and Peric, 2002). It is therefore clear that the rate of reduction of the local truncation error with mesh refinement should produce corresponding reductions in the discretization error (notwithstanding the caveats discussed in Section 5.4.5). Applying this continuous error transport equation to the example for (linear) 1D unsteady heat conduction in Equation (5.8), results in:

$$\frac{\partial \varepsilon_h}{\partial t} - \alpha \frac{\partial^2 \varepsilon_h}{\partial x^2} = -\left[\frac{1}{2}\frac{\partial^2 T_h}{\partial t^2}\right]\Delta t - \left[-\frac{\alpha}{12}\frac{\partial^4 T_h}{\partial x^4}\right](\Delta x)^2 + O\left(\Delta t^2, \Delta x^4\right).$$

Having tied the rate of reduction of the discretization error to the truncation error, we are now in the position to define the formal order of accuracy of a numerical scheme as the smallest exponent which acts upon a discretization parameter in the truncation error, since this will dominate the limiting behavior as $h \to 0$. For problems in space and time, it is sometimes helpful to refer to the formal order of accuracy in time separately from the formal order of accuracy in space. For the 1D unsteady heat conduction example stated earlier, the simple explicit finite difference discretization is formally first-order accurate in time and second-order accurate in space. While it is not uncommon to use different order-of-accuracy discretizations for different terms in the equations (e.g., third-order convection and second-order diffusion), the formal order of accuracy of such a mixed-order scheme is simply equal to the *lowest order* discretization employed.

The truncation error can usually be derived for even complicated, nonlinear discretization methods (e.g., see Grinstein et al., 2007). For cases where the formal order of accuracy has not been determined from a truncation error analysis, there are three approaches that can be used to estimate the formal order of accuracy. The actual formal order of accuracy will be at least as high as the estimated order of accuracy obtained from these three methods. While this is a much weaker statement on the code correctness than performing a true truncation error analysis, it does allow one to make a judgment call as to whether sufficient accuracy has been obtained.

The first approach is to approximate the truncation error by inserting the exact solution to the mathematical model into the discrete equations. Since the exact solution to the mathematical model will not satisfy the discrete equation, the remainder (i.e., the discrete residual) will approximate the truncation error as shown in Equation (5.13). By evaluating the discrete residual on successively finer meshes (i.e., as $h \rightarrow 0$), the rate of reduction of the truncation error can be determined, which may, however, be lower than the true formal order of accuracy. This first approach is called the residual method and is discussed in detail in Section 5.7.3. The second approach is to compute the observed order of accuracy for a series of meshes as $h \rightarrow 0$, and this approach is addressed in the next section. In this case, if the observed order of accuracy is found to be two, then it is generally safe to assume that the formal order of accuracy is *at least* second order. The final approach is to simply assume the theoretical order of accuracy from the numerical approximations employed in the discretization. For example, a linear basis function used with the finite element method generally results in a second-order accurate scheme for the dependent variables, as does linear interpolation/extrapolation when used to determine the interface fluxes in the finite volume method.

In the development of the truncation error, a certain degree of solution smoothness was assumed. As such, the formal order of accuracy can be reduced in the presence of discontinuities and singularities in the solution. For example, the formal order of accuracy for problems with nonlinear discontinuities (such as shock waves in inviscid gas dynamics) has been shown to reduce to first order, *regardless* of the formal order of accuracy of the scheme on smooth problems (Banks et al., 2008). Even for viscous flows, which do not admit discontinuities, the mesh is rarely refined enough at a shock wave to resolve the viscous effects in the shock. For linear discontinuities (e.g., contact discontinuities in inviscid gas dynamics and material interfaces in thermal or structural mechanics problems), the formal order generally reduces to $p/(p+1)$ (i.e., less than one) for methods which have a formal order p for smooth problems (Banks et al., 2008). As a result, the formal order of accuracy of a numerical method typically refers to order of accuracy for smooth solutions.

5.5.2 Observed Order of Accuracy

The *observed order of accuracy* is the actual order of accuracy obtained from the software package on a series of systematically-refined meshes. For now, we only consider the case where the exact solution to the mathematical model is known. In this case, the discretization error can be evaluated exactly (or at least within round-off and iterative error) and only two mesh levels are required to compute the observed order of accuracy. The more difficult case of computing the observed order of accuracy when the exact solution to the mathematical model in not known is addressed in Chapter 9 as part of the solution verification process.

5.5.2.1 Spatial Discretization

Consider a power series expansion of the solution to the discrete equations u_h in terms of the mesh spacing h in the limit as $h \to 0$ for steady-state (i.e., time-independent) problems:

$$u_h = u_{h=0} + \left.\frac{\partial u}{\partial h}\right|_{h=0} h + \left.\frac{\partial^2 u}{\partial h^2}\right|_{h=0} \frac{h^2}{2} + \left.\frac{\partial^3 u}{\partial h^3}\right|_{h=0} \frac{h^3}{6} + O(h^4). \tag{5.16}$$

If a convergent numerical scheme is employed (i.e., it is consistent and stable), then we should have $u_{h=0} = \tilde{u}$. Furthermore, for a formally second-order accurate scheme, by definition we will have $\left.\frac{\partial u}{\partial h}\right|_{h=0} = 0$ since terms of order h do not appear in the truncation error (recall Equation [5.15]). Employing the definition of the discretization error from Equation (5.14), we find that for a general second-order accurate numerical scheme:

$$\varepsilon_h = g_2 h^2 + O(h^3), \tag{5.17}$$

where the coefficient g_2 is only a function of (x, y, z). Note that for discretizations that exclusively employ second-order accurate central-type differencing, the truncation error only contains even powers of h, and thus the higher order terms in Equation (5.17) would be $O(h^4)$. For a more general pth-order accurate scheme, we have

$$\varepsilon_h = g_p h^p + O(h^{p+1}). \tag{5.18}$$

As just mentioned, if central-type differencing is used, the higher order terms will be $O(h^{p+2})$.

Equation (5.18) provides an appropriate theoretical starting point for computing the observed order of accuracy. In the limit as $h \to 0$, the higher-order terms in Equation (5.18) will become small relative to the leading term and can be neglected. Consider now two discrete solutions, one computed on a fine mesh with spacing h and another computed on a coarse mesh with spacing $2h$ found by eliminating every other node, cell, or element from a structured mesh. Neglecting the higher-order terms (whether they are small or not), Equation (5.18) can be written for the two solutions as

$$\varepsilon_{2h} = g_p (2h)^{\hat{p}},$$
$$\varepsilon_h = g_p h^{\hat{p}}.$$

Dividing the first equation by the second one, then taking the natural log, we can solve for the observed order of accuracy to give

$$\hat{p} = \frac{\ln\left(\frac{\varepsilon_{2h}}{\varepsilon_h}\right)}{\ln(2)}. \tag{5.19}$$

Here the diacritical carat "^" is used to account for the fact that we have neglected higher-order terms in the development of Equation (5.19). Thus $\hat{p}$ no longer represents the formal order of accuracy p, but instead gives the actual behavior of the discretization error, i.e., the *observed order of accuracy*. The observed order can be computed regardless of whether or not the higher-order terms in Equation (5.18) are small; however, it will only match the formal order of accuracy when they are small, i.e., in the limiting sense as $h \to 0$.

A more general expression for the observed order of accuracy can be found that applies to meshes that are systematically refined by an arbitrary factor. We now introduce the *grid refinement factor* r which is defined as the ratio of coarse to fine grid mesh spacing

$$r \equiv \frac{h_{coarse}}{h_{fine}} \tag{5.20}$$

(where $r > 1$). Again, neglecting the higher order terms, the discretization error expansions for the two mesh levels become

$$\varepsilon_{rh} = g_p (rh)^{\hat{p}},$$
$$\varepsilon_h = g_p h^{\hat{p}}.$$

Dividing the first equation by the second and taking the natural log, we find the more general expression for the observed order of accuracy

$$\hat{p} = \frac{\ln\left(\frac{\varepsilon_{rh}}{\varepsilon_h}\right)}{\ln(r)}. \tag{5.21}$$

At this point, it is important to mention that the exact solution to the discretized equations u_h is known to within the accuracy of round-off and iterative convergence errors in the numerical solutions. While round-off and iterative errors are discussed in detail in Chapters 7 and 8, respectively, their impact on the observed order of accuracy will be addressed in Section 5.5.2.4.

The observed order of accuracy can be evaluated using the discretization error in the dependent variables, norms of those errors, or the discretization error in any system response quantities that can be derived from the solution (usually via integration over a surface). When applied to norms of the discretization error, the relationship for the observed order of accuracy becomes

$$\hat{p} = \frac{\ln\left(\frac{\|\varepsilon_{rh}\|}{\|\varepsilon_h\|}\right)}{\ln(r)}, \tag{5.22}$$

where any of the norms discussed in Section 5.3 can be used.

5.5.2.2 Temporal Discretization

The order verification procedure for temporal problems with no spatial dependency is essentially the same as for spatial problems described above. The only difference is that the time step Δt is refined rather than a spatial mesh, thus mesh quality is not a concern. If adaptive time-stepping algorithms are employed, the adaptive algorithm should be disabled to provide explicit control over the size of the time step during the order of accuracy computation. For unsteady problems, the temporal discretization error can sometimes be quite small, thus round-off error should be estimated (see Chapter 7) to ensure that it does not adversely impact the observed order-of-accuracy computation.

5.5.2.3 Spatial and Temporal Discretization

It is more difficult to apply the order verification procedure to problems that involve both spatial and temporal discretization, especially for the case where the formal spatial order of accuracy is different from the temporal order. In addition, temporal discretization errors can, in some cases, be much smaller than spatial errors, especially when explicit time marching schemes are used due to their strict stability limits, thus making it more difficult to verify the temporal order of accuracy since the temporal errors can be masked by the spatial errors.

For numerical schemes involving spatial and temporal discretization, it is helpful to rewrite the discretization error expansion given in Equation (5.18) by separating out the spatial and temporal terms. This expression is given as

$$\varepsilon_h = g_x h_x^p + g_t h_t^q + O\left(h_x^{p+1}\right) + O\left(h_t^{q+1}\right), \tag{5.23}$$

where h_x denotes the spatial discretization ($h_x = \Delta x/\Delta x_{ref} = \Delta y/\Delta y_{ref}$, etc.), h_t the temporal discretization (i.e., $h_t = \Delta t/\Delta t_{ref}$), p is the spatial order of accuracy, and q the temporal order of accuracy. Neglecting the higher-order terms in Equation (5.23) results in

$$\varepsilon_h = g_x h_x^{\hat{p}} + g_t h_t^{\hat{q}}, \tag{5.24}$$

where again the "^" indicates observed orders of accuracy. Equation (5.24) can also be formulated in terms of the error norms as:

$$\|\varepsilon_h\| = g_x h_x^{\hat{p}} + g_t h_t^{\hat{q}}. \tag{5.25}$$

Some numerical schemes may also have a cross term present (Kamm et al., 2003) that takes the form

$$\varepsilon_h = g_x h_x^{\hat{p}} + g_t h_t^{\hat{q}} + g_{xt} h_x^{\hat{r}} h_t^{\hat{s}} \tag{5.26}$$

such as the Lax–Wendroff scheme which has $p = q = 2$ and $r = s = 1$. Again, adaptive time-stepping algorithms should be avoided during code order of accuracy verification. Procedures similar to those described in this section can be used for independent refinement in the spatial coordinates.

The next subsections describe spatial and temporal order verification procedures that can either be conducted separately or in a combined manner.

Separate Order Analysis

The simplest approach for performing order verification on a code with both spatial and temporal discretization is to first verify the spatial discretization on a steady-state problem. Once the spatial order of accuracy has been verified, then the temporal order can be investigated separately. The temporal order verification can employ a problem with no spatial discretization such as an unsteady zero-dimensional case or a case with linear spatial variations which can generally be resolved by second-order and higher spatial discretizations to within round-off error. Alternatively, a problem can be chosen which includes spatial discretization errors, but computed on a highly-refined spatial mesh (Knupp and Salari, 2003). In the latter approach, the use of a highly-refined spatial mesh is often required to reduce the spatial errors to negligible levels, thus allowing the spatial discretization error term $g_x h_x^p$ to be neglected in Equation (5.23) relative to the temporal error term $g_t h_t^q$. In practice, this can be difficult to achieve due to stability constraints (especially for explicit methods) or the expense of computing solutions on highly-refined spatial meshes. An alternative is to reduce the spatial dimensionality of the problem in order to allow a highly-refined spatial mesh to be used. In addition to their expense when very fine meshes are required, the other significant drawback to using separate order analysis is that it will *not* uncover issues related to the interaction between the spatial and temporal discretization.

Combined Order Analysis

Combined spatial and temporal order-of-accuracy analysis is significantly more complicated than separate order analysis. To simplify this process as much as possible, we have decomposed the combined order analysis into three steps, with the third step having two possible options.

Step 1: Establish Formal Order of Accuracy

For problems involving both spatial and temporal discretization, establishing the formal order of accuracy includes both determining the form of the discretization error expansion, usually Equation (5.24) or Equation (5.26), as well as identifying the values for p and q (and possibly r and s). For cases where the formal order of accuracy is not known, or where the cross term is present, the next step will be more difficult.

Step 2: Ensure Terms Are of Similar Magnitude

It is important to be certain that the different terms in the discretization error power series expansion are of the same order of magnitude. If the discretization error terms are of drastically different magnitude, then extremely refined meshes and/or time steps may be needed to detect coding mistakes that impact the smaller term. When the cross term is present in the series expansion, as in Equation (5.26), then the following approach can still provide some useful initial guidance on obtaining error terms of similar magnitude. The simplest approach is to perform a spatial mesh refinement study with very small time steps, then using two solutions, estimate both $\hat{p}$ and g_x as:

$$\hat{p} = \frac{\ln\left(\frac{\varepsilon_{rh}}{\varepsilon_h}\right)}{\ln(r_x)}, \tag{5.27}$$

$$g_x = \frac{\varepsilon_h}{h_x^{\hat{p}}}. \tag{5.28}$$

Next, a temporal refinement study is performed with a very fine grid to estimate $\hat{q}$ and g_t as:

$$\hat{q} = \frac{\ln\left(\frac{\varepsilon_{rh}}{\varepsilon_h}\right)}{\ln(r_t)}, \tag{5.29}$$

$$g_t = \frac{\varepsilon_h}{h_t^{\hat{q}}}. \tag{5.30}$$

Knowing both the order of accuracy and the coefficient, an appropriate spatial discretization size and time step can be chosen (within numerical stability constraints) so that the terms in the error expansion have roughly similar magnitudes.

An alternative approach based on the discretization error expansions of Equations (5.24) and (5.25) that does not require extremely fine meshes can be summarized briefly as follows. First, a spatial mesh refinement study using three meshes is performed with a fixed time step to obtain $\hat{p}$ and g_x. Note that this will introduce a fixed temporal discretization error (i.e., $g_t h_t^q$) for all computations, thus the standard observed order-of-accuracy relationship from Equation (5.21) cannot be used. Considering only the discretization error norms for now (the same analysis will also apply to the discretization error of the system response quantities), Equation (5.25) can be rewritten as

$$\left\| \varepsilon_{h_x}^{h_t} \right\| = \phi + g_x h_x^{\hat{p}}, \tag{5.31}$$

where $\phi = g_t h_t^{\hat{q}}$ is the fixed temporal error term. To calculate $\hat{p}$, the observed order-of-accuracy expression from Chapter 9 given by Equation (9.59) can be used which requires three mesh solutions, e.g., coarse ($r_x^2 h_x$), medium ($r_x h_x$), and fine (h_x). In terms of the error norms, this equation is written as

$$\hat{p} = \frac{\ln\left(\dfrac{\left\| \varepsilon_{r_x^2 h_x}^{h_t} \right\| - \left\| \varepsilon_{r_x h_x}^{h_t} \right\|}{\left\| \varepsilon_{r_x h_x}^{h_t} \right\| - \left\| \varepsilon_{h_x}^{h_t} \right\|} \right)}{\ln(r_x)}. \tag{5.32}$$

By calculating $\hat{p}$ in this manner, the constant temporal error term ϕ will cancel out. The spatial discretization error coefficient g_x can then be found from

$$g_x = \frac{\left\| \varepsilon_{r_x h_x}^{h_t} \right\| - \left\| \varepsilon_{h_x}^{h_t} \right\|}{h_x^{\hat{p}} \left(r_x^{\hat{p}} - 1 \right)}. \tag{5.33}$$

A similar analysis is then performed by refining the time step with a fixed spatial mesh to obtain $\hat{q}$ and g_t. The coarsest spatial mesh is a good candidate since it will have the lowest computational cost. Once these four unknowns have been estimated, the spatial step size h_x and the temporal step size h_t can be chosen such that the spatial discretization error term ($g_x h_x^{\hat{p}}$) has the same order of magnitude as the temporal error term ($g_t h_t^{\hat{q}}$), subject again to any numerical stability restrictions.

Example: Spatial and Temporal Discretization Error Magnitudes

The above approach was implemented in a graphical manner by Xue et al. (2021) for a numerical scheme that is formally second order in both space and time. Figure 5.1 shows the norms of the discretization error in the solution plotted versus the time step (d_t). These results are shown for six different mesh levels, from a coarse mesh ($d_x = 1/8$) to a fine mesh ($d_x = 1/256$). On the coarsest

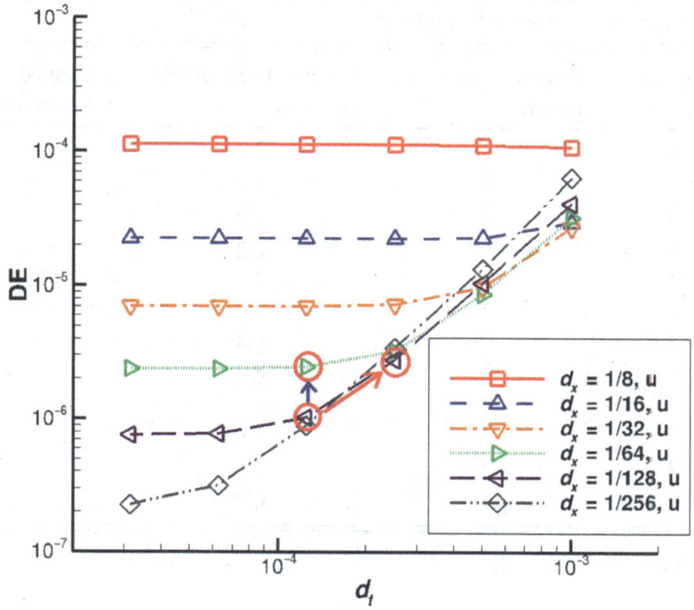

Figure 5.1 Example of a separate temporal and spatial refinement study to obtain error terms of similar magnitude.
(adapted from Xue et al., 2021)

mesh, the errors are dominated by the spatial discretization error since changing the time step has no effect on the error norms. Conversely, the finest mesh ($d_x = 1/256$) is dominated by the temporal error for all but the very smallest time step since coarsening the mesh has little or no impact on the error norms. Any spatial mesh and time step combination near the "bend" in the curves where the behavior changes from horizontal (no change in error level with a change in the time step d_t) to diagonal would serve as an appropriate starting mesh. For example, at $d_x = 1/128$ and $d_x = 1.25 \times 10^{-4}$, coarsening the spatial mesh (blue arrow) would result in a similar increase in error as coarsening the time step (red arrow). Again, the goal is to have the spatial and temporal error components be of similar magnitudes. Conversely, at $d_x = 1/128$ and $d_t = 3.125 \times 10^{-5}$, coarsening the spatial grid increases the error by a factor of four, while coarsening the time step results in no change in the error, indicating that the spatial error term is much larger than the temporal error term.

Step 3a: Simple Combined Spatial and Temporal Analysis
For the first case, it is assumed that (a) the spatial and temporal discretization error terms are of similar magnitude, (b) the formal spatial and temporal orders of accuracy are the same (i.e., if $p = q$), and (c) the spatial and temporal mesh are refined/coarsened by the same factor. As a result, the standard order verification procedure with only two mesh levels can be applied using Equations (5.21) or (5.22) since $r_x = r_t$. For the more complicated case when $p \neq q$, temporal refinement can be conducted by choosing the temporal refinement factor r_t according to Equation (5.34) following Richards (1997):

Table 5.1 *Temporal refinement factors required to conduct combined spatial and temporal order verification using only two numerical solutions*

Spatial order, p	Temporal order, q	Spatial refinement factor, r_x	Temporal refinement factor, r_t	Expected error reduction ratio (coarse/fine)
1	1	2	2	2
1	2	2	$\sqrt{2}$	2
1	3	2	$\sqrt[3]{2}$	2
1	4	2	$\sqrt[4]{2}$	2
2	1	2	4	4
2	2	2	2	4
2	3	2	$\sqrt[3]{4}$	4
2	4	2	$\sqrt[4]{4}$	4
3	1	2	8	8
3	2	2	$\sqrt{8}$	8
3	3	2	2	8
3	4	2	$\sqrt[4]{8}$	8

$$r_t = (r_x)^{p/q}. \tag{5.34}$$

This choice for r_t will ensure that the spatial and temporal discretization error terms are reduced by the same factor with refinement when the solutions are in the asymptotic range.

Example values of r_t for the most commonly used spatial refinement factor $r_x = 2$ are given in Table 5.1 according to Equation (5.34) for various formal spatial and temporal orders of accuracy. The observed orders of accuracy in space and time are then computed using two mesh levels according to

$$\hat{p} = \frac{\ln\left(\frac{\|\varepsilon_{r_x h_x}^{r_t h_t}\|}{\|\varepsilon_{h_x}^{h_t}\|}\right)}{\ln(r_x)} \quad \text{and} \quad \hat{q} = \frac{\ln\left(\frac{\|\varepsilon_{r_x h_x}^{r_t h_t}\|}{\|\varepsilon_{h_x}^{h_t}\|}\right)}{\ln(r_t)}. \tag{5.35}$$

The analysis using system response quantities from Equation (5.24) is exactly the same as given above in Equation (5.35), only with the norm notation omitted.

Step 3b: Complicated Combined Spatial and Temporal Analysis

A more complicated combined spatial and temporal order verification method was developed by Kamm et al. (2003). Their procedure begins with a general formulation of the discretization error, which for a global system response quantity can be written in the form of Equation (5.26). Repeating the equation here for convenience,

$$\varepsilon_{h_x}^{h_t} = g_x h_x^{\hat{p}} + g_t h_t^{\hat{q}} + g_{xt} h_x^{\hat{r}} h_t^{\hat{s}}, \tag{5.27}$$

where the observed orders of accuracy $\hat{p}, \hat{q}, \hat{r}$, and $\hat{s}$ are to be solved for along with the three coefficients g_x, g_t, and g_{xt}. In addition, for the norms of the discretization error, a similar expansion can be employed

Table 5.2 *Mesh levels used in the spatial and temporal code verification study of Kamm et al. (2003)*

Spatial step, h_x	Temporal step, h_t
Δx	Δt
$\Delta x/r_x$	Δt
$\Delta x/r_{x^2}$	Δt
Δx	$\Delta t/r_t$
Δx	$\Delta t/r_{t^2}$
$\Delta x/r_x$	$\Delta t/r_t$
$\Delta x/r_x$	$\Delta t/r_{t^2}$

$$\left\|\varepsilon_{h_x}^{h_t}\right\| = g_x h_x^{\hat{p}} + g_t h_t^{\hat{q}} + g_{xt} h_x^{\hat{r}} h_t^{\hat{s}}, \tag{5.36}$$

where the coefficients (and possibly even the observed orders) will be different.

In order to solve for these seven unknowns, at least seven independent levels of spatial and/or temporal refinement are required. Beginning with an initial coarse mesh with Δx and Δt, Kamm et al. (2003) successively refined in space (by r_x) and time (by r_t) to obtain the seven mesh levels shown in Table 5.2. By computing these seven different numerical solutions, the discretization error expressions for all seven mesh levels result in a coupled, nonlinear set of algebraic equations. The authors solved this nonlinear system of equations using a Newton-type iterative procedure. Since good initial guesses for the unknowns are usually available from the formal orders of accuracy and the sizing study from *Step 2* above, this process should converge quickly if the solutions are near the asymptotic range. If convergence becomes an issue, then one of the following approaches can be used. First, a constant error term can be added to Equations (5.26) or (5.36) to account for the fact that the numerical solutions might not be converging to the exact solution to the mathematical model. Second, additional mesh levels can be computed in order to solve for the unknowns using a least squares process (e.g., see Eca et al., 2016).

An advantage of this more complicated combined spatial and temporal analysis is that it does not require that the terms in the discretization error expression be the same order of magnitude (although it may help the nonlinear iterations converge). The main drawbacks are the possibility of poor convergence of the iterative solution process and the computational expense since a single calculation of the observed orders of accuracy requires seven different numerical solutions. Additional solutions should also be computed to ensure the asymptotic behavior of the observed order of accuracy has been achieved, further increasing the computational expense.

For many spatial/temporal discretization approaches, the mixed spatial/temporal term does not appear in the truncation error, thereby reducing the unknowns and the required number of independent mesh levels down to four. The resulting error expansions were given earlier as Equations (5.24) and (5.25). Although not unique, candidate mesh levels to employ for the simpler error expansions are given in Table 5.3. The resulting four nonlinear algebraic

Table 5.3 *Mesh levels recommended for the simpler error expansion of Equations (5.24) and (5.25) which omit the mixed spatial/temporal term*

Spatial step, h_x	Temporal step, h_t
Δx	Δt
$\Delta x/r_x$	Δt
Δx	$\Delta t/r_t$
$\Delta x/r_x$	$\Delta t/r_t$

equations also have no closed form solution and thus must be solved numerically (e.g., using Newton's method) for the observed orders of accuracy $\hat{p}$ and $\hat{q}$ and the coefficients g_x and g_t.

5.5.2.4 Additional Considerations

The observed order of accuracy can fail to match the nominal formal order due to a number of reasons including:

(1) use of an inconsistent numerical algorithm (see Section 5.4.3),
(2) mistakes in the computer code,
(3) discrete solutions which are not in the asymptotic range (i.e., when the higher-order terms in the truncation error are not small),
(4) the presence of round-off and iterative error, and
(5) when the observed order is computed locally (e.g., element-by-element) in the domain.

Identifying the first two issues is the main goal of code verification activities. The last three topics merit further discussion in the following sub-sections.

Asymptotic Range

The *asymptotic range* is defined as the range of discretization sizes ($\Delta x, \Delta y, \Delta t$, and so on, denoted here collectively by the parameter h) where the lowest-order terms in the truncation error and discretization error expansions dominate. It is only in the asymptotic range that the limiting behavior of these errors can be observed. For code verification purposes, the order of accuracy test will only be successful when the solutions are in this asymptotic range. In practice, the *asymptotic range is surprisingly difficult to identify and achieve* for all but the simplest scientific computing applications. Even an experienced computational analyst with a good intuition on the mesh resolution required to obtain a "good" solution will generally underestimate the resolution required to obtain the asymptotic range. However, many of the cases considered for code verification tend to be much simpler than the actual applications, and thus achieving the asymptotic range on moderately-sized meshes is feasible. Finally, as we shall see in Chapter 9, the approaches for estimating discretization error also rely on the solution (or even multiple solutions) being in the asymptotic range.

Effects of Iterative and Round-off Error

Recall that the underlying theory used to develop the general observed order-of-accuracy expression given in Equation (5.21) made use of the exact solution to the discrete equation u_h. In practice, u_h is only known within some tolerance determined by the number of digits

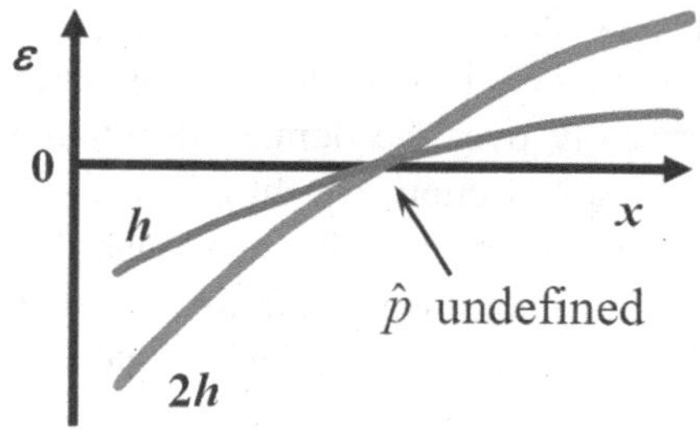

Figure 5.2 Qualitative plot of local discretization error on a coarse mesh ($2h$) and a fine mesh (h) showing that the observed order of accuracy from Equation (5.19) can be undefined when examined locally.

used in the computations (i.e., round-off error) and the criterion used for iterative convergence. The discrete equations can generally be iteratively solved to machine zero (i.e., the level at which round-off errors dominate the iterative errors). However, in practice, the iterative procedure is usually terminated earlier to reduce computational effort. Round-off and iterative error are discussed in detail in Chapters 7 and 8, respectively. To ensure that the computed solutions are accurate approximations of the exact solution to the discrete equations u_h, both round-off and iterative error should be estimated and shown to be at least 100 times smaller than the discretization error on the finest mesh (and time step) employed (i.e., $\leq 0.01 \times \varepsilon_h$).

Evaluation for Local (Field) Solution Quantities

Care must be taken when computing the observed order of accuracy locally since unrealistic orders can be produced. The most common reason for this anomaly is that it requires much finer meshes to achieve the asymptotic region for local quantities of interest. But other complicating numerical phenomena occur, such as the existence of a *crossover point*. For example, when the discrete solutions approach the exact solution to the mathematical model from below in some region of the domain and from above in another, the observed order of accuracy will be undefined at the crossover point. Figure 5.2 gives an example of such a case and shows the discretization error versus a spatial coordinate x. Applying Equation (5.21) would likely produce $\hat{p} \approx 1$ almost everywhere except at the crossover point, where if $\varepsilon_h = \varepsilon_{2h} = 0$, the observed order of accuracy from Equation (5.21) is undefined. These effects can be even more pronounced when one considers other numerical error sources such and round-off and iterative error. For this reason, global quantities and solution error norms are recommended rather than local quantities for code verification purposes.

5.6 Systematic Mesh Refinement

Until this point, the asymptotic behavior of the truncation and discretization error (and thus the formal and observed orders of accuracy) has been addressed by the somewhat vague notion of taking the limit as the discretization parameters ($\Delta x, \Delta y, \Delta t$, etc.) go to zero. For time-dependent problems, refinement in time is straightforward since the time step is fixed over the spatial domain and can be coarsened or refined by an arbitrary factor, subject of course to numerical stability constraints. For problems involving the discretization of a spatial domain, the refinement process is more challenging since the spatial mesh resolution and quality can vary significantly over the domain depending on the complexity of the geometry and the solution. In this section, we introduce the concept of systematic mesh refinement which requires uniformity of the refinement over the spatial domain and consistency of the refinement as $h \to 0$. These two requirements are discussed in detail, as well as additional issues related to the use of local and/or global mesh transformations and mesh topology.

5.6.1 Uniform Mesh Refinement

Local refinement of the mesh in selected regions of interest, while often useful for reducing the discretization error, is *not* appropriate for assessing the asymptotic behavior of discrete solutions. The reason is that the series expansion for the discretization error given in Equation (5.16) is in terms of a single parameter h which is assumed to apply over the entire domain. When the mesh is refined locally in one region but not in another, then the refinement can no longer be described by a single parameter. This same concept holds when the mesh is refined by a different factor in different spatial coordinate directions, which requires special procedures when used for evaluating the observed order of accuracy or for estimating the discretization error. Refining by different refinement factors in different directions should be treated similarly to refinement in space and time (see Section 5.5.2.3). The requirement that the mesh refinement be uniform is *not* the same as requiring that the mesh itself be uniform, only that it be refined in a uniform manner. Note that the requirement of uniform refinement is not restricted to integer refinement factors between meshes. Assuming that a coarse and fine mesh are in fact related through *uniform refinement*, the grid refinement factor can be computed as

$$r_{12} = \left(\frac{N_1}{N_2}\right)^{1/d}, \tag{5.37}$$

where N_1 and N_2 are the number of nodes/cells/elements on the fine and coarse meshes, respectively, and d is the number of spatial dimensions.

An example of uniform and nonuniform mesh refinement is presented in Figure 5.3. The initial coarse mesh (Figure 5.3a) has 4×4 cells, and when this mesh is refined by a factor of two in each direction, the resulting uniformly refined mesh (Figure 5.3b) has 8×8 cells. The mesh shown in Figure 5.3c also has 8×8 cells, but it has been selectively refined to the x-axis and in the middle of the two bounding arcs. While the average cell length scale is refined by a factor of two, the local cell length scale varies over the domain, thus the mesh in Figure 5.3c has not been uniformly refined. Stated differently, the mesh refinement factor from Figure 5.3a to Figure 5.3c cannot be described by a single parameter.

5.6.2 Consistent Mesh Refinement

In general, one should not expect to obtain convergent solutions with mesh refinement when poor quality meshes are used. This is obviously true for the extreme example of a mesh with

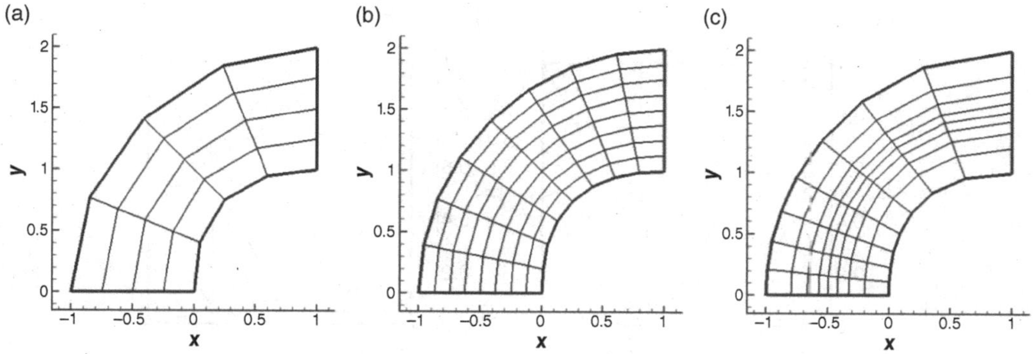

Figure 5.3 Example of uniform and nonuniform mesh refinement: (a) coarse mesh with 4×4 cells, (b) uniformly refined mesh with 8×8 cells, and (c) nonuniformly refined mesh with 8×8 cells.

degenerate cells, such as those involving mesh crossover, that persist with uniform refinement. We now introduce the concept of *consistent mesh refinement* which requires that mesh quality must either stay constant or improve in the limit as $h \to 0$. Examples of mesh quality metrics include cell aspect ratio, skewness, and stretching rate (i.e., the rate at which the mesh transitions from coarse to fine spacing).

To further illustrate the concept of consistent mesh refinement, consider the simple 2D triangular mesh over the square domain given in Figure 5.4a. While this initial coarse mesh certainly has poor quality, it is the approach used for refining this mesh that will determine the consistency. Consider now three cases where this initial coarse mesh is uniformly refined. In the first case, the midpoints of each edge are connected so that each coarse mesh cell is decomposed into four finer cells with similar shape, as shown in Figure 5.4b. Clearly, if this refinement procedure is performed repeatedly, then in the limit as $h \to 0$ even very fine meshes will retain the same mesh qualities related to cell skewness, cell volume variation, and stretching rate. Another refinement approach might allow for different connectivity of the cells and also provide more flexibility in choosing new node locations (i.e., not necessarily at edge mid-points) as shown in Figure 5.4c. As this refinement approach is applied in the limit as $h \to 0$, the mesh quality will generally improve. A third refinement strategy might employ arbitrary edge node placement while not allowing changes in the cell-to-cell connectivity (Figure 5.4d). Considering Figure 5.4, refinement strategy (b) employs fixed quality meshes, and strategy (c) employs meshes with improving quality as $h \to 0$. Thus, both mesh refinement strategies are considered consistent. Strategy (d) is an inconsistent mesh refinement approach, since the quality of the mesh degrades with refinement.

Borrowing concepts from Knupp (2003), we assume the existence of a global mesh quality metric σ which varies between 0 and 1, with $\sigma = 1$ denoting an isotropic mesh, i.e., one with "ideal" mesh quality (square quadrilaterals, cubic hexahedrals, equilateral triangles, etc.). Smaller values of σ would denote anisotropic meshes with lower quality (skewness, stretching, curvature, etc). Techniques exist for quantifying this anisotropy such as the algebraic theory of mesh quality metrics (see Knupp, 2003 for details). Consistent mesh refinement can thus be defined by requiring that $\sigma_{fine} \geq \sigma_{coarse}$ during refinement. Consistent refinement with $\sigma \to 1$ as $h \to 0$ can place significant burdens on the mesh generation and refinement procedure

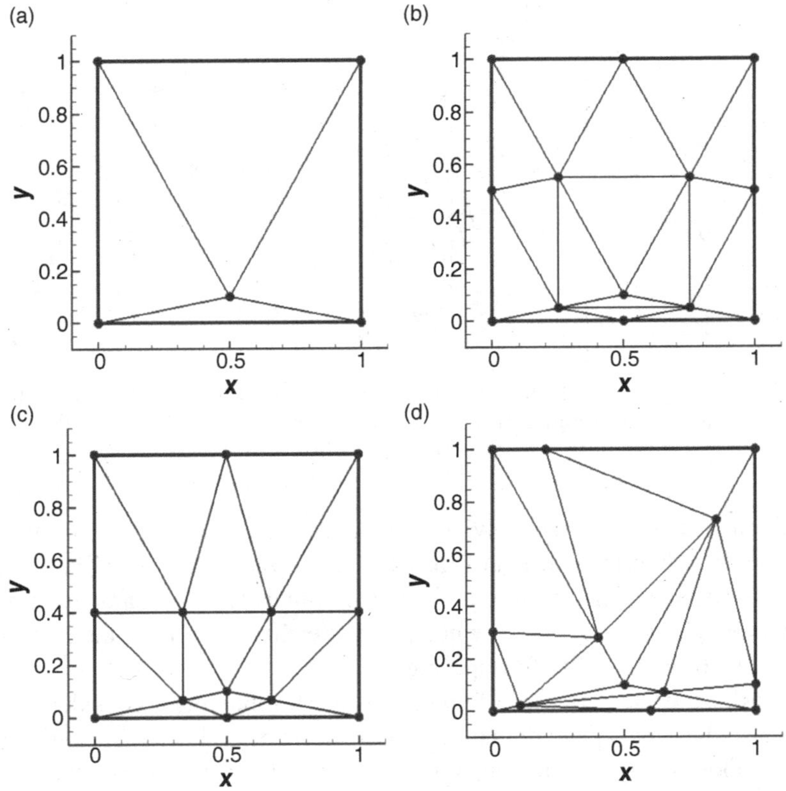

Figure 5.4 Example showing consistent and inconsistent mesh refinement: (a) poor quality coarse mesh with four unstructured triangular cells, (b) uniformly refined mesh that retains a fixed mesh quality, (c) uniformly refined mesh with improved mesh quality, and (d) uniformly refined mesh with inconsistent refinement.

(especially for unstructured meshes). Requiring $\sigma_{fine} \geq \sigma_{coarse}$ as the mesh is refined provides the easiest criteria for discretization schemes to satisfy since meshes become more isotropic with refinement. A more difficult mesh quality requirement to satisfy from the code verification point of view is to require convergence of the numerical solutions for meshes with a fixed quality measure σ as $h \to 0$. Such nuances relating the numerical scheme behavior to the mesh quality fall under the heading of solution verification and are addressed in more detail in Chapter 9. For code verification purposes, it is important to document the asymptotic behavior of the quality of the meshes used for the code verification study.

5.6.3 Other Mesh-Related Issues

5.6.3.1 Mesh Transformations

Mesh transformations are sometimes used to handle complex geometries and to allow local mesh refinement. For discretization methods employing body-fitted structured (i.e., curvilinear) meshes, these transformations can take the form of global transformations of the governing equations. For finite-volume and finite-element methods on structured or

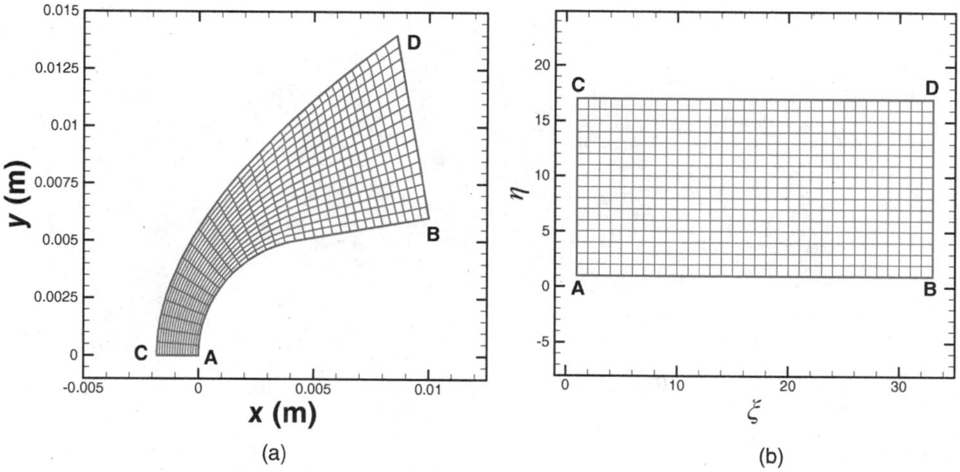

Figure 5.5 Example of a global transformation of a 2D body-fitted structured mesh: (a) mesh in physical (x, y) coordinates and (b) mesh in the transformed computational (ξ, η) coordinates.

unstructured meshes, these transformations usually take the form of local mesh transformations centered about each cell or element. An example of a global transformation for a body-fitted structured grid in 2D is presented in Figure 5.5. The transformation must ensure a one-to-one mapping between the grid line intersections in physical space (a) and those in computational coordinates (b). For additional discussion on mesh transformations, see Thompson et al. (1985) or Oberkampf and Roy (2010).

5.6.3.2 Mesh Topology

There are many different mesh topologies used in scientific computing. When conducting code verification studies, it is recommended that the most general mesh topology that will be employed for solving the problems of interest be used for the code verification studies. For example, if simulations will only be performed on Cartesian meshes, then it is sufficient to conduct the code verification studies on Cartesian meshes. However, if simulations will be performed on non-isotropic meshes consisting of a combination of hexahedral, prismatic, and tetrahedral cells, then those mesh topologies should be used during code verification.

Meshes in 1D consist of an order set of nodes, cells, or elements that may either be uniformly or nonuniformly distributed. For 2D meshes, the nodes/cells/elements may either be structured quadrilaterals, unstructured triangles, unstructured polygons with an arbitrary number of sides, or some hybrid combination of these. Examples of 2D mesh topologies that can be used for code verification are given in Figure 5.6 (Veluri et al., 2008).

In 3D, structured meshes can either be Cartesian, stretched Cartesian, or curvilinear (i.e., body fitted). The 3D unstructured meshes can contain cells that are tetrahedral (four-sided pyramids), pyramidal (five-sided pyramids), prismatic (any 2D cell type extruded in the third direction), hexahedral, general polyhedral, or hybrid combinations of these. An example of a 3D hybrid mesh topology that has been employed for performing code verification on a scientific computing code with general unstructured mesh capabilities is given in Figure 5.7. This mesh consists of hexahedral and prismatic triangular cells extruded from the curved $y_{\min}$ and $y_{\max}$ boundaries and

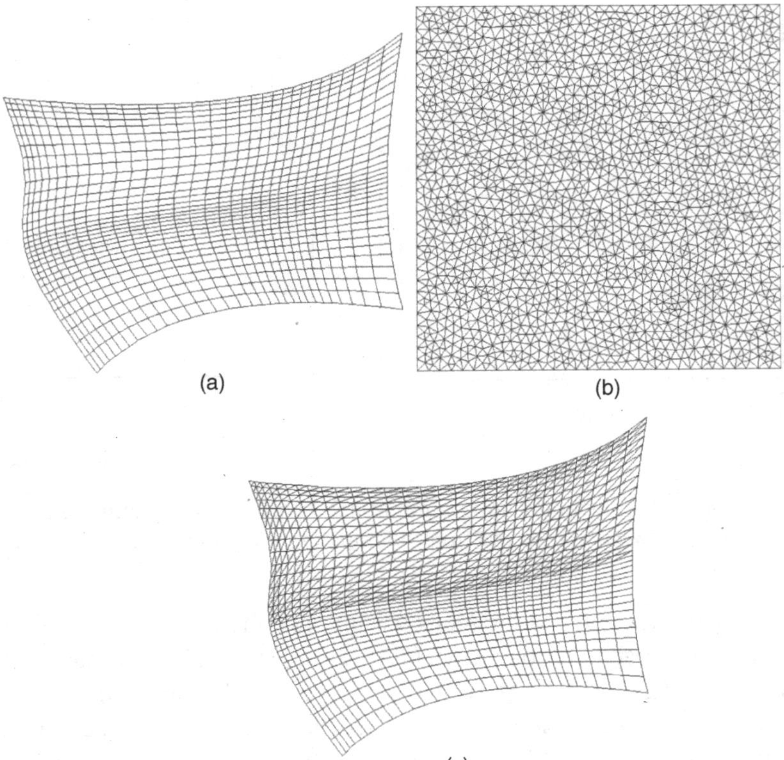

Figure 5.6 Example mesh topologies in 2D: (a) structured curvilinear, (b) unstructured triangles, and (c) hybrid structured/unstructured curvilinear.
(adapted from Veluri et al., 2008)

joined together with a region of tetrahedral cells in the middle. In order to create a family of systematically refined hybrid unstructured grids for code verification, an initial structured grid fine mesh was generated. The resulting hexahedral cells were then further decomposed into tetrahedral cells in the middle layer and prisms in the front layer. To generate systematically coarsened grids, every other point was eliminated in each direction from the underlying structured grid, then the tetrahedral and prismatic layers were added as before to each mesh level. See Veluri et al. (2012) for additional details. When structured grids are employed, it is generally easiest to first generate the finest mesh, then eliminate every other grid line in order to obtain the coarser meshes. In order to ensure systematic coarsening, the number of grid points N for each coordinate direction of the finest mesh should be chosen as

$$N = M \cdot 2^L + 1, \tag{5.38}$$

where M is an integer and L is the desired number of coarse grid levels.

5.6.3.3 Departures from Systematic Mesh Refinement

There are numerous reasons that can cause a family of meshes to deviate from systematic mesh refinement. For structured grids, eliminating every other grid line from a fine mesh

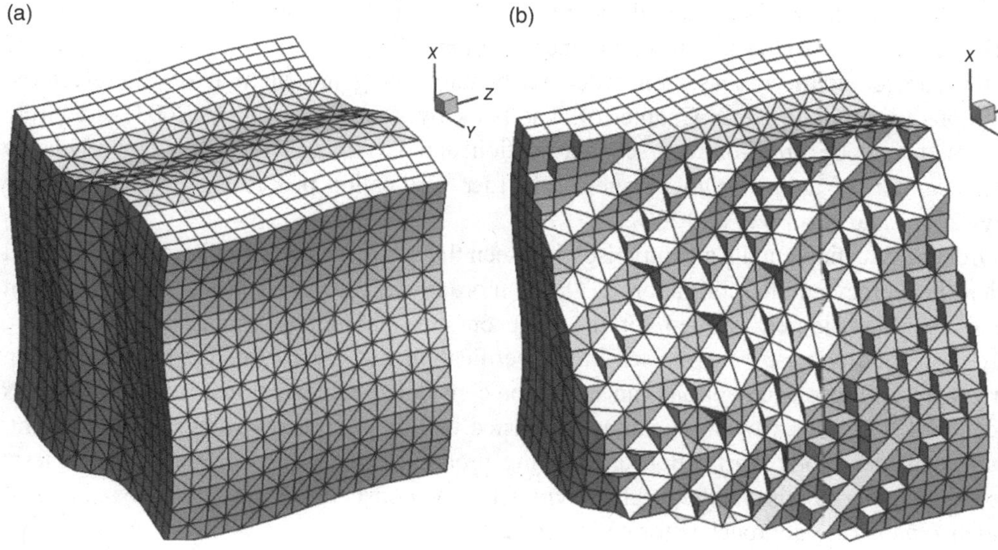

Figure 5.7 A general hybrid mesh topology in 3D: (a) full 3D mesh and (b) internal view showing hexahedral cells (back layer), tetrahedral cells (middle layer), and prismatic cells (front layer).

(adapted from Veluri et al., 2012)

results in a grid refinement factor of exactly two. Refinement factors other than two (e.g., a factor of $r = \sqrt{2}$) can be generated either via interpolation from the fine grid or even the use of a mesh generation script with different mesh sizing parameters. This process often results in refinement that is not quite systematic (i.e., the refinement is not perfectly uniform and/or consistent). Deviations from systematic refinement are even more common on unstructured meshes where systematic refinement is usually attempted by sub-dividing existing cells or by using a mesh generation script. Such departures from systematic refinement can give rise to "noisy" values of the observed order of accuracy. In such cases, the observed order of accuracy can be computed in a least squares sense to filter out the noise due to the non-systematic refinement. See Eca et al. (2016) for additional details. While the least squares approach can aid in filtering out noise, it is even better to simply avoid deviations from systematic mesh refinement in the first place. Since meshes used for code verification generally involve domains with simple geometries, the approach of starting with a structured fine mesh (discussed in the previous section) can be used.

5.7 Order-of-Accuracy Verification Procedure

This section addresses the order verification procedure applicable for discretizations in space and/or time. The procedure can be invaluable for identifying the presence of coding mistakes (i.e., bugs) and problems with the numerical algorithms. Techniques are also discussed to aid in the debugging process once a coding mistake is found to exist. Limitations of the order verification procedure are then described, as well as different variants of the standard order verification procedure.

There are four books which deal extensively with the subject of order-of-accuracy verification. Roache (1998, 2009) and Oberkampf and Roy (2010) all provide an overview of the subject, with emphasis on order verification using the Method of Manufactured Solutions (discussed here in Chapter 6). The book by Knupp and Salari (2003) is entirely dedicated to code order-of-accuracy verification and is one of the most comprehensive references on the subject. Other reviews of order verification procedures are provided by Roy (2005) and Knupp et al. (2007).

Order verification entails a comparison between the convergence behavior of the observed order of accuracy and the formal order. Once an order verification test has been passed, then the code is considered verified for the code options (submodels, numerical algorithms, boundary conditions, etc.) exercised in the verification test. Any further order verification performed for those code options can simply be considered confirmation of code correctness (Roache, 1998). This viewpoint can be contrasted with the "evidence gathering" viewpoint, which considers code verification an ongoing process. The simple code verification tests described earlier in this chapter, such as symmetry and conservation checks, are examples of evidence gathering approaches for code verification.

The order verification procedure discussed here is adapted from the procedure recommended by Knupp and Salari (2003). In brief, this procedure is used to determine whether or not the code output (numerical solution and other system response quantities) converges to the exact solution to the mathematical model at the formal rate with systematic mesh refinement. If the formal order of accuracy is observed in an asymptotic sense, then the code is considered verified for the coding options exercised. Failure to achieve the formal order of accuracy for smooth solutions indicates the presence of a coding mistake or a problem with the numerical algorithm. The steps in the order verification procedure for steady-state problems are presented in Figure 5.8, and these steps are discussed in detail below.

1. **Define Mathematical Model**
   The governing equations (i.e., the mathematical model) generally occur in partial differential or integral form and must be specified unambiguously along with all required initial conditions, boundary conditions, auxiliary equations, and submodels. Small errors in defining the mathematical model and/or its parameters can easily cause the order verification test to fail. For example, an error in the fourth significant digit of the thermal conductivity caused an order verification test to fail for a computational fluid dynamics code used to solve the Navier–Stokes equations (Roy et al., 2007).
2. **Choose Numerical Algorithm**
   A discretization scheme, or numerical algorithm, must be chosen. This includes both the general discretization approach (finite difference, finite volume, finite element, etc.) and the specific approaches to spatial quadrature. Discretization of any boundary or initial conditions involving spatial derivatives (e.g., Neumann-type boundary conditions) must also be considered.
3. **Establish Formal Order of Accuracy**
   The formal order of accuracy of the numerical scheme should be established, ideally by an analysis of the truncation error (see Section 5.5.1). For cases in which the formal order of accuracy has not been established via truncation error analysis, one of the less-rigorous

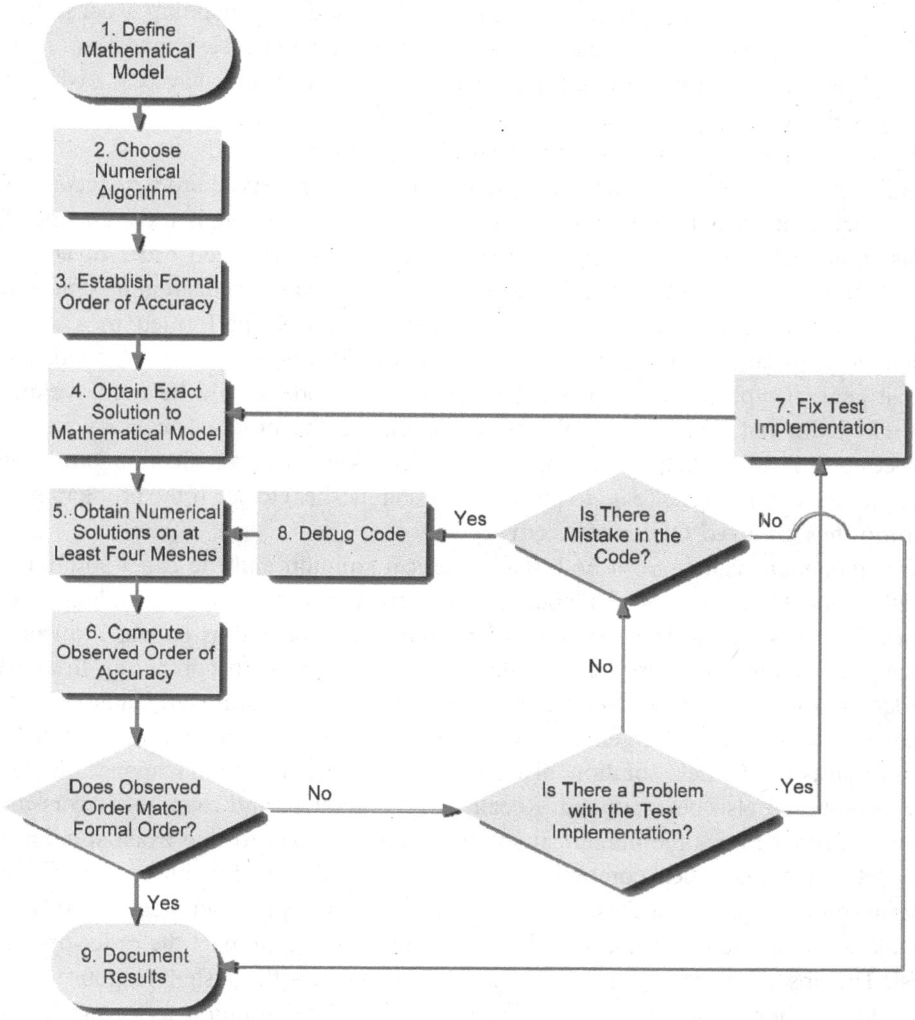

Figure 5.8 Flowchart showing the order verification procedure.
(adapted from Knupp and Salari, 2003)

approaches to approximating the formal order (discussed in Section 5.5.1) may be used. These approaches include the residual method, the order verification procedure itself (i.e., the observed order of accuracy), or the expected order of accuracy.

4. **Obtain Exact Solution to Mathematical Model**
The exact solution to the mathematical model must be obtained including both the solution (i.e., the dependent variables) and the system response quantities of interest. The same exact solution can be reused if the only changes in the code relate to the numerical scheme (e.g., a different flux function or a different numerical parameter is employed). A key point is that actual numerical values for the exact solution must be computed, which may reduce the utility of an infinite series solution because the convergence rate of the series is different for different locations in the solution domain.

See Chapter 6 for a description of recommended methods for obtaining exact solutions to mathematical models for scientific computing applications. New exact solutions are needed any time the governing equations are changed in any way (e.g., when a new submodel is examined or boundary conditions are changed).

5. **Obtain Numerical Solutions on at Least Four Meshes**
While only two mesh levels are required to compute the observed order of accuracy when the exact solution to the mathematical model is known, it is strongly recommended that at least four mesh levels be used to demonstrate that the observed order of accuracy is asymptotically approaching the formal order as the mesh discretization parameters (e.g., $\Delta x, \Delta y, \Delta t$) approach zero. The mesh refinement must be performed in a systematic manner as discussed in Section 5.6. If only one grid topology is to be tested, then the most general type of grid that will be run with the code should be used. Veluri et al. (2012) point out that when problems occur during the order-of-accuracy verification process, it is often helpful to apply the order-of-accuracy verification procedures to simpler mesh topologies (e.g., to purely Cartesian meshes) to see if the problem reoccurs.

6. **Compute Observed Order of Accuracy**
With numerical values from both the numerical solution and the exact solution to the mathematical model now available, the discretization error can be evaluated. Global norms of the solution discretization error should be computed as opposed to examining local values. Not only are there subtle issues associated with computing the observed order for a local value (see Section 5.5.2.4), but it is often computationally expensive to achieve asymptotic convergence at every point in the solution domain. In addition to the error norms for the field solution, discretization errors in all system response quantities of interest should also be examined. Recall that the iterative and round-off errors must be small in order to use the numerical solution as a surrogate for the exact solution to the discrete equations when computing the discretization error. For higher-order schemes, highly-refined spatial meshes, and/or small time steps, the discretization error can be small and thus round-off and iterative error can adversely impact the order-of-accuracy test. The observed order of accuracy can be computed for the system response quantities and for the norms of the discretization error in the field solution using the techniques discussed in Section 5.5.2.

Note that only for simple scientific computing cases (e.g., linear elliptic problems) will the observed order of accuracy match the formal order to more than approximately two significant figures during a successful order verification test. For complex scientific computing codes, it is more common that the observed order of accuracy will approach the formal order with increasing mesh refinement. Thus, it is the asymptotic behavior of the observed order of accuracy that is of interest. In addition, observed orders of accuracy that converge to a value higher than the formal order can either indicate the presence of unforeseen error cancellation (which should not be cause for concern) or mistakes in establishing the formal order of accuracy.

If the observed order of accuracy does not match the formal order in an asymptotic sense, then one should first troubleshoot the test implementation (see Step 7) and then debug the code (Step 8), if necessary. If the formal order of accuracy is still not achieved, then the code verification test is considered to have failed, and the user must then decide if the code has achieved "sufficient" accuracy to justify its use. If the observed order does match the formal order, then the verification test is considered successful, and the user can proceed to Step 9 to document the test results.

7. **Fix Test Implementation**
   When the observed order of accuracy does not match the formal order, the first step is to make sure that the test was implemented correctly. Common examples of incorrect test implementations include mistakes in constructing or evaluating the exact solution to the mathematical model, comparing the numerical and exact solutions, and computing the error norms. If there is significant variation of the observed order of accuracy using different pairs of meshes, it is likely that the asymptotic region has not been achieved. If the test implementation is found to be flawed, then the test should be fixed and then repeated. If the test was correctly implemented, then a coding mistake or algorithm inconsistency is likely present, and one should proceed to Step 8 to debug the code.
8. **Debug the Code**
   When an order-of-accuracy test fails, it indicates either a mistake in the programming of the discrete algorithm, or worse, an inconsistency in the discrete algorithm itself. See Section 5.7.1 for a discussion of approaches to aid in debugging the scientific computing code.
9. **Document Results**
   All code verification results should be documented so that a subsequent user understands the code's verification status and does not duplicate the effort. In addition to documenting the observed order of accuracy, the magnitude of the discretization error should also be reported for both system response quantities and the numerical solution (i.e., the norms of the discretization error). It is further recommended that the meshes and solutions used in the code verification test be added to one of the less-frequently run dynamic software test suites (e.g., a monthly test suite), thus allowing the order-of-accuracy verification test to be repeated on a regular basis. Coarse grid cases, which can typically be executed rapidly, can be added as system-level regression or defect tests to a test suite that is run more frequently as discussed in Chapter 4.

5.7.1 Recommendations for Debugging

Order verification provides a highly-sensitive test as to whether there are mistakes in the computer code and/or inconsistencies in the discrete algorithm. In addition, aspects of the order verification procedure can be extremely useful for tracking down the mistakes (i.e., debugging the code) once they have been found to exist. Once an order verification test fails, and assuming the test was properly implemented, the local variation of discretization error in the domain should be examined. For example, accumulation of error near a boundary or a corner cell generally indicates that the errors are in the boundary conditions, usually due to a lower-order accurate or even an inconsistent boundary condition. Errors in regions of mild grid clustering or skewness can indicate mistakes in the mesh transformations or spatial quadratures/reconstruction. Mesh quality problems could be due to the use of a poor quality mesh, or worse, due to a discrete algorithm that is overly-sensitive to mesh irregularities.

5.7.2 Limitations of Order Verification

A limitation of the order verification process is that the formal order of accuracy can change due to the level of smoothness of the solution as discussed at the end of Section 5.5.1. In addition, since order verification can only detect problems in the solution itself, it generally cannot be used to detect coding mistakes affecting the efficiency of the code. For example, a mistake that causes an iterative scheme to converge in 500 iterations when it should converge in 10 iterations will not be detected by order verification. Note that this type of mistake would be found with an appropriate component-level test fixture for the iterative scheme (as discussed in Chapter 4). Similarly, mistakes which affect the robustness of the numerical algorithm will not be detected, since these mistakes also do not affect the converged numerical solution.

Finally, the standard order verification procedure is not able to verify individual terms in the discretization. Thus, a numerical scheme which is formally first-order accurate for convection and second-order accurate for diffusion results in a numerical scheme with first-order accuracy. Mistakes reducing the order of accuracy of the diffusion term to first order would therefore *not* be detected. This limitation can be addressed by selectively turning terms on and off, which is most easily accomplished when the Method of Manufactured Solutions is employed (see Chapter 6).

5.7.3 Alternative Approaches for Order Verification

A number of variants of the order verification procedure have been proposed. Most of these alternative approaches were developed in order to avoid the high cost of generating and computing numerical solutions on highly-refined 3D meshes. All of these approaches do reduce the cost of conducting the order verification test relative to the standard approach of systematic mesh refinement. However, each approach is also accompanied by drawbacks which are discussed in Oberkampf and Roy (2010) in detail and summarized briefly here.

One simplified approach is called the residual method. In general, when the discrete operator $L_h(\bullet)$ operates on anything but the exact solution to the discrete equations u_h, the nonzero result is referred to as the *discrete residual*, or simply the *residual*. This discrete residual is not to be confused with the iterative residual which is found by inserting an approximate iterative solution into the discrete equations (see Chapter 8). Recall that the truncation error can be evaluated by inserting the exact solution to the mathematical model $\tilde{u}$ into the discrete operator as given previously in Equation (5.13). Assuming an exact solution to the mathematical model is available, the truncation error (i.e., the discrete residual) can be evaluated directly on a given mesh without actually solving for the numerical solution on this grid. Since no iterations are needed, this truncation error evaluation is usually very inexpensive. The truncation error found by inserting the exact solution to the mathematical model into the discrete operator can be evaluated on a series of systematically refined meshes. The observed order of accuracy is then computed by Equation (5.22) but using norms of the truncation error rather than the discretization error. There are a number of drawbacks to the residual form of order verification that are related to the fact that the residual does not incorporate all aspects of the code. See Oberkampf and Roy (2010) for more details.

There are two other simplified approaches to order verification that have been proposed. The *statistical form of order verification* was proposed by Hebert and Luke (2005) and uses only a

Table 5.4 *Comparison of different order verification approaches showing the cost and the type of order-of-accuracy estimate produced (adapted from Thomas et al., 2008)*

Verification method	Cost	Type of order estimate
Standard order verification with systematic mesh refinement	High*	Precise order of accuracy
Downscaling method	Moderate to low	Admits false positives
Statistical method	Moderate (due to sampling)	Admits false positives
Residual method	Very low	Admits false positives and false negatives

**Note: While the cost of standard order verification is relatively high, it is not as high as typical applications since the solutions are generally simpler and the asymptotic range is usually relatively easy to achieve.*

single mesh, which is successively scaled down in size and used to sample over the domain covered by the original mesh. The sampling is performed randomly, and norms of the volume-weighted discretization error are examined. The main advantage of the statistical method is that it is relatively inexpensive because it does not require refinement of the mesh. The final approach to order verification is *downscaling* (Diskin and Thomas, 2007; Thomas et al., 2008), which shares a number of attributes with the statistical method. The major difference is that instead of statistically sampling the smaller meshes in the domain, the mesh is scaled down about a single point in the domain, which eliminates the statistical convergence issues associated with statistical order verification. The focal point to which the grids are scaled can be chosen to emphasize the internal discretization, the boundary discretization, or singularities (Thomas et al., 2008). A major benefit of the downscaling method is that it allows for boundary condition verification in the presence of complex geometries. For more details on these two simplified order verification approaches, see Oberkampf and Roy (2010).

In order to summarize the characteristics of the different order verification approaches, it is first helpful to categorize the results of an order verification test. Here we will define a positive result as one in which the observed order of accuracy is found to match the formal order in an asymptotic sense, while a negative result is one where the observed order is less than the formal order. We define a false positive as a case where a less-rigorous order verification test achieves a positive result, but the more rigorous order verification procedure with systematic mesh refinement produces a negative result. Similarly, a false negative occurs when the test result is negative, but standard order verification with systematic mesh refinement is positive. The characteristics of the four approaches for order-of-accuracy verification are given in Table 5.4. As one might expect, the cost of conducting an order verification study varies inversely with the reliability of the observed order-of-accuracy estimate.

5.8 Responsibility for Code Verification

The ultimate responsibility for ensuring that rigorous code verification has been performed lies with the *user* of the scientific computing code (AIAA, 2021). The user of the code produces the simulation results that are acted upon either by the user or the person to whom the results are given. This holds true whether the code was developed by the user, by someone else in the user's organization, or by an independent organization (government laboratory, commercial software company, etc.). It is *not sufficient* for the code user to simply assume that code verification studies have been adequately performed.

In the ideal case, code verification should be performed during, and as an integrated part of, the software development process. While code verification studies are most often conducted by the code developers, a higher level of independence can be achieved when they are conducted by a separate group, by the code customer, or even by an independent regulatory agency (recall the discussion of independence of the verification and validation process in Chapter 2). The argument is commonly made that fewer coding mistakes exist in a scientific computing code that has a long history of use. This is actually a weaker argument than it appears because the longer a software package is used, the more modifications, extensions, and options are added to the package. Attempting to conduct rigorous code verification on large legacy software packages can be difficult and frustrating.

Commercial companies that produce scientific computing software rarely perform rigorous code verification studies, or if they do, they do not make the results public. Most code verification efforts that are documented for commercial codes are limited to simple benchmark examples that demonstrate "engineering accuracy" rather than verifying the order of accuracy of the code (Oberkampf and Trucano, 2008). Abanto et al. (2005) performed order verification studies on three different commercial computational fluid dynamics codes which were formally at least second-order accurate. Most tests resulted in either first-order accuracy or non-convergent behavior with mesh refinement. It is our opinion that code users should be aware that commercial software companies are unlikely to perform rigorous code verification studies unless users request it. An increasing impetus for rigorous code verification is due to regulatory authorities allowing the use of modeling and simulation-based evidence as part of the regulatory filings for certification of the safety and reliability of their systems or processes.

In the absence of rigorous, documented code verification evidence, there are code verification activities that can be performed by the user. In addition to the simple code verification activities discussed in Section 5.2, order verification tests can also be conducted when an exact solution (or a verifiably accurate surrogate solution) to the mathematical model is available. While the traditional approach for finding exact solutions can be used, the more general Method of Manufactured Solutions procedure requires the ability to employ user-defined boundary conditions, initial conditions, and source terms. Even if a user does not have access to the source code, the Method of Manufactured Solutions can be used as long as source terms can be added to the software, i.e., the right-hand side of the equations for the mathematical model. The next chapter focuses on different approaches for obtaining exact solutions to the mathematical model including the Method of Manufactured Solutions.

5.9 Problems

1. True/False: Convergence addresses whether or not the discrete equations approach the partial differential equations as the mesh and time step are systematically refined.
2. Consistent mesh refinement requires that the mesh quality ________ or __________ with mesh refinement.
3. Briefly explain the difference between the formal and observed orders of accuracy.

4. List four distinct reasons why the observed order of accuracy might not equal the formal order of accuracy for a code verification exercise.
5. You are performing a code verification study of a formally third-order accurate code where the exact solution to the partial differential equation produces a system response quantity (SRQ) of $\tilde{f} = 100$. Performing the simulations produces SRQs of $f_1 = 104.3$ on the fine grid and $f_2 = 108.8$ on the coarse grid. Compute the observed order of accuracy for the case where these grids are systematically-refined by a factor of $r = 1.3$.
6. You are performing a code verification study of a formally second-order accurate code where the exact solution to the partial differential equation produces a system response quantity (SRQ) of $\tilde{f} = 10$. Performing the simulations produces SRQs of $f_1 = 10.2$ on the fine grid, $f_2 = 10.46$ on the medium grid, and $f_3 = 11.0$ on the coarse grid. Compute the observed order of accuracy for the case where these grids are systematically-refined by a factor of $r = 1.5$. Did the code pass this code verification test? Briefly explain why.
7. Find the formal order of accuracy in both space and time for the following finite-difference discretization of the first-order linear wave equation $\frac{\partial u}{\partial t} + c\frac{\partial u}{\partial x} = 0$ with $c > 0$ using Taylor series expansions:

$$\frac{3u_i^n - 4u_i^{n-1} + u_i^{n-2}}{2\Delta t} + c\frac{u_i^n - u_{i-1}^n}{\Delta x} = 0.$$

5.10 References

AIAA (2021). Recommended Practice: Code Verification in Computational Fluid Dynamics. AIAA, AIAA R-141-2021.

Abanto, J., D. Pelletier, A. Garon, J.-Y. Trepanier, and M. Reggio (2005). Verification of Some Commercial CFD Codes on Atypical CFD Problems. AIAA Paper 2005-682.

Banks, J. W., T. Aslam, and W. J. Rider (2008). On Sub-linear Convergence for Linearly Degenerate Waves in Capturing Schemes. *Journal of Computational Physics*, **227**(14), 6985–7002.

Crank, J., and P. A. Nicolson (1947). Practical Method for Numerical Evaluation of Solutions of Partial Differential Equations of the Heat-Conduction Type. *Proceedings of the Cambridge Philosophical Society*, **43**, 50–67.

Despres, B. (2004). Lax Theorem and Finite Volume Schemes. *Mathematics of Computation*, **73**(247), 1203–1234.

Diskin, B., and J. L. Thomas (2007). Accuracy Analysis for Mixed-Element Finite Volume Discretization Schemes. National Institute of Aerospace Technical Report, TR 2007-8, Hampton, VA.

Eca, L., C. M. Klaij, G. Vaz, M. Hoekstra, and F. S. Pereira (2016). On Code Verification of RANS Solvers. *Journal of Computational Physics*, **310**, 418–439.

Ferziger, J. H., and M. Peric (2002). *Computational Methods for Fluid Dynamics*, 3rd ed. Berlin: Springer-Verlag.

Grinstein, F. F., L. G. Margolin, and W. J. Rider (2007). *Implicit Large Eddy Simulation: Computing Turbulent Fluid Dynamics*. Cambridge: Cambridge University Press.

Hebert, S., and E. A. Luke (2005). Honey, I Shrunk the Grids! A New Approach to CFD Verification Studies. AIAA Paper 2005-685.

Hirsch, C. (2007). *Numerical Computation of Internal and External Flows (Vol. 1)*, 2nd ed. Burlington, MA: Elsevier.

Kamm, J. R., W. J. Rider, and J. S. Brock (2003). Combined Space and Time Convergence Analyses of a Compressible Flow Algorithm. AIAA Paper 2003-4241.

Knupp, P. M. (2003). Algebraic Mesh Quality Metrics for Unstructured Initial Meshes. *Finite Elements in Analysis and Design*, **39**(3), 217–241.

Knupp, P., C. Ober, and R. Bond (2007). Measuring Progress in Order-Verification within Software Development Projects. *Engineering with Computers*, **23**(4), 271–282.

Knupp, P. M., and K. Salari (2003). *Verification of Computer Codes in Computational Science and Engineering*, K. H. Rosen (ed.). Boca Raton, FL: Chapman & Hall/CRC.

Morton, K., and D. Mayers (2005). *Numerical Solution of Partial Differential Equations: An Introduction*, 2nd ed. Cambridge: Cambridge University Press.

Nakao, M. T., M. Plum, and Y. Watanabe (2019). *Numerical Verification Methods and Computer-Assisted Proofs for Partial Differential Equations*. New York, NY: Springer.

Oberkampf, W. L., and C. J. Roy (2010). *Verification and Validation in Scientific Computing*. Cambridge: Cambridge University Press.

Oberkampf, W. L., and T. G. Trucano (2008). Verification and Validation Benchmarks. *Nuclear Engineering and Design*, **238**(3), 716–743.

Richards, S. A. (1997). Completed Richardson Extrapolation in Space and Time. *Communications in Numerical Methods in Engineering*, **13**(7), 573–582.

Richtmyer, R. D., and K. W. Morton (1967). *Difference Methods for Initial-value Problems*, 2nd ed. New York, NY: John Wiley & Sons.

Rider, W. J. (2019). The Foundations of Verification in Modeling and Simulation. In C. Beisbart and N. J. Saam (eds.), *Computer Simulation Validation – Fundamental Concepts, Methodological Frameworks, and Philosophical Perspectives*, 271–293. Berlin: Springer.

Roache, P. J. (1998). *Verification and Validation in Computational Science and Engineering*. Albuquerque, NM: Hermosa Publishers.

Roache, P. J. (2009). *Fundamentals of Verification and Validation*. Albuquerque, NM: Hermosa Publishers.

Roy, C. J. (2005). Review of Code and Solution Verification Procedures for Computational Simulation. *Journal of Computational Physics*, **205**(1), 131–156.

Roy, C. J. (2009). Strategies for Driving Mesh Adaptation in CFD (Invited). AIAA Paper 2009-1302.

Roy, C. J., E. Tendean, S. P. Veluri, R. Rifki, E. A. Luke, and S. Hebert (2007). Verification of RANS Turbulence Models in Loci-CHEM Using the Method of Manufactured Solutions. AIAA Paper 2007-4203.

Tannehill, J. C., R. H. Pletcher, V. Shankar, and D. A. Anderson (2021). *Computational Fluid Mechanics and Heat Transfer*, 4th ed. Boca Raton, FL: CRC Press.

Thomas, J. L., B. Diskin, and C. L. Rumsey (2008). Toward Verification of Unstructured-Grid Solvers. *AIAA Journal*, **46**(12), 3070–3079.

Thompson, J. F., Z. U. A. Warsi, and C. W. Mastin (1985). *Numerical Grid Generation: Foundations and Applications*. New York, NY: Elsevier.

Trucano, T. G., M. M. Pilch, and W. L. Oberkampf (2003). On the Role of Code Comparisons in Verification and Validation. *Sandia National Laboratories*, SAND 2003-2752, Albuquerque, NM.

Veluri, S. P., C. J. Roy, S. Hebert, and E. A. Luke (2008). Verification of the Loci-CHEM CFD Code Using the Method of Manufactured Solutions. AIAA Paper 2008-661.

Veluri, S. P., C. J. Roy, and E. A. Luke (2012). Comprehensive Code Verification Techniques for Finite Volume CFD Codes. *Computers and Fluids*, **70**, 59–72.

Xue, X., H. Wang, and C. J. Roy (2021). Code Verification for 2D Unsteady Flows in SENSEI. *AIAA SciTech Forum*, AIAA Paper 2021-1742, January 11–21, 2021, Virtual Conference.

Yee, H. C. (2004). Building Blocks for Reliable Complex Numerical Simulations. In D. Drikakis and B. Geurts (eds.), *Turbulent Flow Computations*, 199–236. New York, NY: Springer.

Yee, H. C., and P. K. Sweby (1998). Aspects of Numerical Uncertainties in Time Marching to Steady-State Numerical Solutions. *AIAA Journal*, **36**(5), 712–724.

6

Exact and Manufactured Solutions

A man who does not plan long ahead will find trouble at his door. —Confucius

Contents of Chapter 6

The focus of this chapter is on exact solutions to mathematical models for use with code order-of-accuracy verification testing. In scientific computing, "correct" code output depends on the chosen spatial mesh, time step, iterative convergence tolerance, machine precision, adequacy of code verification testing, and so on. We are thus forced to rely on indirect methods for assessing code correctness. In Chapter 5, the order of accuracy test was shown to be the most rigorous approach for code verification testing. To conduct order of accuracy testing, an exact solution to the underlying mathematical model is required. When used for code verification, the ability of this exact solution to exercise all terms in the mathematical model is more important than any physical realism of the solution. In fact, realistic exact solutions are often avoided for code verification due to the presence of singularities, discontinuities, and/or geometric complexity. Numerous examples will be given in this chapter of exact solutions and their use with the order verification test. The final example given in this chapter employs the less rigorous convergence test with benchmark numerical solutions. In addition to code verification applications, exact solutions to mathematical models are extremely valuable for evaluating the accuracy of numerical schemes, determining solution sensitivity to mesh quality and topology, evaluating the reliability of discretization error estimators, and evaluating solution adaptation schemes.

This chapter is organized as follows. First, an introduction to differential equations is presented, followed by a discussion of traditional exact solutions. Then, significant time is spent on the Method of Manufactured Solutions including the procedures, benefits and limitations; alternative forms; discontinuities; and manufactured solutions in different disciplines. Physically realistic manufactured solutions are then described, followed by a discussion of less-rigorous approximate solution methods.

6.1 Introduction

This chapter discusses methods for obtaining exact solutions to mathematical models used in scientific computing. These mathematical models typically take the form of either integral or differential equations. Because we will have to compute actual numerical values for these exact solutions, we will use a different definition for an exact solution than found in standard mathematics textbooks. The definition used here for an exact solution is a solution to the mathematical model that is in *closed form*, i.e., in terms of elementary functions (trigonometric functions, exponentials, logarithms, powers, etc.) or readily-computed special functions (e.g., Gamma, beta, error, Bessel functions) of the independent variables. Solutions involving infinite series or a reduction of a partial differential equation to a system of ordinary differential equations which do not have exact solutions will be considered as approximate solutions and are addressed at the end of the chapter.

In scientific computing, the mathematical models can take different forms. When a set of physical laws such as conservation of mass, momentum, or energy (i.e., the conceptual model), are formulated for an infinitesimally small region of space, then the resulting mathematical model generally takes the form of differential equations. When applied to a region of space with finite size, the mathematical model takes the form of integral (or integrodifferential) equations. The differential form is called the *strong form* of the equations, whereas the integral form, after application of the divergence theorem (which converts the volume integral of the gradient of a quantity to fluxes through the boundary), is called the *weak form*. The finite difference method employs the strong form of the equations, while the finite element and finite volume methods employ the weak form.

The strong form explicitly requires solutions that are differentiable, whereas the weak form admits solutions that can contain discontinuities while still satisfying the underlying physical laws, and these discontinuous solutions are called *weak solutions*. Weak solutions satisfy the differential equation only in a restricted sense. For much of this chapter, we will focus on the strong (i.e., differential) form of mathematical models; however, since strong solutions also satisfy the weak form of the equations, the finite element and finite volume methods will not be excluded by this assumption. Section 6.4 includes a discussion of a weak form implementation of the Method of Manufactured Solutions along with solutions which include discontinuities.

Because scientific computing often involves complex systems of coupled partial differential equations (PDEs) which have relatively few exact solutions, the organization of this chapter is very different from that of standard mathematics texts. After a short introduction to differential equations in Section 6.2, a discussion of "traditional" exact solutions and solution methods is presented in Section 6.3. The Method of Manufactured Solutions (MMS) is a more general approach for obtaining exact solutions to complicated mathematical models and is discussed in detail in Section 6.4. When physically realistic manufactured solutions are desired, the approaches discussed in Section 6.5 can be used. As discussed above, solutions involving infinite series, reduction of PDEs to ordinary differential equations, or numerical solutions of the underlying mathematical model with established numerical accuracy are relegated to Section 6.6 which deals with approximate solution methods.

6.2 Introduction to Differential Equations

Differential equations are ubiquitous in the study of physical processes in science and engineering. A differential equation is a relation between a variable and its derivatives. When only one independent variable is present, the equation is called an *ordinary differential equation*. A differential equation involving derivatives with respect to two or more independent variables (e.g., x and t, or x, y, and z) is called a *partial differential equation* (PDE). A differential equation can be a single equation with a single dependent variable or a system of equations with multiple dependent variables. The *order* of a differential equation is the largest number of derivatives applied to any dependent variable.

A differential equation is considered to be *linear* when the dependent variables and all of their derivatives occur with powers equal to one and there are no products involving derivatives and/or functions of the dependent variables. Solutions to linear differential equations can be combined to form new solutions using the linear superposition principle. A *quasi-linear* differential equation is one that is linear in the highest derivative, i.e., the highest derivative appears to the power one.

General solutions to PDEs are solutions that satisfy the PDE, but they are not unique in the sense that a particular problem of interest is specified. To find *particular solutions* to PDEs, additional conditions must be supplied on the boundary of the domain of interest (i.e., *boundary conditions*), at an initial data location (i.e., *initial conditions*), or some combination of the two. Boundary conditions generally come in the form of *Dirichlet* boundary conditions which specify values of the dependent variables or *Neumann* boundary conditions which specify the values of derivatives of the dependent variables normal to the boundary. When both Dirichlet and Neumann conditions are applied at a boundary, it is called a *Cauchy* boundary condition, whereas a linear combination of a dependent variable and its normal derivative is called a *Robin* boundary condition. The latter is often confused with *mixed* boundary conditions, which occur when different boundary condition types (Dirichlet, Neumann, or Robin) are applied at different boundaries in a given problem.

The maximum order of the boundary conditions is at least one less than the order of the differential equation. Here *order* refers to the highest number of derivatives applied to any dependent variable as discussed earlier. This requirement on the order of the boundary condition is sometimes erroneously stated as a requirement on the order of accuracy of the discretization of a derivative boundary condition when the PDE is solved numerically. On the contrary, a reduction in the formal order of accuracy of a discretized derivative boundary condition generally leads to a reduction in the observed order of accuracy of the *entire* solution.

Partial differential equations can be classified as *elliptic*, *parabolic*, *hyperbolic*, or a combination of these types. For scalar equations, this classification is fairly straightforward in a manner analogous to determining the character of algebraic equations. For systems of PDEs written in quasi-linear form, the eigenvalues of the coefficient matrices can be used to determine the mathematical character, e.g., see Hirsch (2007). For a detailed introduction to PDEs, see Evans (2010) and Haberman (2018).

6.3 Traditional Exact Solutions

The standard approach to obtaining an exact solution to a mathematical model can be summarized as follows. Given the governing partial differential (or integral) equations on some domain with appropriately specified initial and/or boundary conditions, find the exact solution. The main disadvantage of this approach is that there are only a limited number of exact solutions known for complex equations. Here the complexity of the equations could be due to geometry, nonlinearity, physical models, and/or coupling between multiple physical phenomena such as fluid-structure interaction.

When exact solutions are found for complex equations, they often depend on significant simplifications in dimensionality, geometry, physics, etc. For example, the flow between infinite parallel plates separated by a small gap with one plate moving at a constant speed is called Couette flow and is described by the Navier–Stokes equations, a nonlinear, second-order system of PDEs. In Couette flow, the velocity profile is linear across the gap and this linearity causes the diffusion term, a second derivative of velocity, to be identically zero. In addition, there are no solution variations in the direction that the plate is moving. Thus, the exact solution to Couette flow exercises very few of the terms in the Navier–Stokes equations. It should be noted that exact solutions that exist in only one or two dimensions can be generalized to higher dimensions by using coordinate transformations to rotate the solutions and thus produce variations in all of the coordinate directions (see Horner, 2021).

There are many books available that catalogue a vast number of exact solutions for differential equations found in science and engineering. These texts address ordinary differential equations (e.g., Polyanin and Zaitsev, 2017), linear PDEs (Kevorkian, 2000; Meleshko, 2005; Polyanin, 2016), and nonlinear PDEs (Kevorkian, 2000; Meleshko, 2005; Polyanin and Zaitsev, 2011). In addition, many exact solutions can be found in discipline-specific references such as those for heat conduction (Carslaw and Jaeger, 1959), fluid dynamics (Panton, 2005; White, 2005), linear elasticity (Timoshenko and Goodier, 1969; Slaughter, 2002), elastodynamics (Kausel, 2006), and vibration and buckling

(Elishakoff, 2004). The general Riemann problem involves an exact weak (i.e., discontinuous) solution to the 1D unsteady inviscid equations for gas dynamics (Gottlieb and Groth, 1988).

Example of Exact Solutions: 1D Unsteady Heat Conduction

Consider the 1D unsteady heat equation with constant thermal diffusivity α

$$\frac{\partial T}{\partial t} = \alpha \frac{\partial^2 T}{\partial x^2}, \tag{6.1}$$

where $T(t,x)$ is the temperature. Some general solutions to the 1D unsteady heat conduction equation are presented in Table 6.1, where A, B, C, and μ are arbitrary constants and n is a positive integer. These solutions, as well as many others, can be found in Polyanin (2016). Employing the Hopf–Cole transformation (Polyanin and Zaitsev, 2011), these solutions can also be transformed into solutions to Burgers' equation. See Oberkampf and Roy (2010) for more details.

Table 6.1 *Exact solutions to the 1D unsteady heat conduction equation*

Solutions to $\frac{\partial T}{\partial t} = \alpha \frac{\partial^2 T}{\partial x^2}$
$T(t,x) = A(x^3 + 6\alpha tx) + B$
$T(t,x) = A(x^4 + 12\alpha tx^2 + 12\alpha^2 t^2) + B$
$T(t,x) = x^{2n} + \sum_{k=1}^{n} \frac{(2n)(2n-1)\cdots(2n-2k+1)}{k!} (\alpha t)^k x^{2n-2k}$
$T(t,x) = A\exp(\alpha\mu^2 t \pm \mu x) + B$
$T(t,x) = A\frac{1}{\sqrt{t}}\exp\left(\frac{-x^2}{4\alpha t}\right) + B$
$T(t,x) = A\exp(-\mu x)\cos(\mu x - 2\alpha\mu^2 t + B) + C$
$T(t,x) = A\,\mathrm{erf}\left(\frac{x}{2\sqrt{\alpha t}}\right) + B$

Example with Order Verification: Steady Burgers' Equation

This section describes an exact solution for steady Burgers' equation, which is then employed in an order verification test for a finite difference discretization. Benton and Platzmann (1972) describe 35 exact solutions to viscous Burgers' equation, which is given as:

$$\frac{\partial u}{\partial t} + u\frac{\partial u}{\partial x} = \nu \frac{\partial^2 u}{\partial x^2}. \tag{6.2}$$

The steady-state form of Burgers' equation is found by restricting the solution u to be a function of x only, thus reducing Equation (6.2) to the ordinary differential equation

$$u\frac{du}{dx} = \nu \frac{d^2 u}{dx^2}, \tag{6.3}$$

where u is a velocity and ν is the viscosity. The exact solution for a steady, viscous shock (Benton and Platzmann, 1972) is given in dimensionless form, denoted by primes, as

$$u'(x') = -2\tanh(x'). \tag{6.4}$$

This dimensionless solution for Burgers' equation can be converted to dimensional quantities via transformations given by $x' = x/L$ and $u' = uL/\nu$ where L is a reference length scale. This solution for Burgers' equation is also invariant to scaling by a factor α as follows: $\bar{x} = x/\alpha$ and $\bar{u} = \alpha u$. Finally, one can define a Reynolds number in terms of L, a reference velocity u_{ref}, and the viscosity ν as

$$\mathrm{Re} = \frac{u_{ref}L}{\nu}, \tag{6.5}$$

where the domain is generally chosen as $-L \le x \le L$ and u_{ref} the maximum value of u on the domain.

For this example, steady Burgers' equation is discretized using the simple implicit finite difference scheme given by

$$\bar{u}_i\left(\frac{u_{i+1}^{k+1} - u_{i-1}^{k+1}}{2\Delta x}\right) - \nu\left(\frac{u_{i+1}^{k+1} - 2u_i^{k+1} + u_{i-1}^{k+1}}{\Delta x^2}\right) = 0, \tag{6.6}$$

where a uniform mesh with spacing Δx is used between the spatial nodes which are indexed by i. The formal order of accuracy of this discretization scheme is two and can be found from a truncation error analysis. The above discretization scheme is linearized by setting $\bar{u}_i = u_i^k$ and then iterated until the solution at iteration k satisfies Equation (6.6) within round-off error using double precision computations. This results in the iterative residuals, found by substituting $\bar{u}_i = u_i^{k+1}$ into Equation (6.6), being reduced by approximately 14 orders of magnitude. Thus, round-off and iterative convergence errors can be neglected (see Chapters 7 and 8, respectively), and the numerical solutions effectively contain only discretization error.

The solution to Burgers' equation for a Reynolds number of eight is given in Figure 6.1, which includes both the exact solution (in scaled dimensional variables) and a numerical solution

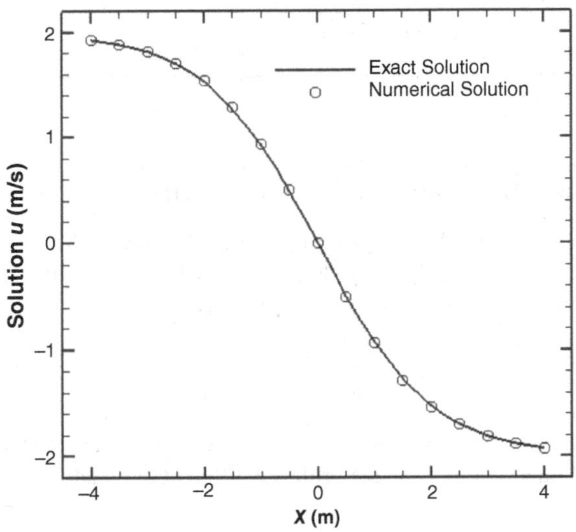

Figure 6.1 Comparison of the numerical solution using 17 nodes with exact solution for steady Burgers' equation with Reynolds number Re = 8.

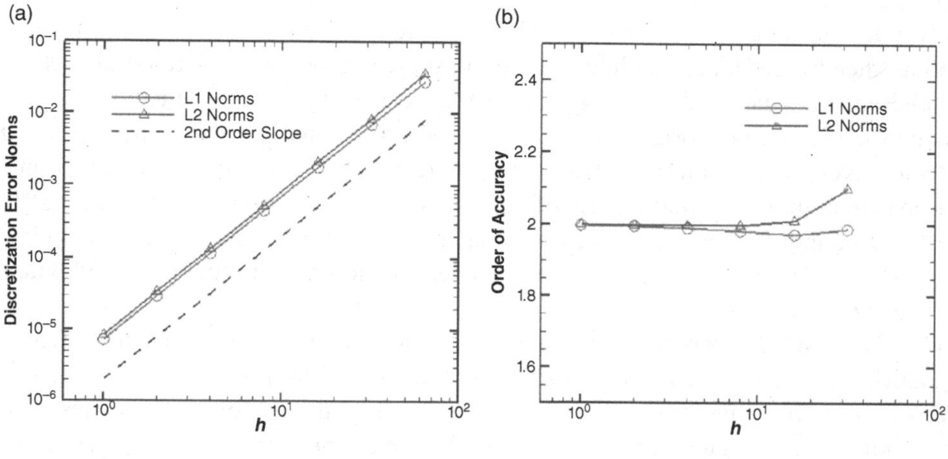

Figure 6.2 Discretization error (a) and observed orders of accuracy (b) for steady Burgers' equation with Reynolds number Re = 8.

obtained using 17 evenly spaced points. The low value for a Reynolds number was chosen for this code verification exercise to ensure that both the convection and the diffusion terms are roughly the same order of magnitude. Note that choosing a large Reynolds number effectively scales down the diffusion term since it is multiplied by 1/Re when written in dimensionless form. For higher Reynolds numbers, extremely fine meshes would be needed to detect coding mistakes in the diffusion term.

Numerical solutions for steady Burgers' equation were obtained on seven uniform meshes from the finest mesh of 513 nodes ($h = 1$) to the coarsest mesh of 9 nodes ($h = 64$). Discrete L_1 and L_2 norms of the discretization error are shown in Figure 6.2a and both norms appear to reduce with mesh refinement at a second-order rate. For reference, in Figure 6.2a, the second-order slope is also shown. The order of accuracy, as computed from Equation (5.22), is given in Figure 6.2b. Both norms rapidly approach the scheme's formal order of accuracy of two with mesh refinement for this simple problem. The code used to compute the numerical solutions to Burgers' equation is thus considered to be verified for the options exercised, namely steady-state solutions on a uniform mesh.

6.4 Method of Manufactured Solutions (MMS)

This section addresses the difficult question of how to create an exact solution for complex PDEs, where complexity refers to characteristics such as nonlinearity, irregular domain shape, higher dimensions, multiple submodels, coupled systems of equations, and coupling of different types of physical processes. The traditional methods for obtaining exact solutions discussed earlier generally cannot handle any of these increases in complexity. The primary need for exact solutions to complex PDEs is for order of accuracy testing during code verification.

The *Method of Manufactured Solutions* (MMS) is a general and very powerful approach for creating exact solutions. Rather than trying to find an exact solution to a PDE with given initial and boundary conditions, the goal is to "manufacture" an exact solution to a slightly modified governing equation. For code verification purposes, it is *not* required that the manufactured solution be related to a physically realistic problem. Recall that code verification deals only with the mathematics of a given problem. The general concept behind MMS is to choose a solution *a priori*, then operate the governing PDEs onto the chosen solution, thereby generating additional analytic source terms that require no discretization. The chosen (manufactured) solution is then the exact solution to the modified governing equations made up of the original equations plus the additional analytic source terms. Thus, MMS involves the solution to the backward problem: given an original set of equations and a chosen solution, find a modified set of equations that the chosen solution will satisfy. The initial and boundary conditions can be found from the chosen solution, i.e., knowing the exact solution over the entire domain of the independent variables, the initial and boundary conditions can be chosen as needed.

While the MMS approach for generating exact solutions was not new (e.g., see Zadunaisky, 1976; Stetter, 1978), Roache and Steinberg (1984) and Steinberg and Roache (1985) appear to be the first to employ these exact solutions for the purposes of code verification. Their original work looked at the asymptotic rate of convergence of the discretization errors with systematic mesh refinement. Shih (1985) independently developed a similar procedure for debugging scientific computing codes, but without employing mesh refinement to assess convergence or order of accuracy.

The concepts behind MMS for the purpose of code verification were later refined by Roache et al. (1990) and Roache (1998). The term "manufactured solution" was coined by Oberkampf et al. (1995) and refers to the fact that the method generates (or manufactures) a related set of governing equations for a chosen analytic solution. An extensive discussion of manufactured solutions for code verification is presented by Knupp and Salari (2003) and includes both details of the method as well as application to a variety of different PDEs. Reviews of the MMS procedure are presented by Roache (2002) and Roy (2005). While it is not uncommon for MMS to be used for verifying computational fluid dynamics codes (e.g., Roache et al., 1990; Pelletier et al., 2004; Roy et al., 2004; Eca et al., 2007; AIAA, 2021), it has also begun to appear in other disciplines (see Section 6.4.6).

MMS allows the generation of exact solutions to PDEs of nearly arbitrary complexity, with notable exceptions discussed in Section 6.4.3. When combined with the order verification procedure described in Chapter 5, MMS provides the most powerful tool for code verification. When physically realistic manufactured solutions are desired, the modified MMS approaches discussed in Section 6.5 can be used.

6.4.1 Procedure

The procedure for creating an exact solution using MMS is fairly straightforward. For a scalar mathematical model, this procedure can be summarized as follows.

Step 1. Establish the mathematical model in the form $L(u) = 0$, where $L(\bullet)$ is the differential operator and u the dependent variable.

Step 2. Choose the analytic form of the manufactured solution $\hat{u}$.

Step 3. Operate the mathematical model $L(\bullet)$ onto the manufactured solution $\hat{u}$ to obtain the analytic manufactured solution source term $s = L(\hat{u})$.

Step 4. Obtain the modified form of the mathematical model by including the analytic source term $L(u) = s$.

Because of the manner in which the analytic source term is obtained, it is a function of the independent variables only and does *not* depend on the solution u or the discretization parameters. The initial and boundary conditions can be obtained directly from the manufactured solution $\hat{u}$. For a system of equations, the manufactured solution $\hat{u}$ and the source term s are simply considered to be vectors, but otherwise the process is unchanged.

An advantage of using MMS to generate exact solutions to general mathematical models is that the procedure is not affected by nonlinearities or coupled sets of equations. However, the approach is conceptually different from the standard training scientists and engineers receive in problem solving. Thus, it is helpful to examine a simple example which highlights the subtle nature of MMS.

Example Manufactured Solution: 1D Unsteady Heat Conduction

Consider the unsteady 1D heat conduction equation given in Equation (6.1). The governing partial differential equation written in the form $L(T) = 0$ is:

$$\frac{\partial T}{\partial t} - \alpha \frac{\partial^2 T}{\partial x^2} = 0. \tag{6.7}$$

Once the governing equation has been specified, the next step is to choose the analytic manufactured solution. A detailed discussion on how to choose the solution will follow in Section 6.4.1.1, but for now, consider a combination of exponential and sinusoidal functions:

$$\hat{T}(x,t) = T_o \exp(t/t_o) \sin(\pi x/L). \tag{6.8}$$

Due to the analytic nature of this chosen solution, the derivatives that appear in the governing equation can be evaluated exactly as:

$$\frac{\partial \hat{T}}{\partial t} = T_o \sin(\pi x/L) \frac{1}{t_o} \exp(t/t_o),$$

$$\frac{\partial^2 \hat{T}}{\partial x^2} = -T_o \exp(t/t_o)(\pi/L)^2 \sin(\pi x/L).$$

We now modify the mathematical model by including the above derivatives, along with the thermal diffusivity α, on the right-hand side of the equation. The modified governing equation that results is:

$$\frac{\partial T}{\partial t} - \alpha \frac{\partial^2 T}{\partial x^2} = \left[\frac{1}{t_o} + \alpha\left(\frac{\pi}{L}\right)^2\right] T_o \exp(t/t_o) \sin(\pi x/L). \tag{6.9}$$

The left-hand side of Equation (6.9) is the same as in the original mathematical model given by Equation (6.7), thus no modifications to the underlying numerical discretization in the code under consideration need to be made. The right-hand side could be thought of in physical terms as a distributed source term. However, from a code verification perspective, the critical fact is that an exact solution in now known to a very closely related mathematical model. The key concept behind MMS is that the exact solution to Equation (6.9) is known and is given by Equation (6.8), the manufactured solution $\hat{T}(x,t)$ that was chosen in the beginning. Again, the initial and boundary conditions can be found by simply querying the manufactured solution given in Equation (6.8).

As the governing equations become more complex, symbolic manipulation tools such as Mathematica, Maple, MuPAD® (MATLAB®'s symbolic engine), or SymPy should be used. These tools have matured greatly over the last two decades and can produce rapid symbolic differentiation and simplification of expressions. Most symbolic manipulation software packages have built-in capabilities to output the solution and the source terms directly as computer source code in most programming languages. This is of great practical importance because it eliminates the risk of an error in the derivation and coding of the source terms that must be added to the code being tested.

Example of MMS Applied to the 2D Steady Euler Equations

This MMS example deals with the Euler equations, which govern the flow of an inviscid (i.e., frictionless) fluid. This example is adapted from Roy et al. (2004) and we will demonstrate the steps of both generating the exact solution with MMS and the order verification procedure. The two-dimensional, steady-state form of the Euler equations is given by

$$\begin{aligned}\frac{\partial(\rho u)}{\partial x}+\frac{\partial(\rho v)}{\partial y}&=s_m(x,y),\\ \frac{\partial(\rho u^2+p)}{\partial x}+\frac{\partial(\rho uv)}{\partial y}&=s_x(x,y),\\ \frac{\partial(\rho vu)}{\partial x}+\frac{\partial(\rho v^2+p)}{\partial y}&=s_y(x,y),\\ \frac{\partial(\rho ue_t+pu)}{\partial x}+\frac{\partial(\rho ve_t+pv)}{\partial y}&=s_e(x,y),\end{aligned} \tag{6.10}$$

where arbitrary source terms $s(x,y)$ are included on the right-hand side for use with MMS. In these equations, u and v are the Cartesian components of velocity in the x- and y-directions, respectively, ρ the density, p the pressure, and e_t is the specific total energy, which for a calorically perfect gas is given by

$$e_t=\frac{1}{\gamma-1}RT+\frac{u^2+v^2}{2}, \tag{6.11}$$

where R is the specific gas constant, T the temperature, and γ the ratio of specific heats. The final relation used to close the set of equations is the perfect gas equation of state:

$$p=\rho RT. \tag{6.12}$$

The manufactured solutions for this case are chosen as simple sinusoidal functions given by:

$$\begin{aligned}\rho(x,y)&=\rho_0+\rho_x\sin\left(\frac{a_{\rho x}\pi x}{L}\right)+\rho_y\cos\left(\frac{a_{\rho y}\pi y}{L}\right),\\ u(x,y)&=u_0+u_x\sin\left(\frac{a_{ux}\pi x}{L}\right)+u_y\cos\left(\frac{a_{uy}\pi y}{L}\right),\\ v(x,y)&=v_0+v_x\cos\left(\frac{a_{vx}\pi x}{L}\right)+v_y\sin\left(\frac{a_{vy}\pi y}{L}\right),\\ p(x,y)&=p_0+p_x\cos\left(\frac{a_{px}\pi x}{L}\right)+p_y\sin\left(\frac{a_{py}\pi y}{L}\right).\end{aligned} \tag{6.13}$$

The subscripts here refer to constants (not differentiation) with the same units as the variable, and the dimensionless constants a are usually chosen between 0.5 and 1.5 to provide low frequency

Table 6.2 *Constants for the supersonic Euler manufactured solution*

Equation, ϕ	ϕ_0	ϕ_x	ϕ_y	$a_{\phi x}$	$a_{\phi y}$
ρ (kg/m^3)	1	0.15	–0.1	1	0.5
u (m/s)	800	50	–30	1.5	0.6
v (m/s)	800	–75	40	0.5	2/3
p (N/m^2)	1×10^5	0.2×10^5	0.5×10^5	2	1

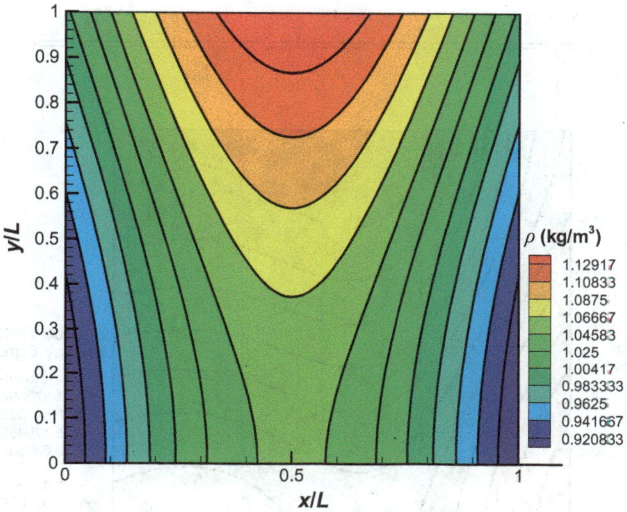

Figure 6.3 Manufactured solution for density for the Euler equations.

solutions over an $L \times L$ square domain. For this example, the constants were chosen to give supersonic flow in both the positive x and positive y directions, thus the bottom and left boundaries are inflow, and the top and right boundaries are outflow. While not necessary, this choice simplifies the inflow boundary conditions to Dirichlet (specified) values at the inflow boundaries, whereas outflow boundary values are simply extrapolated from the interior. The inflow boundary conditions are specified directly from the manufactured solution. The specific constants chosen for this example are shown in Table 6.2, and a plot of the manufactured solution for the density is given in Figure 6.3. The density varies smoothly in both coordinate directions between 0.92 and 1.13 kg/m^3.

Substitution of the chosen manufactured solutions into the governing equations allows the analytic determination of the source terms. For example, the source term for the mass conservation equation is given by:

$$\begin{aligned} s_m(x,y) = {} & \frac{a_{ux}\pi u_x}{L}\cos\left(\frac{a_{ux}\pi x}{L}\right)\left[\rho_0 + \rho_x \sin\left(\frac{a_{\rho x}\pi x}{L}\right) + \rho_y \cos\left(\frac{a_{\rho y}\pi y}{L}\right)\right], \\ & + \frac{a_{vy}\pi v_y}{L}\cos\left(\frac{a_{vy}\pi y}{L}\right)\left[\rho_0 + \rho_x \sin\left(\frac{a_{\rho x}\pi x}{L}\right) + \rho_y \cos\left(\frac{a_{\rho y}\pi y}{L}\right)\right], \\ & + \frac{a_{\rho x}\pi \rho_x}{L}\cos\left(\frac{a_{\rho x}\pi x}{L}\right)\left[u_0 + u_x \sin\left(\frac{a_{ux}\pi x}{L}\right) + u_y \cos\left(\frac{a_{uy}\pi y}{L}\right)\right], \\ & + \frac{a_{\rho y}\pi \rho_y}{L}\sin\left(\frac{a_{\rho y}\pi y}{L}\right)\left[v_0 + v_x \cos\left(\frac{a_{vx}\pi x}{L}\right) + v_y \sin\left(\frac{a_{vy}\pi y}{L}\right)\right]. \end{aligned}$$

Table 6.3 *Cartesian meshes employed in the Euler manufactured solution*

Mesh name	Mesh nodes	Mesh spacing, h
Mesh 1	129×129	1
Mesh 2	65×65	2
Mesh 3	33×33	4
Mesh 4	17×17	8
Mesh 5	9×9	16

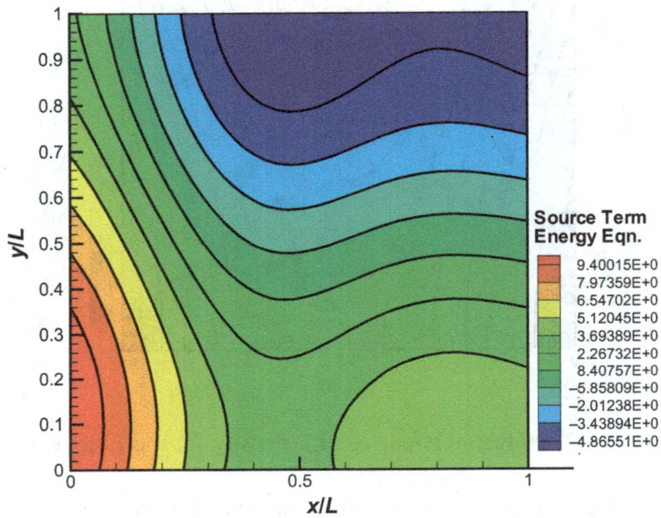

Figure 6.4 Analytic source term for the energy conservation equation.

The source terms for the momentum and energy equations are significantly more complex, and all source terms were obtained using Mathematica. A plot of the source term for the energy conservation equation is given in Figure 6.4. Note the smooth variations of the source term in both coordinate directions.

The governing equations are discretized and solved on multiple meshes. In this case, two different finite-volume computational fluid dynamics codes were employed: Premo, an unstructured grid code, and Wind, a structured grid code (see Roy et al., 2004 for more details). Both codes utilized the second-order Roe upwind scheme for the convective terms (Roe, 1981). The formal order of accuracy of both codes is thus second order for smooth problems.

The five Cartesian meshes employed are summarized in Table 6.3. The coarser meshes were found by successively eliminating every other grid line from the fine mesh (i.e., a grid refinement factor of $r = 2$). It is important to note that while the current example was performed on Cartesian meshes for simplicity, a more general code verification analysis would employ the most general meshes which will be used by the code (e.g., unstructured meshes with significant stretching, skewness, boundary orientations). See Section 5.6 for additional discussion of mesh topology issues.

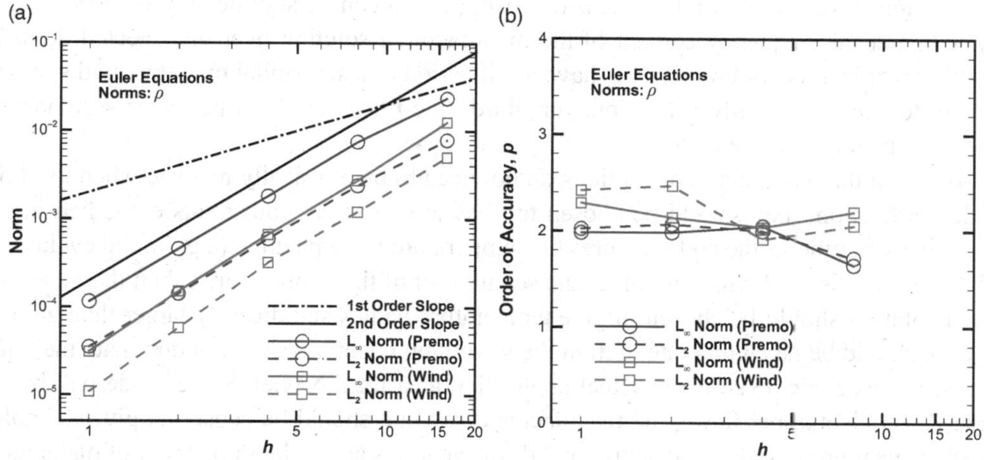

Figure 6.5 Code verification results for the 2D steady Euler equations: (a) norms of the discretization error and (b) observed order of accuracy.

The global discretization error is measured using discrete L_∞ and L_2 norms of the discretization error, where the exact solution comes directly from the chosen manufactured solution given by Equations (6.13). The behavior of these two discretization error norms for the density ρ as a function of the cell size h is given in Figure 6.5a. On the logarithmic scale, a first-order scheme will display a slope of unity, while a second-order scheme will give a slope of two. The discretization error norms for the density appear to converge with second-order accuracy.

A more quantitative method for assessing the observed order of accuracy is to calculate it using the norms of the discretization error. Since the exact solution is known, the relation for the observed order of accuracy of the discretization error norms comes from Equation (5.22). A plot of the observed order of accuracy as a function of the element size h is presented in Figure 6.5b. The Premo code clearly asymptotes to second-order accuracy, while the Wind code appears to asymptote to an order of accuracy that is slightly higher than two. In general, an observed order of accuracy higher than the formal order can occur due to error cancellation and should not be considered as a failure of the order verification test (although it may indicate mistakes is determining the formal order of accuracy of the method). Further grid refinement would possibly provide more definitive results for the Wind code. In this case, the observed order of accuracy for both codes is near two, thus the formal order of accuracy is recovered, and the two codes are considered verified for the options examined.

6.4.1.1 Guidelines for Code Verification

When used for code order-of-accuracy verification studies, manufactured solutions should be chosen to be analytic functions with smooth derivatives. It is important to ensure that no derivatives vanish, including cross-derivatives if these show up in the governing equations. Trigonometric and exponential functions are recommended since they are smooth and infinitely differentiable. Recall that the order verification procedures involve systematically refining the spatial mesh and/or time step, thus obtaining numerical solutions can be expensive for complex 3D applications. In some cases, the high cost of performing order verification studies in multiple dimensions using MMS can be significantly reduced simply by reducing the frequency content of the manufactured solution over the selected domain. In other words, it is not important to have a full period of a sinusoidal manufactured solution in the domain, often only a fraction (one-third, one-fifth, etc.) of a period is sufficient to exercise the terms in the code.

Although the manufactured solutions do not need to be physically realistic when used for code verification, they should be chosen to obey any physical constraints embodied in the code. For example, if the code requires the temperature to be positive (e.g., in the evaluation of the speed of sound which involves the square root of the temperature), then the manufactured solution should be chosen to give temperature values significantly larger than zero.

Care should be taken that one term in the governing equations does not dominate the other terms. For example, even if the actual application area for a Navier–Stokes code will be for high-Reynolds number flows, the manufactured solution should be chosen to give Reynolds numbers near unity so that convective and diffusive terms are of the same order of magnitude. For terms that have small relative magnitudes (e.g., if the term is scaled by a small parameter such as 1/Re), mistakes can still be found through order verification, but possibly only on extremely fine meshes. A more reliable approach to ensuring that all of the terms in the governing equations are roughly the same order of magnitude over some significant region of the domain is to examine ratios of those terms (e.g., see Roy et al., 2007).

6.4.1.2 Boundary and Initial Conditions

The discretization schemes for the PDE and the various submodels comprise a significant fraction of the possible options in most scientific computing codes. When performing code verification studies on these options, mathematically consistent boundary and initial conditions are required according to the mathematical character of the differential equations. For example, if Dirichlet (fixed-value) or Neumann (fixed-gradient) boundary conditions are required, these can be determined directly from the analytic manufactured solution (although these will generally not be constant along the boundary).

In order to verify the implementation of the boundary conditions, the manufactured solution should be tailored to exactly satisfy a given boundary condition on a domain boundary. Bond et al. (2007) present a general approach for tailoring manufactured solutions to ensure that a given boundary condition is satisfied along a general boundary. The method involves multiplying any standard manufactured solution by a function which has values and/or derivatives which are zero over a specified boundary. To modify the standard form of the manufactured solution for a 2D steady-state solution for temperature, one may start by simply writing the manufactured solution as follows:

$$\hat{T}(x,y) = T_0 + \hat{T}_1(x,y), \tag{6.14}$$

where $\hat{T}_1(x,y)$ is any baseline manufactured solution. For example, this manufactured solution could take the form

$$\hat{T}_1(x,y) = T_x f_s\left(\frac{a_x \pi x}{L}\right) + T_y f_s\left(\frac{a_y \pi y}{L}\right), \tag{6.15}$$

where the $f_s(\cdot)$ functions represent a mixture of sines and cosines and T_x, T_y, a_x, and a_y are constants. (Note the subscripts used here do not denote differentiation but instead they denote which constants are associated with solution variations in the x and y directions.)

For 2D problems, a boundary can be represented by the general curve $F(x,y) = C$, where C is a constant. A new manufactured solution appropriate for verifying boundary conditions along this boundary can be found by multiplying the spatially varying portion of $\hat{T}(x,y)$ by the function $[C - F(x,y)]^m$, i.e.,

$$\hat{T}_{BC}(x,y) = T_0 + \hat{T}_1(x,y)[C - F(x,y)]^m. \tag{6.16}$$

This procedure will ensure that the manufactured solution is equal to the constant T_0 along the specified boundary for $m = 1$. Setting $m = 2$, in addition to enforcing the above Dirichlet BC for temperature, will ensure that the gradient of temperature normal to the boundary is equal to zero, i.e., the adiabatic boundary condition for this 2D steady heat conduction example. In practice, the curve $F(x,y) = C$ is used to define the domain boundary where the boundary conditions will be tested.

Example Manufactured Solution for Boundary Conditions

To illustrate this procedure for verifying boundary conditions, we will choose the following simple manufactured solution for temperature:

$$\hat{T}(x,y) = 300 + 25\cos\left(\frac{7}{4}\frac{\pi x}{L}\right) + 40\sin\left(\frac{4}{3}\frac{\pi y}{L}\right), \tag{6.17}$$

where the temperature is assumed to be in units of Kelvin. For the surface where the Dirichlet or Neumann boundary condition will be applied, choose:

$$F(x,y) = \frac{1}{2}\cos\left(\frac{0.4\pi x}{L}\right) - \frac{y}{L} = 0. \tag{6.18}$$

A mesh bounded on the left by $x/L = 0$, on the right by $x/L = 1$, on the top by $y/L = 1$, and on the bottom by the above defined surface, $F(x,y) = 0$, is shown in Figure 6.6a. The standard

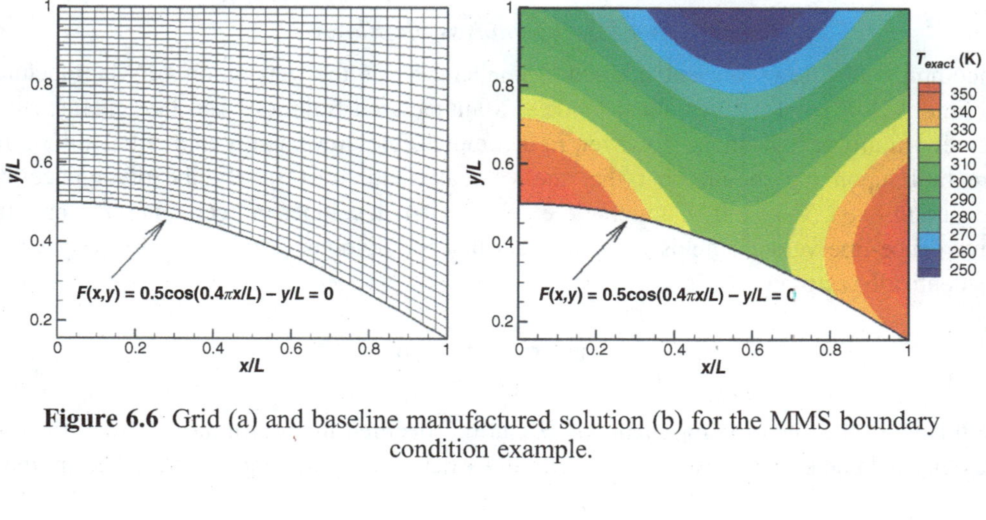

Figure 6.6 Grid (a) and baseline manufactured solution (b) for the MMS boundary condition example.

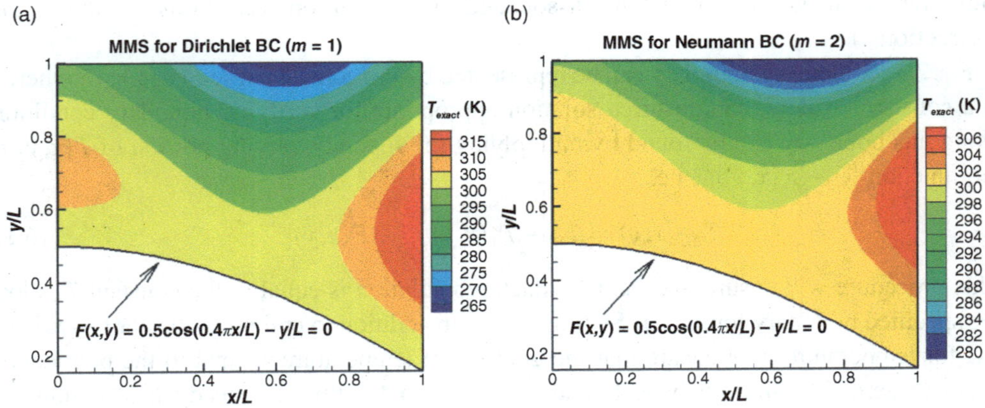

Figure 6.7 Manufactured solutions given by Equation (6.19): (a) fixed temperature boundary condition ($m = 1$) and (b) fixed temperature and zero gradient (adiabatic) boundary condition ($m = 2$).

manufactured solution given by Equation (6.17) is shown in Figure 6.6b, where clearly the constant value and gradient boundary conditions are not satisfied.

Combining the baseline manufactured solution with the boundary specification, as given by Equation (6.16), yields

$$\hat{T}_{BC}(x,y) = 300 + \left[25\cos\left(\frac{7}{4}\frac{\pi x}{L}\right) + 40\sin\left(\frac{4}{3}\frac{\pi y}{L}\right)\right]\left[-\frac{1}{2}\cos\left(\frac{0.4\pi x}{L}\right) + \frac{y}{L}\right]^m. \quad (6.19)$$

When $m = 1$, the curved lower boundary will satisfy the constant temperature condition of 300 K as shown in the manufactured solutions of Figure 6.7a. When $m = 2$ (Figure 6.7b), the zero gradient (i.e., adiabatic) boundary condition is also satisfied along the lower curved boundary. For more details on this approach as well as extensions to 3D, see Bond et al. (2007).

6.4.1.3 Incompressible Flows

Incompressible flows present additional challenges for the generation of Manufactured Solutions. For most mathematical models, Manufactured Solutions can be generated in a somewhat arbitrary fashion. However, in incompressible flows, many numerical algorithms are built upon the requirement that the velocity field must satisfy the divergence-free constraint, i.e., $\vec{\nabla}\cdot\vec{V} = 0$. Choudhary et al. (2016) developed a technique for creating divergence-free velocity fields using the fact that the divergence of the curl of any vector is identically zero, i.e.,

$$\vec{\nabla}\cdot\left(\vec{\nabla}\times\vec{H}\right) \equiv 0. \quad (6.20)$$

In this case, the vector components of $\vec{H}$ can be specified in an arbitrary fashion, then the velocity field determined by $\vec{V} = \vec{\nabla}\times\vec{H}$ will by definition be divergence free. The approach in Choudhary et al. (2016) can be readily extended to unsteady flows by defining the Cartesian vector $\vec{H}$ as:

$$\vec{H} = \begin{bmatrix} H_x(x,y,z,t) \\ H_y(x,y,z,t) \\ H_z(x,y,z,t) \end{bmatrix}.$$

The unsteady Cartesian velocity field can thus be found from $\vec{V} = \vec{\nabla} \times \vec{H}$ and, from the identity given in Equation (6.20), will still satisfy the divergence free condition. The pressure and/or temperature can be found in the usual manner.

6.4.2 Benefits of MMS for Code Verification

There are a number of benefits to using the MMS procedure for code verification. Perhaps the most important benefit is that it handles complex mathematical models without additional difficulty since the procedures described above are readily extendible to non-linear, coupled systems of equations. In addition, MMS can be used to verify most of the coding options available in scientific computing codes, including the consistency/convergence of numerical algorithms. The procedure is not tied to a specific discretization scheme, but works equally well for finite-difference, finite-volume, and finite-element methods.

The use of MMS with order verification for code verification has been shown to be remarkably sensitive to mistakes in the discretization and very small inconsistencies between the solution of the mathematical model and the solution produced by the code. In one particular case of a compressible Navier–Stokes code for unstructured grids (Roy et al., 2007), global norms of the discretization error were found to be non-convergent. Further investigation found a small discrepancy (in the 4th significant figure!) for the constant thermal conductivity between the governing equations used to generate the source terms and the model implementation in the code. When this discrepancy was corrected, the order verification test was passed. The same study uncovered an algorithm inconsistency in the discrete formulation of the diffusion operator that resulted in non-convergent behavior on skewed meshes. This same formulation had been implemented in at least one commercial computational fluid dynamics code (see Roy et al., 2007 for more details).

In addition to its ability to indicate the presence of coding mistakes (i.e., bugs), the MMS procedure combined with order verification is also a powerful tool for finding and removing those mistakes (i.e., debugging). After a failed order of accuracy test, individual terms in the mathematical model and the numerical discretization can be omitted, allowing the user to quickly isolate the terms with the coding mistake. In addition, examining contour plots of the local discretization error can often provide guidance as to whether the problems are occurring in the domain interior or (more likely) at a specific boundary. When combined with the approach for verifying boundary conditions discussed in the previous section and a suite of meshes with different topologies (e.g., hexahedral, prismatic, tetrahedral, and hybrid meshes, as discussed in Chapter 5), the software developer has a comprehensive set of tools to aid in code debugging.

6.4.3 Limitations of MMS for Code Verification

There are some limitations to using the MMS procedure for code verification. The principal disadvantage is that it requires the user to incorporate arbitrary source terms, initial conditions, and boundary conditions in a code. Even when the code provides a framework for including these additional interfaces, the actual coded software changes for each different manufactured solution. The MMS procedure is thus code-intrusive and generally cannot be performed as a black-box testing procedure where the code simply returns some outputs based on a given set of inputs. In other words, new software routines/functions must be added in order to perform the testing. In addition, each code option which changes the mathematical model (even in very minor ways) requires new source terms to be generated. Thus, order verification with MMS can be time consuming when many code options must be verified.

Most manufactured solution development has taken place in the area of fluid dynamics. More recently, manufactured solutions have begun to appear in a wide range of other disciplines, such as structural mechanics, radiation transport, plasma flows, and dynamics (see Section 6.4.6). However, manufactured solutions for coupled physics problems are still quite rare, due in part to the challenges associated with satisfying additional constraints at the multiphysics interface.

Difficulties also arise when applying MMS to mathematical models where the governing equations themselves contain `min`, `max`, or other non-smooth switching functions. These functions generally do not result in continuous manufactured solution source terms. These switching functions can be dealt with by simply turning off different branches of the switching functions (Eca et al., 2007) or by tailoring the manufactured solution so that only one switching branch will be used for a given verification test (Roy et al., 2007). The former is simpler but more code intrusive than the latter approach. We recommend that model developers employ smooth blending functions such as the hyperbolic tangent to both simplify the code verification testing and to possibly make the numerical solution process more robust (see Oberkampf and Roy, 2010).

6.4.4 Alternative Forms of MMS

Since the initial formulation of the Method of Manufactured Solutions was proposed by Roache and Steinberg in 1984, several variations have been proposed. The weak (i.e., integral) form of MMS greatly simplifies the evaluation of MMS source terms, while other approaches seek to avoid the need for including source terms altogether during code verification. Three alternative approaches to MMS are discussed in detail below.

6.4.4.1 Weak Form of MMS

Most implementations of MMS simply use the PDE (i.e., strong) form of the governing equations; however, applying MMS to the integral (i.e., weak) form has some advantages. To set the stage, most scalar mathematical models can be written in conservation form as

$$\frac{\partial u}{\partial t} + \vec{\nabla} \cdot \vec{f} = s, \tag{6.21}$$

where u is the solution, s is the source term, and the vector $\vec{f}$ is a general function of the solution and solution derivatives. In the finite volume and finite element methods, the divergence theorem is used to convert the volume integral of the divergence of quantities (found in the strong form of the equations) to the flux of those quantities through the control surface of the cell/element as:

$$\oiint_{\Omega_i} \left(\vec{\nabla}\cdot\vec{f}\right) dV = \oiint_{\partial\Omega_i} \left(\vec{f}\cdot\hat{n}\right) dS. \tag{6.22}$$

Here, Ω_i is the volume of cell/element i, $\partial\Omega_i$ is the closed control surface about that cell/element, and $\hat{n}$ is an outward-pointing unit vector normal to the face. Derlaga et al. (2013) developed a weak form implementation of MMS and used it to verify a finite volume computational fluid dynamics code.

There are two main advantages to the Manufactured Solutions implementation in the weak form. First, it requires one less derivative to be computed for the Manufactured Solution source term, which can be a significant simplification for complex mathematical models. Second, for problems in three spatial dimensions, volume integrals are converted to area integrals and, similarly, in two dimensions, area integrals are converted to line integrals, thus the spatial dimension of the integration is also reduced by one. The above discussion extends readily to systems of equations by writing the conservation law in Equation (6.21) in vector form (see Hirsch, 2007). Note that the weak form of MMS can be used to generate the source terms, even if the code to be verified employs the governing equations in the strong (i.e., differential) form, such as for finite difference methods. In fact, since smooth exact solutions satisfy both the weak and strong form of the governing equations, either approach can be used to generate the MMS source terms.

6.4.4.2 Adjusted Manufactured Solutions

The requirement to include arbitrary source terms in Manufactured Solutions is generally seen as a drawback for two reasons: one must modify the software in order to perform the order of accuracy testing and one does not always have access to arbitrary source terms, especially for commercial codes. Brglez (2013) developed an approach termed the "Adjusted" Method of Manufactured Solutions as a technique for avoiding the implementation of source terms. Although the approach does have this advantage, it is very restrictive in the types of PDEs that can be addressed. The approach works when one has access to an arbitrary (i.e., user-defined) function that appears as a coefficient in the PDEs, such as viscosity, thermal conductivity, or some other material parameter. It also requires that this function be solvable as a function of the other terms in the governing equation.

For example, consider the 1D unsteady heat equation

$$\frac{\partial T}{\partial t} - \alpha\frac{\partial^2 T}{\partial x^2} = 0$$

with the ability to specify a user-defined thermal diffusivity $\alpha = \alpha(x, t)$. One then chooses the (analytic) Adjusted Manufactured Solution for temperature as $\hat{T} = \hat{T}(x,t)$. If the thermal diffusivity can be analytically specified as

$$\alpha(x,t) = \frac{\frac{\partial\hat{T}}{\partial t}}{\frac{\partial^2\hat{T}}{\partial x^2}}, \tag{6.23}$$

then the chosen Manufactured Solution will satisfy the governing equation without requiring the addition of a source term. This approach only works for systems of equations when each differential equation has an arbitrary function. The Manufactured Solutions must also be chosen carefully to avoid singularities (i.e., ensuring the denominator of Equation (6.23) does not go to zero).

6.4.4.3 External Verification Analysis

Ingraham and Hixon (2013) developed an approach for performing code verification for unsteady mathematical models that also avoids the need for MMS source terms. The External Verification Analysis begins with a Taylor series expansion of the solution at time t_0 for some future time $t_0 + \Delta t$:

$$u(x, y, t_0 + \Delta t) = u(x, y, t_0) + \left.\frac{\partial u}{\partial t}\right|_{x,y,t_0} \Delta t + \left.\frac{\partial^2 u}{\partial t^2}\right|_{x,y,t_0} \frac{(\Delta t)^2}{2!} + \cdots. \tag{6.24}$$

If an arbitrary number of time derivatives are known at time $t = t_0$, then the solution at the later time $t = t_0 + \Delta t$ can be found to arbitrary accuracy (assuming appropriate convergence of the infinite series). The "reference" solution found from the Taylor series expansion at time $t = t_0 + \Delta t$ can then be used to compute the discretization error in the numerical solution computed at time $t = t_0 + \Delta t$. The governing equation is then used to convert spatial derivatives at time $t = t_0$ to time derivatives as follows.

Consider a time-dependent mathematical model written in conservation form, which for a 2D scalar equation is

$$\frac{\partial u}{\partial t} + \frac{\partial f}{\partial x} + \frac{\partial g}{\partial y} = 0, \tag{6.25}$$

where generally $f = f(u)$ and $g = g(u)$. Assuming an arbitrary analytic solution $\hat{u}(x, y, t)$ and initial conditions $\hat{u}(x, y, t_0)$ found by simply setting $t = t_0$, the first time derivative in the Taylor series expansion of Equation (6.24) can be obtained from the governing equation using the chain rule as

$$\left.\frac{\partial \hat{u}}{\partial t}\right|_{t_0} = -\left[\frac{\partial \hat{f}}{\partial x} + \frac{\partial \hat{g}}{\partial y}\right]_{t_0} = -\left[\frac{\partial \hat{f}}{\partial \hat{u}}\frac{\partial \hat{u}}{\partial x} + \frac{\partial \hat{g}}{\partial \hat{u}}\frac{\partial \hat{u}}{\partial y}\right]_{t_0}. \tag{6.26}$$

Second and higher derivatives in time are obtained by repeatedly applying the mathematical model in a manner similar to that for obtaining the modified equation (containing only spatial derivatives) in a truncation error analysis (e.g., see Tannehill et al., 1997). For example, the second derivative in time can be found as

$$\left.\frac{\partial^2 \hat{u}}{\partial t^2}\right|_{t_0} = -\left[\frac{\partial^2 \hat{f}}{\partial x \partial t} + \frac{\partial^2 \hat{g}}{\partial y \partial t}\right]_{t_0}. \tag{6.27}$$

The cross derivatives in Equation (6.27) can be found by repeatedly applying the governing equation along with the chain rule, e.g.,

$$\left.\frac{\partial^2 \hat{f}}{\partial x \partial t}\right|_{t_0} = \left[\frac{\partial^2 \hat{f}}{\partial \hat{u}^2}\frac{\partial \hat{u}}{\partial x}\frac{\partial \hat{u}}{\partial t} + \frac{\partial \hat{f}}{\partial \hat{u}}\frac{\partial^2 \hat{u}}{\partial x \partial t}\right]_{t_0}, \tag{6.28}$$

where $\frac{\partial \hat{u}}{\partial t}\Big|_{t_0}$ is found from Equation (6.26) and $\frac{\partial^2 \hat{u}}{\partial x \partial t}\Big|_{t_0}$ by taking the x derivative of Equation (6.26), and so on. The number of time derivatives can be chosen to obtain an arbitrarily high order of accuracy for the solution at $t_0 + \Delta t$. Just as with the traditional MMS procedure, all of these derivatives should be obtained using symbolic manipulation software. See Ingraham and Hixon (2013) for more details.

6.4.5 Discontinuities

Nearly all of the development in Manufactured Solutions has been for smooth solutions. Two notable exceptions are the works of Grier et al. (2014, 2015) and Woods and Starkey (2015), to be discussed shortly. It should be noted that traditional exact solutions exist for discontinuous problems such as the generalized Riemann problem (Gottlieb and Groth, 1988). More complicated solutions involving shock waves and detonations have been developed that involve infinite series (Powers and Stewart, 1992) or a change of dependent variable (Powers and Aslam, 2006).

Grier et al. (2014) developed a technique for computing MMS source terms for problems with discontinuities. Their approach involves geometrically splitting any cell/element containing a discontinuity, then performing the integration separately on each side of the discontinuity. Although geometrically complex, they demonstrated that their approach resulted in at least second-order accurate source terms, even for curved discontinuities. In Grier et al. (2015), they applied their method to order of accuracy verification with MMS for inviscid flows with shock waves, confirming that the first and second-order finite volume method was indeed only first-order accurate at (nonlinear) shock discontinuities (see Section 5.3.1).

Woods and Starkey (2015) developed a more accurate technique for computing the MMS source terms for discontinuous solutions. They employed the weak form of MMS (see Section 6.4.4.1) in order to reduce the dimension of the required integration. They then used Fubini's theorem to break multidimensional integrals into a series of one-dimensional integrals. Special treatment of the identified discontinuous solution points resulted in fast source term integration that was accurate within machine precision. See Woods and Starkey (2015) for more details. Note that as discussed in Section 5.5.1, the formal order of accuracy will reduce to first order or lower in the presence of discontinuities (Banks et al., 2008).

6.4.6 Manufactured Solutions in Other Disciplines

Nearly all of the early Manufactured Solutions were developed in the area of fluid dynamics. Over the past decade, Manufactured Solutions has seen an explosion of applications over a wide range of disciplines. Selected examples from a number of different disciplines are listed below.

- **Solid mechanics**
  - Aycock, K. I., Rebelo, N., and Craven, B. A. (2020). "Method of Manufactured Solutions Code Verification of Elastostatic Solid Mechanics Problems in a Commercial Finite Element Solver," *Computers and Structures*, Vol. 229, 106175 (13 pages).

- Gfrerer, M. H., and Schanz, M. (2018). "Code Verification Examples Based on the Method of Manufactured Solutions for Kirchhoff–Love and Reissner–Mindlin Shell Analysis," *Engineering with Computers*, Vol. 34, No. 4, pp. 775–785.
- Gong, Z.-X., Lu, C.-J., and Huang, H.-X. (2012). "Effect of Regularized Delta Function on Accuracy of Immersed Boundary Method," *Applied Mathematics and Mechanics*, Vol. 33, No. 11, pp. 1453–1466.
- Hamad, F. (2016). "Formulation of the Axisymmetric CPDI with Application to Pile Driving in Sand," *Computers and Geotechnics*, Vol. 74, pp. 141–150.
- Malaya, N., Estacio-Hiroms, K. C., Stogner, R. H., Schulz, K. W., Bauman, P. T., and Carey, G. F. (2013). "MASA: A Library for Verification Using Manufactured and Analytical Solutions," *Engineering with Computers*, Vol. 29, No. 4, pp. 487–496.
- Roy, C. J. (2017). "Chapter 18: Verification and Validation for the Finite Element Method," *Fundamentals of Finite Element Analysis: Linear Finite Element Analysis*, authored by I. Koutromanos, Wiley, Hoboken, NJ, pp. 615–636.
- Schroder, J. et al. (2021). "A Selection of Benchmark Problems in Solid Mechanics and Applied Mathematics," *Archives of Computational Methods in Engineering*, Vol. 28, pp. 713–751.

Fluid-structure interaction

- Étienne, S., Garon, A., and Pelletier, D. (2012). "Some Manufactured Solutions for Verification of Fluid-Structure Interaction Codes," *Computers and Structures*, Vols. 106–107, pp. 56–67.
- Yu, K. R., Étienne, S., Hay, A., and Pelletier, D. (2015). "Code Verification for Unsteady 3-D Fluid–Solid Interaction Problems," *Theoretical and Computational Fluid Dynamics*, Vol. 29, Nos. 5–6, pp. 455–471.

Electromagnetics

- Dudson, B. D., Madsen, J., Omotani, J., Hill, P., Easy, L., and Løiten, M. (2016). "Verification of BOUT++ by the Method of Manufactured Solutions," *Physics of Plasmas*, Vol. 23, No. 6, pp. 062303-1–062303-12.
- Garcia-Doñoro, D., Garcia-Castillo, L. E., and Ting, S. W. (2016). "Verification Process of Finite-Element Method Code for Electromagnetics: Using the Method of Manufactured Solutions," *IEEE Antennas and Propagation Magazine*, Vol. 58, No. 2, pp. 28–38.
- Neufeld, E., Cassará, A. M., Montanaro, H., Kuster, N., and Kainz, W. (2016). "Functionalized Anatomical Models for EM-Neuron Interaction Modeling," *Physics in Medicine and Biology*, Vol. 61, No. 12, pp. 4390–4401.
- Freno, B. A., and Matula, N. R. (2023). "Code-Verification Techniques for the Method-of-Moments Implementation of the Combined-Field Integral Equation," *Journal of Computational Physics*, Vol. 488, pp. 112231-1–112231-21.

Radiation transport

- Badri, M. A., Jolivet, P., Rousseau, B., Le Corre, S., Digonnet, H., and Favennec, Y. (2018). "Vectorial Finite Elements for Solving the Radiative Transfer Equation," *Journal of Quantitative Spectroscopy and Radiative Transfer*, Vol. 212, pp. 59–74.
- Gleicher, F. N., II, Wang, Y., Gaston, D., and Martineau, R. (2012). "The Method of Manufactured Solutions for RattleSnake, A S_N Radiation Transport Solver inside the

MOOSE Framework," *Transactions of the American Nuclear Society*, Vol. 106, pp. 372–374.
  - Owens, A. R., Welch, J. A., Kópházi, J., and Eaton, M. D. (2016). "Discontinuous Isogeometric Analysis Methods for the First-Order Form of the Neutron Transport Equation with Discrete Ordinate (S_N) Angular Discretisation," *Journal of Computational Physics*, Vol. 315, pp. 501–535.
  - Schunert, S., and Azmy, Y. (2015). "Comparison of Spatial Discretization Methods for Solving the S_N Equations Using a Three-Dimensional Method of Manufactured Solutions Benchmark Suite with Escalating Order of Nonsmoothness," *Nuclear Science and Engineering*, Vol. 180, No. 1, pp. 1–29.
- **Nuclear reactor thermal hydraulics and reactor kinetics**
  - Ganapol, B. D. (2013). "What Is Convergence Acceleration Anyway?" in *Integral Methods in Science and Engineering: Progress in Numerical and Analytic Techniques*, Birkhäuser, New York, NY, pp. 115–136.
  - Merroun, O., Almers, A., El Bardouni, T., El Bakkari, B., and Chakir, E. (2009). "Analytical Benchmarks for Verification of Thermal-Hydraulic Codes Based on Sub-Channel Approach," *Nuclear Engineering and Design*, Vol. 239, No. 4, pp. 735–748.
- **Climate modeling**
  - Bisht, G., and Riley, W. J. (2019). "Development and Verification of a Numerical Library for Solving Global Terrestrial Multiphysics Problems," *Journal of Advances in Modeling Earth Systems*, Vol. 11, pp. 1516–1542.
- **Plasmas**
  - Paruta, P., Ricci, P., Riva, F., Wersal, C., Beadle, C., and Frei, B. (2018). "Simulation of Plasma Turbulence in the Periphery of Diverted Tokamak by using the GBS Code," *Physics of Plasmas*, Vol. 25, No. 11, pp. 112301-1–112301-16.
  - Ricci, P., Riva, F., Theiler, C., Fasoli, A., Furno, I., Halpern, F. D., and Loizu, J. (2015). "Approaching the Investigation of Plasma Turbulence through a Rigorous Verification and Validation Procedure: A Practical Example," *Physics of Plasmas*, Vol. 22, No. 5, pp. 055704-1–055704-12.
- **Dynamics**
  - Atallah, A. M., Woollands, R. M., Elgohary, T. A., and Junkins, J. L. (2020). "Accuracy and Efficiency Comparison of Six Numerical Integrators for Propagating Perturbed Orbits," *Journal of the Astronautical Sciences*, Vol. 67, No. 2, pp. 511–538.
  - Ellis, J. R. and Hall, C. D. (2009). "Model Development and Code Verification for Simulation of Electrodynamic Tether System," *Journal of Guidance, Control, and Dynamics*, Vol. 32, No. 6, pp. 1713–1722.
  - Wu, W., McCue, L. S., and Roy, C. J. (2011). "The Method of Manufactured Solutions Applied to Chaotic Systems," *Nonlinear Dynamics*, Vol. 63, No. 4, pp. 667–679.
- **Conjugate heat transfer**
  - Cai, Z., and Thornber, B. (2018). "An Internal Penalty Discontinuous Galerkin Method for Simulating Conjugate Heat Transfer in a Closed Cavity," *International Journal for Numerical Methods in Fluids*, Vol. 87, No. 3, pp. 134–159.
  - Veeraragavan, A., Beri, J., and Gollan, R. J. (2016). "Use of the Method of Manufactured Solutions for the Verification of Conjugate Heat Transfer Solvers," *Journal of Computational Physics*, Vol. 307, pp. 308–320.

- **Free surface flows**
  - Charlot, L., Etienne, S., Hay, A., Pelletier, D., and Garon, A. (2015). "High-Order Time Integrators for Front-Tracking Finite-Element Analysis of Viscous Free-Surface Flows," *International Journal for Numerical Methods in Fluids*, Vol. 77, No. 11, pp. 668–693.
  - Wang, S. S., and Jia, Y. (2008). "Mathematical Verification Using Prescribed or Manufactured Solutions," in *Verification and Validation of 3D Free-Surface Flow Models,* Wang, S. S., Roache, P. J., Schmalz Jr, R. A., Jia, Y., and Smith, P. E. (eds.). American Society of Civil Engineers, pp. 120–152.
- **Multiphase flows**
  - Brady, P. T., Herrmann, M., and Lopez, J. M. (2012). "Code Verification for Finite Volume Multiphase Scalar Equations Using the Method of Manufactured Solutions," *Journal of Computational Physics*, Vol. 231, No. 7, pp. 2924–2944.
  - Buchheit, K., Altantzis, C., Bakshi, A., Jordan, T., and Van Essendelft, D. (2018). "The BubbleTree Toolset: CFD-Integrated Algorithm for Lagrangian Tracking and Rigorous Statistical Analysis of Bubble Motion and Gas Fluxes for Application to 3d Fluidized Bed Simulations," *Powder Technology*, Vol. 338, pp. 960–974.
  - Choudhary, A., Roy, C. J., Dietiker, J.-F., Shahnam, M., Garg, R., and Musser, J. (2016). "Code Verification for Multiphase Flows Using the Method of Manufactured Solutions," *International Journal of Multiphase Flow*, Vol. 80, pp. 150–163.
  - Vedovoto, J. M., Silveira Neto, A., Mura, A., and Figueira da Silva, L. F. (2011). "Application of the Method of Manufactured Solutions to the Verification of a Pressure-Based Finite-Volume Numerical Scheme," *Computers and Fluids*, Vol. 51, No. 1, pp. 85–99.
- **Chemically-reacting flows**
  - Shunn, L., Ham, F., and Moin, P. (2012). "Verification of Variable-Density Flow Solvers Using Manufactured Solutions," *Journal of Computational Physics*, Vol. 231, No. 9, pp. 3801–3827.
  - Viswanathan, S., Wang, H., and Pope, S. B. (2011). "Numerical Implementation of Mixing and Molecular Transport in LES/PDF Studies of Turbulent Reacting Flows," *Journal of Computational Physics*, Vol. 230, No. 17, pp. 6916–6957.
  - Wang, L., Zhou, W.-J., and Ji, C.-Q. (2015). "Verification of a Chemical Nonequilibrium Flows Solver Using the Method of Manufactured Solutions," *Procedia Engineering*, Vol. 99, pp. 713–722.

6.5 Physically Realistic Manufactured Solutions

The MMS procedure discussed in the previous sections is the most general method for obtaining exact solutions to mathematical models for use in code verification studies. Since physical realism of the solutions is not required during code verification, the solutions are somewhat arbitrary and can be tailored to exercise all terms in the mathematical model. However, there are many cases in which physically realistic exact solutions are desired, such as assessing sensitivity of a numerical scheme to mesh quality, evaluating the reliability of discretization error estimators, and judging the overall effectiveness of solution adaptation schemes (see Chapter 9). There are two main approaches for obtaining physically realistic manufactured solutions to complex equations, and these approaches are discussed below.

6.5.1 Theory-Based Solutions

One approach to obtaining physically realistic manufactured solutions is to use a simplified theoretical model of the physical phenomenon as a basis for the manufactured solution. For example, if a physical process is known to exhibit an exponential decay in the solution with time, then a manufactured solution of the form

$$\alpha \exp(-\beta t)$$

could be employed, where α and β could be chosen to provide physically meaningful solutions.

There are numerous examples of this approach applied to the modeling of fluid turbulence using the Reynolds-Averaged Navier–Stokes (RANS) equations. Pelletier et al. (2004) have verified a 2D incompressible finite element code that employs a k-ε two-equation turbulence model. They constructed manufactured solutions which mimic turbulent shear flows, with the turbulent kinetic energy and the turbulent eddy viscosity as the two quantities specified in the manufactured solution. More recently, Eca and Hoekstra (2006) and Eca et al. (2007) developed physically realistic manufactured solutions mimicking steady, wall-bounded turbulence for 2D incompressible Navier–Stokes codes. They examined both one- and two-equation turbulence models and noted challenges in generating physically realistic solutions in the near-wall region. Additional solutions were developed by Eca et al. (2012) for steady RANS and by Eca et al. (2016) for the unsteady RANS equations. These last two studies include extensive investigations of systematic grid refinement on both structured and unstructured grids, iterative errors, and least squares assessment of the observed order of accuracy.

6.5.2 Reconstructed Numerical Solutions

A second approach for generating physically realistic manufactured solutions is to first compute a highly-accurate numerical solution, then to perform a reconstruction (either global or local) of the numerical solution. This procedure was termed Tuned Test Problems by Sinclair et al. (1997) and Sinclair and Kardak (2022) and applied to solid mechanics problems. A similar method was independently developed under the name Method of Nearby Problems by Roy and Hopkins (2003) and Roy and Sinclair (2009). This general approach involves first computing a highly-refined numerical solution for the problem of interest, then generating an accurate curve fit of that numerical solution. If both the underlying numerical solution and the curve fit are "sufficiently" accurate, then it will result in a physically-realistic manufactured solution which has small source terms. This procedure is described in more detail below.

The steps for generating physically realistic manufactured solutions using the solution reconstruction procedure are:

1. compute the original numerical solution on a highly refined grid,
2. generate an accurate spline or curve fit to this numerical solution, thereby providing an analytic representation of the numerical solution,

3. operate the governing partial differential (or integral) equations on the curve fit to generate analytic source terms (which ideally will be small), and
4. create the nearby problem (or tuned test problem) by appending the analytic source terms to the original mathematical model.

The key point to this approach is that, by definition, the curve fit generated in step 2 is the exact solution to the nearby problem. While the approach is very similar to MMS, the addition of the curve fitting step is designed to provide a physically realistic solution. A 2D example of this approach to generate physically-realistic manufactured solutions was given for the case of viscous, incompressible flow in a lid-driven cavity at a Reynolds number of 100 by Roy and Sinclair (2009). Sinclair and Kardak (2022) provide an example of this procedure applied to stress concentration for finite element analysis.

Simple Example: Numerical Solution Reconstruction

To demonstrate the solution reconstruction procedure, a simple 1D example is presented in Figure 6.8, where the original data used to generate the curve fit are 17 points sampled at equal intervals from the function $\sin(2\pi x)$ (Roy and Sinclair, 2009). The goal of this example is to create a spline fit made up of four spline regions that exhibits C^2 continuity at the spline zone interfaces (i.e., continuity up to the second derivative). The spline fitting is performed in a manner that allows arbitrary levels of continuity at spline boundaries and is readily extendible to multiple dimensions following Junkins et al. (1973).

The first step is to generate five overlapping local fits Z_1 through Z_5, with each of the interior fits spanning two spline regions (see top of Figure 6.8). A least squares method is used to find a best-fit quadratic function in each of the five spline zones:

$$Z_n(\bar{x}) = a_n + b_n\bar{x} + c_n\bar{x}^2. \tag{6.29}$$

The overbars in Equation (6.29) specify that the spatial coordinate x is locally transformed to satisfy $0 \le \bar{x} \le 1$ in each of the interior spline zones. Since each spline zone now has two different local fits, one from the left and the other from the right, these two local fits are combined together with the left and right weighting functions shown in Figure 6.8 (middle). The form of the 1D weighting function used here for C^2 continuity is

$$W_{RIGHT}(\bar{x}) = \bar{x}^3\left(10 - 15\bar{x} + 6\bar{x}^2\right)$$

and the corresponding left weighting function is defined simply as

$$W_{LEFT}(\bar{x}) = W_{RIGHT}(1 - \bar{x}).$$

This weighting function is chosen to provide full weight to the middle of each local fit and goes to zero at the far extent of each local fit, thus ensuring the desired level of solution continuity at spline interfaces. The final fit in each region can be written as

$$F(x, y) = W_{LEFT}Z_{LEFT} + W_{RIGHT}Z_{RIGHT}.$$

For example, for region 2, one would have $Z_{LEFT} = Z_2$ and $Z_{RIGHT} = Z_3$.

Note that in addition to providing the desired level of continuity at spline boundaries, the weighting functions are also useful in reducing the dependence near the boundaries of the local fits

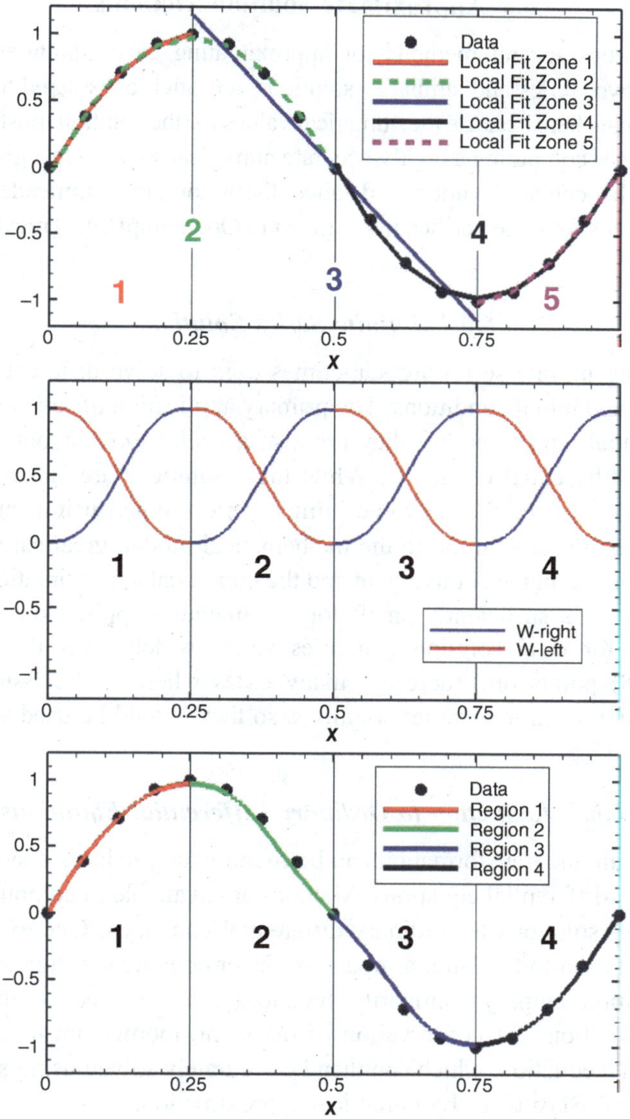

Figure 6.8 Simple one-dimensional example of the weighting function approach for combining local quadratic least squares fits to generate a C^2 continuous spline fit: local fits (top), weighting functions (middle), and resulting C^2 continuous spline fit (bottom). (from Roy and Sinclair, 2009)

where they often exhibit the poorest agreement with the original data. When these final fits are plotted (bottom of Figure 6.8), we see that they are indeed C^2 continuous, maintaining continuity of the function value, slope, and curvature at all three interior spline boundaries. For extension of this technique to two and three dimensions, see Junkins et al. (1973) and Roy and Sinclair (2009).

6.6 Approximate Solution Methods

This section describes several methods for approximating exact solutions to mathematical models. The first two, series and similarity solutions, are often considered to be exact, but are treated as approximate here because the numerical values for the solution must be computed. The third method involves computing a highly-accurate numerical solution to a given problem and is called a numerical benchmark solution. A detailed discussion of recommended requirements for approximate solutions for code verification is given in Oberkampf and Trucano (2008).

6.6.1 Infinite Series Solutions

Solutions involving infinite series are sometimes used to solve differential equations with general boundary and initial conditions. The primary application of series solutions has been for linear differential equations, but they are also a useful tool for obtaining solutions to certain nonlinear differential equations. While these solutions are "analytic," they are not closed form solutions since they involve infinite series. When using infinite series as an approximation of an exact solution to the mathematical model, great care must be taken to ensure that the series is, in fact, convergent and the numerical approximation error created by truncating the series is sufficiently small for the intended application. For example, the convergence rate for local solution quantities varies widely over the solution domain. As Roache (1998) points out, there are many cases where subtle issues arise with the numerical evaluation of infinite series solutions, so they should be used with caution.

6.6.2 Reduction to Ordinary Differential Equations

In some cases, a suitable transformation can be found which reduces a system of PDEs to a system of ordinary differential equations. Methods are available to compute highly-accurate numerical or series solutions for ordinary differential equations. One example is the well-known Blasius solution to the laminar boundary layer equations in fluid dynamics (Schetz, 1993). This solution employs similarity transformations to reduce the incompressible boundary layer equations for conservation of mass and momentum to a single, nonlinear ordinary differential equation, which can then be accurately solved using series solution (the original approach of Blasius) or by numerical approximation.

When using this approach, it must be recognized that *exactly* the same mathematical model used in the solution of the ordinary differential equations must be used in the verification of the PDE code. Consider the situation in which a code based on the full Navier–Stokes equation is used to solve for the laminar boundary layer flow over a flat plate. In this case, the solution from the Naiver–Stokes code would *not* converge to the Blasius solution since these are two different mathematical models. This is because the Navier–Stokes equations contain terms omitted from the boundary layer equations.

6.6.3 Benchmark Numerical Solutions

Another approximate solution method is to compute a *benchmark numerical solution* with a high-degree of numerical accuracy. In order for a numerical solution to a complex PDE to qualify as a benchmark solution, the problem statement, numerical scheme, and numerical solution accuracy should be carefully documented (Oberkampf and Trucano, 2008). Quantifying the numerical accuracy of benchmark numerical solutions is often difficult, and at a minimum should include evidence that (1) the asymptotic convergence range has been achieved for the benchmark problem and (2) the code used to generate the benchmark solution has passed the order of accuracy code verification test for the cptions exercised in the benchmark problem. An additional advantageous feature of a good benchmark numerical solution is that the same solution results are obtained by two different codes with different numerical approximation methods. Benchmark numerical solutions for solid mechanics applications are discussed in Szabo et al. (2010) and Schroder et al. (2021).

Example Benchmark Solution: Turbulent Flow over an Airfoil

The problem under consideration is the Mach 0.15 flow over a 2D NACA 0012 airfoil at a Reynolds number (based on the chord length) of 6 million and an angle of attack of 10 degrees. This problem has been used as a verification and validation test case and is available on the Turbulence Model Resource website (NASA, 2023). The Spalart-Allmaras (1994) turbulence model has been applied to this problem using two computational fluid dynamics codes, FUN3D and CFL3D, both of which have undergone rigorous code verification testing using MMS (Rumsey and Thomas, 2008). Since both of these codes have been demonstrated to converge to the same results, they have been used to establish a benchmark numerical solution for the Spalart-Allmaras turbulence model applied to this case.

Lift and drag coefficients for three codes (FUN3D, CFL3D, and Tau) are shown in Figure 6.9 for a family of systematically refined grids. In terms of grid-converged forces and pitching moment, the following values represent the estimated lower and upper bounds for the benchmark NACA 0012 solution from the CFL3D and FUN3D codes for the SA turbulence model:

- Lift coefficient: $C_L = 1.0909$–1.0911
- Drag coefficient: $C_D = 0.012270$–0.012275

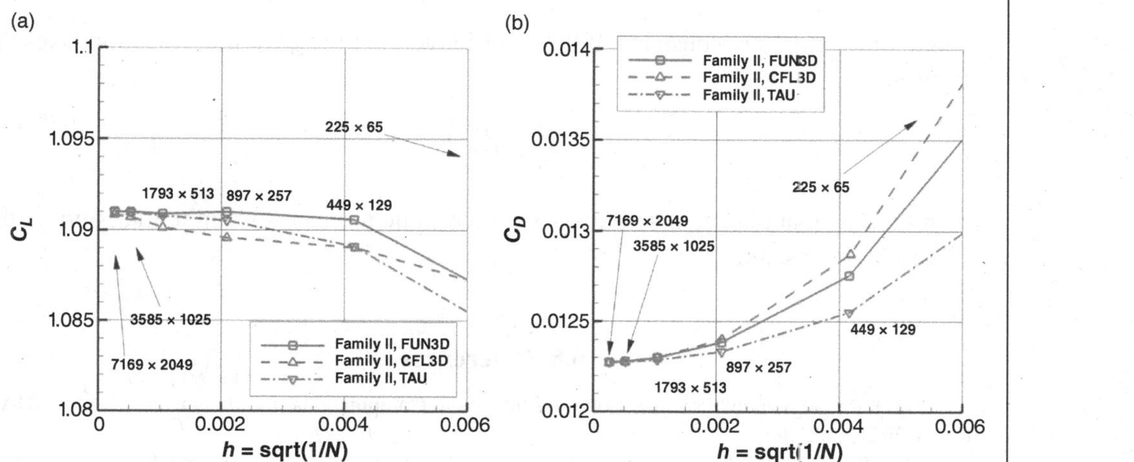

Figure 6.9 Lift and drag coefficients for three different CFD codes.
(adapted from NASA, 2023)

- Pressure drag coefficient: $C_{Dp} = 0.00606\text{–}0.00607$
- Viscous drag coefficient: $C_{Dv} = 0.006205\text{–}0.006206$
- Moment coefficient: $C_{My} = 0.00676\text{–}0.00680$

Thus, for example, the benchmark lift coefficient is "known" to within about 0.02% and drag coefficient to about 0.1%, or 1/10 of a drag count. There is a high degree of confidence in this benchmark solution since both the CFL3D and FUN3D codes have undergone rigorous code order of accuracy testing using MMS. Furthermore, a third independent code (Tau) also produces similar results in the limit of grid refinement. Additional discussion of this case can be found in Roy et al. (2018).

6.7 Problems

1. Multiple choice: Manufactured solutions ______________.
   (a) require the specification of user-defined source terms
   (b) require the specification of user-defined initial/boundary conditions
   (c) should exercise all terms in the governing equations
   (d) should be smoothly varying functions
   (e) all answers are correct
2. Explain the difference between the formal and observed orders of accuracy.
3. Explain the main concepts behind the Method of Manufactured Solutions for generating exact solutions to partial differential equations.
4. Give eight reasons that might prevent the observed order of accuracy from matching the formal order of accuracy during an order of accuracy verification test using the Method of Manufactured Solutions.
5. A manufactured solution for a one-dimensional problem is chosen as:

$$u(x) = A + B\sin\left(\frac{\pi x}{L}\right),$$

where A, B, and L are constants. If the diffusion term in the governing equation takes the form of

$$\frac{\partial}{\partial x}\left(\mu\frac{\partial u}{\partial x}\right),$$

then find the manufactured solution source term due to this diffusion term assuming the viscosity is constant.

6.8 References

AIAA (2021). Recommended Practice: Code Verification in Computational Fluid Dynamics. AIAA, AIAA R-141-2021, 2021.

Banks, J. W., T. Aslam, and W. J. Rider (2008). On Sub-linear Convergence for Linearly Degenerate Waves in Capturing Schemes. *Journal of Computational Physics*, **227**, 6985–7002.

Benton, E. R., and G. W. Platzman (1972). A Table of Solutions of the One-Dimensional Burgers Equation. *Quarterly of Applied Mathematics*, **30**(2), 195–212.

Bond, R. B., C. C. Ober, P. M. Knupp, and S. W. Bova (2007). Manufactured Solution for Computational Fluid Dynamics Boundary Condition Verification. *AIAA Journal*, **45**(9), 2224–2236.

Brglez, S. (2013). Code Verification for Governing Equations with Arbitrary Functions Using Adjusted Method of Manufactured Solutions. *Engineering with Computers*, **30**(4), 669–678.

Carslaw, H. S., and J. C. Jaeger (1959). *Conduction of Heat in Solids*, 2nd ed. Oxford: Clarendon Press.

Choudhary, A., C. J. Roy, E. A. Luke, and S. P. Veluri (2016). Code Verification of Boundary Conditions for Compressible and Incompressible Computational Fluid Dynamics Codes. *Computers and Fluids*, **126** (2016), 153–169.

Derlaga, J. M., T. S. Phillips, and C. J. Roy (2013). SENSEI Computational Fluid Dynamics Code: A Case Study in Modern Fortran Software Development. AIAA Paper 2013-2450.

Eca, L., and M. Hoekstra (2006). Verification of Turbulence Models with a Manufactured Solution. In P. Wesseling, E. Onate, and J. Periaux (eds.), *European Conference on Computational Fluid Dynamics*, ECCOMAS CFD 2006.

Eca, L., M. Hoekstra, A. Hay, and D. Pelletier (2007). On the Construction of Manufactured Solutions for One and Two-Equation Eddy-Viscosity Models. *International Journal for Numerical Methods in Fluids*, **54**(2), 119–154.

Eca, L., M. Hoekstra, and G. Vaz (2012). Manufactured Solutions for Steady-Flow Reynolds-Averaged Navier–Stokes Solvers. *International Journal of Computational Fluid Dynamics*, **26**(5), 313–332.

Eca, L., C. M. Klaij, G. Vaz, M. Hoekstra, and F. S. Pereira (2016). On Code Verification of RANS Solvers. *Journal of Computational Physics*, **310**, 418–439.

Eca, L., G. Vaz., and M. Hoekstra (2016). Code Verification of REFRESCO with a Statistically Periodic Manufactured Solution. *Proceedings of the ASME 2014 33rd International Conference on Ocean, Offshore and Arctic Engineering*, June 8–13, 2014, San Francisco, CA, 11 pages.

Elishakoff, I. (2004). *Eigenvalues of Inhomogeneous Structures: Unusual Closed-Form Solutions*. Boca Raton, FL: CRC Press.

Evans, L. C. (2010). *Partial Differential Equations*, 2nd ed. Providence, RI: American Mathematical Society.

Gottlieb, J. J., and C. P. T. Groth (1988). Assessment of Riemann Solvers for Unsteady One-dimensional Inviscid Flows of Perfect Gases. *Journal of Computational Physics*, **78**(2), 437–458.

Grier, B., E. Alyanak, M. White, J. Camberson and R. Figliola (2014). Numerical Integration Techniques for Discontinuous Manufactured Solutions. *Journal of Computational Physics*, **278**, 193–203.

Grier, B., R. Figliola, E. Alyanak, and J. Camberson (2015). Discontinuous Solutions Using the Method of Manufactured Solutions on Finite Volume Solvers. *AIAA Journal*, **53**(8), 2369–2378.

Haberman, R. (2018). *Applied Partial Differential Equations with Fourier Series and Boundary Value Problems*, 5th ed. Boston, MA: Pearson.

Hirsch, C. (2007). *Numerical Computation of Internal and External Flows: Fundamentals of Computational Fluid Dynamics*, 2nd ed. Oxford: Butterworth-Heinemann.

Horner, M. (2021). The Method of Rotated Solutions: A Highly Efficient Procedure for Code Verification. *Journal of Verification, Validation and Uncertainty Quantification*, **6**(1), 011001, 6 pages.

Ingraham, D., and R. Hixon (2013). External Verification Analysis: A Code-Independent Verification Technique for Unsteady PDE Codes. *Journal of Computational Physics*, **243**, 46–57.

Junkins, J. L., G. W. Miller, and J. R. Jancaitis (1973). A Weighting Function Approach to Modeling of Irregular Surfaces. *Journal of Geophysical Research*, **78**(11), 1794–1803.

Kausel, E. (2006). *Fundamental Solutions in Elastodynamics: A Compendium*. New York, NY: Cambridge University Press.

Kevorkian, J. (2000). *Partial Differential Equations: Analytical Solution Techniques*, 2nd ed., Texts in Applied Mathematics, 35. New York, NY: Springer.

Knupp, P., and K. Salari (2003). *Verification of Computer Codes in Computational Science and Engineering*, K. H. Rosen (ed.). Boca Raton, FL: Chapman & Hall/CRC.

Meleshko, S. V. (2005). *Methods for Constructing Exact Solutions of Partial Differential Equations: Mathematical and Analytical Techniques with Applications to Engineering*. New York, NY: Springer.

NASA Langley Turbulence Modeling Resource website (2023). 2D NACA 0012 Airfoil Validation Case. https://turbmodels.larc.nasa.gov/naca0012numerics_val.html.

Oberkampf, W. L., F. G. Blottner, and D. P. Aeschliman (1995). Methodology for Computational Fluid Dynamics Code Verification/Validation. AIAA Paper 95-2226 (see also Oberkampf, W. L. and Blottner, F. G. (1998). Issues in Computational Fluid Dynamics Code Verification and Validation. *AIAA Journal*, **36**(5), 687–695).

Oberkampf, W. L., and C. J. Roy (2010). *Verification and Validation in Scientific Computing*, 1st ed. Cambridge: Cambridge University Press.

Oberkampf, W. L., and T. G. Trucano (2008). Verification and Validation Benchmarks. *Nuclear Engineering and Design*, **238**(3), 716–743.

Panton, R. L. (2005). *Incompressible Flow*, 3rd ed. New York, NY: John Wiley.

Pelletier, D., E. Turgeon, and D. Tremblay (2004). Verification and Validation of Impinging Round Jet Simulations Using an Adaptive FEM. *International Journal for Numerical Methods in Fluids*, **44**, 737–763.

Polyanin, A. D. (2016). *Handbook of Linear Partial Differential Equations for Engineers and Scientists*, 2nd ed. Boca Raton, FL: Chapman & Hall/CRC.

Polyanin, A. D., and V. F. Zaitsev (2011). *Handbook of Nonlinear Partial Differential Equations*, 2nd ed. Boca Raton, FL: Chapman & Hall/CRC.

Polyanin, A. D., and V. F. Zaitsev (2017). *Handbook of Ordinary Differential Equations: Exact Solutions, Methods, and Problems*, 3rd ed. Boca Raton, FL: Chapman & Hall/CRC.

Powers, J. M., and T. D. Aslam (2006). Exact Solution for Multidimensional Compressible Reactive Flow for Verifying Numerical Algorithms. *AIAA Journal*, **44**(2), 337–344.

Powers, J. M., and D. S. Stewart (1992). Approximate Solutions for Oblique Detonations in the Hypersonic Limit. *AIAA Journal*, **30**(3), 726–736.

Roache, P. J. (1998). *Verification and Validation in Computational Science and Engineering*. Albuquerque, NM: Hermosa Publishers.

Roache, P. J. (2002). Code Verification by the Method of Manufactured Solutions. *Journal of Fluids Engineering*, **124**(1), 4–10.

Roache, P. J., P. M. Knupp, S. Steinberg, and R. L. Blaine (1990). Experience with Benchmark Test Cases for Groundwater Flow. In I. Celik and C. J. Freitas (eds.), *Benchmark Test Cases for Computational Fluid Dynamics*, 49–56, American Society of Mechanical Engineers, Fluids Engineering Division, 93, Book No. H00598.

Roache, P. J., and S. Steinberg (1984). Symbolic Manipulation and Computational Fluid Dynamics. *AIAA Journal*, **22**(10), 1390–1394.

Roe, P. L. (1981). Approximate Riemann Solvers, Parameter Vectors, and Difference Schemes. *Journal of Computational Physics*, **43**, 357–372.

Roy, C. J. (2005). Review of Code and Solution Verification Procedures for Computational Simulation. *Journal of Computational Physics*, **205**(1), 131–156.

Roy, C. J., and M. M. Hopkins (2003). Discretization Error Estimates Using Exact Solutions to Nearby Problems (Invited). AIAA Paper 2003-0629.

Roy, C. J., C. C. Nelson, T. M. Smith, and C. C. Ober (2004). Verification of Euler/Navier–Stokes Codes Using the Method of Manufactured Solutions. *International Journal for Numerical Methods in Fluids*, **44**(6), 599–620.

Roy, C. J., C. L. Rumsey, and E. N. Tinoco (2018). Summary Data from the Sixth AIAA Computational Fluid Dynamics Drag Prediction Workshop: Code Verification. *Journal of Aircraft*, **55** (4), 1338–1351.

Roy, C. J., and A. J. Sinclair (2009). On the Generation of Exact Solutions for Evaluating Numerical Schemes and Estimating Discretization Error. *Journal of Computational Physics*, **228**(5), 1790–1802.

Roy, C. J., E. Tendean, S. P. Veluri, R. Rifki, E. A. Luke, and S. Hebert (2007). Verification of RANS Turbulence Models in Loci-CHEM using the Method of Manufactured Solutions. AIAA Paper 2007-4203.

Rumsey, C. L. and J. L. Thomas (2008). Application of FUN3D and CFL3D to the Third Workshop on CFD Uncertainty Analysis. NASA/TM-2008-215537.

Schetz, J. A. (1993). *Boundary Layer Analysis*. Upper Saddle River, NJ: Prentice Hall.

Schroder, J., T. Wick, S. Reese, P. Wriggers, R. Muller, S., Kollmannsberger, et al. (2021). A Selection of Benchmark Problems in Solid Mechanics and Applied Mathematics. *Archives of Computational Methods in Engineering*, **28**, 713–751.

Shih, T. M. (1985). Procedure to Debug Computer Programs. *International Journal for Numerical Methods in Engineering*, **21**(6), 1027–1037.

Sinclair, G. B., M. Anaya-Dufresne, G. Meda, and M. Okajima (1997). Tuned Test Problems for Numerical Methods in Engineering. *International Journal for Numerical Methods in Engineering*, **40**, 4183–4209.

Sinclair, G. B., and A. A. Kardak (2022). On the Generation of Tuned Test Problems for Stress Concentrations. *ASME Journal of Verification, Validation, and Uncertainty Quantification*, **7**(1), 011003, 6 pages.

Slaughter, W. S. (2002). *The Linearized Theory of Elasticity*. Boston, MA: Birkhauser.

Spalart, P. R., and S. R. Allmaras (1994). A One-Equation Turbulence Model for Aerodynamic Flows. *Recherche Aerospatiale*, **1**, 5–21.

Steinberg, S., and P. J. Roache (1985). Symbolic Manipulation and Computational Fluid Dynamics. *Journal of Computational Physics*, **57**(2), 251–284.

Stetter, H. J. (1978). The Defect Correction Principle and Discretization Methods. *Numerische Mathematik*, **29**(4), 425–443.

Szabo, B. A., I. Babuska, J. Pitkaranta, and S. Nervi (2010). The Problem of Verification with Reference to the Girkmann Problem. *Engineering with Computers*, **26**, 171–183.

Tannehill, J. C., D. A. Anderson, and R. H. Pletcher (1997). *Computational Fluid Mechanics and Heat Transfer*, 2nd ed. Philadelphia, PA: Taylor & Francis.

Timoshenko, S. P., and J. N. Goodier (1969). *Theory of Elasticity*, 3rd ed. New York, NY: McGraw-Hill.

White, F. M. (2005). *Viscous Fluid Flow*, 3rd ed. New York, NY: McGraw Hill.

Woods, C. N., and R. P. Starkey (2015). Verification of Fluid-Dynamic Codes in the Presence of Shocks and Other Discontinuities. *Journal of Computational Physics*, **294**, 312–328.

Zadunaisky, P. E. (1976). On the Estimation of Errors Propagated in the Numerical Integration of Ordinary Differential Equations. *Numerische Mathematik*, **27**(1), 21–39.

Part III

Solution Verification

Solution verification, sometimes referred to as calculation verification, is a critical aspect of ensuring that numerical errors are not corrupting the simulation results. The main focus of solution verification is the estimation of the numerical errors that occur when a mathematical model is discretized and solved on a digital computer. While some of the strategies employed will be similar to those used for code verification, there is an important difference. In solution verification, the exact solution to the mathematical model is *not* known, and thus, the numerical errors must be estimated and not simply evaluated. As a result, estimated numerical errors are characterized as numerical uncertainties.

Chapter 7 discusses three of the sources of numerical errors that arise in scientific computing in detail: computer round-off, statistical sampling, and surrogate modeling error. It closes with a discussion on how these errors, in addition to iterative error and discretization error, are converted into simulation result uncertainties and then combined into total numerical uncertainty. Chapter 8 gives a detailed discussion of various iterative methods. It also addresses the topics of iterative convergence and iterative error and explains how they are related. Chapter 8 closes with a discussion of practical methods for estimating iterative errors suitable for cases when the iterative solution converges in a monotone fashion. Chapter 9 gives a detailed discussion of discretization error and its relationship to truncation error. It is stressed that discretization error is often the largest numerical error source and also the most challenging to estimate. Various discretization error estimators are described, with special attention paid to Richardson extrapolation. The reliability of discretization error estimators and their conversion to uncertainties are also described. Chapter 9 closes with a discussion of solution adaptation methods.

7

Numerical Uncertainty Estimation

On two occasions I have been asked, 'Pray, Mr. Babbage, if you put into the machine wrong figures, will the right answers come out?' I am not able rightly to apprehend the kind of confusion of ideas that could provoke such a question. —Charles Babbage

Contents of Chapter 7

Most *solution verification* activities are focused on estimating the numerical errors in the simulation. This chapter addresses in detail round-off error, statistical sampling error, and surrogate modeling error. Iterative error and discretization error, which are discussed in the following chapters, are usually the largest sources of numerical error and are more difficult to estimate. Iterative errors are related to the iterative tolerances used for iterative solutions to discretized equations and are discussed in detail in Chapter 8. Discretization errors are those associated with the mesh resolution and quality as well as the time step chosen for unsteady problems and are discussed in detail in Chapter 9.

Round-off and discretization errors are always present in scientific computing simulations, while the presence of iterative, statistical, and surrogate modeling errors will depend on the application and the chosen numerical algorithms. Some of the techniques used for solution verification appear to be similar to those used for code verification activities. However, the fundamental difference between code and solution verification is that in code verification, the exact solution is known and thus the various errors can be *evaluated*. In solution verification, the exact solution is *not* known, thus the errors must be *estimated*. It is this estimation process which gives rise to *numerical uncertainty*. This chapter concludes with a discussion of how the various sources of numerical error can be rolled up into a total numerical uncertainty estimate.

7.1 Elements of Solution Verification

Solution verification begins after the mathematical model has been embodied in a verified code; the initial, boundary, and system excitation conditions have been specified; and any other auxiliary relations have been determined. It includes the running of the code on a mesh, or series of meshes, possibly to a specified iterative convergence tolerance. Solution verification ends after all post-processing of the simulation results are completed to provide the final simulation results. There are thus three aspects of solution verification:

1. verification of input data,
2. numerical error estimation, and
3. verification of post-processing tools.

These three aspects form the combined human and numerical errors in the simulation result shown in Figure 7.1.

Verification of input and output data is particularly important when a large number of simulations are performed, such as for an ensemble of simulations that vary different input conditions. Issues associated with the verification of input and output data are discussed in this section. Numerical error estimation is discussed in detail in the remainder of this chapter as well as in Chapters 8 and 9.

Input data is any required information for running a scientific computing code. Common forms of input data include:

- input files describing models, submodels, and numerical algorithms,
- computer-aided drawing (CAD) surface geometry information,
- domain grids,
- boundary and initial conditions,
- excitation of the system of interest (e.g., effect of gravity and magnetic fields),
- data used for submodels (e.g., chemical species properties, reaction rates), and
- information on material properties.

There are various techniques that can aid in the verification of the input data. Checks for consistency between model choices should be made at the beginning of code execution. For

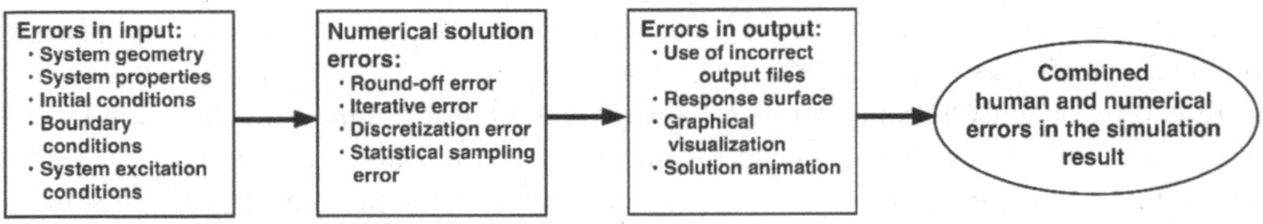

Figure 7.1 Error sources that form the elements of solution verification.

example, the code should not allow a no-slip (viscous) wall boundary condition to be used for an inviscid simulation involving the Euler equations. For more subtle modeling issues, an expert knowledge database could be used to provide the user with warnings when a model is being used outside of its range of applicability (e.g., Stremel et al., 2007). In addition, all input data used for a given simulation should be archived in an output file, such as simulation data management software, so that the correctness of the input data can be confirmed by post-simulation inspection, for example, if needed by regulatory authorities. Verification of input data also includes the verification of any preprocessing software that is used to generate the input data and thus the standard software engineering practices discussed in Chapter 4 can be used.

Post-processing tools are defined as any software that operates on the output from a scientific computing code. If this post-processing involves any type of numerical approximation such as discretization, integration, interpolation, etc., then the order of accuracy of these tools should be verified (e.g., by order verification), otherwise the standard software engineering practices should be followed. If possible, the post-processing steps should be automated, and then verified, to prevent common human errors such as picking the wrong solution for post-processing. If the user of the code is to perform the post-processing, then a checklist should be developed to ensure this process is done correctly.

Numerical errors occur in *every* scientific computing simulation, and thus need to be estimated in order to build confidence in the mathematical accuracy of the solution. In other words, numerical error estimation is performed to ensure that the solution produced by running the code is a sufficiently accurate approximation of the exact solution to the mathematical model. When numerical errors are found to be unacceptably large, then they should either be accounted for in the total uncertainty of the simulation result (see Chapter 13) or reduced to an acceptable level by refining the mesh, reducing the iterative tolerance, etc. The five types of numerical errors are:

1. Round-off error,
2. Statistical sampling error,
3. Surrogate modeling error,
4. Iterative error, and
5. Discretization error.

This chapter discusses the first three numerical error sources in detail, while iterative error and discretization error are addressed separately in Chapters 8 and 9, respectively.

7.2 Round-Off Error

Round-off errors arise due to the use of finite precision arithmetic on digital computers. For example, in a single-precision digital computation, the following result is often obtained

$$3.0 * (1.0/3.0) = 0.9999999,$$

while the true answer using infinite precision is 1.0. Round-off error can be significant for ill-conditioned systems of equations, implicit methods, higher-order methods, as well as time-accurate simulations. Repeated arithmetic operations will degrade the accuracy of a scientific computing simulation, and generally not just in the last significant figure of the solution since these errors are compounded. Furthermore, the addition or subtraction of both large and small numbers will significantly reduce the precision of the computed solution. An example of when such a loss of precision can occur is for the computation

$$\texttt{x} = (\texttt{1.E} - \texttt{9} + \texttt{1.0}) - \texttt{1.0},$$

which will return `x = 0.0` for single precision computations and the correct value of `x = 1.E−9` for double precision computations. In this case, simply rearranging the parentheses would produce the correct result. Thus, the expected magnitude of variables should be considered during the programming of arithmetic operations.

Round-off error can be reduced by using more significant digits in the computation. Although round-off error can be thought of as the truncation of a real number to fit it into computer memory, it should not be confused with truncation error which is a measure of the difference between a partial differential equation and its discrete approximation as defined in Chapter 5.

7.2.1 Floating Point Representation

Scientific computing applications require the processing of real numbers, i.e., numbers expressed in floating point representation. Even when limiting these real numbers to lie within a certain range, say from −1 million to +1 million, there are infinitely many real numbers to be considered. This poses a problem for digital computers, which must store real numbers in a finite amount of computer memory. In order to fit these numbers into memory, both the precision (the number of significant figures) and the range of the exponent must be limited. An efficient way to do this is through an analogy with scientific notation, where both large and small numbers can be represented compactly. For example, 14,000,000 can be represented as 1.4×10^7 and 0.0000014 can be represented by 1.4×10^{-6}. In digital computers, floating point numbers are more generally represented as

$$S \times B^E,$$

where S is the significand (or mantissa), B the base (usually 2 for binary or 10 for decimal), and E is the exponent. The term floating point number comes from the fact that in this notation, the decimal point is allowed to move to represent the significand S more efficiently.

For digital computer hardware and software, the most widely-used standard for floating point numbers is the IEEE Standard 754 (IEEE, 2008). This standard addresses number format, rounding algorithms, arithmetic operations, and exception handling (division by zero, numerical overflow, numerical underflow, etc.). While the IEEE standard addresses both binary and decimal formats, nearly all software and hardware used for scientific computing employ binary storage of floating point numbers. The most commonly used formats are single precision and double precision. Additional standard formats that may or may not be available on a given computer hardware or software system are half precision, extended precision, and quadruple precision.

Single precision employs 32 bits, or 4 bytes, of computer memory. The significand is stored using 24 bits, one of which is used to determine the sign of the number (positive or

Table 7.1 *Summary of floating point number formats in IEEE Standard 754 (IEEE, 2008)*

Precision format	Total # of bits used	Bits used for significand	Bits used for exponent	Maximum number of significant digits (Base 10)	Exponent range (Base 10)
Single	32	24	8	7	±38
Double	64	53	11	15	±308
Half [a]	16	11	5	3	±5
Extended[a]	80	64	16	18	±9,864
Quadruple[a]	128	113	15	34	±4,932

[a] not available in some programming languages and/or compilers

negative). The exponent is then stored in the remaining 8 bits of memory, with 1 bit used to store the sign of the exponent. The significand determines the precision of the floating point number, while the exponent determines the range of numbers that can be represented. Single precision provides approximately seven significant decimal digits when storing a single floating point number and can represent positive or negative numbers with magnitudes as large as $\sim 3.4 \times 10^{38}$ and as small as $\sim 1.2 \times 10^{-38}$. For double precision numbers, 64 bits (8 bytes) of memory are used, with 53 bits assigned to the significand and 11 bits to the exponent, thus providing approximately 15 significant decimal digits for storing floating point numbers. The five standard binary formats are summarized in Table 7.1. The last two columns of Table 7.1 give the maximum and minimum precision and range where, for example, single precision numbers will be represented in base 10 as:

$$1.234567 \times 10^{\pm 38}.$$

Again, it should be emphasized that Table 7.1 shows the maximum possible digits of precision in base 10. Since scientific computing codes often involve millions or billions of floating point operations to obtain a numerical solution, the maximum possible digits of precision is never achievable in practice. The achievable digits of precision will depend strongly on characteristics such as whether the governing equations are solved in an appropriate nondimensional form and whether implicit or explicit solution algorithms are employed.

The use of single (32 bit) and double (64 bit) precision for representing floating point numbers should not be confused with 32-bit and 64-bit computer architectures. The wide availability of 64-bit processors in desktop computers beginning in 2003 was initially driven by the fact that the amount of random access memory (RAM) addressable by a 32-bit integer is only 4 GB (2^{32} bytes or approximately 4.29×10^9 bytes). The 64-bit processors were developed for large databases and applications requiring more than 4 GB of addressable memory, providing a theoretical upper bound of approximately 17 billion GB (17×10^{18} bytes), although in practice the maximum addressable memory was much smaller. In addition to providing more addressable memory, 64-bit processors also generally perform arithmetic faster on double precision (64 bit) floating point numbers, the most common floating point data type used in scientific computing applications. This speed-up occurs because the data path between memory and the processor is more likely to be 64 bits wide rather than 32 bits, so only one memory read instruction is needed to move a double precision floating point number from memory to the processor. However, single precision

Table 7.2 *Data types used to specify the different precision formats in C/C++, Fortran, Python, and MATLAB®*

Precision Format	C/C++	Fortran 95/2003	Python	MATLAB®
Single	float	real, real*4 (default)[a,b]	e	single
Double	double (default)	double precision, real*8[a,b]	float (default)	double (default)
Half	n/a	a, b	e	n/a
Extended	long double[a]	a, b	e	n/a
Quadruple	long double[a]	a, b	e	n/a
Arbitrary	c	c	e	vpa[d]

[a] compiler dependent; [b] accessible via the "`kind`" attribute; [c] accessible via third-party libraries
[d] variable precision arithmetic; [e] accessible via the decimal module

(or more generally, mixed precision) may be of interest when accuracy requirements allow it due to the reduced memory requirements and increased computational speed (Baboulin et al., 2009).

7.2.2 Specifying Precision in a Code

The approach for specifying the precision for floating point numbers generally depends on the programming language, the compiler, and the hardware. The data types used for specifying the precision of real numbers (variables, constants, and functions) in C/C++, Fortran, Python, and MATLAB® are summarized in Table 7.2. In addition, the C/C++ and Fortran programming languages have the capability to employ arbitrary precision floating point numbers through the use of third-party software libraries such as the GNU Multiple Precision (GMP) arithmetic library for C/C++ (GNU, 2009) and FMLIB for Fortran (Smith, 2009). In general, when more digits of precision are employed, program execution will be slower. A more detailed discussion of procedures for specifying the floating point precision for each programming language can be found in Oberkampf and Roy (2010).

7.2.3 Practical Guidelines for Estimating Round-Off Error

While the precision of a number is determined by the number of bits used for the significand (mantissa), the precision of a simulation also depends on all of the arithmetic and functional operations used to obtain the solution. As discussed earlier, some scientific computing applications are more prone to round-off errors than others. A practical approach for assessing the effects of round-off error on a simulation prediction is to run the simulation with the desired precision, then re-run the simulation with higher precision and compare the solutions. For example, if the underlying simulation is performed with single precision floating point arithmetic, then round-off error can be estimated as

$$\bar{\varepsilon}_{\text{Round-Off}} = f_{\text{Single}} - f_{\text{Double}}, \tag{7.1}$$

where f_{Single} is the single precision SRQ, f_{Double} is the double precision SRQ, and the overbar denotes an error estimate (as opposed to the true error). The same mesh and/or time step should be employed in both cases, and when iterative methods are used then *both simulations* should be iteratively converged to within machine precision, i.e., until the iterative residuals can no longer be reduced due to round-off error (see Chapter 8). When a non-deterministic, i.e., stochastic, simulation is computed, this recommended procedure should be used on individual statistical samples. Attempting to use the procedure on statistical response measures, such as the mean of an SRQ, is unworkable because of the extremely large number of samples that are needed to reduce the statistical error to the point where the sensitivity to round-off error is discernable.

7.3 Statistical Sampling Error

In many cases, system response quantities in scientific computing are statistical mean quantities such as average lift and drag coefficients from an aerodynamics simulation or the vibrational modes from a structural dynamics simulation. If the simulations are steady and deterministic, then there will be no statistical sampling error for each solution. Statistical sampling error can occur in scientific computing predictions for many reasons. Certain scientific computing approaches are inherently stochastic in nature (e.g., stochastic differential equations, direct simulation Monte Carlo, lattice Boltzmann) and require time or ensemble averaging to determine mean system response quantities. If a submodel used in a scientific computing prediction is stochastic (e.g., a random walk model for Brownian motion), then a number of realizations may be needed to determine statistical response quantities. In addition, many unsteady simulations require averaging over time to determine mean properties. In some cases, steady simulations which employ iterative techniques can display unsteadiness due to strong physical instabilities in the problem (Ferziger and Peric, 2002). Solution unsteadiness can also occur due to numerical sources such as flux limiters, even when stable numerical schemes are employed. Finally, the use of nondeterministic simulations (e.g., to propagate input uncertainties to the outputs, as discussed in Chapter 3) must also deal with statistical sampling error.

7.3.1 Estimation of Statistical Sampling Error

We are typically interested in estimating the uncertainty due to statistical sampling error in mean values. The approaches discussed here are based on the fact that the estimated mean value from a finite number of samples or time steps (or even experimental measurements) will itself be randomly distributed about the true mean of the entire population of infinite samples. Furthermore, the *central limit theorem* says that under most conditions, the sample mean will be normally-distributed about the true population mean, even when the population itself does not follow a Gaussian (i.e., normal) distribution (Devore, 2009). However, the approach for estimating the uncertainty in the mean value will differ depending on whether the samples are purely random and uncorrelated with each other (i.e., white noise) or whether they come from a time sequence of correlated values such as an unsteady simulation (or a time series of measurements). Both cases are discussed in detail in the following sections.

7.3.1.1 Uncorrelated Samples

When the statistical samples are uncorrelated, as when drawn from a white noise process such a Monte Carlo sampling, then we can rely on standard confidence intervals for the

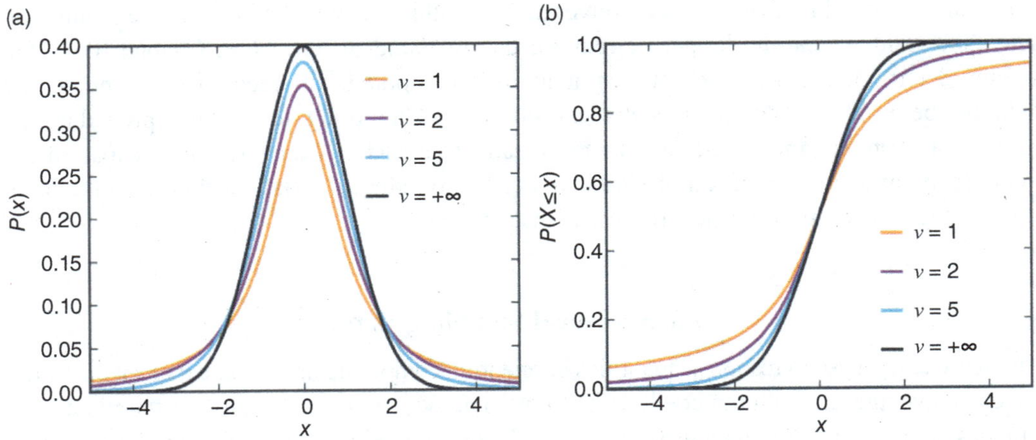

Figure 7.2 Student's *t*-distribution for different degrees of freedom: (a) probability density function and (b) cumulative distribution function.
(reproduced from https://en.wikipedia.org/wiki/Student%27s_t-distribution)

uncertainty in the sample mean. Similar techniques can be used for other statistical measures such as the variance or even quantiles of the distribution (Ross, 2021). Consider a system response quantity f, which in the limit of infinite observations has a true population mean μ and a true variance σ^2. Taking a finite number of observations N from that population allows one to compute the sample mean $\overline{f}$ and the unbiased sample variance S^2 (or standard deviation S) as:

$$\overline{f} = \frac{1}{N}\sum_{i=1}^{N} f_i, \tag{7.2}$$

$$S^2 = \frac{1}{N-1}\sum_{i=1}^{N} \left(f_i - \overline{f}\right)^2. \tag{7.3}$$

If the underlying population follows a Gaussian distribution, then we can construct a new, normalized random variable Z such that

$$Z = \frac{\overline{f} - \mu}{S/\sqrt{N}}, \tag{7.4}$$

which obeys the Student's *t*-distribution shown in Figure 7.2. The parameter ν is the degrees of freedom and is equal to one less than the number of samples, i.e., $\nu = N - 1$. For increasing sample sizes, the Student's *t*-distribution approaches a Gaussian distribution (the black lines in Figure 7.2).

Considering the distribution of Z given in Figure 7.2, and after some rearranging, we find that the true population mean μ falls within the two-sided interval

$$\mu \in \left[\overline{f} - \frac{CS}{\sqrt{N}},\ \ \overline{f} + \frac{CS}{\sqrt{N}}\right], \tag{7.5}$$

Table 7.3 *Values for C in the two-sided Student's t-distribution for different degrees of freedom* ($v = N - 1$)

	Confidence level					
Degrees of freedom, v	50%	70%	90%	95%	99%	99.9%
1	1.000	1.963	6.314	12.71	63.66	636.6
2	0.816	1.386	2.920	4.303	9.925	31.60
3	0.765	1.250	2.353	3.182	5.841	12.92
4	0.741	1.190	2.132	2.776	4.604	8.610
5	0.727	1.156	2.015	2.571	4.032	6.869
6	0.718	1.134	1.943	2.447	3.707	5.959
7	0.711	1.119	1.895	2.365	3.499	5.408
8	0.706	1.108	1.860	2.306	3.355	5.041
9	0.703	1.100	1.833	2.262	3.250	4.781
10	0.700	1.093	1.812	2.228	3.169	4.587
20	0.687	1.064	1.725	2.086	2.845	3.850
100	0.677	1.042	1.660	1.984	2.626	3.390
∞	0.674	1.036	1.645	1.960	2.576	3.291

where the parameter C is a function of both the desired confidence level and the number of observations N (see Table 7.3). For example, for a 95% confidence interval about the true mean and $N = 10$ observations, we have $C = 2.262$ and $\sqrt{N} = 3.162$, thus we can be 95% confident that the true population mean lies in the interval $\mu \in [\overline{f} - 0.715S, \overline{f} + 0.715S]$, where S is the sample standard deviation from Equation (7.3). This approach is again appropriate for uncorrelated samples.

7.3.1.2 Convergence for Time-Varying Signals

When samples are obtained from a single time-varying signal, the approach for estimating the uncertainty in the mean value is more complicated. We still seek to estimate the confidence interval for the true mean about the sample mean, but the correlation between the time samples, represented by the autocorrelation function, provides additional challenges. Here we discuss the approach proposed by Brouwer et al. (2019), which is based on Bendat and Piersol (2010), for determining the random uncertainty of the mean value.

Time-varying signals differ from white noise in that the processes contributing to the variations can occur over many frequencies and lead to correlations between the signal at time t and time $t + \tau$. The highest frequency possible f_{HIGH} is limited by the Nyquist frequency, which is one-half of the sampling frequency, and comes from the time step used in an unsteady simulation or the data acquisition rate for a time-varying measurement. Note that the time step should be small enough (or the sampling rate high enough) to ensure the high frequency variations in the signal are captured. The lowest frequency f_{LOW} is generally dictated by the processes generating the temporal variations, but the simulations (or experiments) should be run long enough so that these low frequency variations are well captured by the time series.

To illustrate the convergence of the uncertainty in the mean from a time-varying signal, a band pass signal is assumed. Assuming all frequency variations are fully captured, the signal is considered to contain *green noise*. If the low frequencies are not fully captured in the time

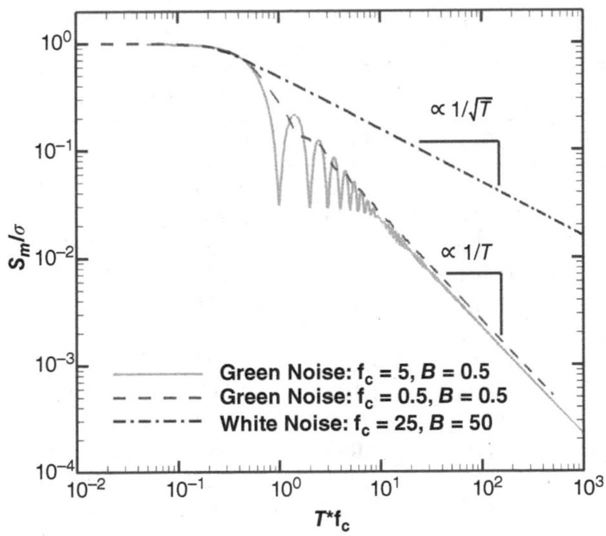

Figure 7.3 Variations of S_m/σ for different central frequencies and frequency widths ($B = \mathrm{f_{HIGH}} - \mathrm{f_{LOW}}$) for green noise and white noise versus the simulation (or measurement) time T times the central frequency $\mathrm{f_C}$.
(adapted from Brouwer et al., 2019)

series, then the signal is considered to contain *red noise*. One can also define a central frequency $\mathrm{f_C}$ which is the average of the low and high frequencies.

Inverting these frequencies gives the longest time period T_{Long} associated with the lowest frequency randomness $\mathrm{f_{LOW}}$ and a central time period T_C associated with the central frequency $\mathrm{f_C}$. Brouwer et al. (2019) focus on estimating S_m, which is the standard deviation of the mean. In other words, S_m is the standard deviation describing how the sample mean ($\bar{f}$) is distributed about the true population mean (μ). The variation of the ratio between S_m and the true signal standard deviation σ for different measurement times T is shown in Figure 7.3 for two different examples of green noise, along with a white noise curve for reference, where B is the bandwidth ($B = \mathrm{f_{HIGH}} - \mathrm{f_{LOW}}$).

When the simulation (or measurement) time T is less than the central time period T_C, then the ratio S_m/σ for green (and red) noise approaches unity; thus, the signal standard deviation itself is the best estimator of the uncertainty in the mean value. This situation is undesirable since it simply says that the true mean lies somewhere within the variations seen in the signal (dependent on the desired confidence level).

Increasing the simulation (or measurement) time to be longer than T_C, but less than the longest period of variation T_{Long} gives:

$$\frac{s_m}{\sigma} = \sqrt{\frac{1}{2(\mathrm{f_{HIGH}} - \mathrm{f_{LOW}})T}}. \tag{7.6}$$

Since, in a discrete sense, $T = N\Delta t$, where N is the number of observations and Δt is the time step, Equation (7.6) gives the expected rate of reduction found in the previous section for uncorrelated white noise processes where $s_m/\sigma \propto \sqrt{\frac{1}{N}}$. In fact, if the number of observations is chosen as $N = 2(\mathrm{f_{HIGH}} - \mathrm{f_{LOW}})T$, or alternatively the time step is chosen as $\Delta t = \frac{1}{2(\mathrm{f_{high}} - \mathrm{f_{low}})}$, then one gets a similar result as in Section 7.3.1.1, namely $s_m/\sigma = \sqrt{1/N}$. (Note that the standard deviation of the mean S_m corresponds to a two-sided confidence level of approximately 68%.)

If the simulation (or measurement) time is chosen to be longer than the longest wavelength of the variations T_{Long}, then one gets:

$$\frac{s_m}{\sigma} = \frac{1}{\sqrt{2\pi T}} \sqrt{\frac{1}{\mathrm{f_{HIGH}} \mathrm{f_{LOW}}}}. \tag{7.7}$$

Again, since $T = N\Delta t$, Equation (7.7) says that the standard uncertainty in the mean scales as $1/N$, and thus reduces *faster* than the $\sqrt{1/N}$ rate seen for uncorrelated processes. In the usual case, when the simulation (or measurement) time is chosen to capture all of the relevant time scales, Equation (7.7) is appropriate to estimate the random uncertainty of the mean value. The key point is that to understand the uncertainty in the mean for a correlated time signal, the analyst must be able to estimate the range of frequencies contained in the signal.

Since the above development focuses on the standard uncertainty, i.e., the uncertainty in the sample mean, it would correspond to a two-sided confidence level of only ~68%. A coverage factor k is needed to achieve other confidence levels, e.g., $k = 2$ for 95% confidence. The actual approach recommended by Brouwer et al. (2019) does not require the estimation of $\mathrm{f_{LOW}}$ and $\mathrm{f_{HIGH}}$, but instead relies on the estimation of the auto-covariance function. This approach works for measurement times long enough so that Equation (7.7) is valid. They also discuss techniques for testing for stationarity of the signal and understanding which frequencies contribute most to the uncertainty of the mean value (see Brouwer et al., 2019 for details). An open source implementation of this approach can be found in Brouwer (2022).

7.4 Surrogate Modeling Error

When simulations are expensive and involve the propagation of uncertainty from the model inputs to the SRQs, surrogate models (or response surface approximations) are often used to approximate the simulation behavior over the range of inputs at a much lower cost (Queipo et al., 2005; Smith, 2014). They are also frequently used for design and optimization. Surrogate models thus provide a low-cost method for approximating the mapping of the uncertain input parameters to the SRQs, but they are usually only feasible for a modest number of input dimensions (on the order of 10). There are numerous approaches to creating surrogate models including polynomial regression, least squares fitting, Gaussian process regression (i.e., Kriging), and machine learning. A sampling strategy, such as Latin hypercube sampling or orthogonal arrays, is needed to obtain the training data from the simulation. Some surrogate models satisfy the training data points exactly (e.g., curve fitting), while other approaches (e.g., least squares) will only match the training data in an approximate sense. The choice of surrogate model will depend strongly on the amount of noise in the data (e.g., from numerical sources such as iterative and discretization error).

Various methods have been developed to estimate the error due to construction of a surrogate model (e.g., Jiang et al., 2020). Cross validation is often used to partition the available simulation results into training data and "validation" data in order to estimate the accuracy of the surrogate model (Geisser, 1993). Although this approach uses the term "validation," in our context it is really a verification strategy since it seeks to characterize the mathematical error between the actual simulation output and the surrogate. One popular cross validation method is the leave-one-out strategy. In this method, one begins with a data set of N points in the input space where the SRQ has been computed using the scientific computing code. The first error estimate is computed by creating the surrogate model with $N-1$ points, then calculating the error at the remaining point that was omitted. This process is repeated N times, each time leaving out a different point and calculating the error. The result is a distribution of errors over the N data points. For example, one may take the average absolute value error or the maximum error over the N points as a measure of the error (or uncertainty) in the surrogate model.

7.5 Numerical Error versus Numerical Uncertainty

All of the sources of numerical error that have been discussed in this chapter are correctly classified as errors. If the value (including the sign) of these errors is known with a high-degree of confidence, then these errors can either be removed from the numerical solutions (similar to the removal of well-characterized bias errors from an experimental measurement) or, if sufficiently small, simply neglected. In cases where numerical errors cannot be estimated with a great deal of reliability (as is generally the case in scientific computing), while they are still truly errors, our knowledge of these errors is uncertain. Thus, when these numerical errors are estimated, they should be treated as epistemic (i.e., lack of knowledge) uncertainties.

7.5.1 Converting Numerical Errors to Uncertainties

One of the simplest methods for converting an error estimate to an uncertainty is to use the magnitude of the error estimate to apply uncertainty bars about the simulation result, possibly with an additional factor of safety included. For example, the round-off error estimate $\bar{\varepsilon}_{\text{Round-Off}}$ discussed above in Equation (7.1) can be represented as a numerical uncertainty $U_{\text{Round-Off}}$ as

$$U_{\text{Round-Off}} = F_s|\bar{\varepsilon}_{\text{Round-Off}}|, \tag{7.8}$$

where $F_s \geq 1$ is the factor of safety (this is, in fact, a generalization of Roache's grid convergence index discussed in Chapter 9). The resulting interval for the numerical solution, accounting for numerical uncertainty, can be approximated by applying this uncertainty symmetrically about the solution, e.g.,

$$f_h \in f_{\text{Single}} \pm U_{\text{Round-Off}} = f_{\text{Single}} \pm F_s|\bar{\varepsilon}_{\text{Round-Off}}|. \tag{7.9}$$

These concepts are shown graphically in Figure 7.4 with a factor of safety of approximately $F_s = 1.5$. The figure shows the single-precision numerical solution f_{Single} as a signed error estimate $\bar{\varepsilon}_{\text{Round-Off}}$ and the uncertainty band that is created by taking plus/minus the absolute value of $\bar{\varepsilon}_{\text{Round-Off}}$ centered on the numerical solution. The factor of safety is needed

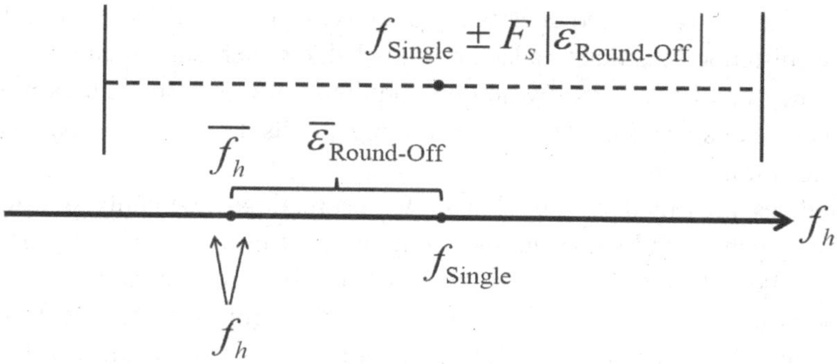

Figure 7.4 Example of converting a round-off error estimate into a numerical uncertainty centered about a single-precision numerical solution.
(adapted from Roy and Balch, 2012)

because $\bar{f}_h$ is only an approximation of the true numerical solution without round-off error, f_h. In other words, even when the error estimate is reasonably accurate, the true numerical solution f_h could still be slightly larger or slightly smaller than the estimated numerical solution $\bar{f}_h$, hence the two arrows in the figure. When the error estimate $\bar{\varepsilon}_{\text{Round-Off}}$ is poor, this heuristic approach is designed to still potentially provide conservative numerical uncertainty estimates, depending of course on the chosen factor of safety. We generally recommend that this uncertainty be centered about the actual numerical solution f_{Single} rather than the estimated exact solution $\bar{f}_h$.

7.5.2 *Estimating Total Numerical Uncertainty*

When the uncertainties due to all numerical error sources have been estimated, a quantitative assessment of the total numerical uncertainty "budget" can provide guidance as to the efficient allocation of resources for adjusting the numerical uncertainty. For example, if the total numerical uncertainty budget is estimated for a given class of problems, and the largest contributor to the total uncertainty is the spatial discretization error, then this suggests that an efficient use of resources would be to focus on refining the spatial mesh. Just as important, this analysis will also suggest areas where resources should not be focused (e.g., in reducing round-off errors), or even areas where cost-savings could be made (e.g., performing fewer iterations if the iterative error is already well below the desired level).

When multiple sources of numerical error are present, then one approach is to simply add the symmetric, two-sided numerical uncertainties together (Roy and Oberkampf, 2011). This is written as

$$U_{\text{Numerical}} = U_{\text{Round-Off}} + U_{\text{Statistical}} + U_{\text{Surrogate}} + U_{\text{Iterative}} + U_{\text{Discretization}}, \tag{7.10}$$

where $U_{\text{Round-Off}}$, $U_{\text{Statistical}}$, $U_{\text{Surrogate}}$, $U_{\text{Iterative}}$, and $U_{\text{Discretization}}$ refer to uncertainty due to round-off, statistical convergence, surrogate modeling, iteration, and discretization, respectively. Again, iterative error estimation is addressed in detail in Chapter 8 and discretization error estimation in Chapter 9. Equation (7.10) is guaranteed to be a conservative bound on the numerical uncertainty if the uncertainty estimates for each of the components is conservative (i.e., if the estimated uncertainty bounds the true error). Numerical uncertainties are generally epistemic (i.e., due to a lack of knowledge rather than inherent randomness) in nature since they can be reduced by adding additional information such as more digits of precision, more statistical samples, more training points for the surrogate model, more iterations, finer meshes, etc.

The techniques presented in this chapter for estimating uncertainty due to round-off, statistical convergence, and surrogate modeling are best used by applying the uncertainty symmetrically about the numerical solution. However, for cases where the iterative errors or discretization errors are reducing monotonically and the error estimates are deemed reasonably reliable, then it is easy to argue that the sign of these errors is known. In such cases, these uncertainties can be applied on only one side of the numerical solution. This approach is particularly useful for discretization errors, where techniques exist for assessing the reliability of the error estimates (see Chapter 9) and these uncertainties often dominate the other numerical uncertainty sources. In this case, upper and lower uncertainties can be established separately.

For example, if the solutions appear to be converging monotonically from below with grid refinement, then one can establish an upper bound for uncertainty due to discretization error $U_{\text{Discretization,Upper}}$ and set the lower bound $U_{\text{Discretization,Lower}}$ to zero. This approach now allows the upper and lower bound for the total numerical error to be more generally evaluated as:

$$U_{\text{Numerical,Upper}} = U_{\text{Round-Off}} + U_{\text{Statistical}} + U_{\text{Surrogate}} + U_{\text{Iterative,Upper}} + U_{\text{Discretization,Upper}}, \tag{7.11}$$

$$U_{\text{Numerical,Lower}} = U_{\text{Round-Off}} + U_{\text{Statistical}} + U_{\text{Surrogate}} + U_{\text{Iterative,Lower}} + U_{\text{Discretization,Lower}}. \tag{7.12}$$

Again, the numerical uncertainty due to round off, statistical convergence, and surrogate modeling should generally be applied symmetrically. The true exact solution to the mathematical model $\tilde{f}$ would then be estimated to lie in the range:

$$\tilde{f} \in \left[f_h - U_{\text{Numerical,Lower}},\ f_h + U_{\text{Numerical,Upper}}\right], \tag{7.13}$$

where f_h is the numerical solution.

7.6 Problems

1. Discuss procedures that could be used for verification of the various types of input data needed for a simulation.
2. Discuss procedures that could be used for verification of post-processing tools for simulation output data.
3. Discuss the two types of error that occur in the representation of floating point numbers on a digital computer.
4. Compute the confidence interval about the true mean for 90% confidence given that five statistically uncorrelated samples have been computed.

5. Explain why the numerical error in constructing a surrogate model rapidly increases as the number of simulation input dimensions increases.
6. Explain why a factor of safety greater than unity should be included in the estimation of numerical errors.

7.7 References

Baboulin, M., A. Buttari, J. Dongarra, J. Kurzak, J. Langou, J. Langou, et al. (2009). Accelerating Scientific Computations with Mixed Precision Algorithms. *Computer Physics Communications*, **180**(12), 2526–2533.

Bendat, J. S., and A. G. Piersol (2010). *Random Data: Analysis and Measurement Procedures*, 4th ed. New York, NY: John Wiley & Sons.

Brouwer, J. (2022). Open Source Python Code for Determining the Random Uncertainty in the mean. https://github.com/blueOceanSustainableSolutions/pyTST.

Brouwer, J., J. Tukker, Y. Klinkenberg, and M. Van Rijsbergen (2019). Random Uncertainty of Statistical Moments in Testing: Mean. *Ocean Engineering*, **182**, 563–576.

Devore, J. (2009). *Probability and Statistics for Engineering and the Sciences*, 7th ed. Belmont, CA: Brooks/Cole.

Ferziger, J. H., and M. Peric (2002). *Computational Methods for Fluid Dynamics*, 3rd ed. Berlin, Germany: Springer.

Geisser, S. (1993). *Predictive Inference*. New York, NY: Chapman & Hall.

GNU (2009). GNU Multiple Precision Arithmetic Library. http://gmplib.org.

IEEE (2008). *IEEE Standard for Floating-Point Arithmetic*. New York: Microprocessor Standards Committee, Institute of Electrical and Electronics Engineers Computer Society.

Jiang, P., Q. Zhou, and X. Shao (2020). *Surrogate Model-Based Engineering Design and Optimization*. Singapore: Springer.

Oberkampf, W. L., and C. J. Roy (2010). *Verification and Validation in Scientific Computing*, 1st ed. Cambridge: Cambridge University Press.

Queipo, N. V., R. T. Haftka, W. Shyy, T. Goel, R. Vaidyanathan, and P. K. Tucker (2005). Surrogate-Based Analysis and Optimization. *Progress in Aerospace Sciences*, **41**, 1–28.

Ross, S. M. (2021). *Introduction to Probability and Statistics for Engineers and Scientists*, 6th ed. London: Elsevier Academic Press.

Roy, C. J., and M. S. Balch (2012). A Holistic Approach to Uncertainty Quantification with Application to Supersonic Nozzle Thrust. *International Journal for Uncertainty Quantification*, **2**(4), 363–381.

Roy, C. J., and W. L. Oberkampf (2011). A Comprehensive Framework for Verification, Validation, and Uncertainty Quantification in Scientific Computing. *Computer Methods in Applied Mechanics and Engineering*, **200**(25–28), 2131–2144.

Smith, D. M. (2009). FMLIB, Fortran Library. http://myweb.lmu.edu/dmsmith/FMLIB.html.

Smith, R. C. (2014). *Uncertainty Quantification: Theory, Implementation, and Applications*. Philadelphia, PA: Society for Industrial and Applied Mathematics.

Stremel, P. M., M. R. Mendenhall, and M. C. Hegedus (2007). BPX – A Best Practices Expert System for CFD. AIAA Paper 2007-974.

8

Iterative Error

The truth is much too complicated to allow anything but approximations.

—John von Neumann

Contents of Chapter 8

Iterative error is the difference between the current approximate solution to an equation or system of equations and the exact solution. It occurs any time an iterative (or relaxation) method is used to solve algebraic equations. In scientific computing, the most common use of iterative methods is for solving the system of algebraic equations resulting from the discretization of a mathematical model. When this system of equations is nonlinear and implicit schemes are employed, then a linearization step must first be used. In most cases, each of these coupled algebraic equations contains only a few of the unknowns, and this type of system is referred to as a sparse system. For cases where the discretized equations are linear, direct solution methods may be used where the algebraic equations are solved exactly in a finite number of steps, which will recover the exact solution to the discrete equations (within round-off error). However, for nonlinear discretizations employing implicit algorithms, iterative methods are required. Even in the case of linear systems, for a large number of unknowns (which generally occurs for 3D scientific computing applications), the resulting system of algebraic equations can be solved to a sufficient level of iterative error much more efficiently using iterative methods than with direct solution methods.

This chapter begins with an introduction to iterative methods for linear and nonlinear systems of equations. The monitoring of iterative convergence for linear and nonlinear systems is discussed, and different types of iterative convergence are described. Different approaches to estimating the iterative error are then presented, followed by a discussion of the relationship between iterative residuals (usually used to monitor iterative convergence) and the iterative error. Finally, a practical approach to estimating iterative error is given, followed by some sample problems.

8.1 Iterative Methods

This section provides an overview of iterative methods for both single equations and systems of equations. The subject of iterative methods for systems of equations, especially those arising from the discretization of mathematical models, is an extremely broad subject. Here we provide only a limited overview of this topic, with some additional details provided for the simpler stationary iterative methods as well as for Newton's method. Further information on iterative methods can be found in the textbooks by Hageman and Young (1981), Hackbush (1994), Kelley (1995), and Saad (2003), as well as on the world-wide-web by Barrett, et al. (1994) and Saad (1996). A review of basic linear algebra and matrix analysis can be found in Golub and Van Loan (2013) and Meyer (2000).

Iterative methods can be used to solve both single equations and systems of equations. The iteration process itself seeks to find a series of approximate solutions to the equations by starting with an initial guess. The difference between these approximate solutions and the exact solution to the equations is the iterative error. This section begins by first discussing iterative methods applicable to a single equation with a single unknown, which sometimes occurs in scientific computing. Systems of equations which arise from the discretization of mathematical models are then discussed. The basic ideas behind Newton's method for implicitly solving systems of nonlinear equations via linearization are outlined. Next are considered linear systems of algebraic equations which can arise from either linear mathematical models or from a linearization of a nonlinear mathematical model. The techniques discussed for solving these linear systems include direct solution methods, stationary methods, Krylov-subspace methods, and hybrid methods. An example of a simple iterative method arising from a linear scientific computing application is then given.

8.1.1 Equations with a Single Unknown

For single equations, iterative methods are often used to find solutions (i.e., roots) when the equation cannot be solved algebraically. This situation commonly occurs for transcendental equations (i.e., those involving transcendental functions such as exponentials, logarithms, or trigonometric functions) and most polynomials of fifth order (quintic) and higher. For such equations, the simplest iterative solution approach is the direct substitution method (e.g., Chandra, 2003). Consider the equation $F(x) = 0$ which we will assume cannot be solved algebraically. This equation is rearranged, possibly by adding x to both sides, to obtain

$$x = g(x), \tag{8.1}$$

where $g(x)$ is some new function of x. Given a suitable initial guess, Equation (8.1) can be applied in an iterative fashion as

$$x^{k+1} = g\left(x^k\right),\ k = 1, 2, 3, \ldots, \tag{8.2}$$

where k is the *iteration number*. The non-zero result obtained by inserting the approximate solution x^k into the function

$$\mathfrak{R}^k = F\left(x^k\right) \tag{8.3}$$

is the *iterative residual* for a single equation, not to be confused with the continuous or discrete residuals discussed in Chapter 5. While extremely simple, the drawback to the direct substitution approach is that it will only converge when $|dg/dx| < 1$.

More sophisticated methods for iteratively solving a single algebraic equation are based on either bracketing a solution (root) of the equation or employing information on the function gradient dF/dx. The most basic bracketing approach is the bisection method, whereupon one starts with two initial guesses x_1 and x_2 for which $F(x_1) \cdot F(x_2) < 0$. This inequality ensures a solution will be within the interval; if the product is equal to zero then one of the endpoints itself is a solution. The interval is then repeatedly sub-divided using $x_3 = (x_1 + x_2)/2$, with the new interval containing the solution being the one for which the product is again less than or equal to zero. This process is repeated (iterated) until a solution is found which provides for $F(x)$ sufficiently close to zero, i.e., until sufficient convergence of the iterative residual. Convergence of this method is linear, meaning that the iterative error will reduce by a constant factor (in this case 0.5) each iteration. While this convergence rate is quite slow, the bisection method is guaranteed to converge to a solution.

There are more advanced iterative methods that incorporate function gradient information to allow faster convergence, but at the cost of a decrease in the robustness of the iteration (i.e., convergence is no longer guaranteed) (Süli and Mayers, 2006). In order of increasing efficiency, and therefore *decreasing* robustness, these methods can be summarized as:

1. The *false position method* – similar to bisection but employing gradient information to aid in bracketing the solution,
2. The *secant method* $x^{k+1} = x^k - \dfrac{x^k - x^{k-1}}{F(x^k) - F(x^{k-1})} F(x^k)$, and
3. *Newton's method* $x^{k+1} = x^k - \dfrac{F(x^k)}{\dfrac{dF}{dx}(x^k)}$.

The secant method is preferred for cases where the analytic gradient dF/dx is not well behaved or is undefined. When sufficiently close to a solution, Newton's method (also called Newton–Raphson) converges quadratically (see Section 8.2.3.1) while the secant method converges somewhat slower. However, far from a solution these two methods may not converge at all. For all of the above methods, convergence of the iterative procedure is monitored with the iterative residual given in Equation (8.3). A more detailed discussion on the convergence of iterative methods is presented in Section 8.2.

8.1.2 Systems of Equations

Systems of equations often arise from the discretization of mathematical models in scientific computing applications. Iterative methods are often, but not always, used when such systems of equations are large and/or nonlinear. For moderate size scientific computing applications in one or two spatial dimensions, the number of unknowns is usually between a thousand and several million. For larger 3D applications, computations are routinely performed on spatial grids with 500 million nodes/cells/elements or more. For mathematical models that employ multiple, coupled differential equations, it is not uncommon to have five or more unknowns per grid point, resulting in over 2 billion total unknowns. The largest scientific computing applications today often have up to 100 billion total unknowns.

Scientific computing applications do not always employ iterative methods. In some cases, direct solution methods can be used; in addition, when explicit methods are used to solve marching problems then no iterations are involved. To understand when iterative methods may be used, and therefore where iterative error may occur, one must first understand the linearity of the mathematical model (discussed in Chapter 6), the type of numerical algorithm employed (explicit or implicit), and the mathematical character of the discrete algorithm. For explicit algorithms (e.g., Euler explicit, explicit Runge–Kutta, Adams–Bashforth), each unknown can be computed as a function of known values, either from initial conditions or from a previous iteration/time step. Implicit algorithms require the solution of a simultaneous set of algebraic equations. The mathematical character of the discrete algorithm often matches that of the underlying mathematical model (see Section 6.1 for a more detailed discussion). In some cases, however, steady elliptic or mixed elliptic-hyperbolic problems are solved by marching in pseudo-time to the steady-state solution, thus converting them into a hyperbolic/parabolic system in time.

Mathematical models that are elliptic in nature are called boundary-value problems. Regardless of whether the algorithms involved are explicit or implicit, boundary-value problems are usually solved using iterative methods. The only exception is when linear

Table 8.1 *Use of iterative and direct solution methods in scientific computing based on the characteristics of the discrete model*

Mathematical character	Linearity	Explicit algorithm	Implicit algorithm
Initial-boundary value problem (hyperbolic/ parabolic)	**Linear**	No iteration	Iteration or direct
	Nonlinear	No iteration	Iteration
Boundary-value problem (elliptic and mixed)	**Linear**	Iteration	Iteration or direct
	Nonlinear	Iteration	Iteration

elliptic problems are solved implicitly, in which case direct solution methods may be used, although these tend to be expensive for even moderate-sized systems. Hyperbolic and parabolic mathematical models are called initial-boundary value problems and are solved via marching techniques, and the marching direction is usually (but not always) time. When explicit algorithms are used to solve initial-boundary value problems, then no iteration is needed. Initial-boundary value problems solved with implicit algorithms require iterative methods unless the problem is linear, in which case a direct solution approach can also be used. Implicit algorithms applied to time dependent problems almost always require the use of iterations within each time step, usually referred to as sub-iterations. Eca et al. (2019) provide a discussion of the iterative errors associated with these sub-iterations and suggest that the convergence tolerances seen in the literature are often insufficient to guarantee accurate simulations. The situations where iterative methods can be used, and thus where iterative error will be present, are summarized in Table 8.1.

8.1.2.1 Nonlinear Systems of Equations

The most common situation for scientific computing applications is for the discretization of the mathematical model to result in a nonlinear system of equations to be solved. As shown in Table 8.1, when explicit algorithms are applied to nonlinear boundary value problems, then iterations are needed to approach the exact solution to the discrete equations, which will never be reached due to the presence of round-off error. For implicit algorithms to be applied to nonlinear systems of equations (regardless of the mathematical character), then a linearization step is needed (e.g., from a Picard iteration or Newton's method). We will address the fundamentals of Newton's method and follow closely the development of Kelley (2003).

Referring to the operator notation from Chapter 5, the (nonlinear) set of discrete equations is $L_h(u) = 0$ which is solved exactly by u_h. If we consider u_h^k to be the kth iterate of the numerical solution, then performing an expansion of our discrete operator $L_h(\cdot)$ at the $k+1$ iterate about the kth iterate gives:

$$L_h\left(u_h^{k+1}\right) = L_h\left(u_h^k\right) + \left.\frac{\partial L_h}{\partial u}\right|_{u_h^k}\left(u_h^{k+1} - u_h^k\right) + O\left[\left(u_h^{k+1} - u_h^k\right)^2\right]. \tag{8.4}$$

The term $\left.\frac{\partial L_h}{\partial u}\right|_{u_h^k}$ is the Jacobian of the discrete equations and is a (usually large) matrix which linearizes the discrete operator about the current iterative solution. Newton's method seeks updated solutions which, when neglecting the higher-order terms, will drive the right-hand side of Equation (8.4) to zero, i.e.,

$$\left.\frac{\partial L_h}{\partial u}\right|_{u_h^k} \Delta u_h^{k+1} = -L_h\left(u_h^k\right). \tag{8.5}$$

For the full Newton step $\Delta u_h^{k+1} = u_h^{k+1} - u_h^k$ is the update to be applied to the solution at iteration k to get it to the $k+1$ level. A benefit of using Newton's method with exact, analytic Jacobians is that one may achieve quadratic convergence (see Section 8.2). Note that applying Newton's method to a nonlinear operator results in a linear system to be solved at each nonlinear iteration, thus the approaches for linear systems discussed in the next section also apply.

In practice, the step size ($\Delta u_h^{k+1} = u_h^{k+1} - u_h^k$) should be controlled to ensure that the norm of the nonlinear residual at iteration $k+1$ is smaller than the value at k, i.e.,

$$\left\|L_h\left(u_h^{k+1}\right)\right\| < \left\|L_h\left(u_h^k\right)\right\|. \tag{8.6}$$

When the inequality in Equation (8.6) is violated for the full Newton step, a simple method to control the step size is to update the iterative solution at $k+1$ according to

$$u_h^{k+1} = u_h^k + \lambda \Delta u_h^{k+1}, \tag{8.7}$$

where λ starts at unity and is reduced by a factor of two until the inequality in (8.6) is satisfied. More sophisticated methods of step size control (such as line searches) are presented by Kelley (2003).

8.1.2.2 Linear Systems of Equations

The remainder of this subsection will provide an overview of the different methods for solving linear systems of algebraic equations. The linear systems can arise from either linear or nonlinear mathematical models, with the latter requiring an additional linearization step as discussed earlier. A basic understanding of linear algebra is assumed for the following discussion (e.g., see Meyer, 2000).

Consider a linear system of algebraic equations of the form

$$A\vec{x} = \vec{b}, \tag{8.8}$$

where A is an $N \times N$ matrix of scalars, $\vec{b}$ a column vector of length N, $\vec{x}$ the desired solution vector of length N, and N is the total number of unknowns. Note that when discussing linear systems, we will employ the vector sign, whereas for the general discretized equations $L_h(u) = 0$, the vector sign is implied. In general, linear systems arising from the discretization of mathematical models will be sparse, meaning that most of the entries in the A matrix are zeroes. When the total number of unknowns N is sufficiently small, then direct solution techniques can be used (see Section 8.1.2.2.1). For large linear systems, iterative methods are much more efficient. The basic concept behind iterative methods is to first make an initial guess at the solution to Equation (8.8) and then make successive approximations to improve the solution until satisfactory iterative convergence is reached. Because the approximate solutions will not exactly satisfy Equation (8.8), convergence of these iterative methods is generally determined by monitoring the difference between the right- and left-hand sides of the equation. The iterative residual associated with the approximate solution to the linear equations (again with the vector sign included) at iteration k is thus defined as

$$\vec{\mathfrak{R}}^k = \vec{b} - A\vec{x}^k. \tag{8.9}$$

Stationary iterative methods are discussed in some detail in Section 8.1.2.2.2, whereas only a broad overview of the more advanced Krylov subspace methods is given in Section 8.1.2.2.3. In addition, hybrid approaches that combine the stationary and Krylov subspace methods are also mentioned in Section 8.1.2.2.4. This section concludes with an example of some stationary iterative methods applied to a linear mathematical model.

8.1.2.2.1 Direct Solution Methods

Direct solution methods are those which, neglecting round-off error, will produce the exact solution to the linear system given by Equation (8.8) in a finite number of steps. In general, direct methods require on the order of N^3 operations (additions, subtractions, multiplications, and divisions) to find the solution, where N is the total number of equations to be solved. Unless the sparse structure of the linear system can be used to reduce this operation count, direct solution methods are prohibitively expensive for large scientific computing applications and iterative methods should be employed instead. The use of iterative methods is further supported by the fact that one is generally not required to drive the iterative error to zero, but only down to an acceptably small level.

An example of a direct solution method which makes efficient use of the sparse matrix structure is the *Thomas algorithm*. When the matrix A contains nonzero entries only along the three diagonals L_T, D, and U_T, e.g.,

$$A = \begin{bmatrix} D & U_T & 0 & \cdots & \cdots & \cdots & 0 \\ L_T & D & U_T & \ddots & & & \vdots \\ 0 & L_T & D & U_T & \ddots & & \vdots \\ \vdots & \ddots & \ddots & \ddots & \ddots & \ddots & \vdots \\ \vdots & & \ddots & L_T & D & U_T & 0 \\ \vdots & & & \ddots & L_T & D & U_T \\ 0 & \cdots & \cdots & \cdots & 0 & L_T & D \end{bmatrix} \tag{8.10}$$

then a simplification of the standard direct solution technique of Gaussian elimination followed by back substitution can be used. The Thomas algorithm only requires on the order of N operations, an extraordinary improvement in computational efficiency versus N^3 operations for other direct solution approaches. This tridiagonal matrix structure commonly arises during the discretization of 1D mathematical models using second-order accurate methods. A similar approach called the block tridiagonal solver can be used when the L_T, D, and U_T entries themselves are square matrices (Saad, 2003).

The approximate number of arithmetic operations (i.e., the operation count) for some common direct solution techniques is given in Table 8.2. To get an idea of the computational

Table 8.2 *Operation count for some direct solution techniques*

Direct solution technique	Approximate operation count	Approximate operations for $N = 1{,}000$	Approximate operations for $N = 1$ million
Thomas algorithm[a]	$8N$	8×10^3	8×10^5
Gaussian elimination	$\frac{2}{3}N^3$	7×10^8	7×10^{17}
LU decomposition	$\frac{2}{3}N^3$	7×10^8	7×10^{17}
Gauss-Jordan	N^3	1×10^9	1×10^{18}
Matrix inversion	$2N^3$	2×10^9	2×10^{18}
Cramer's rule	$(N+1)!$	4.0×10^{2570}	8.3×10^{5565714}

[a] Essentially Gaussian elimination, but optimized for tridiagonal systems

time for these direct solution methods, today's modern desktop processors can perform computations at around 100 gigaflops, meaning roughly 10^{11} floating point operations per second. Thus, with 1 million unknowns, the Thomas algorithm would require approximately 0.00008 seconds while Gaussian elimination would require 11.5 weeks. It is clear that as the total number of unknowns, N, becomes large, the direct solution techniques are no longer feasible unless a sparse matrix structure can be exploited.

8.1.2.2.2 Stationary Iterative Methods

Stationary iterative methods use approximations of the full linear matrix operator A to successively improve the solution. By splitting the matrix as $A = M - N$, these methods can be written in the form

$$M\vec{x}^{k+1} = N\vec{x}^{k} + \vec{b}, \tag{8.11}$$

where k refers to the iteration number and the matrices M and N and the vector $\vec{b}$ are generally constant (i.e., they do not depend on the iteration number). In the limit as the iterations converge, $\vec{x}^{k} = \vec{x}^{k+1}$, and thus the original linear system given in Equation (8.8), is recovered. Note that for some approximations to nonlinear systems, the matrices M and N and the vector $\vec{b}$ may in fact be updated during the iteration process. Examples of stationary iterative methods include Jacobi iteration, Gauss–Seidel iteration, and multigrid methods.

Left-multiplying Equation (8.11) by the inverse of M results in

$$\vec{x}^{k+1} = M^{-1}N\vec{x}^{k} + M^{-1}\vec{b} \tag{8.12}$$

or simply

$$\vec{x}^{k+1} = G\vec{x}^{k} + M^{-1}\vec{b}, \tag{8.13}$$

where $G = M^{-1}N$ is called the iteration matrix. As will be discussed in Section 8.2, the convergence rate of stationary iterative methods depends on the properties of this iteration matrix.

Some of the simpler stationary iterative methods can be developed by splitting the matrix A into its diagonal D, a lower-triangular matrix L, and an upper triangular matrix U:

$$A = \begin{bmatrix} D & U & \cdots & \cdots & U \\ L & D & \ddots & & \vdots \\ \vdots & \ddots & \ddots & \ddots & \vdots \\ \vdots & & \ddots & D & U \\ L & \cdots & \cdots & L & D \end{bmatrix} = \begin{bmatrix} D & 0 & \cdots & \cdots & 0 \\ 0 & D & \ddots & & \vdots \\ \vdots & \ddots & \ddots & \ddots & \vdots \\ \vdots & & \ddots & D & 0 \\ 0 & \cdots & \cdots & 0 & D \end{bmatrix} + \begin{bmatrix} 0 & 0 & \cdots & \cdots & 0 \\ L & 0 & \ddots & & \vdots \\ \vdots & \ddots & \ddots & \ddots & \vdots \\ \vdots & & \ddots & 0 & 0 \\ L & \cdots & \cdots & L & 0 \end{bmatrix}$$

$$+ \begin{bmatrix} 0 & U & \cdots & \cdots & U \\ 0 & 0 & \ddots & & \vdots \\ \vdots & \ddots & \ddots & \ddots & \vdots \\ \vdots & & \ddots & 0 & U \\ 0 & \cdots & \cdots & 0 & 0 \end{bmatrix}. \tag{8.14}$$

With this splitting, the Jacobi method can be defined with $M = D$ and $N = -L - U$, resulting in the following iterative scheme:

$$D\vec{x}^{k+1} = -(L + U)\vec{x}^{k} + \vec{b}. \tag{8.15}$$

The choice of $M = D$ decouples the equations at each iteration, thus resulting in an explicit iterative scheme where each component of the solution vector x_i^{k+1} depends only on known values from the previous iteration x_i^k.

The Gauss–Siedel method is similar; however, it uses the most recently updated information available during the solution. For a forward sweep through the unknowns, the Gauss–Siedel method uses the splitting $M = D + L$ and $N = -U$ to give

$$(D + L)\vec{x}^{k+1} = -U\vec{x}^{k} + \vec{b}. \tag{8.16}$$

A symmetric Gauss–Seidel method would employ a forward sweep with

$$(D + L)\vec{x}^{k+1/2} = -U\vec{x}^{k} + \vec{b} \tag{8.17}$$

followed by a backward sweep with

$$(D + U)\vec{x}^{k+1} = -L\vec{x}^{k+1/2} + \vec{b}, \tag{8.18}$$

where $k = 1/2$ indicates an intermediate iteration step. All three of these Gauss–Seidel iteration methods can be performed explicitly, i.e., without solving for multiple unknowns at the same time.

Another technique that should be mentioned since it has an impact on the convergence of stationary iterative methods is over/under-relaxation. Rather than employing the stationary iterative methods as discussed previously, one instead replaces the solution components found by applying the standard iterative method x_i^{k+1} with the following

$$\hat{x}_i^{k+1} = x_i^k + \omega\left(x_i^{k+1} - x_i^k\right), \tag{8.19}$$

where $0 < \omega < 2$. Setting the relaxation parameter ω to one recovers the baseline iterative methods, whereas values greater than one or less than one result in over-relaxation or under-relaxation, respectively. When implemented with the forward Gauss–Seidel approach and with $\omega > 1$, this approach is called successive over-relaxation (SOR). The splitting for the SOR method is $M = D + \omega L$ and $N = -\omega U + (1 - \omega)D$ which can be written in matrix form as:

$$(D + \omega L)\vec{x}^{k+1} = [\omega U + (1 - \omega D)]\vec{x}^{k} + \omega\vec{b}. \tag{8.20}$$

In certain cases, the optimal relaxation factor can be estimated from properties of the iteration matrix (Saad, 1996).

8.1.2.2.3 Krylov Subspace Methods

In order to understand the basic ideas behind Krylov subspace methods, a quick review of matrix theory (Meyer, 2000) is first required. A vector space is a mathematical construct which includes definitions of vector addition and scalar multiplication. A vector subspace is any non-empty subset of a vector space. A basis for a subspace is a linearly independent set of vectors that can be linearly combined to form any vector in the subspace. The span of a set of vectors is the subspace formed by all linear combinations of those vectors. When a set of spanning vectors are linearly independent, they also form a basis for that subspace.

Krylov subspace methods, also called nonstationary iterative methods, are popular for large sparse systems (Saad, 2003). They work by minimizing the iterative residual defined by Equation (8.9) over the Krylov subspace. This subspace is the vector space spanned by the set of vectors formed by premultiplying an initial iterative residual vector by the matrix A to various powers. For any positive integer $m < N$ (where N is the number of unknowns), this space is defined as

$$K_m\left(A, \vec{\mathfrak{R}}^{k=0}\right) = span\left\{\vec{\mathfrak{R}}^{k=0}, A\vec{\mathfrak{R}}^{k=0}, A^2\vec{\mathfrak{R}}^{k=0}, \ldots, A^{m-1}\vec{\mathfrak{R}}^{k=0}\right\}. \tag{8.21}$$

An advantage of the Krylov subspace methods is that they can be implemented such that they only require matrix–vector multiplication and avoid expensive matrix–matrix multiplication.

The most commonly used Krylov subspace methods are the conjugate gradient method (for symmetric matrices only), the biconjugate gradient method, the stabilized biconjugate gradient method, and the generalized minimum residual method (GMRES) (Saad, 2003). Because the sequence of vectors given in Equation (8.21) tends to become nearly linearly dependent, Krylov subspace methods generally employ an orthogonalization procedure such as Arnoldi orthogonalization (for the conjugate gradient and GMRES methods) or a Lanczos biorthogonalization (for the biconjugate gradient and stabilized biconjugate gradient methods) (Saad, 2003). Krylov subspace methods are technically direct solution methods since they form a basis in the vector space which contains the solution vector $\vec{x}$. As a result, they converge (within round-off error) in N steps. However, since the number of unknowns N is usually large for scientific computing applications, the process is almost always terminated early when the iterative residuals are sufficiently small, resulting in an iterative method.

8.1.2.2.4 Hybrid Methods

Stationary and nonstationary iterative methods are often combined to take advantage of the strengths of each method. Preconditioned Krylov subspace methods premultiply the standard linear system given in Equation (8.8) by an approximate inverse of the A matrix. For example, a Jacobi preconditioner is $P = D$ (the diagonal of matrix A), and the preconditioned system becomes

$$P^{-1}A\vec{x} = P^{-1}\vec{b}. \tag{8.22}$$

The preconditioner should be chosen such that P^{-1} is inexpensive to evaluate and the resulting system is easier to solve (Saad, 2003). Alternatively, the basic concepts behind Krylov subspace methods can be applied to the vector basis formed by linearly combining the sequence of iterative solutions $\vec{x}^k$ produced by a stationary iterative method. Hageman and Young (1981) refer to this process as polynomial acceleration.

Example: Direct and Iterative Methods for Steady Heat Conduction

Consider the steady-state form of the 2D heat conduction equation, which results in the following elliptic boundary value problem:

$$\frac{\partial^2 T}{\partial x^2} + \frac{\partial^2 T}{\partial y^2} = 0. \tag{8.23}$$

Discretizing with second-order accurate spatial finite differences on a Cartesian mesh results in

$$\frac{T_{i+1,j} - 2T_{i,j} + T_{i-1,j}}{\Delta x^2} + \frac{T_{i,j+1} - 2T_{i,j} + T_{i,j-1}}{\Delta y^2} = 0. \tag{8.24}$$

When this discrete equation is applied over each of the grid nodes, a linear matrix system results with a pentadiagonal structure, i.e.,

$$A = \begin{bmatrix} D & U_T & 0 & \cdots & 0 & U_P & 0 & \cdots & 0 \\ L_T & D & U_T & \ddots & & \ddots & U_P & \ddots & \vdots \\ 0 & L_T & D & U_T & \ddots & & \ddots & \ddots & 0 \\ \vdots & \ddots & \ddots & \ddots & \ddots & \ddots & & \ddots & U_P \\ 0 & & \ddots & \ddots & \ddots & \ddots & \ddots & & 0 \\ L_P & \ddots & & \ddots & \ddots & \ddots & \ddots & \ddots & \vdots \\ 0 & \ddots & \ddots & & \ddots & L_T & D & U_T & 0 \\ \vdots & \ddots & L_P & \ddots & & \ddots & L_T & D & U_T \\ 0 & \cdots & 0 & L_P & 0 & \cdots & 0 & L_T & D \end{bmatrix}. \tag{8.25}$$

While this pentadiagonal matrix could be solved directly, and therefore with no iterative error, it is usually more efficient to use an iterative technique, especially for large systems. Introducing the superscript k to denote the iteration number, a line implicit discretization can be defined in the i direction:

$$\frac{T_{i+1,j}^{k+1} - 2T_{i,j}^{k+1} + T_{i-1,j}^{k+1}}{\Delta x^2} + \frac{T_{i,j+1}^{k} - 2T_{i,j}^{k+1} + T_{i,j-1}^{k}}{\Delta y^2} = 0. \tag{8.26}$$

It is observed in this equation that the pentadiagonal contributions (L_p and U_p) to the matrix structure (at $j \pm 1$) are evaluated at the known iteration level k. This discretization results in a tridiagonal system of equations for the unknowns at iteration level $k + 1$ with the same structure as the matrix given in Equation (8.10), which can again be solved efficiently with the Thomas algorithm.

This line implicit iterative scheme can be written in matrix form as follows. Consider a splitting of the full matrix A from Equation (8.25) as

$$A = L_P + L_T + D + U_T + U_P, \tag{8.27}$$

where D represents the diagonal terms (i, j), L_T the $(i-1, j)$ terms, L_P the $(i, j-1)$ terms, U_T the $(i+1, j)$ terms, and U_P the $(i, j+1)$ terms. Equation (8.26) is thus equivalent to splitting the A matrix by separating out the pentadiagonal terms, i.e.,

$$M = (L_T + D + U_T), \quad N = -(L_P + U_P). \tag{8.28}$$

The iteration scheme can then be written as

$$(L_T + D + U_T)\vec{T}^{k+1} = -(L_P + U_P)\vec{T}^{k} + \vec{b} \tag{8.29}$$

(where here $\vec{b} = 0$) or simply

$$\vec{T}^{k+1} = (L_T + D + U_T)^{-1}\left[-(L_P + U_P)\vec{T}^{k} + \vec{b}\right] = M^{-1}\left[N\vec{T}^{k} + \vec{b}\right]. \tag{8.30}$$

As discussed in Section 8.2, the convergence of this iterative scheme is governed by the eigenvalues of the iteration matrix $G = M^{-1}N$.

Note that even for steady-state problems, it is common in many scientific computing algorithms to include a temporal (or pseudo-time) term in order to help stabilize the iteration process. This stabilization occurs because the temporal terms tend to increase the diagonal dominance of the resulting iteration matrix. These temporal terms effectively serve as an under-relaxation technique, with the under-relaxation factor approaching unity as the time step is increased. When temporal terms are included for steady-state problems, one must keep in mind that it is only the steady-state portion of the iterative residuals that should be monitored for iterative convergence. In other words, the iterative residuals (discussed in detail in the next section) should omit any temporal terms for steady problems.

8.2 Iterative Convergence

Convergence of an iterative method is generally assessed by inserting the current iterative solution into the discrete equations, with the resulting imbalance called the iterative residual. Recall from Chapter 5 that the discrete equations can be written in the form

$$L_h(u_h) = 0, \tag{8.31}$$

where L_h is the linear or nonlinear discrete operator and u_h is the exact solution to the discrete equations. The iterative residual is found by substituting the current iterative solution u_h^k into Equation (8.31), i.e.,

$$\mathfrak{R}_h^k = L_h\left(u_h^k\right), \tag{8.32}$$

where $\mathfrak{R}_h^k \to 0$ as $u_h^k \to u_h$. This iterative residual is a vector quantity (although we have omitted the vector sign) and applies for both linear and nonlinear discretizations. Note that although it appears to be evaluated in a similar fashion, this iterative residual is *completely different* than the discrete and continuous residuals discussed in Chapter 5, and this can be a source of much confusion. The residuals discussed in Chapter 5 were due to discretization of the PDE, whereas the iterative residual discussed here deals with the convergence of the iterative solution to the discrete equations.

In scientific computing, one typically assesses iterative convergence by monitoring the norm of the iterative residual vector over all of the nodes/cells/elements relative to its initial value, i.e.,

$$\frac{\left\|\mathfrak{R}_h^k\right\|}{\left\|\mathfrak{R}_h^{k=0}\right\|}. \tag{8.33}$$

When multiple governing equations are being solved (e.g., conservation of mass, momentum, and energy), then this residual norm ratio should be monitored separately for each governing equation. The iterative residuals can typically be reduced as much as 6–7 orders of magnitude for single precision and 12–13 orders of magnitude for double precision before they level off due to the presence of round-off error. However, if the equations are stiff and/or the Jacobian is poorly conditioned, this leveling off (or "stagnation") can occur much sooner (Kelley, 2003).

8.2.1 Linear Systems

This subsection provides a brief discussion of the convergence of stationary iterative methods. See Saad (2003) for a discussion of the convergence of Krylov subspace methods. For stationary iterative methods applied to linear systems, iterative convergence is governed by the eigenvalues of the iteration matrix. Recall that M and N come from the splitting of the full matrix A in Equation (8.8) into $A = M - N$. The iteration matrix G comes from left-multiplying Equation (8.11) by M^{-1}, i.e.,

$$\vec{x}^{k+1} = M^{-1}N\vec{x}^{k} + M^{-1}\vec{b}, \tag{8.34}$$

and can thus be expressed as $G = M^{-1}N$. Specifically, convergence of the iterative scheme is related to the eigenvalue of largest magnitude λ_G for the iteration matrix G. The absolute value of this eigenvalue is also called the *spectral radius* of the matrix

$$\rho(G) = |\lambda_G|. \tag{8.35}$$

Convergence of the iterative method requires that the spectral radius be less than one, and the closer to unity the slower the convergence rate of the method (Golub and Van Loan, 2013). A crude approximation of the spectral radius of a matrix (Meyer, 2000) can be found by using the fact that the spectral radius is bounded from above by any valid matrix norm, i.e.,

$$\rho(G) \le \|G\|. \tag{8.36}$$

For the Jacobi and Gauss–Seidel methods, convergence is guaranteed if the matrix A is strictly diagonally dominant, meaning that the magnitude of the diagonal entry of each row is larger than the sum of the magnitudes of the other entries in the row, i.e.,

$$|A_{ii}| > \sum_{i \neq j} |A_{ij}|. \tag{8.37}$$

For linear problems, when the maximum eigenvalue λ_G of the iteration matrix G is real, then the limiting iterative convergence behavior will be monotone. When λ_G is complex, however, the limiting iterative convergence behavior will generally be oscillatory (Ferziger and Peric, 1996). Different iterative convergence behaviors are discussed in Section 8.2.3.

8.2.2 Nonlinear Systems

Convergence for most approaches for solving nonlinear systems of equations is determined by local convergence theory. We consider here the convergence theory for Newton's method. Assuming the initial iterate $u_h^{k=0}$ is sufficiently close to the solution u_h, Newton's method will converge under the standard assumptions, namely that the discrete solution u_h exists, that the Jacobian $\frac{\partial L_h}{\partial u}\big|_{u_h}$ is Lipschitz continuous near u_h, and that the Jacobian is non-singular (see Kelley, 2003). In addition, it is generally assumed that the resulting linear system is solved to sufficient accuracy during each Newton step, although this requirement is often relaxed in practice.

8.2.3 Types of Iterative Convergence

This section describes the different types of iterative convergence that may be observed. For simple scientific computing applications, iterative convergence is often found to be predominantly monotone. For more complex applications, iterative convergence can also be oscillatory or a complex mixture of monotone and oscillatory.

8.2.3.1 Monotone Convergence

Monotone iterative convergence occurs when subsequent iterative solutions approach the exact solution to the discrete equations in a monotone fashion with no local minima or maxima in the iterative error as a function of the iteration number. The monotone convergence rate is usually either linear or quadratic. Linear convergence occurs when the ratio of iterative error (defined formally in Section 8.3) from one iteration to the next is approximately constant, i.e.,

$$\varepsilon_h^{k+1} = C\varepsilon_h^k \text{ where } C < 1. \tag{8.38}$$

Thus, the iterative error is reduced by the same fraction C for every iteration. Most scientific computing codes converge at best at a linear rate. Quadratic convergence occurs when the iterative errors drop with the square of the iterative errors from the previous iterations (Süli and Mayers, 2006), i.e.,

$$\varepsilon_h^{k+1} = C\left(\varepsilon_h^k\right)^2 \text{ where } C < 1. \tag{8.39}$$

For nonlinear governing equations, quadratic convergence can be achieved for advanced iterative techniques (e.g., full Newton's method) as long as the current iterative solution is in

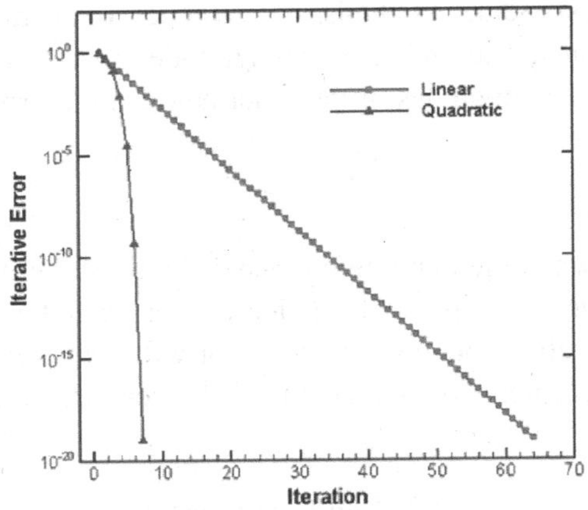

Figure 8.1 Example of monotone iterative convergence showing both linear and quadratic behavior.

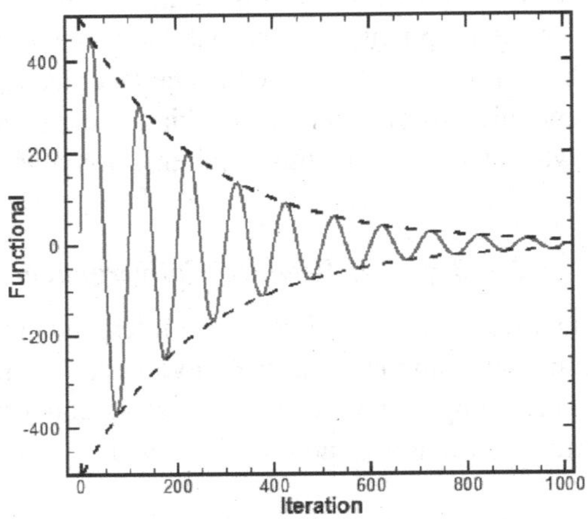

Figure 8.2 Example of oscillatory iterative convergence of a solution functional versus iteration number.

the vicinity of the exact solution to the discrete equations and the standard assumptions given in Section 8.2.2 hold (Kelley, 2003). Examples of linear and quadratic monotone convergence are given in Figure 8.1, which shows the iterative error on a logarithmic scale as a function of the number of iterations. Although clearly desirable, quadratic convergence is difficult to obtain for practical scientific computing problems. In some cases, monotone iterative convergence can be accelerated by using over-relaxation ($\omega > 1$) as discussed for Equation (8.19).

8.2.3.2 Oscillatory Convergence

Oscillatory iterative convergence occurs when the iterative solution converges in an oscillatory manner to the exact solution to the discrete equations. An example of oscillatory iterative convergence is given in Figure 8.2, which shows the iterative convergence error in a general solution functional reducing in an oscillatory but convergent manner. Oscillatory

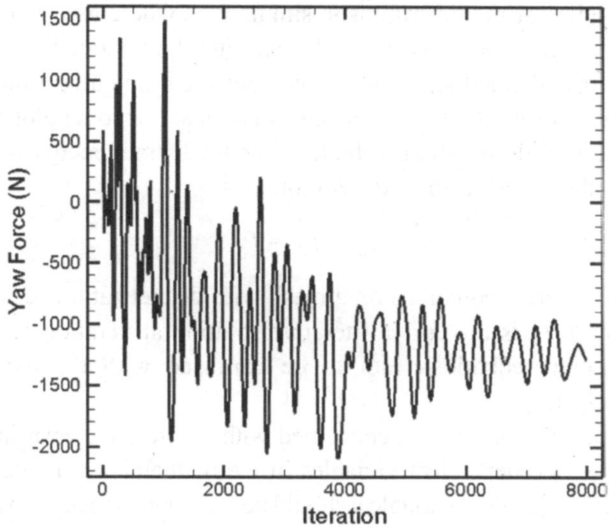

Figure 8.3 Example of general iterative convergence for the yaw force in an inviscid CFD simulation of the flow over a missile.
(adapted from Roy et al., 2007)

convergence occurs when the largest eigenvalues of the iteration matrix $G = M^{-1}N$ are complex (Ferziger and Peric, 1996). In some cases, oscillatory iterative convergence can be accelerated by using under-relaxation ($\omega < 1$) as discussed for Equation (8.19).

8.2.3.3 General Convergence

General iterative convergence occurs when the iterative solutions converge in a complicated manner and exhibit a mixture of monotone and oscillatory convergence. An example of general convergence is given in Figure 8.3 which shows the iteration history for the yaw force on a missile geometry from an inviscid CFD simulation. This complex type of iterative convergence history is not uncommon in many scientific computing simulations. The source of the oscillations for general convergence may be physical (e.g., strong physical instabilities or unsteadiness in a problem) or numerical (e.g., the activation and deactivation of a numerical flux or slope limiter).

8.3 Iterative Error Estimation

While monitoring the iterative residuals often serves as an adequate indication as to whether iterative convergence has been achieved, it does not by itself provide any guidance as to the size of the iterative error in an SRQ of interest. In this section, we discuss approaches for estimating the iterative error in arbitrary solution quantities. These approaches can be used to obtain a quantitative estimate of the iterative error. If the numerical solution is to be used in a code order verification study (Chapter 5) or a discretization error study (Chapter 9), the recommended iterative error levels should be 100 times smaller than the discretization error levels so as not to adversely impact the discretization error evaluation/estimation process. For cases where the solution will not be used for discretization error evaluation/estimation (e.g., parametric studies or nondeterministic simulations where a large number of solutions are required), then larger iterative error levels can often be tolerated.

The iterative error is defined as the difference between the current numerical solution at iteration k and the exact solution to the discrete equations. For some global solution property f, for example, the total lift on a flight vehicle or the total strain energy in a loaded structure, we can thus define the iterative error at iteration k as

$$\varepsilon_h^k = f_h^k - f_h. \tag{8.40}$$

Here h refers to the discrete equations on a mesh with discretization parameters (Δx, Δy, Δt, etc.) represented by h, f_h^k is the current iterative solution at iteration k, and f_h is the exact solution to the discrete equations (not to be confused with the exact solution to the mathematical model $\tilde{f}$).

In some cases, we might instead be concerned with the iterative error in the entire solution over the domain (i.e., the dependent variables in the mathematical model). In this case, the iterative error for *each* dependent variable u should be measured as a norm over the domain, e.g.,

$$\varepsilon_h^k = \left\| u_h^k - u_h \right\|, \tag{8.41}$$

where u_h^k and u_h are column vectors composed of a single dependent variable at each node/cell/element in the domain. When multiple dependent variables are included (e.g., x-velocity, y-velocity, and pressure in a fluid dynamics simulation), then these iterative errors should be monitored separately. The common norms employed include discrete L_1, L_2, and L_∞ norms (see Chapter 5 for norm definitions), but iterative residuals are most commonly monitored with L_2 norms for smooth solutions and L_1 norms for solutions with discontinuities. Note that the standard texts on iterative methods often do not differentiate between the different dependent variables, and thus lump all dependent variables at all domain locations into the unknown variable vector $\vec{x}$. We will switch between the scientific computing notation u_h and the linear systems notation $\vec{x}$ as necessary.

8.3.1 Theoretical Convergence for Newton's Method

Kelley (2003) provides a relationship between the iterative error and the nonlinear iterative residual for Newton's method. If the standard assumptions (given in Section 8.2.2) hold and the initial solution is sufficiently near the exact solution to the (nonlinear) discrete equations u_h, then:

$$\frac{\left\|\varepsilon_h^k\right\|}{4\kappa\left\|\varepsilon_h^{k=0}\right\|} \le \frac{\left\|\mathfrak{R}_h^k\right\|}{\left\|\mathfrak{R}_h^{k=0}\right\|} \le \frac{4\kappa\left\|\varepsilon_h^k\right\|}{\left\|\varepsilon_h^{k=0}\right\|}. \tag{8.42}$$

Here κ is the condition number of the Jacobian $\frac{\partial L_h}{\partial u_h}$ for the given norm, defined as:

$$\kappa = \left\|\frac{\partial L_h}{\partial u_h}\right\| \cdot \left\|\left(\frac{\partial L_h}{\partial u_h}\right)^{-1}\right\|.$$

Thus, for a well-conditioned system where the condition number of the Jacobian is near unity, reductions in the iterative residual should track very well with reductions in the iterative error.

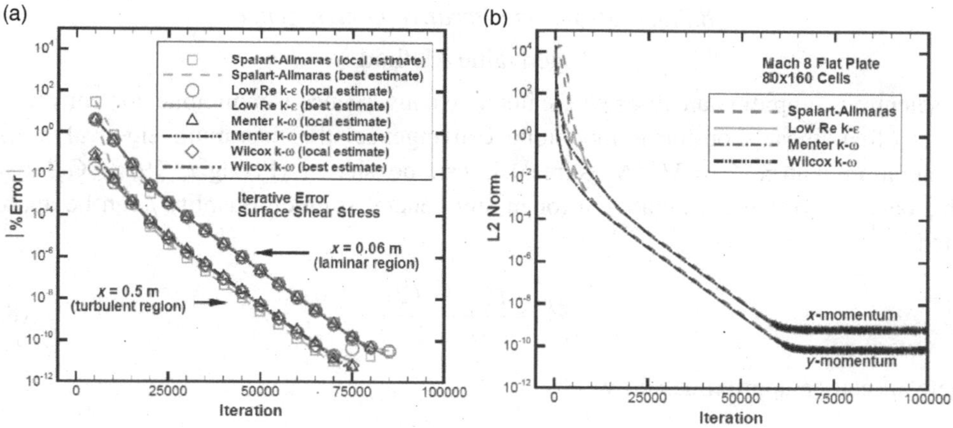

Figure 8.4 Mach 8 flow over a flat plate using the Navier–Stokes equations: (a) iterative error estimates from Blottner's method (local estimate) along with the machine zero method (best estimate) and (b) L_2 norms of the iterative residuals for the x- and y-momentum equations. (reproduced from Roy and Blottner, 2003)

8.3.2 Machine Zero Method

Recall that the iterative error is defined as the difference between the current iterative solution and the exact solution to the discrete equations. We can obtain a very good estimate of the iterative error at any of the prior iterations by first converging the solution down to machine zero, i.e., the iterations where the residual convergence stalls out (or stagnates) due to round-off error. The iterative error in a solution function f at iteration k is then approximated as

$$\varepsilon_h^k = f_h^k - f_h \cong f_h^k - f_h^{k\to\infty}, \tag{8.43}$$

where $f_h^{k\to\infty}$ is the machine zero solution, i.e., the iterative solution found in the limit as the number of iterations goes to infinity. This solution state can be identified as the point in the iteration history where the residuals no longer reduce but instead display seemingly random oscillations about some value (for an example, this stagnation occurs at approximately iteration number 65,000 in Figure 8.4b). This leveling off generally occurs near residual reductions of approximately 7 orders of magnitude for single precision computations and 15 orders of magnitude for double precision computations (see Chapter 7). Due to the expense of converging complex scientific computing codes to machine zero, this approach is only practical when applied to a small subset of cases in a large parametric study.

8.3.3 Local Convergence Rate

Rather than converging the iterative residuals to machine zero, a more efficient approach is to use the current iterative solution and its neighboring iterates to estimate the iterative error during the solution process. This approach requires some knowledge of the type of iterative convergence displayed by the numerical solution. These iterative error estimates are most useful for larger scientific computing applications where the convergence is slow.

8.3.3.1 Monotone Iterative Convergence

Eigenvalue Method

For scientific computing analyses where iterative convergence is monotone, the convergence rate is often linear. For linear monotone convergence, the maximum eigenvalue λ_G of the iteration matrix $G = M^{-1}N$ is real. It can be shown (Ferziger, 1988; Golub and Van Loan, 2013) that the iterative error in any general solution quantity f can be approximated by

$$\varepsilon_h^k \cong \frac{f_h^{k+1} - f_h^k}{\lambda_G - 1}, \tag{8.44}$$

where λ_G can be approximated by

$$\lambda_G \cong \frac{f_h^{k+1} - f_h^k}{f_h^k - f_h^{k-1}}. \tag{8.45}$$

Solving instead for the *estimate* of the exact solution to the discrete equations $\hat{f}_h$, one obtains:

$$\hat{f}_h \cong \frac{f^{k+1} - \lambda_G f^k}{1 - \lambda_G}. \tag{8.46}$$

See Ferziger (1988) or Golub and Van Loan (2013) for additional details.

Blottner's Method

A similar approach was developed by Blottner for cases where linear monotone convergence is observed in the iterative residuals and/or the iterative error (Roy and Blottner, 2001). When linear monotone convergence occurs, then iterative solutions at three different iteration levels can be used to estimate the iterative error. Blottner's approach can be summarized as follows. Recall the iterative error in any quantity f from Equation (8.40)

$$\varepsilon_h^k = f_h^k - f_h, \tag{8.47}$$

where f_h is the exact solution to the discrete equations. If convergence is linear monotonic, then the error will decrease exponentially as a function of iteration number

$$\varepsilon_h^k = \alpha e^{-\beta k}, \tag{8.48}$$

where α and β are constants which are independent of the iteration number. Equations (8.47) and (8.48) can be combined and rewritten as

$$\beta k = \ln \alpha - \ln\left(f_h^k - f_h\right). \tag{8.49}$$

Equation (8.49) is then evaluated at three different iteration levels, for example, $(k-1)$, k, and $(k+1)$, and these relationships are then used to eliminate α and obtain

$$\begin{aligned} \beta(k-(k-1)) &= \ln\left[\left(f_h^{k-1} - f_h\right)/\left(f_h^k - f_h\right)\right], \\ \beta(k+1-(k)) &= \ln\left[\left(f_h^k - f_h\right)/\left(f_h^{k+1} - f_h\right)\right]. \end{aligned} \tag{8.50}$$

Assuming the increments between iterations are equal (as in the current example), then the left-hand sides of Equation (8.50) are equivalent. Equating the right hand sides gives

$$\ln\left[\left(f_h^{k-1}-f_h\right)/\left(f_h^k-f_h\right)\right]=\ln\left[\left(f_h^k-f_h\right)/\left(f_h^{k+1}-f_h\right)\right]$$

or simply

$$\left(f_h^{k-1}-f_h\right)/\left(f_h^k-f_h\right)=\left(f_h^k-f_h\right)/\left(f_h^{k+1}-f_h\right). \tag{8.51}$$

An estimate of the exact solution to the discretized equations $\hat{f}_h$ is then given by

$$\hat{f}_h=\frac{f_h^k-\Lambda^k f_h^{k-1}}{1-\Lambda^k} \quad \text{where } \Lambda^k=\frac{f_h^{k+1}-f_h^k}{f_h^k-f_h^{k-1}}. \tag{8.52}$$

This final result is closely related to the eigenvalue approach given above in Equation (8.46), where here Λ^k plays the role of the estimated eigenvalues with the largest magnitude λ_G. When convergence is slow, it is often advantageous to use nonconsecutive iterations (e.g., $k-10$, k, and $k+10$) due to the relative importance of round-off error when the iterative solutions change very little from iteration to iteration.

Example: Blottner's Method for Iterative Error Estimation

Blottner's method has been used to estimate the iterative error for a Navier–Stokes CFD code by Roy and Blottner (2003). They examined the iterative error in the surface shear stress for the hypersonic flow over a flat plate aligned with the flow. Figure 8.4a shows iterative error estimates for two different regions of the plate: one with laminar flow and the other with turbulent flow. The symbols labeled "local estimate" are from Blottner's iterative error estimation procedure, while the lines labeled "best estimate" were found from the machine zero method discussed in Section 8.3.2. Machine zero is reached at an iterative residual reduction of approximately 14 orders of magnitude for these double precision computations as shown in Figure 8.4b. Blottner's method agrees quite well with the machine zero method and has the advantage of allowing early termination of the iterations once the iterative error is sufficiently small. However, Blottner's method should only be applied for cases where the iterative convergence is known (or demonstrated) to be monotone and linear.

8.3.3.2 Oscillatory Iterative Convergence

Although less common, Ferziger and Peric (1996) have developed an iterative error estimator which addresses oscillatory iterative convergence. Recall that in the eigenvalue approach for monotone solutions (discussed in the last section), the eigenvalue of the iteration matrix with the largest magnitude λ_G is used to estimate the iterative error. For oscillatory convergence, the eigenvalues with the largest magnitude are complex (and occur as conjugate pairs) and can also be used to estimate the iterative error. See Ferziger and Peric (1996) and the references cited therein for details.

8.4 Relation between Iterative Residuals and Iterative Error

Section 8.3.1 provided theoretical results relating the iterative residual reductions to reductions in the iterative error for Newton's method. Furthermore, a number of applied studies

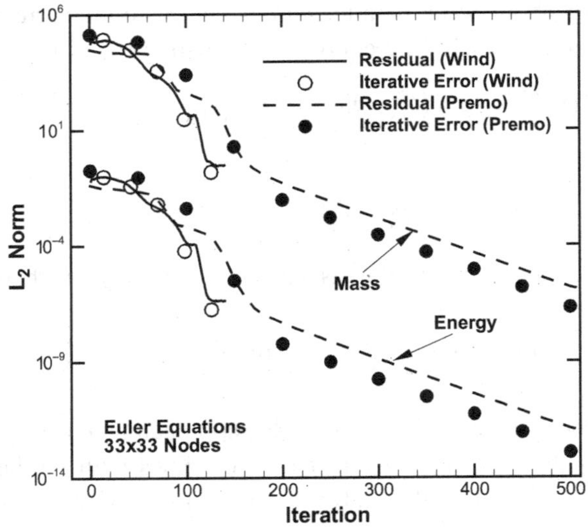

Figure 8.5 Discrete L_2 norms of the iterative convergence error and iterative residuals for manufactured solutions of the Euler equations.
(reproduced from Roy et al., 2004)

have shown that the iterative residuals track well with iterative error for a wide range of linear and nonlinear problems in scientific computing (e.g., Ferziger and Peric, 2002; Roy and Blottner, 2003). This behavior was demonstrated earlier in Figure 8.4 for a compressible Navier–Stokes example, which shows that the slope of the iterative error (Figure 8.4a) corresponds to the slope of the iterative residuals (Figure 8.4b), and thus they converge at the same rate. This observation supports the common practice of assessing iterative convergence indirectly by examining the iterative residuals. The key is then to find the appropriate scaling factor for a given class of problems to relate the norms of the iterative residuals to the iterative error in the quantity of interest.

Another example showing the relationship between iterative error and iterative residuals for the Euler equations is presented in Figure 8.5. These solutions were obtained for a manufactured solution computation with both a double precision code (Premo) and a single precision code (Wind) (Roy et al., 2004). The system response quantity (SRQ) of interest is the full flowfield solution, so L_2 norms of the iterative error over the domain for the mass and total energy per unit volume are found from the machine zero method (Section 8.3.2). These iterative error norms are then compared with L_2 norms of the iterative residuals of the mass and total energy conservation equations. For both codes, the iterative residual norms closely follow the actual iterative error norms from the machine zero method. Note that for these cases, the iterative residuals were scaled by a constant factor so as to line up more closely with the iterative errors, which will not affect the slopes. This scaling is required because the iterative residuals have the same general behavior, but not the same magnitude (or units), as the iterative error.

8.5 Practical Approach for Estimating Iterative Error

This section describes a practical and computationally efficient approach for estimating the iterative error in SRQs. This approach is particularly helpful when a large number of computational predictions must be made, e.g., for a parametric study or for conducting nondeterministic simulations. Most scientific computing codes monitor iterative convergence by examining norms of the iterative residuals. Since the iterative residual norms have been shown to follow closely with actual iterative errors for many problems (see Section 8.4), a small number of computations should be sufficient to determine how the iterative errors in the SRQ scale with the iterative residuals for the cases of interest.

The steps in this practical approach to iterative error estimation are as follows. First, the iterative residuals for all equations are converged to a value that the analyst expects to be more than sufficient to be iteratively converged. Next, the value of the SRQ at this final iteration is taken as an approximation to the exact solution to the discrete equations $\bar{f}_h$. The iterative error for all previous iterations, k, is then approximated using this estimate of f_h as $\bar{\varepsilon}_h^k = f_h^k - \bar{f}_h$. With the iterative error history now available as a function of the iteration number, the iterative residual convergence level to achieve any given iterative error level can be found.

Example: Practical Iterative Error Estimation Approach

An example of this practical approach to iterative error estimation is given in Figure 8.6 for laminar viscous flow through a packed bed of spherical particles (Duggirala et al., 2008). The SRQ of interest is the average pressure gradient across the bed, and the desired iterative error level in the pressure gradient is 0.01%. This low iterative error level is typical of what might be needed for a

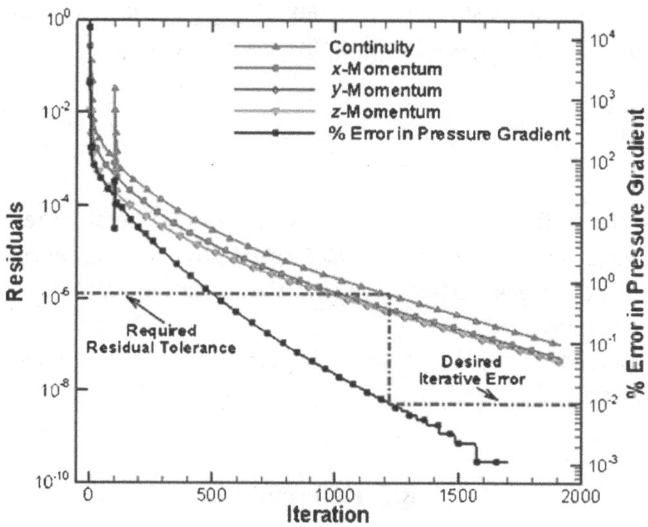

Figure 8.6 Norms of the iterative residuals (left axis) and percentage error in pressure gradient (right axis) for laminar flow through a packed bed.
(reproduced from Duggirala et al., 2008)

grid convergence study (see Chapter 9). Here the iterative residuals are first converged down seven orders of magnitude below their initial levels for each governing equation, which takes just over 1,900 iterations. The SRQ at this final iteration is taken as an estimate of the exact solution to the discrete equations and used to compute the iterative error in the SRQ at each prior iteration. Figure 8.6 shows that for the desired iterative error level in the pressure gradient of 0.01%, the iterative residual norms (shown here normalized by their initial values) only need to be converged down to approximately 10^{-6}, which takes only approximately 1,200 iterations (~40% reduction in computational effort). Simulations for similar problems can be expected to require approximately the same level of iterative residual convergence in order to achieve the desired iterative error tolerance in the pressure gradient.

For a looser iterative error tolerance of say 2%, typical for a set of parametric or nondeterministic simulations, Figure 8.6 indicates that a relative iterative residual reduction of less than five orders of magnitude (down to 2×10^{-5}) would be sufficient. This second example highlights the use of iterative error estimation to make an ensemble of simulations more efficient by reducing the number of iterations from over 1,900 down to ~350, thus requiring less than 20% of the original computing resources (a factor of five improvement in computational efficiency).

8.6 Problems

1. A nonlinear scientific computing problem is solved via Newton's method and has a condition number of $\kappa = 2$. If the standard assumptions hold, then the normalized iterative error norm $\frac{\left\| \varepsilon_h^k \right\|}{\left\| \varepsilon_h^{k=0} \right\|}$ should be within a factor of ________ of the normalized iterative residual norm $\frac{\left\| \mathfrak{R}_h^k \right\|}{\left\| \mathfrak{R}_h^{k=0} \right\|}$.
2. Multiple Choice (pick the best answer). Iterative ________ is/are the imbalance found by inserting the current iterative solution into the discrete equations.
   a. convergence
   b. error
   c. residuals
   d. tolerance
3. Briefly explain the difference between iterative errors and iterative residuals.
4. For linear systems, *monotone* and *oscillatory* iterative convergence occurs when the eigenvalue of the largest magnitude of the iteration matrix is ________ and ________, respectively.
5. Provide two specific examples of scientific computing simulations which might not contain iterative error.

8.7 References

Barrett, R., M. Berry, T. F. Chan, J. Demmel, J. M. Donato, J. Dongarra, et al. (1994). *Templates for the Solution of Linear Systems: Building Blocks for Iterative Methods*. Philadelphia, PA: SIAM. www.netlib .org/linalg/html_templates/Templates.html.

Chandra, S. (2003). *Computer Applications in Physics: With Fortran and Basic*. Pangbourne: Alpha Science International.

Duggirala, R., C. J. Roy, S. M. Saeidi, J. Khodadadi, D. Cahela, and B. Tatarchuck (2008). Pressure Drop Predictions for Microfibrous Flows Using CFD. *Journal of Fluids Engineering*, **130**, 13 pages. DOI: 10.1115/1.2948363.

Eca, L, G. Vaz, S. L. Toxopeus, and M. Hoekstra (2019). Numerical Errors in Unsteady Flow Simulations. *Journal of Verification, Validation and Uncertainty Quantification*, **4**, 021001-1–021001-10.

Ferziger, J. H. (1988). A Note on Numerical Accuracy. *International Journal for Numerical Methods in Fluids*, **8**, 995–996.

Ferziger, J. H., and M. Peric (1996). Further Discussion of Numerical Errors in CFD. *International Journal for Numerical Methods in Fluids*, **23**(12), 1263–1274.

Ferziger, J. H., and M. Peric (2002). *Computational Methods for Fluid Dynamics*, 3rd ed. Berlin, Germany: Springer.

Golub, G. H., and C. F. Van Loan (2013). *Matrix Computations*, 4th ed. Baltimore, MD: The Johns Hopkins University Press.

Hackbush, W. (1994). *Iterative Solution of Large Sparse Systems of Equations*. New York, NY: Springer-Verlag.

Hageman, L. A., and D. M. Young (1981). *Applied Iterative Methods*. London: Academic Press.

Kelley, C. T. (1995). *Iterative Methods for Linear and Nonlinear Equations*. Philadelphia, PA: SIAM Frontiers in Applied Mathematics Series, Society for Industrial and Applied Mathematics.

Kelley, C. T. (2003). *Solving Nonlinear Equations with Newton's Method*. Philadelphia, PA: SIAM Fundamentals of Algorithms Series, Society for Industrial and Applied Mathematics.

Meyer, C. D. (2000). *Matrix Analysis and Applied Linear Algebra*. Philadelphia, PA: Society for Industrial and Applied Mathematics.

Roy, C. J., and F. G. Blottner (2001). Assessment of One- and Two-Equation Turbulence Models for Hypersonic Transitional Flows. *Journal of Spacecraft and Rockets*, **38**(5), 699–710 (see also Roy, C. J. and Blottner, F. G. (2000). Assessment of One- and Two-Equation Turbulence Models for Hypersonic Transitional Flows. AIAA Paper 2000-0132.

Roy, C. J., and F. G. Blottner (2003). Methodology for Turbulence Model Validation: Application to Hypersonic Flows. *Journal of Spacecraft and Rockets*, **40**(3), 313–325.

Roy, C. J., C. J. Heintzelman, and S. J. Roberts (2007). Estimation of Numerical Error for 3D Inviscid Flows on Cartesian Grids. AIAA Paper 2007-0102.

Roy, C. J., C. C. Nelson, T. M. Smith, and C. C. Ober (2004). "Verification of Euler/Navier–Stokes Codes Using the Method of Manufactured Solutions. *International Journal for Numerical Methods in Fluids*, **44**(6), 599–620.

Saad, Y. (1996). *Iterative Methods for Sparse Linear Systems*, 1st ed. Boston, MA: PWS Publishing. Updated manuscript available online at: www-users.cs.umn.edu/~saad/books.html.

Saad, Y. (2003). *Iterative Methods for Sparse Linear Systems*, 2nd ed. Philadelphia, PA: Society for Industrial and Applied Mathematics.

Süli, E., and D. F. Mayers (2006). *An Introduction to Numerical Analysis*. Cambridge: Cambridge University Press.

9

Discretization Error

Mathematics takes us still further from what is human, into the region of absolute necessity, to which not only the world, but every possible world, must conform.

—Bertrand Russell

Contents of Chapter 9

Recall that our definition of a mathematical model is a governing partial differential or integral equation along with its associated initial and boundary conditions. In scientific computing, one is concerned with finding approximate solutions to this mathematical model, a process that involves the discretization of both the mathematical model and the domain. The approximation errors associated with this discretization process are called discretization errors, and they occur in nearly every scientific computing simulation. The *discretization error* can be formally defined as the difference between the exact solution to the discrete equations and the exact solution to the mathematical model:

$$\varepsilon_h = u_h - \tilde{u}. \tag{9.1}$$

In this chapter, it will be shown that the discretization error arises out of the interplay between the chosen discretization scheme for the mathematical model, the mesh resolution, the mesh quality, and the solution behavior itself. Discretization error is the most difficult type of numerical error to estimate accurately and is usually the largest of the numerical error sources identified in Chapter 7.

This chapter begins with an overview of the discretization process, including discretization of the governing equations and of the domain, followed by a discussion of the sources of discretization error. An overview and classification of several approaches for estimating discretization error is then presented. Particular attention is focused on Richardson extrapolation, which involves post-processing discrete solutions on two (or more) meshes to estimate the discretization error in either local or global solution quantities. Techniques are then given for estimating the reliability of the discretization error estimators based on the observed order of accuracy. In nearly all practical cases, the discretization error is more appropriately characterized as an (epistemic) uncertainty due to uncertainties in the error estimation process itself. Roache's Grid Convergence Index (GCI) is discussed for converting discretization error estimates into uncertainties (note: a general approach for converting numerical errors into numerical uncertainties was discussed in Chapter 7). Issues associated with systematic mesh refinement are discussed, along with approaches for assessing just how systematic the refinement is. This chapter concludes with a discussion of solution adaptation to either reduce or appropriately size the discretization error for scientific computing applications.

9.1 Elements of the Discretization Process

The goal of the discretization process is to convert the mathematical model into a set of algebraic equations which can be readily solved on a digital computer. There are two distinct steps in this process: discretization of the mathematical model and discretization of the domain. While most applied scientific computing codes employ both aspects of discretization, some discretization schemes do not. For example, the method of weighted residuals does not include a domain discretization step (Heubner et al., 2001).

9.1.1 Discretization of the Mathematical Model

Throughout this book, we have made numerous references to the mathematical model being composed of partial differential equations. However, this is not the only form that the governing equations can take in scientific computing. The partial differential equation form arises when the conceptual model (e.g., the conservation of energy principle) is applied to an infinitesimal element. If instead a control volume of finite size is used, then the divergence theorem can be used to convert the volume integral of the gradient of a quantity to fluxes of that quantity through the closed control surface bounding the control volume. The resulting integral (or integrodifferential) form of the governing equations is called the weak form of the equations since it admits solutions with discontinuities (i.e., weak solutions) that satisfy the differential equation only in a restricted sense. The weak form is more fundamental since it admits discontinuous (weak) solutions, whereas the differential form relies on the dependent variables being differentiable and therefore continuous (i.e., strong solutions).

In order to find the solution to a mathematical model, initial and/or boundary conditions must be supplied. In some cases, these initial and boundary conditions also take the form of differential equations, such as the Neumann (or gradient) boundary condition, where the gradient of a property normal to a boundary is specified. For example, the adiabatic (or zero heat flux) boundary condition for a heat conduction problem has the form

$$\left.\frac{\partial T}{\partial n}\right|_{\Gamma} = 0 \tag{9.2}$$

on the boundary Γ, where n is defined to be normal to the boundary. To ensure that the boundary condition discretization does not have an adverse impact on the overall order of accuracy of the code, the formal order of accuracy of the boundary condition discretization must be the same as (or higher than) that of the interior discretization scheme. This is not to be confused with the order of the boundary condition itself (i.e., the highest derivative in the boundary condition), which must be at most one order less than the governing equation.

While there are a number of different discretization approaches, scientific computing software generally utilizes one of the following three approaches. The finite difference method employs the strong form of mathematical model, whereas the finite volume method and the finite element method employ the weak form of the equations. An introduction to these three main discretization approaches is given in chapter 8 of Oberkampf and Roy (2010). Other approaches which can also result in consistent and convergent discretizations include boundary element methods (Hall, 1994), spectral element methods (Van de Vosse and Minev, 1996), and pseudo-spectral methods (Fornberg, 1998).

9.1.2 Discretization of the Domain

For steady problems, discretization of the domain involves decomposing the spatial domain into nodes, cells, or elements where the discrete equations will be applied. For unsteady problems, the temporal domain must also be discretized by selecting an appropriate time step Δt to advance the simulation. This time step must be chosen to resolve the physical phenomena in question, to satisfy any stability criteria in the problem (recall the discussion of numerical stability in Chapter 5), and to achieve the desired level of temporal accuracy. The terms "mesh" and "grid" are often used to refer to the partitioned domain in space; however, for unsteady simulations the term "mesh" will also refer to the discretized temporal domain.

For complex, 3D geometries, the generation of an appropriate spatial mesh can require significant time and effort. Two mesh-related issues that have a significant impact on the discretization error are mesh resolution and mesh quality. The effects of mesh resolution on the discretization error are clear when one considers that the truncation error serves as the local source for generating discretization error. (This relationship was initially shown in Chapter 5 and is further developed in this chapter during the discussion of error transport equations in Section 9.3.2.1.) The truncation error contains the discretization parameters (Δx, Δy, etc.) to various powers, with the lowest power determining the formal order of accuracy of the discretization scheme. Thus, the discretization error can be reduced by simply refining the mesh. Mesh quality addresses factors such as cell aspect ratio, cell stretching ratio, cell skewness, and orthogonality of mesh lines at a boundary. Poor quality spatial meshes can result in a reduction in the order of accuracy and even non-convergence of the discrete solution in extreme cases. The relationship between mesh quality and the discretization error is discussed further in Section 9.8.

There are two basic types of spatial meshes: structured and unstructured. In 3D *structured meshes*, each node or element is uniquely defined by an ordered set of indices (i, j, and k in 3D). Structured grids can either be Cartesian, stretched Cartesian, skewed Cartesian, or curvilinear. Cartesian grids have fixed spacing (Δx, Δy, and Δz) between orthogonal grid

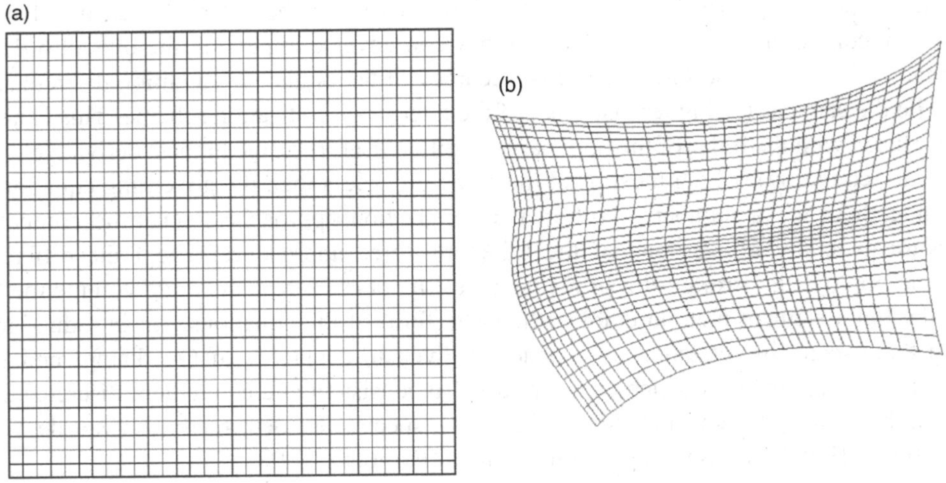

Figure 9.1 Structured mesh examples in 2D: (a) Cartesian and (b) curvilinear.

lines, whereas stretched Cartesian and skewed Cartesian relax the requirements for fixed spacing and orthogonality, respectively. Curvilinear structured grids allow the mesh to wrap around a surface and the spacing can vary over the domain. Examples of structured Cartesian and curvilinear meshes in 2D are shown in Figure 9.1a and Figure 9.1b, respectively. The regular ordering of the structured grid generally makes it more efficient for computing numerical solutions than unstructured meshes, but at the cost of increased difficulty during mesh generation for complicated geometries.

Unstructured meshes do not employ regular ordering of the nodes or elements. They can be composed of triangles, quadrilaterals, or arbitrary elements in 2D, and pyramids, tetrahedrons, hexahedrons, and arbitrary elements in 3D. There is no equivalent unstructured element type in 1D as the nodes/elements in a 1D mesh can always be indexed in an ordered manner. Two examples of unstructured meshes in 2D are given in Figure 9.2. A general unstructured triangular mesh is shown in Figure 9.2a, while a hybrid mesh containing structured quadrilateral and unstructured triangular elements is shown in Figure 9.2b.

Cartesian meshes can also be used for general geometries. For Cartesian grid methods, curved boundaries are usually handled with special boundary cells called cut-cells. An example of a Cartesian grid where the underlying mesh is structured is given in Figure 9.3a. This type of mesh could be handled by a structured grid code with special treatment of the curved boundaries. The more common example of a Cartesian grid where the underlying mesh is unstructured is given in Figure 9.3b. The unstructured nature of the grid allows clustering by subdividing selected cells, thus allowing more resolution of the surface geometry and/or solution features.

Finally, there are a number of mesh-free methods in the area of fluid mechanics which generally do not require a volume mesh in the domain but may require surface meshes. These approaches thus have the advantage of greatly simplifying the grid generation process by reducing the dimensionality of the mesh to be generated by one. These methods take a Lagrangian approach, thus allowing individual vortices or fluid particles to travel through the domain and interact with each other. Some examples of grid-free methods include vortex

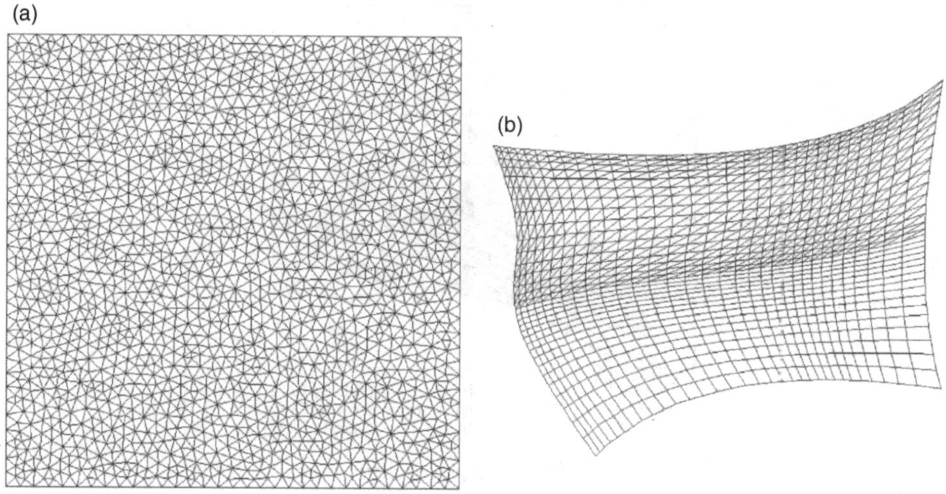

Figure 9.2 Unstructured meshes in 2D: (a) general triangular and (b) hybrid structured/unstructured.

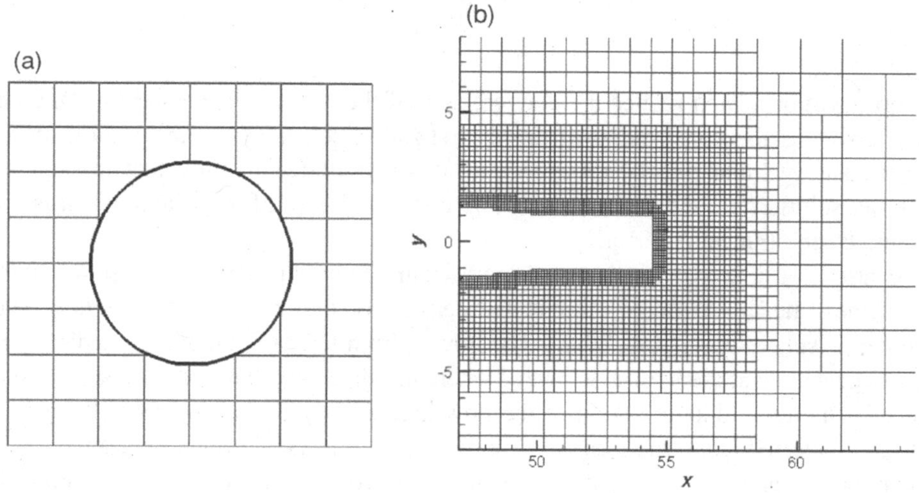

Figure 9.3 Cartesian grids in 2D: (a) structured Cartesian and (b) unstructured Cartesian.
(Figure 9.3b is reproduced from Roy et al., 2007)

methods, smoothed particle hydrodynamics, and lattice Boltzmann. The actual form of the underlying governing equations and the consistency/convergence of these methods with increasing number of vortices, particles, etc. has generally not been adequately addressed.

9.2 Sources of Discretization Error

The discretization error has two components: one that is locally-generated and one that is transported from other parts of the domain. The transported component is also called pollution error in the finite element literature (e.g., Babuska et al., 1997). The existence of these two components was shown mathematically in Chapter 5 using the continuous error

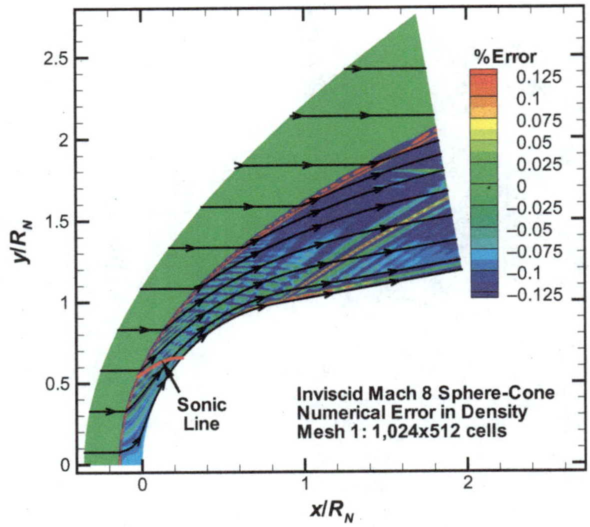

Figure 9.4 Contours of total estimated discretization error in density for the flow over an inviscid hypersonic sphere cone.
(from Roy, 2003)

transport equation which relates the convergence of the numerical method (i.e., the discretization error) to the consistency of the discretization scheme (i.e., the truncation error). There we showed that the discretization error is transported in the same manner as the underlying solution properties (e.g., it can be convected and diffused) and that it is locally generated by the truncation error.

An example of discretization error transport for the Euler equations is shown in Figure 9.4, which depicts the discretization error in the density for the inviscid, Mach 8 flow over an axisymmetric sphere-cone (Roy, 2003). The flow is from left to right, and large discretization errors are generated at the bow shock wave where the shock and the grid lines are misaligned. In the subsonic (i.e., elliptic) region of the flow immediately behind the normal shock, these errors are convected along the local streamlines. In the supersonic (hyperbolic) regions these errors propagate along characteristic Mach lines and reflect off the surface. Additional discretization error is generated at the sphere-cone tangency point (i.e., the point where the spherical nose joins the conical body), which represents a singularity due to the discontinuity in the surface curvature. Errors from this region also propagate downstream along the characteristic Mach line. An adaptation process which is driven by the global error levels would adapt to the characteristic line emanating from the sphere-cone tangency point, which is not desired. An adaptation process driven by the local contribution to the error should adapt to the sphere-cone tangency point itself, thus obviating the need for adaption to the characteristic line that emanates from it. Targeted mesh adaptation is discussed in detail in Section 9.8.

9.3 Approaches for Estimating Discretization Error

There are many approaches available for estimating discretization error. These methods can be broadly categorized as *a priori* methods and *a posteriori* methods. The *a priori* methods are those that allow a bound to be placed on the discretization error before any numerical solution is even computed. In general, one looks to bound the discretization error by an equation of the form

$$\varepsilon_h \leq C(u)h^p, \tag{9.3}$$

where ε_h is the discretization error, the function $C(u)$ usually depends on various derivatives of the exact solution, h is a measure of the element size (e.g., Δx), and p is the formal order of accuracy of the method. One approach to developing an *a priori* discretization error estimator is to perform a truncation error analysis for the scheme, relate the truncation error to the discretization error (as in Chapter 5), then develop some approximate bounds on the solution derivatives that comprise $C(u)$. The main failing of *a priori* error estimators is that $C(u)$ is extremely difficult to bound and even when this is possible for simple problems, the resulting error estimate often greatly overestimates the true discretization error. *A priori* methods are generally only useful for assessing the formal order of accuracy of a discretization scheme. Current efforts in estimating the discretization error are focused on *a posteriori* methods. These methods provide an error estimate *only after* a numerical solution has been computed. They use a computed solution to the discrete equations (possibly including information from the problem being solved) to estimate the discretization error relative to the exact solution to the mathematical model.

The mathematical formalism that underlies the finite element method makes it fertile ground for the rigorous estimation of discretization error. Beginning with the pioneering work of Babuska and Rheinboldt (1978a), a tremendous amount of work has been done over the last 40 years on *a posteriori* estimation of discretization error by the finite element community (Ainsworth and Oden, 2000). The initial developments up to the early-1990s were concentrated on linear, elliptic, scalar mathematical models and focused on the h-version of finite elements. Early extensions of the *a posteriori* methods to parabolic and hyperbolic mathematical models were done by Eriksson and Johnson (1987) and Johnson and Hansbo (1992), respectively. Up to this point, *a posteriori* error estimation in finite elements was limited to analysis of the energy norm of the discretization error, which for Poisson's equation can be written on element k as:

$$|||\varepsilon_h|||_k = \left[\int_{\Omega_k} \left| \vec{\nabla} u_h - \vec{\nabla} \tilde{u} \right|^2 d\Omega \right]^{1/2}. \tag{9.4}$$

The finite element method produces the numerical solution from the chosen set of basic functions that minimizes the energy norm of the discretization error (Szabo and Babuska, 2021). Taken locally, the energy norm can provide guidance on where adaptive refinement should occur. Taken globally, the energy norm provides a global measure of the overall optimality of the finite element solution. In the early 1990s, important extensions of *a posteriori* error estimators to system response quantities were found that require the solution to an adjoint, or dual, problem (e.g., Johnson and Hansbo, 1992). For additional information on *a posteriori* error estimation in finite element methods, see Babuska et al. (1986), Whiteman (1994), Ainsworth and Oden (1997, 2000), and Estep et al. (2000). A more introductory discussion of error estimation in finite element analysis is presented by Akin (2005).

In general, the level of reliability of *a posteriori* error estimation methods is strongly problem dependent. All of the discretization error estimators to be discussed here were originally developed for elliptic problems. As a result, they tend to work well for elliptic problems, but are not generally effective for mathematical models that are parabolic or hyperbolic in nature. The level of complexity of the problem is also an important issue. The error estimators work well for smooth, linear problems with simple physics and geometries; however, strong nonlinearities, discontinuities, singularities, and physical and geometric complexity can significantly reduce the reliability and applicability of *a posteriori* discretization error estimation methods.

One approach for measuring the accuracy and reliability of a discretization error estimator is through the effectivity index θ. Here we will define the *effectivity index* as a general norm of the estimated discretization error $\bar{\varepsilon}_h$ divided by the norm of the true error ε_h. The effectivity index can be computed locally by evaluating it on element or cell k

$$\theta_k = \frac{\|\bar{\varepsilon}_h\|_k}{\|\varepsilon_h\|_k} \tag{9.5}$$

or globally over the entire domain Ω

$$\theta k = \frac{\|\bar{\varepsilon}_h\|_\Omega}{\|\varepsilon_h\|_\Omega}. \tag{9.6}$$

One important property that is desired in a discretization error estimator is consistency (i.e., asymptotic exactness) of the estimate, which means the estimated error must approach the true error (i.e., $\theta \to 1$) as the mesh is refined ($h \to 0$) or as the order of the scheme is increased ($p \to \infty$). The practical utility of a given error estimator can also be gauged by examining its effectivity index on coarse meshes. Finally, if asymptotic exactness of the discretization error estimator cannot be shown, then it is recommended to have error estimates that are conservative (i.e., they overestimate the error). Thus, a discretization error estimator that overestimates the error is more desirable than one which underestimates it.

There are two types of discretization error estimators discussed in this section. In *recovery methods*, an estimate of the exact solution to the mathematical model (or possibly its gradient) is obtained by post-processing the numerical solution. This estimate of the exact solution will provide good discretization error estimates when it has a higher order of accuracy than the underlying numerical solution. For some of the recovery-based methods, higher-order estimates can be easily obtained for system response quantities as well. *Residual-based methods*, by contrast, also incorporate information on the specific problem being solved into the error estimate. This additional information comes either from the underlying mathematical model or from the discrete equations. While the implementation of residual-based methods in a scientific computing code can be more difficult and code-intrusive than for recovery methods, they have the potential to provide more detailed information on the discretization error and its various sources. Some residual-based methods require the solution to an adjoint (or dual) problem in order to provide discretization error estimates in solution functionals, which are usually of more interest to the scientific computing analyst.

9.3.1 Recovery Methods

One approach to error estimation is to compare the discrete solution to a higher-order estimate of the exact solution to the mathematical model. While this approach uses only information from the discrete solution itself, in some cases more than one discrete solution is needed, with the additional solutions being obtained either on systematically-refined/ coarsened meshes or with different formal orders of accuracy.

9.3.1.1 Mesh Refinement Methods

Mesh refinement methods are based on the general concept of Richardson extrapolation (Richardson, 1911, 1927). Recall that in Chapter 5 we used a power series expansion to relate the discretization error in a system response quantity $\varepsilon_h = f_h - \tilde{f}$ to the mesh spacing parameter h, which for a convergent, pth-order discretization method can be written as

$$f_h - \tilde{f} = g_p h^p + O(h^{p+1}), \tag{9.7}$$

where f_h is the discrete solution and $\tilde{f}$ the exact solution to the mathematical model.

The basic premise behind mesh refinement methods is to compute two discrete solutions on systematically-refined meshes, then use Equation (9.7) to solve for an estimate of the exact solution to the mathematical model $\bar{f}$. As with all discretization error estimators, the estimate will only be reliable in the asymptotic range where the terms of order h^{p+1} and higher can be neglected. In the asymptotic range, the estimate $\bar{f}$ is generally accurate to within order p+1. This higher-order estimate of $\tilde{f}$ can then be used to estimate the error in the discrete solutions. The main advantages of mesh refinement-based error estimators are that they can be applied as a post-processing step to *any type of discretization* and that they can be applied both to the field solution and to any system response quantities. A more detailed discussion of Richardson extrapolation and its use as a discretization error estimator is presented in Section 9.4.

9.3.1.2 Order Refinement Methods

Order refinement methods are those which employ two or more discrete solutions on the same mesh but with differing formal orders of accuracy. The results from the two numerical solutions are then combined to produce a discretization error estimate. An early example of order refinement methods for error estimation is the Runge–Kutta–Fehlberg method (Fehlberg, 1969) for adaptive step size control in the solution of ordinary differential equations. This approach combines a basic fourth-order Runge–Kutta integration of the differential equations with an inexpensive fifth-order estimate of the error. Order refinement can be difficult to implement in finite difference and finite volume discretizations due to difficulties formulating higher-order accurate gradients and boundary conditions. Order refinement methods have been implemented within the context of finite elements under the name *hierarchical bases* (e.g., see Bank, 1996; Szabo and Babuska, 2021).

9.3.1.3 Finite Element Recovery Methods

Recovery methods for estimating the discretization error were developed by the finite element community (e.g., Zienkiewicz and Zhu, 1987, 1992a). For the standard h-version of finite elements with linear basis functions the solution is piece-wise linear; therefore, the gradients are only piece-wise constant and are discontinuous across the element faces. The user of a finite element code is often more interested in gradient quantities such as stresses than the solution itself, so most finite element codes provide for post-processing of these discontinuous gradients into piece-wise linear gradients using existing finite element infrastructure.

In some cases (see the following discussion of superconvergence), this reconstructed gradient is of a higher order of accuracy than the gradient found in the underlying finite element solution. Recall the definition of the energy norm of the discretization error given in Equation (9.4). If the true gradient from the mathematical model is available, then this important error measure can be computed exactly. For the case where a reconstructed gradient is higher-order accurate than the finite element gradient, then it can be used to approximate the true gradient in the energy norm. In addition to providing estimates of the discretization error in the solution gradients, due to their local nature, recovery methods are also often used as *indicators* of where solution refinement is needed in adaptive solutions (see Section 9.8).

In order to justify the use of the recovered gradient in the energy norm, it must in fact be higher-order accurate than the gradient from the finite element solution. This so-called *superconvergence property* can occur when certain regularity conditions on the mesh and the solution are met (Wahlbin, 1995) and results in gradients that are up to one order higher in accuracy than the underlying finite element gradients. For linear finite elements, the superconvergence points occur at the element centroids, whereas for quadratic finite elements, the location of the superconvergence points depends on the element topology. If the reconstructed gradient is superconvergent, and if certain consistency conditions are met by the gradient reconstruction operator itself, then error estimators based on this recovered gradient can be shown to be asymptotically exact with effectivity indices approaching unity (Ainsworth and Oden, 2000). While the superconvergence property appears to be a difficult one to attain for complex scientific computing applications, the discretization error estimates from some recovery methods tend to be "astonishingly good" for reasons that are not well understood (Ainsworth and Oden, 2000).

Recovery methods have been shown to be most effective when the reconstruction step employs solution gradients rather than solution values. The *superconvergent patch recovery* (SPR) method (Zienkiewicz and Zhu, 1992a) is the most widely-used recovery method in finite element analysis. Assuming the underlying finite element method is of order p, the SPR approach is based on a local least squares fitting of the solution gradient values at the superconvergence points using polynomials of degree p. The SPR recovery method was found to perform extremely well in an extensive comparison of *a posteriori* finite element error estimators (Babuska et al., 1994). A more recent approach called *polynomial preserving recovery* (PPR) was proposed by Zhang and Naga (2005). In their approach, they use polynomials of degree $p + 1$ to fit the solution values at the superconvergence points, then take derivatives of this fit to recover the gradient. Both the SPR and PPR gradient reconstruction methods can be used to obtain error estimates in the global energy norm and in the local solution gradients. Estimates of the discretization error in system response quantities must be computed heuristically. For example, a 5% error in the global energy norm may be found to correspond to a 20% error in a global system response quantity for a given class of problems.

9.3.2 Residual-Based Methods

Residual-based methods use the discrete solution along with additional information from the problem being solved such as the mathematical model, the discrete equations, and/or the local sources of discretization error. Residual-based discretization error estimation methods include error transport equations (both continuous and discrete), defect correction, finite element residual methods, and adjoint methods. As is shown in the next section, all of these residual-based methods employ estimates of the truncation error. The truncation error can be approximated either by inserting the exact solution to the mathematical model (or an approximation thereof) into the discrete equation or by inserting the discrete solution into the continuous mathematical model (or a higher-order discretization). The former is the discrete residual, while the latter is the continuous residual (e.g., the finite element residual). The use of adjoint methods to extend discretization error estimation methods to provide error estimates in system response quantities is also discussed.

9.3.2.1 Error Transport Equations

Discretization error is transported through the domain in a similar fashion as the solution variables in the underlying mathematical model. For example, if a mathematical model governs the convection and diffusion of a scalar variable, then a discrete solution to the mathematical model will contain discretization error that is also convected and diffused. Babuska and Rheinboldt (1978a) appear to be the first to develop such a discretization error transport equation within the context of the finite element method. However, rather than solve this transport equation directly for the discretization error, the typical approach used in finite elements is to use this equation to either indirectly bound the error (explicit residual methods) or approximate its solution (implicit residual methods). The main advantages of error transport equations are that they provide estimates of the local error in the field variables for the problem and that they are generally solved on the same mesh as the original (primal) problem. These local error estimates can be applied to the numerical solution to (1) estimate the error in the solution functionals (i.e., global system response quantities such as the integrated force on a boundary) and/or (2) iteratively update the numerical solution and thus provide more accurate discretization and truncation error estimates. Examples of the use of error transport equations for finite volume schemes can be found in Zhang et al. (2000), Shih and Williams (2009), Banks et al. (2012), Yan and Ollivier-Gooch (2017), Tyson and Roy (2019), and Tyson et al. (2019).

Continuous Discretization Error Transport Equation

The following development is applicable to any discretization approach and is based on the *generalized truncation error expression* (Roy, 2009) discussed in Chapter 5 as Equation (5.12). Recall that the original (possibly nonlinear) governing equation operator $L(\cdot)$ and the discrete equation operator $L_h(\cdot)$ are exactly solved by $\tilde{u}$ (the exact solution to the original mathematical model) and u_h (the exact solution to the discrete equations), respectively. Thus, we can write:

$$L(\tilde{u}) = 0 \tag{9.8}$$

and

$$L_h(u_h) = 0. \tag{9.9}$$

Furthermore, the partial differential equation and the discretized equation are related through the *generalized truncation error expression* (Equation 5.12, repeated here for convenience) as

$$L_h(u) = L(u) + TE_h(u). \tag{9.10}$$

Equation (9.10) assumes some suitable mapping of the operators and the solutions onto either a continuous or discrete space for finite difference and finite volume methods (see Tyson and Roy, 2019), and onto function spaces of different order (e.g., different polynomial orders) for finite element methods (e.g., see Wang and Roy, 2023). Substituting u_h into Equation (9.10) and noting that $L_h(u_h) = 0$, gives:

$$L(u_h) = -TE_h(u_h). \tag{9.11}$$

Inserting the definition of the discretization error

$$\varepsilon_h = u_h - \tilde{u}, \tag{9.12}$$

we can thus rewrite Equation (9.11) as

$$L(\varepsilon_h + \tilde{u}) = -TE_h(u_h). \tag{9.13}$$

If the governing mathematical model is linear, then $L(\varepsilon_h + \tilde{u}) = L(\varepsilon_h) + L(\tilde{u})$. Since $L(\tilde{u}) = 0$, we now have simply

$$L(\varepsilon_h) = -TE_h(u_h). \tag{9.14}$$

Equations (9.13) and (9.14) are the continuous error transport equations and are the nonlinear and linear mathematical models, respectively, that govern the transport of the discretization error ε_h through the domain. Furthermore, the truncation error acting upon the discrete solution serves as a source term which governs the local production or destruction of discretization error and is a function of the local discretization parameters (Δx, Δy, Δt, etc.). These *continuous discretization error transport equations* can be discretized and solved for the discretization error in the solution variables assuming that the truncation error is known or can be estimated.

Discrete Discretization Error Transport Equation

A more useful discrete version of the discretization error transport equation can be derived as follows (Tyson and Roy, 2019). First, the exact solution to the mathematical model $\tilde{u}$ is substituted into Equation (9.10), then noting that $L(\tilde{u}) = 0$ and using the definition of the discretization error Equation (9.12), we obtain

$$L_h(u_h - \varepsilon_h) = TE_h(\tilde{u}). \tag{9.15}$$

This is the nonlinear form of the discrete error transport equations. Assuming that the original discrete problem has been solved for u_h, Equation (9.15) can be solved for the discretization error ε_h. However, it requires either an estimate of the truncation error (since $\tilde{u}$ is not known) or else the definition of the discretization error can be substituted and then iteratively corrected towards $\tilde{u}$.

Consider the following approximate form of Equation (9.12)

$$\bar{\varepsilon}_h = u_h - \bar{u}, \tag{9.16}$$

where the overbars denote estimates of the discretization error or the exact solution. Once the error transport equations have been solved for the error estimates in the field variables, the corrected solution can be computed as

$$\bar{u} = u_h - \bar{\varepsilon}_h. \tag{9.17}$$

This corrected solution is a better approximation of $\tilde{u}$ than u_h, and thus provides a better estimate of the truncation error. This approach can be applied in an iterative fashion to successively improve the truncation error estimate, the linearization (for the linearized form discussed below), and therefore, the discretization error estimates (see Tyson et al., 2019 and Wang and Roy, 2023 for details).

A linearized form of the discrete error transport equations can be found by linearizing the discrete equations at $\tilde{u}$ about u_h:

$$L_h(\tilde{u}) = L_h(u_h) + \left.\frac{\partial L_h}{\partial u}\right|_{u_h} (\tilde{u} - u_h) + \left.\frac{\partial^2 L_h}{\partial u^2}\right|_{u_h} \frac{(\tilde{u} - u_h)^2}{2} + \dots. \tag{9.18}$$

Since $L_h(u_h) = 0$, inserting the definition of the discretization error and dropping the higher-order terms of order ε_h^2 and higher, results in

$$\left.\frac{\partial L_h}{\partial u}\right|_{u_h} \bar{\varepsilon}_h = -L_h(\tilde{u}) = -TE_h(\tilde{u}), \tag{9.19}$$

where the overbar denotes an error estimate since the higher-order terms in Equation (9.18) were omitted. For example, if the underlying numerical scheme is second-order accurate (i.e., $\varepsilon_h \propto h^2$), then the omitted terms are $\propto \varepsilon_h^2 \propto h^4$ and the corrected solution should be fourth-order accurate. The matrix on the left-hand side $\left.\frac{\partial L_h}{\partial u}\right|_{u_h}$ is the Jacobian matrix and is generally available (possibly in an approximate sense) when implicit solution methods such as Newton's method are used. If the Jacobian is available, then this linearized form can be extremely inexpensive to compute for steady-state problems. When implemented in this linearized form, the truncation error and the Jacobian matrix can both be iteratively updated as mentioned above.

Equations (9.15) and (9.19) are the nonlinear and linearized forms of the *discrete discretization error transport equation*, respectively. They govern the transport of the discretization error ε_h through the domain. These equations can be solved for the discretization error if the numerical solution u_h has been computed and an estimate of the truncation error is available.

Approximating the Truncation Error

While the development of discretization error transport equations is relatively straightforward, questions remain as to the treatment of the truncation error, which acts as the local source term. The truncation error can be difficult to derive for complex, nonlinear discretization schemes such as those used for the solution to the compressible Euler and Navier–Stokes equations in fluid dynamics. However, if the truncation error can be reliably approximated, then this approximation can be used as the source term for the error transport equation.

Here we present two approaches for approximating the truncation error based on the *generalized truncation error expression* given by Equation (9.10) (Roy, 2009). In the first approach, the exact solution to the mathematical model $\tilde{u}$ is inserted into Equation (9.10). Since this exact solution will exactly solve the mathematical model, because the term $L(\tilde{u}) = 0$, this allows the truncation error to be approximated as

$$TE_h(\tilde{u}) = L_h(\tilde{u}). \tag{9.20}$$

Since this exact solution is generally not known, it could be approximated by substituting an estimate of the exact solution, for example from any local discretization error estimator using Equation (9.17), into the discrete operator:

$$TE_h(\tilde{u}) \approx L_h(\overline{u}_h). \tag{9.21}$$

Alternatively, the solution from a fine grid solution u_h could be inserted into the discrete operator for a coarse grid $L_{rh}(\cdot)$:

$$TE_h(\tilde{u}) = \frac{1}{r^p} TE_{rh}(\tilde{u}) \approx \frac{1}{r^p} TE_{rh}(u_h) = \frac{1}{r^p} L_{rh}(u_h). \tag{9.22}$$

Note that the subscript rh denotes the discrete operator on a grid that is a factor of r coarser in each direction than the fine grid. For example, $r = 2$ when the coarse mesh is formed by eliminating every other point in each direction of a structured mesh. This approach was used by Shih and Qin (2007) to estimate the truncation error for use with a discrete discretization error transport equation.

A second approach for estimating the truncation error is to insert the exact solution to the discrete equations u_h into Equation (9.10). Since this solution exactly solves the discrete equations $L_h(u_h) = 0$, we have

$$TE_h(u_h) = -L(u_h). \tag{9.23}$$

If a continuous representation of the solution is available, then this evaluation is straightforward. In fact, the right-hand side of Equation (9.23) is the definition of the finite element residual that is given in the next section. For other numerical methods (e.g., finite difference and finite volume), a continuous projection of the numerical solution must be made in order to estimate the truncation error. It is often sufficient to replace the continuous mathematical model with a higher-order discrete operator since this higher order operator only needs to be applied to the lower-order numerical solution u_h and not solved. As mentioned earlier, since error transport equations provide local error estimates, the numerical solution can be corrected (with $\overline{u} = u_h - \overline{\varepsilon}_h$) in order to improve the truncation error estimate (Tyson et al., 2019).

Discretization Error Estimation

Error transport equations provide local error estimates in the field variables by solving a similar set of equations on the same grid using the truncation error as a source term. For example, in a linear elasticity problem they provide error estimates in the displacements over the domain, while for an incompressible fluid dynamics problem they provide error estimates in the velocities and pressure over the domain. The only other discretization error estimation approach discussed herein that provides local error estimates is Richardson extrapolation, which requires multiple, systematically-refined grids. Note that Richardson extrapolation presents some additional challenges when applied to local quantities (Oberkampf and Roy, 2010).

Error estimates for solution functionals (such as the integrated force or net heat transfer through a boundary) derived from the local quantities can be obtained as follows. First, the numerical solution is corrected to higher order using the error estimate, $\overline{u} = u_h - \overline{\varepsilon}_h$. The functional then can be recomputed using this higher-order corrected solution. Note that any necessary post-processing of the corrected solution should also be performed with a higher order of accuracy. For example, if the original (primal) solution produces a second-order accurate solution for the pressure on a surface, and error transport equations are used to correct this surface pressure to fourth order, then integration of the surface pressure to determine the corrected net force on the surface should also be performed with fourth order accuracy. The error in the net force in the original (primal) solution is then $\overline{\varepsilon}_h = f_h - \overline{f}$ where f_h is the primal solution and $\overline{f}$ is the higher-order integration of the corrected solution.

9.3.2.2 Finite Element Residual Methods

In a broad mathematical sense, a residual refers to what is left over when an approximate solution is inserted into an equation. In Chapter 5 we discussed the residual method for code order of accuracy verification where the residual was found by substituting the exact solution to the mathematical model into the discrete equations. In Chapter 8 we discussed iterative convergence in terms of the iterative residuals which are found by substituting an approximate iterative solution into the discrete equations. Consider now the general mathematical operator $L(u) = 0$ which is solved exactly by $\tilde{u}$. Because the finite element method provides for a continuous representation of the numerical solution u_h, it is natural to define the finite element residual in a continuous sense over the domain as

$$\mathfrak{R}(u_h) = L(u_h). \tag{9.24}$$

In a manner analogous to the development of the previous section, a continuous discretization error transport equation can be derived within the finite element framework (Babuska and Rheinboldt, 1978a). This so-called residual equation has three different types of terms: (1) interior residuals that determine how well the finite element solution satisfies the mathematical model in the solution domain, (2) terms associated with any discretized boundary conditions on the domain boundary (e.g., Neumann boundary conditions), and (3) inter-element residuals which are functions of the discontinuities in normal fluxes across element–element boundaries (Ainsworth and Oden, 2000). It is the treatment of these three terms that differentiates between explicit and implicit residual methods. While the use of explicit and implicit residual methods for finite elements has reached a certain level of maturity for elliptic problems (Ainsworth and Oden, 2000), the drawback to these methods is that they only provide error estimates in the energy norm of the discretization error given in Equation (9.4). While the energy norm is a natural quantity by which to judge the overall goodness of a finite element solution, most scientific computing is used to make an engineering decision with regards to a specific system response quantity (called "quantities of interest" by the finite element community). Extension of both the explicit and implicit residual methods to provide error estimates in a system response quantity generally requires the solution to the adjoint system (see Section 9.3.2.3).

Explicit Residual Methods

Explicit residual methods are those which employ information available from the finite element solution along with the finite element residuals to directly compute the error estimate. First developed by Babuska and Rheinboldt (1978b), explicit residual methods lump all three types of residual terms under a single, unknown constant. The analysis requires the use of the triangle inequality, which does not allow for cancellation between the different residual types. Due both to the use of the triangle inequality and the methods for estimating the unknown constant, explicit residual methods tend to be conservative estimates of the discretization error. They provide an element-wise estimate of the *local contribution* to the bound for the global energy norm of the error, but *not* a local estimate of the true error, which would include both local and transported components. Since explicit residual methods deal only with local contributions to the error, they can also be used for solution adaptation procedures. Stewart and Hughes (1998) have provided a tutorial on explicit residual methods and also discuss their relationship to *a priori* error estimation.

Implicit Residual Methods

Implicit residual methods avoid the approximations required in explicit residual methods by seeking solutions to the residual equation which governs the transport and generation of the discretization error. In order to achieve non-trivial solutions to the global residual equation, the mesh would either have to be refined or the order of the finite element basis functions increased. Both of these approaches would be significantly more expensive than obtaining the original finite element solution and, therefore, are not considered practical. Instead, the global residual equation is decomposed into a series of uncoupled, local boundary value problems approximating the global equation. These local problems can be solved over a single element using the element residual method (Demkowicz et al., 1984; Bank and Weiser, 1985) or over a small patch of elements using the subdomain residual method (Babuska and Rheinboldt, 1978a, 1978b). The solution to the local boundary value problems provides the local discretization error estimate, while the global error estimate is simply summed over the domain. By directly treating all three types of terms that show up in the residual equation, implicit residual methods retain more of the structure of the residual equation than do the explicit methods, and thus should, in theory, provide tighter discretization error bounds.

9.3.2.3 Adjoint Methods for System Response Quantities

The analyst employing scientific computing is often interested in system response quantities that can be post-processed from the solution. These system response quantities can take the form of integrated quantities (e.g., net flux through or force acting on a boundary), local solution quantities (e.g., maximum stress or maximum temperature), or even an average of the solution over some region.

Adjoint methods in scientific computing were initially used for determining solution sensitivities for design optimization problems (e.g., see Jameson, 1988). In the optimization setting, the adjoint (or dual) problem can be solved for sensitivities of a solution functional (e.g., a system response quantity) that one wishes to optimize relative to some chosen design parameters. The strength of the adjoint method is that it is efficient even when a large number of design parameters are involved. In the context of optimization in scientific computing, adjoint methods can be thought of as constrained optimization problems where a chosen solution functional is to be optimized subject to the constraint that the solution must also satisfy the mathematical model (or possibly the discrete equations).

Adjoint methods can also be used for estimating the discretization error in a system response quantity in scientific computing applications. Consider a scalar solution functional $f_h(u_h)$ evaluated on mesh h. The true discretization error in the functional is

$$\varepsilon_{f_h} = f_h(u_h) - f(\tilde{u}). \tag{9.25}$$

An approximation of the discretization error in this functional is given by

$$\varepsilon_{f_h} \cong f_h(u_h) - f_h(\tilde{u}). \tag{9.26}$$

If the functional involves integration (e.g., integrated lift force on an aerodynamic body), then the higher-order functional denoted by $f_h(\tilde{u})$ should be computed with a corresponding higher-order integration. Performing a Taylor series expansion of $f_h(\tilde{u})$ about the discrete solution gives

$$f_h(\tilde{u}) = f_h(u_h) + \left.\frac{\partial f_h}{\partial u}\right|_u (\tilde{u} - u_h) + O(\varepsilon_h^2). \tag{9.27}$$

Here $\left.\frac{\partial f_h}{\partial u}\right|_{u_h}$ is a row vector denoting the derivative of the functional with respect to the discrete solution at every node/cell and $(u_h - \tilde{u})$ is the column vector of local solution errors over every node/cell. Next, an expansion of the discrete operator $L_h(\cdot)$ is performed at $\tilde{u}$ about u_h:

$$L_h(\tilde{u}) = L_h(u_h) + \left.\frac{\partial L_h}{\partial u}\right|_{u_h} (\tilde{u} - u_h) + O(\varepsilon_h^2), \tag{9.28}$$

where $L_h(\tilde{u})$ is the discrete residual, an approximation of the truncation error from Equation (9.20), and $\left.\frac{\partial L_h}{\partial u}\right|_{u_h}$ is the Jacobian which linearizes the discrete equations with respect to the solution. This Jacobian may already be available since it can also be used to formulate implicit solutions to the discrete equations and for design optimization. Since $L_h(u_h) = 0$, if the higher-order terms are dropped, then Equation (9.28) can be rearranged to obtain

$$(\tilde{u} - u_h) = \left[\left.\frac{\partial L_h}{\partial u}\right|_{u_h}\right]^{-1} L_h(\tilde{u}). \tag{9.29}$$

Substituting this equation into Equation (9.27) to eliminate the column vector of local errors gives

$$f_h(\tilde{u}) \cong f_h(u_h) + \left.\frac{\partial f_h}{\partial u}\right|_{u_h} \left[\left.\frac{\partial L_h}{\partial u}\right|_{u_h}\right]^{-1} L_h(\tilde{u}) \tag{9.30}$$

or

$$f_h(\tilde{u}) \cong f_h(u_h) + \Psi^T L_h(\tilde{u}), \tag{9.31}$$

where Ψ is the row vector of discrete adjoint sensitivities and the superscript "T" is the transpose which converts this column vector into a row vector. The adjoint sensitivities are found by solving

$$\Psi^T = \left.\frac{\partial f_h}{\partial u}\right|_{u_h} \left[\left.\frac{\partial L_h}{\partial u}\right|_{u_h}\right]^{-1}. \tag{9.32}$$

This equation can be put into the standard linear equation form by transposing both sides of Equation (9.32) and noting that for a column vector b and matrix A, $\left(b^T A^{-1}\right)^T = \left(A^{-1}\right)^T \left(b^T\right)^T = \left(A^T\right)^{-1} b$, thus

$$\left[\left.\frac{\partial L_h}{\partial u}\right|_{u_h}\right]^T \Psi = \left[\left.\frac{\partial f_h}{\partial u}\right|_{u_h}\right]^T. \tag{9.33}$$

The linearizations of the functional and the discrete operator both omit terms of order $\varepsilon_{f_h}^2$. The discretization error estimate in the functional can thus be computed by inserting Equation (9.31) into Equation (9.26) as:

$$\bar{\varepsilon}_{f_h} = -\Psi^T L_h(\tilde{u}) = -\Psi^T TE_h(\tilde{u}). \tag{9.34}$$

For example, if the numerical method for the original (primal) problem is third-order accurate, then these functional error estimates will be sixth-order accurate. The relationship between the linearized form of the error transport equations given by Equation (9.19) and adjoint methods can be shown as follows. Inserting Equation (9.32) into Equation (9.34) gives the following expression for the discretization error estimate in the solution functional:

$$\bar{\varepsilon}_{f_h} = -\left.\frac{\partial f_h}{\partial u}\right|_{u_h} \left[\left.\frac{\partial L_h}{\partial u}\right|_{u_h}\right]^{-1} TE_h(\tilde{u}). \tag{9.35}$$

The product of the first two factors in Equation (9.35) is equal to the adjoint sensitivities $(-\Psi^T)$, while the product of the last two factors is equal to the local error estimate $(-\bar{\varepsilon}_h)$ in the field solution. Thus, the functional error estimate can be computed as either the (linearized) adjoint sensitivities times the local truncation error or as the functional linearization times the local field solution error estimates. In fact, when the same exact linearizations and truncation error estimates are used, the two approaches produce identical functional error estimates within round-off error (Tyson and Roy, 2019).

The adjoint solution provides the linearized sensitivities of the solution functional f_h to perturbations in the discrete operator $L_h(\cdot)$. As such, the adjoint solution vector components are often referred to as the adjoint sensitivities. Equation (9.31) shows that the adjoint solution provides the sensitivity of the discretization error in the solution functional $f(\cdot)$ to the local sources of discretization error (i.e., the truncation error) in the domain. This observation can be used as the basis for providing solution adaptation targeted for solution functionals (see Section 9.8). Because the discrete operator $L_h(\cdot)$ is used above, this approach is called the *discrete adjoint method.* A similar analysis using expansions of the continuous mathematical operator $L(\cdot)$ and functional $f(\cdot)$ can be performed to obtain discretization error estimates using the *continuous adjoint method.* Both continuous and discrete adjoint methods also require appropriate formulations of initial and boundary conditions.

Adjoint methods have been used extensively for both finite element methods (see Johnson and Hansbo, 1992; Paraschivoiu et al., 1997; Rannacher and Suttmeier, 1997; Estep et al., 2000; Cheng and Paraschivoiu, 2004) and finite volume methods (see Venditti and Darmofal, 2002, 2003). The main drawbacks to adjoint methods are that (a) they can only be implemented in a linearized fashion, (b) they do not provide local discretization error estimates for the field quantities, and (c) for unsteady problems, the adjoint problem must be solved backwards in time (after the original/primal problem is solved and stored forwards in time).

9.4 Richardson Extrapolation

As mentioned earlier, Richardson extrapolation is a recovery method that utilizes mesh refinement to obtain a more accurate numerical solution. It relies on discrete solutions on two (or more) meshes to estimate the discretization error. We place special emphasis on Richardson extrapolation because it is the only discretization error estimator that can be applied as a post-processing step to both local solution variables and global system response quantities from *any* discretization approach. The main drawback is the expense of generating and then computing multiple discrete solutions on systematically-refined meshes.

The basic concept behind Richardson extrapolation (Richardson, 1911, 1927) is as follows. If one knows the formal rate of convergence of a discretization method with mesh refinement, and if discrete solutions on two systematically refined meshes are available, then one can use this information to obtain an estimate of the exact solution to the mathematical model. Depending on the level of confidence one has in this estimate, it can be used to either correct the fine mesh solution or to provide a discretization error estimate for it. While Richardson's original work applied the approach locally over the domain to the dependent variables in the mathematical model (e.g., u_h), it can be readily applied to any solution functional/system response quantity (f_h). There is, however, the additional requirement that the numerical approximations (integration, differentiation, etc.) used to obtain the system response quantity be at least of the same order of accuracy as the underlying discrete solutions. While the following development is for global system response quantities, e.g., the lift or drag on a flight vehicle, the application to local field variables is identical.

Recall the definition of the discretization error in some general global solution variable f on a mesh with spacing h

$$\varepsilon_h = f_h - \tilde{f}, \tag{9.36}$$

where f_h is the exact solution to the discrete equations and $\tilde{f}$ is the exact solution to the original partial differential equations. We can expand the numerical solution f_h in a power series in h about the exact solution

$$f_h = f_{h=0} + g_1 h + g_2 h^2 + g_3 h^3 + O(h^4), \tag{9.37}$$

where $O(h^4)$ denotes a leading error term on the order of h to the fourth power and, for a convergent numerical method, $f_{h=0} = \tilde{f}$. Making this substitution and moving $\tilde{f}$ to the left-hand side allows us to write the discretization error for a mesh with spacing h as

$$\varepsilon_h = f_h - \tilde{f} = g_1 h + g_2 h^2 + g_3 h^3 + O(h^4). \tag{9.38}$$

In general, we require numerical methods that are higher than first-order accurate, and thus discretization methods are chosen which cancel out selected lower-order terms. For example, if a formally second-order accurate numerical scheme is chosen, then the general discretization error expansion becomes

$$\varepsilon_h = f_h - \tilde{f} = g_2 h^2 + g_3 h^3 + O(h^4). \tag{9.39}$$

Equation (9.39) forms the basis for standard Richardson extrapolation, which is described next.

9.4.1 Standard Richardson Extrapolation

The *standard Richardson extrapolation* procedure, as originally formulated by Richardson (1911), applies exclusively to cases where the numerical scheme is formally second-order accurate and the mesh is systematically refined (or coarsened) by a factor of two. Consider a second-order discretization scheme that is used to produce numerical solutions on two meshes: a fine mesh with spacing h and a coarse mesh with spacing $2h$. Since the scheme is second-order accurate, the g_1 coefficient is zero and the discretization error equation (Equation [9.39]) can be rewritten as

$$f_h = \tilde{f} + g_2 h^2 + g_3 h^3 + O(h^4). \tag{9.40}$$

Applying this equation on the two mesh levels h and $2h$ gives

$$\begin{aligned} f_h &= \tilde{f} + g_2 h^2 + g_3 h^3 + O(h^4), \\ f_{2h} &= \tilde{f} + g_2 (2h)^2 + g_3 h^3 + O(h^4). \end{aligned} \tag{9.41}$$

Eliminating $\tilde{f}$ from these equations and solving for g_2 yields

$$g_2 = \frac{f_{2h} - f_h}{3h^2} - \frac{7}{3} g_3 h + O(h^2). \tag{9.42}$$

Substituting Equation (9.42) into the fine grid expansion of Equation (9.41) and solving for $\tilde{f}$ gives

$$\tilde{f} = f_h + \frac{f_h - f_{2h}}{3} + \frac{4}{3} g_3 h^3 + O(h^4). \tag{9.43}$$

Combining the terms of order h^3 and higher in Equation (9.43) with the exact solution yields

$$\bar{f} = \tilde{f} - \frac{4}{3} g_3 h^3 + O(h^4). \tag{9.44}$$

This Richardson extrapolated solution $\bar{f}$ is thus a higher-order accurate estimate of the exact solution $\tilde{f}$. The Richardson extrapolated solution can thus be computed by inserting Equation (9.44) in Equation (9.43) to give

$$\tilde{f} = f_h + \frac{f_h - f_{2h}}{3}. \tag{9.45}$$

Equation (9.45) is the standard Richardson extrapolation relationship and provides an estimate $\bar{f}$ of the exact solution $\tilde{f}$ which is higher-order accurate than the underlying numerical solution f_h. This estimate will, therefore, converge to the exact solution faster than the numerical solutions themselves as the mesh is refined. A discretization error estimate derived from the Richardson extrapolation procedure will thus be asymptotically consistent.

There is often confusion as to the order of accuracy of the Richardson extrapolation estimate. As shown in Equation (9.44), the estimates from standard Richardson extrapolation are generally third-order accurate. In Richardson's original work (Richardson, 1911), he used this extrapolation procedure to obtain a higher-order accurate solution for the stresses in a masonry dam based on two second-order accurate numerical solutions. The original partial differential equation was Poisson's equation and he employed central differences which cancelled out the odd powers in the truncation error (i.e., $g_3 = 0$). His estimate for the exact solution was thus fourth-order accurate as is clearly shown in Equation (9.44) when $g_3 = 0$.

9.4.2 Generalized Richardson Extrapolation

Richardson extrapolation can be generalized to pth-order accurate schemes and for two meshes systematically refined by an arbitrary factor. First, consider the general discretization error expansion for a pth-order scheme:

$$\varepsilon_h = f_h - \tilde{f} = g_p h^p + g_{p+1} h^{p+1} + g_{p+2} h^{p+2} + \ldots. \tag{9.46}$$

Introducing the grid refinement factor as the ratio of the coarse to fine grid spacing, we have

$$r = \frac{h_{coarse}}{h_{fine}} > 1, \tag{9.47}$$

and the coarse grid spacing can thus be written as $h_{coarse} = r h_{fine}$. Choosing $h_{fine} = h$ for simplicity, the discretization error equations on the two meshes can be written as

$$\begin{aligned} f_h &= \tilde{f} + g_p h^p + g_{p+1} h^{p+1} + O(h^{p+2}), \\ f_{rh} &= \tilde{f} + g_p (rh)^p + g_{p+1} (rh)^{p+1} + O(h^{p+2}). \end{aligned} \tag{9.48}$$

As before, these equations can be used to eliminate the g_p coefficient and solve for $\tilde{f}$ to give

$$\tilde{f} = f_h + \frac{f_h - f_{rh}}{r^p - 1} + g_{p+1} h^{p+1} \frac{r^p (r-1)}{r^p - 1} + O(h^{p+2}). \tag{9.49}$$

Again, combining terms of order h^{p+1} and higher with the exact solution $\tilde{f}$ gives

$$\bar{f} = \tilde{f} - g_{p+1} \frac{r^p (r-1)}{r^p - 1} h^{p+1} + O(h^{p+2}). \tag{9.50}$$

Substituting this expression into Equation (9.49) results in the generalized Richardson extrapolation estimate $\bar{f}$:

$$\bar{f} = f_h + \frac{f_h - f_{rh}}{r^p - 1}. \tag{9.51}$$

As is shown by Equation (9.50), this estimate of the exact solution is generally $(p+1)$-order accurate estimate of the exact solution to the mathematical model $\tilde{f}$ unless additional error cancellation occurs in the higher-order terms (e.g., if the g_{p+1} coefficient is zero).

There have been numerous extensions to the original Richardson extrapolation approach. These extensions address methods for obtaining the Richardson extrapolation estimate on all fine grid spatial points, extending it to all fine mesh points in space and time, and combining it with a discrete residual minimization (resulting in a hybrid extrapolation/residual method). See chapter 8 of Oberkampf and Roy (2010) for more details.

9.4.3 Assumptions for Richardson Extrapolation

There are five basic assumptions required for Richardson extrapolation to provide reliable estimates of the exact solution to the mathematical model: (1) *both* discrete solutions are in the asymptotic range, (2) the meshes have a uniform (Cartesian) spacing over the domain, (3) the coarse and fine meshes are related through systematic refinement, (4) the solutions are smooth, and (5) other sources of numerical error are small. These five assumptions are each discussed in detail below.

9.4.3.1 Asymptotic Range

The formal order of accuracy of a discretization scheme is the theoretical rate at which the discretization error is reduced as the mesh is refined. Recall that in Chapter 5 we used a continuous error transport equation to relate the formal order of accuracy to the lowest order term in the truncation error. This lowest order term will necessarily dominate the higher-order terms in the limit as the mesh spacing parameter h goes to zero. The dependent solution variables generally converge at the formal order of accuracy in the asymptotic range. It must be stressed, however, that the mesh associated with the asymptotic range for one system response quantity can be *very different* than the mesh required for the asymptotic range of a different system response quantity. For example, the mesh required for achieving the asymptotic range of a local quantity, such as local shear stress, may be much finer than that for a global quantity, such as aerodynamic lift. One should keep in mind that this asymptotic requirement applies not just to the fine mesh solution but to the coarse mesh solution as well. Procedures for confirming that the asymptotic range has been reached will be given in Section 9.5.2.

9.4.3.2 Uniform Mesh Spacing

The discretization error expansion given in Equation (9.46) is in terms of a single mesh spacing parameter h. This parameter is a measure of the size of the discretization, and thus has units of length for spatial discretizations and time for temporal discretizations. This could be strictly interpreted as allowing only Cartesian meshes with spacing h in each of the spatial coordinate directions. While this restriction seemingly prohibits the use of Richardson extrapolation for practical scientific computing applications, this is in fact not the case. Recall our previous discussions on the use of local or global transformations and their impact on the order of accuracy in Section 5.6. Those findings also apply to the use of Richardson extrapolation. The spatial mesh quality can affect the formal order of accuracy of the method through interactions with the mesh resolution and the local solution behavior; this relationship is explored in detail in Section 9.8. Discrete transformations will not adversely impact the Richardson extrapolation procedure as long as they are of the same order of accuracy as the discretization scheme or higher. For example, if the underlying discretization is second-order accurate, then the numerical approximations used to determine grid metrics and element geometry should also maintain second-order accuracy. Thompson et al. (1985) note that there may be accuracy advantages to evaluating the discrete transformation metrics with the same underlying discretization method used for the solution derivatives.

9.4.3.3 Systematic Mesh Refinement

An often-overlooked requirement for the use of Richardson extrapolation is that the two mesh levels be systematically refined. Recall the definition of systematic mesh refinement given in Section 5.6, which requires that the mesh refinement be both uniform and consistent. Uniform refinement means that the mesh is refined by the same factor over the entire domain, which *precludes* the use of local refinement or adaptation during the Richardson extrapolation procedure. Consistent refinement requires that the mesh quality must either remain constant or improve with mesh refinement. Examples of mesh quality metrics include cell aspect ratio, cell skewness, and cell-to-cell stretching factor.

For structured meshes, systematic mesh refinement can be performed by simply removing every other grid node from a fine mesh in each coordinate direction. For general unstructured meshes, the most common technique is to start with a coarse mesh, then systematically refine each element, usually with a refinement factor of two. However, such refinement can lead to a reduction in the quality of the refined elements, e.g., when generating eight new tetrahedra from a single tetrahedron in three dimensions ($r = 2$). For all mesh types, the mesh parameters can by systematically modified to refine/coarsen a mesh within the original mesh generation software, but care must be taken to ensure that the refinement is indeed systematic. Techniques for quantitatively evaluating just how systematic the mesh refinement is are discussed in Section 9.7. For an example of how Richardson extrapolation can fail in the presence of nonuniform mesh refinement, even in the asymptotic range, see Eca and Hoekstra (2009a).

9.4.3.4 Smooth Solutions

As discussed earlier, the coefficients g in the discretization expansion given by Equation (9.46) are generally functions of the solution derivatives. As such, the Richardson extrapolation procedure will break down in the presence of discontinuities in any of the dependent variables or their derivatives. This is further complicated by the fact that the observed order of accuracy often reduces to first order or lower in the presence of certain discontinuities and singularities, *regardless* of the formal order of accuracy of the method (see Chapter 5). Examples of discontinuities include shock waves, contact surfaces, free surfaces, and material interfaces, while examples of singularities include stagnation points, surface curvature discontinuities, and surface slope discontinuities (e.g., sharp corners).

9.4.3.5 Other Numerical Errors Sources

Recall that the discretization error is defined as the difference between the exact solution to the discrete equations and the exact solution to the mathematical model. The exact solution to the discrete equations is never known due to round-off error, iterative error, and statistical sampling errors (when present). In practice, the available numerical solutions are used as surrogates for the exact solution to the discretized equations. If the other numerical error sources are too large, then they will adversely impact the Richardson extrapolation procedure since any extrapolation procedure will tend to amplify "noise" (Roache, 1998). A good rule of thumb is to ensure that all other sources of numerical error are at least two orders of magnitude smaller than the discretization error in the fine grid numerical solution (Roy, 2005; Eca and Hoekstra, 2009b). Of course, the accuracy of any error estimate relies on a foundation of rigorous code verification (see Chapters 5 and 6).

9.4.4 Discretization Error Estimation

While it may be tempting to use the Richardson extrapolated value as a more accurate solution than the fine grid numerical solution, this should only be done when there is a high degree of confidence that the five assumptions underlying Richardson extrapolation have indeed been met. In particular, the observed order of accuracy (which generally requires discrete solutions on three systematically-refined meshes as discussed in Section 9.5.2) must first be shown to match the formal order of accuracy of the discretization scheme. In any case, when only two mesh levels are available, the asymptotic nature of the solutions cannot be confirmed, thus one is limited to simply using the Richardson extrapolated value to estimate the discretization error in the numerical solutions.

Substituting the estimated exact solution from the generalized Richardson extrapolation expression (Equation 9.51) into the definition of the discretization error on the fine mesh (Equation 9.36) gives the estimated discretization error for the fine mesh (with spacing h)

$$\bar{\varepsilon}_h = f_h - \bar{f} = f_h - \left(f_h + \frac{f_h - f_{rh}}{r^p - 1} \right)$$

simply

$$\bar{\varepsilon}_h = -\frac{f_h - f_{rh}}{r^p - 1}. \tag{9.52}$$

The equivalent discretization error estimate for the coarse grid solution is

$$\bar{\varepsilon}_{rh} = -r^p \frac{f_h - f_{rh}}{r^p - 1}. \tag{9.53}$$

While these expressions do provide consistent discretization error estimate as the mesh is systematically refined, there is no guarantee that the estimate will be reliable for any given fine mesh (h) and coarse mesh (rh) discrete solutions. Therefore, if only two discrete solutions are available, then this numerical error estimate should be converted into a numerical uncertainty as discussed in Chapter 7.

9.4.5 Advantages and Disadvantages

The primary advantage that Richardson extrapolation holds over other discretization error estimation methods is that it can be used as a post-processing technique applied to any discretization scheme (finite difference, finite volume, finite element, etc.). In addition, it gives estimates in the total error, which includes both locally generated errors and those transported from other regions of the domain. Finally, it can be used for any quantity of interest including both local solution quantities and derived system response quantities.

There are, however, some disadvantages to using discretization error estimators based on Richardson extrapolation. First and foremost, they rely on having multiple numerical solutions in the asymptotic grid convergence range. This can place significant burdens on the grid generation process, which is already a bottleneck in many scientific computing applications. Furthermore, these additional solutions can be extremely expensive to compute. Consider the case where one starts with a 3D mesh consisting of 1 million elements. Performing a mesh refinement with a refinement factor of two thus requires a solution on a mesh with 8 million elements. When one also accounts for the additional time steps or iterations required for this finer mesh, the solution cost easily increases by an order of magnitude with each refinement (note that integer refinement is not necessarily required as discussed in Oberkampf and Roy, 2010).

The underlying theory of Richardson extrapolation requires smooth solutions, thus reducing the reliability of these error estimators for problems with discontinuities or singularities. In addition, the extrapolation procedure tends to amplify other sources of error such as round-off and iterative convergence error (Roache, 1998). Finally, the extrapolated quantities will not satisfy the same governing and auxiliary equations as either the numerical solutions or the exact solution. For example, if an equation of state is used to relate the density, pressure, and temperature in a gas, there is no guarantee that extrapolated values for density, pressure, and temperature will also satisfy this equation.

9.5 Reliability of Discretization Error Estimators

One of the key requirements for reliability (i.e., accuracy) of any of the discretization error estimators discussed in this chapter is that the solution (or solutions) must be in the asymptotic range. This section provides a discussion of just what this asymptotic range means for discretization approaches involving both mesh (h) refinement and order (p) refinement. Regardless of whether h- or p-refinement is used, the *demonstration* that the asymptotic range has been achieved generally requires three discrete solutions be computed. Demonstrating that the asymptotic range has been reached can be surprisingly difficult for complex scientific computing applications involving nonlinear, hyperbolic, coupled systems of equations. It is unlikely that the asymptotic range will be reached in such cases without using some type of solution adaptation strategy (see Section 9.8).

9.5.1 Asymptotic Range

The asymptotic range is defined differently depending on whether one is varying the mesh resolution or the formal order of accuracy of the discretization scheme. When mesh refinement is employed, then the asymptotic range is defined as the sequence of systematically-refined meshes over which the discretization error reduces at the formal order of accuracy of the discretization scheme. Examining the discretization error expansion for a pth-order accurate scheme given by Equation (9.46), the asymptotic range is achieved when h is sufficiently small that the h^p term is much larger than any of the higher-order terms (h^{p+1}, h^{p+2}, etc.). Due to possible differences in the signs for the higher-order terms, the behavior of the discretization error outside of the asymptotic range is *unpredictable*. Demonstrating that the asymptotic range has been reached using systematic mesh refinement is achieved by evaluating the observed order of accuracy. The observed order assesses the behavior of the discrete solutions over a range of meshes and its evaluation is discussed in the next section.

For discretization methods involving order refinement, the asymptotic range is determined by examining the behavior of the numerical solutions with successively refined basis functions, all on the same mesh. As the basis functions are refined and the physical phenomena in the problem are resolved, the discrete solutions will eventually become better approximations of the exact solution to the mathematical model. Convergence is best monitored with error norms since convergence can be oscillatory with increased basis order. An example of hierarchical basis functions used within the finite element method is given by Bank (1996).

9.5.2 Observed Order of Accuracy

The observed order of accuracy is the measure that is used to assess the confidence in a discretization error estimate. When the observed order of accuracy is shown to match the formal order, then one can have a high degree of confidence that the error estimate is reliable. In the discussion of the observed order of accuracy from Chapter 5, only two numerical solutions were required for its calculation because the exact solution was known. When the exact solution is not known, which is the case for solution verification, three numerical solutions on systematically refined meshes are usually required to calculate the observed order of accuracy. For this observed order of accuracy to match the formal order of the discretization scheme, the requirements are the same as those given in Section 9.4.3 for Richardson extrapolation. When any of these requirements fail to be met, inaccurate values for the observed order of accuracy can be obtained (e.g., see Roy, 2003; Salas, 2006; Eca and Hoekstra, 2009a).

9.5.2.1 Constant Grid Refinement Factor

Consider a pth-order accurate scheme with numerical solutions on a fine mesh (h_1), a medium mesh (h_2), and a coarse mesh (h_3). For the case of a constant grid refinement factor, i.e.,

$$r = \frac{h_2}{h_1} = \frac{h_3}{h_2} > 1,$$

we can write

$$h_1 = h, \quad h_2 = rh, \quad h_3 = r^2 h.$$

Using the discretization error expansion from Equation (9.46), we can now write for the three discrete solutions:

$$\begin{aligned} f_1 &= \tilde{f} + g_p h^p + g_{p+1} h^{p+1} + O(h^{p+2}), \\ f_2 &= \tilde{f} + g_p (rh)^p + g_{p+1} (rh)^{p+1} + O(h^{p+2}), \\ f_3 &= \tilde{f} + g_p (r^2 h)^p + g_{p+1} (r^2 h)^{p+1} + O(h^{p+2}). \end{aligned} \tag{9.54}$$

Neglecting terms of order h^{p+1} and higher allows us to recast these three equations in terms of the observed order of accuracy $\hat{p}$ and the estimated exact solution $\bar{f}$:

$$\begin{aligned} f_1 &= \bar{f} + g_p h^{\hat{p}}, \\ f_2 &= \bar{f} + g_p (rh)^{\hat{p}}, \\ f_3 &= \bar{f} + g_p (r^2 h)^{\hat{p}}. \end{aligned} \tag{9.55}$$

The observed order of accuracy $\hat{p}$ will only match the formal order if the higher order terms are indeed small. Subtracting f_2 from f_3 and f_1 from f_2 yields

$$f_3 - f_2 = g_p (r^2 h)^p - g_p (rh)^{\hat{p}} = g_p r^{\hat{p}} h^{\hat{p}} (r^{\hat{p}} - 1) \tag{9.56}$$

and

$$f_2 - f_1 = g_p (rh)^{\hat{p}} - g_p h^{\hat{p}} = g_p h^{\hat{p}} (r^{\hat{p}} - 1). \tag{9.57}$$

Dividing Equation (9.56) by Equation (9.57) gives

$$\frac{f_3 - f_2}{f_2 - f_1} = r^{\hat{p}}. \tag{9.58}$$

Taking the natural log of both sides and solving for the observed order of accuracy $\hat{p}$ gives

$$\hat{p} = \frac{\ln\left(\frac{f_3 - f_2}{f_2 - f_1}\right)}{\ln(r)}. \tag{9.59}$$

Consistent with the development of generalized Richardson extrapolation in Section 9.4.2, the Richardson extrapolated estimate of the exact solution $\bar{f}$ and the leading error term coefficient g_p are now given in terms of the observed order of accuracy $\hat{p}$ by

$$\bar{f} = f_1 + \frac{f_1 - f_2}{r^{\hat{p}} - 1} \tag{9.60}$$

and

$$g_p = \frac{f_1 - \bar{f}}{h^{\hat{p}}}. \tag{9.61}$$

Note that it is only when the observed order of accuracy nearly matches the formal order of the numerical scheme that we can expect the discretization error estimate given by Equation (9.60) to be reliable. This is equivalent to saying that the solutions on all three meshes are in the asymptotic range and the higher-order terms in Equations (9.54) are small for all three meshes. In practice, when this locally observed order of accuracy is used for the extrapolation estimate, it is often restricted to be in the range

$$0.5 \leq \hat{p} \leq p_f,$$

where p_f is the formal order of accuracy of the discretization scheme. Allowing the observed order of accuracy to increase above the formal order can result in discretization error estimates that are not conservative (i.e., they underestimate the error). Furthermore, as $\hat{p}$ approaches zero, the magnitude of the extrapolated estimate grows without bound, providing no practical guidance as to the size of the discretization error.

9.5.2.2 Nonconstant Grid Refinement Factor

For the case of non-constant grid refinement factors

$$r_{12} = \frac{h_2}{h_1} > 1, \qquad r_{23} = \frac{h_3}{h_2} > 1,$$

where $r_{12} \neq r_{23}$, the determination of the observed order of accuracy $\hat{p}$ is more complicated. For this case, the following transcendental equation (Roache, 1998) must be solved for $\hat{p}$:

$$\frac{f_3 - f_2}{r_{23}^{\hat{p}} - 1} = r_{12}^{\hat{p}} \left(\frac{f_2 - f_1}{r_{12}^{\hat{p}} - 1} \right). \tag{9.62}$$

This equation can usually be solved with a simple direct substitution iterative procedure as discussed in Chapter 8 to give

$$\hat{p}^{k+1} = \frac{\ln\left[\left(r_{12}^{\hat{p}^k} - 1\right)\left(\frac{f_3 - f_2}{f_2 - f_1}\right) + r_{12}^{\hat{p}^k}\right]}{\ln(r_{12} r_{23})}, \tag{9.63}$$

where an initial guess of $\hat{p}^k = p_f$ (the formal order of the scheme) can be used. Once the observed order of accuracy is found, the estimate of the exact solution $\bar{f}$ and the leading error term coefficient g_p are given by Equations (9.60) and (9.61) by replacing the constant grid refinement factor with $r = r_{12}$.

9.5.2.3 Residual-Based Methods

When residual-based methods are used to obtain error estimates, only two systematically-refined meshes are needed to compute the observed order of accuracy. The equation is similar to Equation (5.22) used for code verification, but the true discretization error (e.g., ε_h) is replaced with the error estimates (e.g., $\bar{\varepsilon}_h$), resulting in

$$\hat{p} = \frac{\ln\left(\frac{\bar{\varepsilon}_{rh}}{\bar{\varepsilon}_h}\right)}{\ln(r)}, \tag{9.64}$$

or in terms of error norms,

$$\hat{p} = \frac{\ln\left(\frac{\|\bar{\varepsilon}_{rh}\|}{\|\bar{\varepsilon}_{h}\|}\right)}{\ln(r)}. \tag{9.65}$$

The primary advantage of requiring only two systematically refined grids is that it is computationally easier to get two solutions in the asymptotic range than it is to get three.

9.5.2.4 Application to System Response Quantities

Recall that system response quantities are defined as any solution result derived from the solution to the mathematical model or its discrete approximation. Examples of common system response quantities in scientific computing are lift and drag in aerodynamics, heat flux through a surface in heat transfer analysis, and maximum stress in a structural mechanics problem. The observed order of accuracy for system response quantities can be evaluated by the approaches described earlier in this section. For this observed order of accuracy to match the formal order of the discretization scheme, there is an additional requirement beyond those described in Section 9.4.3 for Richardson extrapolation. This additional requirement pertains to the order of accuracy of any numerical approximations used to compute the system response quantity. When the system response quantity is an integral, a derivative, or an average, then the numerical approximations used in its evaluation must be of *at least* the same order of accuracy as the underlying discretization scheme. In most cases, integrated quantities and averages are better behaved and converge more rapidly with mesh refinement than local quantities. However, in some cases the errors due to the numerical quadrature can interact with the numerical errors in the discrete solution and adversely impact the observed order of accuracy computation. An example of these interactions for the computation of drag on an inviscid supersonic blunt body (mixed elliptic/hyperbolic) problem is given by Salas and Atkins (2009).

9.6 Discretization Uncertainty and the Grid Convergence Index

As discussed previously, the calculation of the observed order of accuracy requires at least three systematically refined meshes. It is common that the observed order of accuracy differs from the formal order of accuracy. In this case, the error estimate is much less reliable and should generally be converted into a numerical uncertainty. While the difference between the discrete solution and the (unknown) exact solution to the mathematical model is still truly a numerical error, because we do not know the true value of this error, it should be represented as an epistemic uncertainty. Uncertainties due to lack of knowledge are called epistemic uncertainties and are distinct from aleatory (or random) uncertainties. Epistemic uncertainties can be reduced by providing more information, in this case, additional computations on more refined or even adapted meshes. This subsection describes a method for converting the discretization error estimate from Richardson extrapolation into an uncertainty.

Roache (1994) proposed the *Grid Convergence Index*, or GCI, as a method for uniform reporting of grid refinement studies. Roache's stated goal in formulating the GCI is to achieve a 95% certainty (i.e., that it provide conservative uncertainty estimates in 19 out of 20 cases) over a wide range of applications. The original GCI is based on the often reported relative difference between two discrete solutions, but also properly accounts for the amount of grid refinement and the order of accuracy. The GCI takes the further step of converting the discretization error estimate into an uncertainty estimate using absolute values. The GCI for the fine grid numerical solution is defined as

$$GCI = \frac{F_s}{r^p - 1}\left|\frac{f_2 - f_1}{f_1}\right|, \tag{9.66}$$

where F_s is a factor of safety. When only two discrete solutions are available, then the formal order of accuracy of the discretization scheme is used in the definition of the GCI along with a factor of safety of $F_s = 3$. However, when three discrete solutions are available then the observed order of accuracy can be computed and used in Equation (9.66). When the observed order matches the formal order of accuracy, then a less conservative factor of safety of $F_s = 1.25$ can be used (Roache, 1998).

If the GCI is re-derived in terms of the Richardson extrapolation estimate, then f_1 in the denominator of Equation (9.66) will be replaced with the Richardson extrapolation estimate $\bar{f}$; however, this modification will only significantly affect the uncertainty estimate when the uncertainties themselves are large (Roy, 2001). When the implementation of Equation (9.66) is used, the GCI returns a fractional estimate of the relative uncertainty in the fine grid solution. For example, a GCI value of 0.15 indicates an uncertainty due to the discretization process of 15% in the fine grid solution. As with all extrapolation-based approaches, systematic mesh refinement must be used.

It is important to include the factor of safety in the GCI, primarily because it is implicitly assumed that the two solutions are near the asymptotic region. Even when three solutions are used to show that the observed order of accuracy matches the formal order, the GCI is based on the Richardson extrapolated value, which is itself an ordered estimate of the exact solution to the mathematical model (recall Equation 9.50). Thus, we do not know *a priori* whether the estimated exact solution is above or below the true exact solution to the mathematical model. Consider Figure 9.5, which shows two numerical solutions (f_1 and f_2), the estimated exact solution from Richardson extrapolation $\bar{f}$, and the true exact solution $\tilde{f}$. In general, there is an

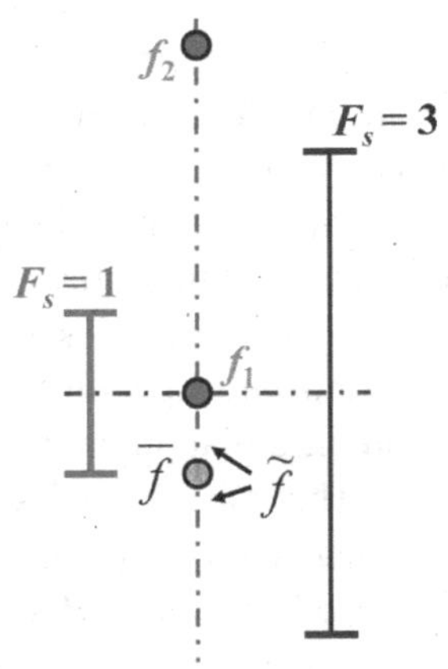

Figure 9.5 Factor of safety for extrapolation-based error bands.
(from Roy, 2005)

equal chance that the true exact solution is above or below the estimated value. Thus, a factor of safety of $F_s = 1$ centered about the fine grid numerical solution f_1 will only provide a 50% chance that the true error $\tilde{f}$ is within the uncertainty band. Increasing the factor of safety should increase the likelihood that the true error will fall within the band. This argument is made simply to support the use of a factor of safety in converting the discretization error estimate into an epistemic uncertainty. It in *no way* implies that the resulting uncertainty is a randomly distributed variable (i.e., an aleatory uncertainty).

To reiterate the main point made in Section 9.5, the reliability of any discretization error or uncertainty estimate can only be determined by assessing the asymptotic nature of the discrete solutions. When the solutions are found to be far outside the asymptotic range, the reliability of the error/uncertainty estimate will likely be poor and its behavior erratic. In this case, no value for the factor of safety is assured to be conservative.

9.6.1 GCI Implementation

In order to avoid numerical problems when the solution values are near zero, most modern implementations of the GCI omit the normalization by the fine grid solution in the denominator, i.e., the GCI is redefined as

$$GCI = \frac{F_s}{r^p - 1}|f_2 - f_1|. \tag{9.67}$$

This implementation thus provides an uncertainty estimate for the fine grid solution in the same units as the solution. Roache (1998) provides clear guidelines for choosing the factor of safety when:

1. solutions on only two grids are available ($F_s = 3$), or
2. solutions on three grids are available and the observed order of accuracy is calculated and shown to match the formal order of the scheme ($F_s = 1.25$).

In the first case, when only two grid levels are available, GCI estimates should be used with caution since there is no information on whether the solutions are in (or even near) the asymptotic range. Far from the asymptotic range, all approaches for estimating discretization error or uncertainty will be unreliable (see Section 9.5). For the second case, when three solutions are available, if the observed order of accuracy matches the formal order, then $F_s = 1.25$ is used and either the formal or observed order of accuracy can be employed (since they match, this choice will have little effect on the estimate). Difficulties arise when the observed and formal orders of accuracy do not agree. Furthermore, exactly how does one define agreement? Ideally, one would simply continue to systematically refine the mesh until the solutions are demonstrably asymptotic, possibly incorporating adaptive solution strategies to achieve the asymptotic range faster (see Section 9.8). When no additional solutions are possible due to time or resource limitations, Roache (1998) provides anecdotal examples of how to apply the GCI to a range of situations.

We recommend the following procedure for calculating the GCI when solutions on *three* systematically refined meshes are available. (If more than three systematically refined meshes are available, then a least squares form of GCI is recommended as discussed in Section 9.6.3.) In all cases, the non-normalized GCI given by Equation (9.67) is used. When the observed order

Table 9.1 *Recommended implementation of the GCI for solutions on three systematically-refined grids using Equation (9.67)*

$\left\lvert\frac{\hat{p}-p_f}{p_f}\right\rvert$	F_s	p
≤ 0.1	1.25	p_f
> 0.1	3.0	$\min\left(\max(0.5,\hat{p}),p_f\right)$

of accuracy $\hat{p}$ agrees with the formal order p_f within 10%, then the formal order of accuracy along with a factor of safety of 1.25 is used in the GCI calculation. When the observed order of accuracy does not agree within 10%, then a factor of safety of three ($F_s = 3.0$) is used. Furthermore, the order of accuracy is limited between 0.5 and the formal order. Allowing the order of accuracy to be much larger than the formal order causes the uncertainty estimates to be unreasonably small since the GCI goes to zero as $p \rightarrow \infty$. This situation should be avoided since it is an artifact of the solutions not being in the asymptotic range. Allowing the order of accuracy to be too small causes the uncertainty estimate to get extremely large, which does not provide useful information to the analyst. These recommendations are summarized in Table 9.1.

9.6.2 Reliability of the GCI

Three studies have carefully examined the reliability of the GCI. Eca and Hoekstra (2002) examined a wide range of fluid flow problems using a large number of non-integer refined meshes (at least 16 different grids for each case). They employed a locally-evaluated observed order of accuracy and found that the GCI with a factor of safety of $F_s = 1.25$ worked well on the finer meshes. Cadafalch et al. (2002) examined five test cases for fluid flow (some with heat transfer) using four to seven meshes for each case with a grid refinement factor of two along with a global averaging method for the observed order of accuracy. They also found good results for the GCI with a factor of safety of 1.25 on finer meshes. Finally, Eca et al. (2004) looked at potential flow solutions around 2D airfoils where exact solutions were available. They achieved good uncertainty estimates using the GCI when the locally evaluated, observed order of accuracy was not significantly larger than the formal order of the method. See Pelletier and Roache (2006) for additional discussion of the effectiveness of the GCI.

The above assessments of the reliability of the GCI are mainly in the context of solutions that are in or near the asymptotic range. For solutions that are far outside the asymptotic range, no uncertainty estimator (including the GCI) will be reliable. Only limited studies have addressed the case with solutions that are "moderately" outside the asymptotic range (e.g., see Phillips and Roy, 2014 and Eca et al., 2021).

9.6.3 Least Squares GCI

The calculation of the observed order of accuracy often exhibits large departures from the formal order of the discretization scheme. These variations can be due to a number of sources including the discrete solutions not being asymptotic, errors transported from other regions (especially for hyperbolic problems), iterative error, round-off error, interpolation of solutions onto a common grid, and nonuniformity of the mesh refinement. Eca and Hoekstra (2002) developed a method for filtering out the "noise" from the observed order of accuracy calculation using a least squares fitting approach over four or more mesh levels. Recall the series expansion from generalized Richardson extrapolation from Equation (9.48), which can be written for a general mesh level k as:

$$f_k = \bar{f} + g_p h_k^{\hat{p}}.$$

In their approach, they minimize the function

$$S\left(\bar{f}, g_p, \hat{p}\right) = \left\{ \sum_{k=1}^{NG} \left[f_k - \left(\bar{f} + g_p h_k^{\hat{p}} \right) \right]^2 \right\}^{1/2}, \tag{9.68}$$

where k refers to a mesh level and NG is the total number of mesh levels ($NG > 3$). This minimization is accomplished by setting the derivatives of S with respect to $\bar{f}$, g_p, and $\hat{p}$ to zero, which results in the following expressions:

$$\bar{f} = \frac{\sum_{k=1}^{NG} f_k - g_p \sum_{k=1}^{NG} h_k^{\hat{p}}}{NG}, \tag{9.69}$$

$$g_p = \frac{NG \sum_{k=1}^{NG} f_k h_k^{\hat{p}} - \left(\sum_{k=1}^{NG} f_k\right)\left(\sum_{k=1}^{NG} h_k^{\hat{p}}\right)}{NG \sum_{k=1}^{NG} h_k^{2\hat{p}} - \left(\sum_{k=1}^{NG} h_k^{\hat{p}}\right)\left(\sum_{k=1}^{NG} h_k^{\hat{p}}\right)}, \tag{9.70}$$

$$\sum_{k=1}^{NG} f_k h_k^{\hat{p}} \ln(h_k) - \bar{f} \sum_{k=1}^{NG} h_k^{\hat{p}} \ln(h_k) - g_p \sum_{k=1}^{NG} h_k^{2\hat{p}} \ln(h_k) = 0. \tag{9.71}$$

Eca and Hoekstra (2002) employed a false position method to iteratively solve Equation (9.71) for $\hat{p}$.

The main drawback to the least squares method is that it requires solutions on four or more systematically-refined grids. While their original approach actually applies the uncertainty estimate to the extrapolated value $\bar{f}$ (Eca and Hoekstra, 2002), subsequent studies have all applied the uncertainty estimate to the fine grid solution by using the order of accuracy found by solving Equation (9.71) directly in the GCI calculation of Equation (9.67) (e.g., see Eca et al., 2005). Applying the uncertainty estimate to the fine grid solution is considered to be more conservative since it avoids the variability/noise that can occur in the extrapolated solution and generally provides a more physically consistent solution than the extrapolated solution. A more recent update by Eca and Hoekstra (2014) employs four different power series expansions and uses the one that best fits the data on a series of systematically refined meshes in a least squares sense. Their approach provides the numerical uncertainty in both integral and local flow quantities and is applied to four different example applications. The recently-developed *robust verification analysis* also provides a statistical estimate of the exact solution to the mathematical model with grid refinement (see Rider et al., 2016 for details).

Example: Richardson Extrapolation and the GCI

An example of discretization error and uncertainty estimation is now given using Richardson extrapolation and the Grid Convergence Index (GCI). Consider computed solutions for a solution functional (or system response quantity) f found from running scientific computing simulations on nine systematically refined grid levels as shown in Table 9.2. The finest grid is 2048 × 2048 cells and has the mesh spacing parameter h arbitrarily set to unity. The coarsest grid is 8 × 8 cells with $h = 256$. These functional values are plotted in Figure 9.6a versus the mesh parameter h. Since the mesh is systematically refined with a refinement factor of $r = 2$, any three meshes in the sequence can be used to compute the observed order of accuracy using Equation (9.59). (Note that the observed order for a grid triplet is typically plotted at an h value corresponding to the finest of the three meshes, as indicated in Figure 9.6a.) In this case, due to the non-monotone behavior of the

Table 9.2 *Solution functional values* f *computed on nine systematically refined grids*

Mesh cells	h	f
8 × 8	256	24.20175
16 × 16	128	10.30191
32 × 32	64	11.58527
64 × 64	32	9.070761
128 × 128	16	8.135920
256 × 256	8	7.858402
512 × 512	4	7.723745
1024 × 1024	2	7.658460
2048 × 2048	1	7.629082

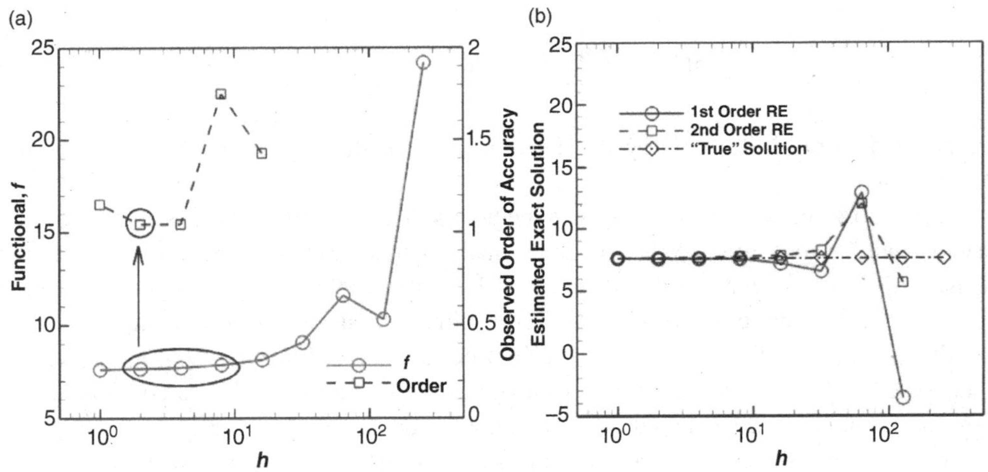

Figure 9.6 Example of a functional f computed on a sequence of systematically refined meshes: (a) the computed value f along with the observed order of accuracy found from each grid triplet and (b) the estimated exact solution computed with Richardson extrapolation assuming both first-order and second-order accuracy (the "true" solution is also given).

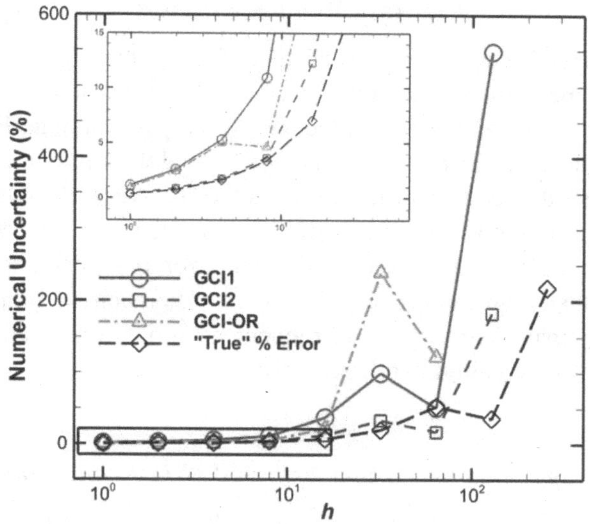

Figure 9.7 Estimated percentage uncertainty in the functional using the GCI with first-order accuracy (GCI1), second-order accuracy (GCI2), and the approach recommended in Section 9.6.1 (GCI-OR) along with the "true" percentage error.

solution on the coarser meshes, the observed order of accuracy is undefined until the three meshes at $h = 16$, 32, and 64. The three coarsest meshes can thus be considered to be well outside the asymptotic grid convergence range. The three finest meshes are in the asymptotic range and show an observed order of accuracy near unity, suggesting that the formal order for this application is likely only first order. The middle set of meshes could be considered to be near the asymptotic range. Richardson extrapolation is then used to estimate the exact solution using pairs of adjacent meshes according to Equation (9.51) assuming an order of accuracy of $p = 1$ and $p = 2$ as shown in Figure 9.6b. Also given in the figure is an estimate of the "true" exact solution computed using the two finest mesh solutions and the mixed-order extrapolation technique found in Roy (2003). It is clear from the figure that the extrapolated solutions are not close to the "true" exact solution on the coarser grids, but they do approach this solution on the finer meshes.

The estimated percentage numerical uncertainty is given in Figure 9.7 using three different versions of the GCI. The first approach (GCI1) uses two solutions and assumes first-order accuracy, the second (GCI2) uses two solutions and assumes second-order accuracy, and the third (GCI-OR) comes from Equation (9.67) using the implementation given in Section 9.6.1. In all cases, a factor of safety of $F_s = 3.0$ is used, although one could argue that the smaller value of $F_s = 1.25$ could be used on the finer meshes if the formal order of accuracy was known to be one. Also shown in the figure is the "true" percentage error. The numerical uncertainty estimates generally bound the true error except for GCI2 evaluated at $h = 64$; however, the uncertainty estimates are generally overly-conservative on the coarser meshes. On the finer meshes (see inset of Figure 9.7), GCI2 provides the tightest uncertainty estimates, but this is simply due to the use of the factor of safety of $F_s = 3.0$. If the smaller factor of safety of $F_s = 1.25$ was used, then GCI1 and GCI-OR would tightly bound the true error, while GCI2 would no longer be conservative (i.e., the uncertainty estimates would be smaller than the true error). While this example is given for a large number of grid levels, in practice it is much more common to only have three or four systematically refined grids available.

9.7 Mesh Refinement Issues

Second only to the inability to achieve the asymptotic grid convergence range, the failure of extrapolation-based methods for estimating discretization error can often be tied to problems achieving systematic mesh refinement on the meshes in question (e.g., Baker, 2005a; Salas, 2006). The use of locally refined meshes, meshes refined in only one coordinate direction, and refinement where the refinement factor varies locally over the domain are all examples of improper approaches to systematic mesh refinement. When combined with the extrapolation-based methods discussed earlier for error or uncertainty estimation, these approaches are virtually guaranteed to fail. For a discussion of non-integer refinement factors, refinement versus coarsening, and refinement in only one spatial coordinate direction, see chapter 8 of Oberkampf and Roy (2010).

9.7.1 Mesh Spacing Parameter h and Grid Refinement Factor

Recall we have used the mesh spacing parameter h as a way to characterize a reference element length associated with the mesh. When the mesh is equally spaced, or isotropic, the determination of this parameter is straightforward. However, most scientific computing applications require complex mesh topologies with significant variations in local element size and quality. Following closely the development of Eca et al. (2013), we define the mesh spacing parameter h for mesh level k as

$$h_k = (\Lambda_k)^{1/d}, \tag{9.72}$$

where d is the dimensionality of the problem and Λ is a representative length in 1D ($d=1$), an area in 2D ($d=2$), and a volume in 3D ($d=3$). The simplest way to determine Λ on mesh level k is to sum the values of Λ over each cell/element and then divide by the number of cells:

$$\Lambda_k = \frac{1}{N_k}\sum_{i=1}^{N_k}\lambda_i = \frac{\Lambda_{Total}}{N_k}, \tag{9.73}$$

where λ_i is the length, area, or volume of cell i in 1D, 2D, or 3D, respectively. (Note that for finite difference methods, one must deal with the spaces/partitions between nodes, or one minus the number of nodes in each coordinate direction.) The grid refinement factor, when normalized to $r = 1$ on the fine grid ($k = 1$), becomes

$$r_{1,k} = \left(\frac{\Lambda_k}{\Lambda_{k=1}}\right)^{1/d} = \left(\frac{N_{k=1}}{N_k}\right)^{1/d}. \tag{9.74}$$

Equation (9.74) is a global measure of the refinement and is based solely on the number of cells/elements on each mesh. By itself, $r_{1,k}$ cannot be used to ensure the uniformity of the grid refinement, it is simply a measure of how many cells/elements are used in each mesh level (see the right-hand side of Equation [9.74]). Techniques for assessing the level of uniformity of the refinement are discussed in the next section, which can be especially important for unstructured meshes. When significant local mesh refinement is applied from one mesh to another, the *local* grid refinement factor will vary widely over the domain and the numerical solutions should *not* be used in an extrapolation procedure or for the computation of the observed order of accuracy.

9.7.2 Assessing Uniformity of the Mesh Refinement

Recall the definition of *systematic mesh refinement* given in Section 5.6, which requires that the mesh refinement be both uniform and consistent. Uniform refinement requires that the mesh be refined by the same factor over the entire domain. This does *not* mean that the mesh itself must be uniform over the domain, only that the ratio of grid refinement from one mesh to another must not vary over the domain. Consistent refinement requires that the meshes should maintain the same grid quality (skewness, aspect ratio, stretching factor, etc.) or possibly provide for improved grid quality with mesh refinement.

To assess the level of uniformity of the mesh refinement, additional statistical assessments of the local reference cell/element size λ_i (discussed in the previous section) can be used (Eca et al., 2013). The goal is to determine if λ_i is varying appropriately from one refinement level to another. The simplest approach is to look at the RMS (or L_2 norm) of the cell size, given by

$$\Lambda_{2,k} = \left(\frac{1}{N_k} \sum_{i=1}^{N_k} \lambda_i^2 \right)^{1/2}, \tag{9.75}$$

where again N_k is the total number of cells on mesh k and λ_i is the length/area/volume of cell i. The grid refinement factor associated with this RMS measure, when normalized to $r = 1$ on the fine grid ($k = 1$), is

$$r_{2,k} = \left(\frac{\Lambda_{2,k}}{\Lambda_{2,k=1}} \right)^{1/d}, \tag{9.76}$$

where $d = 1$, 2, or 3 in 1D, 2D, or 3D, respectively. When the sequence of meshes are uniformly refined, then the refinement factors computed from Equations (9.74) and (9.76) will be equal, i.e.,

$$r_{1,k} = r_{2,k}. \tag{9.77}$$

In other words, when the refinement factor based on the L_2 norm of the local cell/element size $r_{2,k}$ equals the refinement factor based solely on the change in the number of cells/elements $r_{1,k}$, then there is strong evidence that the grids have been uniformly refined. Conversely, when the sequence of meshes is not uniformly refined, these two measures of the grid refinement factor will almost always differ. For 2D and 3D problems, additional reference cell measures and refinement factors can be computed related to the lower-dimensional curves and areas, respectively, of the various domain boundaries. See Eca et al. (2013) for more details.

Another technique to ensure that two meshes retain a constant local refinement factor over the domain was used by Roy et al. (2007) and applied to 3D unstructured Cartesian meshes used for aerodynamic force and moment computations. The level of uniformity of the mesh refinement was evaluated by comparing cell volume ratios between the two successive mesh

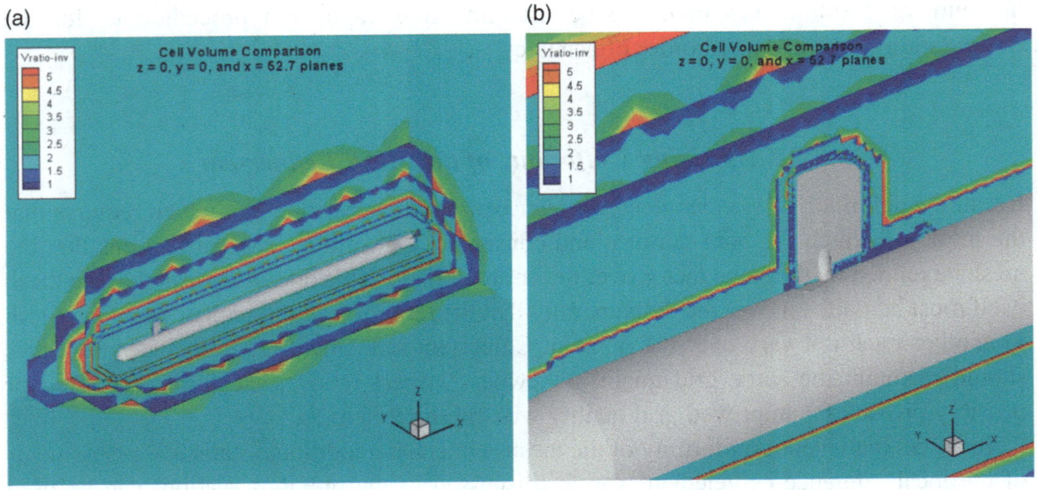

Figure 9.8 Contour plots of the cell volume ratio between the two unstructured Cartesian mesh levels for (a) the entire missile geometry and (b) a close-up of the canard (from Roy et al., 2007); the cell volume ratios are near 1.6 except in the regions where the cells change resolution.

levels as follows. First, the local cell volumes were calculated and stored at the nodes. Next, the fine grid volume distribution was interpolated onto the coarse grid using an inverse distance function. Now that the coarse grid contains both fine and coarse grid cell volume information, the coarse to fine grid volume ratios can be calculated and examined locally. An example of the Cartesian mesh used here was given earlier in Figure 9.3b showing different layers of resolution, with smaller cells near the surface. The volume ratios between two mesh levels are shown in Figure 9.8 and are fairly constant in the domain around a value of 1.6 (light blue shading), with the exception of the regions where the Cartesian grid layers transition from one cell size to another (dark blue and red shading). The departures from uniform refinement, where the contours vary from the nominal value of approximately 1.6, are likely to degrade the reliability of the error/uncertainty estimates. This approach could be extended to assess mesh quality parameters such as skewness, aspect ratio, stretching factor, etc., thus addressing both the uniformity and the consistency of the refinement.

9.8 Solution Adaptation

Up to now, this chapter has focused on the estimation of numerical errors and numerical uncertainties due to the discretization. In addition to estimating the discretization error, we also desire methods for reducing it either when it is found to be too large or when the solutions are not yet in the asymptotic range and, therefore, the error estimates are not reliable. Applying systematic mesh refinement, although required for assessing the reliability of all discretization error estimation approaches, is not the most efficient method for reducing the discretization error. Since systematic refinement, by definition, refines by the same factor over the entire domain, it generally results in meshes with highly-refined cells or elements in regions where they are not needed. Recall that for 3D scientific computing applications, each time the mesh is refined using grid halving (refinement factor of two) the number of cells/

elements increases by a factor of eight. Thus, systematic refinement for reducing discretization error can be prohibitively expensive. Targeted, local solution adaptation is a much better strategy for *reducing* the discretization error.

There are three factors that affect the discretization error for uniform meshes: the chosen discretization scheme, the mesh resolution, and the local solution behavior (including solution derivatives). For nonuniform meshes, including unstructured meshes, the mesh quality also plays a role. These factors are discussed in detail below. This subsection also makes an important distinction between the quantity that is chosen to drive the adaptation process (called the *adaptation criteria*) and the specific type of adaptation used (called the *adaptation approach*).

9.8.1 Relating Discretization Error to Truncation Error

Recall that earlier in this chapter we derived both continuous and discrete error transport equations. These error transport equations show that the discretization error can be both locally generated by the truncation error and also transported from other regions of the domain. (This transported error is also referred to as pollution error in the finite element community.) Since the truncation error serves as the local source in the discretization error transport equations, reducing the truncation error should result in a commensurate reduction in the discretization error. Furthermore, any solution adaptation scheme should adapt only on the locally generated error and *not* on the error transported from other regions in the domain.

9.8.2 1D Truncation Error Analysis on Uniform Meshes

The process of examining the truncation error first requires that a mathematical model be defined along with its discrete approximation. As an example, consider the steady-state form of Burgers' equation given by

$$L(\tilde{u}) = \tilde{u}\frac{d\tilde{u}}{dx} - \nu\frac{d^2\tilde{u}}{dx^2} = 0, \tag{9.78}$$

where the first term is a nonlinear convection term, the second term is a diffusion term (multiplied by the constant viscosity ν), and $\tilde{u}$ is the exact solution to this differential equation.

A simple second-order accurate finite difference discretization for the steady-state Burgers' equation is

$$L_h(u_h) = u_i\left(\frac{u_{i+1} - u_{i-1}}{2\Delta x}\right) - \nu\left(\frac{u_{i+1} - 2u_i + u_{i-1}}{\Delta x^2}\right) = 0, \tag{9.79}$$

where u_h denotes the exact solution to this discrete equation. We have assumed a Cartesian mesh with constant node spacing $h = \Delta x$. To find the truncation error associated with this discretization method, we first find the Taylor series expansions of u_{i+1} and u_{i-1} expanded about u_i:

$$u_{i+1} = u_i + \left.\frac{du}{dx}\right|\Delta x + \left.\frac{d^2u}{dx^2}\right|_i\frac{\Delta x^2}{2} + \left.\frac{d^3u}{dx^3}\right|_i\frac{\Delta x^3}{6} + \left.\frac{d^4u}{dx^4}\right|_i\frac{\Delta x^4}{24} + O\left[\Delta x^5\right]. \tag{9.80}$$

$$u_{i-1} = u_i - \frac{du}{dx}\bigg|\Delta x + \frac{d^2u}{dx^2}\bigg|_i \frac{\Delta x^2}{2} - \frac{d^3u}{dx^3}\bigg|_i \frac{\Delta x^3}{6} + \frac{d^4u}{dx^4}\bigg|_i \frac{\Delta x^4}{24} + O\left[\Delta x^5\right]. \tag{9.81}$$

Substituting these expansions into the discrete operator $L_h(\cdot)$ and rearranging gives

$$L_h(u) = \underbrace{u_i \frac{du}{dx}\bigg|_i - \nu \frac{d^2u}{dx^2}\bigg|_i}_{L(u)} + \underbrace{u_i \frac{d^3u}{dx^3}\bigg|_i \frac{\Delta x^2}{6} - \nu \frac{d^4u}{dx^4}\bigg|_i \frac{\Delta x^2}{12} + O\left[\Delta x^4\right]}_{TE_h(u)}. \tag{9.82}$$

The right-hand side of Equation (9.82) can be thought of as the actual differential equation that is solved by the discretization scheme. It contains the original partial differential equation, higher-order derivatives of the solution, and coefficients which are functions of the grid spacing to different powers. Since the leading truncation error terms are on the order of Δx^2, we find that the formal order of accuracy of this method is second order. Equation (9.82) also tells us that this discretization is consistent since the discrete equations approach the partial differential equation in the limit as Δx goes to zero. By analogy, Equation (9.82) can be rewritten as the *generalized truncation error expression* (developed in Chapter 5) by employing operator notation:

$$L_h(u) = L(u) + TE_h(u). \tag{9.83}$$

Equation (9.83) states that the discretized equation is equal to the continuous mathematical model (in this case Burgers' equation) plus the truncation error (TE) associated with the discretization $h = \Delta x$. Note that in this form of the equation u can be any smooth (i.e., differentiable) function. We will make extensive use of Equation (9.83) in the following discussion.

9.8.3 1D Truncation Error Analysis on Nonuniform Meshes

One approach for examining the truncation error on nonuniform (anisotropic) meshes is to first perform a global transformation of the governing equation. An additional motivation for performing such a global transformation is that it will provide insight into the role of mesh quality on the solution discretization and truncation errors. Following Thompson et al. (1985) and Roy (2009), the first derivative in physical space can be transformed into a uniform computational space using the transformation $\xi = \xi(x)$ and the truncation error for the centered central difference method can be found as follows:

$$\underbrace{\frac{du}{dx}}_{L(u)} = \frac{1}{x_\xi}\frac{du}{d\xi} = \underbrace{\frac{1}{x_\xi}\left(\frac{u_{i+1}-u_{i-1}}{2\Delta\xi}\right)}_{L_h(u)} - \underbrace{\frac{1}{6}\frac{x_{\xi\xi\xi}}{x_\xi}\frac{du}{dx}\Delta\xi^2}_{Small} - \underbrace{\frac{1}{2}x_{\xi\xi}\frac{d^2u}{dx^2}\Delta\xi^2}_{Stretching} - \underbrace{\frac{1}{6}x_\xi^2\frac{d^3u}{dx^3}\Delta\xi^2}_{Standard\ TE} + O(\Delta\xi^4). \tag{9.84}$$

Here we have assumed a fixed transformation $\xi = \xi(x)$ and thus $\Delta\xi \to 0$ with systematic mesh refinement. The derivatives of x with respect to ξ (e.g., x_ξ, $x_{\xi\xi}$) are called metrics of the transformation and are functions only of the mesh geometry, not the solution. On the right-hand side of this equation, the first term is the finite difference equation for the first derivative in transformed coordinates, the second term involving $x_{\xi\xi\xi}$ is zero when the discrete form of the metrics is used with the same central difference approximation (Mastin, 1999), the third term involving $x_{\xi\xi}$ is a grid stretching term, and the fourth term involving the square of x_ξ is the standard leading truncation error term that appears when the equations are discretized on a uniform mesh. The higher order terms (order $\Delta\xi^4$ and higher) can be neglected in the limit as the mesh spacing goes to zero. The grid stretching term is zero for a uniformly spaced mesh, but gets large as the mesh is stretched (e.g., from coarse to fine spacing).

This approach for examining the truncation error of the transformed equation illustrates some important features of the truncation error for anisotropic meshes. First, the truncation error is affected by the mesh resolution ($\Delta\xi$), the mesh quality ($x_{\xi\xi}$), and the local solution derivatives. Second, the quality of a mesh can *only* be assessed in the context of the local solution behavior and the mesh resolution, since the grid stretching term $x_{\xi\xi}$ is multiplied by the second derivative of the solution and $\Delta\xi^2$. When the solution varies linearly, this term will go to zero.

In a similar manner, the truncation error for the second-order accurate centered second derivative in non-conservative form is:

$$\begin{aligned}
\frac{d^2u}{dx^2} &= \overbrace{\frac{1}{x_\xi^2}\frac{d^2u}{d\xi^2} - \frac{x_{\xi\xi}}{x_\xi^3}\frac{du}{d\xi}}^{L(u)} = \overbrace{\frac{1}{x_\xi^2}\left(\frac{u_{i+1}-2u_i+u_{i-1}}{\Delta\xi^2}\right) - \frac{x_{\xi\xi}}{x_\xi^3}\left(\frac{u_{i+1}-u_{i-1}}{2\Delta\xi}\right)}^{L_h(u)} + \overbrace{\frac{1}{12}\left(\frac{2x_{\xi\xi}x_{\xi\xi\xi}-x_\xi x_{\xi\xi\xi\xi}}{x_\xi^3}\right)\frac{du}{dx}\Delta\xi^2}^{Small} \\
&\quad + \underbrace{\frac{1}{12}\left(\frac{3x_{\xi\xi}^2-4x_\xi x_{\xi\xi\xi}}{x_\xi^2}\right)\frac{d^2u}{dx^2}\Delta\xi^2}_{Small} - \underbrace{\frac{1}{3}x_{\xi\xi}\frac{d^3u}{dx^3}\Delta\xi^2}_{Stretching} - \underbrace{\frac{1}{12}x_\xi^2\frac{d^4u}{dx^4}\Delta\xi^2}_{Standard\ TE} + O(\Delta\xi^4).
\end{aligned} \tag{9.85}$$

The third and fourth terms on the right-hand side will be small when the discrete form of the metrics and smooth transformations are used. The fifth term involving $x_{\xi\xi}$ is a grid stretching term and the sixth term is the standard leading truncation error term that appears when the equations are discretized in physical space on a uniform grid.

Combining these two truncation error expressions and neglecting the small terms results in the following truncation error expression for the transformed Burgers equation:

$$\begin{aligned}
u\frac{du}{dx} - \nu\frac{d^2u}{dx^2} &= \overbrace{\left(\frac{u}{x_\xi}+\frac{\nu x_{\xi\xi}}{x_\xi^3}\right)\frac{du}{d\xi} - \frac{\nu}{x_\xi^2}\frac{d^2u}{d\xi^2}}^{L(u)} = \overbrace{\left(\frac{u_i}{x_\xi}+\frac{\nu x_{\xi\xi}}{x_\xi^3}\right)\left(\frac{u_{i+1}-u_{i-1}}{2\Delta\xi}\right) - \frac{\nu}{x_\xi^2}\left(\frac{u_{i+1}-2u_i+u_{i-1}}{\Delta\xi^2}\right)}^{L_h(u)} \\
&\quad + \underbrace{x_{\xi\xi}\left(\frac{\nu}{3}\frac{d^3u}{dx^3}-\frac{u_i}{2}\frac{d^2u}{dx^2}\right)\Delta\xi^2}_{Stretching} + \underbrace{x_\xi^2\left(\frac{\nu}{12}\frac{d^4u}{dx^4}-\frac{u_i}{6}\frac{d^3u}{dx^3}\right)\Delta\xi^2}_{Standard\ TE} + O(\Delta\xi^4).
\end{aligned} \tag{9.86}$$

Thus, the truncation error for second-order accurate central differences applied to Burgers' equation contains two types of terms. The first is due to mesh quality (stretching in 1D) and the second is due to the mesh resolution. For a uniform grid, the mesh stretching term is zero and the standard truncation error terms are exactly equal to the truncation error for Burgers' equation discretized in physical space on a uniform grid as given in Equation (9.82).

Extending these mesh transformation procedures to higher dimensions results in additional truncation error terms related to the mesh quality such as mesh skewness, aspect ratio, etc. (Mastin, 1999). While the above procedures rely on a global transformation of the mathematical model, similar procedures could be developed for unstructured grid methods which rely on local transformations centered about each cell or element.

9.8.4 Isotropic versus Anisotropic Mesh Adaptation

As shown above, the mesh quality plays a role in the truncation error which acts as the local source term for the discretization error. Some argue that isotropic or nearly uniform meshes should be used in order to minimize the impact of the mesh quality on the solution. However, examination of Equation (9.86), developed for steady Burgers' equation, shows that the mesh quality (mesh stretching in this case) appears in the truncation error multiplied by second and third derivatives of the solution. Therefore, in regions where these solution derivatives are small, large variations in mesh stretching can be tolerated with little impact on the discretization error of the solution. Alternatively, the magnitude of the mesh stretching term can be reduced by increasing the mesh resolution. These two observations provide a sound argument for allowing significant mesh anisotropy in regions where either the solution is well behaved or there is significant mesh clustering.

Two examples are now given for hyperbolic problems that employ mesh adaptation with large anisotropy. The first example is from Laflin (1997) who examined laminar hypersonic flow over a compression corner using structured grids. Figure 9.9a shows the adapted mesh and Figure 9.9b the Mach number contours. The hyperbolic freestream region, where no solution gradients occur, is contained by a single layer of large, high-aspect ratio, skewed

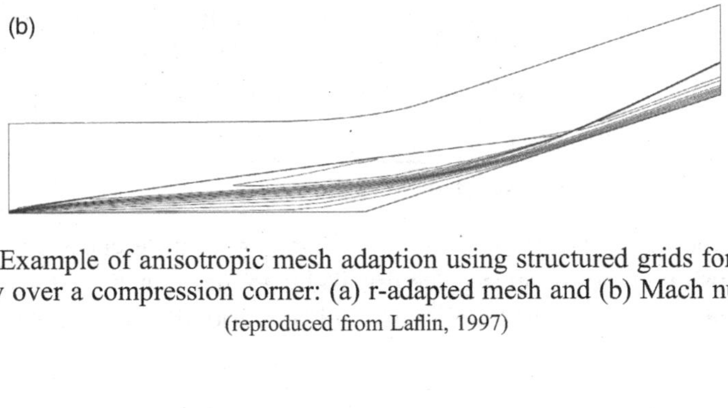

Figure 9.9 Example of anisotropic mesh adaption using structured grids for the laminar hypersonic flow over a compression corner: (a) r-adapted mesh and (b) Mach number contours. (reproduced from Laflin, 1997)

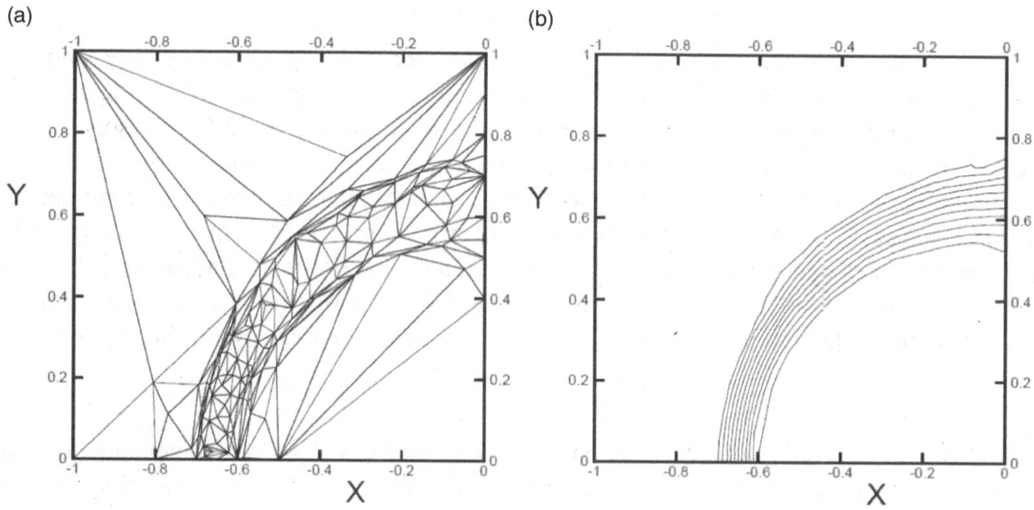

Figure 9.10 Example of anisotropic mesh adaption using unstructured grids for a 2D linear advection equation: (a) r-adapted mesh and (b) velocity contours.
(reproduced from Wood and Kleb, 1999)

cells. While these cells have very poor mesh quality, the fact that there are no gradients in this region means that no discretization errors are introduced.

The second example of anisotropic mesh adaptation is from Wood and Kleb (1999) who examined the 2D linear wave equation using unstructured mesh adaptation. The adapted mesh is given in Figure 9.10a and the corresponding solution in Figure 9.10b. Again, the mesh is highly anisotropic away from the gradient regions and clustered only in the regions of the solution variation.

9.8.5 Adaptation Criteria

One of the most difficult aspects of solution adaptation is finding a good criterion for driving the adaptation process. The *least* defensible methods for adaptation are based on either solution features or estimates of the discretization error. A better approach is based on higher-order reconstructions of solution values or gradients (e.g., finite element recovery methods). The most rigorous methods are based on evaluating or estimating the local truncation error, or equivalently the residual (e.g., from finite elements). We refer to these latter approaches as residual-based methods, and they account for the local cell/element contributions to the discretization error. In their basic form, residual-based methods, and to a lesser extent, the reconstruction methods, can provide some reduction in the overall numerical error; however, they do not target specific solution functionals that are often of interest to the analyst. For adaptation targeted to a specific system response quantity, such as the aerodynamic lift or drag force on a vehicle, an adjoint problem must also be solved in order to find the sensitivity of the discretization error in the system response quantity to the local errors generated at each cell/element.

9.8.5.1 Solution Features

A commonly-used method for driving the solution adaptation process is to use solution features such as solution gradients, solution curvature, or other specific solution features to drive the adaptation process. When there is only one dominant feature to be resolved in a problem, then feature-based adaptation can often improve the solution. However, when multiple features are present (e.g., shock waves, expansion fans, contact discontinuities, and boundary layers in a compressible flow problem), feature-based adaptation often results in some features being over-refined while other features are not refined adequately. In such cases, feature-based adaptation can fail "disastrously" (Ainsworth and Oden, 2000). An example of the failure of solution gradient-based adaptation is given in Section 9.8.6.

9.8.5.2 Discretization Error

Since it is the discretization error that one wishes to reduce with solution adaptation, on the surface it might appear the discretization error would serve as an appropriate driver for the adaption process. However, as will be shown later, the total discretization error is not an appropriate solution adaption criterion. One should instead adapt based on the *local contributions* to the discretization error, i.e., the truncation error. Due to the fact that the discretization error is also transported (i.e., convected and diffused) from other regions of the domain, it is possible that the total discretization error may be large in a region where the local contributions to the discretization error are small (e.g., near, but not at, a singularity). Similarly, it is possible for the total discretization error to be small even when the local contribution to the discretization error is large. For these reasons, adaptation based on the total discretization error is not recommended.

An example of the failure of mesh adaptation based on the total discretization error was given by Gu and Shih (2001) who examined the flow in an incompressible lid-driven cavity. The cavity aspect ratio (width to height) was two, and the Reynolds number based on the cavity width was 2000. Figure 9.11 shows the changes in the L_1 norm of the discretization error in velocity with the total number of cells. For the case of uniform refinement (Figure 9.11a), the error was reduced at the expected formal order of accuracy of two with refinement. In the case of mesh adaptation (Figure 9.11b), the solutions were non-convergent when the relative discretization error was used, whereas only a small error reduction occurred when the absolute discretization error was used. This example highlights the dangers of adapting based on the total discretization error which also includes errors transported from other regions.

9.8.5.3 Recovery Methods

An error indicator that is frequently used for solution adaptation for finite element methods is gradient recovery or reconstruction. In this approach, gradients of the finite element solution are compared to the gradients found from post-processing patches of neighboring elements. Larger mismatches between these two gradient computations serve as an indicator of larger local errors. Recall that recovery-based methods in finite elements, such as the superconvergent patch recovery (SPR) method (Zienkiewicz and Zhu, 1992a), were discussed earlier in this chapter. The SPR method can provide higher-order accurate estimates of the solution gradients when the superconvergence property is attained. In this case, the SPR error

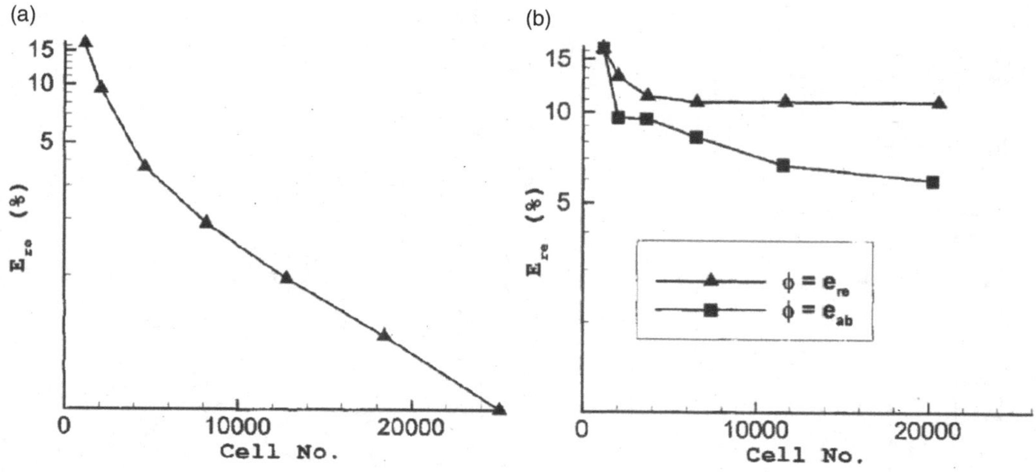

Figure 9.11 Example of Cartesian grid-based h-adaptation for a lid-driven cavity: (a) uniform refinement and (b) refinement based on the total discretization error.
(reproduced from Gu and Shih, 2001)

estimator will act like an estimator of the total discretization error in the solution gradients. One would thus expect the SPR method to be a poor criterion on which to base a solution adaptation strategy. However, as was the case for discretization error estimation, the SPR method has been shown to be effective for driving solution adaptation, at least for elliptic problems (Ainsworth and Oden, 2000).

An example of mesh adaptation based on the SPR method in a linear elasticity problem was given by Zienkiewicz and Zhu (1992b). They examined a uniformly loaded L-shaped plate in a plane stress condition as shown in Figure 9.12a. For this problem, the SPR method will provide local estimates of the discretization error in the stresses. By using adaptive remeshing with h-refinement (discussed in Section 9.8.7), an adapted mesh with 229 elements is shown in Figure 9.12b with maximum discretization errors in the local stress of under 1%. For an application of the SPR method to compressible turbulent flows with adaptive remeshing, see Ilinca et al. (1998).

Laflin (1997) and McRae and Laflin (1999) have developed a similar driver for adaptation based on solution interpolation error. The implementation along with r-adaptation (see Section 9.8.7.2) on structured meshes has been successful for a wide range of 2D problems. The basic idea behind the approach is that the solution values are compared to values interpolated from a patch of neighboring cells. The main difference between their approach and finite element-based recovery methods described above is that they deal with solution values rather than solution gradients. Examples of meshes and solutions found using their solution interpolation error approach along with r-adaptation were given in Figure 9.9; an additional example will be presented later in Figure 9.18. Both examples show a significant amount of anisotropic mesh adaptation.

More recently, anisotropic adaptation of unstructured meshes has been examined using metric fields. The approach of Loseille and Alauzet (2011) uses the Hessian (i.e., matrix of second derivatives) of a scalar field such as Mach number to drive the adaptation. This approach can be shown to target the average (L_1) error in the linear reconstruction of the

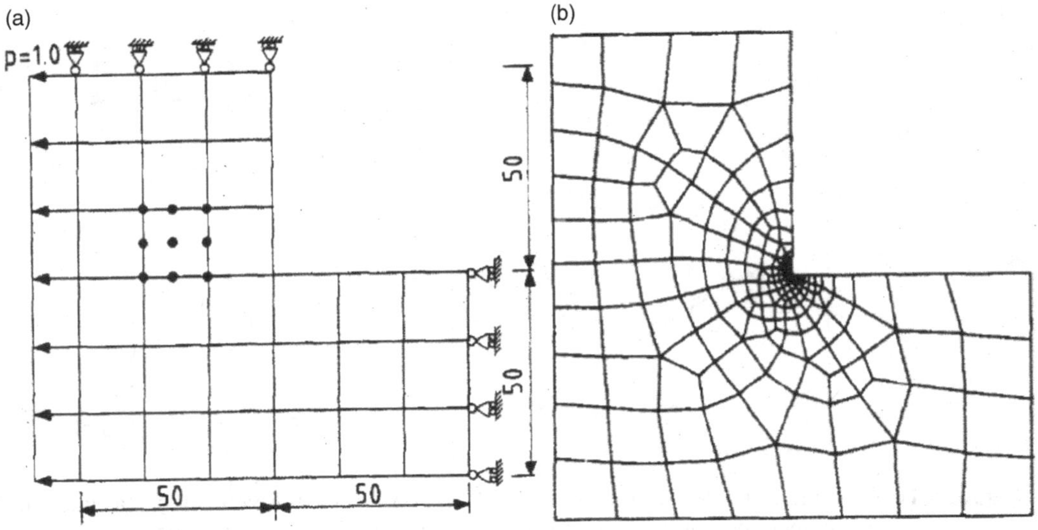

Figure 9.12 Application of the SPR recovery-based error estimator for estimating plane stress in an L-shaped linear elastic domain: (a) problem set up and initial mesh and (b) final adapted mesh with 229 elements.
(reproduced from Zienciewicz and Zhu, 1992b)

function and can thus be considered a recovery-based method. Metric-based adaptation provides guidance as to both where the adaptation should occur, and which direction should be targeted for adaptation. Thus, it is ideal for unstructured meshes which allow cells/elements to be subdivided in different directions. An example application of the metric-based approach was given by Balan et al. (2020) for transonic, turbulent flow over a 3D wing using different unstructured grid computational fluid dynamics solvers. They demonstrated that the adapted mesh forces and moments converged to fine-mesh values at a faster rate than that found using systematic mesh refinement. While the adapted mesh surface pressures also converged well, the skin friction distributions along the surface tended to be noisy in the adapted meshes.

9.8.5.4 Truncation Error/Residuals

The formal relationship between the discretization error and the truncation error can be shown by the discretization error transport equations discussed earlier in Section 9.3.2.1. It was shown that the truncation error provides the *local* element's contribution to the discretization error, to be contrasted with discretization errors that are transported from other regions of the domain. As such, the truncation error is a good indicator of where mesh adaptation should occur. The general concept behind truncation error-based adaptation is to reduce the truncation error where its magnitude is large in order to reduce the total discretization error. Baker (1997) notes that although the idea of truncation error-based adaptation is fundamentally sound, it has "surprisingly" seen little use. In our experience, adaptation can fail when certain regions of the domain, such as inflow boundaries, can have a dominant impact on the discretization error. Adjoint-based adaptation (discussed next) can mitigate this problem. The factors affecting the truncation error for a given discretization scheme include the element size, mesh quality, and the local solution behavior.

9.8.6 Adjoint-Based Adaptation

A promising approach for solution adaptation to reduce the discretization error in a system response quantity is the adjoint method. Most adjoint methods for mesh adaptation use the truncation error/residual weighted by the adjoint sensitivities as indicators of where mesh adaptation should occur. The truncation error/residual indicates where the discretization errors are *locally* being generated. The adjoint method provides the *sensitivity* of the chosen system response quantity (i.e., solution functional) to those local truncation errors. As a result, adjoint methods provide targeted mesh adaption for the chosen system response quantity. The main drawbacks to adjoint methods for adaptation are (1) their complexity and code intrusiveness, (2) the fact that they only apply to a single system response quantity/solution functional, and (3) for unsteady flows the adjoint must be marched backwards in time (thus requiring storage of the primal solution in space and time). This section provides examples of adjoint-based adaptation methods based on the truncation error/residual or approximations thereof.

An example of a truncation error approximation method combined with an adjoint method for system response quantities is given by Dwight (2008), who considered the inviscid transonic flow over an airfoil using an unstructured finite-volume discretization. The discretization scheme employed a central-type flux quadrature combined with an artificial viscosity method for numerical stability. The adjoint method was formulated to provide the sensitivities of the total drag force to the numerical parameters used in the artificial viscosity stabilization technique. This artificial viscosity term can be considered as the leading truncation term since it is only added in the discrete equations and does not show up in the original mathematical model (the Euler equations).

Figure 9.13 shows the discretization error in the drag coefficient as a function of the number of nodes. Uniform refinement (squares) shows second-order convergence on the

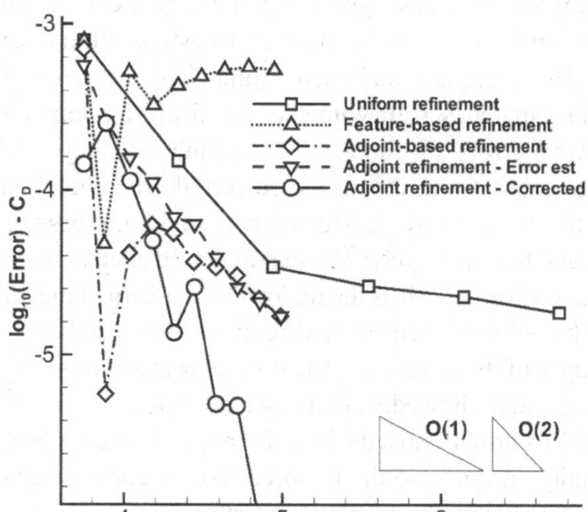

Figure 9.13 Discretization error in the drag coefficient for transonic flow over an airfoil comparing uniform refinement, local gradient-based adaptation, and adjoint, artificial dissipation-based adaptation.

(reproduced from Dwight, 2008)

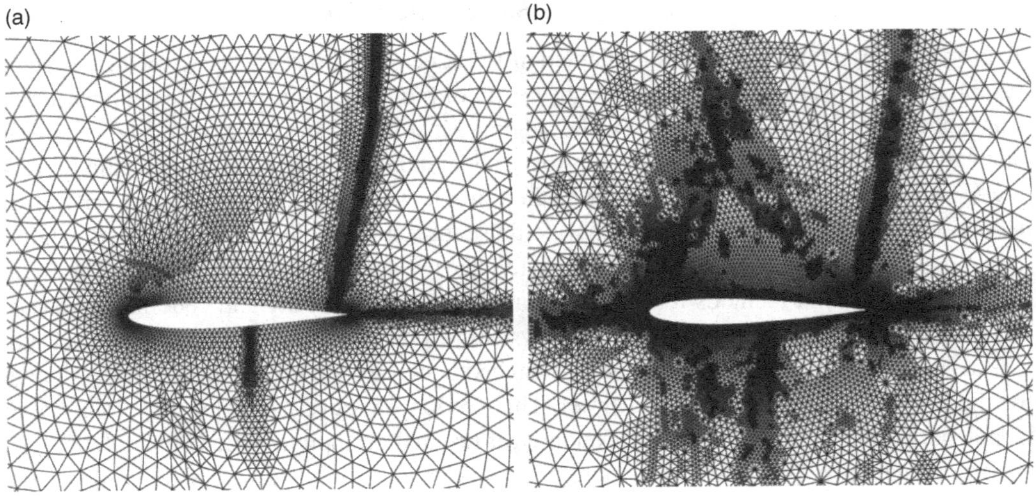

Figure 9.14 Adapted grids for transonic flow over an airfoil: (a) solution gradient-based adaptation and (b) adaptation with an adjoint-based artificial dissipation estimator for the drag force.
(reproduced from Dwight, 2008)

coarser grids, then a reduction to first order on the finer grids, likely due to the presence of the shock discontinuities or the singularities at the leading and trailing edge of the airfoil (e.g., see Banks et al., 2008). Feature-based adaptation using solution gradients (triangles) initially shows a reduction in the discretization error for the drag coefficient, but then subsequent adaptation steps show an increase in the error. Adjoint-based refinement (denoted by the diamond symbols) gives significantly better results. As discussed in Section 9.3.2.3, adjoint methods can also be used to estimate the discretization error in the solution functionals. Here, the adjoint-based estimate of the error in the drag coefficient (upside-down triangles) matches well with the actual error (diamonds), especially on the finer meshes. The lowest errors are found by correcting the drag coefficient solutions with the adjoint-based error estimates to obtain the corrected drag coefficient (circles).

The adapted grids for the feature-based (gradient) and adjoint-based adaption strategies are given in Figure 9.14a and b, respectively. The feature-based adaptation refines the shock waves on the upper and lower surface as well as the region downstream of the trailing edge. In addition to the shock waves, the adjoint-based method refines along the stagnation streamlines at the leading and trailing edges of the airfoil, as well as other regions of the mesh where errors could propagate to the leading and trailing edge. In many cases, the mesh resulting from adjoint-based adaptation can be used to provide insight into the physical mechanisms that impact a given system response quantity, in this case the importance of accurately resolving the leading and trailing edge stagnation points in determining the airfoil drag.

Another example of adjoint-based adaptation used to reduce discretization errors in a viscous, compressible flow problem was given by Venditti and Darmofal (2003). They used an unstructured grid finite volume scheme and examined both the laminar and turbulent viscous flow over an airfoil. Their study also included a comparison to feature-based adaptation using the solution curvature (Hessian). A plot of the drag versus the number of

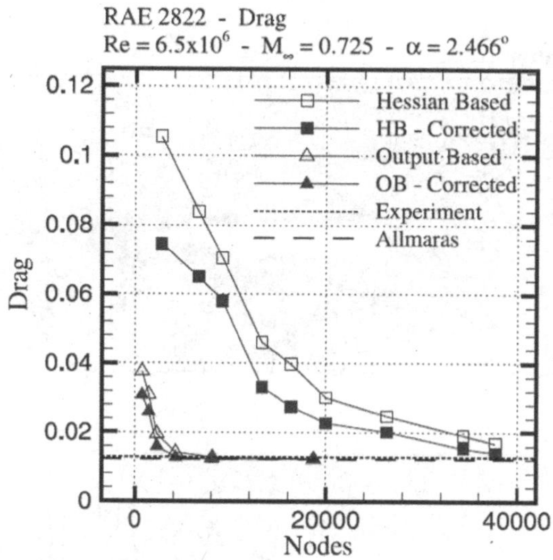

Figure 9.15 Drag for viscous turbulent transonic flow over an airfoil using solution curvature-based (Hessian) and adjoint-based adaptation.
(reproduced from Venditti and Darmofal, 2003)

mesh nodes is presented in Figure 9.15. The adjoint-based adaption (labeled "Output Based" in the figure) converges much faster with mesh adaptation than the solution curvature-based (Hessian) method. Also shown on this figure are "Corrected" results where the adjoint-based error estimate in the drag was used to correct the adapted solution. The adapted grids for this case are presented in Figure 9.16 and show that adjoint-based adaptation results in significantly different final adapted meshes.

Rannacher and Suttmeier (1997) examined the case of a square elastic disc with a crack subjected to a constant boundary traction acting along a portion of the upper boundary as shown in Figure 9.17a. They then employed an explicit residual method both alone and combined with an adjoint method to drive a mesh adaptation routine. The system response quantity for the example shown here was the average normal stress over the clamped bottom and right boundaries. Final adapted meshes for the residual-based adaption and the adjoint method are given in Figure 9.17b and c, respectively. While both approaches adapted to the crack tip (near the center), the adjoint method also adapted to the lower boundary. The adjoint-based adaption was found to provide much lower discretization error levels in the average normal stress than the residual-based adaption. In fact, the adjoint-based method achieved similar error levels as the 65,000 elements residual-based method using under 9,000 elements.

9.8.7 Adaptation Approaches

Once the adaptation criteria has been selected (see Section 9.8.5), the end result is typically a weighting function that varies over the spatial domain. In some cases, the weighting function may be a vector quantity indicating directionality preferences for the adaptation. With the weighting function in hand, local solution adaptation can be achieved in a variety of different ways. Adaptive remeshing generally starts with a coarse mesh and then a mesh generation

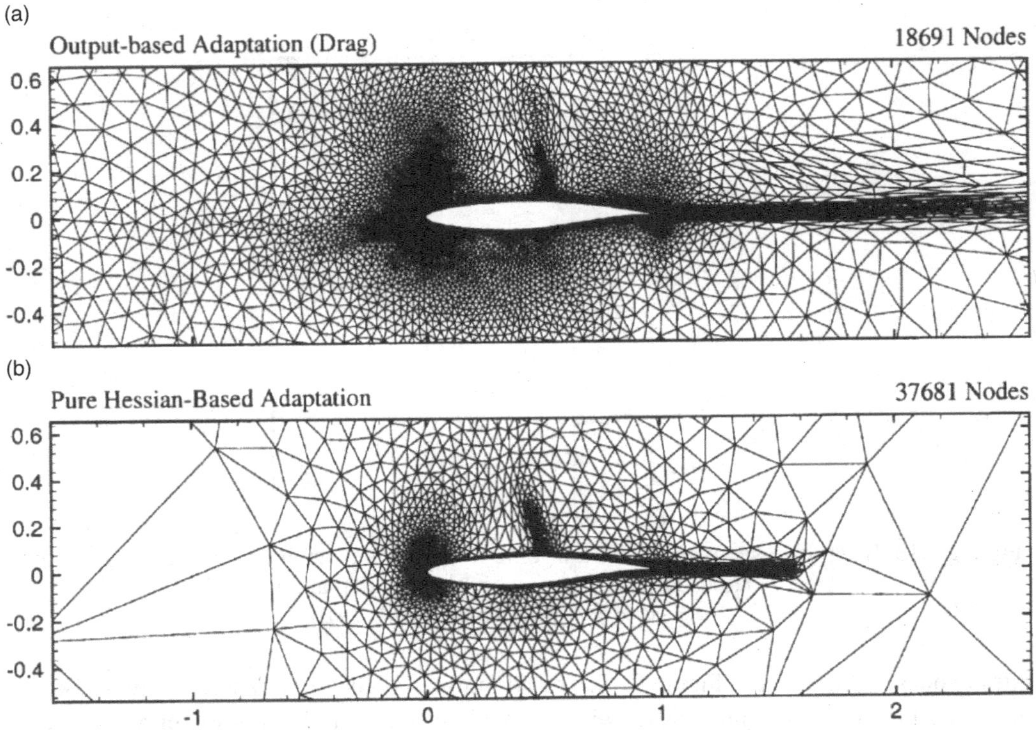

Figure 9.16 Adapted grids for subsonic turbulent flow over an airfoil: (a) adjoint-based adaptation for drag and (b) solution curvature-based (Hessian) adaptation.
(reproduced from Venditti and Darmofal, 2003)

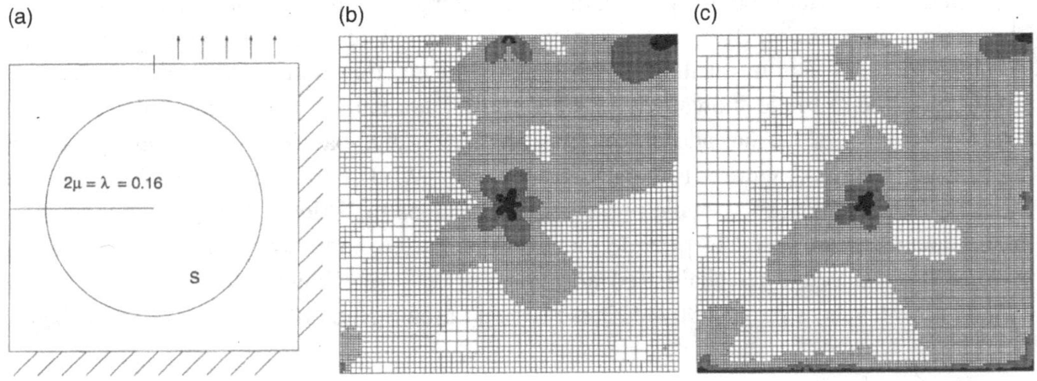

Figure 9.17 Linear elasticity considering the average normal stress over the clamped boundaries: (a) problem set up showing the crack emanating from the center and extending to the left boundary (note that S is simply a circular path used in their analysis), (b) final mesh for residual-based adaption, and (c) final mesh for adjoint-based adaption.
(reproduced from Rannacher and Suttmeier, 1997).

tool is employed in a recursive manner to improve the mesh. Alternatively, one could seek to adapt the mesh by adding additional cells/elements (h-adaptation) or moving them from one region to another while maintaining the original mesh connectivity (r-adaptation). Order refinement (p-adaptation) increases the formal order of accuracy of the discretization scheme in regions of high values of the weighting function. Hybrid combinations of these approaches are also used such as mixed h and r adaption (mesh movement and refinement) and mixed h and p adaption (e.g., the hp-version of finite elements). For general unstructured grid methods, adaptive remeshing and h-adaptation are the most popular. For structured grid methods, the r-adaption approach is the most common since h-adaptation cannot be performed locally due to the required *i, j, k* ordering of the mesh. In addition to mesh refinement, other issues that should be considered when adapting a mesh are mesh quality, local solution behavior, and the alignment of the element faces.

9.8.7.1 Adaptive Remeshing

Initial approaches to adaptation were based on adaptive remeshing, where a grid generation tool (often the same one used to create the initial mesh) is employed to adaptively improve the mesh. The weighting functions are usually accounted for as constraints in the remeshing process. An early example of adaptive remeshing for compressible fluid flow problems was given by Peraire et al. (1987), while a more recent application for solid mechanics is given by Bugeda (2006).

9.8.7.2 Mesh Adaptation

There are two approaches for adapting a mesh that are not based on remeshing. First, mesh refinement (h-adaptation) selectively subdivides cells where the weighting function is large, and possibly coarsening cells where it is small. Second, mesh movement (r-adaptation) maintains the same number of nodes and node-to-node connectivity but moves elements toward regions targeted for adaptation. In some cases, both procedures are used together in a mixed h- and r-adaptation process. These approaches to mesh adaptation are discussed below.

Local Mesh Refinement/Coarsening (h-adaptation)

Local mesh refinement, or h-adaptation, is achieved by subdividing cells that have been targeted for refinement. One usually starts with an initial coarse mesh, and then begins refining, and possibly coarsening (a process known as agglomeration) the mesh in an iterative manner. The total number of local mesh refinement levels is usually specified, for example, by allowing a coarse grid cell to be refined up to a maximum of six times. Another constraint that is often used is that neighboring cells should be within one level of refinement of each other, which prevents the possibility of having a highly-refined cell adjacent to an extremely coarse cell. In some cases, h-adaptation employs edge swapping where the edge connecting two cells is moved to either improve mesh quality or to better align the faces with the solution features. One drawback of using pure h-adaptation is that the mesh quality is adversely affected at the interface between adapted and unadapted regions, since the ratio of adjacent cell length scales (i.e., stretching factor) approaches two.

Mesh Movement (r-adaptation)

Mesh movement, or r-adaptation, is more common in structured grids since local mesh refinement (h-adaptation) by subdividing cells results in an unstructured mesh. When mesh movement is used, the current number of cells and the mesh connectivity is retained, but cells/elements are moved in space according to the weighting function. The most common approaches for achieving mesh movement (Burg, 2006) are:

1. the Laplacian and modified Laplacian (i.e., elliptic adaptation),
2. the linear elasticity equations,
3. the linear spring analogy,
4. the torsional spring analogy, and
5. variational methods.

An example of mesh movement is given in Figure 9.18 for laminar flow in a supersonic inlet (McRae, 2000). An elliptic adaption scheme is used based on the solution interpolation error, which is then modified to include grid quality measures. The figure shows both the solution adapted mesh and the density contours. The shock wave formed by the compression ramp separates the laminar boundary layer, leading to another shock forming upstream. The impingement of these combined shocks on the upper wall also causes the upper-wall boundary layer to separate, producing another shock that emanates from the top wall. It is interesting to note that the shock waves and the boundary layers are clearly evident simply by examining the mesh. The mesh used in this example consists of only 121×91 nodes and the solution was found to provide similar solution quality to one computed on a mesh with approximately 700×900 evenly spaced (i.e., nearly isotropic) cells.

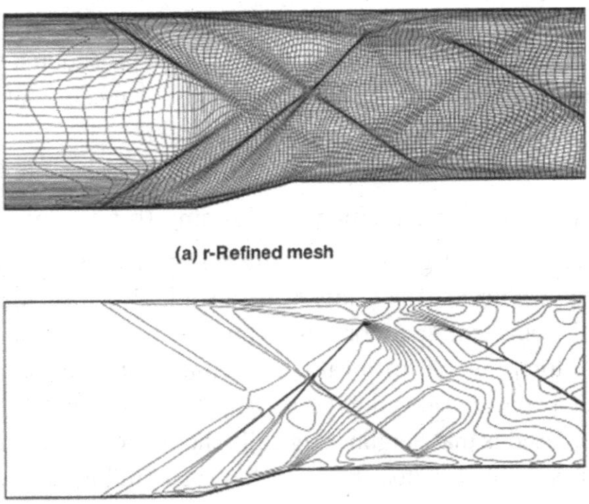

(b) Density contours

Figure 9.18 Structured grid adaptation for laminar flow through a supersonic inlet: (a) r-adapted mesh and (b) density contours.
(reproduced from McRae, 2000)

Mixed Mesh Refinement (r- and h-adaptation)

A number of researchers have found that a combination of different solution adaptation strategies provides the best approach for reducing discretization error. For example, Baker (2005b) found a combination of local mesh refinement/coarsening and mesh movement to be effective for temporally varying (i.e., unsteady) problems on unstructured meshes.

9.8.7.3 Order Refinement (p-adaptation)

Order refinement (or p-adaptation) can also be used to adaptively improve the solution. Order refinement consists of increasing the formal order of accuracy of the discretization scheme in regions targeted for refinement. While this approach is natural to implement with the p-version of finite elements, applying it to finite difference and finite volume discretizations is more difficult due to the difficulties of formulating higher-order discretizations for both the governing equations and the boundary conditions. In the finite element method, p-adaption and h-adaption are often combined in what is known as hp-adaptive finite elements. For a review of hp-adaptive finite elements, see Patra and Oden (1997) and Mitchell and McClain (2014).

9.9 Problems

1. True or False: The Grid Convergence Index (GCI) provides a method for converting a discretization error estimate based on Richardson extrapolation into an epistemic uncertainty.
2. List four distinct reasons why the observed order of accuracy might not equal the formal order of accuracy for a solution verification exercise.
3. List the four assumptions that are needed for Richardson extrapolation to provide reliable estimates of the exact solution to the mathematical model.
4. In order to apply Richardson extrapolation, the iterative and round-off errors should be at least ________ times smaller than the fine grid discretization error.
5. Assume that the numerical solution has been computed on coarse and fine grid levels, where the grid refinement is by a factor of two in each coordinate direction, using a formally third-order accurate discretization. If the numerical solutions on the coarse and fine grid are 20 and 23.5, respectively, use Richardson extrapolation to estimate the exact solution to the differential equation. What major assumption must you make?
6. Numerical solutions are obtained on three systematically refined grids resulting in solutions of 2.19, 2.50, and 2.96 on the fine, medium, and coarse grids, respectively. However, the grid refinement factor between these three grid levels is not the same. The grid refinement factor between the two finest meshes is $r_{12} = 2$, while the refinement factor between the two coarsest meshes is $r_{23} = 1.6$. Using the direct substitution iterative procedure given in Equation (9.63) with a starting guess of $p = 2$, determine the observed order of accuracy. (The iterative procedure should converge within about 2.5% after 10 iterations). In addition, estimate the exact solution to the mathematical model. Hint: Use the two finest mesh levels and the observed order of accuracy that you computed.
7. You are performing a solution verification study by running three simulations on systematically refined grids with a grid refinement factor of $r = 2$. The formal order

of accuracy of the method is second order. The fine, medium, and coarse grid solutions are $f_1 = 1.20, f_2 = 1.26$, and $f_3 = 1.43$, respectively. Find the following:

a. The observed order of accuracy
b. The Richardson extrapolated value (using the observed order of accuracy)
c. The Grid Convergence Index (GCI)
d. The interval about the fine grid solution within which you expect the true exact solution to the model to be found with 95% confidence: $[f_1 - \mathrm{U}, f_1 + \mathrm{U}]$

8. Consider that you are performing a grid refinement study using a formally second-order accurate code to estimate the numerical uncertainty (due to discretization error) in a system response quantity (SRQ) due to discretization error. Neglecting iterative and round-off errors, the three SRQ values you obtain are 5.1, 5.4, and 6.35 for the fine, medium, and coarse grids, respectively. If the meshes are systematically refined by a factor of two, estimate the numerical uncertainty in the fine grid solution by computing the Grid Convergence Index (GCI).
9. The most rigorous solution adaptation criterion is to adapt based on:
   a. solution gradients
   b. solution curvature
   c. discretization error
   d. interpolation error
   e. truncation error (possibly with adjoint weighting)
10. Briefly, explain the difference between h-adaptation, r-adaptation, and p-adaptation.

9.10 References

Ainsworth, M., and J. T. Oden (1997). A Posteriori Error Estimation in Finite Element Analysis. *Computer Methods in Applied Mechanics and Engineering*, **142**(1–2), 1–88.

Ainsworth, M., and J. T. Oden (2000). *A Posteriori Error Estimation in Finite Element Analysis*. New York, NY: Wiley Interscience.

Akin, J. E. (2005). *Finite Element Analysis with Error Estimators*. Burlington, VT: Elsevier.

Babuska, I., and W. C. Rheinboldt (1978a). A Posteriori Error Estimates for the Finite Element Method. *International Journal for Numerical Methods in Engineering*, **12**, 1597–1615.

Babuska, I., and W. C. Rheinboldt (1978b). Error Estimates for Adaptive Finite Element Computations. *SIAM Journal of Numerical Analysis*, **15**(4), 736–754.

Babuska, I., T. Strouboulis, S. K. Gangaraj, and C. S. Upadhyay (1997). Pollution Error in the h-Version of the Finite Element Method and Local Quality of the Recovered Derivative. *Computer Methods in Applied Mechanics and Engineering*, **140**(1–2), 1–37.

Babuska, I., T. Strouboulis, and C. S. Upadhyay (1994). A Model Study of the Quality of A Posteriori Error Estimators for Linear Elliptic Problems. Error Estimation in the Interior of Patchwise Uniform Grids of Triangles. *Computer Methods in Applied Mechanics and Engineering*, **114**(3–4), 307–378.

Babuska, I., O. C. Zienkiewicz, J. Gago, E. R. Oliveira (1986). *Accuracy Estimates and Adaptive Refinements in Finite Element Computations*. Chichester: Wiley.

Baker, T. J. (1997). Mesh Adaptation Strategies for Problems in Fluid Dynamics. *Finite Elements in Analysis and Design*, **25**, 243–273.

Baker, T. J. (2005a). On the Relationship between Mesh Refinement and Solution Accuracy. AIAA Paper 2005-4875.

Baker, T. J. (2005b). Adaptive Modification of Time Evolving Meshes. *Computer Methods in Applied Mechanics and Engineering*, **194**, 4977–5001.

Balan, A., M. A. Park, W. K. Anderson, D. S. Kamenetskiy, J. A. Krakos, T. Michal, et al. (2020). Verification of Anisotropic Mesh Adaptation for Turbulent Simulations over ONERA M6 Wing. *AIAA Journal*, **58**(4), 1550–1565.

Bank, R. E. (1996). Hierarchical Bases and the Finite Element Method. *Acta Numerica*, **5**, 1–45.

Bank, R. R., and A. Weiser (1985). Some A Posteriori Error Estimators for Elliptic Partial Differential Equations. *Mathematics of Computation*, **44**, 283–301.

Banks, J. W., T. Aslam, and W. J. Rider (2008). On Sub-linear Convergence for Linearly Degenerate Waves in Capturing Schemes. *Journal of Computational Physics*, **227**, 6985–7002.

Banks, J., J. Hittinger, J. Connors, and C. Woodward (2012). Numerical Error Estimation for Nonlinear Hyperbolic PDEs via Nonlinear Error Transport. *Computer Methods in Applied Mechanics and Engineering*, **213**, 1–15.

Bugeda, G. (2006). A New Adaptive Remeshing Scheme Based on the Sensitivity Analysis of the SPR Point Wise Error Estimation. *Computer Methods in Applied Mechanics and Engineering*, **195**(4–6), 462–78.

Burg, C. (2006). Analytic Study of 2D and 3D Grid Motion using Modified Laplacian. *International Journal for Numerical Methods in Fluids*, **52**, 163–197.

Cadafalch, J., C. D. Perez-Segarra, R. Consul, and A. Oliva (2002). Verification of Finite Volume Computations on Steady-State Fluid Flow and Heat Transfer. *Journal of Fluids Engineering*, **24**, 11–21.

Cheng, Z., and M. Paraschivoiu (2004). A Posteriori Finite Element Bounds to Linear Functional Outputs of the Three-Dimensional Navier–Stokes Equations. *International Journal for Numerical Methods in Engineering*, **61**(11), 1835–1859.

Demkowicz, L., J. T. Oden, and T. Strouboulis (1984). Adaptive Finite Elements for Flow Problems with Moving Boundaries. Part I: Variational Principles and A Posteriori Estimates. *Computer Methods in Applied Mechanics and Engineering*, **46**(2), 217–251.

Dwight, R.P. (2008). Heuristic A Posteriori Estimation of Error due to Dissipation in Finite Volume Schemes and Application to Mesh Adaptation. *Journal of Computational Physics*, **227**(5), 2845–2863.

Eca, L., and M. Hoekstra (2002). An Evaluation of Verification Procedures for CFD Applications. *24th Symposium on Naval Hydrodynamics*, Fukuoka, Japan, July 8–13, 2002.

Eca, L., and M. Hoekstra (2009a). Error Estimation Based on Grid Refinement Studies: A Challenge for Grid Generation. *Congress on Numerical Methods in Engineering*, Barcelona, Spain, June 29–July 2, 2009.

Eca, L., and M. Hoekstra (2009b). Evaluation of Numerical Error Estimation Based on Grid Refinement Studies with the Method of the Manufactured Solutions. *Computers and Fluids*, **38**, 1580–1591.

Eca, L., and M. Hoekstra (2014). A Procedure for the Estimation of the Numerical Uncertainty of CFD Calculations Based on Grid Refinement Studies. *Journal of Computational Physics*, **262**, 104–130.

Eca, L., M. Hoekstra, J. F. Beja Pedro, and J. A. C. Falcao de Campos (2013). On the Characterization of Grid Density in Grid Refinement Studies for Discretization Error Estimation. *International Journal for Numerical Methods in Fluids*, **72**, 119–134.

Eca, L., M. Hoekstra, and P. Roache (2005). Verification of Calculations: An Overview of the Lisbon Workshop. AIAA Paper 2005-4728.

Eca, L., G. B. Vaz, J. A. C. Falcao de Campos, and M. Hoekstra (2004). Verification of Calculations of the Potential Flow around Two-Dimensional Foils. *AIAA Journal*, **42**(12), 2401–2407.

Eca, L., G. Vaz, M. Hoekstra, S. Doebling, R. Singleton, G. Srinivasan, et al. (2021). Assessment of Discretization Uncertainty Estimators Based on Grid Refinement Studies. *ASME Journal of Verification, Validation and Uncertainty Quantification*, **6**(4), 13 pages.

Eriksson, K., and C. Johnson (1987). Error-Estimates and Automatic Time Step Control for Nonlinear Parabolic Problems. Part 1. *SIAM Journal of Numerical Analysis*, **24**(1), 12–23.

Estep, D., M. Larson, and R. Williams (2000). *Estimating the Error of Numerical Solutions of Systems of Nonlinear Reaction-Diffusion Equations*. Memoirs of the American Mathematical Society, **146**(696). Providence, RI: American Mathematical Society.

Fehlberg, E. (1969). Low-Order Classical Runge-Kutta Formulas with Step Size Control and Their Application to Some Heat Transfer Problems. NASA Technical Report 315, National Aeronautics and Space Administration, July 1969.

Fornberg, B. (1998). *A Practical Guide to Pseudospectral Methods*. Cambridge: Cambridge University Press.

Gu, X., and T. I-P. Shih (2001). Differentiating between Source and Location of Error for Solution-Adaptive Mesh Refinement. AIAA Paper 2001-2660.

Hall, W. S. (1994). *Boundary Element Method*. Netherlands: Springer.

Huebner, K. H., D. L. Dewhirst, D. E. Smith, and T. G. Byrom (2001). *The Finite Element Method for Engineers*, 4th ed. New York, NY: John Wiley & Sons.

Ilinca, F., D. Pelletier, and L. Ignat (1998). Adaptive Finite Element Solution of Compressible Turbulent Flows. *AIAA Journal*, **36**(12), 2187–2194.

Jameson, A. (1988). Aerodynamic Design via Control Theory. *Journal of Scientific Computing*, **3**(3), 233–260.

Johnson, C., and P. Hansbo (1992). Adaptive Finite Element Methods in Computational Mechanics. *Computer Methods in Applied Mechanics and Engineering*, **101**(1–3), 143–181.

Laflin, K. R. (1997). Solver-Independent r-Refinement Adaptation for Dynamic Numerical Simulations. Doctoral Thesis, North Carolina State University, 1997.

Loseille, A., and F. Alauzet (2011). Continuous Mesh Framework Part I: Well-Posed Continuous Interpolation Error. *SIAM Journal on Numerical Analysis*, **49**(1), 38–60.

Mastin, C. W. (1999). Truncation Error on Structured Grids. in *Handbook of Grid Generation*, J. F. Thompson, B. K. Soni, and N. P. Weatherill (eds.) Boca Raton, FL: CRC Press.

McRae, D. S. (2000). r-Refinement Grid Adaptation Algorithms and Issues. *Computer Methods in Applied Mechanics and Engineering*, **189**, 1161–1182.

McRae, D. S., and K. R. Laflin (1999). Dynamic Grid Adaptation and Grid Quality. in *Handbook of Grid Generation*, J. F. Thompson, B. K. Soni, and N. P. Weatherill (eds.), 34-1–34-33 Boca Raton, FL: CRC.

Mitchell, W. F., and M. A. McClain (2014). A Comparison of hp-Adaptive Strategies for Elliptic Partial Differential Equation. *ACM Transactions on Mathematical Software*, **41**(1), 1–39.

Oberkampf, W. L., and C. J. Roy (2010). *Verification and Validation in Scientific Computing*. New York, NY: Cambridge University Press.

Paraschivoiu, M., J. Peraire, and A. T. Patera (1997). A Posteriori Finite Element Bounds for Linear Functional Outputs of Elliptic Partial Differential Equations. *Computer Methods in Applied Mechanics and Engineering*, **150**(1–4), 289–312.

Patra, A., and J. T. Oden (1997). Computational Techniques for Adaptive hp Finite Element Methods. *Finite Elements in Analysis and Design*, **25**(1–2), 27–39.

Pelletier, D., and P. J. Roache (2006). Chapter 13: Verification and Validation of Computational Heat Transfer. *Handbook of Numerical Heat Transfer*, 2nd ed., Minkowycz, Sparrow, and Murthy (eds.) Hoboken, NJ: Wiley.

Peraire, J., M. Vahdati, K. Morgan, and O. Zienkiewicz (1987). Adaptive Remeshing for Compressible Flow Computations. *Journal of Computational Physics*, **72**(2), 449–466.

Phillips, T. S., and C. J. Roy (2014). Richardson Extrapolation-based Discretization Uncertainty Estimation for Computational Fluid Dynamics. *Journal of Fluids Engineering*, **136**, 121401-1–121401-10.

Rannacher, R., and F. T. Suttmeier (1997). A Feed-Back Approach to Error Control in Finite Element Methods: Application to Linear Elasticity. *Computational Mechanics*, **19**(5), pp. 434–446.

Richardson, L. F. (1911). The Approximate Arithmetical Solution by Finite Differences of Physical Problems Involving Differential Equations, with an Application to the Stresses in a Masonry Dam. *Philosophical Transactions of the Royal Society of London*. Series A, Containing Papers of a Mathematical or Physical Character, **210**, 307–357.

Richardson, L. F. (1927). The Deferred Approach to the Limit. Part I. Single Lattice. *Philosophical Transaction of the Royal Society of London*. Series A, Containing Papers of a Mathematical or Physical Character, **226**, 299–349.

Rider, W., W. Witkowski, J. R. Kamm, and T. Wildey (2016). Robust Verification Analysis. *Journal of Computational Physics*, **307**, 146–163.

Roache, P. J. (1994). Perspective: A Method for Uniform Reporting of Grid Refinement Studies. *Journal of Fluids Engineering*, **116**, 405–413.

Roache, P. J. (1998). *Verification and Validation in Computational Science and Engineering*. NM: Hermosa Publishers.

Roy, C. J. (2001). Grid Convergence Error Analysis for Mixed-Order Numerical Schemes. AIAA Paper 2001-2606.

Roy, C. J. (2003). Grid Convergence Error Analysis for Mixed-Order Numerical Schemes. *AIAA Journal*, **41**(4), 595–604.

Roy, C. J. (2005). Review of Code and Solution Verification Procedures for Computational Simulation. *Journal of Computational Physics*, **205**(1), 131–156.

Roy, C. J. (2009). Strategies for Driving Mesh Adaptation in CFD (Invited). AIAA Paper 2009-1302.

Roy, C. J., C. J. Heintzelman, and S. J. Roberts (2007). Estimation of Numerical Error for 3D Inviscid Flows on Cartesian Grids. AIAA Paper 2007-0102.

Salas, M. D. (2006). Some Observations on Grid Convergence. *Computers and Fluids*, **35**, 688–692.

Salas, M. D., and H. L. Atkins (2009). On Problems Associated with Grid Convergence of Functionals. *Computers and Fluids*, **38**, 1445–1454.

Shih, T. I-P., and Y. C. Qin (2007). A Posteriori Method for Estimating and Correcting Grid-Induced Errors in CFD Solutions Part 1: Theory and Method. AIAA Paper 2007-100.

Shih, T. I-P., and B. R. Williams (2009). Development and Evaluation of an A Posteriori Method for Estimating and Correcting Grid-Induced Errors in Solutions of the Navier–Stokes Equations. AIAA Paper 2009-1499.

Stewart, J. R., and T. J. R. Hughes (1998). A Tutorial in Elementary Finite Element Error Analysis: A Systematic Presentation of A Priori and A Posteriori Error Estimates. *Computer Methods in Applied Mechanics and Engineering*, **158**(1–2), 1–22.

Szabo, B. A., and I. Babuska (2021). *Finite Element Analysis: Method, Verification and Validation*, 2nd ed. Hoboken, NJ: Wiley.

Thompson, J. F., Z. U. A. Warsi, and C. W. Mastin (1985). *Numerical Grid Generation: Foundations and Applications*. New York, NY: Elsevier.

Tyson, W. C., and C. J. Roy (2019). A Higher-Order Error Estimation Framework for Finite-Volume CFD. *Journal of Computational Physics*, **394**, 632–657.

Tyson, W. C., G. K. Yan, C. J. Roy, and C. F. Ollivier-Gooch (2019). Relinearization of the Error Transport Equations for Arbitrarily High-Order Error Estimates. *Journal of Computational Physics*, **397**, 1–22.

Van de Vosse, F. N., and P. D. Minev (1996). Spectral Element Methods: Theory and Applications. EUT Report, Dept. of Mechanical Engineering; Vol. 96-W-001, Eindhoven University of Technology.

Venditti, D. A., and D. L. Darmofal (2002). Grid Adaptation for Functional Outputs: Application to Two-Dimensional Inviscid Flows. *Journal of Computational Physics*, **176**, 40–69.

Venditti, D. A., and D. L. Darmofal (2003). Anisotropic Grid Adaptation for Functional Outputs: Application to Two-Dimensional Viscous Flows. *Journal of Computational Physics*, **187**, 22–46.

Wahlbin, L. B. (1995). *Superconvergence in Galerkin Finite Element Methods*. Volume 1605 of Lecture Notes in Mathematics. Berlin, Germany: Springer-Verlag.

Wang, H., and C. J. Roy (2023). Error Transport Equations Implementation for Discontinuous Galerkin Methods. *Journal of Computational Physics*, **474**, 111760-1–111760-18.

Whiteman, J. R. (1994). *The Mathematics of Finite Element and Applications: Highlights 1993*. New York, NY: Wiley.

Wood, W. A., and W. L. Kleb (1999). On Multi-dimensional Unstructured Mesh Adaption. AIAA Paper 1999-3254.

Yan, G., and C. F. Ollivier-Gooch (2017). Towards Higher Order Discretization Error Estimation by Error Transport using Unstructured Finite-Volume Methods for Unsteady Problems. *Computers and Fluids*, **154**, 245–255.

Zhang, X. D., J.-Y. Trepanier, and R. Camarero (2000). A Posteriori Error Estimation for Finite-Volume Solutions of Hyperbolic Conservation Laws. *Computer Methods in Applied Mechanics and Engineering*, **185**(1), 1–19.

Zhang, Z., and A. Naga (2005). A New Finite Element Gradient Recovery Method: Superconvergence Property. *SIAM Journal of Scientific Computing*, **26**(4), 1192–1213.

Zienkiewicz, O. C., and J. Z. Zhu (1987). A Simple Error Estimator and Adaptive Procedure for Practical Engineering Analysis. *International Journal for Numerical Methods in Engineering*, **24**, 337–357.

Zienkiewicz, O. C., and J. Z. Zhu (1992a). The Superconvergent Patch Recovery and A Posteriori Error Estimates, Part 2: Error Estimates and Adaptivity. *International Journal for Numerical Methods in Engineering*, **33**, 1365–1382.

Zienkiewicz, O. C., and J. Z. Zhu (1992b). Superconvergent Patch Recovery (SPR) and Adaptive Finite Element Refinement. *Computer Methods in Applied Mechanics and Engineering*, **101**, 207–224.

Part IV

Model Validation and Predictive Capability

Chapter 10 discusses the basic concepts of model validation from the perspective of model accuracy assessment based on experimental measurements. This chapter explains that model validation forms the basis for predictive capability of models and how model validation accuracy requirements impact simulation accuracy requirements. Chapter 11 stresses how model validation experiments differ from traditional experiments. Six primary guidelines for the design and execution of model validation experiments are discussed, along with technical and practical issues of conducting and funding model validation experiments. Chapter 12 discusses various methods for estimating model accuracy when simulation results are compared to experimental measurements, and also addresses the topic of model calibration. We discuss how model validation metrics are constructed, as well as how aleatory and epistemic uncertainty in both the model predictions and the experimental measurements make the accuracy assessment less precise.

Chapter 13 synthesizes the key results of verification, validation, and uncertainty quantification from the previous chapters and incorporates them into an approach for nondeterministic predictions. This chapter deals with the complex issue of model-based extrapolation to conditions where no experimental data are available. The discussion of predictive capability stresses the importance of estimating the total uncertainty in model predictions due to all possible sources. Eight procedural steps are explained for nondeterministic predictions applicable to any type of system of interest.

Model Validation and Predictive Capability

10

Model Validation Fundamentals

It doesn't matter how beautiful your theory is, it doesn't matter how smart you are. If it doesn't agree with experiment, it's wrong. —Richard Feynman

Contents of Chapter 10

This chapter discusses the fundamental philosophy and concepts of model validation. As discussed earlier in this text, model validation focuses on the assessment of the accuracy of the mathematical model for the conditions of the validation experiment. Model validation experiments differ from various types of traditional experiments, and this is discussed in detail. An analysis of the sources of error in experiments and simulations is presented in this chapter to help the reader better understand the issues involved in comparing the measurements of system response quantities (SRQs) from a model validation experiment with simulation results. When model validation is directed toward an engineered system of interest, the concept of a validation hierarchy has proven to be most effective. This chapter presents the features and construction of a model validation hierarchy and gives a detailed example. The chapter ends with a discussion of the practical and conceptual difficulties commonly encountered in model validation, particularly in the presence of conducting model input parameter calibration.

10.1 Philosophy of Model Validation Experiments

Because of the well-developed methodologies and techniques established in metrology, experimental measurements are usually viewed as the best way to estimate the true value of a physical attribute in nature. This level of trustworthiness in experimental measurements has been built over at least four millennia of experience, both from mistakes and successes. This is not meant to imply that experimental measurements *always* yield an accurate estimate of the true value. Experimental measurements can be inaccurate, or downright wrong, for many reasons. However, the trustworthiness of an experimental measurement is strengthened by the fact that techniques for investigating its limitations, weaknesses, and uncertainties are generally well understood. When new experimental diagnostic techniques are developed, they must be carefully investigated and understood. Where possible, comparisons are made with measurements from existing techniques that are better understood, so that measurement uncertainty in the new technique can be estimated. As science and technology progress, new measurement techniques increase the confidence and precision of measurements and also begin to measure physical quantities that were previously unmeasurable.

An enlightening way of thinking about the trustworthiness of experimental measurements is to think of an experimental measurement as "asking a question of nature" (Hornung and Perry, 1998). When a measurement result is obtained, the result can be thought of as nature's answer to a question that was asked. Our natural assumption is to believe that the answer obtained is the answer to the question asked. However, this is not actually the case because there are always assumptions and preconceptions involved concerning the formulation of the question and the interpretation of the answer. For example, when measuring the fluid velocity in a flow field, the experimentalist believes the question asked was: Given the flow field of interest, what is the velocity of the fluid at a certain point, without any influence of the measurement itself? The intent is to ask nature the question that excludes or minimizes the effect of random or systematic errors in the measurement. However, if there is significant random measurement error or if there is an unknown systematic error either in the measurement itself or in the data reduction procedure, then the question asked of nature is different from what was thought to be asked. That is, nature answered the question that *includes all* the random and systematic errors, whether large or small, as part of the answer.

By asking nature a question, experimentalists can deceive themselves because either they do not recognize that the question is vague, or because their preconceived notions or agendas led them astray. For example, suppose a group of experimentalists are interested in measuring some local quantity in a physical domain, e.g., the local strain in a solid, the local velocity in a flow field, or the total energy of a material at a point. They frame their question of nature within a certain mindset, such as the strain is elastic, the flow field is steady, or certain components of energy are unimportant, respectively. Then they make their measurements and interpret the results within their mindset. Often the interpretation makes perfect sense and is fully consistent within their framework, but their interpretation could be *completely* wrong. The mindset always involves assumptions concerning the relevant physical phenomena and the way the measurement technique works. Any disagreements or oddities in the measurements that don't quite fit the mindset are commonly relegated to random measurement uncertainty. Human nature places great value on success, but the natural world has no agenda.

From a model validation perspective, the experimentalist must ask the *same* question of the model that is asked of nature in the experimental measurement. However, there is a critical dichotomy. From an experimental measurement perspective, the experimentalist must strive for precision in the question of nature while minimizing the impact of (a) the assumptions in the diagnostic and measurement technique and (b) the systematic and random measurement errors. From a simulation perspective, there are two completely different issues: (a) the assumptions and approximations in the formulation of the mathematical model and (b) the input data needed for the model to produce a simulation result. In model validation, the goal is to determine the impact of the assumptions and approximation in the formulation of the model, as opposed to the impact of the input data needed for the model. To achieve this distinction, the experimentalist must measure all the input data needed for the simulation of the experiment. In this way, the repercussions of the modeling assumptions and approximations on the SRQs that are measured in the experiment can be critically tested.

10.1.1 Validation Experiments vs. Traditional Experiments

Traditional experiments can be grouped into three general categories (Oberkampf and Trucano, 2002). The first category comprises experiments that are conducted primarily to improve the fundamental understanding of some physical process. Sometimes these are referred to as physical discovery or phenomena discovery experiments. Examples are: (a) experiments that measure characteristics of fluid turbulence; (b) experiments that investigate crack propagation in solids; (c) experiments in high-energy density physics; and (d) experiments probing the onset and stability of phase changes in solids, liquids, and gases.

The second category of traditional experiments consists of those conducted primarily for determining or improving parameters in an existing mathematical model of a physical process. Sometimes these are referred to as model-calibration or parameter-estimation experiments. Examples are: (a) experiments to measure chemical reaction rate parameters in reacting or detonating flows; (b) experiments to measure thermal emissivity of material surfaces; (c) experiments to calibrate parameters in a model for predicting large plastic deformation of a material; and (d) experiments to calibrate the mass-diffusion rate parameters in a mass-transport chemistry model.

The third category of traditional experiments includes those that determine the reliability, performance, or safety of components, subsystems, or complete systems. These experiments are sometimes referred to as functionality tests, acceptance tests, or qualification tests of components, subsystems, or systems. Examples are: (a) certification tests of a new combustor design in gas turbine engines; (b) proof tests of a new design of a filament-wound composite pressure vessel; (c) safety tests of the emergency cooling system in a nuclear power reactor; and (d) confirmation of the thrust profile of a solid-propellant rocket motor.

Model validation experiments constitute a new type of experiment. (See Oberkampf and Aeschliman, 1992; Marvin, 1995; Aeschliman and Oberkampf, 1998, for a discussion of the early concepts of a model validation experiment.) A model validation experiment is conducted for the primary purpose of determining the accuracy of a mathematical model relative to the measurements of SRQs of interest. Hence, a validation experiment is designed, executed, and analyzed for the purpose of quantitatively determining the ability of the model, and its embodiment in a computer code, to simulate a physical process. In a model validation experiment, *the computational analyst is the customer*, as opposed to a researcher building a model or a customer interested in the performance or safety of the system. Scientific computing has only matured to the point where it could even be considered as a separate customer in experimental activities during the last few decades. As modern technology increasingly moves toward engineering systems that are designed, certified, and even fielded based primarily on scientific computing, then scientific computing itself will increasingly become the customer of experiments.

One other aspect of model validation experiments that is different from traditional experiments is that traditional experiments place a great deal of emphasis on measurements in a controlled environment. Only with a controlled environment can measurements of physical processes be reliably repeated by other experimentalists and models carefully calibrated. In model validation experiments, however, *characterization* of the experiment is the more important goal. Here *characterization* means measuring all of the important input data that are needed for the simulation, both within the system and in the surroundings. For example, if certain input conditions cannot be tightly controlled, their random variability should be carefully characterized. Thus, precise control and repeatability of the experiment are less important in a model validation experiment than *measuring the variability of the simulation input data* of an experiment. Variability in the system and surroundings data of a model validation experiment does not, in itself, cause a problem, as long as the variability of the system and surroundings are precisely measured. For experiments with uncontrolled conditions, a number of experimental realizations are necessary to carefully characterize the variability of the system and surroundings so that this information can be provided to the computational analyst. For example, suppose the system of interest is exposed to the variability of outdoor weather conditions. The important measurements for a model validation experiment in this environment are all of the relevant system and surroundings data (as they actually occurred) that are needed for input data for the simulation. Furthermore, the measurement of the SRQs of interest should be related to either the time-dependent nature or the variability of the input data.

10.1.2 Goals and Strategy of Model Validation

In Section 2.2.3, three aspects of model validation were discussed regarding Figure 2.8, duplicated as Figure 10.1. Aspect 1 involves the quantitative comparison of computational

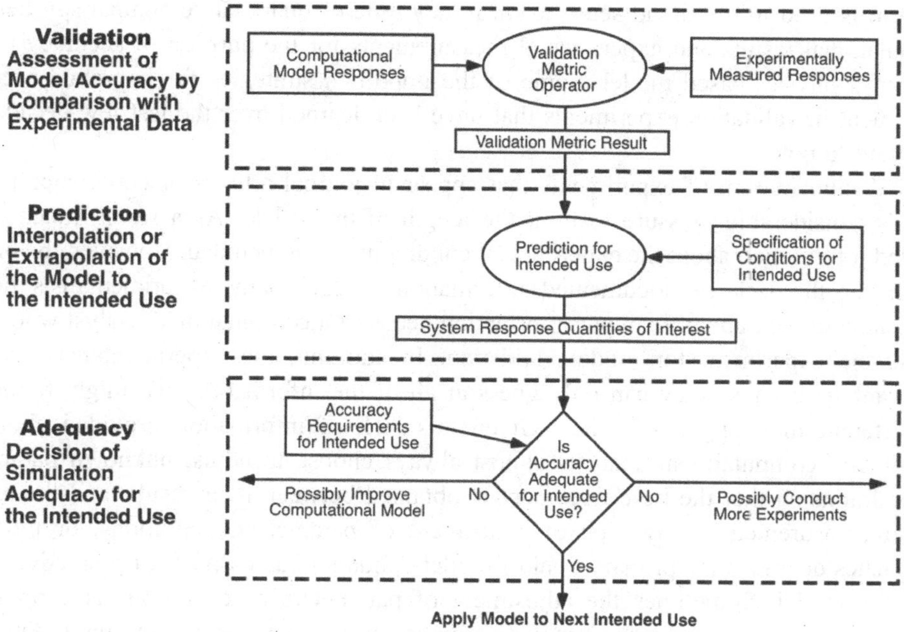

Figure 10.1 Three aspects of model validation.
(Oberkampf and Trucano, 2008)

responses and experimentally measured responses. The mathematical operator used to make the comparisons is a *validation metric operator*. The operator is usually formulated as a difference operator so that the validation metric result is a measure of the disagreement between the simulation and experimental responses. Aspect 2 deals with the use of the model to make predictions, in the sense of interpolation or extrapolation beyond existing data, for the conditions of the intended use of the model. Aspect 3 deals with (a) the comparison of the estimated accuracy of the model relative to the accuracy requirements of the model for the model's intended use domain, and (b) the decision of adequacy/inadequacy of the model over the model's intended use domain. Aspect 1 can be thought of in two ways: scientific model validation and project-oriented model validation. *Scientific model validation* is a quantitative assessment of model accuracy without regard to any specific accuracy requirements or engineering project needs. This is the most common type of validation activity that is published in the research literature. *Project-oriented model validation* is a quantitative assessment of model accuracy with the needs of the project taking priority. The following sections discuss each type of model validation activity in detail. The reader is reminded that this book uses the restricted view of the term *model validation*, meaning model accuracy assessment as depicted in Aspect 1.

10.1.2.1 Scientific Model Validation

Most published work dealing with comparing simulation and experimental results is directed toward what is referred to here as *scientific model validation*. The use of this term is *not* meant to contrast validation in scientific endeavors versus engineering endeavors. Instead, the term is used in the broad sense to mean any type of quantitative comparison between computational results and experimental measurements for the purpose of accuracy assessment of a physics-based model. Some of the important strategies for improved accuracy assessment in validation experiments that have been learned over the last few decades are important to note.

For documentation of experiments that appear in journal articles or conference papers, there is considerable pressure to limit the length of the article. As a result, much of the detailed information about the experimental conditions is not included. However, the biggest reason for the lack of documented information is that many experimentalists do not understand or care about the input information needs of a computational analyst who wants to use the experiment for model validation. In addition, the experimentalists are at a disadvantage because they can only guess at all of the information that might be needed for different modeling approaches. Given this lack of information provided about the experiment, computational analysts almost always choose to adjust unknown parameters or conditions so that the best agreement is obtained between their results and the experimental measurements. Any type of adjustment of parameters, conditions, or modeling approaches degrades the primary goal of model validation: assessment of predictive accuracy of a model. Sometimes the adjustment of parameters or conditions is explicit; for example, using calibration or parameter estimation procedures for missing input data. This directly damages the critical assessment of model predictive accuracy. Sometimes, the adjustment procedures are not explicitly mentioned or explained, either intentionally or unintentionally. For example, certain modeling approaches may have been attempted, may have given poor agreement with the experimental data, and then were abandoned. The experience, however, was used to guide the analysts toward new modeling assumptions that yielded modeling results with improved agreement with the data.

For future validation experiments, an experiment design collaboration between the computational analysts and the experimentalists is a much more constructive procedure. Many of the difficulties just mentioned can be reduced or eliminated in a joint effort. This topic will be discussed in depth in Chapter 11 "Design and Execution of Model Validation Experiments." Here, two aspects of a joint effort are mentioned. First, in a joint effort, the analyst must inform the experimentalist of the input information required by the model. In addition, the analyst must communicate to the experimentalist what SRQs are of particular interest so that they can be measured. Second, the experimental measurements of the SRQs should be withheld from the analysts so that a blind-test prediction can be made. Although the value of blind predictions in scientific computing is not uniformly accepted, it is crucial to withhold the measured SRQs so as to critically assess the predictive accuracy of a model. Some fields of science, such as drug testing and clinical trials in health care, have long recognized that without blind or double-blind testing, conclusions are commonly distorted and misleading.

In comparing simulation and experimental results, emphasis should be placed on quantitative comparison methods. Color contour levels of an SRQ over a two-dimensional region, one for a simulation result and one for an experimental result, can be qualitatively useful. However, little quantitative information can be gained, especially if the quantitative values of the color scale are not defined or carefully chosen. The most common comparison method is a graph where simulation results and experimentally measured SRQs are plotted over a range of an input or control parameter. However, there are two quantitative shortcomings of this method. First, agreement or disagreement between computational and experimental results is not quantified. The focus is on the trends of the simulation relative to the measured SRQ, not quantifying the computation-experiment disagreement. Observations and conclusions are commonly made about the claimed agreement, such as "good" or "excellent" or "the model is validated," but all of these are subjective.

Second, these types of graphs rarely have any type of information concerning the effect of the uncertainties on the measured or computed results. For example, the simulation results are commonly deterministic results. Additionally, no information is shown concerning the influence of the numerical solution errors on the results, specifically discretization or iterative errors. For the experimental results, it is still common to see experimental results without any experimental uncertainty estimates shown or discussion given on how the estimates were obtained. Significant improvement in quantitative comparisons is made if the effects of the uncertainties on both the simulation and experimental results are displayed. One method is to show the mean value and the uncertainty bars with plus or minus two standard deviations for both the simulation results and experimental measurements. The topic of quantitative comparison of simulation and experimental results will be discussed in detail in Chapter 12 "Model Accuracy Assessment."

Simulation results and experimental results can produce many different SRQs. Model validation is concerned with the comparison of the measured SRQ with the same simulation SRQ under the same conditions as the experiment. There is a range of difficulty in predicting and experimentally measuring different SRQs. *Prediction difficulty* means aspects such as (a) the fidelity of the physics model that is required to accurately predict an SRQ; (b) the range of spatial and/or temporal scales exhibited by an SRQ; and (c) the spatial, temporal, and statistical convergence characteristics required to computationally resolve multiple physical scales and physical phenomena. *Experimental difficulty* means that there is a wide range of difficulty in measuring different SRQs. In experiments, this range of difficulty is primarily due to large differences in spatial and/or temporal scales for various SRQs. With modern digital electronic equipment, temporal scales do not usually generate difficulty compared to measurement of a wide range of spatial scales. Difficulty in experimental measurements commonly translates to increased experimental uncertainty for smaller spatial scales, both in bias and random uncertainty.

Figure 10.2 depicts a mathematical operator perspective of the range of difficulty in simulating and measuring different SRQs. The spectrum could have a different type of ordering for some complex physics processes, such as those strongly dependent on statistical features of the SRQ. The scale of difficulty is ordered in terms of derivatives and integrals of the dependent variables in the partial differential equations (PDEs) of the model. For example, consider steady-state heat conduction through a homogeneous solid with constant thermal conductivity. The PDE for the temperature distribution through a two-dimensional solid is given by Laplace's equations:

$$\frac{\partial^2 T}{\partial x^2} + \frac{\partial^2 T}{\partial y^2} = 0. \tag{10.1}$$

The dependent variable in the PDE is the temperature $T(x,y)$. This quantity is represented in the middle of the range in Figure 10.2. If one is interested in predicting and measuring the heat flux through a vertical line at $x =$ constant, the heat flux q as a function of y is given by

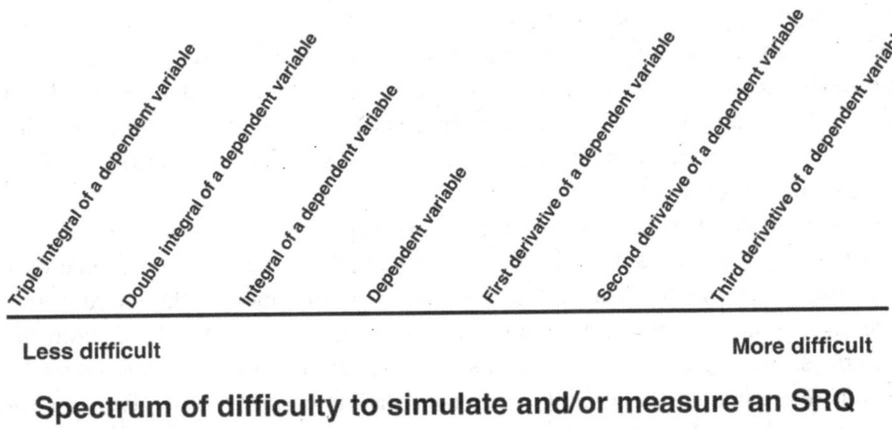

Figure 10.2 Spectrum of various types of SRQs and the difficulty to simulate and measure each.

$$q(y) = -k\left(\frac{\partial T}{\partial x}\right)_{x=\text{constant}} . \tag{10.2}$$

From this equation, it is seen that the heat flux is an SRQ that is dependent on the first derivative of the dependent variable in the PDE. This quantity is represented just to the right of the dependent variable in Figure 10.2. Various types of integrals of dependent variables are also of interest as SRQs. These are referred to as functionals because they operate on a function and produce a scalar quantity. These are shown to the left of the middle in Figure 10.2 and they are ordered in terms of the number of integrals of the dependent variable of the PDE. For example, lift and drag on an object is the double integral of the normal and shear stresses over the surface of the object.

In model validation activities, computational analysts will commonly compare their computational results with experimental measurements at one level of difficulty shown in the spectrum in Figure 10.2. If the computational results are accurate, then it is implied that the model is accurate at all levels in the spectrum. Because a spectrum of difficulty exists in simulation, a validation claim at one level of difficulty does *not necessarily* translate into accuracy at higher levels of difficulty. For example, if good agreement was shown for the temperature distribution through a solid, it is more demanding of the model to accurately simulate the heat flux because of the derivative operator in the heat flux. However, demonstrated accuracy at high levels of difficulty *does imply* accuracy at lower levels of simulation difficulty because of the integral operator. The integral operator is a strong smoothing operator which can act, for example, over space, time, angle, and energy. For instance, if the analyst is interested in the dynamic motion of a system, it is more difficult to simulate time-dependent acceleration than velocity and position.

10.1.2.2 Project-Oriented Model Validation

Project-oriented model validation is similar to scientific model validation in that one is still interested in a quantitative assessment of model accuracy as measured with respect to experimental data. However, in project-oriented model validation, the focus shifts to how validation activities contribute to assessing the predictive accuracy of the model when the model is applied to a specific system of interest. Project-oriented model validation is interested in evaluating the model accuracy using project-relevant experiments, which includes experiments using portions of the complete systems. As discussed in Chapter 2, the validation domain may or may not overlap the intended use domain of the system. For regions of overlap, interpolation is required to estimate the model accuracy for conditions that do not match the validation data conditions. For nonoverlapping regions, extrapolation of model accuracy is required for conditions outside the validation domain, but inside the intended use domain. (See Figure 2.9 for a discussion of overlapping and non-overlapping regions of the validation and the intended use domains.) Chapter 13 "Predictive Capability," deals with the issue of model-based extrapolation incorporating validation metrics as a measure of observed model accuracy.

During the last two decades, large scientific computing projects, such as the US National Nuclear Security Administration (NNSA) sponsored Advanced Simulation and Computing (ASC) program (NNSA, 2019), have developed methods to improve the design, planning, and execution of project-oriented validation experiments. More informative and effective model validation experiments were found to be critical for managing the allocation of time, money, facilities, and personnel to optimally advance the predictive capability of project-oriented simulation. Project-oriented validation experiments, referred to here as *directed experiments*, are those designed purposefully with specific goals linked to the project or the objectives of the system of interest. Occasionally, experimental data directly relevant to the system of interest can be found in the published technical literature or proprietary company reports, but this is rare because of the specificity of project needs. Because directed experiments are required to obtain quantitative comparisons of simulations with experimental data, important requirements are placed on directed experiments to create the greatest value for performing these comparisons. In addition, directed experiments should be designed to assist computational analysts and mathematical-model builders, if needed, to understand *why* the model may have performed poorly.

Figure 10.3 shows a two-dimensional input-parameter space defined by α and β, each one characterizing some feature of the system or the conditions imposed by the surroundings. The validation domain is shown as the interior region in which various validation experiments have been conducted, denoted by V. The validation domain is defined to be the *convex hull* of the set of input conditions that describe the conditions of the validation experiments. (The convex hull of a set of points, also known as the minimum convex polygon or as the intersection of all convex sets of the points, is the envelope of all points such that the enclosure is convex everywhere [Leonard and Lewis (2016)].) The intended use domain (also referred to as the application domain, the system operating condition domain, or the context of use domain) shows the region of interest in which the predictive capability of the model is needed to assist in the design and prediction of the system performance, safety, and reliability. The convex hull of the intended use domain can also be defined based on the specified operating conditions of the system. Typical of operating conditions of a system, the operating envelope is given as a polygon with the corners specified in terms of pairs of coordinates (α_i, β_i), $i = 1, 2, \ldots, 5$. It is defined in terms of the convex hull of the set of points (α_i, β_i). As shown in Figure 10.3, the validation domain is commonly *a small subset of the intended use domain*. For many types of complex systems, the validation domain is completely disconnected from the intended use domain, i.e., they are disjoint sets.

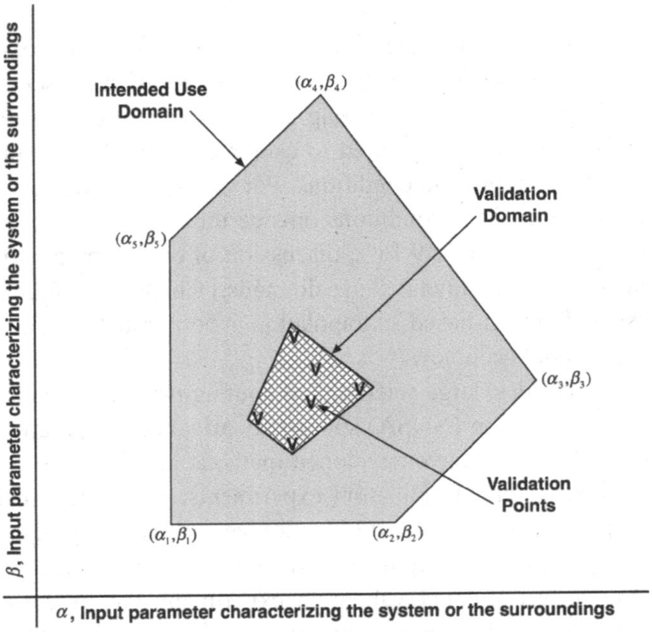

Figure 10.3 Example of an intended use domain and the validation domain.
(adapted from Trucano et al., 2002)

The difference between the simulation result and the experimental measurements for an SRQ can be quantitatively estimated at each of the conditions marked with a V using a validation metric operator. If all of the relevant input data needed for the mathematical model are measured by the experimentalist, as well as provided to the computational analyst, then the difference between the simulation and the experiment is the *model form uncertainty.* That is, the model form uncertainty (MFU) is the uncertainty due to the assumptions and approximations in the formulation of the mathematical model. An estimate of the MFU could be made over the validation domain, but estimating this uncertainty is difficult for the following reasons:

1. The estimate depends on the uncertainty in the experimental measurements. Uncertainty in the experimental measurements depends not only on the measurement uncertainty, but also on the variability in the input data for the conditions of the experiment. For example, if important input data to the model have high variability (because they cannot be controlled), this variability occurs in addition to the experimental measurement uncertainty.
2. Any significant computational errors incurred in the simulation will negatively impact the ability to interpret the difference between the simulation and the experimental results as the MFU. That is, computational errors pollute the difference between the simulation and the experiment such that the difference is not just due to assumptions and approximations in the formulation of the mathematical model. As discussed in previous chapters, computational errors can be due to coding errors and numerical solution errors; computer round-off error; iterative error; discretization error; and statistical sampling error.

3. MFU is *not, by nature, a stochastic quantity.* It is fundamentally a *bias error* due to the formulation of the mathematical model. However, because of the experimental uncertainty and the simulation uncertainty, if a nondeterministic simulation is computed, then the MFU is a mixture of both stochastic uncertainty and bias error.

Although it can be difficult to estimate MFU between the experimental conditions within the validation domain, there should be reasonable confidence in the result because the estimate can be thought of as a type of functional interpolation. Conceptualization of the MFU within the validation domain as interpolation does *not* assume that interpolant is smooth or that its magnitude is small. It only assumes that the limits of the MFU can be reasonably estimated based on the experimental data and the physics fidelity captured in the mathematical model. Estimation of MFU outside of the validation domain is much more difficult because there are no anchor points based on experimental measurements. One approach has been to construct a regression fit of MFU over the validation domain and then extrapolate this fit, including the additional uncertainty from a prediction interval, into the intended use domain. The weakness of this approach, as just mentioned, is that MFU is a mixture of stochastic uncertainty and bias error. Therefore, MFU depends on how the weaknesses in the assumptions and approximations in the formulation of the model expose themselves in terms of the SRQs outside the validation domain. A detailed discussion of MFU and its extrapolation is discussed in Chapter 12 "Model Accuracy Assessment," and Chapter 13 "Predictive Capability."

With directed experiments, system designers and project managers tend to focus on model accuracy assessment initially, instead of how good the model accuracy needs to be for application in the successful design of their system. This is understandable because they are focused on their system meeting design and performance requirements. From a project perspective, MFU increases simulation uncertainty beyond the contribution due to uncertainty of the system and the surroundings within the intended use domain. For example, large variability may be present due to BCs, system excitation, and ICs over certain portions of the intended use domain. These uncertainties must be considered as well as MFU in the successful design of the system of interest. All of the sources of uncertainty in the simulation should be considered, including numerical uncertainties, when estimating the total uncertainty of the SRQs of interest. If MFU is a dominant contributor, then the project needs will guide what conditions are needed for future validation experiments to better estimate this uncertainty. If there are multiple conditions for potential validation experiments, these should be prioritized. Chapter 14 "Planning and Prioritization in Modeling and Simulation," provides a discussion of various methods for planning and prioritization of not only directed experiments, but also other activities that are needed for simulation credibility.

In a directed experiment, there must be a balance between the project needs and the needs of a high-quality validation experiment. The primary purpose of a validation experiment is determining the MFU. Because the project and the validation experiment have different goals, the tension between the two is exacerbated if the project funds the design, execution, and analysis of the experiment. For example, in the design of the directed experiment, the project leader typically would like the system geometry and functionality of the hardware to be the same, or similar, to the actual system being designed. However, the computational analysts would like the experiment to be focused on the physical processes in the modeling vis-à-vis those occurring in the experiment. The viewpoints and goals of the experimentalists are to obtain the highest accuracy measurements possible, given the time and resources available. There are no straightforward cures to this three-way tension. The only general recommendation that can be made is to openly discuss the reasoning and priorities of each perspective so that a logical compromise can be attained. Experience has shown that the project perspective commonly dominates the discussion concerning design trade-offs of the validation experiment to the detriment of the validation goals of the experiment. Project managers who have training and experience in computational simulation can provide more enlightened decision making on needed compromises.

Example: Earthquake Engineering

Consider the design of a building that is intended for command, control, and communications for first responders dealing with a natural disaster. The building and all equipment inside should be resilient and continue functioning during and after an extreme earthquake. In terms of the Mercalli scale, the acceleration could be greater than 1.2 g and the ground motion velocity greater than 100 cm/s, in three orthogonal directions. The building is intended to have two floors with a total usable floor space of 1,000 m^2 for offices, personnel, furnishings, and equipment.

Modern public works projects are designed based on both computational simulation and previous experience. For abnormal environments, computational simulation bears an especially heavy responsibility for the prediction of the system's performance because there is so little experimental data available. Due to the size and mass of most public works structures, validation experiments can assess the accuracy of the simulation only for scale models of the structure, portions of the structure, or only specific subsystems. However, the vibration of a scale-model structure is very different from a full-scale structure.

Figure 10.4 gives an example of a validation domain in terms of two important input parameters to a mathematical model of the vibration of a test article. Acceleration of the structure, in three orthogonal coordinates, is a characteristic of the excitation due to the surroundings. Mass of the structure is a characteristic of the structure itself. The experimental capability shown is representative of the University of Buffalo's Structural Engineering and Earthquake Simulation Laboratory. A mathematical model for the vibration of the structure would have tens to hundreds of input data

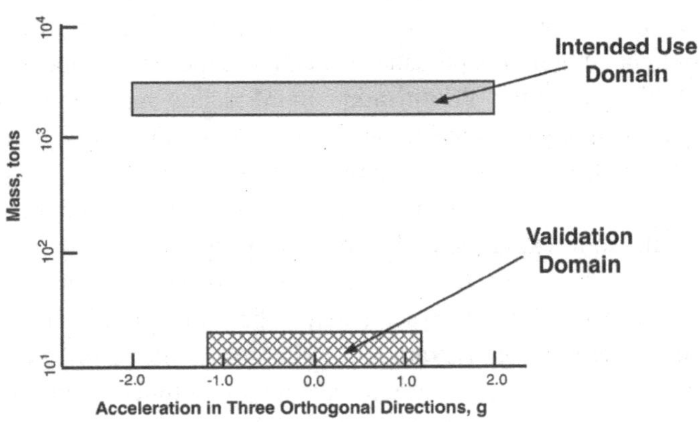

Figure 10.4 Example of the validation domain and intended use domain for response of large structures to earthquakes.

parameters, depending on the complexity of the model. Since graphical visualization of a high-dimension space is not possible, it is helpful to pick a few of the most important input parameters for visualization in two or three dimensions. These important parameters should also reflect the limitations in experimental facility capabilities. Figure 10.4 uses two of these input parameters to clearly show that the validation domain is far, and completely disconnected, from the intended use domain for the structure. As a result, very large model-based extrapolation is required based on the fidelity of the physics in the model to predict the response of the structure.

10.2 Model Validation Using Data from Existing Experiments

Since conducting new experiments is time consuming and expensive, many organizations face great pressure to use existing experimental data for model validation. This is especially true for directed experiments where time schedules and budgetary constraints are immense. The resistance is usually presented as an inquiry to why the experimental data that has been collected for decades on similar systems can't be used as existing data to assess the current models. A technically sound and defensible answer to this recurring request is so important that a detailed discussion of possible responses is presented in this chapter. Although these responses do not apply in some situations, they give the reader an idea of typical difficulties and challenges when using data from existing experiments.

10.2.1 Unmeasured Model Input Data

The most common reason for not being able to use existing experimental data for model validation is that important input data for the simulation is not available or documented as part of the description of the experiment. Examples of categories of information not documented, or poorly quantified, are system characteristics, BCs, ICs, and system excitation. Because the estimation of MFU *has not been an important issue* in the past, the measurement and documentation of these data are not typically available. When input data were missing, the usual goal of comparing simulation and experiment was to obtain the best agreement between simulation and experiment by calibrating unknown input parameters. This is an appropriate goal in many situations, but it is a distinctly different goal than model validation in the sense discussed in this text. The level of detailed information needed from the experimentalist in order to estimate MFU is much more demanding than is commonly recognized.

Examples of system characteristics not commonly measured or documented are: (a) mechanical, electrical, thermal, chemical, magnetic, optical, acoustical, and radiological properties of materials, components, or assemblies; (b) detailed geometric features of the system, such as material gaps, gaskets, and as-tested geometric inspection data; and (c) system assembly details, such as the preload torque on bolts, inspection of welds and rivets, and detailed information of friction-fit assemblies. Examples of BCs not measured are: (a) Dirichlet, Neumann, Robin, mixed, periodic, and Cauchy; and (b) information needed by the PDEs concerning how the surroundings interact with the boundaries of the system, including possible time dependence of the BCs. Examples of ICs not measured are: (a) the spatial distribution of all dependent variables of the PDEs over the domain of interest, and (b) temporal derivatives of the dependent variables over the domain of the PDE. And finally, examples of system excitation not measured are: (a) variation of the excitation over the spatial domain of the PDEs, (b) how excitation may change over time, and (c) how excitation may change when the system is deformed or an altered state due to response to the excitation occurs.

The level of uncertainty characterization of the input data needed, listed from most precise to least precise, for a simulation can range from:

- a precisely known (deterministic) value or function;
- a precisely known random variable, i.e., a pure aleatory uncertainty;
- a random variable characterized by a family of probability distributions, but the parameters of the distribution are imprecisely known, i.e., a mixed aleatory and epistemic uncertainty; and/or
- an opinion or a recollection of a quantity that results in an interval, i.e., a pure epistemic uncertainty.

The first category would only require a deterministic simulation, whereas the remaining three categories would require a nondeterministic simulation. Not all input data needed for a model is of the same level of importance relative to the prediction of the SRQs of interest. Global sensitivity analysis can be used to determine the order of importance of input data for each of the SRQs (discussed in Chapter 13 "Predictive Capability"). The less knowledge of important input information, the poorer the ability to quantitatively estimate the MFU.

Example: Unmeasured Model Input Data

Consider an example where an experiment is conducted and some of the input data needed for a simulation are measured. Some of the input data are characterized as deterministic quantities and some are not measured. The experimentalist is only able to provide an interval-valued estimate for the unmeasured input. That is, the experimentalist could not justify that the unmeasured input is either a precisely known random variable or an imprecisely known random variable. There is only one SRQ of interest, y, and the experimental measurement uncertainty is negligibly small. Using all of the information provided by the experimentalist for the input data, a nondeterministic simulation of the experiment is computed.

Figure 10.5 shows the experimental measurements made at four values of x and the interval-valued simulation result for y shown as the vertical bars. As discussed in Chapter 3 "Modeling and Computational Simulation," interval-valued uncertain input will produce an interval-valued uncertain response. As can be seen in Figure 10.5, a great deal of imprecision exists in the comparison of the interval-valued simulation and experiment. There is some evidence that the simulation does not agree quite as well with the experiment for x_3 and x_4 compared to x_1 and x_2. However, this is a weak argument because the uncertainty in the simulation result contains no explicit or implied likelihood within the interval. The only justified conclusion concerning the comparison is that there is *no evidence of disagreement* between the simulation and the experiment.

Contrast this example with the traditionally used procedure. The traditional approach considers the unmeasured inputs as adjustable parameters in the model, either scalar parameters or random variables. As discussed in Chapter 3, the solution to the inverse model is used to optimize the unmeasured parameters to obtain the best agreement with experimental measurements. Hence, the

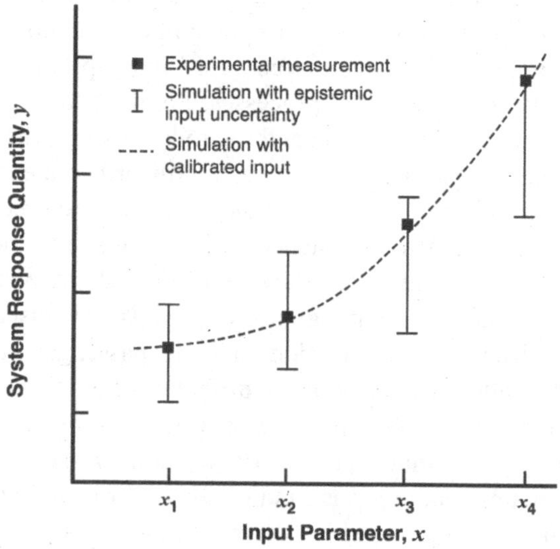

Figure 10.5 Example of comparison of experimental measurements with different approaches to input uncertainty.

unmeasured parameters are optimized or updated, given the constraint of the model, to obtain the greatest agreement with the measured SRQ. Using this calibration approach, a representative simulation result is shown in Figure 10.5 as the dashed line. This approach has little to do with assessing model accuracy, but everything to do with improving agreement between simulation and experiment. Whether this approach improves the predictive capability of the model is a matter of intense debate; however, the procedure clearly produces better agreement of the simulation and the measurements. For the present example, there are completely different sources of uncertainty in play; uncertainty in model inputs and MFU. Generally, there are many different sources of uncertainty in a complex model, both physical and numerical. Because of the flexibility of the model, one source of uncertainty can be counteracted with other sources of uncertainty, thus achieving the measured result for the wrong reasons.

10.2.2 Insufficient Coverage of SRQs

The second most common difficulty in attempting to use data from traditional experiments for validation is that a small number of SRQs are measured and few, if any, replicate experiments are conducted. With limited experimental measurements, only limited and/or weak conclusions can be made concerning the quantitative accuracy of the model, leaving two common situations. First, experimental measurements are made of high-level SRQs, but few lower-level SRQs are measured, such as local quantities. This situation is very common when experiments are funded and conducted by product-driven projects. As discussed earlier with directed experiments, project managers are focused on one goal: designing and certifying the performance of their system of interest on schedule and on budget. As a result, they concentrate on conducting experiments that directly contribute to ensuring that the system meets high-level performance and functionality requirements specified for their system.

Consider the example of an automobile manufacturer interested in improving the crashworthiness of an existing product line. Suppose a significant number of crash experiments have been conducted on the existing line of vehicles. Crush-up of a vehicle is an extremely complex convolution of diverse physical processes. Some of the processes that occur are large plastic deformation; material recontact and friction; buckling; shear banding; brittle and ductile fracture; and material tearing; all with an extraordinarily wide range of materials. The SRQs of primary interest to a manufacturer are those that are required by the governmental regulatory authority, such as crush-up distance and intrusion into the passenger compartment. Although these are important SRQs, they are of limited value in an assessment of the accuracy of a model because they are high-level SRQs. As discussed with regard to Figure 10.2, crush-up distance and intrusion into the passenger compartment represent multiple integrals of dependent variables of the differential equations being solved.

The second example of insufficient coverage of SRQs includes experiments where local SRQs are made over a very limited portion of the domain of the PDE. This situation commonly occurs when a project group is funding the experiment because they are dealing with an unexpected failure of a system either during the prototype testing phase or during in-service operations of a system. These types of investigations are under extreme time constraints, and they tend to concentrate on the local nature of the failure, as opposed to a broader lack of knowledge of the failure mechanism or failure in the modeling of the system. For example, suppose that significant cavitation occurs in a fluid pump or turbine such that the surface is being pitted and eroded. Experiments are conducted to measure the local static pressure in the region where surface erosion occurs. Suppose that the computational fluid dynamics (CFD) simulation shows that, indeed, the measured pressure is significantly lower than the predicted pressure. CFD studies are conducted to investigate the usual suspect: turbulence modeling. If a turbulence model is found that agrees with the measured pressure in the cavitation region, then it is usually assumed that this must be a "better" simulation. The CFD model would then be used to modify the design of the pump or alter operating conditions such that the simulation shows no cavitation in the region. This is an expedient solution to the immediate problem, but it certainly lends little credibility that the simulation obtained the correct answer for the right reason.

10.2.3 Insufficient Information on Experimental Uncertainty

The third most common deficiency in trying to use traditional experiments in model validation is that little or no estimation of experimental uncertainty is given for either the measured SRQs or important model input data. Even though uncertainty estimation has been a long tradition in experimental measurements, it is surprising how often measurements are reported without an uncertainty estimate. If an estimate of measurement uncertainty is provided, many times it underestimates the true uncertainty. Some reasons for an underestimate are: (a) the uncertainty estimate provided is an estimate of the repeatability of the measurement, not an estimate of a broader class of random and systematic uncertainties in the experiment; (b) only one experimental realization is measured so the estimate is simply based on experience or what *appears to look right*; and (c) systematic uncertainties due to particular diagnostic techniques, specific experimental procedures, or experimental facilities have not been quantified by conducting experiments using different diagnostics, procedures, or experimental facilities, respectively.

The widespread situation of underestimation of experimental measurement uncertainty was highlighted in a famous article by metrologist William Youden, "Enduring Values" (Youden, 1972). In his article, Youden states, "Why do results obtained by different investigators characteristically disagree by more than would be expected by their estimates of uncertainty?" Youden stresses that *everything* gets changed in another laboratory, whereas the investigator can (or does) make only minor changes within their own laboratory. Youden gives two examples of systematic uncertainties in measurements of fundamental physical constants. He shows 15 values of the Astronomical Unit measured by investigators over the period 1895 to 1961. He points out that each investigator's reported value of uncertainty is *outside the limits* reported by the immediate predecessor. Youden also references the article by McNish (1962) concerning measurements of the speed of light.

McNish and Youden show 24 measurements of the speed of light along with each investigator's estimate of their experimental uncertainty. The graph of the measurements shows that the estimated value (the experimenter's best estimate for the speed of light) is changing at a rate of about 1 km/s for every seven investigators who measure it! For the 24 investigators, the estimated value has changed 3.5 km/s; however, half of the investigators estimated their experimental uncertainty as *under* 0.5 km/s. Youden points out that the problem is that the experimenters significantly underestimated unknown systematic uncertainties in their measurements. Even though Youden and other metrologists have called attention to the recurring problem of measurement uncertainty commonly being underestimated, human nature, saving face among investigators, and competition between laboratories, seldom change. The topic of improved experimental uncertainty estimation is discussed in Chapter 11 "Design and Execution of Model Validation Experiments."

10.3 Sources of Error in Experiments and Simulations

The fundamental sources of error in experimental measurements and simulation are presented in the following section. This discussion focuses on errors instead of uncertainties because the concept of a true value, in either the experiment or the simulation, is advantageous. This discussion expands on the development in Oberkampf et al. (2004). Let y_{sim} be an SRQ that results from a simulation. As discussed in Section 3.5, y_{sim} is a final result of the various mappings shown in Figure 10.6. Let y_{nature} be the true value of the SRQ from nature that, clearly, can never be known exactly. Using the definition of error discussed in Section 2.3, the error in the simulation, E_{sim}, is defined as

$$E_{sim} = y_{sim} - y_{nature}. \tag{10.3}$$

Both terms on the right-hand side of Equation (10.3) can be separated into additional terms in order to explicitly identify various contributors to error. Equation (10.3) is written as

$$E_{sim} = \left(y_{sim} - y_{Pcomputer}\right) + \left(y_{Pcomputer} - y_{model}\right) + \left(y_{model} - y_{exp}\right) + \left(y_{exp} - y_{nature}\right). \tag{10.4}$$

The term $y_{Pcomputer}$ is the SRQ that could be theoretically computed on a perfect computer, meaning a computer with infinite speed, precision, and memory such that all arithmetic error sources approach zero in the limit. Note that the same mathematical model, the same

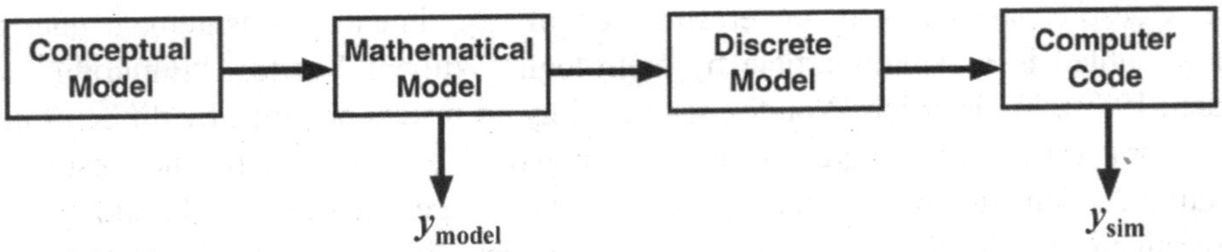

Figure 10.6 Sequence of model mappings to produce a simulation result.

algorithms, the same discretization, and the same computer code used in y_{sim} are also used in $y_{Pcomputer}$. The term y_{model} is the SRQ resulting from the exact solution to the mathematical model, i.e., the calculus problem; the model given by the PDEs, BCs, ICs, system excitation, geometric features, and all other input data needed for the simulation. The term y_{exp} is the value of the SRQ that is measured in an experiment.

A more compact form of Equation (10.4) is written as

$$E_{sim} = E_1 + E_2 + E_3 + E_4, \tag{10.5}$$

where

$$\begin{aligned} E_1 &= \left(y_{sim} - y_{Pcomputer}\right), \\ E_2 &= \left(y_{Pcomputer} - y_{model}\right), \\ E_3 &= \left(y_{model} - y_{exp}\right), \\ E_4 &= \left(y_{exp} - y_{nature}\right). \end{aligned} \tag{10.6}$$

E_1 through E_4 represent all of the fundamental sources of error in a comparison of a simulation result and an experimental measurement. The only two quantities in Equation (10.6) that are known are y_{sim} and y_{exp}. Writing the simulation error in this way clearly demonstrates the increased likelihood of error cancellation when the focus is on obtaining agreement between simulation and experimental measurement, i.e., model calibration, as opposed to attempting to estimate each source of error independently. The processes of verification and validation attempt to estimate each error contributor in Equation (10.5) so that increased confidence can be gained concerning the magnitude of the sum of all the contributors. Conversely, model calibration attempts *to achieve a zero sum of all of the terms* in Equation (10.5). Each of the processes in verification and validation are depicted in Figure 10.7 with respect to each of the error terms identified in Equation (10.5).

The term E_1 represents all numerical errors resulting from the difference between the approximate discrete solution, y_{sim}, and the exact solution to the discrete equations, $y_{Pcomputer}$, which is obtained on a perfect computer and all of the discretization errors approach zero. To be more precise, y_{sim} is the discrete solution obtained using a finite discretization size, finite iterative convergence, finite number of statistical samples (if the solution is nondeterministic), obtained on a finite precision computer. E_1 is referred to as the

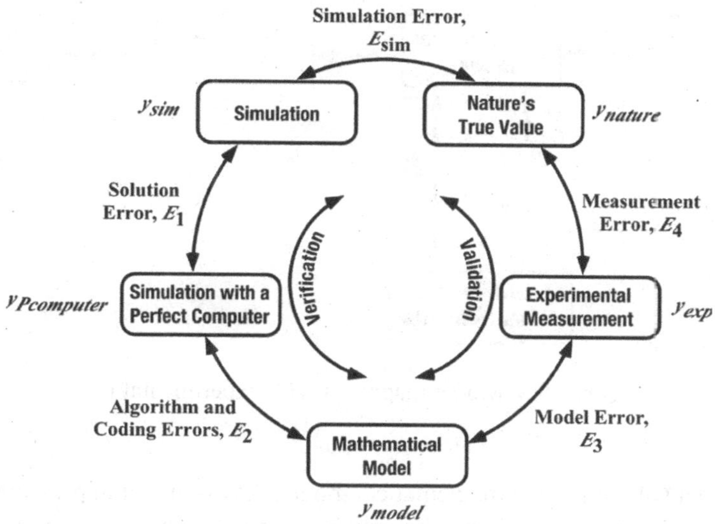

Figure 10.7 Error sources and verification and validation.

solution or calculation error and is estimated by solution verification procedures. Chapters 7 through 9 discussed a number of methods for estimating the magnitude of each of these solution error contributors. In the limit, as all of these errors approach zero, the same numerical algorithms and computer code are used for $y_{Pcomputer}$ as are used for y_{sim}. If the numerical algorithm is deficient in some way, or the computer code contains programming errors, E_1 would be contingent on these errors, but they would not technically be present in E_1. For example, suppose an algorithm error or a coding mistake was present in the computer code. Further, suppose that these errors caused the numerical solution to converge to an incorrect solution. E_1 would still represent the numerical solution error due to a finite discretization size, finite iterative convergence, finite number of statistical samples, and a finite precision computer, exclusive of the algorithm and coding errors.

The term E_2 represents all errors resulting from the difference between the exact solution to the discrete equations as the discretization size approaches zero, with perfect iterative convergence, an infinite number of statistical samples, solved on a perfect computer, $y_{Pcomputer}$, and the exact solution of the mathematical model, y_{model}. That is, E_2 comprises the errors due to algorithm and coding errors in the software. Chapters 4 through 6 discussed a number of methods for detecting and removing these errors. Since the exact solution of the discrete equations in the limit is not computable, we must rely on a discretization convergence study with highly converged iterative solutions, and a large number of statistical samples. That is, the strategy is to achieve adequate refinement in the discretized quantities to compute the observed order of accuracy in the asymptotic region, where these errors are well behaved and ordered. If an algorithm error (or deficiency) or a coding error causes an unexpected observed order of accuracy, we can be assured that something is amiss. However, the reverse is not true, i.e., if the observed order of error matches the expected order of error, it is not a proof that the algorithms are perfect and the coding is without error. The pivotal issue to computing the observed order of accuracy is the availability of a

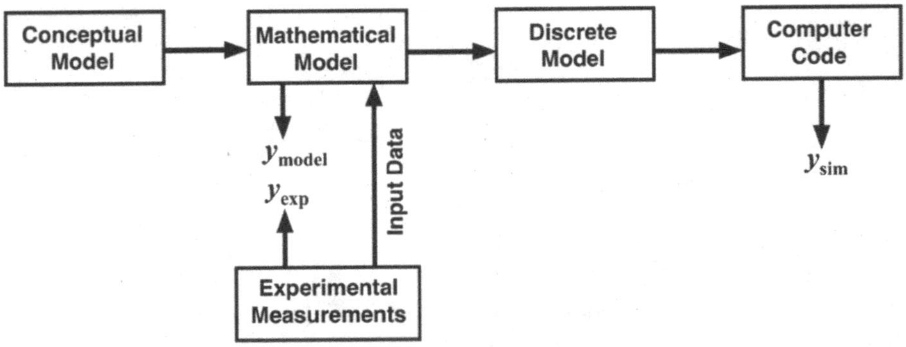

Figure 10.8 Model mappings with experimental data.

demanding exact solution to the mathematical model. As discussed in previous chapters, the most comprehensive exact solutions are manufactured solutions, i.e., solutions obtained by picking y_{model}, so that all of the terms in the differential equations show themselves in the determination of the observed order of accuracy.

The term E_3 represents all errors resulting from the difference between the exact solution to the mathematical model, y_{model}, and the experimental measurement, y_{exp}. The term E_3 is due to approximations and assumptions made during the formulation of the mathematical model, i.e., the model form error. Estimating model error is conceptually and mathematically more difficult that estimating E_1 and E_2, for two reasons. First, experimental measurements provide not only y_{exp}, but also the necessary input data for the mathematical model to compute a result, y_{model}. That is, the evaluation of the model form uncertainty depends on experimental measurements in two different ways; the computation of the difference between y_{model} and y_{exp}, and the computation of the model result, y_{model}. Therefore, the assessment of the model form error can be depicted as a closed-feedback loop, Figure 10.8, using experimental measurements in two different ways. If model input data are not measured by the experiment, but are determined by calibration of the input data, then it is clear how the evaluation of model form uncertainty is corrupted by the use of the model itself. The only way to break this feedback mechanism is by independent, experimental, measurement of the input data; but this is commonly not possible. This issue of coupling model form error and model input parameter calibration was discussed earlier in this chapter. However, Figure 10.8 makes the coupling explicit in terms of the flow of information.

The second reason for the conceptual and mathematical difficulty in assessing the model form error is that experimental measurements always contain uncertainty, because the true value from nature is never known. Consider how the two types of measurement uncertainty, random and systematic, impact the SRQ measurement and the input data measurement shown in Figure 10.8. The impact of random and systematic error on the SRQ measurement, y_{exp}, is *not* an issue in the discussion of E_3 because these two errors are accounted for in the discussion of E_4; discussed in the next paragraph. However, in the measurement of the model input data they *are* an issue. The existence of these two error sources can impact the assessment of model form error by way of the closed feedback loop shown in Figure 10.8, just as in calibration of model input parameters. For example, if a systematic error exists in the measurement of the model input data, then it will also corrupt the estimate of the model form uncertainty; even though there is no systematic error in the measurement of y_{exp}.

The term E_4 represents all errors due to the difference between the true, but unknown value of nature, y_{nature}, and the measurement of the SRQ, y_{exp}. The true value from nature can be viewed as either a deterministic quantity or a nondeterministic quantity. If y_{nature} is a deterministic quantity, then it means that for a fixed set of physical conditions in an experiment, the same SRQ is always produced by nature. Only in very simple and perfectly controlled experiments can y_{nature} be considered as a deterministic quantity. An example of such an experiment is the measurement of the mass of a fixed object. In general, viewing y_{nature} as a deterministic quantity is not helpful and can be misleading because in most experiments there are uncontrollable factors that change the true value. It is more useful to consider y_{nature} as a nondeterministic quantity because exactly the same physical conditions cannot be reproduced from one physical realization to the next. For example, even in a very well-controlled experiment using the same physical system, it is common to have slightly different BCs, ICs, or system excitation. The term y_{exp} is *always* nondeterministic because of random and systematic uncertainties in the experimental measurement. As a result, E_4 should be considered as an uncertain quantity composed of both aleatory and epistemic uncertainties.

10.4 Model Validation Hierarchy

During the 1980s and 1990s, a number of researchers in fluid dynamics were struggling with the fundamental issue: How should model validation be undertaken for complex engineering systems? The *building block approach* that is used today was developed by Lin et al. (1992), Marvin (1995), Sindir et al. (1996), and Rizzi and Vos (1998). Before this time, hierarchical thinking for a complex system was also developed in the nuclear reactor safety community. However, this community did not broaden its focus beyond improving the understanding of mass, momentum, and energy exchange in an accident environment (Zuber et al., 1998).

The building block or hierarchical approach is shown in Figure 10.9, adapted from the AIAA Guide (AIAA, 1998). This approach divides the complex engineering system of interest into an arbitrary number of progressively simpler tiers. If the complex system is divided into three tiers, they are usually referred to as *subsystem*, *benchmark*, and *unit problem*. The strategy of the tiered approach encourages assessment of the accuracy of the model at multiple levels of complexity and physics coupling. The approach is clearly informative in that it (a) recognizes there is a hierarchy of diverse forms of complexity in real systems and simulations and (b) recognizes that the quantity and accuracy of information that is obtained from experiments vary radically over the various tiers of the hierarchy. The arrows from a higher tier to a lower tier indicate the primary impact of that element on a lower-tier element.

The hierarchical view of model validation is distinctly a system-oriented perspective and, as a result, can also be referred to as a system hierarchy. The system of interest is placed at the pinnacle of the validation hierarchy. The purpose of the validation hierarchy is to help identify a range of lower-level tiers where experiments can be conducted for assessment of model accuracy for simpler systems and physics. A validation hierarchy is typically constructed by large-scale projects, with their system of interest at the top. When passing from

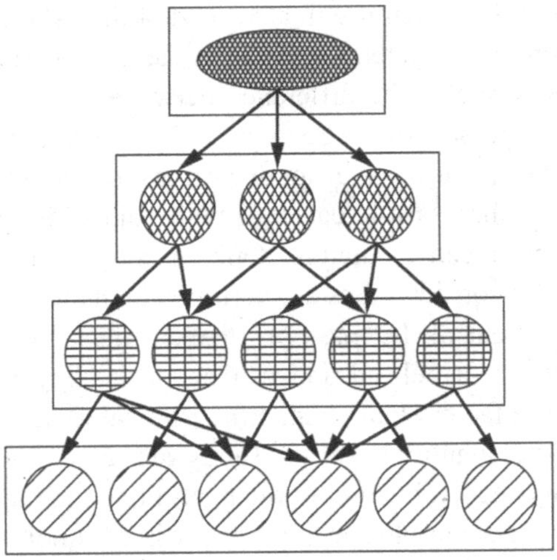

Figure 10.9 Model validation hierarchy tiers.
(adapted from AIAA, 1998)

the higher tiers to the lower tiers of the hierarchy, the emphasis moves from system-oriented engineering to simpler assemblies, and to a physics-based model assessment. As a result, most validation hierarchies differ at the higher tiers because the system or subsystems of interest are different. However, the lower tiers share many common elements, even though the top tier system of interest is completely different. For the common elements at the lower tiers, scientific model validation, as discussed earlier, can have a widespread impact on many different systems of interest.

Characteristics of each of the four tiers are discussed below. Additional tiers could be added to the validation hierarchy, but it would not significantly alter the discussion nor the recommended methodology.

10.4.1 Characteristics of the Complete System Tier

The *complete system tier* consists of the actual engineering hardware or system of interest (Figure 10.10). Thus, by definition, it is operational, functioning hardware with all the geometric characteristics, materials, and features associated with manufacturing and assembly of the system. For typical, complex engineering systems such as a gas turbine engine, multidisciplinary, coupled physical phenomena occur in the system. The BCs, ICs, and system excitation would commonly correspond to those that are of interest for realistic operating conditions of the system. One may also be interested in poorly-controlled or unusual environments, such as abnormal or hostile environments. Abnormal and hostile environments require a separate validation hierarchy because they are typically very different from a normal operating environment.

Experimental data are measured on the engineering hardware under operating conditions. However, the quantity and quality of these measurements are always very limited because (a) the diagnostic and instrumentation systems must have minimal impact on the operation of the system; and (b) the project-oriented test programs are typically conducted on a rigid

Complete System Tier	
Physical Characteristics	**Measured Data for Validation**
Actual system hardware Actual geometry, materials, and features Complete physics and chemistry Actual BCs, ICs, and system excitation	Very limited measurement of model inputs Very limited measurement of model outputs Very few experimental realizations Little or no estimate of experimental uncertainty

Figure 10.10 Characteristics of model validation at the system tier.
(adapted from AIAA, 1998)

schedule and with a tightly constrained budget. It is difficult, and sometimes impossible, for complex systems' tests to quantify even a small percentage of the input data required for simulation. As a result, little or no information is available for most of the input data needed for simulation, including the uncertainty of the input data. Such tests generally provide only measurements that are related to high-level system performance measures, such as those related to system design interest, system functionality, or system certification. Referring to Figure 10.2, these high-level measures would correspond to multiple integrals of dependent variables appearing in the PDEs of the model. Sometimes, the performance measure of a complete system test is simply: Did it work? Did it meet contract specifications? Did it fail safely, as designed?

Experimental data from complete systems are always specific to existing operational hardware and are available mainly through large-scale test programs. Existing data from these tests have traditionally focused on issues such as the performance, certification, safety, or reliability of the system. Although these expensive tests are focused on goals other than model validation, it is commonly possible to broaden the goals of the experiment to obtain valuable model validation data. For example, additional SRQs and model input data could be measured that would be extremely beneficial in assessing model predictive accuracy, although not necessarily needed for a certification test. The test programs typically require expensive ground-test facilities, full-scale flight testing, or testing under dangerous or adverse conditions, such as unpredictable weather or specified failure of certain subsystems. Also, there are certain situations where it is not possible to conduct a validation experiment of the complete system. Such situations could involve public safety or environmental safety hazards, unattainable experimental testing requirements, or international treaty restrictions.

10.4.2 Characteristics of the Subsystem Tier

The *subsystem tier* represents the first decomposition of the actual system hardware into subsystems or subassemblies (Figure 10.11). Each of the subsystems or subassemblies is composed of actual functional hardware from the complete system. Subsystems usually exhibit three or more types of physics that are coupled. The physical processes of the complete system are partially represented by the subsystem tier, but the degree of coupling between various physical phenomena in the subsystem tier is reduced. For example, there is reduced coupling between subsystems, as compared to operation of the complete system.

Subsystem Tier	
Physical Characteristics	**Measured Data for Validation**
Functional subsystem hardware Most geometry, materials, and features Some physics and chemistry coupled Simplified BCs, ICs, and system excitation	Some measurement of model inputs Some measurement of model outputs Few experimental realizations Experimental uncertainty given on some quantities

Figure 10.11 Characteristics of model validation at the subsystem tier.
(adapted from AIAA, 1998)

Most geometric characteristics are restricted to the particular subsystem and its attachment or simplified connection to the complete system. Essentially all of the materials, features, and capabilities of the subsystems are present. During subsystem testing, the BCs, ICs, and excitation are usually simplified compared to the operation of the complete system.

In subsystem tests, there is commonly a significantly increased opportunity for all types of experimental measurements. Project managers are usually more willing to install instrumentation and there is more interest in better understanding the details and operating conditions of the subsystem. In addition, there is usually less pressure from the project schedule, cost constraints, and high-level management attention. A much wider range of test facilities is possible for subsystems, with an accompanying improvement in control of the test conditions. There is an increased percentage of measured input data for the simulation, as well as more SRQs measured, relative to complete system tests. There are commonly more experimental realizations than complete system tests so that improved statistics of measurements are obtained. Experimental uncertainty estimates are given on some of the measured SRQs, but few uncertainty estimates are commonly measured for model input data. It must also be stressed that at the subsystem tier, as well as at every lower tier in the hierarchy, the conceptual model for *each element* in the hierarchy is simplified compared to the conceptual model of higher tier elements. That is, as you progress down the hierarchy the conceptual model for each element can be thought of as smaller and smaller subsets of the conceptual model at the higher tiers in the hierarchy.

10.4.3 Characteristics of the Benchmark Tier

The *benchmark tier*, Figure 10.12, sometimes referred to as the component or assembly tier, represents the next level of decomposition and simplification beyond the subsystem tier. For the benchmark tier, special hardware is fabricated to represent the main geometric features of each subsystem. Special hardware is hardware that is specially fabricated with simplified materials, properties, and features. For example, benchmark hardware is normally *not functional or production hardware*, nor is it fabricated with the same materials as actual subsystems. For example, if the simulation of living biological systems is of interest, surrogates of the system of interest such as rats or mice could be used. For benchmark cases, typically only two or three types of physical phenomena are considered. The benchmark tier hardware is simpler geometrically than the hardware at the subsystem level. The

Benchmark Tier	
Physical Characteristics	**Measured Data for Validation**
Special, nonfunctional hardware fabricated Simplified geometry, materials, and features Little coupling of physics and chemistry Very simple BCs, ICs, and system excitation	Most model inputs measured Many model outputs measured Several experimental realizations Experimental uncertainty given on most quantities

Figure 10.12 Characteristics of model validation at the benchmark tier.
(adapted from AIAA, 1998)

only geometric features that are retained from the subsystem tier are those critical to the types of physical phenomenon that are considered at the benchmark tier. In addition, at this tier there is a distinct shift from project-focused goals and schedules to those that are aimed at improved understanding of the physics involved and assessment of the accuracy of the mathematical models that are used.

For this tier, most of the input data needed for simulation are measured, or at least most of the important inputs are measured. Some of the model SRQs that are measured correspond to dependent variables in the PDEs, or possibly one integral removed from the dependent variables. Since the benchmark tier uses nonfunctional hardware and special materials, the ability to instrument the hardware is significantly improved. Most of the experimental data obtained have associated estimates of measurement uncertainties. The experimental data, both model input data and output data, are usually documented in moderate detail. Examples of important experimental data that are documented include: (a) detailed inspection of all hardware used in the experiment, (b) characterization of the variability of materials used in the experiment, (c) detailed information concerning assembly of the hardware, and (d) detailed measurement of BCs and excitation that were produced by the experimental apparatus or testing equipment.

10.4.4 Characteristics of the Unit Problem Tier

Unit problems represent the total decomposition of the complete system into isolated physical processes that are amenable to high-quality model validation experiments (Figure 10.13). At this level, high precision, special-purpose hardware is fabricated and inspected. This hardware may only vaguely resemble some features of the subsystem or benchmark tier, especially in the view of the system project manager. Unit problems are characterized by simple geometries that are accurately characterized for the specific conditions during the experiment. The geometry features are commonly two-dimensional, either planar or axisymmetric, or they are simple three-dimensional geometries with important geometric symmetry features. One element of complex physics is allowed to occur in each of the unit cases that are examined. The purpose of these cases is to isolate elements of complex physics so that critical evaluations of mathematical models or submodels can be evaluated. For example, in fluid dynamics, unit problems could individually involve (a) fluid turbulence for single phase flow, (b) fluid turbulence for two phase flow, (c) unsteady laminar flows, or (d) laminar diffusion flames. When examining turbulent reacting flow, which combines

Unit Problem Tier	
Physical Characteristics	**Measured Data for Validation**
Very simple, nonfunctional hardware fabricated Very simple geometry and features No coupled physics Very simple BCs, ICs, and system excitation	All model inputs measured Most model outputs measured Many experimental realizations Experimental uncertainty given on all quantities

Figure 10.13 Characteristics of model validation at the unit problem tier.
(adapted from AIAA, 1998)

turbulence and chemical reactions, it is recommended to consider this as one tier above the unit problem tier since it combines two aspects of complex, coupled physics.

For this tier, all of the important model inputs needed for a simulation must be measured in the experiment or they must be characterized by an appropriate combination of input parameter calibration followed by model validation. If important input parameters cannot be independently measured during the experiment, they can be calibrated using the following four-step procedure:

1. A large number of experiments are conducted so that well-characterized probability distributions can be determined for the calibrated parameters. As discussed earlier, this involves the solution of the inverse model to determine the input parameters.
2. A number of new experiments are conducted, such that the calibrated input parameters *remain the same*, but new physical conditions are introduced in the experiment. These new conditions should test whether the mathematical model, which must include the physics-based impact of these new conditions on the calibrated parameters, can accurately predict how these calibrated parameters change due to the new physical conditions.
3. A number of nondeterministic simulations are computed for the new conditions so that new, well-characterized probability distributions are estimated for the SRQs of interest.
4. Quantitative comparisons are made between the new simulation and experimental results using a validation metric operator. If the new experimental results are made available to the computational analyst, and *any* modifications are made to any feature of the model, the entire activity has changed from model validation to model calibration. If the new experimental results are not made available to the analyst, i.e., a blind prediction is made, then the activity is not only a model validation activity, but also an assessment of predictive capability of the model. The relationship between validation, calibration, and predictive capability is discussed in more detail at the end of this chapter.

In the unit problem tier, highly-instrumented, highly-accurate experimental data are obtained, and an extensive uncertainty analysis of the experimental data is conducted. If possible, experiments on unit problems should be repeated using different diagnostic techniques, and possibly in separate experimental facilities, to ensure that systematic uncertainties (bias) in the experimental data are quantified. These types of independent experiments could be conducted in any organization that has the experimental facilities, the technical expertise, *and* the willingness to see how their measurements compare with others. Commonly, these experiments are conducted at universities or research laboratories.

Example: Model Validation in Structural Dynamics

Consider the four steps just described as applied to the simulation of the vibrational modes of bolted-frame structure. The simulation model advertises that, given adequate experimental information to calibrate the stiffness and damping due to the bolted joints in the structure, it can predict the change in vibrational modes due to changes in bolt preloads. First, a large number of bolted-frame structures are assembled using a fixed specification for the pre-load torque on all of the bolts in the structure. Experimental measurements are made on all of the assembled structures to determine the vibrational modes of interest. Using the mathematical model and the experimental measurements for the vibrational modes, the probability distributions for the stiffness and damping in each of the bolted joints is estimated using Bayesian inference. (Bayesian inference is discussed in Chapters 12 and 13.) The probability distributions represent the effect of variability in the manufactured beams and the assembly of the structures.

Second, a large number of new structures are assembled, but for these new structures the preload torque specification on all the bolts is changed. The torque on half of the bolts is increased by 50%, and the torque on the remaining half is decreased by 50%. The new structures are assembled using the same production lot of frame components and the same technicians assembling the structures. Experiments are conducted on the new structures to measure the vibrational modes of interest. Each set of structures is tested in order to construct a well-characterized probability distribution for the vibrational modes of interest. In addition, the new measurements are *not* shown to the computational analysts.

Third, simulations are computed using the new pre-load information on the bolt torques for each set of new structures. Using the new-preload torque information, a large number of simulations is computed so that a well-characterized probability distribution is computed for the vibrational modes of interest for each set of new structures. If no other alterations are made to the new simulations, then these simulations are blind predictions for the new measurements.

Fourth, using the new simulations and the new experimental data, a validation metric operator is used to compute the quantitative difference between the frequencies and amplitudes of the vibrational modes of interest. The validation metric operator must be able to accept probability distributions as input from the simulation and the experiment, as well as properly filter out the signal noise due to higher vibrational modes in the structure and the experimental measurement noise. The magnitude of the validation metric result for each set of new structures can be computed. The magnitude of the result for each set of new structures can then be compared to the required predictive accuracy for the application. As discussed in Chapter 2, the accuracy requirements for the intended application of the model determines if there is adequate or inadequate agreement between simulation and experiment.

10.4.5 Construction of a Model Validation Hierarchy

System engineering is the underlying concept in the construction of the validation hierarchy. The goal of the hierarchy, however, is the identification of simpler and more affordable subsystems and components to assess simulation accuracy using experimental data. Hierarchical model validation is *application driven*, i.e., driven by the system of interest for the conditions to which the system is exposed. The coupled physics perspective of the system becomes important at the lower tiers of the hierarchy. The construction of an effective and productive validation hierarchy is a formidable challenge. There are many ways of constructing the tiers; no single construction is best for all cases. In addition, one validation hierarchy would not be appropriate for systems in different environments, i.e., normal, abnormal, and hostile. In fact, it is common to construct different mid and lower tiers of the hierarchy for different scenarios within a given environment. For example, the lower tiers of a hierarchy for the abnormal environment of a gas turbine engine would be different, considering the inflight loss of a fan blade versus the loss of a turbine blade.

A good hierarchical tier construction is one that accomplishes two tasks. First, the construction carefully disassembles the complete system into tiers in which each lower-level tier has one fewer level of physical complexity. For complex engineered systems, this will require many more than the four tiers shown in Figure 10.9. The types of physical complexity that could be uncoupled from one tier to the next are spatial dimensionality, temporal nature, geometric complexity, and physical process coupling, including multi-scale coupling. The most important of these types of physical complexity to decouple, from one tier to the next, is physical process coupling. Physical process coupling generally produces the largest nonlinear response due to the various contributors. It is important to recognize the nonlinear nature of all of the contributors in the construction of the tiers because the philosophy of the tier construction rests heavily on linear system thinking. That is, the underlying theme in the hierarchy is that confidence in predictive capability at one tier can be built from assessment of predictive accuracy at the lower tiers. The system of interest clearly does not have to be linear, but the effectiveness of the hierarchical validation approach decreases when strong nonlinear coupling occurs from one tier to the next lower tier.

The second task accomplished by a good validation hierarchy is the identification of individual experiments that are practically attainable and able to produce validation-quality data. In other words, the individual experiments should be (a) practically achievable given the experimental test facilities, budget, and schedule, and (b) they should be able to measure all the important input data and multiple SRQs. As discussed earlier, the ability to conduct a true validation experiment at the complete system tier is extremely difficult, if not impossible, for complex systems such as commercial aircraft or complete power plants. At the subsystem tier, it is feasible to conduct validation experiments, but it can still be quite difficult and expensive. Furthermore, as validation experiments are conducted at various tiers there will be new insights gained about the system, as well as unexpected weaknesses in the mathematical modeling of these experiments. The new insights can then be used to update various portions of the validation hierarchy and show, for example, physics coupling that may not have been recognized earlier.

At the benchmark or assembly tier, validation experiments are much more feasible because nonfunctional, nonrealistic hardware is fabricated. This allows considerable flexibility in choosing geometries, materials, experimental facilities, and what coupled physics will be considered. The benchmark tier is probably the most difficult to formulate because it represents the transition from a focus on hardware and system/subsystem functionality in the top two tiers to a physics-based focus in the lower tiers of the hierarchy.

At the lower tiers, for example the unit problem tier, component tier, or single-physics tier, much simpler and affordable experiments can be identified. At the bottom tier of the hierarchy, there should be one type of physical process occurring so that a rigorous test of the modeling can be assessed. For example, if fracture in ductile materials is of interest, there could be three separate bottom tier elements: opening mode tensile fracture, sliding mode shear fracture, and out-of-plane shear fracture. For bottom-tier validation experiments, one must be able to provide a rigorous estimate of (a) the uncertainty in the important input data, especially inherent variability, and (b) the uncertainty in the SRQs of interest. As discussed earlier in this chapter, systematic uncertainties are the most difficult to estimate because they are hidden in the test procedures, test equipment, diagnostic sensors, human element, and even in the test facility itself.

The concept of the model validation hierarchy can also be applied to the manufacture of a part or component. For example, suppose a manufactured part required five processes: metal rolling, metal stamping, arc welding, drilling, and final machining. A sequential-process diagram could be constructed of the five manufacturing processes. Under each of these processes, a validation hierarchy could be constructed. Given the specificity and independence of each of the processes, the validation hierarchy for each process would effectively begin at the benchmark tier. This tier could then be divided into various physical processes that occur during each of the manufacturing processes. Validation experiments could then be identified that could assess the accuracy of the mathematical model of individual process elements.

As a final comment, the concept of how to construct a model validation hierarchy has recently been extended. The system engineering and digital engineering communities have recognized the importance of a model validation hierarchy. As pointed out by Price et al. (2006), they see the concept as a method to address "four key issues . . . 1) analysis fidelity; 2) reductionist vs. holistic system design; 3) simulation driven design environments; 4) real vs. perceived system characteristics." They see the issue as identifying interfaces in the context of a mathematical model of subsystems and subassemblies that maps model input data to model output data. Luckring et al. (2024) define the concept of a *transition tier* in the model validation hierarchy. Above the transition tier, any experiments that are conducted use actual system hardware for subsystems, assemblies, subassemblies, etc. Below the transition tier, experiments use nonfunctional hardware that are representative of certain physical features of the actual system hardware. Once a model validation hierarchy has been formulated, Shaw et al. (2024) address the important practical question: Given a mathematical model for potentially every element in the model validation hierarchy, how can model validation experiments be prioritized to identify those mathematical models that produce the largest uncertainty to the predicted SRQs of interest. The procedure is based on the Phenomena Identification and Ranking Table (PIRT) and an M&S gap analysis, both of which are discussed in Chapter 14 "Planning and Prioritization in Modeling and Simulation."

10.5 Example: Hypersonic Cruise Missile

Consider the construction of a validation hierarchy for a hypersonic cruise missile (Oberkampf and Trucano, 2002). Assume the following characteristics of the missile system: (a) it is launched by an aircraft toward a ground target; (b) it is powered by a conventional gas turbine propulsion system, without a ducted fan; (c) it has an autonomous guidance, navigation, and control (GNC) system with an on-board millimeter-wave target seeker; and (d) it is functioning in a normal operating environment. Figure 10.14 shows a five-tier validation hierarchy for the hypersonic cruise missile that forms the basis of this example.

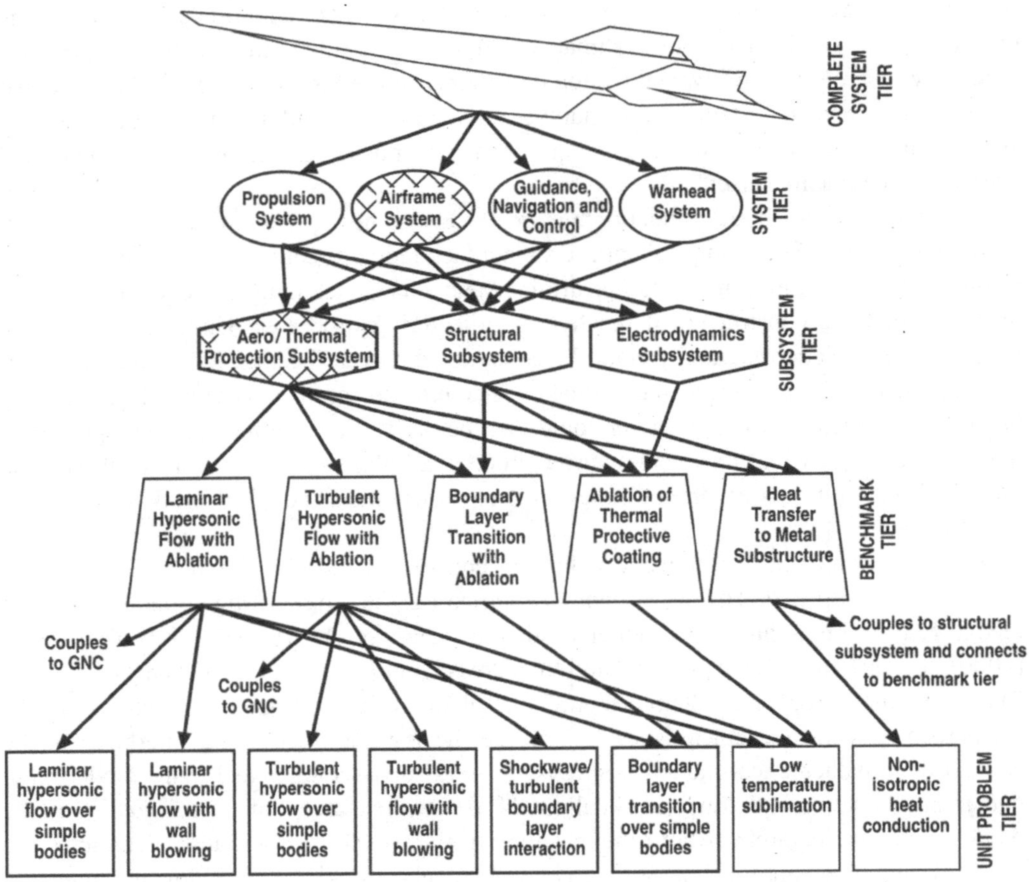

Figure 10.14 Example of a validation hierarchy for a hypersonic cruise missile. (Oberkampf and Trucano, 2002)

10.5.1 System Tier

The entire missile is referred to as the *complete system* and the following elements are identified as part of the system tier: propulsion, airframe, GNC, and warhead. The first three elements would be expected in the validation hierarchy of any flight vehicle. The launch aircraft is not included, but it could be included if a higher-level system view was needed.

10.5.2 Subsystem Tier

At the subsystem tier, the following are identified as elements of the airframe system: aero/thermal protection, structural, and electrodynamics. These three elements are identified at the subsystem tier because they are the primary functional elements of the airframe system. The electrodynamics subsystem deals with the aspects of electromagnetic detectability of the cruise missile. This would range from radio frequencies used to detect the missile by radar, to detection in the visible spectrum. This element is included in the hierarchy because of its importance to effectiveness of the weapon system. For example, if a commercial transport aircraft was of interest, this element would not be included.

Arrows drawn from the system-tier elements to the subsystem-tier elements indicate the primary elements that influence the subsystem tier. At the subsystem tier, each element contains functional hardware of the system of interest. For example, the aero/thermal protection subsystem would contain actual, functioning, subsystem hardware of the cruise missile. If the system of interest has never been built, then this would be prototype hardware as opposed to manufactured hardware for an operational system. Note that the same test hardware is commonly used in multiple subsystem-tier elements. For example, essentially all of the hardware in the aero/thermal protection and the electrodynamics subsystem elements are also used in the structural subsystem. However, the types of experiments and measurements would be very different for validation experiments in each of these subsystems because the mathematical models are entirely different.

Consider the types of experiments that would be conducted in the structural subsystem element. If an experimentalist was interested in the static deflection and local stress in the various structural components, then the mechanical coupling with attached components, including the aerodynamic loading and thermally induced loads would need to be considered. For example, if interested in the deflection of the horizontal tail, the vertical tail and the attached mechanical components of the fuselage would need to be included in this experiment. If interested in the structural dynamic response of the various components, then essentially every piece of hardware from the missile would need to be considered because every part of the structure is dynamically coupled to every other part of the structure. However, certain simplifications of the hardware would be appropriate. For example, an experimentalist could substitute mass mock-ups for certain components, such as the completely functional propulsion system and the warhead, with little loss in fidelity. The structural dynamic and acoustic excitation caused by the propulsion system would be quite important in the validation of the structural dynamics of the subsystem.

10.5.3 Benchmark Tier

At the benchmark tier, Figure 10.14 shows only five representative elements that would be functionally related to the aero/thermal protection subsystem, although additional benchmark-tier elements could be shown. The arrows drawn from the structural and electrodynamics subsystems to the benchmark tier show important coupling from these two subsystems to elements depicted at the benchmark tier.

At the benchmark tier, specialized, nonfunctional hardware is fabricated. For example, the laminar, turbulent, and boundary-layer-transition elements may not contain the as-manufactured ablative coating of the missile. Instead, a simpler material might be used – one that would produce wall blowing and possibly gases or particles that may react within the boundary layer, yet simpler than the typically complex gas and particle chemistry that results from actual ablative materials. The element for ablation of the thermal protective coating may use the actual material on the missile, but the validation experiment may be conducted, for example, at conditions that are attainable in arc-jet tunnels. Explanations for particular arrows shown from the subsystem or benchmark tiers are noted below:

- The arrow from the structural subsystem to the boundary-layer-transition element is drawn to show that structural vibration modes of the vehicle surface can influence transition.
- An arrow is drawn from the two elements labeled "Flow with Ablation" (on the left side of the diagram) that are marked "Couples to GNC." These arrows indicate a coupling of the boundary layer flow field to the millimeter-wave seeker in the GNC hierarchy (not shown here). These arrows are drawn because the ablation products from the heat shield can influence the performance of the millimeter-wave seeker.
- The element labeled "Heat Transfer to Metal Substructure" shows an arrow that would connect to elements at the benchmark tier in the structural subsystem (not shown here). This arrow indicates the coupling that will result in thermally-induced (a) stresses and cause temperature-dependent material properties to be considered in the static deflection of the structure and (b) changes in the vibrational modes of the structure.

10.5.4 Unit Problem Tier

Eight elements, primarily focused on external fluid dynamics, are identified at the unit-problem tier in Figure 10.14. Many other elements could be identified at this tier, but these are representative of the types of validation experiments that should be conducted at the unit-problem tier. Elements related to the fluid dynamics and heat transfer in the inlet of the engine would be located in the unit-problem tier under the propulsion system element. The identification of elements at the unit-problem tier is easier than at the benchmark tier because unit-problem elements are more closely related to traditional mathematical model building experiments and model calibration experiments in fluid dynamics and heat transfer.

A point of clarification should be made concerning experiments at the lower tiers of the hierarchy, particularly at the unit-problem tier. Some project managers and system designers refer to experiments at the lower tiers, such as laminar hypersonic flow in a wind tunnel, as a "simulation" of the flight vehicle in the atmosphere. From the perspective of a project engineer interested in performance of the *real* vehicle, this view about experimental "simulation" is understandable. From the perspective of conducting a validation experiment, however, this view causes significant confusion concerning the issue of simulation versus physical experiments. That is, any experiment conducted at any tier is a physical realization of a process of interest, i.e., *any experiment is reality*. However, from the perspective of a project engineer, experiments that are not conducted at the actual operational conditions and using operational hardware, are of peripheral interest. This lack of appreciation of the value of validation experiments, particularly at the unit-problem tier, is a serious concern when attempting to obtain funding for validation experiments.

10.5.5 Model Validation Pyramid

To better visualize the validation hierarchy of a complex, coupled system, a three-dimensional perspective is helpful. Figure 10.15 shows the top three tiers of the validation hierarchy of the hypersonic cruise missile. The four systems (propulsion, airframe, GNC, and warhead) could be viewed as faces of a four-sided pyramid. In the earlier discussion, the airframe face was divided into three additional facets, each representing the three subsystems: aero/thermal protection, structural, and electrodynamics. The propulsion system face

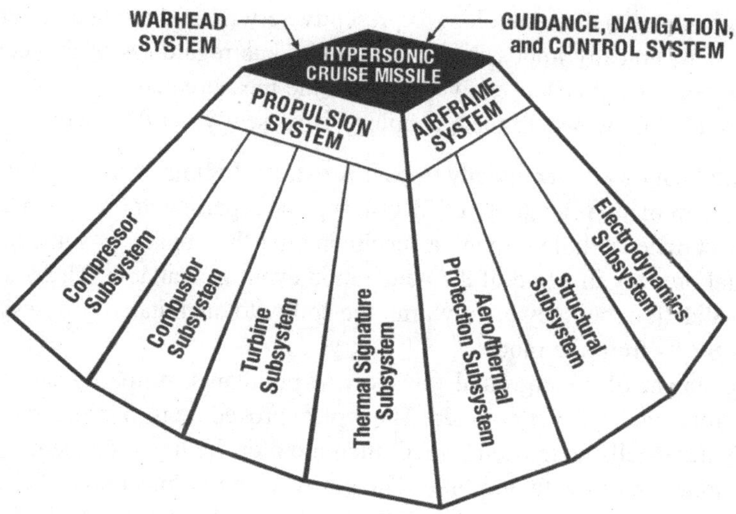

Figure 10.15 Example of a model validation pyramid for a hypersonic cruise missile. (Oberkampf and Trucano, 2002)

could be divided into four additional facets to represent its subsystems: compressor, combustor, turbine, and thermal signature. Similarly, the GNC and the warhead systems could be divided into subsystems appropriate to each. On the surface of this multifaceted pyramid, one could more clearly and easily indicate the coupling from one subsystem to another. For example, the coupling of laminar and hypersonic flow with ablation to the millimeter-wave seeker of the GNC system can be shown. This coupling would be shown by an arrow connecting these hypersonic flow elements to appropriate elements on the GNC facet of the pyramid. To show additional facets at the lower tiers of the hierarchy would require graphics software with animation capability that could rotate the pyramid and zoom in on particular elements of interest. The coupling of one element to any other elements could be displayed if the cursor is placed above an element.

The validation pyramid stresses the system engineering viewpoint at the top tiers, but then the emphasis shifts to physics-based aspects of experiments at the lower tiers. Each facet of the pyramid is devoted to identifying validation experiments for that particular facet. At the higher tiers, the coupling of the simulation models reflects the strong interactions occurring during the operation of the entire system. By traversing around the lower tiers of the pyramid, the number of facets is equal to the total number of elements that have been identified around the pyramid. For a complex system, the total number of facets would be over a hundred. For the hypersonic cruise missile, the aero/thermal protection subsystem divided into five elements, and these subsequently divided into eight elements. However, each facet from the other subsystems would divide into many more facets moving down the hierarchy.

10.6 Practical and Conceptual Difficulties of Model Validation

10.6.1 Practical Difficulties

There are situations in which practical difficulties in obtaining experimental measurements either hinder or eliminate the possibility of model validation. These difficulties can be generally grouped as: (a) the technology is presently very limited to make the measurements needed; (b) it is technically impossible to obtain the data regardless of the technology; and (c) it is impractical or not cost effective to obtain the measurements.

Three examples follow, where the technology is presently very limited:

1. Experimental data on hypervelocity impact is extremely limited in terms of spatial detail and the measurement of a variety of SRQs. Typical experimental results are photographs of the impact crater or hole through a specimen after the impact. In some facilities, high-speed visual or x-ray imaging of the penetration event is available. These data are useful in model validation, but they greatly limit the ability to quantitatively assess the accuracy of various SRQs from the model.
2. The measurement of underground transport of pollutants, primarily due to water flow through porous media, is very limited. The typical procedure is to inject tracer materials at locations where wells have been drilled, then monitor the tracer concentration at nearby wells as a function of depth and time. The porosity and permeability of the surrounding subsurface material that appear in the PDEs are then calibrated so that the observed tracer concentration record is matched. That is, the solution of an inverse problem is used to calibrate the porosity and permeability input parameters over the entire solution domain to obtain the improved agreement with the measured results.
3. The ability of measuring both simulation input quantities and SRQs at the micrometer and smaller scale is very limited. As mathematical models continue to develop for spatial scales in these ranges, the ability to assess the accuracy of these models will be a pacing item. As a result, confidence in predictions at these spatial scales for materials science, biochemistry, and biophysics will be significantly impeded.

An example of a situation in which it is technically impossible to obtain the needed experimental data for model validation is in modeling physical phenomena with (a) very long time scales, on the order of centuries or tens of centuries, or (b) very large physical scales. Two examples of very long-time scales are the long-term prediction of the underground storage of toxic or nuclear wastes, and long-term prediction of the effects of various natural and human contributors to global warming. Two examples of very large physical scale events are the response of the global environment to a large-scale atmospheric event, such as the impact of a sizeable asteroid, and the eruption of a large volcano causing a massive tsunami.

There are also situations in which it is not cost effective, allowed, or is impractical to obtain experimental data for model validation, even though experiments are technically feasible. Some examples are: (a) conducting an experiment on the explosive failure of a full-scale reactor containment building, (b) conducting an experiment on the response of a large-scale public-works structure to an explosive either inside or outside of the structure, (c) obtaining certain experimental data for the physiological response of humans to toxic chemicals or substances, and (d) hazardous or environmentally damaging tests that are banned by international treaties.

10.6.2 Conceptual Difficulties

The restrictive concept of model validation used in this book and shown in the top portion of Figure 10.1 is a very clear and precise assessment of model accuracy by comparison of simulation results with experimental data. Conceptual difficulties or issues arise when the relationship of validation and calibration is explored in more depth. As discussed in Chapter 2, the issue is *not* with regard to the goals of each. The goal of model validation is the assessment of model accuracy. The goal of model calibration is the improvement of agreement between simulation and experimental results by way of updating model input parameters or altering modeling assumptions. The complexity of the relationship between validation and calibration is exemplified in two different ways. First, when *both goals are combined* in the same mathematical process, the question becomes: Is this process model validation? *or* Is this parameter calibration? The very popular Bayesian approach of Kennedy and O'Hagan (2001), as well as variants of their approach, does exactly this, it combines both of these goals. Second, a portion of the available experimental data for the SRQs is used to calibrate the model input parameters, and then the remaining data is used to "validate" the model. This type of "validation," is usually referred to as *cross-validation*.

When the Kennedy and O'Hagan or similar approach is used, it should be recognized by the computational analysts and the decision maker using the updated model that *both goals* are being pursued in the procedure. When simulation results are compared with experimental measurements that were also used in part for parameter calibration, then this would *not* be viewed as a model validation. The counter argument can be made that a comparison of simulation results *is* being made with previously unused experimental measurements; therefore, it should be referred to as model validation. But, suppose a set of experimental measurements of the SRQs is randomly sampled and a portion is used for parameter calibration and the remaining portion is used for model accuracy assessment. This test of model accuracy assessment is extremely weak because of the random sampling of data used and unused for calibration versus accuracy assessment, respectively. That is, there is no difference in the physics occurring for each set of data; they are only different because of random sampling. The test of model accuracy assessment is so weak that it is a myth, at best; and, at worst, it is deceitful. The relationship between validation and calibration is discussed in more detail in Chapter 13 "Predictive Capability."

As discussed in Chapter 2, the issue of calibration versus validation has been hotly debated in the field of hydrology for over three decades. Hydrology has been the epicenter of the debate because of the extremely flexible nature of hydrological models (Anderson and Bates, 2001). The simulation results of these models are *dominated* by parameters that are calibrated based on available measurements of SRQs. The calibrated parameters typically are not just a few scalar quantities, but entire scalar fields in two and three dimensions, as well as tensor fields. As a result, there is astounding adaptability in hydrological models to match the measurements. With this adaptability, the hydrology community recognizes that their models live in the world of calibrated models, as opposed to validated models. In other fields of complex simulations where experimental data are available for a range of SRQs, such as weather forecasting and global climate modeling, model calibration is the reality, as opposed to model validation.

Regardless of any perspective on this issue, a crucial question remains: can a calibrated model be "validated" in the sense of genuinely testing the accuracy of the results from the model? The answer to this question is best understood in the context of predictive capability. For an excellent modern review of various types of conceptual difficulties involved in modeling and computational simulation, see Winsberg (2019).

10.7 Predictive Capability

10.7.1 Predictive Capability and Concepts of Model Validity

Recall the definition of *predictive capability* given in Section 2.2.4:

Use of a computational model to foretell the state of a physical system, including relevant uncertainties, under conditions for which the computational model has not been validated.

This definition is similar to ASME (2022), but the present definition specifically requires that an estimate of the uncertainty in the prediction is provided. The crux of the definition is that the state of the system has *not* been seen before, whether by model validation or calibration. Furthermore, the model does not even have to be a physics-based model. For example, the model can be constructed using any method, including machine learning, regression-based methods, data mining, artificial neural networks, and genetic algorithms. Predictive capability cares nothing about how the model attained its present state. It is entirely focused on foretelling previously unseen physical processes, as opposed to retrodiction or postdiction.

Although this specific concept of predictive capability is relatively new, the basis for predicting the state of a physical system has been debated by philosophers for millennia. It is beyond the scope of this text to address the issue of foretelling the response of a system from a philosophical perspective. Engineering is about adequacy and decision making, not about absolute truth. It is appropriate, however, to discuss further how the concept of predictive capability relates to the concepts of model validation as developed over several decades in the theory of modeling and simulation. Zeigler et al. (2018, p. 31) explain model validation in terms of a sequence of more demanding requirements for the model. At the first level, they define *replicative validity* to mean, "for all of the experiments possible within the experimental frame, the behavior of the model and system agree within acceptable tolerance." At the second level, *predictive validity*, they "require not only replicative validity, but also the ability to predict as yet unseen system behavior." Finally, at the third level, *structural validity*, they require "that the model not only is capable of replicating the data observed from the system, but also mimics in step-by-step, component-by-component fashion the way in which the system does its transitions."

Bossel (1994) addresses the basis for predicting the state of a system by defining two types of models: descriptive models and explanatory models. *Descriptive models* are those that can imitate the behavior or responses of the system based on previous observations of the system. *Explanatory models* are those that represent a system's structure and its components, and their connections and coupling so that one can understand the future system behavior or responses even under conditions never before experienced.

From this discussion, it is seen that predictive capability is closely aligned with the classical concepts of structural validity and explanatory models.

10.7.2 Predictive Capability and Calibrated Models

There are two different ways to conceptualize predictive capability with regard to calibrated models. First, consider the challenge of making an accurate prediction, given previously observed and/or calibrated model parameters, such as from a machine learning model. Stated differently, what is the perilousness, or the potential surprise level, or the estimated uncertainty of a prediction? Machine learning techniques and software packages have become very popular because they seem to always show improvements in predictive capability compared to more traditional parameter calibration techniques. However, it must be remembered that machine learning techniques excel at minimization of global stochastic uncertainty, given the structure of the model and the available data. If practitioners of machine learning techniques are uninterested or unwilling to make predictions for conditions dissimilar to the training data, this is a form of conformation bias.

Consider the example of a machine learning model that is calibrated for low speed, fully developed flow inside a small diameter pipe. Suppose experimental data, even a very large amount of data, have been obtained for the pressure drop in a pipe for various small pipe diameters, various lengths, flow rates, and fluids of differing viscosity. A perilous test of predictive capability is whether a model can predict the pressure drop in a larger diameter pipe, or a large flow rate, or low viscosity fluid, for flow in a pipe. A calibrated model that has never been exposed to, or trained for, the existence of turbulent flow cannot possibly make an accurate prediction for this previously unseen type of process. A physics-based model, which includes both physics-based knowledge and models for the existence of laminar and turbulent flow, does have the potential to make an accurate prediction for this previously unseen process.

The second conceptualization of predictive capability examines the influence of the calibrated parameters in determining the simulation results. To explain this perspective, a more precise explanation is needed for "calibrated parameters" and "parameters influential in determining results." Calibrated parameters are those that are determined using both the mathematical model of the process under consideration and experimental measurements of the SRQs. That is, given the experimental measurements, the inverse model is solved in order to infer the input parameters in the model. Sometimes these are referred to as *free parameters* in the sense that they *cannot* be independently measured separate from the mathematical model used for the system of interest. These parameters can be fixed scalars, probability distributions, or they can vary over the range of the dependent variables of the differential equations. If input parameters are determined *independent* of the mathematical model of the process under consideration, even using experimental measurements, then they are *not* considered calibrated parameters in the present sense. For example, parameters such as Young's modulus, fluid viscosity, and thermal conductivity are not considered calibrated parameters because they are determined *separately* from the complex model for structural dynamics, fluid flow, and heat transfer, respectively, using simple independent models.

The influential input parameters in a simulation can be determined using global sensitivity analysis. As discussed in Chapter 3, global sensitivity analysis is used to determine which input parameters have the most influence on the trends of the SRQs of interest. For example, suppose there are one hundred input parameters required for the solution of the mathematical model. These parameters could be associated with the system of interest (i.e., the geometry, initial conditions, or the physical-modeling parameters) or the surroundings (i.e., the boundary conditions or the system excitation). A global sensitivity analysis can be conducted to determine the importance ranking of the input parameters, in order of the most influential to least influential on the global response of each SRQ. The influence ranking of the input parameters can vary from one SRQ to another SRQ. Chapter 13 discuss sensitivity analyses in more detail.

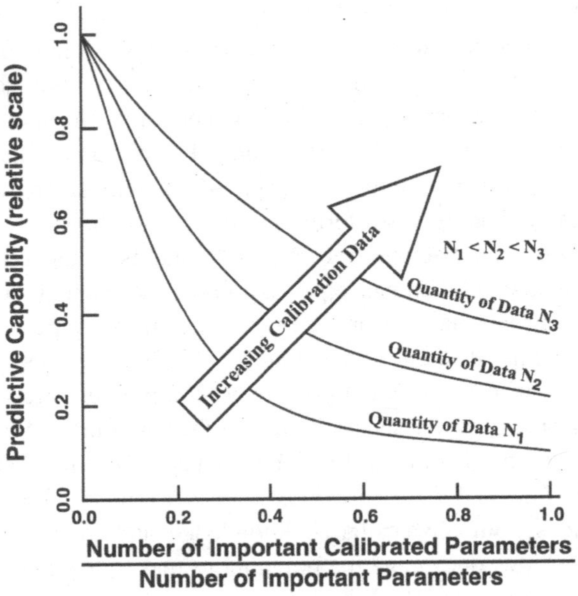

Figure 10.16 Impact of calibrated parameters on predictive capability of a model.

Figure 10.16 depicts the impact of calibrated parameters on the predictive capability of a model. The abscissa in the graph is the ratio of the number of important calibrated parameters in the model divided by the total number of important parameters in the model. That is, the abscissa is the percent of important parameters that are calibrated as part of the mathematical model. For example, suppose that a global sensitivity analysis determines the 10 most important input parameters for a given SRQ. If four of these are calibrated parameters and the remainder can be independently determined from the model, then the value of the abscissa is 0.4. The ordinate is the notional concept of predictive capability of a model in the sense of ability of the model to make accurate predictions for input conditions other than (a) what have been previously assessed in model validation, or (b) what have been previously used to calibrate the model. As is seen in the figure, for a given amount of calibration data, there is a decrease in predictive capability as the percent of important calibrated parameters increases. For example, if there are no calibrated parameters in the model, then there is no loss in predictive capability due to these parameters. That is *not to say* that the model has perfect accuracy to foretell unseen physical processes. It simply says that the ability to make accurate predictions is not influenced by calibrated parameters.

Figure 10.16 displays a truth of paramount importance in *any type* of modeling and simulation. The predictive capability of a calibrated model rapidly decreases when little calibration data are available. This situation is exacerbated as the percentage of calibrated parameters increases. This truth has been well known in statistical modeling for over a century. For a model with a high percentage of important calibrated parameters, *the parameters themselves* become the essence of the model, as opposed to posited knowledge in the formulation of the model, such as physical principles. Hydrological models, for example, are clearly in this domain. Furthermore, machine learning, data mining, and artificial neural network models (that are constructed with little or no influence of physical principles) are also in this domain. It is well known that the reliability of these types of model is entirely dependent on the quantity and diversity of the training data available. These are descriptive models that can have some level of predictive capability "near" the training data, including stochastic features of the SRQs. Models that are developed to simulate any type of human behavior in the social sciences, such as economics, linguistics, psychology, and sociology, are also descriptive models. For a critical review of big data models, see O'Neil (2016). For a critical discussion how parameter calibration, model validation, and numerical errors are intertwined in simulation, see Lenhard (2018) and Pronskikh (2019).

Finally, Figure 10.16 also displays an important message for physics-based models which are the focus of this book. At a global level, these models are supposed to conserve mass, momentum, and energy, and satisfy other fundamental physical principles, such as the second law of thermodynamics. Calibration of parameters in these models can be viewed as an optimization problem, given the constraints of the physics-based model. If a global sensitivity analysis is conducted, and it is found that 100% of the important parameters are calibrated parameters, is it appropriate to refer to the model as a physics-based model? It could be argued that the model retains its high-level physics basis, but the extent of the physics-basis should be questioned with regard to the important submodels that contain the calibrated parameters. For example, in fluid dynamics, what is the physics fidelity of the turbulence submodel? In solid mechanics of ductile materials, what is the physics fidelity of the plastic deformation submodel? Multi-scale modeling can provide some answers to these types of questions.

Example: Validation, Calibration, and Predictive Capability

Consider a physics-based, deterministic mathematical model with a large number of input parameters. Three experiments are conducted at dispersed conditions of input data. As a result, the validation domain is a triangle in a high-dimensional input space with one experiment at each vertex of the triangle, somewhat similar to Figure 10.3. The simulation prediction for each of the experimental conditions is a blind prediction, i.e., no knowledge of the system response is available to the computational analyst, other than the measured input data for the model. Comparing the simulation results with the experimental measurements of the SRQs showed large disagreement, even though the measurement uncertainty is considered negligible. To take advantage of these valuable new experiments, the computational analyst uses the measurements of the SRQs to calibrate the input parameters of the model. It is found that the model has sufficient flexibility such that the input parameters can be updated to achieve perfect agreement of the simulation with the three experimental measurements. A global sensitivity analysis finds that the parameters that were updated to match the experimental measurements constituted 100% of the important parameters of the model.

The predictive capability of the model can be questioned not only within the validation domain, but also outside of the validation domain. More specifically, what is the model uncertainty due to updating the model parameters? Recall that this is a determinist model, i.e., there is no explicit uncertainty in any input parameter. Within the validation domain, estimating the model uncertainty could be thought of as analogous to interpolation, whereas outside of the validation domain it could be viewed as extrapolation. To inform a system designer who intends to use the model for future simulation-informed decision making, an estimate of model uncertainty should be provided. One method of estimating the predictive uncertainty within the validation domain is to compute the difference between the predictions made before the parameters were calibrated and those after the parameters were calibrated. This error measure could be scaled according to bilinear interpolation such that the error is zero at the three experimental points, where the updated parameters yield perfect agreement, but is nonzero in between the three points.

Outside of the validation domain, a similar method of estimating model uncertainty is used. The same bilinear interpolation procedure is used outside of the validation domain. This procedure is less defensible outside of the validation domain because of the extrapolation of the interpolation function into a region where there is no measure of model error based on experimental measurements. In fact, the procedure is not even a physics-based estimate of model uncertainty because the error measure relies partially on a linear extrapolation for the estimated error. However, the procedure does inform the designer of some level of uncertainty in using the model. Without the computational analyst providing some estimate of uncertainty, a designer has little capability, interest, or time to investigate model uncertainty. If the designer is not satisfied with this estimate of uncertainty, then the age old response can be given: "We need more data."

10.8 Further Reading

For recent research on approaches to model validation, see: Sewall et al., 2006; Trucano et al., 2006; Oberkampf and Trucano, 2008; NaRC, 2012; Stewart et al., 2012; Bose et al., 2013; Kutluay and Winner, 2014; Morrison et al., 2014; Kwasniewski and Bojanowski, 2015; Murray-Smith, 2015; Scott et al., 2015; Arifin et al., 2016; Schroeder et al., 2016; Pereira et al., 2017.

For recent research on planning and conducting model validation experiments, see: Blanchat and Jernigan, 2010; Kearney et al., 2010; Blanchat and Suo-Anttila, 2011; Hariharan et al., 2011; Leishear et al., 2012; Boyce et al., 2014; Boyce et al., 2016; Harris et al., 2016; McGrattan et al., 2016; Ao et al., 2017; Oberkampf and Smith, 2017.

10.9 Problems

1. In your field of interest, discuss and contrast an experiment that is conducted for scientific model validation versus project-oriented model validation.
2. In your field of interest, identify a simple system of interest and specify its intended use domain. Define a realistic model validation domain, given generally available experimental facilities and measurement capabilities. Assume the dimensionality of the input data space is two and there is one SRQ of interest. Graphically show each of these domains and discuss the rationale for your specifications of the intended use domain and the validation domain.
3. In your field of interest, identify a simple system of interest and specify its intended use domain. Define a realistic model validation domain, given generally available experimental facilities and measurement capabilities. Assume the dimensionality of the input data space is three and there is one SRQ of interest. Graphically show each of these domains and discuss the rational for your specifications of the intended use domain and the validation domain.

4. In your field of interest, pick a research journal and an article of interest that is focused on experimental measurements of some physical process. Analyze and discuss the article from the perspective of using the experimental data given in the article to conduct validation of a mathematical model of the physical process described in the article.
5. Construct a model validation hierarchy of a simple system or component of interest within your field of interest. Pick one element from each tier of the hierarchy and discuss how a model validation experiment could be conducted for that element.
6. In your field of interest, pick a system of interest for which the operational system cannot be tested because the instrumentation technology is presently not available to make the measurements needed for model validation.
7. In your field of interest, pick a system of interest for which the operational system cannot be tested because this type of test is not allowed for practical, ethical, or governmental reasons. Suggest an alternative experiment that would be allowed, such that the experiment would provide valuable measurements that would assist in model validation. Discuss what model input data and what SRQs could be measured for model validation.
8. In your field of interest, describe how an experiment on a system of interest could be designed and executed that would comprise the separate activities of input parameter calibration, model validation, and assessment of predictive capability. Explain the steps that would need to be taken such that each of the goals of parameter calibration, model validation, and predictive capability are achieved.

10.10 References

Aeschliman, D. P., and W. L. Oberkampf (1998). Experimental Methodology for Computational Fluid Dynamics Code Validation. *AIAA Journal*, **36**(5), 733–741.

AIAA (1998). Guide for the Verification and Validation of Computational Fluid Dynamics Simulations. *American Institute of Aeronautics and Astronautics*, AIAA-G-077-1998, Reston, VA.

Anderson, M. G., and P. D. Bates (eds.) (2001). *Model Validation: Perspectives in Hydrological Science*. New York, NY: John Wiley & Sons.

Ao, D., Z. Hu, and S. Mahadevan (2017). Design of Validation Experiments for Life Prediction Models. *Reliability Engineering and System Safety*, **165**, 22–33.

Arifin, S. M. N., G. R. Madey, and F. H. Collins (2016). Chapter 7: Verification, Validation, Replication, and Reproducibility, in *Spatial Agent-Based Simulation Modeling in Public Health: Design, Implementation, and Applications for Malaria Epidemiology*. New York, NY: John Wiley.

ASME (2022). Verification, Validation, and Uncertainty Quantification Terminology in Computational Modeling and Simulation. *American Society of Mechanical Engineers*, ASME VVUQ 1-2022.

Blanchat, T., and D. Jernigan (2010). Temperature and Heat Flux Datasets of a Complex Object in a Fire Plume for the Validation of Fire and Thermal Response Codes. *Sandia National Laboratories*, SAND2009-6944, Albuquerque, NM.

Blanchat, T. K., and J. Suo-Anttila (2011). Hydrocarbon Characterization Experiments in Fully Turbulent Fires – Results and Data Analysis. *Sandia National Laboratories*, SAND2010-6377, Albuquerque, NM.

Bose, D., J. L. Brown, D. K. Prabhu, P. Gnoffo, C. O. Johnston, and B. Hollis (2013). Uncertainty Assessment of Hypersonic Aerothermodynamics Prediction Capability. *Journal of Spacecraft and Rockets*, **50**(1), 12–18.

Bossel, H. (1994). *Modeling and Simulation*. Wellesley, MA: A. K. Peters.

Boyce, B. L., S. L. B. Kramer, T. R. Bosiljevac, E. Corona, J. A. Moore, K. Elkhodary, et al. (2016). The Second Sandia Fracture Challenge: Predictions of Ductile Failure Under Quasi-Static and Moderate-Rate Dynamic Loading. *International Journal of Fracture*, **198**(1–2), 5–100.

Boyce, B. L., S. L. B. Kramer, H. E. Fang, T. E. Cordova, and others (2014). The Sandia Fracture Challenge: Blind Round Robin Predictions of Ductile Tearing. *International Journal of Fracture*, **186** (1–2), 5–68.

Hariharan, P., M. Giarra, V. Reddy, S. W. Day, K. B. Manning, S. Deutsch, et al. (2011). Multilaboratory Particle Image Velocimetry Analysis of the FDA Benchmark Nozzle Model to Support Validation of Computational Fluid Dynamics Simulations. *Journal of Biomechanical Engineering*, **133**, 041002–1–041002-14.

Harris, J. R., B. W. Lance, and B. L. Smith (2016). Experimental Validation Data for Computational Fluid Dynamics of Forced Convection on a Vertical Flat Plate. *Journal of Fluids Engineering*, **138**, 011401-1–011401-14.

Hornung, H. G., and A. E. Perry (1998). [Personal Communication].

Kearney, S., T. K. Blanchat, A. Ricks, T. Grasser, and D. Jernigan (2010). Test Plan for Validation of the Radiative Transfer Equation. *Sandia National Laboratories*, SAND2010-4800, Albuquerque, NM.

Kennedy, M. C., and A. O'Hagan (2001). Bayesian Calibration of Computer Models. *Journal of the Royal Statistical Society Series B-Statistical Methodology*, **63**(3), 425–450.

Kutluay, E., and H. Winner (2014). Validation of Vehicle Dynamics Simulation Models – A Review. *Vehicle System Dynamics*, **52**(2), 186–200.

Kwasniewski, L., and C. Bojanowski (2015). Principles of Verification and Validation. *Journal of Structural Fire Engineering*, **6**(1), 29–40.

Leishear, R. A., S. Y. Lee, M. D. Fowley, M. R. Poirier, and T. J. Steeper (2012). Comparison of Experiments to Computational Fluid Dynamics Models for Mixing Using Dual Opposing Jets in Tanks with and Without Internal Obstruction. *Journal of Fluids Engineering*, **134**, 111102-1–111102-21.

Lenhard, J. (2018). Holism, or the Erosion of Modularity – A Methodological Challenge for Validation. *Philosophy of Science*, **85**(5), 832–844.

Leonard, I. E., and J. E. Lewis (2016). *Geometry of Convex Sets*. New York, NY: John Wiley.

Lin, S. J., S. L. Barson, and M. M. Sindir. (1992). Development of Evaluation Criteria and a Procedure for Assessing Predictive Capability and Code Performance. *Advanced Earth-to-Orbit Propulsion Technology Conference*, **3174**, 154–164. Marshall Space Flight Center, Huntsville, AL.

Luckring, J. M., S. Shaw, W. L. Oberkampf, and R. Graves (2024). Development of a Model Validation Hierarchy for Modeling and Simulation. In J. H. Morrison, E. I. Walker, and M. Nikbay (eds.), *Approaches to Model Validation*. NATO – Science and Technology Organization.

Marvin, J. G. (1995). Perspective on Computational Fluid Dynamics Validation. *AIAA Journal*, **33**(10), 1778–1787.

McGrattan, K., R. Peacock, and K. Overholt (2016). Validation of Fire Models Applied to Nuclear Power Plant Safety. *Fire Technology*, **52**, 5–24.

McNish, A. G. (1962). The Speed of Light. *Institute of Radio Engineers, Transactions on Instrumentation*, **I-11** (3–4), 138–148.

Morrison, J. H., B. Kleb, and J. C. Vassberg (2014). Observations on CFD Verification and Validation from the AIAA Drag Prediction Workshops. *AIAA SciTech*, AIAA Paper No. 2014-0202, National Harbor, MD.

Murray-Smith, D. J. (2015). *Testing and Validation of Computer Simulation Models: Principles, Methods and Applications*. New York, NY: Springer.

NaRC (2012). Assessing the Reliability of Complex Models: Mathematical and Statistical Foundations of Verification, Validation, and Uncertainty Quantification. *National Research Council of the National Academies*, ISBN-13: 978-0-309-25634-6. Washington, DC.

NNSA (2019). National Nuclear Security Administration: Maintaining the Stockpile. www.energy.gov/nnsa/missions/maintaining-stockpile.

O'Neil, C. (2016). *Weapons of Math Destruction: How Big Data Increases Inequality and Threatens Democracy*. New York, NY: Crown Publishers.

Oberkampf, W. L., and D. P. Aeschliman (1992). Joint Computational/Experimental Aerodynamics Research on a Hypersonic Vehicle: Part 1, Experimental Results. *AIAA Journal*, **30**(8), 2000–2009.

Oberkampf, W. L., and B. L. Smith (2017). Assessment Criteria for Computational Fluid Dynamics Model Validation Experiments. *Journal of Verification, Validation and Uncertainty Quantification*, **2**(3), 031002-1–031002-14.

Oberkampf, W. L., and T. G. Trucano (2002). Verification and Validation in Computational Fluid Dynamics. *Progress in Aerospace Sciences*, **38**(3), 209–272.

Oberkampf, W. L., and T. G. Trucano (2008). Verification and Validation Benchmarks. *Nuclear Engineering and Design*, **238**(3), 716–743.

Oberkampf, W. L., T. G. Trucano, and C. Hirsch (2004). Verification, Validation, and Predictive Capability in Computational Engineering and Physics. *Applied Mechanics Reviews*, **57**(5), 345–384.

Pereira, F. S., L. Eca, and G. Vaz (2017). Verification and Validation Exercises for the Flow around the KVLCC2 Tanker at Model and Full-Scale Reynolds Numbers. *Ocean Engineering*, **129**, 133–148.

Price, M., J. M. Early, R. Curran, E. Benard, and S. Raghunathan (2006). Identifying Interfaces in Engineering Systems. *AIAA Journal*, **44**(3), 529–540.

Pronskikh, V. (2019). Computer Modeling and Simulation: Increasing Reliability by Disentangling Verification and Validation. *Minds and Machines*, **29**, 169–186.

Rizzi, A., and J. Vos (1998). Toward Establishing Credibility in Computational Fluid Dynamics Simulations. *AIAA Journal*, **36**(5), 668–675.

Schroeder, B. B., K. T. Hu, J. G. Mullins, and J. G. Winokur (2016). Summary of the 2014 Sandia Verification and Validation Challenge Workshop. *Journal of Verification, Validation and Uncertainty Quantification*, **1**(1), 015501-1–015501-9.

Scott, S. N., J. A. Templeton, P. D. Hough, J. R. Ruthruff, M. V. Rosario, and J. P. Peterson (2015). Statistical Validation for Heat Transfer Problems: A Case Study. *International Journal of Computational Methods and Experimental Measurements*, **3**(2), 101–120.

Sewall, E. A., D. K. Tafti, A. B. Graham, and K. A. Thole (2006). Experimental Validation of Large Eddy Simulations of Flow and Heat Transfer in a Stationary Ribbed Duct. *International Journal of Heat and Fluid Flow*, **27**, 243–258.

Shaw, S., J. M. Luckring, W. L. Oberkampf, and R. Graves (2024). Exploitation of a Model Validation Hierarchy for Modeling and Simulation. In J. H. Morrison, E. I. Walker, and M. Nikbay (eds.), *Approaches to Model Validation*. NATO – Science and Technology Organization.

Sindir, M. M., S. L. Barson, D. C. Chan, and W. H. Lin (1996). On the Development and Demonstration of a Code Validation Process for Industrial Applications. *27th AIAA Fluid Dynamics Conference*, AIAA Paper 96-2032, New Orleans, LA.

Stewart, S. F. C., E. G. Paterson, G. W. Burgreen, P. Hariharan, M. Giarra, V. Reddy, et al. (2012). Assessment of CFD Performance in Simulations of an Idealized Medical Device: Results of FDA's First Computational Interlaboratory Study. *Cardiovascular Engineering Technology*, **3**(2), 139–160.

Trucano, T., C. Garasi, and T. Mehlhorn (2006). ALEGRA-HEDP Validation Strategy. *Sandia National Laboratories*, SAND2005-6890, Albuquerque, NM.

Trucano, T. G., M. Pilch, and W. L. Oberkampf (2002). General Concepts for Experimental Validation of ASCI Code Applications. *Sandia National Laboratories*, SAND2002-0341, Albuquerque, NM.

Winsberg, E. (2019). Computer Simulations in Science. In E. N. Zalta (ed.), *Stanford Encyclopedia of Philosophy*. Stanford, CA: Stanford University: The Metaphysics Research Lab.

Youden, W. J. (1972). Enduring Values. *Technometrics*, **14**(1), 1–11.

Zeigler, B. P., A. Muzy, and E. Kofman (2018). *Theory of Modeling and Simulation: Discrete Event & Iterative System Computational Foundations*, 3rd ed. Cambridge, MA: Academic Press.

Zuber, N., G. E. Wilson, M. Ishii, W. Wulff, B. E. Boyack, A. E. Dukler, et al. (1998). An Integrated Structure and Scaling Methodology for Severe Accident Technical Issue Resolution: Development of Methodology. *Nuclear Engineering and Design*, **186**(1–2), 1–21.

11

Design and Execution of Model Validation Experiments

To consult the statistician after an experiment is finished is often merely to ask him to conduct a post mortem examination. He can perhaps say what the experiment died of.

—Ronald Fisher

Contents of Chapter 11

This chapter discusses the six primary guidelines for the design and execution of model validation experiments. As discussed in Chapter 10 "Model Validation Fundamentals," the goal of model validation is to quantify the impact of the assumptions and approximations in the formulation of the mathematical model describing the physical process of interest. The quantitative impact is assessed in terms of comparing the measured system response quantities (SRQs) from the experiment and the corresponding SRQs resulting from the simulation. The SRQs from the model may be a prediction, such as a blind prediction, or they may be a retrodiction, where measured SRQs are known to the computational analyst before the comparison. This chapter does not discuss the mathematical operators that quantify the comparisons between experimental measurement and simulation. That topic is discussed in Chapter 12 "Model Accuracy Assessment."

Chapter 11 gives a detailed discussion of a high-quality validation experiment that was conducted by the authors and other contributors in order to demonstrate how the six guidelines for the design and execution of validation experiments are implemented in practice. The experiment is referred to as the Joint Computational-Experimental Aerodynamics Program (JCEAP). From the beginning of the project, it was designed to synergistically couple computational fluid dynamics development and experimental techniques in a hypersonic wind tunnel. The program helped develop the six guidelines for conducting high-quality model validation experiments discussed here. As part of the discussion, technical and practical issues that commonly cause difficulty in the design, execution, and funding of validation experiments are addressed.

11.1 Guidelines for Validation Experiments

The six guidelines were developed over a period of several years and were originally reported in Aeschliman and Oberkampf (1998) and Oberkampf and Trucano (2002). Although JCEAP greatly assisted in refining these ideas, most of these ideas were presented individually by computational and experimental researchers in the past. Some of the key contributors to developing these ideas were Marvin (1995), Cosner (1995), Porter (1996), Barber (1998), Benek et al. (1998), and Roache (1998).

Some of the guidelines to be discussed have the character of a task to be completed during the design and execution of a validation experiment, and some have the character of the quality of the overall activity, both experiment and associated simulation. In the discussion, no distinction is made between the task and character features of the guidelines. Two points should be noted in the description of the guidelines. First, the guidelines refer to *a validation experiment*. This term is used simply for convenience and brevity. A high-quality validation experiment is always a *suite* or *ensemble* of many experiments, i.e., multiple individual experiments, sometimes numbering into the hundreds. Multiple experiments are required to (a) characterize nondeterministic input and response features, and (b) quantify experimental measurement uncertainties. All of these experiments share a common goal: critical evaluation of the accuracy of a mathematical model of a physical process, or processes, occurring during the experiments.

Second, the term *simulation* always refers to the activity of producing a computational result using computer software that embodies a mathematical model of the processes of interest. Some people, particularly engineering project managers, refer to an experiment that is not fully representative of the engineering system of interest as a *simulation*. For example, they will refer to a wind tunnel experiment or an experiment of a subscale system or subsystem as a simulation. This text rejects this terminology because it is diametrically opposed to the distinction between a mathematical model that produces a simulation result and a physical experiment that produces an observation or measurement in nature. Thus, *any* physical experiment *is* reality, but it may not be the reality that is of particular interest with regard to a complex system.

11.1.1 Joint Effort between Analysts and Experimentalists

Guideline 1: *A validation experiment should be jointly designed by experimentalists, model developers, code developers, and code users working closely together throughout the program, from inception to documentation, with complete candor about the strengths and weaknesses of the simulation and the experiment.*

Applied researchers who build the mathematical models incorporated into simulation codes are referred to in this text as *model developers*. The term *code developers* is used to describe those who write the computer software or implement the mathematical models and numerical solution techniques into the software. Those who use the code for analysis of physical processes and engineering systems are known as *code users*. For the purposes of this text, these various types of people are grouped into the broad term *analysts*.

To some, Guideline 1 may sound relatively easy to accomplish. However, experience has shown that it is extraordinarily difficult to accomplish in practice. There are several reasons why this is true; some are practical, and some are related to human nature and business realities. In order to form a team of experimentalists and analysts for the design and execution of a validation experiment, funding must be available for both at the same time. Yet, experimentalists and analysts have traditionally sought funding from separate sources. If these two groups wanted to design and conduct a validation experiment, they would typically each go their own way to seek funding. It would be rare for both to obtain support for their work during the same time frame, if at all. Thus, there remains an unfaltering tradition (particularly in government) that one source is not interested in, nor responsible for, funding both experiments and simulations. While it is understandable why these traditions are in place, if they continue, high-quality validation experiments will remain rare.

If funding does become available for both experimentalists and analysts at the same time, then project leadership must deal with the technical synchronization of both activities. For example, the experimental facility as well as the diagnostic instrumentation needed for the validation experiment must be available and ready for use and not in the construction or development phase. Similarly, the code and the physics modeling options needed for the validation experiment must be operational and have been adequately tested using code verification activities. Some analysts will question the need for this last requirement. These analysts feel that code debugging and testing can be conducted in parallel with validation. This is a gross misjudgment, discussed in previous chapters, that can lead to false positives and false negatives concerning the capability of the simulation. Having both the experimental and simulation capabilities fully prepared, as well as the needed personnel available at the same time, requires careful planning and preparation.

Another practical reason why forming a unified team of experimentalists and analysts proves difficult is that team members are commonly from different organizations. For example, suppose a group of analysts is from a company that does not have the needed experimental facilities. If the analysts' company obtains funding for the planned validation activities, then the analysts' company would seek bids for the experiments to be conducted. Although technical qualifications of potential bidders for the experimental work are important, there are two business issues that come into play. First, when the analysts' company is seeking a facility in which to conduct the experiment, this is obviously a competitive situation for potential bidders. The potential bidders will not expose any weaknesses, limitations, or deficiencies within their facility or inexperience with conducting validation experiments. Not only would they risk losing the potential contract, but they would also fear that any detrimental information could be used against them in the future by their competitors. Second, the analysts' company, once the contract is placed, is in control of funding and the direction of the project. From the perspective of the experimentalists' company, they are the supplier and the analysts' company is the customer. As a result, the experimentalists are the subordinate team members who do not want to risk poor performance on the present contract or potential future contracts.

Organizations that contract for experimental services to conduct high-quality validation experiments need to understand and be convinced of the importance of Guideline 1. They need to develop ways to implement the guideline and mitigate their risks, as well as those associated with the contractor's perspective. In seeking bids for the experimental services, the analysts' company should request information concerning the experience of the company and its staff in conducting validation experiments. For example, they could request the names and contact information of previous validation experiment customers. Once the contract is executed, leadership should clearly emphasize that candor and forthrightness are critical in both the experimental and simulation activities. Thus, contracting for validation experiment services is difficult for both sides of the business enterprises.

Suppose the experimentalists and the analysts are from different organizations within the same corporation, government organization, or university. If so, the experimentalists and the analysts may be familiar with each other, but they may not have developed a close working relationship. In fact, their history may be that they have been competitors for funding or recognition of achievement in the past. For example, the "Computers versus wind tunnels" mentality (Chapman et al., 1975) that has existed in the past in some areas of the computational fluid dynamics (CFD) community is a perfect example of competition, not the collaboration and synergism stated in Guideline 1. Also, because of the nature of their technical training and professional responsibilities, there is typically a gulf separating computationalists' and experimentalists' technical interests and backgrounds. These factors, both obvious and insidious, will deter the candor and openness needed to forthrightly discuss the strengths and weaknesses in the simulation and experimental approaches during both the design and the execution of the validation experiment. These can be overcome in time, but would require patience, respect, understanding, and, most importantly, management leadership. These management qualities are difficult to identify and attain in a fast-paced, modern, competitive environment.

It may be clear what is needed to overcome these substantive difficulties, but implementing solutions will likely prove difficult. Leadership in funding organizations should begin to recognize the importance of funding both experimental and simulation activities together. Even if this need were recognized, there would probably be structural changes needed within the funding organization to actually accomplish funding joint experimental and simulation activities. Most funding organizations, particularly government, are structured such that responsibilities for simulation and experimental research are entirely separate.

Finally, organizations that conduct validation experiments and simulations would most likely need a change in management perspective. Funding groups within an organization need to learn that experimental and simulation activities need to be funded at the same time to improve the effectiveness of both simulation and experiments. For applied research groups, it is reasonable to imagine this change. For project groups focused on building and delivering hardware and staying on budget and schedule, this is extremely difficult to achieve. Some farsighted project managers may devote some project funds to core competency development such as validation experiments, but this is rare. For organizations that *are* able to fund both experimental and simulation groups in a validation experiment project, management must understand that it is the success or failure of the *joint activity* that should be assessed. A viewpoint of some managers responsible for simulation development is that if the validation activity shows that the simulation performed poorly, then the activity is considered useless or even considered a failure. Thus, some simulation development managers feel that a validation experiment only represents a downside risk, as opposed to a path for improvement of simulation.

11.1.2 Measurement of All Input Data Needed for the Simulation

Guideline 2: *A validation experiment should be designed to capture the essential physics of interest; measure all relevant physical modeling input data, and initial and boundary conditions; and system excitation information required by the simulation.*

A validation experiment should be designed to address the physics of interest with regard to its spatial dimensionality, temporal nature, geometric complexity, and physics coupling. For example, when conducting a two-dimensional (2-D), either planar or axisymmetric, experiment, consider the following question: Are the three-dimensional (3-D) effects in the experiment negligibly small so that a 2-D simulation can be appropriately compared with the experimental data? Since the ability of the experimentalist to detect 3-D effects is very limited, the analyst should attempt to answer this question. Both 2-D and 3-D simulations of the experiment should be computed to predict the spatial dimensionality effect on the system response quantities (SRQs) of interest. If the change in the SRQs between the 2-D and 3-D simulations are (a) comparable to the expected experimental measurement uncertainty in the SRQs, and (b) the 3-D simulations have demonstrably negligible numerical solution error, then a 3-D simulation should be used in the comparison with experiment. If the 3-D simulations are not computed and the disagreement with experiment is large, then the question remains: Is the disagreement due to the 2-D assumption in the model or other physical approximations and assumptions in the model? Recent experience by several researchers in fluid dynamics has shown that some experiments that were thought to be essentially planar 2-D were found to have significant 3-D effects. When 3-D simulations were performed, with no other changes in the modeling, very good agreement was obtained. Yet, an analogous question arises with regard to temporal assumptions in the model: Can a steady-state assumption be made in the model, or is an unsteady simulation required?

Guideline 2 points out the demanding measurement requirements on the experimentalist and the narrow goal of the validation experiment. From the simulation side, the analysts must understand which spatial dimensionality, temporal nature, geometric simplifications, and physics couplings are appropriate in modeling a specific experiment. From the experimental side, the experimentalists must decide what design and operational features should exist regarding the test article, in the experimental facility, and in the needed instrumentation so that the specified modeling assumptions are properly assessed. If the experimental parameters that are initially requested by the analyst for the simulation cannot be satisfied in the proposed experimental facility, it may be feasible to alter the model inputs and still satisfy the primary validation goals. Alternatively, it may be necessary to look for another facility to achieve the primary goals. For example, in a fluid dynamics experiment, how can the location of boundary layer transition over surface of a test article be measured? In a solid mechanics experiment, how can the rigidity of a mechanical attachment to a test article be measured? Conversely, analysts must understand or be informed by the experimentalists concerning the limitations of the experimental equipment, the instrumentation, and the test conditions of the facility.

As part of the requirement for Guideline 2, the experimentalists should measure all important physical modeling input data, initial conditions (ICs), boundary conditions (BCs), and system excitation information needed as input for the simulation. These measurement responsibilities are shown in Figure 11.1 from the perspective of the input data needed for the simulation and the output data produced by the simulation. Any important experimental conditions needed by the simulation, but not measured, will significantly undermine the value of the validation experiment. As discussed in Chapter 10, if key input data are not measured,

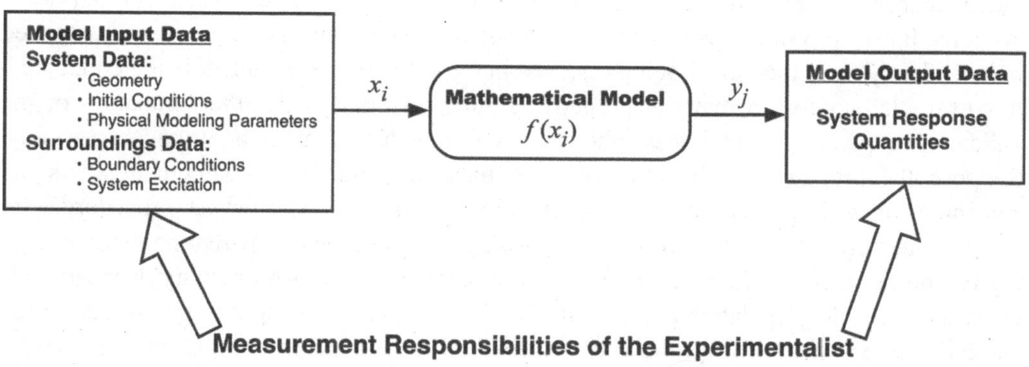

Figure 11.1 Measurement responsibilities of the experimentalist.

the *best result* that can be obtained is that the simulation will yield imprecise results, e.g., probability-boxes (p-boxes), which must be compared with the measurements. The much more likely result is that the analysts will use the unmeasured input quantities as adjustable parameters to calibrate or update their models to obtain the best agreement with the measured SRQs. This destroys the fundamental goal of model validation: critical evaluation of the accuracy of the simulation for the conditions of the validation experiment.

The input quantities that should be used in the simulation are those that *actually occur* and are (hopefully) measured in the validation experiment. Two different reasons for not knowing the input quantities can arise. First, a quantity can be unknown or epistemically uncertain for various reasons. Examples of how this can occur follow:

- a quantity was not measured because the analyst did not request it from the experimentalist, or the experimentalist was not aware that certain details may be needed in a simulation;
- a quantity was not measured, and the experimentalist simply estimated the quantity based on experience, or took the value from an engineering handbook, as opposed to actual measurement; or
- the geometric description of the test article is taken from the design drawings of the hardware, as opposed to determining the as fabricated geometry or the as tested geometry, which may entail deflection due to mechanical and thermal loading during the experiment.

Second, a model input quantity cannot be independently measured before or during the validation experiment. For example:

- a quantity cannot be measured independently before the test because the measurement alters the quantity, such as a destructive test;
- a quantity cannot be measured during the test because the measurement alters the physical processes occurring;
- a quantity cannot be measured independently because it changes during the experiment, such as material deformation and microslip causing heating; and
- a quantity cannot be measured independently because it only exists when the test article is assembled for the experiment, such as any physical feature that only exists when the assembly exists, e.g., thermal and electrical contact resistance.

For this second situation, the quantity should be determined using a procedure that maintains independence between the determination of the quantity and the assessment of model accuracy for the physical experiment being conducted. They can sometimes be characterized independently from the model being assessed in the validation experiment being conducted. For example, suppose a material property is an input quantity for the simulation of the validation experiment, but it has inherent variability due to manufacturing or assembly. Suppose the property can be independently measured, but the measurement alters the specimen, rendering it useless for the validation experiment. The proper procedure is to measure the variability of the property by drawing samples from the parent population, e.g., a given production lot. After a number of samples are measured, a probability distribution is constructed for the population and this distribution is used as the input to the model for the validation experiment. This probability distribution, along with other input uncertainties, is propagated through the model to obtain a probability distribution for the SRQ of interest. A quantitative comparison can then be made between a probability distribution of the experimental measurements and the probability distribution from the simulation. However, the more common situation is that some model input quantities cannot be independently measured separately from the validation experiment. Dealing with this situation is discussed in Chapter 12 "Model Accuracy Assessment."

If the input quantity is uncertain because of both a lack of knowledge (epistemic uncertainty) and a random process (aleatory uncertainty), then the quantity should be characterized as an imprecise probability, e.g., a p-box. Although an imprecise probability distribution is not as desirable in assessing the accuracy of the model for the validation experiment, it is still possible to do so. See Chapter 12 for further discussion.

Example: Wind Tunnel Experiment

A test article representing a flight vehicle geometry is to be used in a validation experiment conducted in a wind tunnel. The following are examples of the level of detail that may be needed for mathematical modeling of the system of interest and the surroundings for the simulation of the experiment:

- Accurate measurements of the actual, as fabricated, test article dimensions, as opposed to the nominal or as requested specifications. These measurements would include any imperfections, surface mismatches, or gaps between components of the test article, as well as hardware needed for mounting the test article in the test section.
- Measurement of the location of boundary layer transition over the surface of the test article for all free-stream conditions, angles of attack, and control surface deflections. If boundary layer trips are used, the location and characterization of the trips should be measured.
- Measurement of free-stream turbulence quantities and steady-state, free-stream, flowfield non-uniformity at the beginning of the test section of the wind tunnel. These measurements can be made without the test article installed.
- Measurement of actual locations of all instrumentation, as opposed to requested locations stated in the fabrication drawings. This may include measurement of the geometric features of the diagnostic sensors, such as pressure orifices or pressure probes, if the flow over the sensors is included in the simulation of the flow over the test article.
- If there is non-negligible deformation of the test article due to aerodynamic loading or aerodynamic heating, the actual deformed geometry should be measured. If this is not possible, then coupled fluid-structure simulations should be conducted for the various test conditions.
- Location of where the free-stream conditions in the test section and the total pressure and temperature conditions are measured in the wind tunnel, especially for subsonic flow in the test section.
- If the flow in the entire test section is included in the simulation, then measurements must be made of the wind tunnel wall geometry upstream of and in the test section. Also, boundary layer profiles and static pressures at various locations on the walls upstream and in the test section should be made before the test article is inserted.

Example: Ductile Material Failure Experiment

A validation experiment is to be conducted concerning the ductile failure of a metal fastener, e.g., a bolt, rivet, or screw, in shear. A single fastener secures two metal plates in a lap joint. Each end of the metal plates is loaded in tension by a test fixture. Various SRQs of interest are to be measured, e.g., elongation of the system (plates and fastener) as a function of load, ultimate load of the system, and deformation geometry of the fastener and plates as a function of load. Because of the large variability known to occur in ductile failure of materials, a large number of experiments is conducted in order to properly estimate statistical features of both the input data needed for the mathematical model and the SRQs of interest. The following are examples of the level of detail that may be needed for the model input data, as well as examples of SRQs that could be measured:

- Detailed geometric measurements of the fastener, each of the metal plates, and the clearance hole for the fastener. The surface finish of each component should be measured, particularly the area where the two plates are in contact.
- If the fastener is a bolt or screw, the preload torque should be measured. If possible, measurements should be made of the friction characteristics of the bolt or screw with the base material. If the fastener is a rivet or a pin, details should be measured concerning the load required for insertion, temperature during insertion, and the deformation generated after insertion of a rivet, but before loading in the test fixture.
- If the fastener is a bolt or a screw, experiments should be conducted for multiple preload torques. The torque on the bolt or screw should be measured as a function of rotational angle when the torque is applied.
- Geometrical description of the gripping mechanism and forces applied for connecting the plates to the test fixture.
- Unless material constitutive submodels of the fasteners are to be assessed as part of the validation activity, separate experiments on the fasteners should be made to determine what constitutive submodels should be used in the simulation. These measurements should include experiments on multiple fasteners and plates to determine the statistical properties of the parameters in the constitutive submodels chosen.
- Load history of the test fixture, including time dependence of the loading.
- Video recording of the system deformation history from two orthogonal directions. Using this video with a suitable calibration method, the measurement of the local deformation of the plates and the fastener should be made for each experiment.
- Detailed geometric measurements of the fastener and plates after failure.

11.1.3 Synergism between Simulation and Experiment

Guideline 3: *A validation experiment should strive to emphasize the inherent synergism that is attainable between simulation and experimental approaches.*

Synergism between simulation and experiment can occur in various ways before, during, and after the execution of the experiment. The three most common are the following.

First, use of simulation to improve the design and execution of the planned validation experiment. Because of the flexibility and responsiveness of simulation, it can be used to improve the design of the experiment to better achieve the goals of assessing the accuracy of the mathematical model. Simulation can be used in the planning stages of a validation experiment to dramatically improve the design, instrumentation, and execution of the experiment. For example, the spatial locations where large changes in the SRQs occur can be computed so that instrumentation can be localized in these regions. Locating experimental instrumentation, however, should also anticipate potential surprises in responses that were not predicted by the simulation. Another example of using simulation in the design of the experiment is advising the experimentalist regarding the sensitivity and frequency response characteristics needed for the instrumentation that is planned for the experiment.

A closely related strategy is having the computational analyst work with the experimentalist to determine how the mathematical model applied to the experiment could be challenged even more than originally envisioned. This strategy is commonly referred to as *design the experiment to break the model.* Optimizing the experiment in this way can be done by (a) modifying the test geometry or the physical modeling parameters to increase the difficulty of prediction, even adding features that do not exist in the system of interest; (b) modifying the boundary conditions or system excitation conditions to extreme conditions that may not exist for the system of interest; and (c) modifying the initial conditions for an initial value problem to increase the likelihood that the mathematical model will fail. It should be noted that analysts commonly do not find it appealing to pursue a strategy of optimizing the experiment to break their models. Project managers, however, who use simulation for design and certification of their system, do find it appealing. Finally, it is stressed that modifying the experimental conditions to achieve a failure of the mathematical model is *completely separate* from a failure of the system design.

Second, use of experimental diagnostic information to improve the assumptions and approximation in the formulation of the mathematical model that is applied to the validation experiment. When analysts formulate the assumptions and approximations for the mathematical model, they focus on the application system of interest, as opposed to the conditions that occur in the validation experiment. For example, the boundary conditions and the system excitation that exist during the validation experiment are critical issues; *not* whether they perfectly match those that occur in the application system of interest. Some analysts view the differences between the validation experiment and the application system of interest as an error or a shortcoming in the validation experiment. This is not only a misunderstanding of the goals of a validation experiment, but also confuses the issue of representativeness of a validation experiment compared to the application system of interest. For example, they tend to think of the validation experiment as a "simulated test of the real system" or a "simulated test of a system component." This line of thinking is deeply embedded in the thinking of system designers and especially project managers. This perspective is certainly understandable, but it fails to recognize how mathematical simulation is fundamentally different in its contribution to the design of a system compared to physical testing of a system or its components. Mathematical simulation is a general purpose, yet unreliable, tool to investigate possible outcomes of a real system. Validation experiments are an exercise in assessment of the *physical reliability of the simulation tool*; *not whether the experiment constitutes the system of interest.* Designers and project managers must be continually reminded that a validation experiment is *not* an approximation of nature, but it *is reality* of a physical process.

As an example, consider a validation experiment to measure the first 10 modes of the vibration of a complex structure. Suppose the structure is fully representative of the actual manufactured system of interest. However, because of the size of the structure and the experimental facilities available, the validation experiment can only be conducted at room temperature. However, the actual operating temperature of the structure of interest is 200°C. Suppose that the validation experiment shows that unexpected vibrational modes occur. That is, these modes are known not to occur in similar systems at operational temperatures because of expected thermal expansion characteristics of certain subassemblies of the system. The system designers and the project manager may regard the validation experiment as uninformative, inaccurate, and even misleading. This view would be mistaken relative to the goal of the validation experiment conducted at room temperature. The goal of the validation experiment is to determine if the mathematical model can accurately predict the vibrational modes at room temperature; *not* whether these modes occur at operating conditions of the system.

Third, simulation (in the sense of a general purpose capability) can be used to improve the understanding and operation of the test facility, as well as the understanding of the experimental diagnostic techniques. This synergism can be used not only to improve the measurement of mathematical model input and system response data for the present validation experiment, but also to improve future validation experiments in the facility. Some experimentalists have so little faith in simulation to represent the operation of their complex facility that they feel that simulation is useless in understanding their facility's operation. They do not recognize that certain simple physical processes, or subsets of processes, can actually be simulated more accurately than they can be measured in an experiment. We refer to this realization as the synergy of using the strength of one approach, either simulation or experiment, to offset the weakness of the other approach. A validation experiment for a complex physical process takes advantage of the confidence in the experimental facility and measurements relative to the lack of confidence in the simulation. But this situation can be reversed such that the experimentalist can have more confidence in the simulation of very simple physical processes compared to the ability to physically produce and measure certain features of an experiment.

Example: Heat Exchanger Experimental Facility

Consider an experimental facility for measuring the thermal efficiency of heat exchangers operating at steady-state conditions. The heat exchangers can be of a wide variety of designs, such as two-fluid shell and tube, plate heat exchangers, and phase-change heat exchangers. The heat exchangers are placed inside a cubical-shaped room that is thermally insulated from the surroundings. The accuracy of measuring the thermal efficiency of the heat exchangers being tested has been questioned regarding (a) the heat conduction through the fluid piping to the heat exchanger, (b) the natural convection between the heat exchanger and the air in the room, (c) the thermal radiation from the heat exchanger to the inside surface of the room, and (d) the heat loss from the room to the surroundings. Each one of these extraneous heat transfer mechanisms can impact the inferred thermal efficiency of the heat exchanger being tested, and each mechanism is very difficult to measure individually. The approach to estimate each of these mechanisms is to use computational simulation in combination with specially conducted experiments and measurements. The specialized experiments are conducted with very simplified experimental test configurations and operating conditions. These test configurations and operating conditions would not be geometrically or physically representative of any actual heat exchangers or their operating conditions; but they would be designed to produce very high confidence in the accuracy of the simulation results. For some complex features of the heat transfer simulations, such as thermal contact resistance and the dependence of emissivity on wave length and angle, a combination of simulation and experimental measurements are used to infer, i.e., calibrate, these heat transfer parameters.

Example: Experiment to Measure Deformation of a Structure

Consider an experimental facility for determining the bending and torsional deformation of a structure under static load. For experiments in which the structure will be configured as a cantilever, the facility must attach the structure of interest to a steel structure which is attached to a steel-reinforced, concrete wall. Although the wall and the attachment structure are very stiff, they cannot be assumed to be perfectly rigid in conducting a model validation experiment to measure the deformation of the structure of interest. To determine the stiffness of the wall and the attachment structure around three orthogonal axes, the following coupled simulation and experiment are conducted.

A solid, steel beam of square cross-section is fabricated, and the geometric and mass properties are measured. In both ends of the beam are sets of threaded holes machined so that the beam can be attached to the wall-mounting structure, as well as attached to a mechanism that can apply a load to the other end of the beam. To determine effective Young's modulus of the entire beam, experiments are conducted where the end points of the beam are simply supported, and the beam is uniformly loaded. The deflected shape of the beam is measured at five equally spaced locations along the beam for a range of loads. The beam is rotated 90° and the set of experiments is repeated. All loads applied to the beam allow the beam to remain clearly in the linear elastic range. Using the set of deflection measurements from these experiments and the solution to the inverse problem of the linear elastic deflection of the beam, the effective Young's modulus is estimated for the entire beam.

The next set of experiments is conducted for the beam attached to the wall-mounting structure. The beam is uniformly loaded as before, in addition to having a pure torsional and lateral moment applied to the end of the beam; all three for a range of loads. The vertical, lateral, and torsional deflection of the beam is measured at five equally spaced locations along the beam. All loads applied to the beam allow the beam and the wall-mounting structure to remain clearly in the linear elastic range. Using the set of deflection measurements from these experiments and the solution to the inverse problem of the linear elastic deflection of the beam, the moment stiffness of the wall and the wall-mounting structure around three orthogonal axes is determined.

11.1.4 Independence and Dependence between Simulation and Experiment

Guideline 4: *Although the experimental design and sharing of model input data should be developed cooperatively, independence should be maintained in obtaining the simulation and experimental system response results.*

The goal of this guideline is to achieve, if possible, a blind test prediction for comparison with the experimentally measured results. There are varying opinions as to the value of blind test predictions contrasted to comparisons made with measurements known to the analysts. Guideline 4 clearly emphasizes the importance of blind test predictions. Most experienced project managers and decision makers who have relied on simulation in the past share this view because they obviously live in the world of unknown and unexpected outcomes. In pharmaceutical drug testing and clinical trials in health care, these fields have learned long ago of the critical importance of blind or double-blind testing. Guideline 4 can be viewed as the essence of what predictive capability is in this text and how it should be experimentally evaluated.

Computational analysts typically have a very different opinion about the value of blind test predictions. Sometimes they view the recommendation of a blind test prediction as a personal affront to their integrity. This is clearly *not* the intention. The intention is to learn from many other fields in the natural and social sciences, and the experiences of project managers as to the impact of knowing the presumed correct (experimental) result when a computation-experiment comparison is to be made. Blind test predictions are very rarely conducted openly and published in the literature. If a blind test prediction exercise is conducted within an industrial organization, then the results are usually considered proprietary, whether the comparisons are good or bad. Analysts, or more precisely their organizations, see very little benefit to be gained by sharing blind test predictions, and could possess a great deal of risk if poor comparison results become public knowledge. Some organizations that have produced poor results may resort to devious means to conceal or limit the distribution of poor results of a blind prediction. There must be an extraordinary incentive for organizations to participate openly in blind test prediction activities. The most common motivator is that if the organization does not participate, then they will be barred from certain future bidding opportunities or government certification of their codes and organization.

It is difficult to accomplish the close cooperation of the analysts and the experimentalists promoted in Guideline 1 and, at the same time, retain the recommended independence of simulation and experimental results in Guideline 4. However, this challenge can be met by careful attention to the procedural and management details. When the experimental measurements are processed and analyzed, the analysts should *not* initially be given the system response results. The analysts should be given the complete details of all of the input data needed for their modeling approach, including the physical modeling parameters, ICs, BCs, and system excitation, *exactly* as it was conducted and measured, including uncertainties. That is, everything that is needed for the analysts to compute simulations must be provided – but not measured responses. The analysts should *not* be shown the experimental measurements of the SRQs. As discussed above, the human tendency is to believe that the experimental measurement is correct, as opposed to the simulation result. This causes a cognitive bias in the analyst, which damages the ability to assess the predictive accuracy of the mathematical model to independently foretell the experimental result. In addition, the analysts should be required to quantify the numerical solution errors and propagate any aleatory or epistemic uncertainties occurring in the input so as to obtain uncertainties in the quantities that will be compared with experimental measurements. Then the comparisons between the simulation and experimental results are made, preferably by the experimentalists or organizers of the comparison effort. Even if the agreement is poorer than hoped for, investigating the causes of differences invariably leads to a deeper understanding of the simulations and experiments.

As a final procedural comment, it is recommended that management *not* be involved in the initial discussion of results and comparisons. The discussions should just involve analysts and experimental staff. It is often found in the initial discussions that miscommunications or misinterpretations between team members will be exposed, and appropriate changes are made in the results of one or both team members. For example, analysts may have not recognized that important model input data has larger uncertainty or was less controlled than originally expected. If management is involved before these necessary discussions or corrections are made, overzealousness on either side of the team can sometimes occur, possibly leading to behavior that is destructive of teamwork.

Guideline 4 stresses the independence of the simulation and experimental validation activities for a project engineering or regulatory environment, not a research code or a mathematical model-building environment. For example, experiments may be conducted for the purpose of building or improving mathematical models of physical processes, e.g., turbulent-reacting flow models or fracture dynamics models. As was pointed out in Chapter 10, experiments conducted to better model physical processes are not validation experiments, but instead are model construction or parameter updating experiments. Model construction experiments require very close cooperation and communication between the model builder and the experimentalist; sometimes they are the same person.

While validation experiments may rely on separate mathematical models to obtain experimental response measurements, those models should *not* depend closely on mathematical models that are being assessed in the validation experiment. The term *closely related models* refers to those that may be used in a data reduction code that shares many of the same physical assumptions as the mathematical model that is being validated. Measurements produced from such experiments do not adequately address the need for independence of experimental measurement results and the mathematical models in question. Experimental data that require closely related model simulations for data processing can never be a desirable outcome for a validation experiment, although this situation may arise in other kinds of experiments, such as scientific experiments for knowledge discovery (see, for example, Lindl et al., 2004).

Example: Shock Wave Physics

Consider the problem of determining the material temperature from experimentally acquired shock hydrodynamic data (density, pressure, and velocity fields) using simulations of shock wave physics rather than some type of diagnostic that directly measures the temperature. The only experimental data that will emerge from shock hydrodynamics experiments without temperature diagnostic measurements are model-inferred shock hydrodynamic data. This problem is relevant, since it arises in investigations of temperature dependence in high-pressure, shock-driven, material response. Such experiments often need simulations to estimate the associated thermal conditions under shock loading. For purposes of scientific discovery, this is permissible though speculative because it is nearly circular logic. That is, such experiments cannot be claimed to provide validation data for the high-pressure thermomechanical response of the material because of the lack of independence between modeling assumptions and experimental results.

11.1.5 Measurement of a Hierarchy of System Response Quantities

Guideline 5: *Experimental measurements should be made of a hierarchy of system response quantities, for example, from globally integrated quantities to local quantities.*

A discussion was given in Chapter 10 concerning the concept of a hierarchy of difficulty to simulate and measure various SRQs. The spectrum of difficulty was ordered in terms of integrals and derivatives of the dependent variables appearing in the mathematical model. The figure discussed in Chapter 10 is repeated here as Figure 11.2. Measurements in a

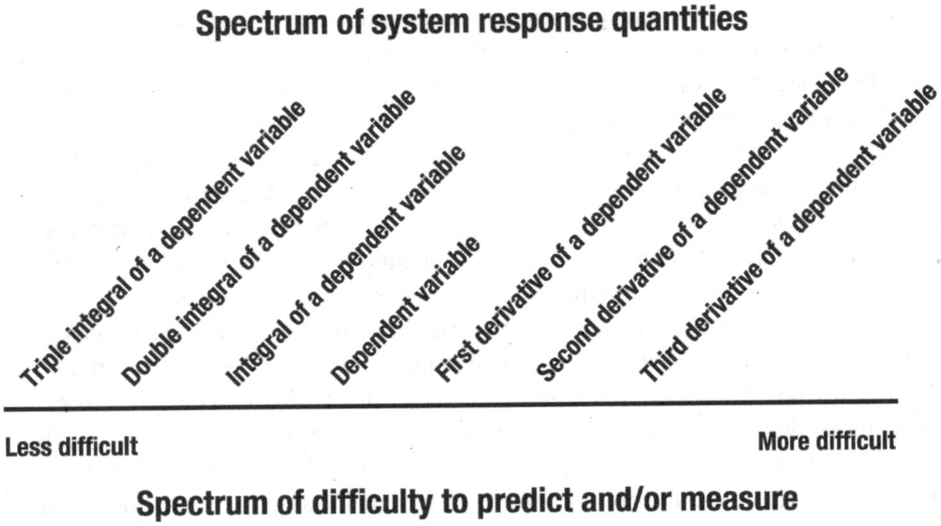

Figure 11.2 Spectrum of SRQs and the difficulty to simulate and measure each.

validation experiment should be made, if possible, over multiple levels of the hierarchy. That is, design the experiment with the strategy that measurements of the SRQs are made over two or three levels of the hierarchy. Experience has shown that a strategy of concentrating on measuring one or two SRQs at the same level can produce inconclusive and misleading model accuracy conclusions. A model with substantive predictive capability should yield robust predictions over at least two, if not more, levels of the hierarchy.

An important, practical advantage also must be noted when taking measurements over a range of difficulty. There are occasions where the model performs much more poorly than expected. If measurements are only taken at one level of difficulty, or only a few SRQs are measured at the same level, then it can be very difficult for the model builders to discover the root causes of the weaknesses in the model. In addition, multiple levels of measurement could also help determine if there are inconsistencies, flaws, or mistakes in the experimental results. On the other hand, if measurements were made over a range of difficulty and a range of different types of SRQs, then this is much more helpful to the model builders in uncovering weaknesses. This is especially important if there is a wide range of spatial and temporal scales occurring in the physics, such as multi-scale modeling, or if there is strong coupling of different types of physics.

There is another advantage in conducting measurements over multiple levels of difficulty. Suppose that measurements were only made at one level of difficulty, but it was a very demanding level for the simulation. Also, suppose that the simulation agreed well, maybe surprisingly well, with the measurements. However, sometime afterwards a similar validation experiment is conducted, but it is found that there is a surprisingly large *disagreement* between the simulation and the measurements. Further investigation shows that the good agreement on the first set of experiments was entirely fortuitous. It was discovered that cancellation of modeling errors, or the fortuitous benefit of numerical solution errors, was the reason for the good agreement, not the high-fidelity physics of the model. If measurements are made at different levels of difficulty and different types of SRQs, fortuitous, good agreement can usually be spotted. For example, suppose an observation of a very good agreement is obtained for fine spatial or temporal detail in the physics, but the model is not able to predict larger scale features of the physics. Something is seriously amiss in either the model or the experiment. The recommendation is that whether the agreement between simulation and experiment is good or bad, computation and experiment each should be subjected to critical examination.

Two pragmatic realities that must be factored into the strategy of measurements at multiple levels of difficulty follow. First, a wider range of experimental diagnostic capability must be available. This is an expense not only for the additional diagnostics and possibly modifications to the experimental facility, but also in the wider range of technical staff capabilities that must be developed and supported. Second, the experimental measurement uncertainty typically increases when moving to higher levels of difficulty in the spectrum shown in Figure 11.2. The rate of increase of measurement uncertainty can be substantial when also moving to higher levels in the spectrum, particularly if new diagnostic measurement techniques are still in their infancy.

One diagnóstic method that can be used to cover two, and possibly three, levels on the measurement hierarchy is the use of modern optical or electro-optical techniques. With the wide range in framing speeds and pixel resolutions available in digital recording cameras, many time-dependent features of a system's response over a large two- or three-dimensional field can be recorded. This breadth of spatial and temporal scales allows quantitative data to be obtained over multiple levels of the measurement hierarchy. For example, in certain types of experiments, it is very difficult or impossible, to place sensors in locations of interest, without disturbing the specimen. Some examples are: (a) flight motion of small objects such as bullets, birds, or insects; (b) impact and penetration of high-speed collisions; (c) interaction of a fluid and a very thin flexible structure; (d) in situ shock wave dynamics in both fluids and solids; and (e) various types of free surface flows. In some situations where the sensors influence the physical phenomena of interest, the physical and geometric characteristics of the sensors can be directly included in the simulations. For example, if the mass properties of the attached accelerometers influence the motion of the object of interest, then the mass and geometric properties of the accelerometers could be included in the simulation of the motion of the object.

Video and audio recording of an experiment or setup of the experiment has proven to be very helpful in detecting experimental setup problems, improving the understanding of the analysts simulating and documenting the experiment. Depending on the time scales in the experiment, one could use inexpensive standard framing rate digital AV equipment or use high-speed digital video recordings. Some examples where standard AV equipment can prove very beneficial are: (a) unexpected interactions of sensors with the phenomenon that is being sensed; (b) unexpected flowfield interactions either within the facility in which the experiment is conducted or between various elements of the geometry being tested; (c) discovering that modeling assumptions concerning BCs or ICs were inappropriate; and (d) discovering that surface characteristics or assembly of test articles were not what was intended or what was modeled. High-speed video recordings can now be coupled with automatic feature recognition algorithms to produce quantitative data for position, velocity, and acceleration. In addition, now that large AV digital records can be stored, new technologies are becoming available that can search AV records for certain classes of objects, features, or words.

Example: Fluid Dynamics Experiment

Consider a wind tunnel experiment in which a range of diagnostic sensors are available for measurement of a range of SRQs. The test article in the wind tunnel is a flight vehicle with forward-mounted canard control surfaces and aft-mounted control fins. The SRQs of primary interest for the engineering application are the forces and moments on the entire flight vehicle, as well as the forces and moments on each of the canards and fins. The following are examples of response features of increasing difficulty to simulate and to measure during the experiment:

- Forces and moments on the entire test article. These could be measured by strain gauges mounted on the support sting within the test article.
- Forces and moments on each of the canards and fins. These could be measured by strain gauges as an integral part of the shafts that hold the canards and fins.
- Location and strength of the vortices shed by the flight vehicle and the canards. These could be measured by a particle image velocimetry (PIV) technique in a laser sheet that is normal to the centerline of the test article.
- Time-averaged surface pressure on the canards and the fins. These could potentially be measured by applying pressure sensitive paint to each of the surfaces and calibrating the color changes to changes in pressure.
- Time-averaged local shear stress on the canards and fins. This could potentially be measured by applying shear-stress sensitive paint to each of the surfaces and calibrating the color changes to the changes in shear stress.

Example: Fracture Mechanics Experiment

Consider an experiment in which fracture is initiated at the end of a thin slot machined through the thickness of a metal plate. The edge of the plate that does not have the slot is firmly clamped all along the edge. On the opposite free edge of the plate, a load is applied equally to each side of the slot such that the load is normal to the plane of the slot. This creates a large tensile stress at the end of the slot, which can initiate local plastic deformation and a Mode I type fracture. The load on each of the free edges of the plate is slowly increased as a function of time. After a specified period of time, and after the initiation of a crack at the end of the slot, the load is slowly deceased to zero. Because of the large variability known to occur in fracture mechanics of materials, a large number of experiments are conducted. The following are examples of SRQs of increasing difficulty to simulate and to measure during the experiment. These SRQs can be measured during the loading and unloading of the plate, as well as after the load is removed:

- Slot width at the free edge of the plate as a function of applied load. This could be measured with a crack-mouth opening extensometer.
- Local strain and stress on both sides of the surface of the plate along the length of the slot, but far from the region of plastic deformation near the crack. This could be measured by traditional strain gauges applied to different locations on the plate.
- Strain in the region of plastic deformation near the fracture. This could be measured by video and laser extensometers viewing the sides of the plate.
- Geometric shape of the fracture. This could possibly be measured by microscopic imagining of the surfaces, X-rays, or ultrasound instruments and imaging.

11.1.6 Rigorous Estimation of Experimental Uncertainty

Guideline 6: *The experimental design should be constructed to analyze and estimate the components of both random (precision) and systematic (bias) experimental uncertainties in both the needed model input data and the system response quantities of interest.*

This guideline is recommended in any type of experimental measurement, but in validation experiments, its importance is crucial. This is critically important because the primary goal of a validation experiment is to estimate the accuracy of a mathematical model that simulates the experiment or process of interest. By contrast, in the field of metrology, the focus is on estimating the uncertainty of a measurement as it relates to the defined *true value*. Granted, these two perspectives are closely related, but the emphasis is different. In validation experiments, it is more important to quantitatively and rigorously estimate the uncertainty of all the measured quantities, both model inputs and outputs, than to infer what the true value might be based on those measurements. This is *not* suggesting that high accuracy measurements are unimportant. However, it must be stressed that accurate estimation of uncertainty of (a) model input data needed for simulation of the experiment and (b) measured SRQs, are of the utmost importance in validation experiments. Thus, in validation experiments, uncertainty is not an indication of shortcomings in the experimental techniques or lack of control in the experimental procedures, diagnostics, or the facility. Experimental uncertainty in a validation experiment is a quantitative measure of the confidence, as well as the skepticism, in the characterization of the experiment from a modeling perspective.

The standard technique for estimating experimental measurement uncertainty is the International Organization for Standardization (ISO) technique described in the *Guide to the Expression of Uncertainty in Measurement* (GUM) (ISO, 1995, 2008; JCGM, 2008). The same procedure is codified in the *US Guide to the Expression of Uncertainty in Measurement* published by the American National Standards Institute (ANSI) (ANSI, 1997). The ISO/ANSI technique has also been described in detail in various disciplines, such as wind tunnel testing (AIAA, 1999, 2003). This technique is the minimum level of effort that should be required for uncertainty estimation in validation experiments. The technique is primarily concerned with the expression of uncertainty in the measurement of a fixed quantity that is referred to as the true value. Uncertainties are not categorized according to the usual terms random and systematic, but according to the *method* which is used to characterize an uncertainty. The ISO/ANSI technique divides uncertainties into:

Type A: uncertainties that are evaluated using a statistical analysis of a collection or ensemble of observations; and

Type B: uncertainties that are evaluated by means other than a statistical analysis of a collection or ensemble of observations.

Each source of uncertainty is mathematically propagated through the sequence of steps required to obtain experimental data; sensors, transducers, data acquisition system, and digital processing steps. To be more precise, the uncertainties are propagated through a statistical model of each contributor, including correlations between contributors, to obtain the final result of an experimental measurement. The combined standard uncertainty of a quantity (the result of a number of contributing measured quantities) is the positive square root of a sum of contributors. The mathematical terms are the variances and covariances of the contributing measured quantities weighted according to how the measurement result varies with changes in these quantities. The fact that Type B uncertainties are supposed to come from *non-statistical means* does not deter the ISO/ANSI approach and practitioners from using statistical methods anyway. As with any approach, there are strengths and weaknesses for specific applications. Although it is outside the scope of this text to discuss the method in detail, see the above cited references for details, as well as Coleman and Steele (2018).

A distinctly different approach for estimating experimental uncertainty has been used by the statistics community for at least the last four decades (Box et al., 2005; Hinkelmann and Kempthorne, 2008; Montgomery, 2017; Moffat and Henk, 2021). A sampling approach is used that is based on statistically comparing multiple measurements of the final engineering quantities of interest. There is *no* statistical propagation model of any of the contributors that are involved in obtaining an experimental measurement. As a result, it is more robust than the ISO/ANSI approach and is able to identify uncertainty sources that are commonly associated with systematic or Type B uncertainties. The approach statistically analyzes the final experimental measurements that were obtained, not intermediate contributors to the final result. It is usually referred to as statistical design of experiments (DOE) because it analyzes the final measured result of the quantities of interest, based on specially designed sampling techniques. DOE is widely used in sampling both computational simulation results and experimental measurements.

In the DOE approach, various classes of sources that are believed to contribute to the uncertainty in the final quantities of interest are identified. Experimental strategies are designed to quantify the statistical contribution that each of these sources contributes to the final quantities of interest. The strategy of DOE is entirely different from the ISO/ANSI approach mentioned above. Specifically, DOE attempts to broaden the sample population as much as possible to identify sources of uncertainty that are latent and difficult to isolate. DOE uses replication, randomization, and blocking techniques in the design of the sample data collected in an experiment. This approach has been widely used in analyzing data from many fields, including production process control, system and component reliability, environmental statistics, biostatistics, medical trials, epidemiology, etc. However, this approach has seen limited use in experiments in engineering and physics. Because of the power of this approach, it was used in the following detailed discussion of a validation experiment example.

For easy reference, the six guidelines discussed in this section are stated in Table 11.1.

11.2 Example: Joint Computational-Experimental Aerodynamics Program (JCEAP)

In 1990, a research team at Sandia National Laboratories initiated a long-term, jointly-funded CFD-experimental research effort, referred to as the Joint Computational-Experimental Aerodynamics Program (JCEAP). The goal of the program was twofold: (a) improve Sandia's hypersonic wind tunnel experimental capabilities and CFD simulation capabilities; and (b) improve the understanding of how simulation and experiment could synergistically work together for a more rapid and efficient development of each. The research program came to a successful close in 1997 and a number of reports, conference

Table 11.1 *General guidelines for model validation experiments*

Number	Guideline
1	*A validation experiment should be jointly designed by experimentalists, model developers, code developers, and code users working closely together throughout the program, from inception to documentation, with complete candor about the strengths and weaknesses of the simulation and the experiment.*
2	*A validation experiment should be designed to capture the essential physics of interest; measure all relevant physical modeling input data, and initial and boundary conditions; and system excitation information required by the simulation.*
3	*A validation experiment should strive to emphasize the inherent synergism that is attainable between simulation and experimental approaches.*
4	*Although the experimental design and sharing of model input data should be developed cooperatively, independence should be maintained in obtaining the simulation and experimental system response results.*
5	*Experimental measurements should be made of a hierarchy of system response quantities, for example, from globally integrated quantities to local quantities.*
6	*The experimental design should be constructed to analyze and estimate the components of both random (precision) and systematic (bias) experimental uncertainties in both the needed model input data and the system response quantities of interest.*

Source: Aeschliman and Oberkampf, 1998;Oberkampf and Trucano, 2002.

papers, and journal articles were produced describing the effort (Oberkampf and Aeschliman, 1992; Walker and Oberkampf, 1992; Oberkampf et al., 1993; Aeschliman et al., 1994; Oberkampf et al., 1995; Aeschliman et al., 1995; Oberkampf et al., 1996; Aeschliman and Oberkampf, 1998; Oberkampf and Blottner, 1998). The program produced an extensive model validation database of force and moment, and surface pressure measurements on a hypersonic vehicle configuration. More importantly, the program developed and expanded upon the fundamental concepts for the design and execution of high-quality validation experiments. The six guidelines discussed earlier evolved from the experiences of JCEAP as well as those of previous experimental researchers interested in model validation. Some of these guidelines were learned the hard way, i.e., they were learned after poor or ineffective procedures were employed. The following discussion concentrates on exemplifying the six guidelines previously discussed. See the references cited for details of the experiment, the model validation database, and the experimental uncertainty quantification procedure.

11.2.1 Joint Planning and Design of the Experiment

The first phase of the program involved the computational and experimental study of the aerodynamic forces and moments on a flight vehicle configuration in a hypersonic flowfield. During this phase, the geometry of the vehicle was refined, the force and moment (F&M) wind tunnel test article was fabricated, the experimental conditions were determined, and the F&M data were obtained. The second phase of the program addressed the next level of difficulty in simulation and experimentation: prediction and measurement of surface pressures on the test article. An identically sized and shaped geometry was fabricated and tested at the same flow conditions as the F&M data.

Figure 11.3 Sandia hypersonic wind tunnel with the aft section retracted from the test section (flow is from left to right). The test article is visible in the open section and the yellow support is part of the rear-mounting fixture to hold the test article.
(Oberkampf et al., 1993)

All of the experiments were conducted in the Sandia National Laboratories long-duration, blowdown, hypersonic wind tunnel (HWT). The tunnel is a blowdown-to-vacuum facility that can operate up to a minute, depending on operating conditions. The nozzles and test sections consist of three contoured axisymmetric nozzles, arranged as a large fixture that can revolve around a common hub. Figure 11.3 shows an overall view of the wind tunnel with the test section open. Depending on which nozzle is used, the test section Mach number is nominally 5, 8, or 14. The diameter of the circular test section of each nozzle is roughly the same, about 0.350 m. Each nozzle is provided with its own electric resistance heater to prevent flow condensation for the low static temperatures that occur in the test section. The total pressure and temperature of the flow is controlled during a run so that a chosen Reynolds number can be obtained within the operating envelope of the facility.

During an experimental run, the total pressure is manually controlled, but all other operating and test article conditions are computer controlled. For these experiments, the run time was about 60 s, during which time the large vacuum spheres downstream of the diffuser section steadily increase in pressure. The turnaround time between runs is typically one hour. Test article pitch angle (angle of attack) is varied during a run using a computer-controlled, arc sector drive, which is located in the diffuser section. Test article roll angle and configuration changes are made between runs by retracting the diffuser section and manually making the needed changes to the test article.

11.2.1.1 Wind Tunnel Conditions

Determining what Mach number condition should be used was the first design decision that needed to be made. In order to achieve the goals of the validation experiment, the primary flow requirement over the test article needed to be entirely laminar flow. This was a very unusual test condition because the normal desire is to have turbulent flow over the vehicle, similar to that of flight in the atmosphere. Laminar flow was the critical driver in the design of the experiment because with laminar flow, it was conceivable that the CFD simulations could *possibly* compute some flow conditions more accurately than the experimental measurements that could be made. With turbulent flow, no such claim could be made, and this is still true three decades after the initiation of JCEAP.

The Reynolds number of the flow over the vehicle is the primary factor in determining if the flow is laminar or turbulent over a vehicle. The Reynolds number is a dimensionless ratio defined as $R_L = \rho VL/\mu$, where ρ is the free-stream density, V is the free-stream velocity, L is a characteristic length scale (usually the total length of the vehicle), and μ is the absolute fluid viscosity in the free-stream. The lower the Reynolds number, the more likely the flow will remain laminar over the entire vehicle. As the Reynolds number increases, the flow will transition to turbulent flow over different portions of the vehicle.

The operating conditions of the HWT facility for the Mach 5 nozzle were examined, given a reasonable estimate on the length of the test article that could be used in the test section. It was found that at the lowest Reynolds number conditions that were producible for Mach 5, it was unavoidable that a mixture of laminar and turbulent flow would occur over the test article. For the Mach 14 nozzle, the range of Reynolds numbers attainable is much lower because of the very low flow density in the test section. As a result, laminar flow could be assured over the test article at any operating condition of the facility. The primary argument against using the Mach 14 nozzle was that the pressures were so low that very accurate measurement of surface pressure would become questionable.

For the range of total pressure and total temperature operating conditions available for the Mach 8 nozzle, Reynolds numbers could be produced that balanced the needs of the experiment. Operating at the lower end of the facility operating envelope would produce the all-laminar flow desired and the surface pressures on the test article were sufficiently large for accurate measurement. For the Mach 8 conditions, dry nitrogen is used for the working fluid. The test section is provided with 0.2×0.38 m Schlieren-grade windows on the top, bottom, and sides. The high-grade windows provide excellent access for optical diagnostics.

11.2.1.2 Test Article Geometry

To achieve the goals of the validation experiment, a geometry was required that could produce a range of flow features. The simplest flowfield would have the possibility of being simulated more accurately than it could be measured. This could possibly be attained with a simple geometry with attached flow at low angles of attack. Complex flowfields could be attained by having an adjustable geometry and using higher angles of attack.

After a number of design iterations for the geometry using CFD, the team settled on a spherically blunted cone with a half-angle of 10° with a sliced region on the aft portion of the cone. A sharp-nosed vehicle was not chosen for three reasons. First, no matter how sharp a machinist fabricates a nose tip, there is always finite nose bluntness. If a sharp tip were used, then a very precise inspection would be required to accurately measure the geometry for the CFD simulations. Second, a very small radius nose tip would change the flow regime from continuum fluid dynamics that exists over the rest of the geometry to rarified flow over the minute nose tip. Stated differently, one of the key assumptions in the CFD model would be violated, damaging the high confidence that is desired for certain flowfield solutions. Third, a sharp tip can be easily damaged over a long period of usage in a wind tunnel test. The damage could be from slight erosion due to impact from dust particles in the free-stream or due to accidental damage during test article installation and modifications. As a result, the nose tip would have to be re-inspected regularly during the course of the experiment to determine if any changes occurred. As a result, a 10% nose bluntness was chosen. Hence, the nose had a radius equal to 0.1 of the base radius of the cone.

Based on CFD simulations, the team decided to begin the slice, which was parallel to the axis of the cone, at 0.7 of the length of the vehicle. See Figure 11.4 for the final vehicle design. Given that the test section diameter of the Mach 8 nozzle was 0.350 m, a base diameter of 0.1016 m was chosen. With this base diameter and the 10° cone half-angle, the resulting length of the test article was 0.2639 m. To increase the difficulty of the CFD simulation for certain geometries, and to make the geometry more appealing to the flight test project group, the test article was designed so that different flaps could be attached to the aft portion of the slice. Flap deflection angles of 0, 10, 20, and 30° were chosen and all of the flaps extended to the baseplane. By designing each flap to extend to the test article baseplane, a substantial simplification became possible in constructing the computational mesh for the body geometry and for the base flow. This also simplifies setting the outflow boundary conditions across the baseplane in the CFD simulation. Note that in an actual flight vehicle, this would not be the case. In a flight vehicle, the flap is hinged near the leading edge of the flap, and as the flap is deflected, the trailing edge of the flap does not extend to the baseplane of the vehicle.

After the vehicle geometry was finalized, the team discussed the geometry of JCEAP with the flight test project group because they were interested in the maneuvering re-entry vehicle geometry and its potential advantages from their perspective. However, when the team explained to the flight test project group that there was only interest in laminar flow, the project group lost interest. As a result, funding for the project was not provided. Fortunately, there was strong support for the project from internal research and development (IR&D) funding. As was briefly discussed in Section 10.1.2, there is commonly conflict between project needs for a validation experiment and the needs for a high-quality validation

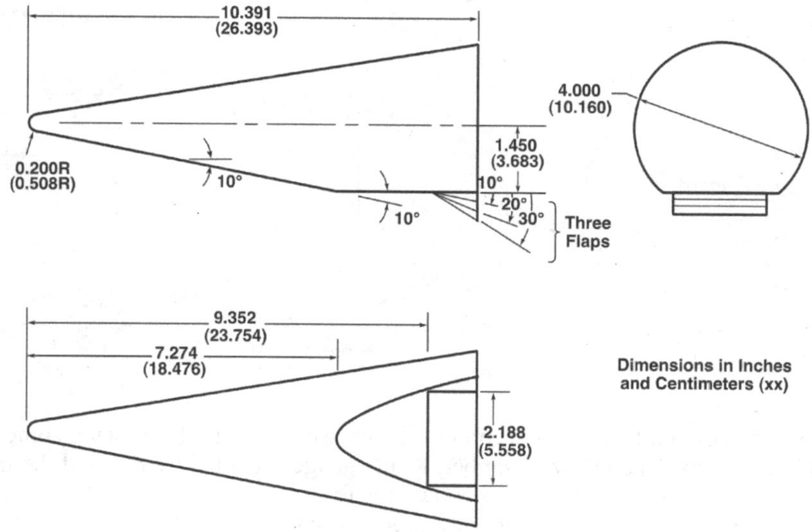

Figure 11.4 JCEAP geometry.
(Aeschliman and Oberkampf, 1998)

experiment. If funding for the experiment does come from the project group, their requirements usually take precedence, and therefore significant compromises must be made from the validation perspective.

11.2.1.3 Test Article Fabrication and Instrumentation

Two different physical test articles were fabricated, but both had the same geometry (Figure 11.4). The F&M test article was used in conjunction with a precision six-component strain gauge balance. The balance was installed internal to the test article geometry, as is typically done to improve the accuracy of the measurements of the aerodynamically induced forces and moments. The pressure test article was significantly more complicated in design and fabrication because of the additional instrumentation required. A total of 96 pressure ports were machined in the test article surface. The orifices were connected to either the 2,482 N/m^2 or the 6,895 N/m^2 electronically scanned pressure module, both of which were located inside the test article. CFD predictions of the pressure level expected determined which ports were attached to which module. The test article also incorporated nine semiconductor bridge Kulite gauges to detect high-frequency surface pressure fluctuations at certain locations along the surface. The test article incorporated four thermocouples mounted on the inside surface of the test article wall at two different axial locations. The thermocouples provided test article surface temperature for input boundary conditions required for the final CFD calculations.

The pressure test article was constructed in four sections (plus detachable nose tips) to permit machining operations and the installation of steel pressure port inserts during fabrication. Figure 11.5 shows a longitudinal cross section of the test article, the adjacent portion of the base cover and sting, and the sting cover. This design greatly facilitated tubing connections between ports and internally mounted pressure modules during subsequent test

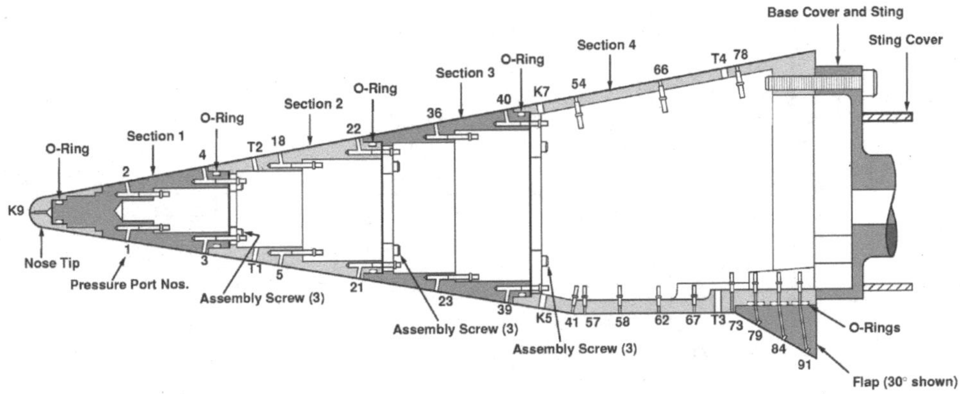

Figure 11.5 Longitudinal cross section of the pressure test article showing some of the surface pressure ports denoted by a number, Kulite gauges denoted by K*n*, and thermocouples denoted by T*n*.
(Oberkampf et al., 1995)

article assembly. A base cover extension, slightly smaller in diameter than the base diameter of the test article, was fabricated as part of the support sting and provided the attachment point to the test article. It also provided a hollow recess with additional volume for the pressure modules, pressure tubing, and electrical wiring. The base cover extension had a flat section on one side to conform to the slice on the test article. The test article was supported in the tunnel by a stiff, thick-walled sting to reduce the deflection of the test article due to aerodynamic loading during a run. The sting connects the test article to the arc sector drive, located in the diffuser section, which pitches the test article to the angle of attack. The sting length is designed so that the center of rotation of the arc sector is at about the mid-length of the test article. In this way, when the test article is moved to a non-zero angle of attack, it remains in the center portion of the test section.

In preliminary tests of the pressure test article, an amalgamation of pressure reference tubing, pressure module electrical cables, wires from the Kulite gauges, and thermocouple wires were taped to the outside of the sting. All of these tubes and cables needed to go from the base of the test article to the diffuser section of the tunnel where they could exit the tunnel. In a project review of the preliminary JCEAP tests, a reviewer noted that this taped bundle of tubing and cables would cause a complex surface geometry for modeling in a CFD analysis. If a CFD analyst chose to simulate the flow in the base region of the test article, along the sting support, then the CFD analysis would need to know the surface geometry of the taped bundle. After this issue was recognized, this complex geometry feature was eliminated because it was of *no interest* to the goals of the validation experiment. A cylindrical section was designed and fabricated to cover the bundle and sting, as shown in Figure 11.5. The cover provided an easily definable and repeatable geometry that could be much more easily meshed in a CFD solution. This simplification in base flowfield geometry exemplifies the model validation design principle: exclude any complex geometric features that *do not* provide value to the goals of the validation experiment.

The overriding consideration in the development of the surface pressure measurement system was the pressure lag time (commonly referred to as settling time) in the small

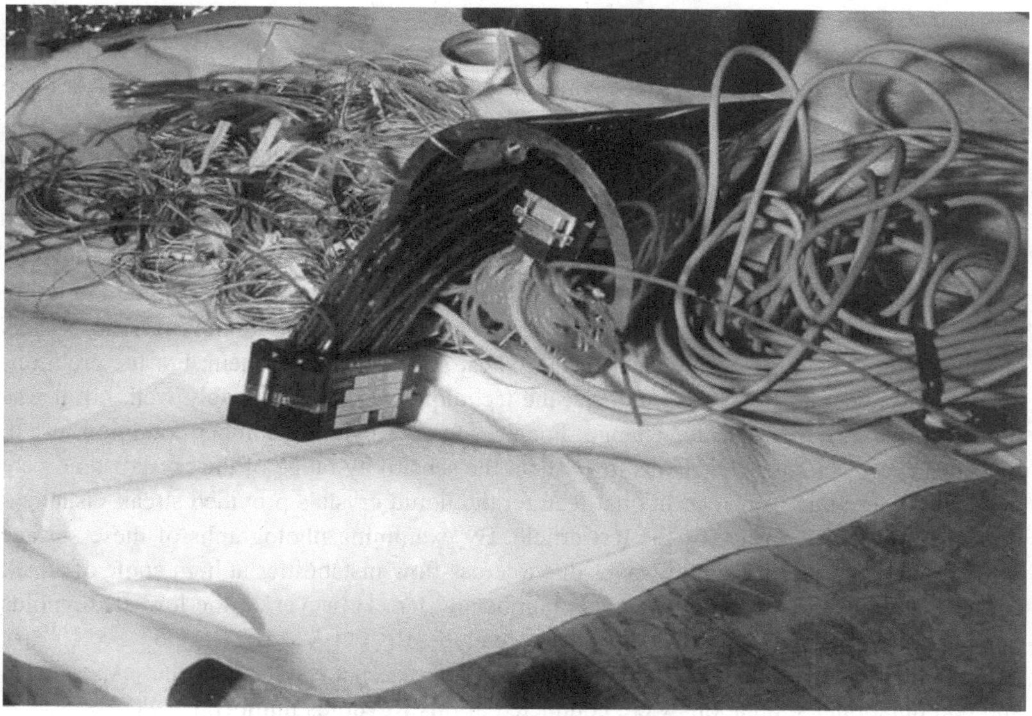

Figure 11.6 Assembly of the pressure test article.
(Oberkampf et al., 1995)

diameter tubing connecting the test article surface pressure ports to the two pressure transducer modules. If a pressure lag error existed, this could introduce a significant systematic uncertainty in the surface pressure measurements. To determine the actual lag characteristic of each port for the tunnel flow conditions of interest, the test article was pitched from a 0° angle of attack to a 10° angle of attack in one continuous sweep, while pressure data were recorded for each of the 96 pressure ports. The same type of sweep was made from 0° to −10° angle of attack. Typical pressure lags were 0.05 – 0.10 s to achieve a stable pressure within 0.1% of the final value. The largest lag time observed was 0.3 s, which occurred at one of the ports on the slice. To be conservative, a delay time of 0.5 s was used prior to recording pressure data following each change to a new angle of attack during data acquisition runs. The key to achieving this response frequency was mounting both pressure modules and the tubing, along with everything else inside the test article. Figure 11.6 suggests the challenge of stuffing everything inside the test article during assembly, without kinking any of the pressure tubing or breaking any electrical connections.

11.2.2 Measurement of the Input Data Needed for the Simulation

After each test article was fabricated, a detailed inspection of the surface geometry of the test article was conducted. Although the test article was precision made, the actual inspected dimensions of the test article were used in future CFD simulations. The location of the pressure orifices in the test article, the Kulite gauges, and the thermocouples were all inspected with regard to the quality of the sensor installation and the precise location of each. As stressed in Guideline 2, the geometry data that should be used in the simulation are the *as tested* geometry. For the F&M test article and the pressure test article, they were built with very thick walls so that there was no deflection due to aerodynamic loads. In addition, for the total temperature conditions available in the HWT, there was no change in geometry due to aerothermal heating and distortion of the test article.

To confirm that laminar flow over the entire test article existed, including at high angles of attack, shear-stress–sensitive liquid crystals were employed. With this technique, the liquid crystals are painted over the surface of the test article before a wind tunnel run. The crystals change color, depending on the local surface shear stress applied to them. For the preliminary tests at moderate Reynolds numbers, the technique worked reasonably well, but it was not definitive in terms of color change. For the lowest Reynolds number condition tested, however, the shear stress level was lower than the sensitivity range of the crystals that were available at the time. As an additional feature, the liquid crystals provided streak visualization of flow on the surface of the test article. By examining photographs of these surface flows, boundary-layer transition waves due to cross-flow instabilities at high angle of attack could be seen at the moderate Reynolds numbers tested. However, at the lowest Reynolds number condition attainable in the operating envelope, $R_L = 1.8 \times 10^6$ ($R = 6.82 \times 10^6$/m), no transition waves could be seen at any angle of attack. As a result, all further experiments and computational simulations were conducted at this Reynolds number.

The wind tunnel flowfield had been calibrated on multiple occasions not only for the Reynolds number and total temperature of interest in JCEAP, but also for many other conditions in the operating envelope. Based on these calibrations, the spatially averaged free-stream Mach number across the central region of the test section was found to be 7.84. As in all wind tunnels, there is a slight variation in free-stream flow characteristics across the central portion of the test section. The spatially averaged conditions from the wind tunnel calibration were used as the inflow boundary conditions in the CFD simulations. The team noted that the only time this Reynolds number and Mach number condition had been exercised before was during calibration of the wind tunnel. This exemplifies the common situation that experimental facilities are commonly used at unusual conditions for validation experiments, relative to the needs of project groups.

The run schedule for the combinations of angle of attack, roll angle, and flap deflection angle was devised very differently from typical wind tunnel tests for project groups. For project groups, the goal is to obtain information at a wide range of operating conditions and test article geometries that are of interest for the system's operational performance. For validation experiments, the goal is to obtain information at a restricted set of conditions so that test time can be used to rigorously estimate experimental uncertainty for all input quantities needed for the simulation, as well as all SRQs that are measured. For JCEAP, the run schedule was constructed so that a large number of replicate samples could be obtained for use in statistical DOE procedure.

During each wind tunnel run the angle of attack was varied from −9 to 18°. The sequence of nominal angle of attack, α, during a run was 0, −9, −6, −3, 0, 3, 6, 9, 12, 15, 18, and 0°. For each of these angles of attack, the roll angle was set at 0 (slice on the windward side), 90, 180, or 270° during different portions of the run schedule. In addition, measurements were

Table 11.2 *Run number schedule for the F&M experiment*

Test article at forward tunnel station				
Roll angle (deg.)	$\delta = 0°$	$\delta = 10°$	$\delta = 20°$	$\delta = 30°$
0	34, 36, 37, 73	63, 72	64, 71	65, 70
90	39	66, 67	69	68
180	40	55	56, 57	58
270	41	62	61	59, 60
Test article at aft tunnel station				
0	74, 75	83	82	81
180	76, 77	78	79	80

Source: Oberkampf and Aeschliman, 1992.

Table 11.3 *Run number schedule for the pressure experiment*

Test article at forward tunnel station				
Roll angle (deg.)	$\delta = 0°$	$\delta = 10°$	$\delta = 20°$	$\delta = 30°$
0	34, 36, 37, 73	63, 72	64, 71	65, 70
90	39	66, 67	69	68
180	40	55	56, 57	58
270	41	62	61	59, 60
Test article at aft tunnel station				
0	74, 75	83	82	81
180	76, 77	78	79	80

Source: Oberkampf et al., 1995.

made with the test article at two different axial locations in the test section by changing the length of the test article sting. The forward axial location places the test article essentially at the center of rotation of the arc sector drive. As a result, the test article stays near the center region of the test section as the angle of attack is changed. At the aft axial station, the test article rotates and translates downward in the test section for positive αs; conversely, it rotates and translates upward for negative αs. Tables 11.2 and 11.3 give the complete run schedule for the F&M experiment and the pressure experiment, respectively.

During the F&M experiment, the aerodynamic load on the test article was measured. This measured load was then used to calibrate the change in the angle of attack due to deflection of the mounting sting. While the tunnel was open, the test article was loaded over the same range of aerodynamic loads and the change in angle of attack was measured. This angular deflection was then used to correct the indicated angle of attack from the arc sector mechanism. For the pressure experiment, the same procedure was used, except because of the stiff sting, the deflections were much smaller than the sting used in the F&M experiment.

Finally, thermocouples were used to measure the surface temperature of the test article during a run. The largest differences in the surface temperatures occur between the windward and leeward sides of the test article at angle of attack. Consequently, the thermocouple measurements were made 180° apart around the test article. The measured temperature was then used as the surface temperature boundary condition in the CFD simulations.

11.2.3 Synergism between Simulation and Experiment

Examples of the synergistic interaction of simulation and experiment were presented earlier regarding the design of the JCEAP experiment. This section includes a more detailed discussion of an interaction that could not have happened without the real-time interaction of the simulation and experimental team members. Thus, if the simulation and experimental team members are diverted from a validation experiment for significant portions of their time, or worse, they do not even work on the effort at the same time, the extraordinary synergism that is possible will not be realized.

After the F&M wind tunnel tests were completed, a detailed experimental uncertainty analysis was conducted. This analysis revealed an anomaly that was most notable in the experimental data for the pitch moment at zero angle of attack for a flap deflection of 0°. The anomaly could also be seen for other flap angles for zero angle of attack, but it was not as obvious. Based on the assumption of a uniform free-stream flowfield, it must be true that at zero angle of attack, the pitch moment for $\phi = 0°$ and $\phi = 180°$ should be the same magnitude for the mean value (based on many statistical samples). It was observed that the difference in pitch moment coefficient between $\phi = 0°$ and 180° was 0.008. Although this may seem small, if this error is converted into an error in measuring the angle of attack, it would convert to an error of 0.6°. This was much larger than what the measurement error in the angle of attack was demonstrated to be.

To resolve this issue, several possible sources of systematic error were investigated. An analysis was conducted to determine if the nonuniform spatial distribution of base pressure could contribute to the pitch moment. Based on five pressure measurements distributed over the base of the test article, this effect was found to be negligible. Another possibility investigated was the misalignment of the strain gauge mounting taper with respect to the test article centerline, which would result in an angle of attack error. Precision test article inspection showed misalignment to be less than one minute; therefore, this was not a significant contribution to the discrepancy.

In parallel with the experimental investigation, a CFD effort was initiated to try and assist with understanding the anomaly. A number of additional CFD solutions were generated with very refined meshes for the zero angle of attack case. These solutions had the highest confidence in simulation accuracy because they were for $\alpha = 0°$ and $\delta = 0°$, where the flowfield is the simplest case of the entire experiment. The computed results for the pitch moment produced a value that did not match experimental results that were in question. The computational results were halfway between each of the experimental results for $\phi = 0°$ and 180°, respectively. However, different members of the team had different levels of confidence in the accuracy of the CFD results.

The experimentalist also suggested that this anomaly could be caused by operating the wind tunnel at a Reynolds number much lower than the design Reynolds number of the nozzle. The nozzle contour included the boundary layer displacement thickness corresponding to operating the tunnel at a Reynolds number of $R = 19.7 \times 10^6$/m. At the low Reynolds number used in JCEAP, the displacement thickness would be larger and consequently, this could produce slight flow nonuniformity in the test section. Therefore, the team decided to conduct another tunnel entry to determine if the anomaly at the JCEAP Reynolds number could be reproduced and also to determine if the anomaly existed at the design Reynolds number of the nozzle. During this second tunnel entry at the JCEAP Reynolds number, the pitch moment anomaly could *not* be reproduced. The mean value of the zero angle of attack pitch moment coefficient for $\phi = 0^\circ$ and $\phi = 180^\circ$ matched in magnitude within 0.001, which was the expected level of uncertainty in the data. Hence, a more intensive investigation to discover what had changed since the previous tunnel runs began.

Some team members suggested that the test section of the tunnel might have moved slightly from the horizontal since the last tunnel entry. This was possible because all of the different Mach number test sections are on a large revolver-type mechanism. The test section was moved, reassembled, and checked several times. It was always found to be horizontal within one minute of angle. However, during the examination of angles, it was found that the test article pitched nose down when the tunnel was closed and locked in place using hydraulic cylinders. Repeated closures and measurements showed this movement to be a constant -0.08°. This consistent systematic error due to tunnel closure had, to the team's knowledge, existed in the facility since it was built. This constant -0.08° error correction in the angle of attack was made to all of the data. The correction, however, was not sufficient in magnitude to explain the anomaly previously identified.

After the second tunnel entry, it was noted that the attempt at reproducing the anomaly did *not* include the base pressure instrumentation. When the base pressure transducers were reinstalled and the pitch moment remeasured, the anomaly *was* precisely reproduced. Each of the five base pressures was connected to a transducer that had a full scale of 0.007 atmospheres; therefore, these transducers were grossly off-scale when the tunnel was opened to atmospheric pressure before and after a run. When all the strain gauge instrumentation was connected into the data collection system and the base pressure transducers were off-scale in response to atmospheric pressure, those analog data channels were saturated. This introduced a systematic error to all data channels through the common signal return line. This systematic error was easily circumvented and eliminated from the data reduction procedure by using the second zero of the F&M output, which was taken *after* the test section and diffuser were evacuated just before a run was initiated.

Two final observations should be reviewed. First, this systematic error due to an electrical ground loop had existed in the past any time base pressures were measured on a test article. While this could not be considered as a major experimental error, the error was definitely detectable by employing synergism between simulation and experiment. These kinds of errors are rarely reported anywhere, but good organizations learn from them and change experimental procedures. Second, the CFD results were exactly correct in predicting the final measured result after the ground loop bias error was removed from the data. Members of the team who had not believed CFD could calculate some flowfield cases more accurately than they could be measured were stunned. They became believers in the power of synergism between simulation and experimental activities.

11.2.4 Independence and Dependence between Simulation and Experiment

The relatively simple physics involved in modeling for JCEAP lent itself to natural independence between simulation and experiment. That is, for laminar flow of an ideal gas, single phase, non-reacting flow, there are no other significant modeling issues remaining. Some of the fine detail assumptions in the submodels could be challenged, but these were not expected to produce significant changes in the simulation results for the flow that was investigated in JCEAP. For complex flows, or any type complex physics, the value of a blind prediction for an experiment cannot be overstated.

As discussed in Guideline 2, the simulation must be provided with the measured input data that characterizes the actual experiment that was conducted. For the test article geometry, these data were measured before the experiment was conducted because there were no changes or flow interactions during the execution of the experiment. The actual operating conditions achieved during the experiment must be measured, recorded, and provided to the CFD analyst. Since the CFD simulations conducted were deterministic simulations, the average value for the operating conditions, including the measured test article wall temperature, were provided to the CFD analyst for the entire run schedule. Since these input data were very similar to what had been assumed in the initial simulations, there was very little change in the simulation results.

11.2.5 Measurement of a Hierarchy of System Response Quantities

The JCEAP experiment was designed to span two levels of the spectrum of difficulty discussed earlier in this chapter (see Figure 11.2). Measurements and predictions were made of a dependent variable in the PDEs (pressure), as well as measurements and predictions of integrated quantities (F&Ms). In addition to these measurements, liquid crystal and Schlieren photographs were taken. The liquid crystal streak images not only provided definitive information in the determination of laminar flow conditions, but they were also very valuable in surface flow visualization.

Figure 11.7 is an example of one of these visualizations of the streak pattern on the surface of the geometry. Figure 11.7 is for $\delta = 0^\circ$ and a flap deflection of 10°. The deflected flap causes a shock wave to form just ahead of the flap, generating an adverse pressure gradient along the slice portion of the geometry. The laminar boundary layer cannot tolerate the adverse pressure gradient, so the boundary layer separates in the slice region. The flow reattaches midway up the length of the flap, as can be seen by the dark gray region on the flap. Near the edges of the flap, the reattachment point moves toward the front of the flap. As a result, a flow separation bubble is formed midway along the slice to roughly midway along the flap. Visualization photographs, such as these, can provide excellent qualitative information for (a) assessing the accuracy of the simulation and (b) assisting in understanding *why* a simulation may have performed poorly.

Schlieren photographs of the flow show the shadow pattern due to the deflection of light by a refractive index gradient through the flowfield. Schlieren photographs were taken at every angle of attack on every wind tunnel run. These photographs were helpful and instructive in understanding the flowfield, particularly the shock waves in the flowfield.

At one point during the F&M testing, it was noticed that for certain test article geometries at large angles of attack (15° to 18°), the normal force and pitch moment measurements deviated from the expected trend with δ. The situations where this trend was particularly evident were large flap deflections for $\phi = 0^\circ$, large angles of attack, and when the test article was at the aft tunnel station. The team suspected that wall interference could be causing the anomaly in the F&M measurements, so the Schlieren photographs were studied during the experiment. In addition, a high-speed Schlieren video was taken of these suspect situations.

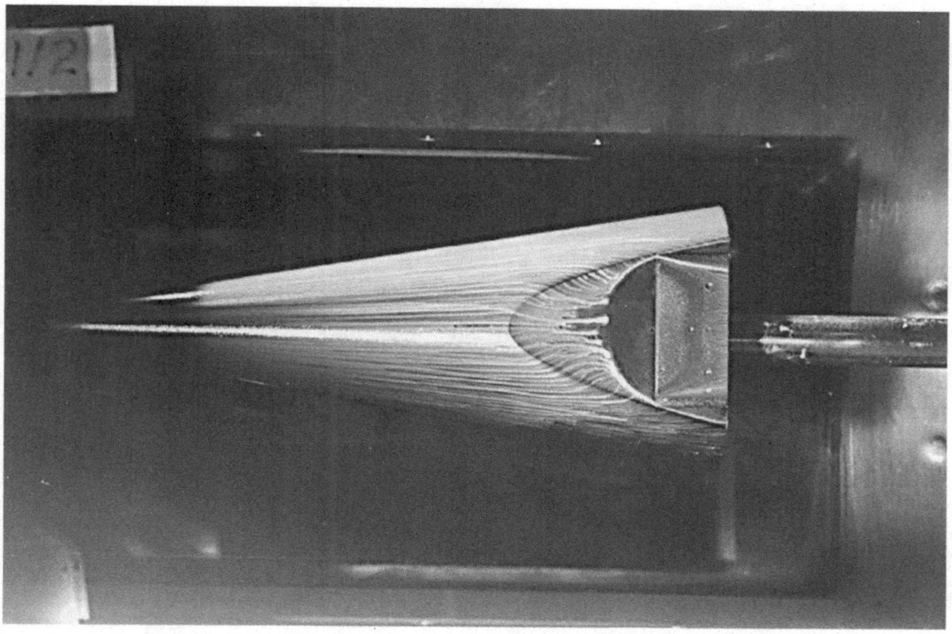

Figure 11.7 Surface flow visualization on JCEAP for $\alpha = 0°$ and $\delta = 10°$.
(Oberkampf and Aeschliman, 1992)

Figure 11.8 is a Schlieren photograph of the test article at $\alpha = 17.9°$, $\phi = 0°$, $\delta = 20°$ at the aft tunnel station (Run 82). This figure graphically shows upstream interference in the form of a shock wave from the lower wall of the test section impinging on the underside of the test article.

After various analyses and tests, including use of the high-speed Schlieren video, the team concluded that the flow mechanism for wall interference upstream of the test article was by way of the window cavity in the wall of the test section. The window cavity is the volume formed between the inside surface of the flat windows and the circular cross section of the wind tunnel test section. At a large angle of attack and large flap deflection, the strong shock wave from the flap merged with the strong bow shock wave and impinged on the window cavity. The high pressure from the shock wave was fed forward in the subsonic separated flow region of the window cavity. In turn, this increased the angle of the free shear layer over the open cavity, which caused a shock wave to be generated at the *front* of the cavity. Since the front of the window cavity is upstream of the test article, this shock wave impinged on the bottom of the test article, as seen in Figure 11.8. This phenomenon was exacerbated at aft tunnel stations because the test article was significantly aft of the center of rotation of the arc sector mechanism. This caused the test article to translate farther downward as the angle of attack increased, bringing it closer to the window cavity.

Methods were investigated to determine at what angle of attack wall interference from the window cavity began to affect the F&M data for any configuration and tunnel station. Using the Schlieren photographs for each angle of attack, the team was able to identify most of the cases when interference occurred. The most sensitive measure was found to be the trend of base pressure with angle of attack. When the wall interference shock wave impinged on the

Figure 11.8 Schlieren photograph showing the shock wave from the window cavity impinging on the lower side of the test article.
(Oberkampf et al., 1993)

separated flow at the base of the test article, the base pressure would begin to rise uncharac teristically. The base pressure for each orifice was plotted versus angle of attack for every wind tunnel run. When the base pressure began to abruptly increase with angle of attack, the conclusion was that the reflected bow shock wave was near the base of the test article. Although this, in itself, does not contaminate the data because the base pressure was eliminated from the data regardless of its value, the team decided not to use any F&M data for the higher angle of attack for that particular run. Once this wall interference was recognized, the windows were removed, and metal blanks were inserted that conformed to the circular inner surface of the test section. This removed any feature of the window with regard to the flow inside the test section. For the pressure experiment, the metal window blanks were installed for the entire experiment.

11.3 Introduction to Statistical Design of Experiments

Techniques for estimating experimental measurement uncertainty can be grouped into (a) the ISO/ANSI method (GUM) described in ISO (1995, 2008); JCGM (2008), and ANSI (1997), (b) recently developed non-Gaussian error estimation procedures (Grabe, 2014; Salicone and Prioli, 2018), and (c) the statistical DOE method (Box et al., 2005; Hinkelmann and

Kempthorne, 2008; Montgomery, 2017; Moffat and Henk, 2021). Each method has strengths and weaknesses and situations where one has advantages over the other. Given the centuries that measurement uncertainty has been studied, many think that the foundational issues have been established. This thinking is far from accurate. There has been an increasing international debate within the metrology community for at least the last three decades concerning fundamental terminology and concepts. Although the ISO/ANSI technique is internationally sanctioned and is the most commonly used in experimental activities, there are a number of well-founded criticisms of the concepts and procedures.

A series of recent monographs has captured most of the criticisms and weaknesses of the ISO/ANSI approach (Drosg, 2009; Grabe, 2014; Rabinovich, 2017; Salicone and Prioli, 2018). The metrology community, like any large technical community, is slowly starting to take notice. But like every entrenched belief system, change only comes "*one funeral at a time.*" – Max Planck. These references recommend more flexible theoretical approaches, such as evidence theory, to better estimate experimental uncertainty than traditional probabilistic methods, that are the basis of the ISO/ANSI approach. Evidence theory, and more generally all imprecise probability theories, allow a distinction to be made between indeterminism and imprecision. Indeterminism results from a random process, whereas imprecision results from lack of knowledge. Although the present authors fully support these new directions in experimental uncertainty estimation, they are beyond the scope of this book. The goal here is to introduce the basic sources of measurement uncertainty and how these are addressed using the DOE method.

11.3.1 Random and Systematic Uncertainties

As discussed in Chapter 2 "Fundamental Concepts and Terminology," this text uses the following definition for error:

Error in a quantity: deviation from the true value of the quantity.

Following the discussion given in Section 2.2,

$$\varepsilon_m = y_m - y_T, \tag{11.1}$$

where ε_m is the error in the measured quantity y, y_m is the measured value of the quantity, and y_T is the true value of the measured quantity. Here it is assumed that y_T is a fixed scalar quantity for a given set of experimental conditions. For a measured quantity, y_T is *never* known; it is only estimated (unless a reference standard is defined to be the true value). Similarly, ε_m is *never* known for a measured quantity, only estimated. Any estimate of ε_m is referred to as an *uncertainty estimate*, as opposed to an error estimate because the true value is never known. Let an estimation of ε_m be given as ε. The estimation ε is always considered a positive quantity, since the sign of ε_m cannot be known. Whether the estimate ε is accurate or not does not change the fundamental concepts or notation. For example, ε_m may or may not be in the interval $[-\varepsilon, +\varepsilon]$, depending on the accuracy of the estimate.

The traditional concept in metrology is to segregate contributors to uncertainty into two types according to their property or nature: random and systematic (Drosg, 2009; Rabinovich, 2017). The following definitions are central to metrology:

Random uncertainties: inaccuracies of a measurement that, in the course of multiple measurements of the same quantity under the same conditions, vary in an unpredictable way.
Systematic uncertainties: inaccuracies of a measurement that, in the course of multiple measurements of the same quantity under the same conditions, remain constant or vary in a predictable or understandable way.

Random and systematic uncertainties are *always* present in any measurement of a physical quantity. The magnitude of the random and systematic uncertainties relative to the true value y_T depends on many details of the entire measurement process.

The characterization of the random uncertainty requires that multiple measurements of the physical quantity be made. This is usually referred to as *replication* of the experimental measurement; but there are, importantly, various types of replication. For example, suppose a fixed mass is being weighed on a mechanical scale with a dial indicator. Each time the mass is weighed, there is a random component to the measurement uncertainty due to the mechanical mechanism, laboratory environment conditions, as well as to the person reading the indicator. If the mass is weighed only once, but several people read the indicator, multiple readings can be recorded because of the variability in the individuals. This is referred to as zeroth-order replication, i.e., there is no repetition in the measurement (Coleman and Steele, 2018; Moffat and Henk, 2021). If multiple measurements are made using the same instrument, using the same procedure, in the same measurement environment, this is referred to as first-order replication. If multiple measurements are made using a *different* instrument, using the same procedure, in the same measurement environment, this is referred to as second order replication. Each higher order of measurement replication includes more diverse sources of uncertainty thereby converting systematic errors into random errors that are sampled. From a statistical sampling perspective, this is referred to as *increasing the size of the sample set.*

To mathematically address random and systematic uncertainties, certain definitions and concepts from probability and statistics need to be introduced. Let y_i, $i = 1, 2, \ldots n$, be the set of individual values obtained in the measurement of a fixed quantity. There is no assumption made concerning how these individual measurements were obtained, e.g., they could be obtained using the same instrument or different instruments, by the same laboratory or even different laboratories, or over a short period of time or a long period of time. Let $\bar{y}$ be defined as the sample mean of these measurements.

$$\bar{y} = \frac{1}{n}\sum_{i=1}^{n} y_i. \tag{11.2}$$

The individual measurements, y_i, are assumed to be independent draws from the parent population. The mean of the parent population, μ, is defined as

$$\mu = \frac{1}{N}\sum_{i=1}^{N} y_i \quad \text{where } N \text{ is countably infinite.} \tag{11.3}$$

The concept of μ as the true mean of the parent population is mathematically useful, but it is unattainable in practice because the number of samples in reality is always finite.

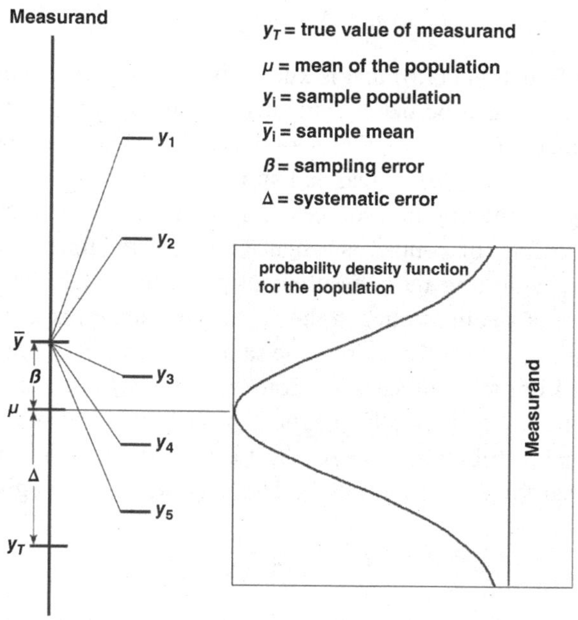

Figure 11.9 Random and systematic contributors to experimental measurement error.

11.3.2 Estimating Random and Systematic Uncertainties

Figure 11.9 depicts the various sources of random and systematic measurement error. In the figure, the sample mean $\bar{y}$ is shown as being computed from five random samples. Let β be defined as the *sampling error* in $\bar{y}$, so that

$$\beta = \bar{y} - \mu. \tag{11.4}$$

Note that here β is referred to as the sampling error, as opposed to the sampling uncertainty, because it is defined with respect to the true mean of the parent population, μ. β is the error in $\bar{y}$ due to taking a finite number of samples from the parent population. Note that μ is *not* assumed to be the true value of the measured quantity of interest because it is *not* assumed that the measurement procedure is free of all systematic errors. By *not* assuming that $\mu = y_T$, the modern metrology literature (Grabe, 2014; Salicone and Prioli, 2018) refers to this either as *non-Gaussian error estimation* or *generalized Gaussian error estimation*. The ISO/ANSI approach does *not* make a distinction between the true value, y_T, and μ. (See Grabe, 2014, for a detailed discussion about why this is done in the ISO/ANSI approach.) Quoting the ISO/ANSI Guide, "the terms 'value of a measurand' (or a quantity) and 'true value of a measurand' (or of a quantity) are viewed as equivalent."

Let Δ be defined as the *systematic error* in the experimental measurement procedure, so that

$$\Delta = \mu - y_T. \tag{11.5}$$

Eliminating μ from Equations (11.4) and (11.5), the fundamental error equation can be written as

$$\beta + \Delta = \bar{y} - y_T. \tag{11.6}$$

The only quantity in Equation (11.6) that is known is $\bar{y}$. Using DOE techniques, the attempt is to estimate Δ so that y_T can be better estimated. Even though β and Δ are convolved in every measurement made, it is known that $\beta \to 0$ as $n \to \infty$, that is, the sampling error must decrease as the number of samples increases. Estimating Δ is much more difficult because of the limited knowledge of the possible sources of systematic error in a measurement procedure. The exception to this, of course, is when there is a reference standard so that y_T is known. For this case, adjusting the measurement procedure such that Δ is set to zero by adjusting the measure procedure to match the reference standard is appropriate.

Consider the sampling error β. To address the sampling error, it is assumed that the sample mean $\bar{y}$ is computed from random samples from an unknown probability distribution. The most commonly assumed, and usually justified, probability distribution for measurement uncertainty is a normal distribution. As a result, it is well known that a statistical confidence interval can be derived for β (Devore, 2007). The magnitude of β is given as

$$|\beta| \leq t_{\alpha/2,\nu} \frac{s}{\sqrt{n}}, \tag{11.7}$$

where $t_{\alpha/2,n-1}$ is the $1 - \alpha/2$ quantile of the t distribution for $n-1$ degrees of freedom, and s is the sample standard deviation. The sample deviation, s, is given by

$$s = \left[\frac{1}{n-1} \sum_{i=1}^{n} (y_i - \bar{y})^2 \right]^{½}. \tag{11.8}$$

Any level of confidence can be chosen, which is given by $100(1 - \alpha)\%$. The higher the confidence level chosen, the wider the bounds on $|\beta|$ because $t_{\alpha/2,\ n-1}$ increases for a fixed n. For n greater than 16, the cumulative t distribution and the cumulative standard normal distribution differ by less than 0.01 for all quantiles. In the limit as $n \to \infty$, the t distribution is identical to the standard normal distribution.

Now consider the systematic (bias) error Δ. The systematic error is much more difficult to address because it is due to a lack of knowledge of the sources of the error. Hence, systematic error has little or no ordered structure, such as sampling uncertainty, because so little is known about it. There are two strategies for dealing with systematic error. First, and by far the most common procedure, is to identify and attempt to reduce or eliminate systematic error. For example, many sensors change their measurement output due to a change in temperature, even though the quantity being measured does not change. As a result, various techniques are used to eliminate the bias shift in sensor output due to temperature. The second strategy is to estimate or reduce the systematic error by expanding the sample space to include different measurement diagnostics and/or procedures that do not suffer from the same systematic error. The most common method to achieve this is to use independent measurement techniques or to use different laboratories to obtain the measurement results of interest. In the following discussion, the term *systematic uncertainty* is used instead of *estimate of the systematic error.*

Systematic uncertainty is an epistemic uncertainty because it is due to lack of knowledge, specifically, lack of knowledge related to the source of the error. Depending on the situation, it could be viewed as either a recognized epistemic uncertainty or a blind epistemic uncertainty. Recalling the definitions given in Chapter 2, epistemic uncertainty is divided into two groups:

Recognized epistemic uncertainty: an epistemic uncertainty for which a conscious decision has been made to either characterize or deal with it in some way, or to ignore it for practical reasons; and
Blind epistemic uncertainty: an epistemic uncertainty for which it is not recognized that the knowledge is incomplete, and that the knowledge is relevant to modeling the system of interest.

In either case, adding knowledge to the measurement process (e.g., by using higher accuracy equipment, diagnostics with different sensing characteristics, or using different high-quality measurement laboratories) can decrease or better quantify the magnitude of the epistemic uncertainty.

There is a long-held belief in metrology that with very accurate reference standards, careful calibration of equipment against reference standards, and regimented laboratory measurement procedures, systematic uncertainties would become insignificant. For simple measurement processes of the basic physical characteristics of manufactured products, this is generally the case. However, for most experimental measurements in engineering and the physical sciences, systematic uncertainties have *not* disappeared. In fact, they can be the *dominant* contributor to uncertainty because *no* reference standards are directly usable for the measurements being made. This situation is disconcerting to many experimentalists and not believed by some. Only rarely is this disconcerting situation discussed. Probably the most famous article in this regard is by Youden (1972), titled "Enduring Values." He emphatically demonstrated that (a) estimated uncertainties commonly do not capture the true value and (b) the largest contributor to uncertainties are commonly systematic uncertainties.

To directly address systematic uncertainties in the metrology community, non-Gaussian error estimation procedures have recently been developed (Grabe, 2014; Salicone and Prioli, 2018). These methods, however, are mathematically more complex than the GUM procedure and have not yet been widely accepted. An effective alternative procedure to estimating measurement uncertainty is to use statistical design of experiments (DOE) (Box et al., 2005; Hinkelmann and Kempthorne, 2008; Montgomery, 2017; Moffat and Henk, 2021). The strategy in DOE is to expand the measurement sample space so that more diverse sources of uncertainty can be statistically quantified. Expanding and segregating the sample space is done by using randomization and blocking techniques, respectively, in the design of the samples taken in an experiment. *Randomization* is designing the samples so that all factors that could possibly affect the measurement result are randomly included in the sampling. *Blocking* is a design technique used to either (a) improve the precision with which comparisons among the factors of interest can be made or (b) identify the individual factors that contribute to measurement uncertainty.

Randomization and blocking techniques use specially designed sampling to *convert systematic uncertainties into random uncertainties.* Then the power of classical statistics can be used to estimate factors that contribute to systematic uncertainty. As a result, when systematic uncertainties are quantified in this way, they are referred to as *correlated systematic uncertainties*, meaning that they are uncertainties that are correlated with specific sources identified through random sampling. Note that in attempting to identify specified or suspected systematic uncertainty sources, the random, or background, uncertainty is also present. The random uncertainty can be segregated from the systematic sources using blocking techniques and comparing the variances of each.

Using randomization to help quantify systematic uncertainties means devising different measurements that can expose factors that could contribute to systematic uncertainty. There are many different kinds of factors that affect measurements conducted in a complex experiment. Some factors and examples are:

- *Experimental instrumentation*: unknown bias in a sensor, measuring device, or transducer; hysteresis in a strain-gauge; shift in zero-reading due to thermal, pressure, acceleration, or radiation sensitivity; frequency response of a complete measuring system; unknown electrical ground loop; misalignment or miscalibration of optical equipment; drifting time reference; drifting frequency reference.
- *Experimental or laboratory procedures*: calibration procedure; preparation procedure; measuring procedure.
- *Experimental hardware*: imperfections or asymmetry in test hardware; assembly of test hardware; installation of test hardware in the facility.
- *Facility characteristics*: preparation of the facility for a test; calibration of the facility; nonuniformity or asymmetry in facility hardware or characteristics.
- *Data recording and reduction*: analog measurement; analog-to-digital conversion; data reduction procedure, such as a programming error in the software.
- *Experimental personnel*: individual technique in human-controlled instruments or pieces of equipment, such as individual operators of the equipment; individual technique in assembly of the test hardware or facility preparation.
- *Time of experiment*: dependence on time of day, day of week, week of month, or month of year.
- *Weather conditions*: dependence on atmospheric conditions including temperature, pressure, winds, humidity, dust, sunshine, and clouds; influence of weather conditions on laboratory procedures or personnel.

As can be seen from the types of factors and examples listed above, the possible sources of systematic uncertainty are surprisingly long. After an experimentalist compiles a list of all of the possible sources of uncertainty for their experiment, the experimentalist should attempt to determine (a) which ones may be important, and (b) which ones can be varied so that their effect can be randomly sampled. When considering which ones may be important, the experimentalist must resolutely keep an open mind on their possible effect. The common attitude that must be overcome is "We've checked all of these factors before and everything is fine." The recommended attitude is: "What can possibly be changed so that we can compare our existing measurement results with a new technique or procedure?" These ideas must be balanced with (a) the options that are possible within a facility; (b) whether qualified personnel needed to implement a new technique are available; (c) costs for new equipment, diagnostic sensors, personnel, and their training; and (d) time needed for the additional experiments to be conducted.

Sometimes it may be impossible to make any significant changes in techniques or procedures within a facility. For this situation, a different experimental facility should be considered. There is no better sampling technique than to conduct blind test comparisons of measurement results using different facilities. Using different facilities (particularly if they have entirely different instrumentation techniques) addresses most of the types of factors listed above, thereby assessing many of the most troublesome and subtle systematic uncertainties. If conducting relatively standard experiments in commercial facilities, then there is typically a range of facilities that could be used for conducting these experiments. If using specialized facilities, such as one-of-a-kind facilities, highly specialized instrumentation, or specialized personnel, then there are few options available.

11.4 Example: Joint Computational-Experimental Aerodynamics Program (continued)

The DOE method was effectively used in the JCEAP validation experiment to improve the estimation of experimental uncertainty beyond the traditional ISO/ANSI approach. The experience clearly demonstrated that the DOE method gave a much more representative estimate of experimental measurement uncertainty than the traditional ISO/ANSI approach. Given the focus on accurate estimation of experimental uncertainty in validation experiments, this is a major finding.

11.4.1 Estimation of Experimental Uncertainty in Force and Moment Measurements

The primary F&M quantities that were measured on the configuration were the normal force, the pitch moment, the axial center of pressure, and the forebody axial force. The forebody axial force is the total axial force measured on the test article using the six-component strain-gauge balance, minus the axial force due to pressure on the base of the test article. The secondary F&M quantities measured were the yaw force, yaw moment, and roll moment. A detailed experimental uncertainty analysis was conducted on the primary F&M quantities, but not on the secondary F&M quantities because the magnitude of these quantities was much smaller than the primary quantities of interest. The following discussion is taken from Oberkampf and Aeschliman (1992), Walker and Oberkampf (1992), Oberkampf et al., (1993) and Aeschliman and Oberkampf (1998) and deals with the F&M measurement portion of the JCEAP experiment. See these references for a detailed discussion.

11.4.1.1 Application of DOE Principles

In the F&M experiment, it was decided to identify only two blocks of measurements over which randomization of experimental runs would take place. The first block contains a mixture of random uncertainties associated with testing procedures and certain systematic uncertainties. The random uncertainties are those that would normally be considered to contribute to run-to-run repeatability in the measurement of F&M. The systematic uncertainties in this group are those that would only be found through deliberate intent to sample their effect. These would be commonly referred to as end-to-end systematic uncertainties. Here the random and systematic uncertainties are listed together because, depending on perspective, they could be segregated differently:

- Run-to-run variation in the F&M strain gauge output;
- Hysteresis, nonlinearity, thermal sensitivity shift, and thermal zero shift in the F&M strain gauge output;
- Run-to-run variation in base pressure transducers and instrumentation for eliminating the base drag;
- Analog data-reduction system and analog-to-digital conversion;
- Digital data recording system;
- Test article pitch, roll, and yaw alignment in the test section using different qualified technicians to make the measurements;
- Imperfections and nonsymmetric features in the test article geometry;
- Assembly and reassembly of the test article using different qualified technicians;
- Run-to-run variation due to facility operating conditions over days, weeks, and months;
- Run-to-run variation due to atmospheric condition effects on the facility;
- Run-to-run variation in the wind tunnel operator setting free-stream conditions in the test section using different qualified operators.

The combination of all of these uncertainties is commonly referred to as end-to-end repeatability uncertainty in the final experimental result. Here, this is referred to simply as *repeatability*.

The second block of samples was directed at identifying the systematic uncertainty that is specifically due to nonuniformity of the flowfield in the test section of the wind tunnel. It has been long recognized that flowfield nonuniformity, or flowfield quality, is a significant contributor to uncertainty in wind tunnel measurements, but there has been only limited success to quantify this contributor (AGARD, 1994; AIAA, 1999). Flowfield nonuniformity in the test section of any type of wind tunnel can be caused by many sources. A few sources that are primarily associated with supersonic wind tunnels follow:

- Poorly designed wall contours ahead of the test section, including the contraction section ahead of the nozzle throat, the nozzle expansion region, and the test section region;
- Turbulence generated from the heater section of the wind tunnel ahead of the contraction section and the nozzle throat;
- Poorly manufactured wind tunnel section and assembly for an experiment;
- Operation of a fixed nozzle wall wind tunnel at a Reynolds number different from the design condition;
- Slight changes and variability in the location of boundary layer transition on the walls upstream and downstream of the nozzle throat due to slight variability in the wall roughness and temperature, acoustic environment, upstream flow control valves, and vibration of the walls.

All of the systematic uncertainties associated with the flowfield nonuniformity component are referred to simply as *flowfield*. Recalling the discussion given earlier introducing the DOE technique, the flowfield nonuniformity systematic uncertainty is converted into a correlated systematic uncertainty because of the blocking technique used to generate a certain type of samples.

A different approach to explaining how the flowfield nonuniformity uncertainty is captured in JCEAP follows. A sampling procedure was devised to assess the *presumption* of a perfectly uniform free-stream flowfield in the test section. The fundamental idea was to estimate the *increase* in uncertainty, beyond the repeatability uncertainty, due to placing the test article at different locations within the test section. By placing the test article at different locations in the test section and comparing this uncertainty to the uncertainty obtained from just testing at one location, what is the increase in uncertainty on the final measured quantities of interest?

The number of possible physical locations within the test section for comparisons in order to estimate the flowfield uncertainty is greatly expanded by recognizing the following. The F&M measured at a positive angle of attack and $\phi = 0°$ should be the same for a negative angle of attack of the same magnitude and $\phi = 180°$, *if* the flowfield is perfectly uniform. Because of physical and mechanical limitations of the hardware used to support the test article in the test section, there were a limited number of locations that could be attained. All of these locations are within regions of the test section where the flowfield is considered fully acceptable for testing.

11.4.1.2 DOE Analysis and Results

Using the technique of blocking, comparisons are made between different wind tunnel runs and portions of different runs to segregate repeatability and flowfield uncertainty contributors. To calculate the repeatability uncertainty, all possible combinations of F&M measurements that were made for the same physical location in the test section for the same configuration must be compared. This allows for estimation of the end-to-end repeatability uncertainty.

By examining the run number schedule (Table 11.2) an experimentalist can choose run pairs that have the same roll angle ϕ, the same flap deflection angle δ, and the same location in the test section. Then comparisons can be made between the same F&M quantities for the same angles of attack. The total number of run pairs that can be formed in this way for quantifying the repeatability uncertainty is 14. All of these run pairs are shown in the second column of Table 11.4. The number of angle-of-attack pairs for each run pair is shown in the third column of the table. The total number of angle-of-attack pairs constructed in this way is 160 for each F&M quantity.

To calculate the additional uncertainty contribution due to the flowfield nonuniformity, examine Table 11.2 to determine which run pairs place the test article at different locations in the test section. Two methods of attaining different locations must then be found. The first method forms run pairs that have the same ϕ and δ but are at different axial stations in the test section. All of the run pairs that can be constructed in this way are shown in the fourth column in Table 11.4. The total number of run pairs that can be formed in this way is 20. Also shown in the fifth column of the table is the number of angles of attack that are in common for each run pair. The total number of angle-of-attack pairs constructed in this way is 220 for each F&M quantity.

The second method forms run pairs based on mirror symmetry between the test article at a roll angle of 0° and pitched to a positive α, and the test article at $\phi = 180°$ and pitched to negative α, both conditions for the same δ. The mirror symmetry pairs are formed for both runs at the same axial station. As the maximum negative angle of attack was $-10°$, the residuals for individual angles of attack can only be calculated over the range $-10°$ to $+10°$. All run pairs that can be constructed in this way for estimation of this uncertainty component

Table 11.4 *Run Pairs for F&M Uncertainty Analysis*

	Repeatability		Flowfield (Axial Location)		Flowfield (Mirror Symmetry)	
Pair Number	Run Pair	No. of αs	Run Pair	No. of αs	Run Pair	No. of αs
1	(34,36)	12	(34,74)	12	(34,40)	8
2	(34,37)	12	(34,75)	12	(36,40)	8
3	(34,73)	12	(36,74)	12	(37,40)	8
4	(36,37)	12	(36,75)	12	(73,40)	8
5	(36,73)	12	(37,74)	12	(63,55)	8
6	(37,73)	12	(37,75)	12	(72,55)	8
7	(63,72)	12	(73,45)	12	(64,56)	8
8	(64,71)	11	(73,75)	12	(64,57)	8
9	(65,70)	10	(63,83)	11	(71,56)	8
10	(66,67)	11	(72,83)	11	(71,57)	8
11	(56,57)	11	(64,82)	10	(65,58)	8
12	(59,60)	11	(71,82)	12	(70,58)	8
13	(74,75)	12	(65,81)	10	(74,76)	8
14	(76,77)	10	(70,81)	10	(74,77)	8
15	-	-	(40,76)	10	(75,76)	8
16	-	-	(40,77)	10	(75,77)	8
17	-	-	(55,78)	10	(83,78)	8
18	-	-	(56,79)	10	(82,79)	8
19	-	-	(57,79)	10	(81,80)	8
20	-	-	(58,80)	10	-	-

Source: Oberkampf et al., 1993.

are shown in the sixth column of Table 11.4. The total number of run pairs that can be formed in this way is 19. Using interpolation for the angle-of-attack range to obtain four positive and four negative angles of attack, the seventh column in the table shows the total number of angles of attack that are in common for each run pair. The total number of angle-of-attack pairs constructed in this way is 152 for each F&M quantity.

After discussing the types of run pairs that are needed for the comparisons shown in Table 11.4, some comments are needed concerning how the run schedule (Table 11.2) was constructed before the experiment. First, replicate runs are *not* afterthoughts in the design of randomization and blocking techniques. Careful thought must be devoted to the construction of the run schedule *before* the experiment begins to achieve the needed samples. Second, replicate runs should be done as far apart in the run schedule as possible in order to incorporate the most changes of all of the factors listed above. For example, replicate runs for conditions associated with runs 63 and 72 are better than replicate runs for conditions associated with runs 66 and 67. Runs 66 and 67 were conducted on the same day, whereas runs 63 and 72 were conducted weeks apart, during which time the test article was disassembled and reassembled.

For each run pair, an average value, $(\bar{\ })$, of the measured F&M quantity can be defined as

$$(\bar{C}_n)_{\alpha_i} = \left\{ \left[(C_n)_p + (C_n)_q \right]_{\alpha_i} \right\} / 2 \quad i = 1, 2 \cdots I,$$

$$(\bar{C}_m)_{\alpha_i} = \left\{ \left[(C_m)_p + (C_m)_q \right]_{\alpha_i} \right\} / 2 \quad i = 1, 2 \cdots I,$$

$$(\bar{x}_{cp})_{\alpha_i} = \left\{ \left[(x_{cp})_p + (x_{cp})_q \right]_{\alpha_i} \right\} / 2 \quad i = 1, 2 \cdots I, \tag{11.9}$$

$$(\bar{C}_a)_{\alpha_i} = \left\{ \left[(C_a)_p + (C_a)_q \right]_{\alpha_i} \right\} / 2 \quad i = 1, 2 \cdots I.$$

C_n, C_m, x_{cp} and C_a are the normal force coefficient, pitch moment coefficient, axial center of pressure, and forebody axial force coefficient, respectively. The symbols p and q represent the run numbers from which measurements are taken, α_i refers to the angle of attack at which each F&M measurement was made, and I is the maximum number of angles of attack that are in common for both run pairs.

The pitch sector mechanism in the wind tunnel is programmed to yield a specified pitch angle according to a digital counter, given an initial starting angle. However, it lacks a feedback control system for the desired pitch angle. As a result, the pitch increments programmed are not accurately repeatable from run to run because of slight variability in friction in the mechanism. As a result, the non-repeatability in pitch angles ranges up to 0.5°. The actual pitch angle attained is known to within $\pm$ 0.02° based on the digital counter output. Therefore, in the average quantities calculated for each run pair in Equation (11.9), it would be inappropriate to assign to measurement uncertainty the fact that identical pitch angles were not obtained from run to run. That is, the inaccuracy is not in the data itself, but caused by the non-repeatability control mechanism itself. This non-repeatability can be eliminated by using a least squares fit of each F&M quantity for each pitch sequence for each run. Using these data fits (discussed later in the pressure experiment), it is possible to then calculate the F&M measurements at precisely the same angle of attack for each run pair.

Let the difference between an individual F&M quantity and the average measurement at each angle of attack be defined as a local residual, so that

$$(\Delta C_n)_{\alpha_i} = (C_n)_{\alpha_i} - (\bar{C}_n)_{\alpha_i} \quad i = 1, 2 \cdots I,$$

$$(\Delta C_m)_{\alpha_i} = (C_m)_{\alpha_i} - (\bar{C}_m)_{\alpha_i} \quad i = 1, 2 \cdots I,$$

$$(\Delta x_{cp})_{\alpha_i} = (x_{cp})_{\alpha_i} - (\bar{x}_{cp})_{\alpha_i} \quad i = 1, 2 \cdots I, \tag{11.10}$$

$$(\Delta C_a)_{\alpha_i} = (C_a)_{\alpha_i} - (\bar{C}_a)_{\alpha_i} \quad i = 1, 2 \cdots I.$$

The total number of repeatability residuals, i.e., measurements at each pitch angle (based on the second and third columns in Table 11.4) is 320 for each F&M quantity. Because of the definition for the run pair average given in Equation (11.9), 160 of the residuals are positive and 160 are negative, i.e., mirror images. Based on the fourth and fifth columns in Table 11.4, the total number of axial location residuals is 440. Based on the sixth and seventh columns in Table 11.4, the total number of mirror symmetry residuals is 304. The total number of residuals for the F&M experiment is 1,064.

Table 11.5 *Summary of results for F&M uncertainty analysis*

Uncertainty component	C_n		C_m		x_{cp}/L		C_a	
	$\hat{\sigma}$	%	$\hat{\sigma}$	%	$\hat{\sigma}$	%	$\hat{\sigma}$	%
Repeatability	0.474×10^{-3}	20	0.406×10^{-3}	19	0.413×10^{-3}	9	0.426×10^{-3}	63
Flowfield	0.941×10^{-3}	80	0.851×10^{-3}	81	1.322×10^{-3}	91	0.324×10^{-3}	37
Total	1.054×10^{-3}	100	0.943×10^{-3}	100	1.385×10^{-3}	100	0.535×10^{-3}	100

Source: Oberkampf and Aeschliman, 1992.

The sample variance for the repeatability component and the flowfield component is calculated using the equation

$$\hat{\sigma}^2 = \frac{1}{2m}\sum_{j=1}^{m}\left(\Delta_1^2 + \Delta_2^2\right)_j. \tag{11.11}$$

The m is the number of residuals compared and Δ_1 and Δ_2 are the residuals for each run pair from Equation (11.10). (Note that the sample variance is an estimate of a population variance and is indicated by [^].) Because of the independence between repeatability uncertainty and flowfield uncertainty sampling, the sum of the variances can be written as

$$\hat{\sigma}^2_{total} = \hat{\sigma}^2_{repeatability} + \hat{\sigma}^2_{flowfield}. \tag{11.12}$$

The sample variance due to the repeatability component and for the total experiment can each be computed separately. As a result, the component due to the flowfield can be found from

$$\hat{\sigma}_{flowfield} = \sqrt{\hat{\sigma}^2_{total} - \hat{\sigma}^2_{repeatability}}. \tag{11.13}$$

Table 11.5 gives the estimated standard deviation $\hat{\sigma}$, (square root of the variance estimate) due to repeatability uncertainty, flowfield uncertainty, and the total combined uncertainty for each F&M quantity in the experiment. The table also shows the percent contribution due to each of the components identified. In the calculation for the standard deviation for $x_{cp/L}$, all residuals for angles of attack of -1, 0, and $+1°$ were excluded. This was done because it is well known that the uncertainty in x_{cp} becomes indefinite as the normal force approaches zero. If the test article geometry were axisymmetric, then the standard wind tunnel procedure of computing the ratio of $C_{m_\alpha}/C_{n_\alpha}$ for x_{cp} would be appropriate. However, for the present nonsymmetric geometry this method is not applicable because it is not known at what angle of attack, if any, both C_m and C_n attain zero. Except for the case of calculating x_{cp} for $\alpha = 0$, all experimental F&M measurements were included in the results shown in Table 11.5, i.e., *no* measurement outliers were discarded in the entire uncertainty analysis.

The repeatability uncertainty computed here captures the very broadest sense of repeatability of the entire wind tunnel facility, facility operations, model assembly and reassembly, and data acquisition system. The common meaning of repeatability uncertainty in the literature is a small subset of what is reported in Table 11.5. Table 11.5 shows that the repeatability uncertainty in C_n, C_m, and x_{cp} ranges from 9 to 20% of the total. Strikingly, the flowfield nonuniformity ranges from 80 to 91% of the total uncertainty for C_n, C_m, and x_{cp}.

The repeatability component of uncertainty in C_a is 63% and that due to the flowfield is 37%. This reversal of uncertainty contributions compared to the other quantities is believed to be due to (a) the axial force being very *insensitive* to flowfield nonuniformity because it is the sum of the normal stress and shear stress acting along the axis of the body and (b) the difficulty in removing the base pressure component from the total axial force.

The DOE experimental uncertainty estimation technique discussed here is new to wind tunnel testing. This analysis, as well as the following analysis for the surface pressure measurements, clearly demonstrates that for most measured quantities, the dominant contributor is the systematic uncertainty due to the flowfield nonuniformity, *not* the repeatability uncertainty. The established procedure for wind tunnel uncertainty estimation is based on the ISO/ANSI procedure, which is focused on estimating the repeatability component of uncertainty, *not* the flowfield uncertainty. Finally, some readers will quickly jump to the conclusion that the obvious culprit is the poor quality flow in the Sandia hypersonic wind tunnel. As will be seen in the discussion of the JCEAP pressure experiment in the next section, this is *not* the case.

11.4.2 Estimation of Experimental Uncertainty in Surface Pressure Measurements

The DOE technique was applied to the surface pressures measured by the two pressure modules, but not the high frequency measurements from the Kulite gauges. The following discussion is taken from Oberkampf et al. (1995), Oberkampf et al. (1996), and Aeschliman and Oberkampf (1998) and deals with surface pressure measurements in the JCEAP experiment. See these references for more details.

11.4.2.1 Application of DOE Principles

The DOE approach applied to the pressure experiment data quantified three classes of contributors to experimental uncertainty. Two of the three contributors were discussed earlier in the F&M experiment: end-to-end repeatability uncertainty and flowfield nonuniformity uncertainty. The repeatability component for the pressure experiment is nearly the same, except now it includes uncertainties associated with the surface pressure measurement system instead of those associated with the F&M measurement system. The flowfield component for the pressure experiment is the same as discussed previously in this chapter. Some examples of the pressure measurement contributors that are grouped into repeatability uncertainty include:

- Run-to-run variation in the individual pressure transducers within each pressure module;
- Hysteresis, nonlinearity, thermal sensitivity shift, and thermal zero shift in the pressure modules;
- Run-to-run variation in the reference pressure that is fed to the pressure modules;
- Analog data reduction system and analog-to-digital conversion;
- Digital data recording system;
- Test article pitch, roll, and yaw alignment in the test section using different qualified technicians to make the measurements;
- Run-to-run variation due to facility operating conditions over days, weeks, and months;

- Run-to-run variation due to atmospheric condition effects on the facility; and
- Run-to-run variation due to the wind tunnel operator setting free-stream conditions in the test section, which includes different qualified operators.

In addition to the repeatability and flowfield components, local surface pressure measurements allow an additional component of systematic uncertainty to be identified. This component is the uncertainty due to (a) imperfections and irregularities associated with the pressure orifice on the surface of the test article, the connecting tubing between the orifice and the pressure module, and the independent pressure sensors in each module; and (b) imperfections or asymmetry in the surface geometry of the test article. All of the systematic uncertainties associated with this component will be referred to as the *instrumentation* component.

For a general wind tunnel experiment, two points must be noted concerning the instrumentation uncertainty component. First, any experiment that measures local features of the flow with independent sensors can use the statistical DOE procedure described below to estimate the instrumentation uncertainty component. For example, if shear-stress sensitive or pressure sensitive liquid crystals were painted on the surface, the following uncertainty estimation procedure could be used. Second, to form many of the run pairs for comparing independent measurements, the uncertainty estimation procedure requires certain features of geometric symmetry of the test article. Specifically, at least two orthogonal planes of mirror symmetry of the test article must exist. If only one plane of mirror symmetry exists, such as an aircraft, then the comparison of different pressure ports includes the contribution due to the flowfield nonuniformity. This is because different pressure ports cannot be placed at the same location in the test section and exposed to the same induced flowfield of the test article.

Examples of some of the sources of systematic uncertainty that fall into the instrumentation component are:

- A poorly fabricated pressure orifice, such as one containing a burr from machining or one having an irregular cylindrical shape;
- Accidental damage to a pressure orifice, such as a damaged edge of an orifice that occurred any time after the test article was fabricated;
- A pressure leak occurring anywhere between the pressure orifice and the connection of the tubing to the pressure module;
- Any blockage or kink in the connecting tubing anywhere between the pressure orifice and the pressure module;
- Test article fabrication deficiencies that cause the loss of mirror symmetry of the surface contour of the test article;
- Accidental damage to the test article, for example during wind tunnel testing or handling, that causes the loss of mirror symmetry of the surface contour of the test article;
- Any deflection or distortion of the axis of the test article, such as due to fabrication or aerodynamic or aerothermal heating of the test article; and/or
- Any deflection or distortion of the test article surface or lifting surfaces that causes the loss of mirror symmetry of the test article, such as due to fabrication or aerodynamic or aerothermal heating of the test article.

Using blocking techniques for the surface pressure measurements, three sets of sources contributing to experimental uncertainty can be identified: repeatability, flowfield, and instrumentation. The blocking strategy for each of these sources can be summarized as:

- Repeatability: compare pairs of measurements that have the same physical location in the test section, the same pressure port on the test article, and exposure to the same induced flowfield of the test article;
- Flowfield: compare pairs of measurements that have the same pressure port on the test article, exposure to the same induced flowfield of the test article, but are at different physical locations in the test section; and
- Instrumentation: compare pairs of measurements that have the same physical location in the test section, exposure to the same induced flowfield of the test article, but use different pressure ports on the test article and different sensors in the pressure modules.

To calculate the repeatability uncertainty, pressure measurements from different runs should be compared for the same location in the test section, pressure port, angle of attack, roll angle, and flap deflection angle. With these restrictions, each pair of ports compared will have the same location in the vehicle-induced flowfield. When differences in pressure port measurements are computed, the uncertainty due to flowfield and instrumentation cancel out computing the residual.

To calculate the flowfield uncertainty, a two-step process is needed. First, pressure measurements should be compared from the same pressure port and exposure to the same induced flowfield, but at different physical locations in the test section. The same induced flowfield can be generated by having the same angle of attack, roll angle, and flap deflection angle, but it can also be generated by using mirror symmetry features as discussed pertaining to the F&M uncertainty analysis. When the differences in pressure port measurements are computed, they will be a combination of the repeatability and flowfield components, but the component due to the instrumentation cancels out in computing the residual. The second step is to subtract the variance due to the repeatability component from the combined variance due to the repeatability and flowfield components, leaving only the contribution due to the flowfield.

To calculate the instrumentation uncertainty, a similar two-step process is needed. First, pressure measurements should be compared from the same physical location in the test section and exposure to the same induced flowfield but using different pressure ports. Given these constraints, the only possibility for comparisons is to have two or more planes of mirror symmetry of the test article geometry. For the JCEAP geometry, this only exits for ports on the conical portion of the geometry ahead of the slice. Once the slice begins, there is only one plane of symmetry: the vertical plane. This single plane of symmetry feature is also common with most aircraft geometries. As a result, only comparisons of pressures measured on the conical portion of JCEAP are made. Note that for four-finned and four-winged missile geometries, there are two planes of symmetry. As a result, the following procedure could be used for these geometries.

For JCEAP, only comparisons of pressure ports can be made for those that are sufficiently upstream of the slice because of the minimal upstream influence of the slice. In addition, only pressure data for the deflected flap cases can be used where experimentalists are confident there is no influence of the deflected flap upstream of the beginning of the slice. The case of $\delta = 0°$ produces no axially separated flow on the test article, so there is no concern that there is any influence upstream of the slice. For $\delta = 10°$ and $20°$, liquid crystal surface flow visualization showed that no axially separated flow progressed ahead of the beginning of the slice. However, a flap deflection of $30°$ showed a separated flow influence over the entire slice region so this flap deflection was not used in these comparisons. Therefore, only run pairs with 0, 10, and 20° flap deflections were used. To further reduce the possibility of any upstream influence of the slice, only axial stations up to 15.75 cm were considered. This is 2.73 cm ahead of the slice, or about 10 to 20 boundary layer thicknesses ahead of the slice. The second step in the calculation process is to subtract the variance due to the repeatability component from the combined variance due to the repeatability and instrumentation components, leaving only the instrumentation component.

11.4.2.2 DOE Analysis and Results

To calculate the repeatability uncertainty, the run schedule (Table 11.3) should be examined and run pairs chosen that have the same roll and flap angles, as well as the same tunnel location. In addition, the run pairs chosen must allow comparison only between the same pressure ports on the test article. Examples of run pairs that meet these conditions are (20,22), (24,61), (103,112), (42,43), (124,126), and (131,133). It can be shown that a total of 29 run pairs meet the required conditions. Pressure measurements were obtained for a total of 12 angles of attack for each run: 9 distinct angles of attack and 3 measurements at $\alpha = 0$. As a result, there are a total of 18 combinations of α where pressure comparisons can be made (9 distinct α comparisons plus 9 permutations of $\alpha = 0$ measurements). Therefore, the total number of pressure port comparisons is

$$(96 \text{ ports}) \times (29 \text{ run pairs}) \times (18\ \alpha \text{ pairs}) = 50{,}112 \text{ comparisons.}$$

The actual number of comparisons is somewhat less than this estimate because some pressure ports were outside the calibrated range for certain conditions. As a result, the total number of pressure port comparisons for repeatability uncertainty was found to be 48,164.

To make these pressure port comparisons, it is required that the α for each of the two runs is identical. If they are not the same, then part of the difference in the two measurements will be due to slight non-repeatability of α caused by the test article pitch mechanism. As mentioned earlier, the pitch control mechanism does not have a feedback control system, so the repeatability of α from one pitch sweep of the mechanism to another can be as large as $\pm 0.5°$. Since it was known that very similar angles of attack were needed for the pressure comparisons, roughly 10 to 20 runs had to be repeated because the resulting α deviated by more than $\pm 0.3°$ from the nominal values. For the final run numbers shown in Table 11.3, α did not deviate from the nominal value by more than $\pm 0.28°$. The average deviation from the nominal value of α for all runs listed in Table 11.3 was 0.057°.

Deviations in α of up to 0.28° from run-to-run would introduce an additional source of uncertainty, but one that had nothing to do with experimental uncertainty in pressure measurements. To remove this source of uncertainty in the DOE analysis, all of the pressure measurements were interpolated to the nominal angles of attack. To accomplish this, a cubic spline interpolation with adjustable knot locations was computed for each pressure port as a function of α for each run.

Differences in pressure port measurements were computed with the following technique. Let the pressure measurement for port i, and angle of attack j, be denoted as $\left(\frac{p_i}{p_\infty}\right)_j^r$, where the superscript r denotes the run number. Then the average pressure of the port for the two runs, r and s, being compared is given by

$$\left(\frac{\bar{p}_i}{p_\infty}\right)_j^{r,s} = \frac{1}{2}\left[\left(\frac{p_i}{p_\infty}\right)_j^r + \left(\frac{p_i}{p_\infty}\right)_j^s\right], \tag{11.14}$$

where $i = 1, 2, \ldots, 96$ and $j = 1, 2, \ldots, 18$, where 18 is the total number of angles of attack. Let the absolute value of the difference between a pressure measurement and the average pressure be defined as the residual. Then, the residual is given by

$$\left(\frac{\Delta p_i}{p_\infty}\right)_j^{r,s} = \left|\left(\frac{p_i}{p_\infty}\right)_j^r - \left(\frac{\bar{p}_i}{p_\infty}\right)_j^{r,s}\right|. \tag{11.15}$$

Note that the residual can be computed using either the pressure measurement from run r or s.

To calculate the flowfield uncertainty, Table 11.3 is examined and run pairs chosen that have the same induced flowfield and the same flap deflection angle, but the test article has different locations in the test section. In addition, the run pairs chosen must allow a comparison only between the same pressure ports on the test article. Four types of run pairs that produce the types of comparisons desired are found. These are (a) comparisons between measurements made at different axial locations in the test section, (b) comparisons between different roll angles at $\alpha = 0$ at the same tunnel station, (c) comparisons between positive α with a roll angle of 0° and negative α with a roll angle of 180°, and (d) comparisons between positive α with a roll angle of 90° and negative α with a roll angle of 270°. Examples of run pairs for each of these types of comparisons are, respectively, (20,101), (24,32), (35,43), and (46,47).

The total number of pressure port comparisons for these four types, minus the number of comparisons lost due to over-ranged ports, is 101,838 residuals. The residuals for flowfield and repeatability uncertainty are computed by the same equations given above, but the number of angles of attack, j, for each of the types is different.

To calculate the instrumentation uncertainty, Table 11.3 is examined for combinations of roll angle and flap deflection angle, but (a) both runs must be at the same physical location in the tunnel, and (b) both pressure ports must experience the same induced flowfield. Since the comparisons are made between different pressure ports, this also means different sensors on the pressure module are used. In addition, some of the pressure comparisons are made using different pressure modules. This feature exemplifies the DOE feature of random sampling from a larger sample space to better estimate the true uncertainty of a larger sample population.

By examining Table 11.3, six types of run pairs that meet the required conditions are found. All of these pairs are formed by comparing different roll angles, the same flap angle, and the same tunnel location. The total number of pressure port comparisons for these six types, minus the number of comparisons lost due to over-ranged ports, is 24,196. The residuals for instrumentation and repeatability uncertainty are computed using the same equations given above.

All of the pressure residuals computed for repeatability, flowfield, and instrumentation uncertainty are plotted in Figure 11.10. In the entire pressure uncertainty analysis, including the F&M uncertainty analysis, *no* measurement outliers were discarded. Given that some of the residual outliers for large pressure are 5 to 10 times greater than the mean residual, many experimentalists would argue that they should be discarded because they are clearly inaccurate. However, the authors question on what physical basis are they inaccurate? Or, are they

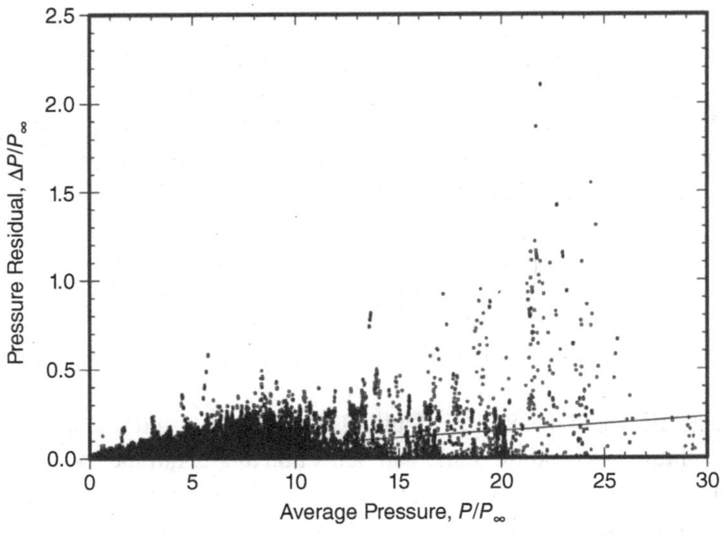

Figure 11.10 All pressure residuals (totaling 174,198) for repeatability, flowfield nonuniformity, and instrumentation uncertainty.
(Oberkampf et al., 1995)

being discarded because they are an embarrassment? They should not be discarded unless there is a clear physical reason identifying what was amiss in the measurement process. Outliers are realities that actually occurred in an experiment, even if the cause is not understood. Their most important impact is on the thickness of the tails of the probability distributions characterizing uncertainty. Outliers tell the experimentalists that the certainty in the measurement results is not as good as what they would like to be believe.

Figure 11.10 shows that the magnitude of the uncertainty steadily increases with the magnitude of the pressure measured. This characteristic is typical of nearly all experimental instrumentation. This trend is accounted for in the residuals by scaling the residuals with the magnitude of measured pressure. A least squares fit of the residuals was computed using a linear function constrained to pass through zero. The resulting fit was computed to be

$$\frac{\Delta p_{ls}}{p_\infty} = 0.00875 \frac{p_s}{p_\infty}, \tag{11.16}$$

where p_s is the surface pressure measured. This linear fit is also shown in Figure 11.10.

The sample variance is now calculated with each local sample scaled according to the least squares fit given above. The equation for estimating each contributing type of variance, normalized by the least squares fit of the residuals, is given by

$$\hat{\sigma}^2 = \frac{1}{N} \sum_{k=1}^{N} \left[\frac{(\Delta p/p_\infty)}{(\Delta p_{ls}/p_\infty)} \right]_k^2. \tag{11.17}$$

N is the total number of residuals (i.e., pressure comparisons) and the subscript k indicates the kth residual. The sample standard deviation due to the flowfield and the instrumentation can then be calculated from

Table 11.6 *Summary of Results for the surface pressure uncertainty analysis*

Component of uncertainty	Number of residuals	Normalized $\hat{\sigma}$ uncertainty	% of total RMS uncertainty
Repeatability	48,164	0.56	12
Flowfield	101,838	1.28	64
Instrumentation	24,196	0.79	24
Total	174,198	1.60	100

Source: Oberkampf et al., 1995.

$$\hat{\sigma}_{flow} = \sqrt{\hat{\sigma}^2_{repeatability\ +\ flow} - \hat{\sigma}^2_{repeatability}}, \tag{11.18}$$

$$\hat{\sigma}_{instrumentation} = \sqrt{\hat{\sigma}^2_{repeatability\ +\ instrumentation} - \hat{\sigma}^2_{repeatability}}, \tag{11.19}$$

respectively. Therefore, the sample standard deviation due to all three sources of uncertainty is given by

$$\hat{\sigma}_{total} = \sqrt{\hat{\sigma}^2_{repeatability} + \hat{\sigma}^2_{flow} + \hat{\sigma}^2_{instrumentation}}. \tag{11.20}$$

Table 11.6 gives the summary statistics for the uncertainty estimates for the pressure measurements. The table shows that the dominant contributor to uncertainty of pressure measurements, as was seen in the F&M measurements, is due to the nonuniformity of the flowfield in the test section. The second most important contributor was the instrumentation. The smallest uncertainty source was the end-to-end repeatability contribution due to the entire measurement system, the wind tunnel facility, and its operation.

The dominant contribution of nonuniform flow to uncertainty begs the question: Is this simply a characteristic of the present wind tunnel or is it characteristic of other supersonic and hypersonic wind tunnels? An earlier application of the DOE uncertainty estimation procedure was applied to surface pressure measurements on a test article in Tunnels A and B of the von Karman Gas Dynamics Facility at the US Air Force Arnold Engineering Development Center, in Tullahoma, Tennessee (Oberkampf et al., 1985). Tunnel A is a variable Mach number facility with a range from Mach 1.5 to 5.5, and Tunnel B can be operated at Mach 6 and 8. The same DOE procedure was applied to pressure data at Mach 3 in Tunnel A and Mach 8 in Tunnel B. The absolute magnitude (no normalization using the free-stream static pressure) of the present results for flowfield uncertainty was compared to those for Mach 8 in Tunnel B of the von Karman Gas Dynamics Facility (Oberkampf et al., 1995). Hypersonic Tunnel B is recognized worldwide for the quality of its flowfield. This comparison showed that the Sandia HWT and Tunnel B are *comparable* in terms of the effect of the flowfield nonuniformity on surface pressure measurements. A similar result concerning the dominance of flowfield nonuniformity compared to repeatability uncertainty was found for Tunnel B.

A similarly relevant question could be raised: How do the uncertainty estimation results using the DOE procedure compare to the traditional ISO/ANSI procedure used in wind tunnels; more specifically, for the *same* experiment? This comparison was made in

Oberkampf et al. (1985). The wind tunnel staff of the von Karman Gas Dynamics Facility used the ISO/ANSI technique to estimate experimental uncertainty in the same experiment. It was found that their uncertainty estimate compared very closely with the repeatability uncertainty computed using the DOE technique. As can be seen from Figure 11.10, this is roughly a factor of 10 less than the total estimated uncertainty in the experiment using the DOE technique.

11.5 Final Thoughts on Validation Experiments

11.5.1 Lessons Learned for Wind Tunnel Experiments

Roughly five years after the completion of the JCEAP project, there was renewed interest at Sandia National Laboratories in the experiment and the simulations. Although there was good agreement between CFD prediction of the force and moment measurements (Walker and Oberkampf, 1992), it was believed that the agreement should be even better for low angle of attack and zero flap deflection. As a result, a new CFD investigation and a reassessment of the experimental data began. The first phase of the investigation was an extensive code verification and solution verification effort. These efforts increased the confidence in the correctness of the CFD code and they reduced the estimated numerical solution error in the CDF simulations (Roy et al., 2000; Roy, McWherter-Payne, et al., 2003). However, there were no significant changes in the simulation results. The second phase of the investigation pursued a number of other possible sources for the disagreement (Roy et al., 2000; Roy, Oberkampf, et al., 2003). The most troublesome disagreement between simulation and experiment was the surface pressure along the conical portion of the test article. This was most troubling because this is the region of the *highest* confidence in the CFD results. As part of the reassessment of the experimental data, the data reduction procedure was independently re-checked. It was found that an error was made in computing the average static pressure for the complete set of experimental runs. The correction was made in the experimental data reduction and the corrected data was used as new input for the CFD simulations. Improved agreement with experimental data was obtained, much to the satisfaction of the experimentalists.

As part of the reassessment of the CFD simulation, two different types of improvements were made in the assumptions and approximations made in the formulation of the mathematical model. First, each of the mathematical submodels were assessed with regard to their accuracy for the conditions of the JCEAP experiment. Those specifically assessed were thermal equilibrium flow assumption in the wind tunnel, ideal gas equation of state assumption, transport models for viscosity and thermal conductivity, continuum flow assumption, axisymmetric flow assumption on the conical portion of the body at zero angle of attack, and supersonic outflow boundary conditions at the end of the conical portion of the body. The accuracy assessment approach taken was to replace each of these submodels with a submodel that is more general, i.e., a submodel that involves more complex physical processes than the one it replaces. When assessing each of the submodels in this way, it was found that there was essentially no change in the level of agreement with experimental measurements.

The second type of improvement in the modeling was to *not assume* uniform flow in the test section of the wind tunnel. This assessment was prompted because the DOE uncertainty estimation procedure showed that the largest contributor to the experimental uncertainty was the nonuniformity of the flow. Relaxing the assumption of uniform flow requires that the flowfield be independently computed using experimental data from the calibration of the empty test section. How this was accomplished is described in Roy, Oberkampf, et al. (2003). Using this computational reconstruction of the nonuniform flow in the test section, this flowfield was then used in the CFD simulation of the flow over JECAP. It was found that significant improvement in agreement was obtained between the revised CFD simulation and the experimental surface pressure measurements at zero angle of attack. Because of both the revised CFD simulations and the DOE uncertainty estimation procedure, it was concluded that flowfield nonuniformity was the dominant contributor to earlier disagreement between simulation and experiment

Lessons learned from reexamination of the JCEAP experimental data and simulations are twofold. First, reevaluation of experimental data, experimental procedures, and data reduction procedures commonly uncovers errors and/or improvements that could be made to the experimental measurements. Second, when comparisons of simulation results and experimental measurements are not as good as expected, there are two paths forward to discover the cause. The first is to reassess the assumptions and approximations made in the formulation of the mathematical model and submodels. The second is to calibrate model input parameters such that improved agreement is attained. Unfortunately, the second path is usually the one taken, without considering that it may be the wrong path.

A major finding from the use of the DOE approach applied to wind tunnel testing is that the flowfield nonuniformity is the dominant contributor to experimental uncertainty in the three facilities in which it was applied. Most, if not all, facility operators would claim that their facility is much better than these and that theirs does not suffer from this deficiency. Since JCEAP was completed, one other facility has used the DOE procedure to quantify the three sources of experimental uncertainty discussed in the example above. It is the high-Mach number test section of the NASA Langley Research Center Unitary Plan Wind Tunnel. This test section measures 1.22 $\times$ 1.22 m in cross-section and can attain Mach numbers in the range 2.3 to 4.63. The test article was a conically blunted cylinder, and the locations of the surface pressure instrumentation were designed to maximize the number of statistical comparisons that could be made using the DOE procedure. The DOE procedure found that the nonuniform flowfield contributor was a factor of 3 to 5.5 times the repeatability uncertainty, depending on the Mach number tested. This result is similar to what was found in three earlier facilities. Details of the experiment and the uncertainty analysis can be found in Rhode and Oberkampf (2017).

The dominant impact of flowfield nonuniformity on wind tunnel experimental uncertainty has also been found by Nagai and Iijima (2011). This finding was obtained for the 1 $\times$ 1 m supersonic wind tunnel of the Japan Aerospace Exploration Agency. Force and moment measurements were made at Mach numbers of 1.4, 1.6, 2.0, and 3.0. Relative to the present discussion, two features of the experiment made it particularly important. First, the primary goal of the experiment was to investigate the various sources of experimental uncertainty, as opposed to the usual situation where uncertainty quantification is a secondary consideration. Second, the approach used to estimate the various sources of uncertainty was the ISO/ANSI procedure. The comprehensive study used two different sized test articles, they moved the test article to various locations in the test section, and they altered the center of rotation so that the test article would be exposed to different portions of the flow in the test section. Purposefully obtaining data at diverse locations and on different sized test articles serves the same purpose as sampling a larger parent population used in the DOE technique discussed earlier.

However, when using the ISO/ANSI technique there is no natural source of uncertainty to which these samples can be assigned because flowfield nonuniformity is a systematic error, *not* a random error. As a result, Nagai and Iijima (2011) defined uncertainty correction factors to the bias error in the test section Mach number uncertainty. Note that these correction factors are in addition to using a coverage factor of 2. This is the only mathematical mechanism available to account for the observed sampling uncertainty in the ISO/ANSI procedure. They found that the correction factors needed to capture the observed uncertainty ranged up to 200! For example, for the smaller test article, the correction factors for pitch moment were 200, 36, 45, and 25 for Mach numbers of 1.4, 1.6, 2.0, and 3.0, respectively. This is convincing, independent confirmation of the dominance of the impact of flowfield nonuniformity, and the extreme to which the ISO/ANSI procedure must go to accommodate this source of uncertainty.

The natural question is: Why hasn't the DOE procedure, or a determined effort such as that conducted by Nagai and Iijima (2011), been pursued by more wind tunnel facilities to investigate the impact of flowfield nonuniformity? There are a number of possible answers, some of which are (a) natural resistance to new procedures and technology in organizations; (b) users of wind tunnels, i.e., paying customers, are more interested in the quantity of data obtained during the experiment than careful uncertainty estimation of the data; and (c) facility owners would rather not deal with or expose a potentially embarrassing weakness in their facility that could negatively impact future business. For this reason, facility operators only see a downside risk, as opposed to a potential upside benefit, in improved uncertainty estimation. A few esteemed metrologists, such as Youden (Youden, 1972; Henrion and Fischhoff, 1986), have called attention to the importance of systematic uncertainties in experimental measurements, but these recommendations have generally been ignored. The customers using these facilities must provide the long-term motivation for improving uncertainty quantification because they are the ones at risk when using measurements from a facility for their systems. They are the ones with the most skin in the game.

As a final comment on validation experiments in wind tunnels, Oberkampf and Smith (2017) have developed a framework for assessing the completeness of model validation experiments with regard to information content concerning the facility and its instrumentation, data completeness for modeling the experiment, and uncertainty quantification of measurements. This framework combines two previously published concepts: strong-sense model validation experiments (Oberkampf et al., 2004) and the modeling maturity assessment procedure referred to as the Predictive Capability Maturity Method (Oberkampf et al., 2007). (PCMM is discussed in detail in Chapter 15 "Maturity Assessment of Modeling and Simulation.") The framework assesses the information completeness of six attributes of the model validation experiment compared to the requirements listed in a table. (Table 11.7 shows the general structure of the model validation experiment completeness table. Each element in the table would contain the requirements for the specified attribute at the specified level of completeness.) The attributes for a wind tunnel experiment are: Experimental Facility, Analog Instrumentation and Signal Processing, Boundary and Initial Conditions, Fluid and Material Properties, Test Conditions, and Measurement of System Response Quantities. Each of the attributes of the experiment are assessed relative to four levels of

Table 11.7 *General structure of a model validation experiment completeness table*

Attribute	Completeness level 0	Completeness level 1	Completeness level 2	Completeness Level 3	Attribute score
Experimental facility					
Instrumentationand signal processing					
System geometry					
System initialconditions					
System physicalparameters					
System boundaryconditions					
System responsequantities					

Adapted from: Oberkampf and Smith, 2017.

information completeness described in the elements of the table. As discussed in Oberkampf and Smith (2017), most existing wind tunnel validation experiments would be assessed at the lower two levels of completeness.

11.5.2 Recommendations for All Validation Experiments

The level of characterization of an experimental facility that is needed for a validation experiment is remarkably higher than what is normally available from an experimental facility. This high level of characterization is difficult and expensive to obtain, for both practical and technical reasons. (Table 11.7 shows the general structure of the model validation experiment information completeness table. This type of table could be constructed for validation experiments in any field of study.) It requires research-oriented personnel who are comfortable questioning accepted assumptions concerning their facility, its operations, and its instrumentation, instead of those who are more oriented toward data production. For a commercial experimental facility, it also requires management who understand the need for critical and detailed investigation of the facility, its instrumentation, and its operating procedures. Lack of interest in seeking out and quantifying systematic errors in facilities and instrumentation is intrinsic in commercial experimental facilities. Their perspective is understandable in the sense that shortcomings in the accuracy of experimental measurements and incomplete understanding of the actual operation of their facility could be detrimental to future business. Only when management of experimental facilities understands that improved knowledge of their facility and experimental uncertainties is to their advantage, as opposed to being a risk, will they attract new business from modeling and simulation (M&S) customers.

It is an open question, concerning the feasibility of conducting validation experiments in high-cost, high-volume data output facilities. These types of facilities usually have testing capabilities that are relevant to actual system and subsystem conditions, commonly well beyond small research facilities. However, these expensive facilities operate with unrelenting pressure to keep data production high for the facility customer, instead of taking time to refine the facility characterization or to investigate possible sources of uncertainty. As a result, conducting validation experiments in high-cost facilities requires a much greater level of planning and preparation before the experiment begins. Furthermore, in these expensive facilities, it is much more likely that a project group for a system or subsystem would provide funding for the experiment, as opposed to an M&S group. Only when M&S matures to the point where it carries more responsibility for predicting system performance, reliability, and safety, will project leadership begin to invest in validation experiments.

The ability to achieve Guidelines 1, 2, and 3 discussed earlier is greatly enhanced if the facility used in the validation experiment is part of the same organization as the M&S group requesting the experiment. These guidelines are (a) achieving a close working relationship and joint planning between experimentalists and simulation analysts; (b) ensuring the measurement of all important model input data; and (c) achieving synergism and constructive teamwork between experimentalists and analysts. Although achieving these three goals in the same organization is difficult in most cases, the authors have seen several successes in organizations. A key ingredient to success is strong management support from both the experimental and simulation groups. This includes the understanding by management that finding weaknesses and limitations in either the experimental facility or the simulation capability is hallmark of a successful model validation project.

When the simulation group contracts with an external facility to conduct a validation experiment, achieving Guidelines 1, 2, and 3 is a constant struggle. If the simulation group has not worked with the facility in the past as part of a validation experiment, then the simulation group should interview staff and management from the facility *before* a contract for the experiment is signed. The simulation group should inquire about the experience of the facility and personnel concerning validation experiments. This inquiry should involve in-depth discussions with staff and management of the facility concerning their knowledge and experience in conducting validation experiments. For example, questions should be raised concerning how they plan to measure the important model input data that are needed for simulation and how the facility has identified and quantified systematic uncertainties in past experiments.

Another key feature that is required to achieve a successful validation experiment is that both technical and practical aspects of the joint experimental-simulation effort must occur simultaneously.

First, the simulation software must be ready for validation activities, the experimental facility must be available, and the required instrumentation must be ready for reliable use. As discussed in Chapters 4 through 9, the software must be under version control; code verification testing should be complete for the simulation options required for the validation experiments; iterative error estimators should be available for use; and solution verification tools should be available for estimating the numerical error on the SRQs being compared with experimental measurements.

Second, the appropriate experimental and simulation personnel must be available to work on the joint effort. As discussed earlier, the validation experience and open-minded perspective of both the experimental and simulation personnel are critical for success.

And third, funding for both the experimental and simulation groups must be available at the same time. If the funding for both the experimental and simulation work come from the same source, this is a major advantage. If the funding comes from different sources, particularly if the sources are outside the primary organization, the changing impulses of these organizations can jeopardize a long-term validation project.

11.6 Further Reading

For recent research on model validation experiments:

- For fluid dynamics experiments, see Wadhams et al., 2008; MacLean et al., 2008; Berry et al., 2014; Bruno et al., 2014; Harris et al., 2016; Andreani et al., 2016; Smith, 2017; Rhode and Oberkampf, 2017; Oberkampf and Smith, 2017; Beresh et al., 2020; Lynch et al., 2020; Lowe et al., 2020; Szoke et al., 2020.
- For solid mechanics experiments, see Thacker et al., 2004; Boyce et al., 2014; Boyce et al., 2016; Hack et al., 2016.

For recent research focused on quantification of systematic uncertainties in experimental diagnostics and data processing, see Pitts et al., 2006; Hariharan et al., 2011; Wilson and Smith, 2013; Sciacchitano et al., 2015; Lampeas et al., 2015; Neal et al., 2015; Ferrar, 2016; Limb et al., 2017; Ao et al., 2017.

11.7 Problems

1. In your field of interest, discuss how you would implement Guideline 1 in a model validation experiment.
2. In your field of interest, discuss how you would implement Guideline 2 in a model validation experiment.
3. In your field of interest, discuss how you would implement Guideline 3 in a model validation experiment.
4. In your field of interest, discuss how you would implement Guideline 4 in a model validation experiment.
5. In your field of interest, discuss how you would implement Guideline 5 in a model validation experiment.
6. In your field of interest, discuss how you would implement Guideline 6 in a model validation experiment.
7. In your field of interest, discuss how you would implement the DOE procedure to estimate experimental uncertainty.
8. In your field of interest, discuss what attributes should be included in a model validation experiment completeness table, like Table 11.7. For one of these attributes, describe the requirements for each of the four levels of completeness, as in Oberkampf and Smith, 2017.

11.8 References

Aeschliman, D. P., and W. L. Oberkampf (1998). Experimental Methodology for Computational Fluid Dynamics Code Validation. *AIAA Journal*, **36**(5), 733–741.

Aeschliman, D. P., W. L. Oberkampf, and F. G. Blottner (1995). A Proposed Methodology for CFD Code Verification, Calibration, and Validation. *16th International Congress on Instrumentation for Aerospace Simulation Facilities*, Paper 95-CH3482–7, Dayton, OH.

Aeschliman, D. P., W. L. Oberkampf, and H. F. Henfling (1994). Fast-Response, Electronically-Scanned Multi-Port Pressure System for Low-Pressure Hypersonic Wind Tunnel Applications. *AIAA Aerospace Ground Testing Conference*, AIAA Paper 94-2580, Colorado Springs, CO.

AGARD (1994). Quality Assessment for Wind Tunnel Testing. *NATO Advisory Group for Aerospace Research & Development (AGARD)*, AGARD-AR-304.

AIAA (1999). Assessment of Experimental Uncertainty with Application to Wind Tunnel Testing. *American Institute of Aeronautics and Astronautics*, S-071A-1999. Reston, VA.

AIAA (2003). Assessing Experimental Uncertainty – Supplement to AIAA S-071A-1999, Vol. G-045-2003. American Institute of Aeronautics and Astronautics, Reston, VA.

Andreani, M., A. Badillo, and R. Kapulla (2016). Synthesis of the OECD/NEA-PSI CFD Benchmark Exercise. *Nuclear Engineering and Design*, **299**, 59–80.

ANSI (1997). U.S. Guide to the Expression of Uncertainty in Measurement. *American National Standards Institute*, ANSI/NCSL Z540–2-1997, Boulder, CO.

Ao, D., Z. Hu, and S. Mahadevan (2017). Design of Validation Experiments for Life Prediction Models. *Reliability Engineering and System Safety*, **165**, 22–33.

Barber, T. J. (1998). Role of Code Validation and Certification in the Design Environment. *AIAA Journal*, **36**(5), 752–758.

Benek, J. A., E. M. Kraft, and R. F. Lauer (1998). Validation Issues for Engine – Airframe Integration. *AIAA Journal*, **36**(5), 759–764.

Beresh, S. J., M. F. Barone, K. J. Dowding, K. P. Lynch, N. E. Miller, and B. W. Lance (2020). A CFD Validation Challenge for Transonic, Shock-Induced Separated Flow: Approach and Metrics. *AIAA SciTech Forum*, Paper No. 2020-1308, Orlando, FL. DOI: 10.2514/6.2020-1308.

Berry, S. A., M. N. Rhode, and K. Edquist (2014). Supersonic Retropropulsion Validation Experiment in the NASA Langley Unitary Plan Wind Tunnel. *Journal of Spacecraft and Rockets*, **51**(3), 664–679.

Box, G. E. P., J. S. Hunter, and W. G. Hunter (2005). *Statistics for Experimenters: Design, Innovation, and Discovery*, 2nd ed. New York, NY: John Wiley.

Boyce, B. L., S. L. B. Kramer, T. R. Bosiljevac, E. Corona, J. A. Moore, K. Elkhodary, et al. (2016). The Second Sandia Fracture Challenge: Predictions of Ductile Failure under Quasi-Static and Moderate-Rate Dynamic Loading. *International Journal of Fracture*, **198**(1–2), 5–100.

Boyce, B. L., S. L. B. Kramer, H. E. Fang, T. E. Cordova, M. K. Neilsen, K. Dion, et al. (2014). The Sandia Fracture Challenge: Blind Round Robin Predictions of Ductile Tearing. *International Journal of Fracture*, **186**(1–2), 5–68.

Bruno, L., M. V. Salvetti, and F. Ricciardelli (2014). Benchmark on the Aerodynamics of a Rectangular 5:1 Cylinder: An Overview after the First Four Years of Activity. *Journal of Wind Engineering and Industrial Aerodynamics*, **126**, 87–106.

Chapman, D. R., H. Mark, and M. W. Pirtle (1975). Computer vs. Wind Tunnels. *Astronautics & Aeronautics*, **13**(4), 22–30.

Coleman, H. W., and W. G. Steele, Jr. (2018). *Experimentation, Validation, and Uncertainty Analysis for Engineers*, 4th ed. New York, NY: John Wiley.

Cosner, R. R. (1995). CFD Validation Requirements for Technology Transition. *26th AIAA Fluid Dynamics Conference*, AIAA Paper 95-2227, San Diego, CA.

Devore, J. L. (2007). *Probability and Statistics for Engineers and the Sciences*, 7th ed. Pacific Grove, CA: Duxbury.

Drosg, M. (2009). *Dealing with Uncertainties: A Guide to Error Analysis*, 2nd ed. New York, NY: Springer.

Ferrar, A. M. (2016). A Framework for Test-to-Test Averaging and Representation of Asynchronous Spatial Measurements. *Journal of Verification, Validation and Uncertainty Quantification*, **1**(3), 031003-1–031003-7.

Grabe, M. (2014). *Measurement Uncertainties in Science and Technology*, 2nd ed. New York, NY: Springer.

Hack, E., G. Lampeas, and E. A. Patterson (2016). An Evaluation of a Protocol for the Validation of Computational Solid Mechanics Models. *Journal of Strain Analysis*, **5**(1), 5–13.

Hariharan, P., M. Giarra, V. Reddy, S. W. Day, K. B. Manning, S. Deutsch, et al. (2011). Multilaboratory Particle Image Velocimetry Analysis of the FDA Benchmark Nozzle Model to Support Validation of Computational Fluid Dynamics Simulations. *Journal of Biomechanical Engineering*, **133**, 041002-1–041002-14.

Harris, J. R., B. W. Lance, and B. L. Smith (2016). Experimental Validation Data for Computational Fluid Dynamics of Forced Convection on a Vertical Flat Plate. *Journal of Fluids Engineering*, **138**, 011401-1–011401-14.

Henrion, M., and B. Fischhoff (1986). Assessing Uncertainty in Physical Constants. *American Journal of Physics*, **54**(9), 791–798.

Hinkelmann, K., and O. Kempthorne (2008). *Design and Analysis of Experiments: Volume 1 – Introduction to Experimental Design*, 2nd ed. New York, NY: John Wiley.

ISO (1995). Guide to the Expression of Uncertainty in Measurement. *International Organization for Standardization*, ISO/IEC Guide 98:1995, ISBN 92-67-10188-9. Geneva, Switzerland.

ISO (2008). Uncertainty of Measurement – Part 3: Guide to the Expression of Uncertainty in Measurement. *International Organization for Standardization*, ISO/IEC Guide 98-3:2008. Geneva, Switzerland.

JCGM (2008). Evaluation of Measurement Data – Guide to the Expression of Uncertainty in Measurement. www.bipm.org/utils/common/documents/jcgm/JCGM_100_2008_E.pdf

Lampeas, G., V. Pasialis, X. Lin, and E. A. Patterson (2015). On the Validation of Solid Mechanics Models Using Optical Measurements and Data Decomposition. *Simulation Modelling Practice and Theory*, **52**, 92–107.

Limb, B. J., D. G. Work, J. Hodson, and B. L. Smith (2017). The Inefficacy of Chauvenet's Criterion for Elimination of Data Points. *Journal of Fluids Engineering*, **139**, 054501-1–054501-3.

Lindl, J. D., P. Amendt, R. L. Berger, S. G. Glendinning, S. H. Glenzer, S. W. Haan, et al. (2004). The Physics Basis for Ignition Using Indirect-Drive Targets on the National Ignition Facility. *Physics of Plasmas*, **11**(2), 339–491.

Lowe, K. T., C. T. Beardsley, A. Borgoltz, W. J. Devenport, J. E. Duetsch-Patel, D. J. Fritsch, et al. (2020). Status of the NASA/Virginia Tech Benchmark Experiments for CFD Validation. *AIAA SciTech Forum*, Paper No. 2020-1584, Orlando, FL. DOI: 10.2514/6.2020-1584.

Lynch, K. P., B. W. Lance, G. S. Lee, J. W. Naughton, N. E. Miller, M. F. Barone, et al. (2020). A CFD Validation Challenge for Transonic, Shock-Induced Separated Flow: Experimental Characterization. *AIAA SciTech Forum*, Paper No. 2020-1309, Orlando, FL. DOI: 10.2514/6.2020-1309.

MacLean, M. G., T. P. Wadhams, M. S. Holden, and H. Johnson (2008). Ground Test Studies of the HIFiRE-1 Transition Experiment, Part 2: Computational Analysis. *Journal of Spacecraft and Rockets*, **45**(6), 1149–1164.

Marvin, J. G. (1995). Perspective on Computational Fluid Dynamics Validation. *AIAA Journal*, **33**(10), 1778–1787.

Moffat, R. J., and R. W. Henk (2021). *Planning and Executing Credible Experiments*. New York, NY: John Wiley.

Montgomery, D. C. (2017). *Design and Analysis of Experiments*, 9th ed. New York, NY: John Wiley.

Nagai, S., and H. Iijima (2011). Uncertainty Identification of Supersonic Wind-Tunnel Testing. *Journal of Aircraft*, **48**(2), 567–577.

Neal, D. R., A. Sciacchitano, B. L. Smith, and F. Scarano (2015). Collaborative Framework for PIV Uncertainty Quantification: The Experimental Database. *Measurement Science and Technology*, **26**, 074003-1–074003-17.

Oberkampf, W. L., and D. P. Aeschliman (1992). Joint Computational/Experimental Aerodynamics Research on a Hypersonic Vehicle: Part 1, Experimental Results. *AIAA Journal*, **30**(8), 2000–2009.

Oberkampf, W. L., D. P. Aeschliman, J. F. Henfling, and D. E. Larson (1995). Surface Pressure Measurements for CFD Code Validation in Hypersonic Flow. *26th AIAA Fluid Dynamics Conference*, AIAA Paper 95-2273, San Diego, CA.

Oberkampf, W. L., D. P. Aeschliman, J. F. Henfling, D. E. Larson, and J. L. Payne (1996). Surface Pressure Measurements on a Hypersonic Vehicle. *34th Aerospace Sciences Meeting*, AIAA Paper 96-0669, Reno, NV.

Oberkampf, W. L., and F. G. Blottner (1998). Issues in Computational Fluid Dynamics Code Verification and Validation. *AIAA Journal*, **36**(5), 687–695.

Oberkampf, W. L., A. Martellucci, and P. C. Kaestner (1985). SWERVE Surface Pressure Measurements at Mach Numbers 3 and 8. *Sandia National Laboratories*, SAND84-2149, SECRET Formerly Restricted Data, Albuquerque, NM.

Oberkampf, W. L., M. Pilch, and T. G. Trucano (2007). Predictive Capability Maturity Model for Computational Modeling and Simulation. *Sandia National Laboratories*, SAND2007-5948, Albuquerque, NM.

Oberkampf, W. L., and B. L. Smith (2017). Assessment Criteria for Computational Fluid Dynamics Model Validation Experiments. *Journal of Verification, Validation and Uncertainty Quantification*, **2**(3), 031002-1–031002-14.

Oberkampf, W. L., and T. G. Trucano (2002). Verification and Validation in Computational Fluid Dynamics. *Progress in Aerospace Sciences*, **38**(3), 209–272.

Oberkampf, W. L., D. P. Aeschliman, R. E. Tate, and J. F. Henfling (1993). Experimental Aerodynamics Research on a Hypersonic Vehicle. *Sandia National Laboratories*, SAND92-1411, Albuquerque, NM.

Oberkampf, W. L., T. G. Trucano, and C. Hirsch (2004). Verification, Validation, and Predictive Capability in Computational Engineering and Physics. *Applied Mechanics Reviews*, **57**(5), 345–384.

Pitts, W. M., A. V. Murthy, J. L. de Ris, J.-R. Filtz, K. Nygard, D. Smith, et al. (2006). Round Robin Study of Total Heat Flux Gauge Calibration at Fire Laboratories. *Fire Safety Journal*, **41**, 459–475.

Porter, J. L. (1996). A Summary/Overview of Selected Computational Fluid Dynamics (CFD) Code Validation/Calibration Activities. *27th AIAA Fluid Dynamics Conference*, AIAA Paper 96-2053, New Orleans, LA.

Rabinovich, S. G. (2017). *Evaluating Measurement Accuracy: A Practical Approach*, 3rd ed. New York, NY: Springer.

Rhode, M. N., and W. L. Oberkampf (2017). Estimation of Uncertainties for a Model Validation Experiment in a Wind Tunnel. *Journal of Spacecraft and Rockets*, **54**(1), 155–168.

Roache, P. J. (1998). *Verification and Validation in Computational Science and Engineering*. Albuquerque, NM: Hermosa Publishers.

Roy, C. J., W. L. Oberkampf, and M. A. McWherter-Payne (2003). Verification and Validation for Laminar Hypersonic Flowfields, Part 2: Validation. *AIAA Journal*, **41**(10), 1944–1954.

Roy, C. J., M. A. McWherter-Payne, and W. L. Oberkampf (2000). Verification and Validation for Laminar Hypersonic Flowfields. *Fluids 2000 Conference*, AIAA Paper 2000-2550, Denver, CO.

Roy, C. J., M. A. McWherter-Payne, and W. L. Oberkampf (2003). Verification and Validation for Laminar Hypersonic Flowfields, Part 1: Verification. *AIAA Journal*, **41**(10), 1934–1943.

Salicone, S., and M. Prioli (2018). *Measuring Uncertainty within the Theory of Evidence*. New York, NY: Springer.

Sciacchitano, A., D. R. Neal, B. L. Smith, S. O. Warner, P. P. Vlachos, B. Wieneke, et al. (2015). Collaborative Framework for PIV Uncertainty Quantification: Comparative Assessment of Methods. *Measurement Science and Technology*, **26**(7), 074004-1–074004-16.

Smith, B. L. (2017). The Difference between Traditional Experiments and CFD Validation Benchmark Experiments. *Nuclear Engineering and Design*, **312**, 42–47.

Szoke, M., V. Vishwanathan, J. T. Loeschen, A. Gargiulo, D. J. Fritsch, J. E. Duetsch-Patel, et al. (2020). Developing a Numerical Model of the Virginia Tech Stability Wind Tunnel for Uncertainty Quantification Based on Real-World Geometry. *AIAA SciTech Forum*, Paper No. 2020-0343, Orlando, FL. DOI: 10.2514/6.2020-0343.

Thacker, B. H., J. E. Pepin, and P. C. McKeighan (2004). Model Validation for the Collapse of Spherical Shells. *22nd IMAC Conference and Exposition on Structural Dynamics*, 1, 435–444, Dearborn, Michigan.

Wadhams, T. P., E. Mundy, M. G. MacLean, and M. S. Holden (2008). Ground Test Studies of the HIFiRE-1 Transition Experiment, Part 1: Experimental Results. *Journal of Spacecraft and Rockets*, **45**(6), 1134–1148.

Walker, M. A., and W. L. Oberkampf (1992). Joint Computational/Experimental Aerodynamics Research on a Hypersonic Vehicle: Part 2, Computational Results. *AIAA Journal*, **30**(8), 2010–2016.

Wilson, B. M., and B. L. Smith (2013). Uncertainty on PIV Mean and Fluctuating Velocity Due to Bias and Random Errors. *Measurement Science and Technology*, **24**(3). DOI: 10.1088/0957-0233/24/3/035302.

Youden, W. J. (1972). Enduring Values. *Technometrics*, **14**(1), 1–11.

12

Model Accuracy Assessment

Essentially, all models are wrong, but some are useful. —George E. P. Box

Contents of Chapter 12

As has been discussed in a number of chapters, particularly Chapter 10 "Model Validation Fundamentals" and Chapter 11 "Design and Execution of Model Validation Experiments," model accuracy assessment is the core issue of model validation. Our intent in model accuracy assessment is to critically and quantitatively determine the ability of a mathematical model and its embodiment in a computer code to simulate a well-characterized physical process. Well-characterized physical experiments are the most useful for model validation because of the higher precision in assessing the accuracy of the mathematical model. How critical and quantitative the model accuracy assessment is will depend on (a) how extensive the experimental data set in exploring the important model input quantities that affect the system response quantities (SRQs) of interest is, (b) how well characterized the important model input quantities based on measurements in the experiments are, (c) how well characterized the experimental measurements and the model predictions of the SRQs of interest are, (d) if the experimental measurements of the SRQs available to the computational analyst before the model accuracy assessment were conducted, and (e) if the experimental measurements were available to the computational analysts, and they were used for model updating or model calibration. This chapter will explore these intricate issues both conceptually and quantitatively.

We begin the chapter by discussing the fundamental elements of model accuracy assessment. As part of this discussion, we review traditional and recent methods for comparing model results and experimental measurements, and we explore the relationship between model accuracy assessment, model calibration, and model prediction. Beginning with the definitions of terms given in Chapter 2 "Fundamental Concepts and Terminology," the perspective of this book is to segregate, as best as possible, each of these activities. There is, however, an alternative approach in the published literature that believes all of these activities should be combined. We briefly review this alternative approach and contrast it with our perspective.

During the last decade, model accuracy assessment has focused on constructing mathematical operators that compute the difference between the experimentally measured results and simulation results. These operators are referred to as *validation metrics*. We review recommendations that have been proposed for the optimum construction of validation metrics. We then discuss in some detail two approaches to model validation that have been developed. The first approach can be used within a Bayesian framework and allows for the estimation of the discrepancy between the model prediction and experimental data. The second approach computes the area between the cumulative distribution functions and/or probability-boxes (p-boxes) resulting from the measurements and the predictions. Some illustrative examples are given for both. This chapter concludes with a discussion of u-pooling, an approach for combining physically unrelated quantities into a single probability distribution.

12.1 Elements of Model Accuracy Assessment

The task of model accuracy assessment might, at first, seem fairly simple: compare the prediction supplied by the modeler to the observation(s) made by the empiricist and see whether they match. They might match perfectly, or the model might be somewhat in error, or the model could be totally incorrect. However, in reality there are several issues that arise to complicate this comparison. Some of these may impact the model's ultimate use in predictions for which no experimental data are available. Some of the important questions are:

- How should one deal with experiment-to-experiment variability in the measured data?
- What should be done if the experimental data are given as a probability distribution versus a sequence of interval-valued quantities?
- What if there are statistical trends in the experimental data that do not appear in the simulation?
- What if the prediction is a probability distribution rather than a point value, i.e., a deterministic value?
- Do model accuracy requirements have any place in model accuracy assessment?
- Should the measure of model accuracy assessment represent evidence for *agreement* between experiment and simulation, or evidence for *disagreement*?
- What should be done in a comparison with experimental data if important information is *not* available to make the prediction?
- How should model accuracy assessment be influenced by limited experimental data obtained over a high-dimensional model input space?
- What could be claimed concerning model accuracy if there is only one experiment for comparison?
- What could be done to synthesize comparisons about different outputs from the model and experimental data?
- How should aleatory and epistemic uncertainty in either the simulation or the experiment be handled?
- How should an accuracy measure be constructed to penalize or reward a very precise model prediction versus a very imprecise prediction?
- What if the predictions from the model are extremely expensive to compute and only a few simulations can be computed for a comparison with measurements?
- What should be done differently in model accuracy assessment if the parameters in the model are first calibrated using related experimental data?

Some of these questions were dealt with in previous chapters, but they will be revisited here in more detail.

In many research situations in scientific simulations, the emphasis in validation is on deciding whether a model is right or wrong *per se*. In most engineering situations, however, this is *not* the case. In engineering, the emphasis is about estimating the accuracy of the model, based on comparisons with experimental data, and determining if the model is adequate for the intended use. As George Box famously asserted over three decades ago, “All models are wrong. Some are useful” (Box and Draper, 1987). For example, a model could be consistently inaccurate and yet close enough for its intended use. Likewise, even a highly accurate model might not be good enough if the accuracy requirements are more demanding for modeling some high-performance system.

12.1.1 Methods of Comparing Simulations and Experiments

There are a wide variety of methods for comparing simulations and experimental measurements. Figure 12.1 summarizes the diverse methods for comparisons and identifies the direction in which more quantitative methods need to progress. This figure illustrates the conceptual increase of quantification in performing validation comparisons as increased

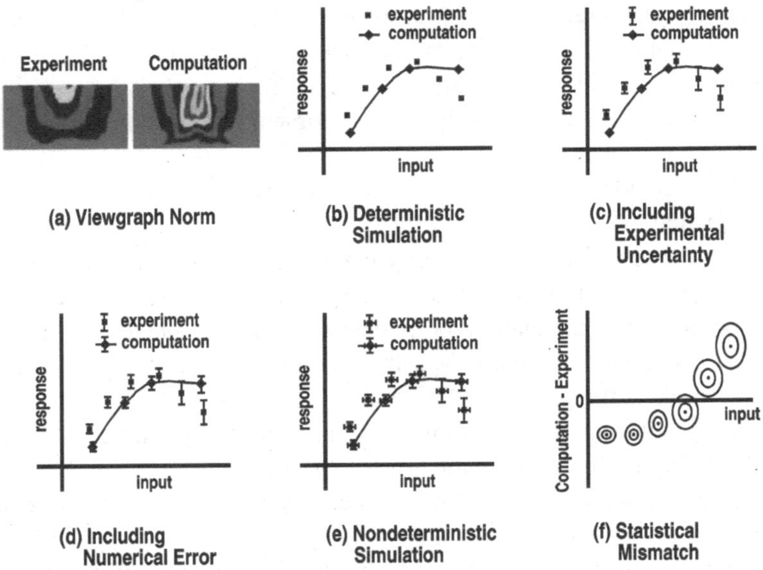

Figure 12.1 Increasing precision for comparing simulations and experiments.
(adapted from Oberkampf et al., 2004)

attention is paid to experimental uncertainty, numerical uncertainty, and nondeterministic simulations. Let us consider each panel of this figure in some detail.

Figure 12.1a depicts a qualitative comparison referred to as a *viewgraph norm* comparison (Rider, 1998) of computational and experimental data often seen in practice. This comparison is typically one picture or contour plot next to another and sometimes no legend is given showing the scales involved. Or similarly, the scales are adjusted so as to present the most flattering agreement between computation and experiment. Clearly, in such a case no quantitative statement about the comparison between computation and experiment can be made, but it does provide some feel of comparison at an intuitive level. Feel and intuition about agreement between computation and experiment are, of course, in the eye of the beholder. This type of comparison is commonly seen in advertising and marketing material for scientific computing software, or in proposals for potential funding.

The plot in Figure 12.1b portrays the most common type of comparison between computational results and experimental data. It shows the system response as a function of the input, or control, parameter in the experiment. While discrete experimental and computational points are shown in this plot, the style of comparison also encompasses the display of curves for experiment and computation without any points shown. The key problem with comparisons implemented at the level of Figure 12.1b is that there is no recognition of uncertainty in the experimental or computational results, or in the quantitative comparison of the two. Conclusions drawn from this type of comparison are really only qualitative, such as, "fair agreement" or "generally good agreement."

Figure 12.1c suggests that the next step for improving the method of comparison is to place estimated uncertainty bars around the experimental data. Occasionally, the meaning of the uncertainty bars is carefully justified and clearly explained. The much more common situation is one in which the uncertainty bars are ambiguous, in the sense that: (a) they are not rigorously justified concerning what is included in terms of random and systematic experimental uncertainty, (b) a statement is made such as "Essentially all of the experimental data fell within the uncertainty bars shown," or (c) the effect of experimental uncertainty in the input/control quantity is not addressed with regard to the response quantity uncertainty. An increasing number of technical journals are requiring some type of statement of experimental uncertainty, such as Figure 12.1c.

Figure 12.1d represents the case in which there is a more quantitative estimate of experimental uncertainty and there is an estimate of numerical solution uncertainty. For example, concerning the experimental uncertainty, multiple experimental realizations could have been obtained so the experimental data point shown would represent the mean of all of the samples. In addition, it might be clarified if any "outlier" measurements were discarded, and that the uncertainty bar would represent two standard deviations of an assumed normal probability distribution. Concerning the numerical solution uncertainty, an *a posteriori* numerical uncertainty estimate from the computation would be given for the specific response quantity that was plotted in the graph.

Figure 12.1e suggests a further improvement in the estimation of experimental uncertainty and now at this level, nondeterministic simulations are included. The information concerning experimental uncertainty could be improved, for example, in two ways. First, one could use a statistical design of experiments (DOE) approach with randomization and blocking to better quantify certain systematic uncertainties. Also, one may have conducted the same experiment at separate facilities, possibly using different diagnostic techniques. Second, an estimate of the experimental uncertainty in the measured input quantity is also obtained and is shown as the lateral uncertainty bar. The uncertainty bars for the input quantity and the measured response, for instance, could represent two standard deviations for a normal probability distribution. Concerning the nondeterministic computational simulations, we are referring to an ensemble of computations sampled for each of the experimental conditions. For example, multiple simulations would be made using the experimentally estimated probability distribution for the input quantity. As a result, the computational data point symbol would be the mean of the nondeterministic simulations for *both* the input data and the response quantity. Note that to use the experimental uncertainty distributions for the input quantity, computations would need to be sampled over the distribution of the measured input quantity.

Figure 12.1f shows a genuine quantitative measure of the comparison between the nondeterministic simulations and the experimental measurements over the range of the input quantity. What is depicted in Figure 12.1f is the *mismatch* between the simulation and the experiment at the mean value of each of the experimental data points. In terms of information content, one could have the same data as contained in Figure 12.1e, but now the statistical differences in the simulation and experiment are displayed. Assuming that probability distributions for both computational and experimental data are well characterized, as discussed regarding Figure 12.1e, comparing computations and experiments will require a difference, or more properly, a convolution of pairs of probability distributions. The elliptical symbols in Figure 12.1f are meant to signify one and two standard deviation contours of the convolutions of the simulation and experimental probability distributions. The "dot" in the center of each contour is the difference in the mean, or expected value, of the simulation and experimental distributions.

We will refer to quantitative comparisons between simulation and experiment, similar to Figure 12.1f, as a *validation metric operator*. We will formally use the definition:

Validation metric operator: a mathematical operator that measures the difference between a system response quantity (SRQ) obtained from a simulation result and one obtained from experimental measurements.

The validation metric should be an objective measure of the distance, in some sense, between simulation and experimental data. The distance measure should have the characteristic that any difference between the simulation and experimental data is a positive quantity. For example, whether the simulation is less than or greater than the experimental data, the metric is positive and additive. Objectiveness means that, given a collection of predictions and observations, a validation metric will produce the same assessment no matter what analyst uses the metric. This is a basic tenet of scientific and engineering practice; that the conclusion is reproducible and that it does not depend on the attitudes or predilections of the analysts. If some inescapable subjectivity must enter the evaluation of the metric, it would be good to keep this intrusion as small and as limited as possible so as to emphasize the objectiveness of the method and minimize the elements subject to dispute.

In general, the validation metric result is a function of all of the input quantities to the model. Commonly, however, the SRQ is primarily dependent on a handful of dominant input quantities. The validation metric should be viewed as a statistical operator because, in general, the simulation and experimental results are not single numbers but are functions. Specifically, they are commonly either a probability distribution or a *probability box* (p-box). A p-box is a special type of cumulative distribution function (CDF) that represents the set of possible CDFs that fall within prescribed bounds (Ferson et al., 2002; Ferson et al., 2004; Aughenbaugh and Paredis, 2006). The p-boxes were first introduced in Chapter 3 "Modeling and Computational Simulation." A p-box can express both epistemic and aleatory uncertainty in either the simulation or the experimental results, or both. The p-boxes will be discussed in more detail later in this chapter.

12.1.2 Uncertainty in Model Accuracy Assessment

Model accuracy assessment would be fairly straightforward if we did not have to deal with uncertainties clouding the issue in the experiment and in the simulation. In earlier chapters, e.g., Chapter 10 and Chapter 11, we have discussed uncertainties primarily from an estimation point of view. Here we discuss them with regard to where they occur, how they are propagated, and how they affect validation metrics results.

Figure 12.2 shows a validation experiment and the corresponding computational model, the sources of uncertainty in experimental measurements and numerical simulations, and the validation metric as a difference operator. First, we will discuss the experimental sources of uncertainty. In Chapter 10, we discussed the two fundamental sources of experimental uncertainty: measurement uncertainty and characterization uncertainty. Measurement uncertainty is due to random and systematic (bias) uncertainties in every experimental measurement that is made. Measurement uncertainty is primarily dependent on (a) the diagnostic techniques used, (b) the instrument calibration techniques used, and (c) the experimental measurement procedures used, e.g., replication, randomization, and blocking techniques to estimate random and systematic measurement uncertainties. Characterization uncertainty is

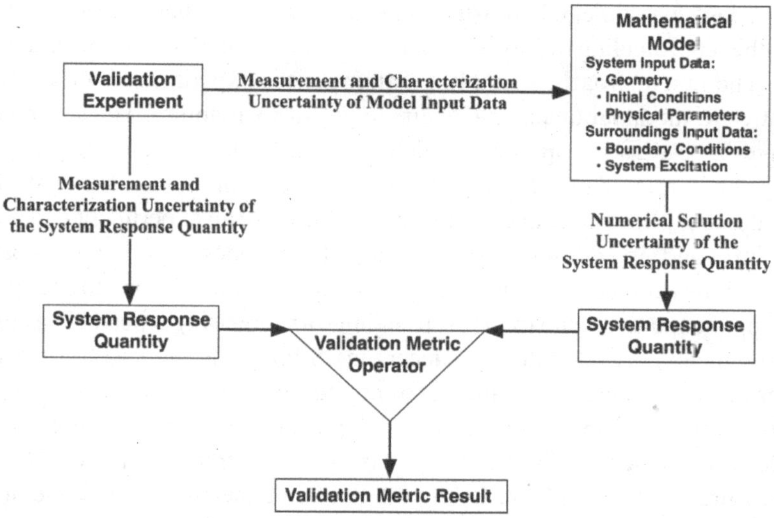

Figure 12.2 Sources of uncertainty in model accuracy assessment.

due to a limited number of measurements of a quantity that is a random variable. Two examples are (a) an input quantity needed for the computational model was not measured in the experiment, so the experimentalists recommend an interval in which they believe the true value lies, and (b) an SRQ is known to be a random variable in the experiment, but too few samples of the quantity are measured because of time and cost constraints.

Also discussed in Chapter 10 were the three sources of uncertainty in the simulation: approximations and formulation of the mathematical model (i.e., resulting in model form uncertainty), mapping of the mathematical model to the discrete model expressed in software, and numerical solution of the discrete model on a computer. On the right-hand side of Figure 12.2, we simply combine the second and third sources into "Numerical Solution Uncertainty." The estimation of the model form uncertainty is, of course, the primary goal of model accuracy assessment, which is indicated as the "Validation Metric Result."

It is seen in Figure 12.2 that the experimental measurement and characterization uncertainties directly affect both the input data to the model and the SRQ measured in the experiment. Noting that the experimental measurement input uncertainties are statistically confounded with the inherent variability of the input quantities in the experiment, an important issue in model accuracy assessment is exposed. We are requiring that the model correctly map the experimental measurement input uncertainties to the SRQ computed, even though there is *no physics variability* that corresponds to these in the model. Stated differently, the experimental measurement uncertainties *do not* represent physics uncertainties in the input but are purely an artifact of the measurements made in the experiment. Contrast these experimental uncertainties with true physics uncertainties in the experiment, such as: (a) a parameter in the experiment that is a random variable, e.g., the material used in an experiment that is a random draw from a production-lot population, or (b) the boundary conditions are uncontrolled parameters in the experiment, e.g., weather conditions in multiple flight experiments. These later examples are *true physics uncertainties* that we expect the model to correctly propagate, whereas we are also requiring the model to correctly propagate non-physics uncertainties, i.e., experimental measurement uncertainties. There are two approaches to addressing the confounding of experimental measurement uncertainty and model input uncertainty. The first approach is simply to minimize the experimental measurement uncertainties relative to the input uncertainties due to physics variability. The second approach is to use a Bayesian estimation approach which is discussed later in this chapter.

To clarify the confounding of the measurement uncertainty and the model input uncertainty, consider the following conceptual example. Suppose an experiment is repeated a large number of times and, in addition, every aspect of the physics is perfectly repeatable as the ensemble of all measurements. That is, all (uncertain) input conditions to the system and all (uncertain) input conditions due to the surroundings are precisely repeatable as the ensemble of all measurements. As a result, the (uncertain) SRQ is also precisely repeatable as the ensemble of all measurements. Further, assume that the experimental measurement uncertainty is *zero* for the measurement of the model inputs, but random measurement uncertainty exists in the measurement of the SRQ, i.e., the system output. As a result, (a) the experimentally measured SRQ would consist of both random uncertainty due to the uncertainty in the random inputs in the experiment, as well as randomly distributed measurement uncertainty, and (b) the experimental characterization error would be zero because all input quantities and the SRQ are perfectly characterized (i.e., known) because of the large number of random samples. In addition, suppose the model *perfectly* represents the relevant physics in the validation experiment, which means the model *perfectly maps* the input uncertainties to the uncertainty in the SRQ. Finally, assume that the numerical solution uncertainty is zero.

Given these assumptions in order to clarify the issue at hand, Figure 12.3 gives an example of the cumulative distribution functions of the SRQ obtained from the experimental measurements and the CDF from the simulation model. As shown in Figure 12.3, these two CDFs will be different because the CDF from the experiment represents both the variability due to the input uncertainty and the random measurement error in the SRQ, whereas the

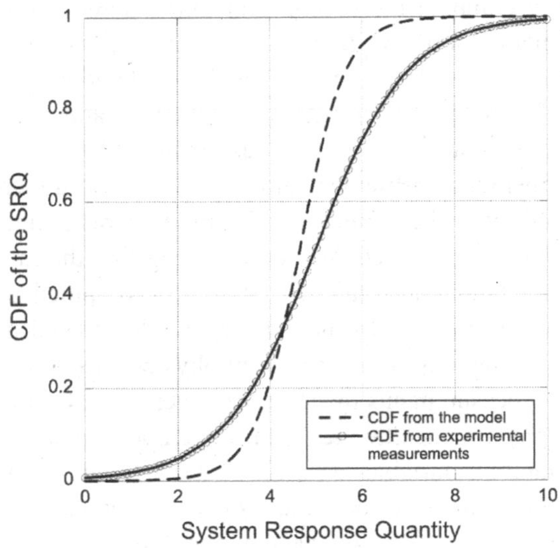

Figure 12.3 Example of cumulative distribution functions for a system response quantity from experimental measurements and from a model.

CDF from the model only represents the variability in the SRQ due to the variability in the input quantities. That is, the perfect physics model, with zero numerical uncertainty, will perfectly propagate the experimental input uncertainty to the SRQ. When the validation metric operator measures the difference between the experimentally measured CDF and the CDF from the model, the difference will *not* be zero. That is, the model would be judged to be "imperfect," when in fact it is *perfect*. The random measurement uncertainty in measuring the SRQ shows itself as increased variability (i.e., dispersion) relative to a measurement that contained *no* random measurement uncertainty. Note that a Bayesian estimation approach would be able to identify this discrepancy between the two CDFs as an experimental measurement uncertainty, as opposed to a model form uncertainty.

12.1.3 Relationship between Model Accuracy Assessment, Calibration, and Prediction

Figure 12.4 depicts several important aspects of model validation, as well as features of calibration and prediction using a simulation model. The left-center portion of Figure 12.4 shows the first step in model validation. The figure illustrates that the same SRQ must be obtained from both the model and the physical experiment. The SRQ can be any type of physically measurable quantity, or it can be a quantity that is based on, or inferred from, measurements. For example, the SRQ can involve derivatives, integrals, or more complex data processing of computed or measured quantities such as the maximum or minimum of functionals over a domain. When significant data processing is required to obtain an SRQ, it is important to process both the computational results and the experimentally measured quantities in *the same manner.* The computational and experimental SRQs are input to the validation metric operator to compute a validation metric result. The SRQs are commonly one of three mathematical forms: (a) a deterministic quantity, e.g., a single (scalar) value, or a

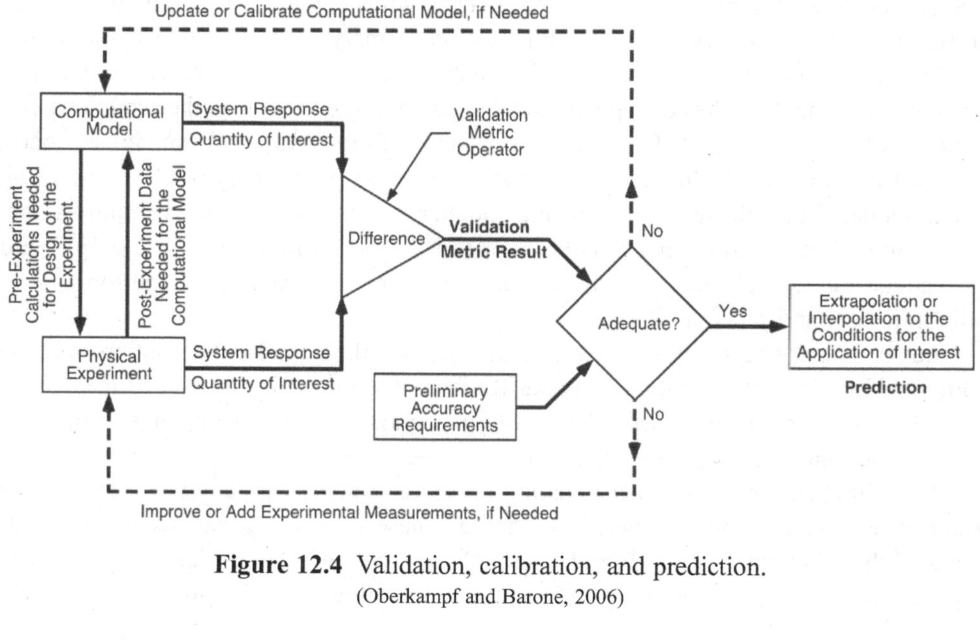

Figure 12.4 Validation, calibration, and prediction.
(Oberkampf and Barone, 2006)

mean value, or a minimum or maximum value over a domain; (b) a probability distribution; or (c) a p-box. Each of these forms can be functions of a parameter or multiple parameters in the model, such as a temperature or pressure; a function of spatial coordinates, such as Cartesian coordinates (x, y, z); or a function of both space and time. If both the computational and experimental SRQs are deterministic quantities, the validation metric will also be a deterministic quantity. If either of the SRQs is a probability distribution or p-box, the result of the validation metric could be a scalar, a probability distribution, or a p-box; depending on the construction of the validation metric operator.

As discussed in Chapters 10 and 11, Figure 12.4 suggests the desired interaction between computation and experimentation that should occur in a validation experiment. To achieve the most value from the validation experiment, there should be in-depth, forthright, and frequent communication between analysts and experimentalists during the planning, design, and execution of the experiment. Also, after the experiment has been completed, the experimentalists should measure and provide to the analysts all the important *input quantities* needed to conduct the computational simulation. What should *not* be provided to the analysts in a rigorous validation activity is the *measured SRQ*. Stated differently, a blind-test prediction should be compared with experimental results so that a true measure of predictive capability can be assessed in the validation metric.

The right-center portion of Figure 12.4 shows the comparison of the validation metric result with the preliminary accuracy requirement for the application of interest. This is referred to as a *preliminary* requirement because the conditions of the application of interest, i.e., the application domain, may be different from the conditions of the validation experiment. If they are different, as is the usual case, then only preliminary accuracy requirements can be applied. If the model passes these preliminary accuracy requirements, then the model prediction would be made using extrapolation or interpolation of the model. Accuracy requirements for the application, i.e., final accuracy requirements, would be applied after the model has been used to extrapolate or interpolate to the conditions of the application of interest. Setting preliminary accuracy requirements, which should be more demanding than for the application of interest, is also useful if the preliminary model accuracy is not satisfied. The following gives two examples. First, it may be concluded that the assumptions and approximations made in the conceptual model are inadequate to achieve the accuracy needed for the validation experiment. This may necessitate a reformulation of the model, as opposed to calibration of model parameters at a later stage. Second, preliminary accuracy assessment provides a natural mechanism for adjusting and adapting the final accuracy requirements to the cost and schedule requirements of the engineering system of interest. For example, trade-offs can be made between required final accuracy, system design, and performance, in addition to cost and schedule.

If the preliminary accuracy requirements are not met, then one has two options as shown in Figure 12.4. First, the dashed-line upper feedback loop provides for any of the activities referred to as model updating, model calibrating, or model tuning. Depending on the ability to determine model input parameters, either by direct measurement or by inference from a model, we have divided parameter updating into three categories: parameter measurement, parameter estimation, and parameter calibration. These three categories were discussed in Chapter 2 "Fundamental Concepts and Terminology," and they will be discussed later in this chapter. If the model parameters are updated, one could proceed again through the validation metric operator. It is now very likely that the validation metric result would be significantly reduced, implying much better agreement between the model and the experimental results. The parameter updating may be physically justifiable, or it may simply be expedient. How to determine the scientific soundness of the updating and its impact on predictive capability is a very difficult issue. This issue will be discussed in more detail in the next chapter. Second, the lower feedback loop could be taken if: (a) improvements or changes are needed in the experimental measurements, such as improved diagnostic techniques; (b) additional experiments are needed to reduce the sampling uncertainty for the experimental measurements; or (c) additional experiments are needed that are more representative of the application of interest.

The right portion of the figure deals with the issue of use of the model to make a prediction for the conditions of the application of interest. It should be noted that Figure 12.4 is completely consistent with the diagram showing the three aspects of model validation (Figure 2.8) in Chapter 2. Also, Figure 12.4 is conceptually consistent with the ASME Standard diagram shown in Figure 2.5 of Chapter 2. The ASME diagram, however, does not explicitly show the extrapolation or interpolation of the model to conditions where no experimental data are available. It is important stress that each of these figures can be applied at *any* level of the validation hierarchy.

Two examples of applications requiring different types of extrapolation are given. First, suppose the physical experiment conducted is on the complete system of interest, i.e., the actual operational system hardware. Suppose the experiment, however, is conducted with a set of conditions that do not correspond to the actual operational conditions of interest because of physical size constraints, safety concerns, or environmental restrictions. The SRQs are measured in the experiment, a model prediction is made, and a validation metric result is computed. Suppose further that the model satisfies the preliminary accuracy requirements. Then an extrapolation is required, using the model, to predict the SRQs of interest corresponding to the actual operational conditions of interest. This type of extrapolation is the most common in engineering applications.

Second, suppose now that the experiments in Figure 12.4 are conducted on all of the subsystems of the complete system of interest, but no validation experiments were conducted on the complete system. Suppose further that all of the validation metrics of all of the SRQs of interest for each subsystem meet the preliminary accuracy requirements. Even if the subsystems are tested at the same operational conditions as the complete system, any real engineering system functions as a coupled, highly interacting group of subsystems. As a result, one could characterize this as a large extrapolation because the model of the complete system is required to predict the interactions of each of the models for each of the subsystems. This extrapolation, of course, has no parameter associated with it, such as the first example, and the ability of the model to predict the interactions of the subsystems has in no way been assessed.

12.2 Approaches to Parameter Estimation and Validation Metrics

During the last two decades, model validation (in the broad sense of the term) and validation metrics have received increased attention. Model validation has been primarily pursued using traditional approaches, specifically, parameter estimation, hypothesis testing, and

Bayesian updating. These approaches, particularly hypothesis testing, are related to the concepts embodied in validation metrics as discussed here. Although we refer to parameter estimation and Bayesian updating as model calibration, many researchers and practitioners in the field refer to these approaches as model validation. In this section, parameter estimation, hypothesis testing, Bayesian updating, and newer approaches to validation metrics will be briefly discussed. It should be stressed that the development of quantitative measures for model accuracy assessment is a new field of research, and various perspectives are being vigorously debated. A special workshop was held in May 2006 to bring together a number of researchers in the field to discuss and debate various approaches to model validation. For this workshop, three challenge problems were proposed, each one in a different engineering field: heat transfer, solid mechanics, and solid dynamics. All of the contributions to the workshop were published in a special issue of *Computer Methods in Applied Mechanics and Engineering* (Pilch, 2008; Dowding, Red-Horse, et al., 2008). The reader should consult this special issue for an excellent survey of the wide-ranging approaches to model validation.

12.2.1 Parameter Estimation

In the 1960s, the structural dynamics community began developing sophisticated techniques for comparing computational and experimental results, as well as techniques for improving agreement between the results using parameter estimation techniques (Wirsching et al., 1995). In many analyses in structural dynamics, certain model input parameters are considered as deterministic (but poorly known) quantities that are estimated using experimental modal data on the structures of interest. A numerical optimization procedure is used so that the best agreement between computational and experimental results can be obtained for a single SRQ or a group of SRQs. A large number of solutions of the model are required to evaluate the effect of different values of the model parameters on the SRQ. Although these techniques are used to compare computational and experimental results, their primary goal is to improve agreement based on available experimental data. During the last decade, more sophisticated and reliable methods have been developed for optimizing parameters in a wider range of stochastic systems (van den Bos, 2007; Crassidis and Junkins, 2012; Aster et al., 2019). In this more recent work, the parameters are considered as random variables represented as probability distributions with unknown parameters or represented by nonparametric distributions. The experimental data can come from multiple experiments on the same structure but under different excitation and loading conditions, or from experiments on physically different structures, but all with the same design and assembly features. A similar, but more complex optimization procedure is used to determine the probability distributions for each of the uncertain parameters that maximize agreement between the model and the experimental results.

Although the terminology used in this text refers to this type of procedure as parameter estimation, as opposed to model validation, we mention it here because the process begins with estimating the difference between the model results and the experimental results.

12.2.2 Hypothesis Testing

Statistical hypothesis testing, also referred to as significance testing, is used throughout the experimental sciences (Devore, 2016; Ross, 2020; Lehmann and Romano, 2022).

Hypothesis testing is a highly developed statistical method of choosing between two competing models of an experimental outcome by using probability theory to minimize the risk of an incorrect decision. In hypothesis testing, the validation-quantification measure is formulated as a "decision problem" to determine whether or not the hypothesized model is consistent with the experimental data. This technique is regularly used in the operations research (OR) community for comparing mutually exclusive models, i.e., the model is either true or false. For example, suppose the hypothesis is made that a coin is fair, i.e., in the toss of the coin it is equally likely that "heads" will appear as often as "tails." The competing hypothesis is that the coin is unfair. Experimental data are then obtained by tossing the coin a number of times, say *N*, and recording what percentage of the outcomes is heads and what percentage is tails. Hypothesis testing then allows one to statistically determine the confidence of a fair coin. The confidence in the determination will depend on *N*, that is, as *N* increases, the confidence in the conclusion increases.

Hypothesis testing has been employed in model validation studies (Li et al., 2014; Nannapaneni and Mahadevan, 2016; Deng et al., 2018; Vanslette et al., 2020). In this approach, the validation assessment is formulated as a decision problem of whether or not the model's predictions are consistent with the available empirical information. Typically, empirical observations are collectively compared to a distribution of realizations computed by the model, such as by Monte Carlo sampling, which together constitutes a prediction about some SRQ. The consistency between the model and the data is characterized as a probability value, with low probability values denoting a mismatch of such magnitude that it is unlikely to have occurred by chance. This probability is often considered meaningful only within the hypothesis-testing context and does not serve as a stand-alone validation metric that indicates the degree of agreement or disagreement between model and empirical results. Instead, the approach focuses on answering the yes-no question about whether a model is accurate to within a specified margin of error, or sometimes on the either-or question choosing between mutually exclusive models.

Despite its use as a tool for model validation, hypothesis testing is not ideally suited to the task. In general, the purpose of hypothesis testing is to identify statements for which there is compelling evidence of truth or falsehood. But this is a different goal than that of model validation, which is more pragmatic and more focused on assessing the quantitative accuracy of a model. The model could be relatively "poor," but the question in model validation should be stated as "How poor?" In hypothesis testing, an accuracy requirement could be stated either as a preliminary or final accuracy statement, and then the result would be the probability that the model met the required accuracy. Two practical difficulties arise with this type of result. First, how should a builder of physics models, a project manager, or a decision maker interpret the result? It is not intuitive how to interpret probability as an accuracy measure. The natural perspective of design engineers and project managers is to ask, "What is the absolute or relative error of the model?" Second, no matter what level of accuracy is specified, the model can be proven false at that level with more experimental data. That is, any model can be proven false, given enough data. As will be discussed shortly, a key difficulty with hypothesis testing is that the requirement of accuracy is built *directly into* the mismatch measure.

Hypothesis-testing has developed the concept of error types for incorrect conclusions which is informative for model validation (Law, 2014; Lehmann and Romano, 2022). They are actually types of errors in logic. A type 1 error, also referred to as *model builder's risk*, is the error in rejecting the validity of a model when the model is actually valid. This can be caused by errors on both the computational side and the experimental side. On the computational side, for example, if a mesh is not sufficiently converged and the computational result is contaminated by numerical error, then an adverse comparison with experimental data is misleading. That is, a poor comparison leads one to conclude that a model needs to be improved or recalibrated when the source of the poor comparison could simply be an under-resolved mesh. On the experimental side, the model builder's risk is most commonly caused by a poor comparison of computational results and experimental data that is due to an unknown bias error in the experimental data. We believe that unknown bias errors in experimental results are the *most damaging* in model validation because if the experimental measurement is accepted as correct, then it is concluded that the computational result is in error, whereas in reality, the experimental data is the culprit. If the error is believed to be in the simulation, then a great deal of effort will be expended trying to find the source of the error. Or worse, a computational model will be recalibrated using the biased experimental data. This results in *transferring* the experimental bias into the mathematical model and then biasing all future computations using the model.

The type 2 error, also referred to as *model user's risk*, is the error in accepting the validity of a model when the model is actually invalid. As with the type 1 error, this can be caused by errors on both the computational side and the experimental side. On the computational side, the logical reverse of the type 1 error described above can occur. That is, if a mesh is not sufficiently converged and the computational result agrees well with the experiment, then the favorable comparison is also misleading. For example, if a finer mesh is used, one can find that the favorable agreement can disappear. This shows that the original favorable agreement has compensating, or canceling, errors in the comparison. We believe that compensating errors in complex simulations is a ubiquitous phenomenon in complex simulations (Lenhard, 2018). Only the tenacious user of the model may dig deep enough to uncover compensating errors. In a competitive, time-constrained, or commercial code-development environment, such users or model developers as these can be very unpopular, and even muffled by coworkers and project management. On the experimental side, the logical reverse of the type 1 error described above can occur. That is, if an unknown bias error exists in the experiment, and a favorable comparison between computational results and experimental data is obtained, the implication of good simulation accuracy is actually incorrect. Similar to the type 2 error on the computational side, only the self-critical and determined experimentalist will continue to examine the experiment in an attempt to find any experimental bias errors. For further discussion of type 1 and type 2 errors in model validation, see Chapter 12 of Oberkampf and Roy (2010).

12.2.3 Bayesian Updating

Bayesian updating, or Bayesian statistical inference, has received a great deal of attention during the last four decades from statisticians, risk analysts, and some physicists and structural dynamicists. For a modern treatment of the topic from a Bayesian perspective, see Gelman et al. (2013), Smith (2014), Martin and Liu (2016), and Santner et al. (2018). Although the process is quite involved, Bayesian analysis can be summarized in three steps.

Step 1 is to construct, or assume, a probability distribution for each input quantity in the model that is chosen to be a tunable, random variable. Step 2 involves conditioning or updating the previously chosen probability models for the input quantities based on comparison of the computational and experimental results. To update the probability models, one must first propagate input probability distributions through the model to obtain probability distributions for the SRQs commensurate with those measured in the experiment. The updating of the input probability distributions, using Bayes Theorem to obtain posterior distributions, commonly assumes that the computational model is structurally correct. That is, the updating is done assuming that all of the disagreement between the model and the experimental data is due to deficient probability distributions of the input parameters. Step 3 involves comparing new computational results with the existing experimental data and quantifying the changes in the updated probability distributions. The new computational results are obtained by propagating the updated input probability distributions through the model. If any new experimental data becomes available, then the entire process is repeated. Much of the theoretical development in Bayesian estimation has been directed toward optimum methods for updating statistical models of uncertain input parameters and in reducing the computational resources required for propagating the uncertainties through the model.

During the last two decades, the Bayesian approach has taken a different direction in development. In 2001, Kennedy and O'Hagan (2001) devised an approach to estimate the model form uncertainty, which could be used in combination with updating the model input parameters using experimental measurements. They referred to the model form uncertainty as the *model discrepancy*. The model discrepancy term was explicitly added to the formulation so that a bias, or systematic, error in the assumed model could be included in the estimation process. The model discrepancy term is characterized as a random variable, but the mean of the random variable is *not* required to be zero. As a result, the model discrepancy is very close to the concept of estimating the accuracy of a mathematical model with respect to experimental measurements of the response of a system. The approach of Kennedy and O'Hagan (2001) is discussed in detail in Section 12.4.

The Bayesian perspective always prefers to place any analytical question in the context of a decision problem. Model accuracy assessment per se need not be a part of decision making. Asking whether, or to what degree, a model is supported or contradicted by available evidence is surely a legitimate question by itself. Assessing the value of a metric that measures the conformity of the model structure with observations should be recognized as a reasonable question to ask. What one does with the knowledge that different models have different discrepancy metrics is not formally part of the decision problem. Bayesians might argue that it should be because making good decisions requires such inclusive considerations. For various practical reasons discussed in this chapter, model accuracy assessments are sometimes needed before the decision context has even been specified.

12.2.4 Comparison of Mean Values

A distinctly different approach, and one that is conceptually and mathematically simpler, is to compare the estimated mean (or expected) values from both the experiment and the simulation. For the case where there is only one experiment and only a deterministic simulation result, the comparison approach is equivalent. For comparing a single value of the experiment and a single value of the simulation, one would intuitively think of using traditional vector norms. Let $S(x_i)$ be the simulation result as a function of the control parameter x_i and let $E(x_i)$ be the experimental measurement, where $i = 1,2,3...N$. The parameter x_i could be an input parameter or it could be time in a time dependent system. The L_1 and L_2 vector norms have been used as validation metrics, which are given by

$$\|S - E\|_p = \left(\sum_{i=1}^{N} |S(x_i) - E(x_i)|^p \right)^{1/p}, \tag{12.1}$$

where $p = 1$ or 2.

Several researchers have constructed validation metrics based on the L_1 and L_2 norms (Coleman and Stern, 1997; Stern et al., 2001; Oberkampf and Barone, 2006; Oberkampf and Roy, 2010). For strongly time dependent responses or system responses with a wide spectrum of frequencies, metrics have been constructed that summed the amplitude and phase differences between the simulation and the experiment (Geers, 1984; Russell, 1997a, 1997b; Sprague and Geers, 1999, 2004). Although the vector norm metrics and those that combine amplitude and phase errors have different perspectives, their common theme is a more engineering-oriented, less statistical approach than those discussed above. They focus only on comparing a deterministic result of the SRQ from the model with the estimated statistical mean (or a single time history) of the experimental data. Most of these approaches do not propagate uncertain input parameters through the model to obtain a probability distribution for the SRQs of interest. Rather, the approximation is made that the mean of any uncertain input quantities maps to the mean of the SRQs of interest.

The primary advantage in deterministic computational results, as opposed to propagating input uncertainties to determine output uncertainties, is the much lower computational costs involved in deterministic simulations. Many computational analysts argue that computational resources are not available to provide both spatially and temporally mesh-resolved solutions, as well as a large number of nondeterministic solutions, for complex simulations. Risk assessment of high-consequence systems, e.g., safety of nuclear power reactors and underground storage of nuclear waste, has shown that with an adequate, but not excessive, level of physical modeling detail, one *can* afford the computational costs of nondeterministic simulations. However, we recognize that there is substantial resistance in many fields to attain both mesh-resolved, nondeterministic simulations. Consequently, there is a need to construct validation metrics that can be computed with only deterministic computational results. The perspective of many computational analysts is that physics modeling fidelity is *much more* important than uncertainty quantification in engineering analyses. As a result, they construct models of such physical complexity that they consume essentially all of the computer resources available for a single solution, leaving no computer resources for nondeterministic simulations. This tradition is deeply embedded in many scientific and engineering fields. As a result, it will be very slow and difficult to change this culture.

An alternative approach to model validation that involves comparing mean (or expected) values was developed by the ASME Committee on Verification and Validation in Computational Fluid Dynamics and Heat Transfer (ASME, 2009), referred to herein as V&V 20. The implementation of V&V 20 involves the computation of two quantities. The first is the error E between the mean values of the simulation S and the experiment D:

$$E = S - D.$$

The second quantity is the standard validation uncertainty, u_{val}, which is given by

$$u_{val} = \sqrt{u_{num}^2 + u_{input}^2 + u_D^2},$$

where u_{num} is the numerical uncertainty in the model SRQs, u_{input} is the effect of propagating the input uncertainties through the simulation to obtain the uncertainty in the SRQs, and u_D is the uncertainty in the experimental data. When applying this metric to estimate where the true (but unknown) value exists, the true model error δ_{model} is said to lie in the interval

$$\delta_{model} = [E - ku_{val}, E + ku_{val}],$$

where k is a statistical coverage factor. When $|E| >> ku_{val}$, $\delta_{model} \sim E$ and there is a clear bias error in the model. In cases where E has a similar magnitude to ku_{val}, then the interval equation for δ_{model} should be used. When $|E| >> ku_{val}$, then the model bias error cannot be determined due to inadequacies in the model validation experiment or the validation simulations. In this latter case, the model form uncertainty is expected to be less than or equal to ku_{val}. Since u_{val} represents a standard uncertainty (approximately 68% confidence for a normal distribution), a coverage factor must be included to achieve other confidence levels. For 95% confidence when the distribution is not known, the V&V 20 standard recommends coverage factors between 2 and 3. For more details on this approach, see ASME (2009) and Eca, Dowding, and Roache (2022).

12.2.5 Comparison of Probability Distributions and P-Boxes

The difference between the probability distribution of the experiment and the distribution of the simulation result can be characterized in many ways. Suppose X and Y are the CDFs from the experiment and the simulation, respectively. One could consider using the following types of differences between X and Y: the convolved distribution $X - Y$, some type of average of $X - Y$, or focus on some type of difference between the shapes of X and Y. The characterization that seems to be the most useful and understandable in the context of model validation is based on comparing the shape of the two CDFs. Random variables whose distribution functions are identical are said to be "equal in distribution." As discussed earlier, this cannot occur even with a perfect model of the physics because of random measurement uncertainty. If the distributions are not quite identical in shape, the discrepancy can be measured with many possible measures that have been proposed in traditional statistics (Huber-Carol et al., 2002; Thas, 2010; Berry et al., 2021).

The difficulty with these traditional measures is that they are only well developed for random variables, i.e., purely aleatory uncertainty. We are interested in determining the difference between experiment and simulation when either or both results characterize aleatory and epistemic uncertainty. As discussed in Chapter 3 "Modeling and Computational Simulation," a mixture of aleatory and epistemic uncertainty is referred to as an imprecise probability. In this text we focus on one characterization of an imprecise probability referred to as a p-box. None of the approaches discussed earlier are able to address a combination of aleatory and epistemic uncertainty in the simulation or experimental results. Later in this chapter, we will discuss an approach for computing a model validation metric result using probability distributions and/or p-boxes as inputs (Ferson et al., 2008; Ferson and Oberkampf, 2009; Whiting et al., 2023).

12.3 Recommended Features for Validation Metrics

A model validation metric is a formal measure of the mismatch between model results (usually embodied in a simulation) and experimental data. A low value of the metric means there is a good match, and a higher value means that simulation and measurements disagree more. It should be possible to apply the validation metric when simulations are either deterministic or nondeterministic. The metric should be mathematically well behaved and intuitively understandable to engineers, project managers, and decision makers. In the following discussion we will recommend seven desirable properties of validation metrics that would be useful in assessing the accuracy of models used in science and engineering simulations (Oberkampf and Trucano, 2002; Oberkampf and Barone, 2006; Ferson et al., 2008).

12.3.1 Influence of Numerical Solution Uncertainty

A metric should either (a) explicitly include an estimate of the numerical uncertainty in the SRQ of interest resulting from the computational simulation, or (b) exclude the numerical uncertainty in the SRQ of interest only if the numerical uncertainty was quantitatively estimated to be negligible. The primary numerical uncertainty of concern here is the uncertainty due to lack of spatial and/or temporal resolution in the discrete solution, and secondarily the uncertainty due to iterative solution error. As discussed in Chapter 7, numerical uncertainty could be explicitly included in the predicted SRQ by including an upper and a lower estimated bound on the simulation outcome, i.e., an epistemic uncertainty in the SRQ that would be represented by an interval. Recall that an interval is a special case of a p-box. An explicit inclusion of the numerical uncertainty in the simulation result would be appealing because it would clearly identify the numerical uncertainty contribution, exclusive of any other uncertainty, either in the experimental data or the simulation input data. However, it would add complexity to the theoretical derivation and calculation of the metric because it would require the metric to use p-boxes as input.

A simpler approach would be to quantitatively show that the numerical uncertainty is negligible before the metric is computed. The numerical uncertainty should be judged small in comparison to the magnitude of the experimental measurement uncertainty in the measurement of the SRQ of interest. For two-dimensional or three-dimensional unsteady PDEs, it is a formidable task to get the numerical uncertainty due to the mesh and time-step small enough to be ignored. If this can be done, however, one can eliminate this confounding issue from the calculation and interpretation of the validation metric.

12.3.2 Assessment of the Physics Modeling Assumptions

A metric should be a quantitative evaluation of predictive accuracy of the SRQ of interest, including all of the combined modeling assumptions, physics approximations, and previously obtained physical parameters embodied in the model. Stated differently, the metric evaluates the aggregate accuracy of the model, including all of its submodels, and all of the physical parameters associated with the models. Consequently, there could be offsetting errors or widely ranging sensitivities in the model that could show very accurate results for one SRQ, but poor accuracy for a different SRQ. If there is interest in evaluating the accuracy of individual submodels or the effect of the accuracy of individual input parameters within the model, one should conduct a sensitivity analysis of the SRQ to these effects. However, sensitivity analysis is a separate issue from constructing a validation metric. As stated earlier, a blind-test prediction should be compared with experimental results so that a true measure of predictive capability can be assessed in the validation metric.

12.3.3 Inclusion of Experimental Data Post-Processing

A metric should include, either implicitly or explicitly, an estimate of the approximation error or uncertainty resulting from post-processing of the experimental and computational data to obtain the SRQ of interest. Examples of the types of post-processing of experimental data are as follows: (a) the construction of a regression function, e.g., least-squares fit, of the data to obtain a continuous function over a range of an input (or control) quantity; (b) the processing of experimental data obtained on a very different spatial or temporal scale than what is assumed in the model; and (c) the use of complex mathematical models of the physically measured quantities to process the experimental data into a useable SRQ. A case in which the post-processing described in example (b) would be necessary occurs when there are localized underground measurements of a pollutant concentration and the model contains a large-scale, spatially averaged permeability model. One might require the type of post-processing defined in example (c) when additional mathematical models of the physical features measured are also needed to process and interpret the experimental data into a useable form for comparison with the computational results. Some examples of experimental data processing are (a) image processing, (b) processing the motion of markers in a sequence of experimental images, (c) time-averaging of measurements, and (d) pattern recognition.

It should be noted that this recommendation is completely separate from the features discussed in Section 12.3.1. Any numerical uncertainty associated with the post-processing of the numerical solution of PDEs should be considered as part of the uncertainty in the model. As discussed in Section 12.3.1, the numerical uncertainty should be either quantified using an interval or it should be demonstrated that it is negligible compared to the experimental measurement uncertainty.

12.3.4 Inclusion of Experimental Measurement Uncertainty

A metric should incorporate, or include in some explicit way, an estimate of the aleatory and epistemic measurement uncertainties in the experimental data for the SRQ. The possible sources for measurement uncertainties depend on a very wide range of factors, some of which were discussed in Chapter 11. For a comprehensive discussion of experimental measurement uncertainty, see Drosg (2009), Grabe (2014), and Rabinovich (2017). At a minimum, a validation metric should include an estimate of random measurement uncertainties, i.e., uncertainties due to random measurement errors. To the extent possible, the metric should also include an estimate of the systematic uncertainties in the experiment.

Epistemic measurement uncertainties are usually segregated into two parts: (a) characterization uncertainty due to a limited number of measurements of the SRQ, and (b) epistemic uncertainty that can exist in the measurement itself. The characterization uncertainty is due to limited knowledge of the stochastic nature of the random error in the measurements and is usually referred to as the sampling error. Epistemic uncertainty in the measurements themselves can be due to (a) uncertainty in a measurement that is reported as a plus-or-minus value, (b) uncertainty implied by specifying a certain number of significant digits in a quantity, (c) uncertainty due to intermittent measurements of a process that is known to be periodic, (d) uncertainty in measurements when the quantity being measured is less than the detectable limit of the diagnostic method, i.e., non-detects, (e) uncertainty due to statistical censoring, e.g., when the data is only known with specified ranges or bins, and (f) uncertainty due to important data that is known to be missing, such as interval-valued measurements, from a random variable (Manski, 2003; Gioia and Lauro, 2005; Ferson et al., 2007). When a collection of such intervals comprises a data set, one can think of the breadths of the intervals as representing epistemic uncertainty, while the scatter among the intervals represents aleatory uncertainty. The characterization uncertainty can be directly represented by using the individual measurements themselves, i.e., using an empirical distribution function (EDF). The epistemic uncertainty in the measurements would require a validation metric that could use a p-box as input to compute the difference between the experiment and the simulation.

If possible, the experimental uncertainty estimation method should use measurements from replications of experiments, as opposed to the traditional propagation of uncertainty technique (ISO, 1995). If replications of the experiment are conducted, then the uncertainty in the metric should depend on the number of replicated measurements of a given SRQ of interest. Replications should be used, along with blocking and randomization, to vigorously attempt to quantify random and systematic uncertainties in measurements. This approach is referred to as statistical design of experiments (DOE) and it was discussed in detail in Chapter 11. If an SRQ is measured for different conditions of an input or control parameter, then techniques should be used to reduce the experimental uncertainty of the SRQ over the range of the input parameter.

12.3.5 Inclusion of Aleatory and Epistemic Uncertainties

A metric should, in a mathematically rigorous way, be able to compute the difference between the computational and experimental results for the SRQ when these results exhibit aleatory and epistemic uncertainty. This uncertainty, for example, could be due to uncertain input quantities or an estimate of numerical solution uncertainty. If the computational and experimental results both exhibit aleatory and epistemic uncertainty, then they could both be characterized as p-boxes or some other imprecise probability structure (Bernardini and Tonon, 2010; Augustin et al., 2014).

Consider the case in which the computational and experimental results are each characterized as precise probability distributions, i.e., only aleatory uncertainty is present. If the variance in each distribution approaches zero, then the difference between the distributions should approach the difference in the mean of each distribution. As another example, suppose the computational and experimental results are each characterized as p-boxes, i.e., a combination of aleatory and epistemic uncertainty. If the aleatory and epistemic uncertainty in each distribution approaches zero, then the difference between each p-box should reduce to the difference in the two point values, i.e., the difference between two scalar quantities.

12.3.6 Exclusion of Any Type of Adequacy Implication

A validation metric should *exclude* any indications, either explicit or implicit, of the level of adequacy in agreement, or satisfaction of accuracy requirements, between computational and experimental results. That is, the metric should only measure the mismatch between the computational and experimental results, separate from *any* other features or characteristics of the computational or experimental results. If any other features or characteristics are combined in the mismatch measure, then one defeats the goal of independently setting accuracy requirements, as set by the intended use of the model (see Figure 12.4, as well as Figures 2.5 and 2.8). Some examples of inappropriately combining the mismatch feature of the comparison with other features, are the following: (a) comparisons of computational and experimental results that yield or imply value judgments, such as "adequate," "good," or "excellent"; (b) computational results judged to be adequate if they lie within some stated uncertainty band or observed range of the experimental measurements; (c) a comparison that combines the mismatch measure and the confidence or probability of the mismatch; and (d) a comparison that combines the mismatch measure and an estimate of the experimentally observed uncertainty.

This last example of constructing an inappropriate validation metric (combining the metric with the experimentally observed uncertainty) is hotly debated. The most common method of constructing an inappropriate metric such as this is to scale the mismatch between the computational and experimental results by the standard deviation of the experimental measurements. For example, suppose the L_1 vector norm is used as the mismatch measure between computation and experiment and this norm is normalized by the sample standard deviation of the measurements, s. For the validation metric one would have $\|S - E\|_1/s$. We argue that this metric would be inappropriate because the metric has explicitly mixed two different types of measures, a difference measure between simulation and experiment SRQ, and a measure of the scatter of the experimental data. The scatter in the experimental data should not be mixed with the first measure because the experimental scatter is controlled by sources that have nothing to do with the ability of the model to predict the observed responses of the system. It simply does not make sense for the validation metric to improve (i.e., get smaller) when the standard deviation of the measurements gets larger (arguably a lower quality experiment). Two examples of these sources are uncertainty in the response due to uncertainty in input data to the model and experimental measurement uncertainty. What would be acceptable, in our view, would be a validation metric that measures the mismatch between the predicted standard deviation and the measured standard deviation.

12.3.7 Properties of a Mathematical Metric

A validation metric should be a *true metric* in the mathematical sense, i.e., a true distance measure. The validation metric would then measure, by some means, the distance between the simulation and experimental results. By definition, a mathematical metric d has four properties (Magnus, 2022):

Non-negativity,	$d(x, y) \geq 0$,
Symmetry,	$d(x, y) = d(y, x)$,
Triangle inequality,	$d(x, y) + d(y, z) \geq d(x, z)$, and
Identity of indiscernibles,	$d(x, y) = 0$ if and only if $x = y$.

12.4 Bayesian Model Discrepancy Approach

The Bayesian model discrepancy approach developed by Kennedy and O'Hagan (2001) is now discussed. This approach represents a significant step forward in that it acknowledges that the model itself is not perfect, and it allows for the discrepancy between the model prediction and experimental data to be estimated as part of the updating process. As with all Bayesian updating procedures, prior probability distributions must be assumed for all inferred quantities. Since these priors often differ from analyst to analyst, this is a significant subjective element of Bayesian inference in most practical situations. In other words, when data are sparse, the model validation results can depend strongly on the choice of the priors.

The Bayesian model discrepancy approach commonly combines model validation, model parameter estimation, and model prediction using Gaussian process regression. When used purely for model validation (i.e., model accuracy assessment), parameter estimation is performed as a separate, independent step. By obtaining experimental results at a range of validation conditions for the model, the observed discrepancy between the experiment and the model at these locations can be estimated. These discrepancies can then be used to update the model discrepancy term to better match the experimental data using Bayesian estimation. The Gaussian process is then used to connect the observations to one another over the validation domain to create the updated model.

For the case where the validation conditions are described by a single variable x, the updated model function is defined as

$$f(x) \sim GP(m(x), c(x, x')),$$

where $f(x)$ is a posterior realization of the Gaussian process, $m(x)$ is the mean posterior function selected to represent the model discrepancy over the validation domain, and $c(x, x')$ is the covariance function (Bailer-Jones, 2017; Lambert, 2018; Santner et al., 2018). The way that the Gaussian process behaves between observation locations is dependent upon the covariance function, which is generally assumed to be of the form

$$c(x, x') = \sigma_{2n} \exp\left(-\frac{|x - x'|}{2\ell^2}\right).$$

Here σ_{2n} is the variance in the observation discrepancies and ℓ is the correlation length scale, which denotes the typical distance over which the function values become decorrelated over the validation domain. Normal distributions are generally assumed for each of these parameters. The length scale can be determined through physical arguments, expert opinion, or by using maximum likelihood estimation (Smith, 2014; Bailer-Jones, 2017). Typically, a large value is chosen for the prior of the variance, while the initial mean function is set to zero; both of these distributions are then updated as experimental data become available. The

Gaussian process must be sampled in such a way that the samples will satisfy the chosen mean function and covariance function over the domain.

Example: Model Discrepancy Estimation Using a Gaussian Process

Figure 12.5 shows a simple example of the Bayesian model discrepancy approach applied to estimate the model error (δ) along with its uncertainty as a function of an angle (α) (Whiting et al., 2023). Here, the initial assumption (i.e., prior) is that the mean model discrepancy is zero and the variance is large. Using the experimental data, along with its associated uncertainty, a series of posterior Gaussian process realizations are made (Figure 12.5a). The mean value of the model discrepancy, along with the 95% confidence intervals showing the remaining model form uncertainty, are shown in Figure 12.5b. Note the growth of the uncertainty in the model discrepancy between the experimental observation locations, as well as the large growth of this uncertainty at the outer bounds of the experimental data set. This behavior is expected, and in fact desired, as one changes from interpolation within the validation domain to extrapolation beyond it.

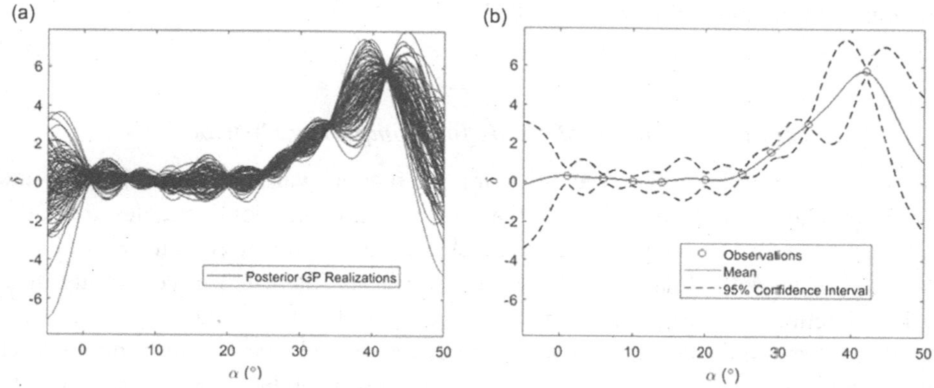

Figure 12.5 Display of updated posterior model using observational data using the Kennedy and O'Hagan approach.
(Whiting et al., 2023)

12.5 Validation Metrics Using CDFs and P-Boxes

When computational predictions are probability distributions, there is a great deal more information contained in how a system responds than compared to a simple deterministic response of a system. One can think of the model as a mapping of all uncertain inputs to produce a set of uncertain system responses. In this mapping, all of the uncertain inputs are convolved according to the physical processes described by the PDEs of interest. A validation metric could be viewed as asking the question: How well does the model map the inputs to the outputs, compared to the way that nature maps these in a physical experiment?

The simplest method of comparing experimental data and predictions is in terms of their means, variances, covariances, and other distributional characteristics. The main limitation of approaches based on comparing summary statistics is that they consider only the central tendencies or other specific behaviors of data and predictions and not their entire distributions. When predictions are distributions, they contain a considerable amount of detail, and it is not always easy to know what statistics are important for a particular application of interest. While some statistical tests are certainly helpful in comparing experimental data and predictions, most do not directly address the validation metric perspective of interest. In addition, traditional statistical tests, as well as the Bayesian approach to model validation, do not address the issue of epistemic uncertainty in either, or both, the experimental data and the prediction. If epistemic uncertainty exists in either, we will consider the representation to be given by a p-box.

In the following sections, we introduce the notion of comparing probabilistic quantities, including p-boxes; describe some of the desirable properties of p-boxes; and suggest a particular approach that has most of the desirable properties of validation metrics discussed in Section 12.3. Several simple examples are given which display some of the features of this validation metric. Portions of this section were taken from Ferson et al. (2008), Ferson and Oberkampf (2009), and Whiting et al. (2023).

12.5.1 Traditional Methods for Comparing Distributions

There are a variety of standard ways to compare random variables in probability theory. If random numbers X and Y always have the same value, the random variables are said to be "equal," or sometimes "surely equal." A much weaker notion of equality is useful in the construction of model validation metrics because we are interested in comparing distributions, i.e., functions as opposed to numbers. If we can only say that the expectation, i.e., the mean, of the absolute values of the differences between X and Y is zero, the random variables are said to be "equal in mean." We can measure the mismatch between X and Y by defining the mean metric, d_E, as

$$d_E(X, Y) = E(|X - Y|) \neq |E(X) - E(Y)|, \tag{12.2}$$

where E denotes the expectation operator. Note that this difference is *not* the same as the absolute value of the difference between the means. The idea can be generalized to higher-order moments, and equality in a higher-order moment implies equality in all lower-order moments.

The notion of equality for randomly varying quantities can be loosened further still by comparing only the *shapes* of the probability distributions of the random variables. Random variables whose distributions are identical are said to be "equal in distribution." This is often denoted as $X \sim Y$, or sometimes by $X^{\circ} = d^{\circ} Y$. This is really a rather loose kind of equality, because it does not require the individual values of X and Y to ever be equal, or even to ever be close. For instance, suppose X is normally distributed with mean zero and unit variance. If we let $Y = -X$, then X and Y are obviously equal in distribution, but are about as far from equality as can be imagined. Nevertheless, equality in distribution is an important concept because a distribution often represents all that is known about the values of a random variable.

If the distributions are not quite identical in shape, the discrepancy can be measured with many possible measures that have been proposed for various purposes. For instance, a very common such measure is the maximal probability, i.e., vertical, difference between the two cumulative distribution functions

$$d_S(X,Y) = \sup_z |Pr(X \leq z) - Pr(Y \leq z)|. \tag{12.3}$$

Here, d_S is the *Smirnov metric*, and $\sup_z$ is the supremum over the sample space of *X* and *Y*. The d_S defines the Kolmogorov–Smirnov statistical test for comparing distributions (Huber-Carol et al., 2002; Mielke and Berry, 2007; Berry et al., 2021). One of the properties of the Smirnov distance is that it is symmetric, which is to say that $d_S(X,Y)$ always equals $d_S(Y,X)$. The symmetry might be considered unnecessary or even counterintuitive as a feature for model validation. We certainly do not view predictions and observations as exchangeable with each other; in a validation metric it matters which is which. Suppose, for instance, that we inadvertently switched the predicted distribution with the experimental data distribution. One might expect to obtain a different result from having made such a mistake, but the Smirnov distance does not change whether the prediction and data are exchanged or not.

The *Kullback–Leibler divergence* is another very widely used measure of the discrepancy between distributions that is not symmetric (Huber-Carol et al., 2002; Mielke and Berry, 2007; Berry et al., 2021). It is defined, in its discrete formulation, for a probability mass function *p* for *X* and a probability mass function *q* for *Y* as

$$\sum_z p(z) \log_2 \frac{p(z)}{q(z)}, \tag{12.4}$$

where *z* takes on all values in the common range of *X* and *Y*. The *p* distribution summarizes the observations, and the *q* distribution summarizes the model prediction. The continuous formulation is similar except that the summation is replaced by integration. The term *divergence* in this metric may be misleading because this quantity has nothing to do with notions of divergence familiar from calculus as the inner product of partial derivatives or the flux per unit volume. Instead, the term is used here in its other meaning as deviation from a standard. The Kullback–Leibler divergence is commonly used in information theory and also in certain fields of physics. It is interpreted as the relative entropy between *p* and *q*, i.e., the entropy of the distribution *p* with respect to the distribution *q*.

There are many other measures that could be used to compare data and prediction distributions. But, given the broad acceptance and ubiquity of the Smirnov and Kullback–Leibler measures in probability and physics, it is perhaps necessary to explain why we simply do not use one of these as our validation metric. This question is addressed in the next section.

12.5.2 Area Validation Metrics

12.5.2.1 Discussion of P-Boxes

A model validation metric was proposed by Ferson et al. (2008) and Ferson and Oberkampf (2009) to measure the mismatch between a prediction and empirical observations. The prediction and the experimental measurements can be given by CDFs or p-boxes. A CDF is typically used when only aleatory uncertainty is present (see Section 12.1.2). A p-box expresses both epistemic and aleatory uncertainty in a way that does not confound the two (see Chapter 3 "Modeling and Computational Simulation"). For a more complete discussion of p-boxes, see Ferson et al. (2002), Ferson et al. (2004), Aughenbaugh and Paredis (2006), and Augustin et al. (2014).

Figure 12.6 shows examples of three p-boxes with varying degrees of aleatory and epistemic uncertainty for a predicted SRQ. Figure 12.6a shows a p-box for an interval, i.e., there is no aleatory uncertainty; there is purely epistemic uncertainty. For values of the SRQ less than the minimum value of the interval, the cumulative probability is zero. That is, there is certainty that all possible values of the SRQ are greater than the minimum value of the interval. For values of the SRQ over the range of the interval, the cumulative probability is in the interval [0, 1]. This interval-valued probability is, of course, a nearly vacuous statement in the sense that the SRQ probability could be *anything* over the range of zero to unity. For values of the SRQ greater than the maximum value of the interval, the cumulative probability is unity. That is, there is certainty that all possible values of the SRQ are less than or equal to the maximum value of the interval.

Figure 12.6b shows a p-box for a mixture of aleatory and epistemic uncertainty in the SRQ. The p-box shows the fraction, expressed as interval-valued quantity, of the sampled population that would have a value less than, or equal to, a particular value of the SRQ. Some examples are (a) for an SRQ = 32, the fraction of the population that would have a value of 32 or less would range from [0.00, 0.02], and (b) for an SRQ = 34, the fraction of the population that would have a value of 34 or less would range from [0.12, 0.52]. The interval-valued fraction is due to lack of knowledge from any source in the simulation. As is typical of a mixture of aleatory and epistemic uncertainty, the range of the interval-valued probability approaches zero for rare events of the SRQ. These rare events can occur in either tail of the probability distribution. It should also be noted that the horizontal breadth of the p-box expresses the magnitude of the epistemic uncertainty in terms of the SRQ.

Figure 12.6c shows a degenerate p-box, i.e., a precise CDF, for the SRQ. The p-box is degenerate in the sense that the epistemic uncertainty is negligible compared to the aleatory uncertainty. Precise probabilities are, of course, the foundation of traditional probability theory and its application. The p-boxes are also referred to as imprecise probabilities because the probability is not necessarily a unique quantity but can be interval valued. During the last few decades, extensions of traditional probability theory have identified a number of types of imprecise probabilities, i.e., more complex than the p-box discussed here. For a discussion of these extensions, some recent texts are Molchanov (2005), Klir (2006), Bernardini and Tonon (2010), and Augustin et al. (2014).

12.5.2.2 Area Validation Metric

Any nondeterministic prediction from the model that contains only aleatory uncertainty can be characterized as a single (precise) cumulative distribution function $F(x)$. In this notation, x is the predicted variable, i.e., the SRQ. The observation(s), on the other hand, is(are) usually provided as a collection of point values in a data set. The distribution function for a data set, which is referred to as an EDF, summarizes the data set as a function suitable for graphical representation. It is a function that maps x to the cumulative probability scale on the interval [0, 1]. It is constructed as a non-decreasing step function with a constant vertical

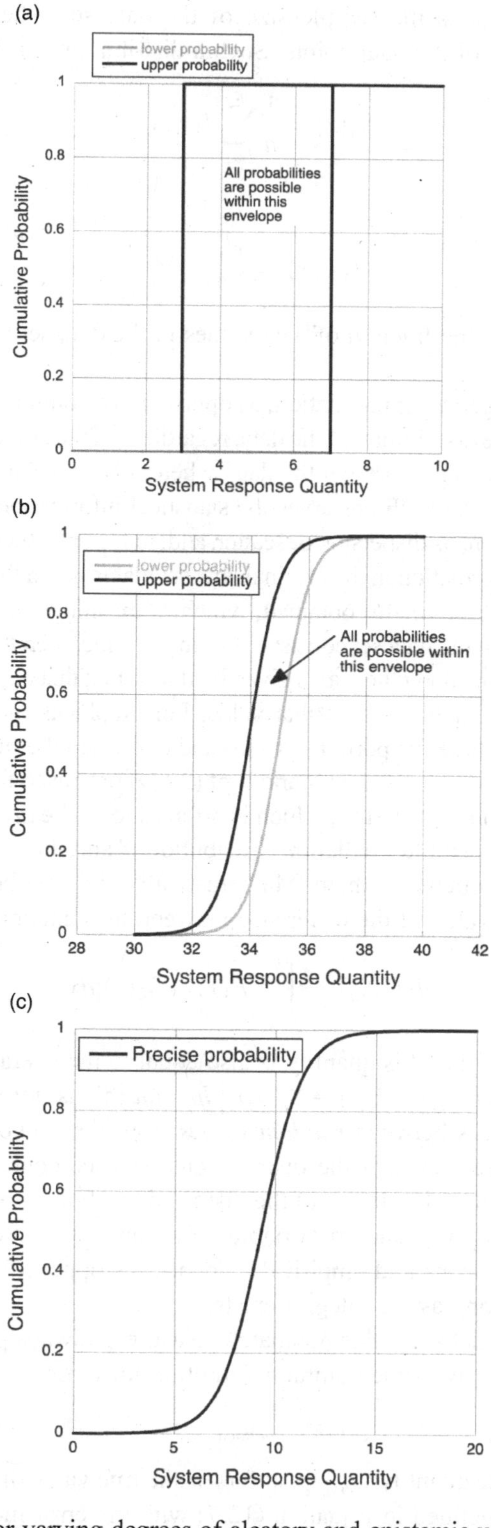

Figure 12.6 P-Boxes for varying degrees of aleatory and epistemic uncertainty: (a) p-box for purely epistemic uncertainty, (b) p-box for a mixture of aleatory and epistemic uncertainty, and (c) degenerate p-box for purely aleatory uncertainty.

step size of $1/n$, where n is the sample size of the data set. The locations of the steps correspond to the values of the data points. Such a distribution for data x_i, $i = 1,\ldots,n$, is

$$S_n(x) = \frac{1}{n}\sum_{i=1}^{n} I(x_i, x), \tag{12.5}$$

where

$$I(x_i, x) .= \begin{cases} 1 & x_i \le x, \\ 0 & x_i > x. \end{cases} \tag{12.6}$$

The term $S_n(x)$ is simply the fraction of data values in the data set that are at or below each magnitude x.

An EDF is an advantageous representation, as opposed to a continuous CDF, because it is an exact representation of the distribution of the data, regardless of the amount of data. An EDF does not require any assumptions to represent the data, whereas bin sizes are required to construct a histogram of the data set. An EDF preserves the statistical information in the data set about its central tendency or location, its dispersion or scatter, and, in fact, all other statistical features of the distribution. The only information in the original data set that is not in the distribution is the order in which the values were originally obtained, which is meaningless whenever the data were sampled at random. When the data set consists of a single value, then the S_n function is a simple step function at the location along the x-axis given by that value; that is, it is zero for all x less than that value and one for all x larger than that value. For graphical clarity, however, it is often convenient not to depict these flat portions at zero and one when the functions are plotted.

Ferson et al. (2008) and Ferson and Oberkampf (2009) proposed the use of the Minkowski L_1 metric as a model validation metric, which is referred to as the *area validation metric*. It is defined as the *area* between the prediction distribution F and the data distribution S_n as the measure of the mismatch between them. Mathematically, the area between the curves is the integral of the absolute value of the difference between the functions

$$d(F, S_n) = \int_{-\infty}^{\infty} |F(x) - S_n(x)| dx. \tag{12.7}$$

It is clear, geometrically, that this quantity is also equal to the average horizontal difference between the two functions, $\int \left|F^{-1}(p) - S_n^{-1}(p)\right| dp$, but this is not the same as the average of the absolute differences between *random values* from these two distributions. (Such an average would not be zero even if the distributions were coincident.) The area validation metric is, thus, a function of the shapes of the distributions, but is not readily interpretable as a function of the underlying random variables. The area validation metric measures the *disagreement* between theory and empirical evidence, as opposed to measuring the agreement. It is a metric so long as the integral exists.

It should be stressed that since this mismatch measure is always positive, it should *not* be interpreted in the same way as the common definition for error,

$$\varepsilon = y_{obtained} - y_T, \tag{12.8}$$

where ε is the error in the quantity $y_{obtained}$ and y_T is the true value of y. For example, contrast the mismatch measure defined in Equation (12.7) with the error measure associated with a confidence interval approach. (See Oberkampf and Barone, 2006; Oberkampf and Roy,

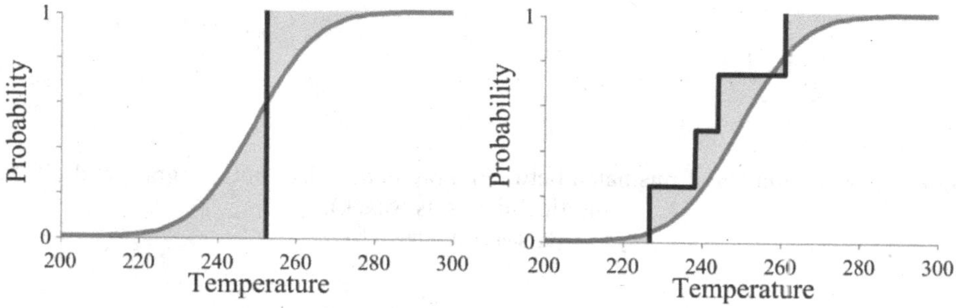

Figure 12.7 Example data sets, with $n = 1$ on the left and $n = 4$ on the right, shown as S_n distributions (black) against a prediction distribution (gray). The validation metric is the areas (shaded) between the prediction distribution and the two data sets.
(Ferson et al., 2008)

2010, for more detail.) In the confidence interval approach, a positive error means the model prediction is *greater* than the experimental measurement. For the area validation metric, there is *no* such indication of how the model differs from the measurement.

Figure 12.7 illustrates this area measure of mismatch for two data sets against a predicted distribution of temperature, which represents an arbitrary SRQ. The predicted distribution, shown as the smooth gray curve, is the same in both graphs. This predicted distribution might be obtained by propagating input uncertainties through a mathematical model by using a large number of samples from a Monte Carlo simulation. Superimposed on these graphs are distribution functions S_n for two hypothetical data sets. On the left graph, the data set consists of the single value 252°C, and on the right, the data set consists of the values {226, 238, 244, 261}°C. In complex engineering systems it is not uncommon to have only one experimental test of the complete system. In such cases, the empirical distribution is a single step function representing the measured point value (i.e., degenerate distributions). Note that the empirical distribution function is zero for all values smaller than the minimum of the data and unity for all values larger than the maximum of the data. Likewise, beyond the range (support) of the predicted distribution, the value of $F(x)$ is either zero or one extending to infinity in both directions. For graphical clarity, however, these flat portions at probability zero or unity are usually not depicted when the distributions are plotted.

The areas measuring the mismatches between the prediction and the two data sets in Figure 12.7 are shaded. In the left graph, the area consists of a region to the left of the datum at 252°C, and a region to right of it. In the right graph, there are four shaded regions composing the total area between the prediction distribution and the data distribution. The area validation metric is a generalization of the deterministic comparisons between scalar values that have no uncertainty. That is, if the prediction and the observation are both scalar point values, the area is simply equal to their difference. It is clear that the area validation metric reflects the difference between the full distribution of both the prediction and the observations. However, the area will tend *not* to be sensitive to minor discrepancies in the distribution tails because there is little probability mass in the tails of the distributions.

Figure 12.8 shows how the area validation metric differs from a validation measure based on merely matching in the mean or matching in both the mean and variance. In each of three cases, the prediction distribution is shown as a smooth gray curve. It is the same in all three

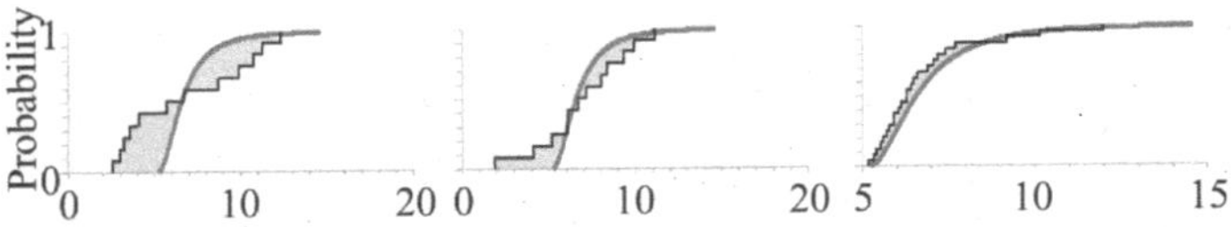

Figure 12.8 Examples of mismatch between a prediction distribution (gray) and different empirical data sets (black).
(Ferson et al., 2008)

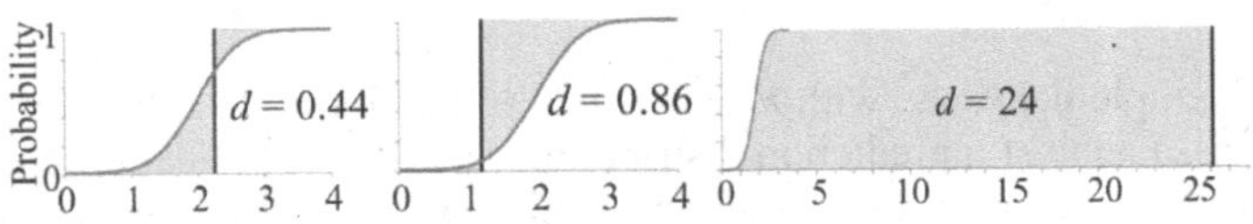

Figure 12.9 Comparisons of a prediction distribution (gray curve) with three different data points (black step functions).
(Ferson et al., 2008)

graphs, although the scale in the third graph is a bit different from the other two. The black step functions represent three different data sets as empirical distribution functions S_n. In the leftmost graph, the prediction distribution and the observed data have the same mean. But, otherwise, the data look rather different from the prediction; the data appear to be mostly in two clusters on either side of the mean. Indeed, so long as the average of the data balances at that mean, those data clusters could be arbitrarily far away from each other. Any validation metric based only on the mean would not detect any discrepancy between the theory and the data, even though the data might bear utterly no resemblance to the prediction, apart from their matching in the mean. In the middle graph, both the mean and the variance of the observed data and the theoretical prediction match. However, one would not claim the comparison between prediction and data were extremely good because of how the prediction deviates from the left tail of the empirical distribution. Smaller values are more prevalent in the experimental data than were predicted. In the third graph, the agreement between the prediction and the data is good overall. This is reflected in the smallness of the area between the prediction distribution and the data distribution. The only way for the area to be small is for the two distributions to match closely over the entire range of each. In each of these cases, the overall mismatch can be measured by the area between the two curves. It measures disagreements that the lower-order moments like the mean and variance cannot address.

Figure 12.9 shows how the area validation metric d varies with different values for a single datum matched against a prediction distribution. In these three examples, the prediction distribution is the same, and it is centered at 2 and ranges between 0 and 4. The most important thing to notice is that a single value can never perfectly match the entire distribution, unless that distribution is itself a degenerate point value. The first and second graphs of the figure compare the prediction distribution to a single observation at 2.25 and 1.18, respectively, yielding corresponding values for the area validation metric of 0.44 and 0.86. The best possible match that a single datum could have occurs when the datum is located at the prediction distribution's median, but, even there, the area validation metric will often be significant. In the case of the prediction distribution depicted in Figure 12.9, the area validation metric will be smallest when the observation is 2, which yields a value of 0.4 for the metric.

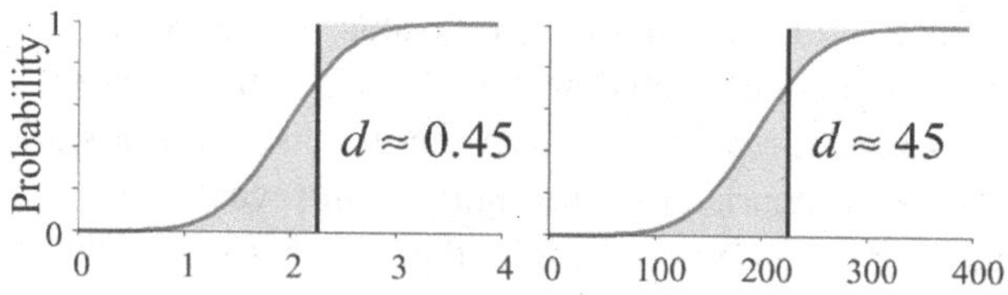

Figure 12.10 Examples showing the metric's dependence on the scale.
(Ferson et al., 2008)

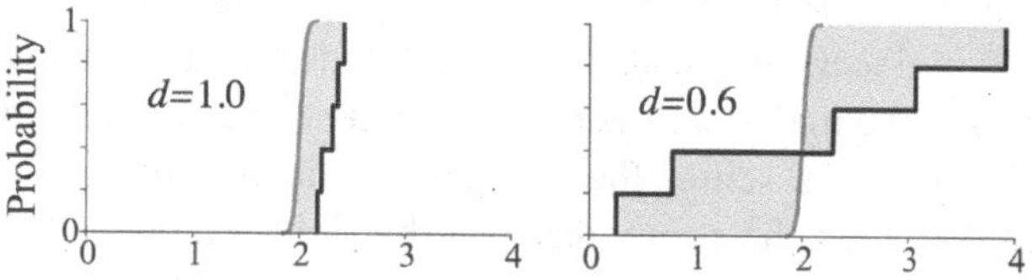

Figure 12.11 Example as to why physical units are important for a validation metric.
(Ferson et al., 2008)

If, for example, the prediction distribution was a uniform probability density function over the range $[a, b]$, in Figure 12.9 this distribution would be a straight line from $(a, 0)$ to $(b, 1)$. For this case, a single observation cannot be any "closer" to it than $(b - a) / 4$, which is the value of the area validation metric if the point is at the median. That is the best match possible with a single data point. Stated differently, the mismatch between theory and experiment *could* be much better if more experimental measurements were available, but, due to sampling error in the experiment, it is the smallest mismatch that can exist. How *bad* could the match be? The match could be very bad; indeed, it can be bad to an arbitrarily large degree. The rightmost graph in Figure 12.9 shows another example of a single datum compared to the same prediction distribution. In this case, the data point is at 26, which means that it is about 24 units away from the distribution. The area validation metric can be arbitrarily large, and it reduces to the simple difference between the experimental datum and the prediction when both are point values. In fact, if the two distributions do not overlap, then the area validation metric also reduces to the difference in the mean values of the distributions. Because probability is dimensionless, the units of the area are always the same as the units of the abscissa, thus *the metric reports disagreement in the same units as the SRQ*.

The area validation metric depends on the scale in which the prediction distribution and experimental data are expressed. The two graphs in Figure 12.10 depict a pair of comparisons in which the corresponding shapes are identical but the scales are different, as though the left graph were expressed in meters and the right graph in centimeters. Although the shapes are the same, the area validation metric is different by 100 fold. It would, of course, be possible to normalize the area measure, perhaps by dividing it by the standard deviation of the prediction distribution, but we do not believe this would be a good idea because the result would no longer be expressed in the physical units of the abscissa. Such normalization would cloud the *physical meaning* of the metric.

Figure 12.11 illustrates why retaining the scale and physical units (degrees, meters, Newtons, Newtons/m^2, etc.) of the data is important for the intuitive appeal of a validation metric. The two graphs are drawn with the same x-axis, and they depict the same prediction distribution as a gray curve concentrated around 2. Two data sets are summarized as the S_n distributions shown as black step functions on the graphs. From a statistical perspective it might be argued that the comparison in the right graph reveals a better match between the theory and the data than the comparison in the left graph. In the left graph, the two distributions do not even overlap, whereas in the right graph the distributions at least overlap and are similar in their means. Using a traditional Kolmogorov–Smirnov test for differences between the two distributions, one would find statistically significant evidence that the distributions in the left graph are different ($d_S = 1.0$, $n = 5$, $p < 0.05$), but would *fail* to find such evidence for the distributions in the right graph ($d_S = 0.6$, $n = 5$, $p > 0.05$). But this is not at all how engineers and scientists would understand these two comparisons. For engineers and scientists, the main focus is on the difference between the two distributions, in units along the x-axis. In this sense, the comparison on the left is a much better match between theory and data than the comparison on the right. Engineers and scientists have a strong intuition that the data-theory comparison on the left might be attributed to a small bias error in the theory or the experimental data, but the theory does a good job of capturing the variability of the physics. The discrepancy on the left is never larger than half a unit along the x-axis, whereas the discrepancy on the right could be larger than two units. It is this physical distance measuring the theory-data disagreement that really matters to engineers and scientists, not some arcane distance measured in terms of a probability. This is the reason why the validation metric should be expressed in the original units, as is the case for the area validation metric.

Finally, consider the behavior of the area validation metric as theory and experimental evidence diverge further and further. Figure 12.12 shows two graphs, each with a prediction distribution drawn in gray and data distribution drawn in black. The traditional and commonly used Smirnov's distance (which is the maximum vertical distance between the two distributions) *cannot* distinguish between these two comparisons. The maximal vertical distance in both cases is unity, so the distributions are both as far apart as they can be according to the Smirnov metric. Under this measure, each data distribution is simply "far" from its prediction distribution. The area validation metric, on the other hand, is about 2 for the left graph and about 40 for the right graph. The area validation metric, therefore, identifies the left graph as having considerably more concordance between data and prediction than the right graph. If the criterion for an acceptably accurate prediction is that it is within 10 units of the actual data, then it might be that the prediction in the left graph is acceptable for the intended purpose, even though the prediction in this case does not overlap

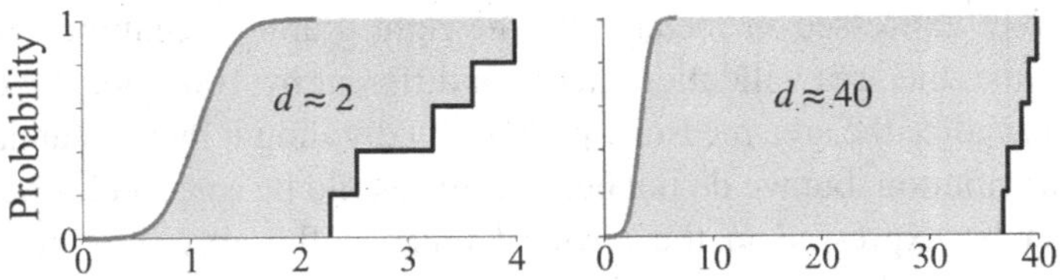

Figure 12.12 Distinguishing nonoverlapping data and prediction distributions.
(Ferson et al., 2008)

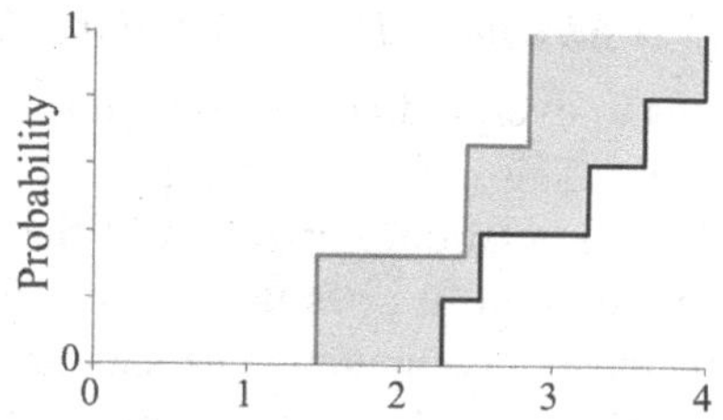

Figure 12.13 The area validation metric when the prediction distribution (gray) is characterized by only three simulations.
(Ferson et al., 2008)

with the data. Likewise, given the same accuracy requirement, the prediction in the right graph is not acceptable for the intended use.

The area validation metric is applicable even when the predictions are sparse. Suppose, for instance, that it is practical to compute only a small number of simulations of a complex model to produce a handful of quantitative predictions. Although these computed results cannot produce the smooth prediction distributions shown earlier, it may be reasonable to consider the values computed to be *samples* from that smooth distribution. If they are random samples (as they would be if inputs are selected randomly), then the "empirical" distribution function formed from these values is an unbiased nonparametric estimator of the true distribution that would emerge with asymptotically many runs.

Figure 12.13 illustrates the idea of constructing S_n functions for both data and predictions, the latter being the values from the sample runs of the model. In this case, there were only three simulations of the model conducted, whose values are random samples from the underlying distributions of the input. They are to be compared against an experimental data set consisting of five values. Having very few simulation runs really means that analysts can construct only a vague picture of what the model is actually predicting. This implies the prediction will have substantial epistemic uncertainty arising from sampling error. When only a few random samples are available from either the model or the measurements, the area validation metric could actually be *much smaller* if a larger number of samples were available. As a result, it is stressed that the present area validation metric should be viewed as the *evidence for the mismatch between the model and the measurements*, instead of the evidence for matching.

As will be discussed in detail in the next chapter, one may also be interested in using the area validation metric to increase the epistemic uncertainty in a model prediction to account for the model form uncertainty. For the situations discussed thus far, one could estimate the model form uncertainty at the conditions where validation data exist by applying the area validation metric symmetrically about the simulation outcome. For example, the true value would be estimated to occur in the interval $[F(Y) - d, F(Y) + d]$, where $F(Y)$ is the simulation outcome, usually a CDF. In this section we have stressed the impact of sampling uncertainty on the area validation metric and its interpretation as *evidence for disagreement*. However, when the primary interest becomes an *estimation of model form uncertainty*, the next section and Section 12.5.3 discuss how the calculation of the area validation metric should be modified.

12.5.2.3 Modified Area Validation Metric

The area validation metric tends to overpredict the model form uncertainty of the simulation result in relation to the true value. This situation occurs due to the model uncertainty being applied symmetrically about the simulation, regardless of the relation between the experiment and simulation CDFs. In addition, it does not account for the additional uncertainty arising due to small experimental (or computational) sample sizes. In order to overcome these weaknesses, Voyles and Roy (2014, 2015) proposed the modified area validation metric. Here we summarize the implementation as described in Whiting et al. (2023). Small sample sizes are accounted for by applying a confidence interval for the mean, assuming a Student's t-distribution, about the entire experimental CDF. This approach can be justified because, when the experimental and simulation CDFs do not overlap, the area validation metric simply defaults to the difference in the mean values of the two distributions. Note that the context of Whiting et al. (2023) is the estimation of model form uncertainty, as opposed to construction of a modified area validation metric. As discussed below, the maximum area between the experimental confidence interval bounds and the simulation CDF are chosen when the goal is to develop a conservative estimate of the model form uncertainty.

The modified area validation metric separately tracks two different areas between the CDFs. These two regions are the area where the right 95% confidence interval of the experimental CDF provides SRQs *larger* than the simulation CDF (d^+), and the area where the left 95% confidence interval of the experimental SRQs are *smaller* than the predicted SRQs from the model (d^-), as shown in Figure 12.14. The interval in which the modified area validation metric estimates the true value is most likely to exist is then $[F(Y) - d^-, F(Y) + d^+]$, where $F(Y)$ is the simulation CDF and d^+ and d^- are defined as:

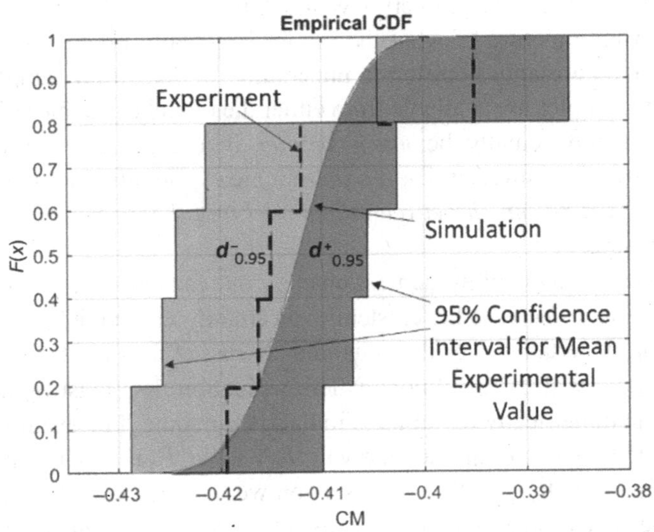

Figure 12.14 Modified area validation metric with confidence intervals for use in estimating the model form uncertainty.
(Whiting et al., 2023)

$$d^{+} = \int_{-\infty}^{\infty} max\left(\left[S_n(Y) + t_{0.025}\frac{s}{\sqrt{N}}\right] - F(Y), 0\right) dY \tag{12.9}$$

$$d^{-} = \int_{-\infty}^{\infty} max\left(F(Y) - \left[S_n(Y) - t_{0.025}\frac{s}{\sqrt{N}}\right], 0\right) dY. \tag{12.10}$$

This form of the modified area validation metric is appropriate for *estimating the model form uncertainty* (see Section 12.5.3.2). The interval-valued estimate of the model form uncertainty $[-d^{-}, d^{+}]$, will yield an asymmetric p-box as opposed to the symmetric p-box found for the area validation metric. Examples of the modified area validation metric are given later in this chapter, as well as in Chapter 13. The larger the bias error in the simulation relative to the experimental data, the more asymmetric the p-box will be when using the modified area validation metric. If instead, one was interested in determining the *evidence for disagreement* between simulation and experiment (see Section 12.5.3.1), then the *max* functions in Equations (12.9) and (12.10) must be changed to *min* functions and the other side of the confidence interval must be used (i.e., the sign must change). The form of the modified area validation metric to determine the *evidence for disagreement* is thus:

$$d^{+} = \int_{-\infty}^{\infty} min\left(\left[S_n(Y) - t_{0.025}\frac{s}{\sqrt{N}}\right] - F(Y), 0\right) dY, \tag{12.11}$$

$$d^{-} = \int_{-\infty}^{\infty} min\left(F(Y) - \left[S_n(Y) + t_{0.025}\frac{s}{\sqrt{N}}\right], 0\right) dY. \tag{12.12}$$

Although not presented in the prior references on the modified area validation metric, small sample sizes in the simulation results could also be accounted for using a confidence interval about the simulation CDF. When used for estimating the model form uncertainty, Equations (12.9) and (12.10) would become

$$d^{+} = \int_{-\infty}^{\infty} max\left(\left[S_n(Y) + t_{0.025}\frac{s_{exp}}{\sqrt{N_{exp}}}\right] - \left[F(Y) - t_{0.025}\frac{s_{sim}}{\sqrt{N_{sim}}}\right], 0\right) dY, \tag{12.13}$$

$$d^{-} = \int_{-\infty}^{\infty} max\left(\left[F(Y) + t_{0.025}\frac{s_{sim}}{\sqrt{N_{sim}}}\right] - \left[S_n(Y) - t_{0.025}\frac{s_{exp}}{\sqrt{N_{exp}}}\right], 0\right) dY, \tag{12.14}$$

where the "*exp*" subscripts refer to the experimental samples and the "*sim*" subscripts refer to the simulation samples. Again, if the analyst is interested in the *evidence for disagreement* rather than an estimate of model form uncertainty, then these equations become

$$d^{+} = \int_{-\infty}^{\infty} min\left(\left[S_n(Y) - t_{0.025}\frac{s_{exp}}{\sqrt{N_{exp}}}\right] - \left[F(Y) + t_{0.025}\frac{s_{sim}}{\sqrt{N_{sim}}}\right], 0\right) dY, \tag{12.15}$$

$$d^{-} = \int_{-\infty}^{\infty} min\left(\left[F(Y) - t_{0.025}\frac{s_{sim}}{\sqrt{N_{sim}}}\right] - \left[S_n(Y) + t_{0.025}\frac{s_{exp}}{\sqrt{N_{exp}}}\right], 0\right) dY. \tag{12.16}$$

12.5.3 Dealing with Epistemic Uncertainty in the Comparisons

As discussed in Chapters 10 and 11, high-quality model validation experiments should minimize, if not eliminate, epistemic uncertainties in the input quantities for the model. There are a number of situations, however, in which they cannot be avoided. Some examples are (a) the experiment was not conducted as a high quality validation experiment so that several important inputs were not measured; (b) information concerning certain inputs was measured, but it was not documented; (c) certain input quantities were quantified as interval-valued quantities based on expert opinion; and (d) the experimentalist never expected that the fidelity of physics models would be developed to the point that extremely detailed information would be needed as input data in the future. For most of these situations, the unknown information should be treated as a lack of knowledge, i.e., as an epistemic uncertainty, in the model inputs for the prediction.

For many of these situations, the lack of knowledge of input data needed for simulations should be represented as an interval. Giving an interval as the representation of a quantity is asserting that the value (or multiple possible values) of the quantity lie somewhere within the interval. Note that the interval-valued quantity can be used to represent an uncertain input quantity, or it can be used to represent the uncertainty in a parameter in a parameterized family of probability distributions. In the latter case, the uncertain quantity would be a mixture of aleatory and epistemic uncertainty that would be represented as a p-box. As discussed earlier, when these intervals and/or p-boxes are propagated through the model, the model prediction is p-box for the SRQ of interest. An example of such a p-box was shown in Figure 12.6b.

Empirical observations can also contain epistemic uncertainty. A number of examples of where these occur in experimental measurements were discussed in Section 12.3.4. Again, the simplest form of this is an interval. When a collection of such intervals comprises a data set, one can think of the breadth of each interval as representing epistemic uncertainty while the scatter among the intervals represents aleatory uncertainty. Ferson et al. (2007), Manski (2003), and Gioia and Lauro (2005) have described how interval uncertainty in data sets also produce p-boxes. When empirical observations have uncertainty of this form that is too large to simply ignore, these elementary techniques can be used to characterize it in a straightforward way.

12.5.3.1 Metrics Used as Evidence for Disagreement

The comparison between two fixed real numbers reduces to the scalar difference between the two. Suppose that, instead of both being scalar numbers, at least one of them is an interval range representing an epistemic uncertainty. If the prediction and the observation overlap, then we should say that the prediction is *correct*, in a specific sense, relative to the observation. If the prediction is an interval, this means that the model, for whatever reason, is making a weaker claim about what is being predicted. For example, the assertion that some system component will achieve a maximum operating temperature between 400 and 800°C is a much weaker claim than saying it will be exactly 600°C. In the extreme case, a prediction with extraordinarily large bounds, while not very useful, is certainly true, if just because it is not claiming anything that might be false. For example, predicting that some probability will be between zero and one does not require any foresight, but at least it is free from contradiction. It is proper that a prediction's express uncertainty be counted toward reducing any measure of mismatch between theory and data in this way because the model is admitting doubt. If it were not so, an uncertainty analysis could otherwise have no epistemological value. From the perspective of model validation, when the uncertainty of prediction encompasses the actual observation, there is *no evidence* of mismatch because *accuracy is distinct from precision.* Both are important in determining the usefulness of a model, for example, for decision-support value, but it is reasonable to distinguish them and give credit where it is due.

A reciprocal consideration applies, by the same token, if the experimental datum is an interval to be compared against a prediction that is a real number. If the datum is an interval, and the prediction falls within the measured interval, there is *no evidence* of mismatch between the two. For instance, if the prediction is, say, 30% and the observation tells us that it was measured to be somewhere between 20% and 50%, then we would have to admit that the prediction might be perfectly correct. If on the other hand the evidence was that it was between 35% and 75%, then we would have to say that the disagreement between the prediction and the observation might be as low as 5%. We could also be interested in how bad the comparison might be, but a validation metric should not penalize the model for the empiricist's imprecision. In most conceptions of the word, the "distance" between two things is the length of the shortest path between them. Thus, the validation metric between a point prediction and an interval datum is the shortest difference between the characterizations of the quantities.

Figure 12.15 gives three examples of how a single point observation might compare with a prediction that is expressed as a p-box. The prediction is the same in all three graphs and is shown as gray bounds representing the p-box for some SRQ. Against this prediction, a single observation is to be compared, and the area between this datum and the prediction is shaded. In the leftmost graph, the datum happens to fall entirely inside the bounds on the prediction. The comparison in this graph evidences no discrepancy at all between the datum and the uncertain prediction. At this value of x, the step function fits entirely inside the graph of the p-box, which tells us that a (degenerate) point distribution at 8 is perfectly consistent with the predictions made by the model. Thus, the area between the datum and the prediction is zero, i.e., there is no evidence for mismatch. In contrast, the data value at 15 in the middle graph is completely outside the range of either bounding distribution. The area between 15 and the prediction is about 4 units, which is the area between the step function and the right bound of the p-box. In the rightmost graph, the observation is located at the intermediate value of 11, which is within the range of the right bound of the prediction distribution. The area of

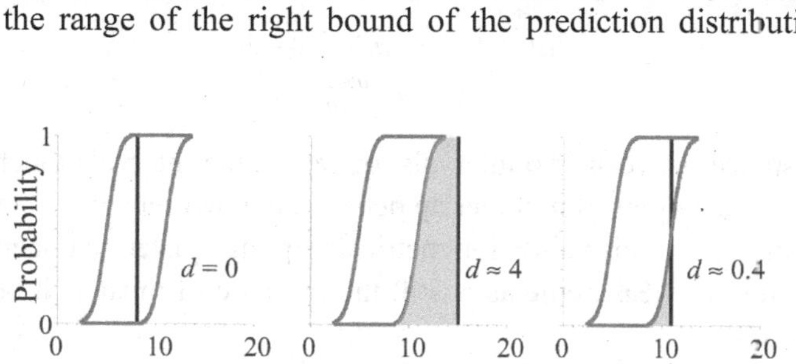

Figure 12.15 Comparison of a prediction characterized as p-boxes (gray bounds) against three separate single observations (black step functions).

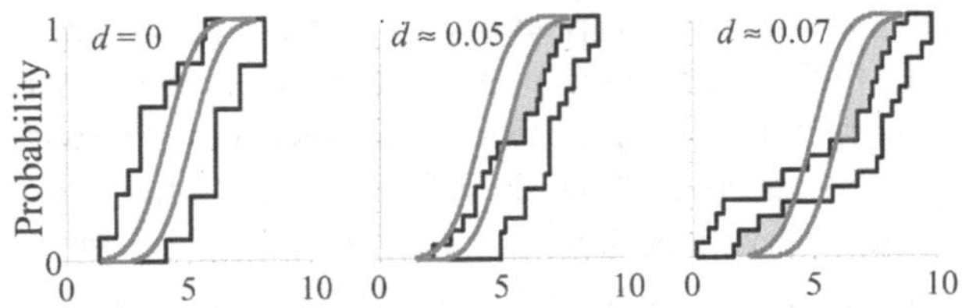

Figure 12.16 Three comparisons of predictions (gray bounds) against empirical observations (black step function bounds), both containing aleatory and epistemic uncertainty.

mismatch in this case is only about 0.4 because the area only includes the small, shaded region between 9 and 11. These comparisons are qualitatively unlike those between a scalar observation and a well-specified probability distribution. We see now that a single observation *can* perfectly match a prediction so long as the prediction has epistemic uncertainty.

Figure 12.16 illustrates three more examples, this time with epistemic uncertainty in both the prediction (shown as gray lines) and the data (black step functions). The leftmost comparison has a distance of zero because there is *at least one distribution* from the prediction that can be drawn simultaneously inside both empirical distribution functions from the measurements. The area of mismatch exists only when there are no possible probability distributions that lie within both the bounds on the prediction distribution and the bounds on the data distribution. For instance, there is no such possible distribution consistent with both data and prediction in the middle or rightmost graph.

It should be noted that the area between the prediction and data no longer constitutes a true mathematical metric when at least one is an interval or a p-box. The reason is that the area can fall to zero without the prediction and data becoming identical (as in the leftmost graph of Figure 12.16), which violates the identity-of-indiscernibles property of a true metric. There may be ways to generalize the area validation metric from probability distributions to p-boxes that are mathematical metrics. However, these have not yet been developed.

In the cases just discussed in Figure 12.15 and Figure 12.16, the shaded regions indicate the mismatch. The validation metric now is defined by the integral

$$\int_{-\infty}^{\infty} \Delta([F_R(x), F_L(x)], [S_{nR}(x), S_{nL}(x)])dx, \tag{12.17}$$

where F and S_n denote the prediction and the data distributions, respectively. The subscripts L and R denote the left and right bounds for those distributions, and

$$\Delta(A, B) = \min_{\substack{a \in A \\ b \in B}} |a - b| \tag{12.18}$$

is the shortest distance between two intervals, or zero if the intervals touch or overlap. This measure integrates the regions of nonoverlap between the two sets of bounds, for every value along the probability axis. This validation metric accepting epistemic uncertainty in either, or both, the model and the measurements is still the measure of mismatch between the model and the measurements.

Figure 12.17 illustrates another feature of this approach to assessing mismatch between prediction and measurement. As before, predictions are shown in gray, and the experimental

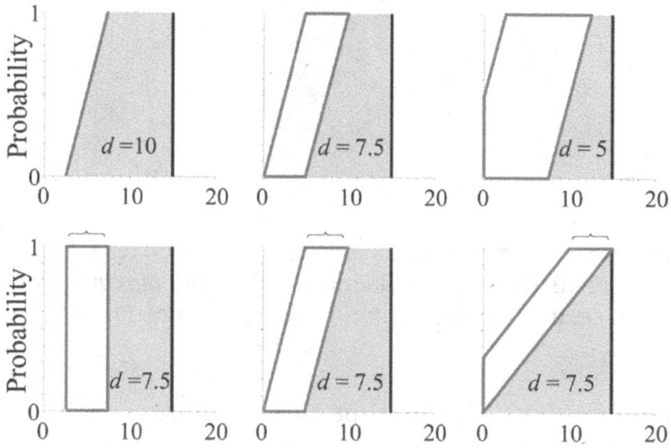

Figure 12.17 Increasing epistemic uncertainty (breadth) of predictions in the top panel and increasing variance (slant) of predictions in the lower panel.

datum is a black step function. Increasing the *breadth* of an uncertain prediction so that it possesses larger epistemic uncertainty and wider bounds can result in lowering the mismatch between the theory and data, as illustrated in the upper panel of three graphs. This breadth is a measure of the epistemic uncertainty in the prediction. It is not the same as the dispersion or variance of a distribution, which measures aleatory uncertainty. By contrast, the lower panel of three graphs in the figure shows that increasing variance in the prediction – reflected in the *slant* of the p-box – does *not* by itself reduce the mismatch.

These behaviors of the area validation metric distinguish it from another commonly used measure of disagreement between an experimental data value and a probability distribution expressed as the datum's displacement in standard deviation units. That measure would suggest that the three graphs in the lower panel of Figure 12.17 depict increasing agreement because the datum is progressively closer to the prediction in terms of standard deviation units. This contrast suggests that accounting for epistemic uncertainty of predictions and observations with p-boxes and measuring their mismatch as the area between those p-boxes is quite different from the common statistical idea of measuring the disagreement as displacement in standard deviation units. The area validation metric has the distinct advantage of distinguishing between aleatory and epistemic uncertainties. These uncertainties are confounded by the validation metric based on displacement in standard deviation units. For example, although the displacement decreases in the three lower graphs of Figure 12.17, the differences between small realizations from the prediction in the rightmost graph on the lower panel are necessarily larger than the corresponding difference in the graphs on the left.

12.5.3.2 Metrics Used to Estimate Model Form Uncertainty

When the goal is to use the validation metric as an estimate of the model form uncertainty, then the mindset must change from trying to determine the evidence for disagreement to trying to estimate, in a conservative manner, just how large the model form uncertainty could be. When epistemic uncertainties are present, this mindset change will have a major effect on how the metric is computed. Rather than seeking the minimum area, as discussed in the previous section, we now focus on determining the maximum area. Consider again the

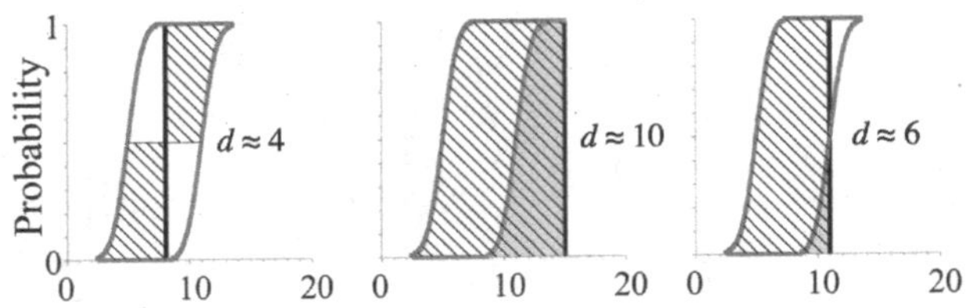

Figure 12.18 Comparison of a prediction characterized as p-boxes (gray bounds) against three separate single observations (black step functions) when the *maximum area* is considered as needed for estimating model form uncertainty.

situation portrayed in Figure 12.15, where the simulation takes the form of a p-box and the experiment the form of a single value (i.e., a step function). The equivalent area validation metrics computed using the maximum area would seek to find the single CDF, from all possible CDFs inside the p-box, which would *maximize* the magnitude of the metric. Such a notional CDF is shown in Figure 12.18, with the shaded area from Figure 12.15 also shown.

Example: Assessing Model Validation/Calibration Approaches

Whiting et al. (2023) examined several different approaches to model validation, calibration, and prediction applied to the turbulent flow over a multielement airfoil. Here we summarize the results of their study pertaining primarily to validation and calibration. The approaches examined include the area validation metric (Section 12.5.2.2), the modified area validation metric (Section 12.5.2.3), the ASME V&V 20 approach (Section 12.2.4), and the Bayesian model discrepancy approach (Section 12.4). Since the V&V 20 standard does not address how it should be used for calibration or for estimating model form uncertainty, we summarize Whiting et al.'s implementation briefly here. When used as a validation approach, the error in the mean E and the validation standard uncertainty u_{val} are combined to give the estimated model form uncertainty applied about the simulation value. The true value is assumed to lie in the nonsymmetric interval:

$$[S - \max(E, 0) - ku_{val}, S - \min(E, 0) + ku_{val}], \tag{12.19}$$

where the coverage factor is taken to be $k = 2$ for 95% confidence. When used as a calibration procedure, the simulation result is first calibrated to give

$$S_{calib} = S - E. \tag{12.20}$$

The true value is assumed to lie in the symmetric interval

$$[S_{calib} - ku_{val}, S_{calib} + ku_{val}], \tag{12.21}$$

which is centered about the calibrated simulation value.

Since Whiting et al.'s goal was not to actually validate or calibrate a model, but instead to assess the usefulness of different model validation and calibration frameworks, they used synthetic experimental data generated using the method of manufactured universes (Stripling et al., 2011) to assess the different techniques. Similar to the method of manufactured solutions used for code verification (see Chapter 6), the method of manufactured universes involves the generation of a manufactured reality, or "universe," from which "experimental" observations can be made. A low-fidelity model is then developed and assessed with uncertain inputs and possibly including numerical solution errors. Since the true behavior of "reality" is known from the manufactured universe, the estimation of model form uncertainty and errors in the lower-fidelity model can be performed in an unambiguous and precise manner.

In this example, the manufactured reality was based on a high-fidelity computational fluid dynamics model for turbulent flow over a multielement airfoil as depicted in Figure 12.19. The low fidelity model to be assessed comes from thin airfoil theory. The SRQs of interest are the 2-D lift (c_l) and quarter-chord moment ($c_{m,c/4}$) coefficients, which vary as a function of the airfoil angle of attack (α) and the rear flap deflection angle (δ). The angle of attack and the flap deflection angle are treated as aleatory input uncertainties assuming normal (Gaussian) distributions with a mean at the nominal value and a standard deviation of 0.5° and 1.0°, respectively. Additionally, measurement uncertainty in the SRQs was also included in the "experiments."

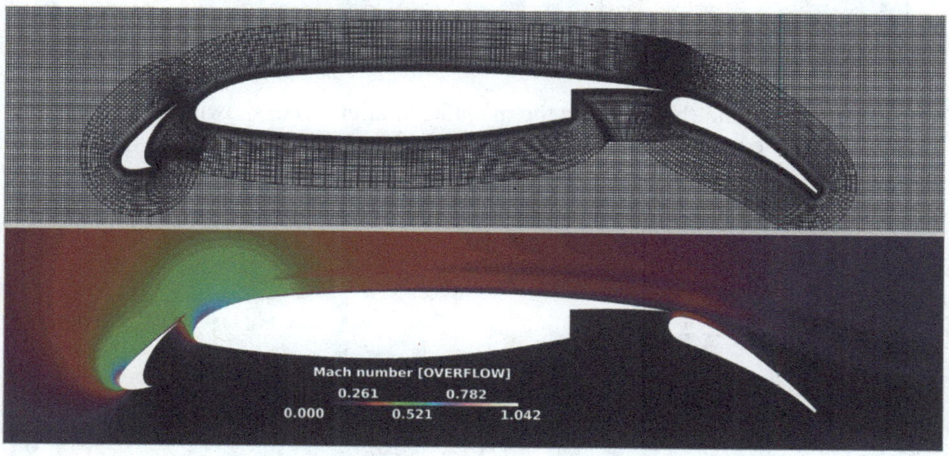

Figure 12.19 Overset grid system and typical flowfield result shown as a coordinate slice colored by local Mach number for the MD 30P/30N airfoil.
(Whiting et al., 2023)

Figure 12.20 shows the lift and moment coefficient provided from the manufactured universe with the red x's representing the locations at which high-fidelity CFD results were computed and where the validation "experiments" are conducted. Bi-harmonic curve fits were used to accurately represent the lift and moment coefficient as a function of angle of attack and flap deflection angle for the "experiments."

The lift and moment coefficients provided from thin airfoil theory are shown in Figure 12.21. Note the linearity of the results for both coefficients because thin airfoil theory does not account for physical phenomena such as flow separation and is thus expected to deviate significantly from the data at high angles of attack and flap deflection angles.

To compare the different validation and calibration methods, two quantities were used to quantify the effectiveness and efficacy of the methods. The first is *conservativeness*, φ_1, which can also be thought of as the reliability of the estimator to capture the true model error. It is measured as the percentage of the time that a respective method encapsulates the true value within its associated uncertainty estimate. In this case, the "true value" is taken as the mean of the synthetic experimental output as the number of input samples becomes very large. Conservativeness values closer to 100% are ideal, with values above 95% considered to be very good. Note that a method can have a high conservativeness, but not be very informative to a

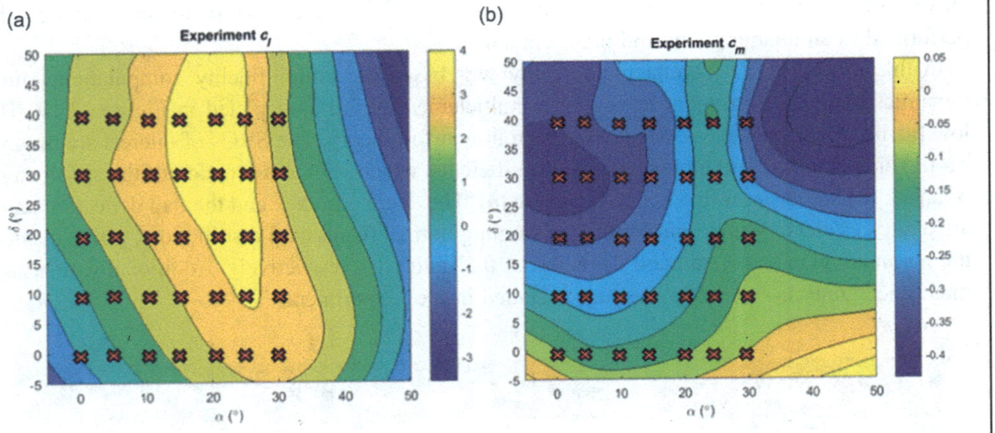

Figure 12.20 (a) c_l and (b) $c_{m,c/4}$ for the manufactured universe showing the 35 CFD runs used to generate the synthetic experimental data curve fits.
(Whiting et al., 2023)

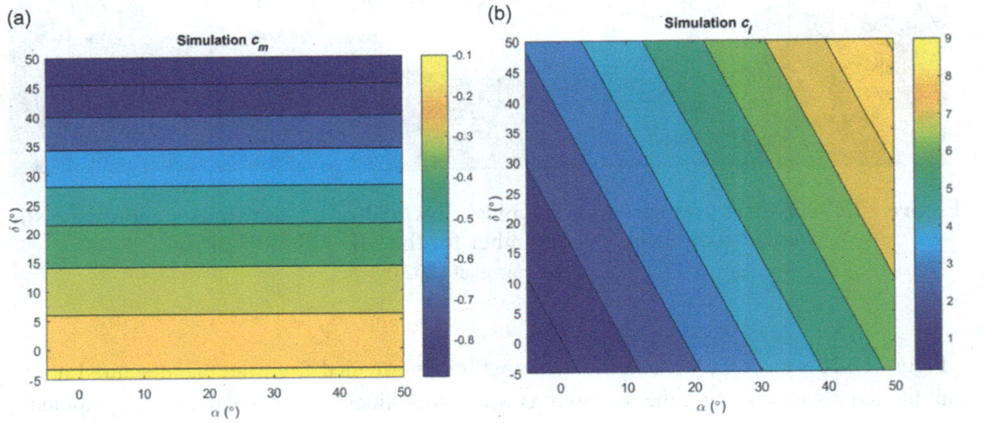

Figure 12.21 (a) c_l and (b) $c_{m,c/4}$ for the low fidelity model.
(Whiting et al., 2023)

decision maker because it is excessively conservative. The second factor taken into consideration is *tightness,* φ_2, which assesses how tightly the uncertainty interval about either the simulated or calibrated value bounds the true value. The tightness definition is implemented slightly differently for validation and calibration methods (see Whiting et al., 2023, for details). As with conservativeness, larger tightness values near 100% are preferred.

Each of the methods previously mentioned in this example were assessed at locations of the input domain where synthetic experimental results were provided (i.e., the red x's in Figure 12.20) and compared with results from the low fidelity model. Uncertainty bounds for each of the results are shown in Figure 12.22 for both 2 and 16 experimental samples for the nominal condition of $\alpha = 30^\circ$ and $\delta = 20^\circ$. For the Bayesian calibration method, the Gaussian process across the entire input

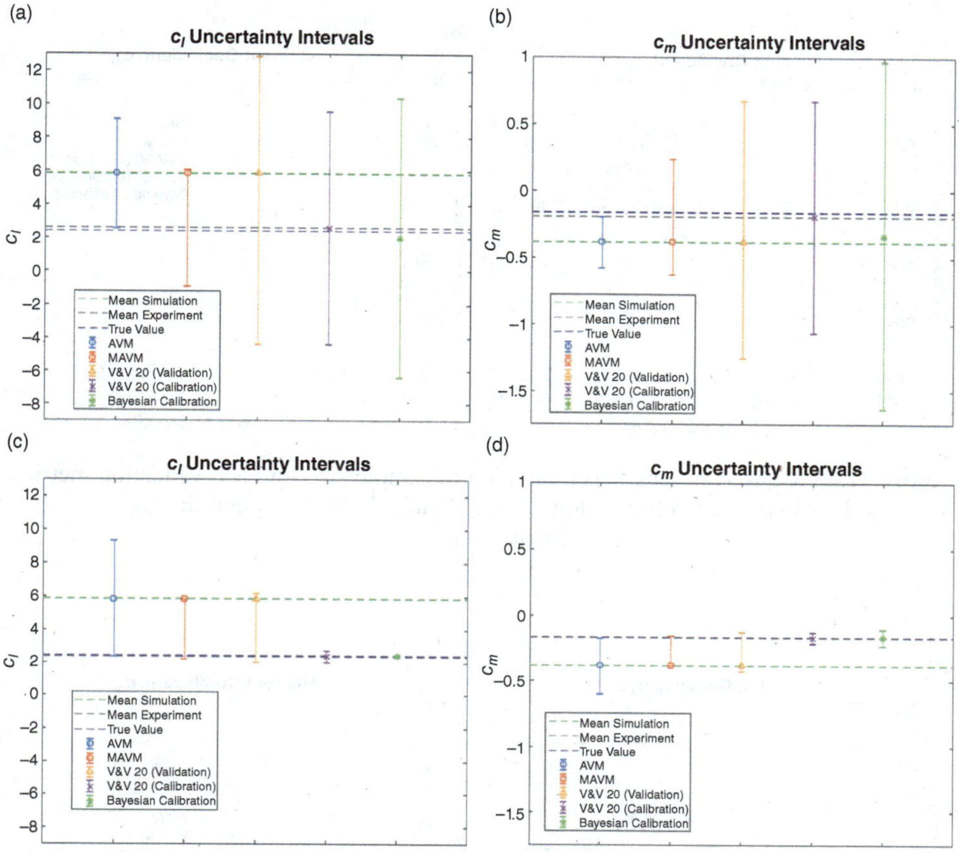

Figure 12.22 (a) Uncertainty intervals for c_l (2 experimental samples), (b) uncertainty intervals for $c_{m,c/4}$ (2 experimental samples), (c) uncertainty intervals for c_l (16 experimental samples), (d) uncertainty intervals for $c_{m,c/4}$ (16 experimental samples) at $\alpha = 30°$ and $\delta = 20°$. (Whiting et al., 2023)

domain was created, but it was only assessed at the single validation location in this section. The green dashed line represents the mean simulation result, and the grey dashed line represents the experimental mean for the given number of samples. As the number of experimental samples increases from 2 to 16, the experimental sample mean (grey line) moves closer to the true value (blue dashed line). It is also seen that as the number of experimental samples increases, the uncertainty bounds about the true value generally become smaller.

In Figure 12.23, the conservativeness of each method aggregated over all of the 35 observation points (red x's in Figure 12.20) as a function of sample size is shown. In addition, since the statistics can vary widely when only a few samples are available, each case with a sample size of 2 is repeated 1,000 times, each case with a sample size of 4 is repeated 500 times, etc. The conservativeness of all of the methods except the original area validation metric is quite good regardless of sample size, suggesting that each of these methods appropriately accounts for small samples sizes. Due to its formulation, when any bias error exists in the model, the area validation metric will only contain the true value 50% of the time, regardless of the number of experimental samples, thus its conservativeness hovers around 50%.

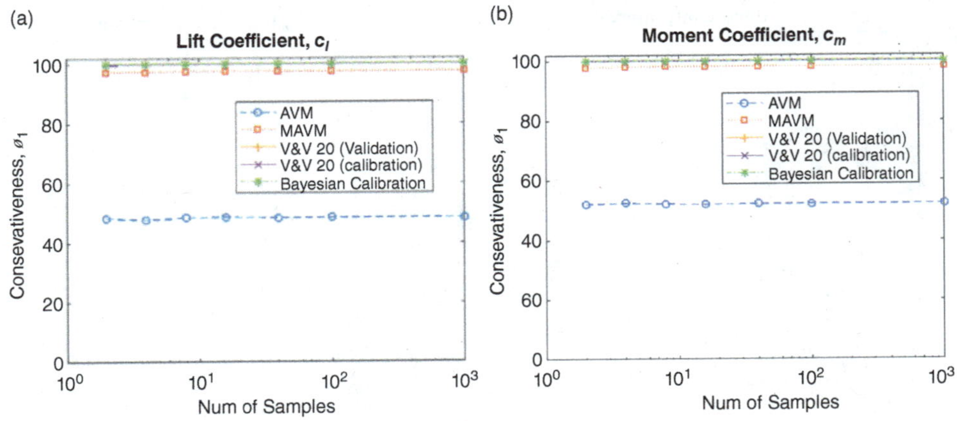

Figure 12.23 Conservativeness as a function of sample size for each validation metric at locations where observation data are available for (a) c_l and (b) $c_{m,c/4}$.
(Whiting et al., 2023)

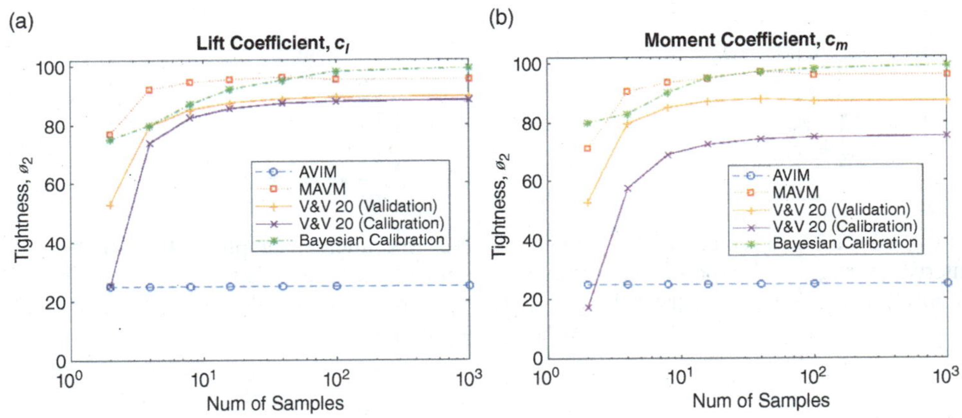

Figure 12.24 Tightness measurements as a function of sample size for locations at which observation data are available for (a) c_l and (b) $c_{m,c/4}$.
(Whiting et al., 2023)

The tightness assessment is shown in Figure 12.24 versus the number of samples. The uncertainty bounds for each approach become tighter around the true value with increasing sample size, excluding the area validation metric. The area validation metric's tightness for these locations converge to about 50% due to the fact that it is applied symmetrically about the simulation outcome, even when there is a clear bias error in one direction. The other four methods' tightness measurements do not quite converge to 100% with increased sample size due to the confidence intervals associated with each method. Thus, at these observation locations, each method slightly overpredicts the associated uncertainty of the model relative to the true value when a large number of experimental samples are available. For a low number of samples, the four approaches beside the area validation metric appear to sacrifice tightness in order to maintain their conservativeness (Figure 12.23).

A final comment should be made on the *(un)fairness* of comparing validation techniques with calibration techniques at conditions where experimental data are available. It is, of course, expected that using the experimental data to calibrate a model will improve the agreement with that very same experimental data. Perhaps a fairer comparison would be to perform the model calibration using the experimental data, then to assess the model at other conditions where experimental data has not been used for calibration. Whiting et al. (2023) addressed this very issue and examined interpolation between, and extrapolation away from, validation conditions, both for cases in which experimental data are sparse and where they are plentiful. See Whiting et al. (2023) for details.

12.6 U-Pooling for Incomparable CDFs

The previous section described how several observations of a physical process can be collected into an empirical distribution S_n for comparison against a single prediction distribution. In practice, however, a model is often used to make predictions of several *different* SRQs. For instance, a single model might predict the uncertainty distribution of temperature, in addition to the uncertainty distributions of electrical resistivity and material stress. This implies that if we want to assess model accuracy for each SRQ, we would have multiple values of the validation metric to compute. One could certainly compute all the area validation metrics separately for each pair of prediction distribution and its observation(s). Each experimental data set must be compared to the prediction distribution to which it corresponds in order to assess the accuracy of the model for each SRQ.

An innovative strategy was introduced, referred to as u-pooling, to deal with this situation (Ferson et al., 2008). Multiple system response quantities, even if they are physically unrelated physical quantities, can be combined by converting them to a universal scale via probability transformations. This strategy allows us to pool fundamentally incomparable data in terms of the relevance of each datum as evidence about the model's mismatch with experimental data. The u-pooling approach is also useful in dealing with small experimental sample sizes. For instance, experimental data replicates at one point in space or time can be transferred to another point in space or time.

To determine whether the experimental data are generally drawn from distributions that are the same as their corresponding prediction distributions, a strategy must overcome the problem of incomparability among data and express the conformance of theory and data on some universal scale. Probability is an appropriate scale for this purpose for probabilistic models. Each datum x_i is transformed by the prediction distribution F_i with which it corresponds to obtain a variate $u_i = F_i(x_i)$ on the universal probability scale, which ranges on the unit interval [0, 1]. Figure 12.25 shows such transformations for three hypothetical cases. Each case is an observation depicted as a step function, i.e., single measurement, and its corresponding prediction distribution shown as a gray curve. The prediction for the first graph is an exponential distribution on a temperature scale. In the middle graph, it is a roughly normal distribution on a scale measuring resistivity in ohms. The third graph depicts an unusual distribution of a dimensionless quantity z. The intersections of the step functions and their respective distribution functions identify values on the probability scale for each u-value. The prediction distributions F_i can be any shape at all, and they need not be the

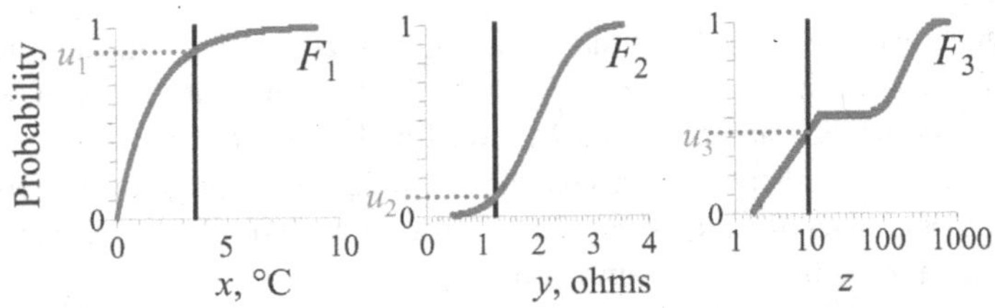

Figure 12.25 Translation of single observations (step functions) through prediction distributions (gray) to a probability scale (ordinates) from three dimensionally inconsistent scales (abscissas). (Ferson et al., 2008)

same for different observations. The u-values are always defined because $F(x) = 1$ for any value of x larger than the largest value in the distribution, and $F(x) = 0$ for any value smaller than the smallest value in distribution.

The various resulting u-values produced by these transformations can then be pooled to obtain an *overall* summary metric for the mismatch of the model's predictions to the data, even when the various individual comparisons are in different dimensional units. Under the assumption that the x_i are distributed according to their respective distributions F_i, these u_i will have a uniform distribution on [0, 1]. This fact is called the *probability integral transform theorem* in statistics (Angus, 1994). This is what it means for a random variable to be "distributed according" to a distribution function. The converse of this fact is perhaps more familiar to engineers and scientists because it is often used to generate random deviates from any specified probability distribution. That is, given a distribution F and a uniform random value u between zero and one, the value $F^{-1}(u)$ will be a random variable distributed according to F. Conversely, as is needed here, if x is distributed according to F, then $u = F(x)$ is distributed according to a uniform distribution over [0, 1]. None of this changes if there happen to be multiple x- and u-values and, in fact, none of it changes if there are multiple distribution functions so long as the x-values are properly matched with their respective distributions. Each u-value tells how the experimental datum from which it was derived compares to its prediction distribution. The x-values are made into compatible u-values by this transformation.

Transforming the observations into a universal probability scale is extremely useful for aggregating incomparable data and comparing evidence collected in incommensurable physical quantities. By itself, however, it has the disadvantage of abandoning the original physical units of the comparisons and retreating to a bounded area validation metric which can be no larger than 0.5. This deficiency can be repaired by back-transforming the u-values through a suitable distribution function G that restores the units, scale, and their interpretation to the u-values. The area validation metric can then be computed in the physically meaningful units after this back-transformation. The justification for this transformation back to SRQs can certainly be questioned for the case where the back-transform u-values were pooled from evidence expressed in different SRQs. However, back-transformation seems very defensible for pooling of the same SRQ that was predicted and measured for different times in a time-dependent simulation, or different spatial locations over the domain of the PDEs.

Figure 12.26 shows how this would work for the three u-values considered in the previous two figures. The result in the left graph is the area between the distribution G and an experimental data distribution of back-transformed data values

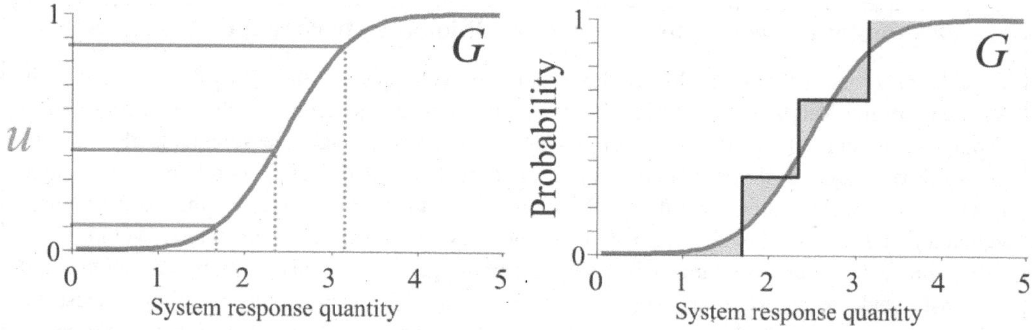

Figure 12.26 Back-transformation from the u-scale to an archetypical scale determined by a distribution G (left) and the area validation metric for the pooled back-transformed values against the G distribution (right).
(Ferson et al., 2008)

$$y_i = G^{-1}(u_i) = G^{-1}(F_i(x_i)). \tag{12.22}$$

All of these y_i have the same units, which are inherited from G. This back-transformation, and any physical meaning associated with the units it reattaches to the u-values, depends on the specification of the G distribution. The right graph in Figure 12.26 also shows, with shading, the area validation metric between the back-transformed y_i and the prediction distribution G. This metric is in the physically meaningful units as a result of the back-transformation.

What distribution should be used to define the back-transformation? In some cases, the scientific or application context of the model accuracy assessment will specify the distribution to use for the back-transformation. This is the distribution, after all, that spells out *where* we are specifically interested in the model's predictive capability. Using the specified prediction distribution as the G distribution allows all the available observations germane to *any* predictions made by the model to be used to characterize the uncertainty about this most important prediction. In general, one would want to use a distribution that expresses where the interest in the model's predictive capability lays. This might be the prediction distribution associated with the prediction that is most important in some sense to the application. Specifying some G distribution to use as the back-transformation allows all the available observations germane to any predictions made by the model to be used to characterize the uncertainty about this most important prediction.

Several researchers have recently investigated the applicability of u-pooling for diverse applications. Li et al. (2014) and Shen et al. (2015) applied u-pooling to the mechanical response of a beam. They extended the u-pooling approach to consider multiple SRQs, as well as addressing the correlation structure between the SRQs. He (2019) developed a method for applying u-pooling to the case where some of the model input data are given by p-boxes. When He (2019) applied the method to the Sandia Validation Challenge problem for heat transfer (Dowding, Pilch, et al., 2008), he found that the uncertainty of the prediction of the temperature response increased significantly.

Example: *U*-Pooling to Generate Additional Experimental Replicates

Choudhary et al. (2016) employed the *u*-pooling technique in the analysis of the Sandia Verification and Validation Challenge Problem for predicting structural failure of a pressurized container (Hu and Orient, 2016). *U*-pooling was used to convert single measurements of stress in a pressurized tank at different spatial locations and at different operating conditions to a single location at new operating conditions. This was done to avoid computing the modified area validation metric (Section 12.5.2.3) with a single experimental replicate, which would likely overestimate the model form uncertainty. Using *u*-pooling, an empirical distribution function could be constructed for the experiment with 10 replicate values. The steps in the *u*-pooling process are depicted in Figure 12.27, showing the simulation CDF being used to transform single experimental measurements of stress to the cumulative probability (u) values at one set of conditions (see panels a–e). Note that only the first 5 experimental locations are shown for brevity. A different simulation CDF is used to transform the 10 u values into 10 experimental replicates for stress at the location and conditions of interest, as shown in panel (*f*). (Here only u1, u2, and u10 are shown for clarity.)

Once the *u*-pooling procedure is completed, the end result is an empirical distribution function for the experiment with 10 replicates, as shown in Figure 12.28 (solid curve). Also shown in the figure is the simulation p-box generated by accounting for both aleatory and epistemic input uncertainty (see Section 12.5). The empirical distribution function was generated from single experimental replicates at 5 different spatial locations and two different operating conditions. The *u*-pooling process can thus provide a much more accurate estimate of the model form uncertainty than a single experimental measurement alone. See Choudhary et al. (2016) for additional details.

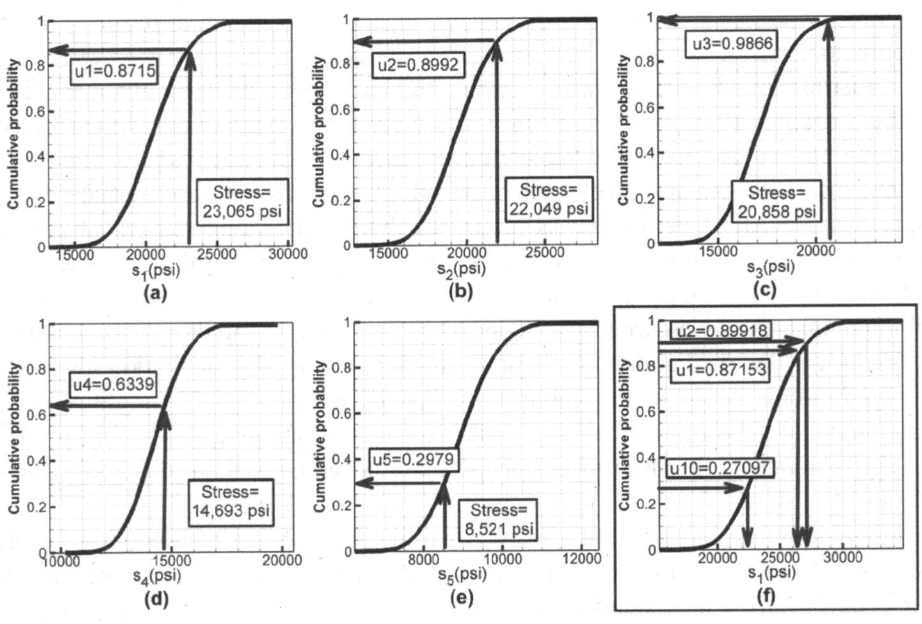

Figure 12.27 (a–e) The *u*-pooling process showing five different spatial locations where single experimental replicate measurements of stress are transformed to cumulative probability (u) space; (f) The u values (only three are shown) are transformed back to stress SRQ values at the desired location and conditions for model assessment.

(Choudhary et al., 2016)

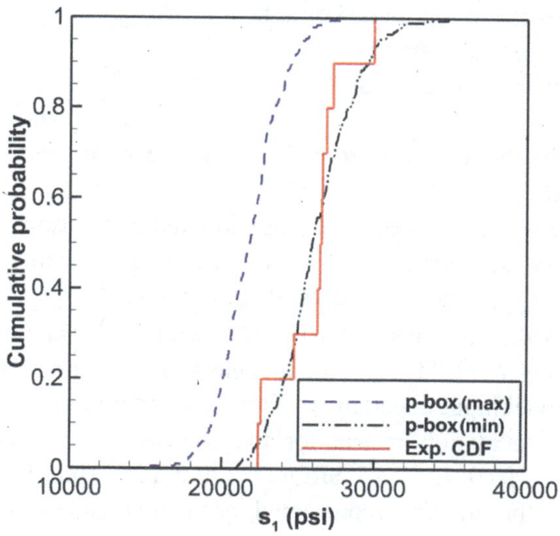

Figure 12.28 Empirical distribution function for the experiment generated by *u*-pooling with 10 experimental replicates along with the p-box for the simulation.
(Choudhary et al., 2016)

12.7 Further Reading

For modern texts on Bayesian methods, see Gelman et al. (2013), Bailer-Jones (2017), Santner et al. (2018), Lambert (2018), and McElreath (2020).

For recent research on model validation using hypothesis testing or Bayesian estimation methods, see Xiong et al. (2009), Liu et al. (2011), Ling and Mahadevan (2013), Sankararaman and Mahadevan (2013), Li et al. (2014), Ling et al. (2014), Nannapaneni and Mahadevan (2016), Deng et al. (2018), Vanslette et al. (2020), and Choo et al. (2022).

For recent research on model validation for both steady-state and time-dependent responses, see Anderson and Papachristodoulou (2009), Sarin et al. (2010), McCusker et al. (2010), Liu et al. (2011), Teferra et al. (2014), Thacker and Paez (2014), Hora and Campos (2015), Wu et al. (2015), Xi et al. (2015), Shen et al. (2015), Allemang et al. (2016), Mullins et al. (2016), Xi and Yang (2016), Ao et al. (2017), Bi et al. (2017), Zhao et al. (2017), Maupin et al. (2018), Wang et al. (2018), Li and Lu (2018), Dvurecenska et al. (2018), He (2019), Gardner et al. (2019), Wilson and Koskelo (2020), Brune et al. (2020), Porter et al. (2021), McKeand et al. (2021), Eca, Dowding, Moorcroft, et al. (2022), Zhang et al. (2022), Hu et al. (2022), Son et al. (2022), White et al. (2022), Deng et al. (2023), and Ledwith et al. (2023).

12.8 Problems

1. True or False: A validation assessment should ideally include the simultaneous calibration of unmeasured model parameters.

2. Multiple Choice: When using existing experimental data for model validation, a model input parameter that is not reported should be ______________.
   a. treated as an aleatory uncertainty
   b. treated as an epistemic uncertainty
   c. ignored
   d. varied to obtain the best agreement with the experimental data for the system response quantity
3. Briefly, explain the difference between validation and calibration.
4. Briefly, explain why the error in a simulation relative to an experimental measurement, $E = SRQ_{Sim} - SRQ_{Exp}$, is not a good validation metric.
5. Compute the area validation metric for the case with 2 replicate experimental measurements of $SRQ_{Exp} = 9.1$ and 11.2 and the (expensive) simulation includes only 2 samples to propagate the model input uncertainty, resulting in computational SRQs of 5.6 and 5.0.
6. Compute the area validation metric for the following experimental and simulation samples: $SRQ_{Exp} = [10.0, 7.0]$ and $SRQ_{Sim} = [9.0, 12.0, 10.0, 10.5]$. Sketch the graph of your results denoting the area represented by the area validation metric.
7. Repeat problem 6 using the modified area validation metric. You can ignore any confidence intervals associated with the small number of simulation samples but include them for the experimental samples.

12.9 References

Allemang, R., M. M. Kolluri, M. Spottswood, and T. Eason (2016). Decomposition-Based Calibration/Validation Metrics for Use with Full-Field Measurement Situations. *Journal of Strain Analysis*, **51**(1), 14–31.

Anderson, J., and A. Papachristodoulou (2009). On Validation and Invalidation of Biological Models. *BMC Bioinformatics*, **10**(Article 132).

Angus, J. E. (1994). The Probability Integral Transform and Related Results. *SIAM Review*, **36**(4), 652–654.

Ao, D., Z. Hu, and S. Mahadevan (2017). Dynamics Model Validation Using Time-Domain Metrics. *Journal of Verification, Validation and Uncertainty Quantification*, **2**(1), 011004-1–011004-15.

ASME (2009). Standard for Verification and Validation in Computational Fluid Dynamics and Heat Transfer. *American Society of Mechanical Engineers*, ASME Standard V&V 20-2009. New York, NY.

Aster, R., B. Borchers, and C. Thurber (2019). *Parameter Estimation and Inverse Problems*, 3rd ed. Burlington, MA: Elsevier.

Aughenbaugh, J. M., and C. J. J. Paredis (2006). The Value of Using Imprecise Probabilities in Engineering Design. *Journal of Mechanical Design*, **128**(4), 969–979.

Augustin, T., F. P. A. Coolen, G. de Cooman, and M. C. M. Troffaes (eds.) (2014). *Introduction to Imprecise Probabilities*. Chichester: Wiley & Sons.

Bailer-Jones, C. A. L. (2017). *Practical Bayesian Inference: A Primer for Physical Scientists*. Cambridge: Cambridge University Press.

Bernardini, A., and F. Tonon (2010). *Bounding Uncertainty in Civil Engineering*. New York, NY: Springer.

Berry, K. J., K. L. Kvamme, J. E. Johnston, and P. W. Mielke, Jr. (2021). *Permutation Statistical Methods with R*. New York, NY: Springer.

Bi, S., S. Prabhu, S. Cogan, and S. Atamturktur (2017). Uncertainty Quantification Metrics with Varying Statistical Information in Model Calibration and Validation. *AIAA Journal*, **55**(10), 3570–3583.

Box, E. P., and N. R. Draper (1987). *Empirical Model-Building and Response Surfaces*. New York, NY: John Wiley.

Brune, A. J., T. K. I. West, and L. M. White (2020). Calibration of Probe Uncertainty and Validation for the Hypersonic Material Environmental Test System. *Journal of Thermophysics and Heat Transfer*, **34**(2), 404–420.

Choo, J., Y. Jung, and I. Lee (2022). A Bayesian Model Calibration under Insufficient Data Environment. *Structural and Multidisciplinary Optimization*, **65**(3), 96-1–96-20.

Choudhary, A., I. T. Voyles, C. J. Roy, W. L. Oberkampf, and M. Patil (2016). Probability Bounds Analysis Applied to the Sandia Verification and Validation Challenge Problem. *Journal of Verification, Validation and Uncertainty Quantification*, **1**(1), 011003-1–011003-13.

Coleman, H. W., and F. Stern (1997). Uncertainties and CFD Code Validation. *Journal of Fluids Engineering*, **119**(4), 795–803.

Crassidis, J. L., and J. L. Junkins (2012). *Optimal Estimation of Dynamics Systems*, 2nd ed. Boca Raton, FL: Chapman & Hall/CRC Press.

Deng, W., X. Lu, and Y. Deng (2018). Evidential Model Validation under Epistemic Uncertainty. *Mathematical Problems in Engineering*, **2018**, 6789635-1–6789635-11.

Deng, Y., M. R. Eden, and S. Cremaschi (2023). Metrics for Evaluating Machine Learning Models Prediction Accuracy and Uncertainty. *Computer Aided Chemical Engineering*, **52**, 1325,1330.

Devore, J. L. (2016). *Probability and Statistics for Engineers and the Sciences*, 9th ed. Pacific Grove, CA: Duxbury.

Dowding, K. J., M. Pilch, and R. G. Hills (2008). Formulation of the Thermal Problem. *Computer Methods in Applied Mechanics and Engineering*, **197**(29–32), 2385–2389.

Dowding, K. J., J. R. Red-Horse, T. L. Paez, I. M. Babuska, R. G. Hills, and R. Tempone (2008). Editorial: Validation Challenge Workshop Summary. *Computer Methods in Applied Mechanics and Engineering*, **197**(29–32), 2381–2384.

Drosg, M. (2009). *Dealing with Uncertainties: A Guide to Error Analysis*, 2nd ed. New York, NY: Springer.

Dvurecenska, K., S. Graham, E. Patelli, and E. A. Patterson (2018). A Probabilistic Metric for the Validation of Computational Models. *Royal Society Open Science*, **5**(11), 180687-1–180687-14.

Eca, L., K. Dowding, D. Moorcroft, and U. Ghia (2022). Comparison of the V&V10.1 and V&V20 Modeling Error Quantification Procedures for the V&V10.1 Example. *Journal of Verification, Validation and Uncertainty Quantification*, **7**(2), 021004-1–021004-7.

Eca, L., K. Dowding, and P. J. Roache (2022). On the Interpretation and Scope of the V&V 20 Standard for Verification and Validation in Computational Fluid Dynamics and Heat Transfer. *Journal of Verification, Validation and Uncertainty Quantification*, **7**(2), 021005-1–021005-9.

Ferson, S., V. Kreinovich, L. Ginzburg, D. S. Myers, and K. Sentz (2002). Constructing Probability Boxes and Dempster-Shafer Structures. *Sandia National Laboratories*, SAND2002-4015, Albuquerque, NM.

Ferson, S., V. Kreinovich, H. Hajagos, W. L. Oberkampf, and L. Ginzburg (2007). Experimental Uncertainty Estimation and Statistics for Data Having Interval Uncertainty. *Sandia National Laboratories*, SAND2007-0939, Albuquerque, NM.

Ferson, S., R. B. Nelsen, J. Hajagos, D. J. Berleant, J. Zhang, W. T. Tucker, et al. (2004). Dependence in Probabilistic Modeling, Dempster-Shafer Theory, and Probability Bounds Analysis. *Sandia National Laboratories*, SAND2004-3072, Albuquerque, NM.

Ferson, S., and W. L. Oberkampf (2009). Validation of Imprecise Probability Models. *International Journal of Reliability and Safety*, **3**(1–3), 3–22.

Ferson, S., W. L. Oberkampf, and L. Ginzburg (2008). Model Validation and Predictive Capability for the Thermal Challenge Problem. *Computer Methods in Applied Mechanics and Engineering*, **197**(29–32), 2408–2430.

Gardner, P., C. Lord, and R. J. Barthorpe (2019). A Unifying Framework for Probabilistic Validation Metrics. *Journal of Verification, Validation and Uncertainty Quantification*, **4**(3), 031005-1–031005-11.

Geers, T. L. (1984). An Objective Error Measure for the Comparison of Calculated and Measured Transient Response Histories. *The Shock and Vibration Bulletin*, **54**(2), 99–107.

Gelman, A. B., J. B. Carlin, H. S. Stern, D. B. Dunson, A. Vehtari, and D. B. Rubin (2013). *Bayesian Data Analysis*, 3rd ed. Boca Raton, FL: CRC Press.

Gioia, F., and C. N. Lauro (2005). Basic Statistical Methods for Interval Data. *Statistica Applicata*, **17**(1), 75–104.

Grabe, M. (2014). *Measurement Uncertainties in Science and Technology*, 2nd ed. New York, NY: Springer.

He, Q. (2019). Model Validation Based on Probability Boxes under Mixed Uncertainties. *Advances in Mechanical Engineering*, **11**(5), 1–9.

Hora, J., and P. Campos (2015). A Review of Performance Criteria to Validate Simulation Models. *Expert Systems*, **32**(5), 578–595.

Hu, K. T., and G. E. Orient (2016). The 2014 Sandia Verification and Validation Challenge: Problem Statement. *Journal of Verification, Validation and Uncertainty Quantification*, **1**(1), 011001-1–011001-10.

Hu, W., T. Zhang, Z. Liu, J. Cheng, Y. Lou, X. Deng, et al. (2022). A New Validation Metric for Models with Correlated Responses Using Limited Experimental and Simulation Data. *Journal of Mechanical Design*, **144**(11), 111702-1–111702012.

Huber-Carol, C., N. Balakrishnan, M. Nikulin, and M. Mesbah (eds.) (2002). *Goodness-of-Fit Tests and Model Validity.* Boston, MA: Birkhauser.

ISO (1995). Guide to the Expression of Uncertainty in Measurement. International Organization for Standardization, ISO/IEC Guide 98:1995, ISBN 92-67-10188-9, Geneva, Switzerland.

Kennedy, M. C., and A. O'Hagan (2001). Bayesian Calibration of Computer Models. *Journal of the Royal Statistical Society Series B-Statistical Methodology*, **63**(3), 425–450.

Klir, G. J. (2006). *Uncertainty and Information: Foundations of Generalized Information Theory.* New York, NY: John Wiley.

Lambert, B. (2018). *A Student's Guide to Bayesian Statistics*. Los Angeles, CA: SAGE Publications.

Law, A. M. (2014). *Simulation Modeling and Analysis*, 5th ed. New York, NY: McGraw-Hill.

Ledwith, M. C., R. R. Hill, L. E. Champagne, and E. D. White (2023). Probabilites of Agreement for Computational Model Validation. *Journal of Verification, Validation and Uncertainty Quantification*, **8** (1), 011003-1–011003-9.

Lehmann, E. L., and J. P. Romano (2022). *Testing Statistical Hypotheses*, 4th ed. New York, NY: Springer.

Lenhard, J. (2018). Holism, or the Erosion of Modularity – A Methodological Challenge for Validation. *Philosophy of Science*, **85**(5), 832–844.

Li, L., and Z. Lu (2018). A New Method for Model Validation with Multivariate Output. *Reliability Engineering and System Safety*, **169**, 579–592.

Li, W., W. Chen, Z. Jiang, Z. Lu, and Y. Liu (2014). New Validation Metrics for Models with Multiple Correlated Responses. *Reliability Engineering and System Safety*, **127**, 1–11.

Ling, Y., and S. Mahadevan (2013). Quantitative Model Validation Techniques: New Insights. *Reliability Engineering and System Safety*, **111**, 217–231.

Ling, Y., J. Mullins, and S. Mahadevan (2014). Selection of Model Discrepancy Priors in Bayesian Calibration. *Journal of Computational Physics*, **276**, 665–680.

Liu, Y., W. Chen, P. Arendt, and H.-Z. Huang (2011). Toward a Better Understanding of Model Validation Metrics. *Journal of Mechanical Design*, **133**(13), 071005: 1–13.

Magnus, R. (2022). *Metric Spaces: A Companion to Analysis*. New York, NY: Springer.

Manski, C. F. (2003). *Partial Identification of Probability Distributions*. New York, NY: Springer.

Martin, R., and C. Liu (2016). *Inferential Models*. Boca Raton, FL: CRC Press.

Maupin, K. A., L. P. Swiler, and N. W. Porter (2018). Validation Metrics for Deterministic and Probabilistic Data. *Journal of Verification, Validation and Uncertainty Quantification*, **3**(3), 031002-1–031002-10.

McCusker, J. R., K. Danai, and D. O. Kazmer (2010). Validation of Dynamic Models in the Time-Scale Domain. *Journal of Dynamic Systems, Measurement, and Control*, **132**(6), 061402-1–061402-9.

McElreath, R. (2020). *Statistical Rethinking: A Bayesian Course with Examples in R and Stan*, 2nd ed. Boca Raton, FL: CRC Press.

McKeand, A. M., R. M. Gorguluarslan, and S.-K. Choi (2021). Stochastic Analysis and Validation under Aleatory and Epistemic Uncertainties. *Reliability Engineering and System Safety*, **205**, 107258-1–107258-14.

Mielke, P. W., Jr., and K. J. Berry (2007). *Permutation Methods: A Distance Function Approach*, 2nd ed. New York, NY: Springer.

Molchanov, I. (2005). *Theory of Random Sets*. New York, NY: Springer.

Mullins, J., Y. Ling, S. Mahadevan, L. Sun, and A. Strachan (2016). Separation of Aleatory and Epistemic Uncertainty in Probabilistic Model Validation. *Reliability Engineering and System Safety*, **147**, 49–59.

Nannapaneni, S., and S. Mahadevan (2016). Reliability Analysis under Epistemic Uncertainty. *Reliability Engineering and System Safety*, **155**, 9–20.

Oberkampf, W. L., and M. F. Barone (2004). Measures of Agreement between Computation and Experiment: Validation Metrics. *34th AIAA Fluid Dynamics Conference*, AIAA Paper 2004-2626, Portland, OR.

Oberkampf, W. L., and M. F. Barone (2006). Measures of Agreement between Computation and Experiment: Validation Metrics. *Journal of Computational Physics*, **217**(1), 5–36.

Oberkampf, W. L., and C. J. Roy (2010). *Verification and Validation in Scientific Computing*. Cambridge: Cambridge University Press.

Oberkampf, W. L., and T. G. Trucano (2002). Verification and Validation in Computational Fluid Dynamics. *Progress in Aerospace Sciences*, **38**(3), 209–272.

Oberkampf, W. L., T. G. Trucano, and C. Hirsch (2004). Verification, Validation, and Predictive Capability in Computational Engineering and Physics. *Applied Mechanics Reviews*, **57**(5), 345–384.

Pilch, M. (2008). Preface: Sandia National Laboratories Validation Challenge Workshop. *Computer Methods in Applied Mechanics and Engineering*, **197**(29–32), 2373–2374.

Porter, N. W., K. A. Maupin, L. P. Swiler, and V. A. Mousseau (2021). Validation Metrics for Fixed Effects and Mixed-Effects Calibration. *Journal of Verification, Validation and Uncertainty Quantification*, **6**(1), 011005-1–011005-11.

Rabinovich, S. G. (2017). *Evaluating Measurement Accuracy: A Practical Approach*, 3rd ed. New York, NY: Springer.

Rider, W. J. (1998). [Personal Communication].

Ross, S. M. (2020). *Introduction to Probability and Statistics for Engineers and Scientists*, 6th ed. London: Elsevier.

Russell, D. M. (1997a). Error Measures for Comparing Transient Data: Part I, Development of a Comprehensive Error Measure. *Proceedings of the 68th Shock and Vibration Symposium*, Hunt Valley, MD.

Russell, D. M. (1997b). Error Measures for Comparing Transient Data: Part II, Error Measures Case Study. *Proceedings of the 68th Shock and Vibration Symposium*, Hunt Valley, MD.

Sankararaman, S., and S. Mahadevan (2013). Distribution Type Uncertainty Due to Sparse and Imprecise Data. *Mechanical Systems and Signal Processing*, **37**(1&2), 182–198.

Santner, T. J., B. J. Williams, and W. Notz (2018). *The Design and Analysis of Computer Experiments*, 2nd ed. New York, NY: Springer.

Sarin, H., M. Kokkolaras, G. Hulbert, P. Papalambros, S. Barbat, and R.-J. Yang (2010). Comparing Time Histories for Validation of Simulation Models: Error Measures and Metrics. *Journal of Dynamic Systems, Measurement, and Control*, **132**, 061401:1–10.

Shen, Z., X. Chen, Q. He, and C. P. Zang (2015). Study on Area Metric Based upon Multiple Correlated System Response Quantities. *SAE 2015 World Congress & Exhibition*, SAE Technical Paper 2015-01-0454.

Smith, R. C. (2014). *Uncertainty Quantification: Theory, Implementation, and Applications*. Philadelphia, PA: Society for Industrial and Applied Mathematics.

Son, H., H. Choi, W. Kim, B. D. Youn, and G. Lee (2022). A Comparative Study of Statistical Validation Metrics with Consideration of Variance to Address Type II Errors in Statistical Model Validation. *Structural and Multidisciplinary Optimization*, **65**(63), 63-1–63-19.

Sprague, M. A., and T. L. Geers (1999). Response of Empty and Fluid-Filled, Submerged Spherical Shells to Plane and Spherical, *Step-Exponential Acoustic Waves. Shock and Vibration*, **6**(3), 147–157.

Sprague, M. A., and T. L. Geers (2004). A Spectral-Element Method for Modeling Cavitation in Transient Fluid-Structure Interaction. *International Journal for Numerical Methods in Engineering*, **60**(15), 2467–2499.

Stern, F., R. V. Wilson, H. W. Coleman, and E. G. Paterson (2001). Comprehensive Approach to Verification and Validation of CFD Simulations-Part 1: Methodology and Procedures. *Journal of Fluids Engineering*, **123**(4), 793–802.

Stripling, H. F., M. L. Adams, R. G. McClarren, and B. K. Mallick (2011). The Method of Manufactured Universes for Validating Uncertainty Quantification Methods. *Reliability Engineering and System Safety*, **96**(9), 1242–1256.

Teferra, K., M. D. Shields, A. Hapij, and R. P. Daddazio (2014). Mapping Model Validation Metrics to Subject Matter Expert Scores for Model Adequacy Assessment. *Reliability Engineering and System Safety*, **132**, 9–19.

Thacker, B. H., and T. L. Paez (2014). A Simple Probabilistic Validation Metric for the Comparison of Uncertain Model and Test Results. *AIAA SciTech Forum*, Paper 2014-0121, National Harbor, MD.

Thas, O. (2010). *Comparing Distributions*. New York, NY: Springer.

van den Bos, A. (2007). *Parameter Estimation for Scientists and Engineers*. New York, NY: John Wiley.

Vanslette, K., T. Tohme, and K. Youcef-Toumi (2020). A General Model Validation and Testing Tool. *Reliability Engineering and System Safety*, **195**, 106684-1–106684-17.

Voyles, I. T., and C. J. Roy (2014). Evaluation of Model Validation Techniques in the Presence of Uncertainty. *American Institute of Aeronautics and Astronautics SciTech Conference*, AIAA Paper-2014-0120, National Harbor, MD.

Voyles, I. T., and C. J. Roy (2015). Evaluation of Model Validation Techniques in the Presence of Aleatory and Epistemic Input Uncertainties. *American Institute of Aeronautics and Astronautics SciTech Conference*, AIAA Paper-2015-1374, Kissimmee, FL.

Wang, C., H. G. Matthies, M. Xu, and Y. Li (2018). Epistemic Uncertainty-Based Model Validation via Interval Propagation and Parameter Calibration. *Computer Methods in Applied Mechanics and Engineering*, **342**, 161–176.

White, A., S. Mahadevan, J. Schmucker, and A. Karl (2022). Multi-Metric Validation Under Uncertainty for Multivariate Model Outputs and Limited Measurements. *Journal of Verification, Validation and Uncertainty Quantification*, **7**(4), 041004-1–041004-15.

Whiting, N. W., C. J. Roy, E. Duque, S. Lawrence, and W. L. Oberkampf (2023). Assessment of Model Validation, Calibration, and Prediction Approaches in the Presence of Uncertainty. *Journal of Verification, Validation and Uncertainty Quantification*, **8**(1), 011001-1–011001-21.

Wilson, B., and A. Koskelo (2020). A Practical Validation Assessment Workflow. *Journal of Verification, Validation and Uncertainty Quantification*, **5**(1), 011004-1–011004-10.

Wirsching, P., T. Paez, and K. Ortiz (1995). *Random Vibrations: Theory and Practice*. New York, NY: John Wiley.

Wu, D., Z. Lu, Y. Wang, and L. Cheng (2015). Model Validation and Calibration Based on Component Functions of Model Output. *Reliability Engineering and System Safety*, **140**(1), 59–70.

Xi, Z., H. Pan, Y. Fu, and R.-J. Yang (2015). Validation Metric for Dynamic System Responses under Uncertainty. *SAE International Journal of Materials and Manufacturing*, **8**(2), 309–314.

Xi, Z., and R.-J. Yang (2016). Reliability Analysis With Model Uncertainty Coupling with Parameter and Experiment Uncertainties: A Case Study of 2014 Verification and Validation Challenge Problem. *Journal of Verification, Validation and Uncertainty Quantification*, **1**(1), 011005-1–011005-11.

Xiong, Y., W. Chen, K.-L. Tsui, and D. W. Apley (2009). A Better Understanding of Model Updating Strategies in Validating Engineering Models. *Computer Methods in Applied Mechanics and Engineering*, **198**(15–16), 1327–1337.

Zhang, H., W. Li, P. Ma, and M. Yang (2022). A Model Validation Method Based on the Orthogonal Polynomial Transformation and Area Metric. *International Journal of Modeling, Simulation, and Scientific Computing*, **13**(2), 2241013.

Zhao, L., Z. Lu, W. Yun, and W. Wang (2017). Validation Metric Based on Mahalanobis Distance for Models with Multiple Correlated Responses. *Reliability Engineering and System Safety*, **159**, 80–89.

13

Predictive Capability

Prediction, not narration, is the real test of our understanding of the world. —Nassim Taleb

Contents of Chapter 13

This chapter synthesizes the key procedures from the previous chapters and incorporates them into the modern concept of predictive capability in scientific computing. This chapter, in contrast to all other chapters, does not just address specific themes in simulation. The theme here is *prediction*, in the sense of foretelling the response of a system to previously unobserved conditions based on a mathematical model of the system of interest. Although this topic has been widely addressed in many applications of simulation and scientific computing, the actual steps in the procedure have not been systematically assembled and discussed in the literature. The framework for predictive capability is organized into eight procedural steps to make a nondeterministic prediction:

1. Formulate the goals of the analysis.
2. Identify all relevant sources of uncertainty.
3. Characterize model input uncertainty.
4. Estimate numerical solution uncertainty.
5. Estimate model form uncertainty.
6. Estimate total uncertainty in the predicted system responses.
7. Update model input parameters.
8. Conduct a sensitivity analysis.

The intent of the chapter is to present all of the critical steps in predictive capability in one chapter; not to provide an in-depth discussion of each of these steps. The analysis approach emphasized for predictive capability is probability bounds analysis (PBA), although Bayesian estimation is also discussed. As discussed in Chapter 3 "Modeling and Computational Simulation," PBA has a number of advantages over the more common approach of Bayesian estimation. Furthermore, PBA is fully consistent with the modern concept of "an abundance of caution" in the sense that PBA attempts include the impact of all identified sources of uncertainty on the simulation result.

The eight steps for predictive capability discussed in this chapter are similar to two topics discussed earlier in this text. The first was the discussion in Chapter 2, Section 2.5 "Integration of Verification, Validation, and Prediction." The second was the discussion in Chapter 3, Section 3.5 "Phases of Computational Simulation." In the eight steps discussed here, the emphasis is on uncertainty quantification (UQ) of all aspects that influence the simulation result. Verification and validation (V&V) should be viewed as important supporting elements in the use of a mathematical model to make a prediction, but they are not the actual goals of a simulation. To demonstrate how the procedural steps are implemented in practice, a detailed example is discussed as a continuous thread through the first five steps of predictive capability.

13.1 Step 1: Formulate the Goals of the Analysis

For a computational simulation that is directed toward informing a decision or aiding in the system of a design, particularly for an engineering system, specific goals of the analysis should be established early in the process. This is in contrast to research simulations where the goals are commonly the discovery of how physical processes interact or the improved understanding of how a system works. The system of interest could be a complex system, such as a rocket launch vehicle for placing satellites into orbit, or it could be an isolated system, such as a fan blade of a gas turbine engine. Keep in mind that the system could also be a manufacturing process of interest, such as metal cutting, additive manufacturing, or mold filling.

As discussed in Chapter 3, the system should be considered in one of three types of environment: normal, abnormal (e.g., in an accident or damaged state), or hostile (e.g., the system is under physical or electronic attack). If multiple environments are of interest, each should be considered in a separate analysis because the goals of the analysis for each environment are usually quite different. For example, in normal environments, goals of the analysis commonly deal with system performance issues, such as fuel consumption, power output, and system range. For abnormal and hostile environments, however, analysis goals usually deal with system safety and the ability to regain control of the system, for instance, how the system can be shut down safely without additional disruption. Whether an environment is considered an unusual normal environment or an abnormal environment usually depends on the potential negative consequences or legal liability of failure or under performance of the system. For example, a somewhat overloaded system or if the operator of the system is confused or distracted, could be considered as either a normal or abnormal environment.

For a given environment, it is also recommended that the analysis address one scenario or sequence of events at a time, as opposed to simultaneously investigating a wide range of scenarios. Focusing on one scenario greatly helps both analysts and customers of the analysis develop and agree on the primary and secondary goals of the analysis. Figure 13.1 depicts a tree structure for a given environment, a set of scenarios identified, and various system response quantities (SRQs) that are of interest in each scenario. Note that the SRQs for each scenario are not necessarily the same. As an example of a system in an abnormal environment, consider an assembled structure in a high vibration environment where multiple dangerous and/or failure scenarios are identified. Examples of possible scenarios are: (a) plastic deformation of portions of the structure, (b) initiation of cracks in the structure, (c) buckling of the structure, and (d) self-loosening of fasteners which could cause a massive failure. Only some of these scenarios may be analyzed in detail, based on (a) the likelihood and possible consequence of each failure mode identified, and (b) the time and resources available for the analysis.

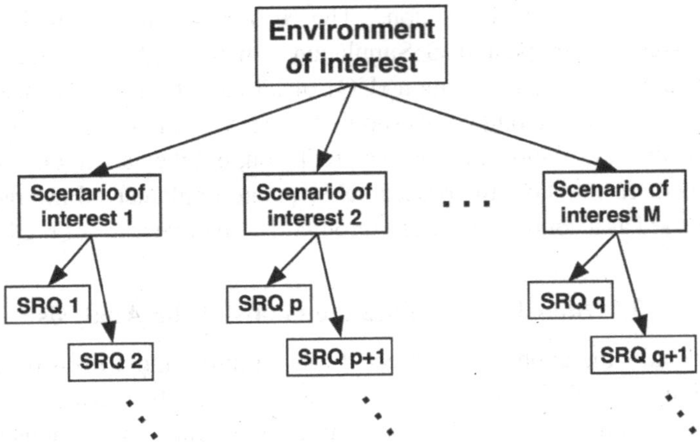

Figure 13.1 Environment-scenario-SRQ tree.

After the environment and the scenario have been established, the SRQs of interest should be discussed in planning the analysis. There are a wide variety of SRQs that could be of interest to the customer, depending on the goals of analysis. For example, the SRQs of interest can be quite different depending on whether t analysis is focused on the design of a fundamentally new system or it is an analysis of an existing system. The design or modification of an existing system usually addresses SRQs related to either localized characteristics such as temperature, pressure, or mechanical stress, or global quantities such as improved performance, safety, or reliability of the system. The analysis of an existing system can also deal with issues related to a possible failure mode of the system, such as failure to meet safety requirements for a newly identified scenario. Failure analyses tend to concentrate on a very small set of SRQs, whereas design analyses address a very wide range of SRQs that cover diverse aspects of performance, safety, and reliability of the system.

There is commonly a desire to set the requirements on the prediction accuracy of the SRQs resulting from the simulation. These requirements are usually associated with requirements that are specified for the performance, safety, and reliability of the system of interest. However, trying to set the simulation accuracy requirements proves to be problematic in most cases. The reason is that by definition, the predicted SRQs are uncertain responses of the system being analyzed. For a given environment and scenario, the uncertainty in the predicted SRQs depends on three diverse sources: (1) uncertainty in the model inputs, (2) numerical solution uncertainties, and (3) model form uncertainty. The impact that these sources can have on the predicted SRQs is *not* known in the planning phase of a simulation because they can only be estimated during the analysis. An appropriate analogy is setting the accuracy requirements on hitting a target, but the target cannot be seen by the shooter. Generally, the best that can be done in the planning stage of an analysis is having a suggested accuracy of key SRQs provided by the customer of the analysis. As an alternative, it may be more constructive in the dialog between the simulation analyst and the customer to discuss an acceptable uncertainty bound in the SRQs, as opposed to an accuracy requirement. For example, it may be more helpful for the analyst to ask, "How inaccurate could the simulation results be, but still satisfy your need for simulation-informed decision making?" The uncertainty of the SRQs in a simulation should be thought of as an interaction and trade-off between each of the three sources given above, plus the time and resources allocated for the simulation.

The final consideration in planning is the alignment of the goals of the analysis with the time, money, and resources available. Necessary resources include the availability of suitably trained and experienced analysis staff, computational facilities, whether existing software capabilities are adequate for the analysis, availability of relevant experimental data needed for the analysis, and whether new experimental data are needed, or new facilities need to be constructed. The magnitude of analysis planning should be proportional to the magnitude and impact of the information produced by the simulation. The value of information produced by a simulation is generally viewed from a positive perspective, i.e., what the value of knowledge gained from a simulation is. For example, what is the value of the simulation toward improving the design of an existing system so that it consumes less fuel, reduces liability by achieving a fail-safe design, or requires less maintenance? The potential negative value, i.e., misinformation, from simulation results should also be considered in design and simulation-informed decision making. For example:

- What is the liability risk to a system supplier if the simulation produces incorrect information for the advertised performance, safety, and reliability of the system?
- What is the damage to an organization's future business, market share, or credibility if a simulation misinforms a system design such that it fails to meet performance, safety, and reliability guarantees?
- What is the damage to public and political confidence in a regulatory authority that approves a product or environmental decision based on misleading, inept, or unreliable simulations?

A number of high-visibility failures of simulation-informed decision making come to mind regarding the negative impact of inadequate simulations and improper use of simulation by management (e.g., the loss of two US Space Shuttles and the Fukushima nuclear disaster) (Petroski, 2012; Huckle and Neckel, 2019).

The critical value of proper planning and setting the goals of an analysis are bluntly pointed out by Vose (2008, p. 35) concerning risk analyses: "The biggest uncertainty in a risk analysis is whether we started off analyzing the right thing and in the right way." Correctly solving the wrong problem is commonly referred to as a *Type III error* (Kimball, 1957).

13.1.1 Example Problem: Step 1 for a Heat Transfer Analysis

There are a number of small, concrete, cinder-block buildings in remote locations that are used to house electrical equipment. The electrical equipment and the walls of the building are mounted securely on a concrete pad. See Figure 13.2 for an example of the type of building under consideration. During a particularly hot summer, inspection and maintenance personnel reported that the temperature inside some of the buildings was very high. As a result, company engineers became concerned that the high temperature inside a building could adversely impact the performance and operational life of the equipment. A heat transfer analysis is requested to estimate the maximum temperature that could occur inside the buildings that are located in the hottest locations during the summer months. The analyst assigned to the task asks, "What are the constraints on the

Figure 13.2 Typical concrete cinder-block building housing electrical equipment.

analysis in terms of cost and schedule?" Management informs the analyst that no budget exists to evaluate and purchase new software, so the analysis must be completed using presently-licensed software, i.e., either MATLAB™ or Mathematica™. Furthermore, management approves the budget and personnel to complete a few experimental measurements on existing buildings to assess the accuracy of the mathematical model.

The electrical cables providing power to the housed equipment and the power distribution cables from the equipment pass through the concrete pad of the building. The four outside walls and roof of the building are exposed to the local weather conditions. The analysis should only examine what is considered to be a worst-case scenario for the hottest conditions at existing locations of the buildings.

13.2 Step 2: Identify All Relevant Sources of Uncertainty

After the goals of an analysis have been established, the conceptual and mathematical models of the system are formulated. This discussion does not address all of the activities that were discussed in Chapter 3, Section 3.5 "Phases of Computational Simulation," but only the activity of *nondeterministic identification*. This activity deals with the identification and specification of all aspects of the model that are considered uncertain, as opposed to those that are considered deterministic or fixed. As mentioned above, the three categories of uncertainty sources are model input, numerical solution, and model form uncertainty. Model input uncertainty and model form uncertainty have their roots in the formulation of the conceptual and mathematical models, whereas numerical solution uncertainty is entirely dependent on the solution of the discrete model that is solved on the computer.

The guiding principle for classifying an aspect of the model as nondeterministic versus deterministic is: *Which is the appropriate choice in order to achieve the goals of the analysis?* Note that the term *aspect* is used to include the assumptions and approximations made in the formulation of the conceptual and mathematical models. For example, aspects of the model are dependent on the specification of the environment, scenario, system, surroundings, and application conditions of interest. The general philosophy that should be used to decide if an aspect should be considered as nondeterministic vs. deterministic is to consider an aspect as uncertain if there is reason to believe that the uncertainty in the aspect could impact the SRQs of interest in a negative way from the perspective of the customer of the analysis. The same philosophy can be stated from the opposite perspective. An aspect should only be considered as deterministic if the potentially detrimental influence of the aspect on the SRQs of interest is *fully understood* to produce what is needed to achieve the goals of the analysis. For example, an aspect may actually be uncertain, but it can be considered as a fixed value if the quantity is fixed at a value such that it *unquestionably* produces the intended impact on the goals of the analysis. This is the philosophy that is used in what is referred to as a *worst-case analysis*. The difficulty with trying to identify the conceptual aspects of a mathematical model formulation, as well as the conditions of model input data, that will produce a worst-case analysis is that a complex system can have interactions between subsystems, coupled physics, humans, and computers that *cannot* be foreseen.

Fixing aspects of the model to produce a worst-case result for an SRQ, from the perspective of the customer of the analysis, can simplify an analysis compared to considering all relevant aspects of the simulation as uncertain. For simple systems, the response can sometimes be sufficiently understood to confidently specify an aspect of the simulation that produces the worst-case system response for the system of interest. However, even single-physics systems can be sufficiently complex such that identifying the aspect beforehand that produces the extreme response is not possible. Great caution should be used in attempting to identify simulation aspects that are believed to produce worst-case results.

Two simple examples follow. First, consider the static loading on a given structure using linear elasticity theory. If the structure is exposed to triaxial loading, it is impossible to determine beforehand what combination of loading conditions will produce the maximum local stress in the structure. Second, consider single phase, incompressible, Newtonian, non-reacting flow through a piping system. It is impossible to determine beforehand what flowrate will produce the lowest local pressure in the piping system. For coupled, multi-physics systems, it is impossible to specify what combination of input data will produce an extreme result for an SRQ because of the complex interactions of physics, subsystems, humans, and computer controls; all of which could be damaged, impaired, or distracted.

In a large-scale simulation, a *screening analysis* or *scoping study* is commonly conducted to obtain an indication of which aspects of the model may be important, or which aspects may be considered as deterministic or even ignored. This screening analysis can only be done after all of the elements of the simulation are computationally operational. A screening analysis is a type of global sensitivity analysis that indicates the sensitivity of the SRQs to each of the possible sources of uncertainty. If the impact of certain aspects of the simulation are sufficiently well understood, the screening analysis can be intentionally biased about being conservative toward the important outcomes of the analysis. This strategy is similar to the strategy used in a worst-case analysis. That is, a screening analysis attempts to err on the side of identifying all possible modeling aspects that *may result* in detrimental outcomes of the SRQs of interest. Sometimes a preliminary screening analysis can be conducted using a

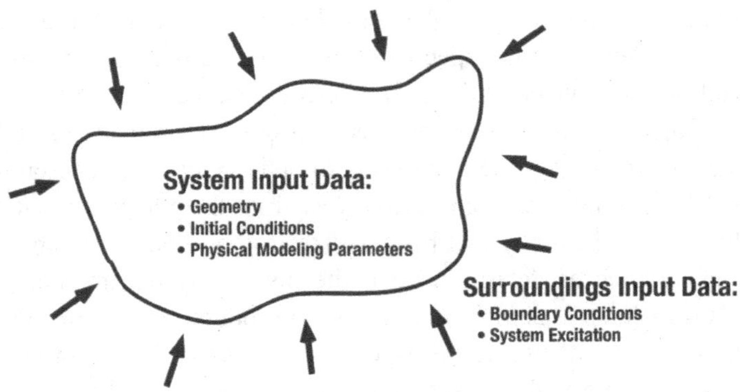

Figure 13.3 System and surroundings input data.

simplified model of the system of interest if there is sufficient confidence in the qualitative reliability of the simplified model.

For example, a complex system can have many subsystems, subassemblies, and components that interact both physically and electronically in several ways. Additionally, there would commonly be many different types of physics occurring, along with different types of interactions, such as human interactions and automatic feedback control systems, which may, or may not, be operating as expected. A screening analysis attempts to identify what aspects of the system, subsystems, components, physics, and interactions should be included in the full analysis, and which could safely be excluded, if any. A screening analysis with a preliminary model formulation can aid analysts and decision makers in directing limited resources toward the more important aspects of the simulation and possibly toward the design of the system. Section 13.8 discusses both screening analyses and sensitivity analyses.

13.2.1 Model Input Uncertainty

Model inputs can enter the simulation in one of two general groups: system input data and surroundings input data. Figure 13.3 depicts these groups, along with the subgroups of each. Quantities in each of the subgroups can be deterministic or uncertain, depending on the goals of the analysis. Each subgroup is briefly discussed to point out what type of uncertainty might need to be considered and what kinds of difficulties commonly arise.

System geometry data can be specified in a number of ways. For example, it could be specified in a commercial software package used in solid mechanics or fluid dynamics, or in a computer-aided design (CAD) software package. In addition, computer-aided manufacturing (CAM) software may be used for more detail on the actual manufacturing and assembly processes, such as final surface preparation and methods, specifications for installing rivets and bolts, welding procedures, installation and assembly of hydraulic lines, location of state sensors, and wiring installation. However, if CAD/CAM software is used, the ability to consider uncertainties in the geometry in an automated way for a nondeterministic simulation can be very limited. By *automated*, we mean that the user of the CAD/CAM package can specify a subset of the geometry features as uncertain, assign values (e.g., random samples) to that subset, and then have all of the remaining geometry features re-computed.

Because of the diverse ways in which most CAD/CAM packages are structured and because of the multitude of ways the geometry within a package can be interrelated, the ability to automate uncertainties for a nondeterministic analysis can be difficult or even impossible. As a result, analysts should be very cautious in choosing CAD/CAM packages so that they have the flexibility that is needed to conduct nondeterministic analyses.

A similar situation can occur when a user constructs a geometry within a commercial software package that is designed for specific types of analyses, e.g., solid mechanics or fluid dynamics. The user usually specifies many of the system geometry parameters through the graphical user interface (GUI) of the software package. Then, if they want to consider some of them as uncertain, it may be possible to individually control these input parameters using a separate code that modifies these parameters. This type of code is sometimes referred to as a *wrapper code*. However, the user must be careful to not accidently over-specify the geometry or cause inconsistencies in the geometry when certain features are considered as uncertain. As a simple example, suppose the user is interested in computing the deflection of a triangular-shaped plate due to a load distribution over the surface of the plate. Because of manufacturing variability, suppose the three internal angles of the plate are considered as continuous random variables. The user can only choose two of the angles as random variables, because choosing three would over-specify the geometry. This example also points out that there is a correlation structure between the three angles. Correlation of input data is discussed in Section 13.3.

Initial conditions (ICs) provide required information for a model of a system that is formulated as an initial value problem. ICs provide required information concerning (a) the initial state of all of the dependent variables in the partial differential equations (PDEs) and (b) the initial state of all other physical modeling parameters, including geometric and material characteristics that could be dependent on time. As a result, the IC data could be a function of the remaining independent variables in the PDEs, such as spatial location. Typically, the most important aspect of the ICs is the state of all of the dependent variables over the domain of the PDE. In addition, the initial state of all of the dependent variables in all of the submodels, e.g., auxiliary PDEs, must be given. If the ICs are considered as uncertain, the uncertainty structure is clearly more complex than input geometry data because the user must not only deal with the correlation structure of the uncertain quantities, but also the correlation structure of one or more independent variables.

Physical modeling parameters provide information describing the physical characteristics of the system, excluding characteristics related to the geometry and the ICs. Physical modeling parameters usually receive the most attention in modeling and simulation because they comprise the independent information related to the physical mechanisms occurring in the system. Some of the physical modeling parameters can describe global characteristics of the system and some can vary as a function of both the independent and dependent variables in the differential equations of the system. Some of the physical parameters can be measured independently of the model of the system. Examples are Young's modulus, material plasticity models, thermal conductivity, fluid viscosity, electrical conductivity, and electromagnetic and fluid permeability. Some physical parameters can only be inferred based on the complete model of the system of interest. Examples are internal dynamic damping of a material, stiffness and damping of assembled joints in a structure, thermal and electrical contact resistance between materials, and effective chemical reaction rates in turbulent reacting flow. The inferential procedure used to update these parameters using the complete model is discussed in Section 13.7.

Surroundings input data consists of two subgroups: boundary conditions (BCs) and excitation of the system. BCs can be dependent on one or more of the independent variables of the PDEs. If the problem is formulated as an initial-boundary value problem, these independent variables are typically other spatial dimensions and time. For example, in a fluid-structure interaction problem, the boundary condition between the structure and the fluid is a compatibility condition. For the boundary condition on the structure, there is a distributed pressure and shear-stress loading imposed by the fluid. For the boundary conditions on the fluid, there is typically no flow through the surface, and the fluid velocity on the boundary must be equal to the local velocity of the boundary. Examples of different types of BCs are: Dirichlet, Neumann, Robin, mixed, periodic, and Cauchy. The character of the uncertainty in BCs considered here is uncertainty in the quantities involved in the BC, e.g., variability of temperature or pressure imposed by the surroundings. The type of BC is considered fixed.

If the BCs are considered as uncertain, the effect on the solution procedure can range from minimal to a situation where the solution procedure must be completely changed. For example, if the uncertainty in the BCs does not require a change in the coupling of the BCs to the solution procedure for the PDEs, then the uncertainty can usually be treated similarly to parametric uncertainty. For instance, the analyst could use a sampling procedure to propagate the effect of the uncertainty in the BCs, such as loading conditions, onto the SRQs. On the other hand, if the uncertainty in the BCs causes a physical or iterative change in the way the BCs are coupled to the solution to the PDEs, then more sophisticated procedures must be used. If the uncertainty in the BCs deforms the boundary to such a degree that the deformation cannot be considered as small, then the analyst must significantly change the numerical solution procedure, and possibly even the mathematical model, to deal with the uncertainty. For example, in fluid dynamics, if a surface is considered to have large roughness, the roughness may be sufficiently large such that two-equation models of turbulence may no longer be appropriate.

System excitation refers to how the surroundings affect the system, *other than* through the BCs. System excitation always results in a change in the PDEs that are being solved. Sometimes system excitation is referred to as a change in the right-hand side of the PDEs to represent the effect of the surroundings on any portion of the system. Common examples of system excitation are (a) a force field acting on the system, such as that due to gravity or an electric or magnetic field and (b) energy deposition distributed through the system, such as by electrical heating, chemical reactions, and ionizing or non-ionizing radiation. System excitation uncertainties are usually treated as an uncertain function over the domain of the PDE. Hence, they can be an uncertain scalar, vector, or tensor field. The uncertainty in the function is usually represented through a set of uncertain parameters in the function. However, similar to large uncertainties in BCs, if large uncertainties occur in the system excitation, then the mathematical model and/or the numerical solution procedure may need to be changed. For example, if the energy deposition within the system is sufficient to cause a phase change in the material, then a complete change in the mathematical model would be required.

A final comment is made regarding the relationship between the number of model input quantities and the dimensionality of the input space for the model. There is no agreed upon method of determining the dimensionality of the input space (DIS) for the model. Some computational analysts define the DIS as equal to the number of uncertain input quantities to the simulation. This is a reasonable perspective, but we believe it is too restrictive in a system design or optimization setting. In this setting, the analyst is interested in varying deterministic input quantities, possibly in addition to nondeterministic quantities, in order to determine how various SRQs change with changes in input data. As a result, we suggest that the DIS should be equal to the total number of input quantities to the model, whether they are uncertain or deterministic. As commonly occurs in system optimization analyses, some of the quantities that had been assumed deterministic are then considered as nondeterministic quantities. Issues related to system design and optimization are briefly addressed in Section 13.8.

13.2.2 Numerical Solution Uncertainty

Numerical solution uncertainty, also referred to as calculation uncertainty, is the uncertainty that is caused by the approximate solution of the mathematical model of the system of interest. Since the mathematical model is almost always given by a differential equation or an integrodifferential equation, the numerical solution to the discrete approximation of the mathematical model results in an approximate solution. The techniques for estimating the error due to discretization and iterative procedures were discussed in detail in Chapters 7, 8, and 9. This chapter addresses sources of numerical solution uncertainty with regard to their impact on the results of a simulation, even a deterministic simulation.

Numerical solution uncertainty can be grouped into four types or sources: (1) computer round-off, (2) iterative solution, (3) spatial and temporal discretization, and (4) statistical sampling. The reader is reminded that each of these sources is a *numerical error*, but they are characterized as an *uncertainty* because the true value is rarely known in practice. Although the methods for estimating the magnitude of each of these numerical errors are quite different, they all rely on the same fundamental approach – comparison of at least two numerical solutions to estimate the magnitude of the error on the SRQs of interest. Procedures for estimating each uncertainty are summarized in Section 13.4 because they can only be estimated and characterized *after* computed solutions are available.

Quantification of numerical solution uncertainty is not commonly addressed in simulations for three reasons. First, in many simulations the numerical solution error is simply ignored or assumed to be small compared to other contributors to uncertainty. Many times, there is little, or no, evidence presented to support the claim that these error sources are small. Second, in some computationally intensive simulations it is recognized, but not necessarily mentioned, that the numerical solution error is important, and possibly even dominant. However, it is argued that various modeling parameters can be adjusted to compensate for the numerical solution error using available experimental data to calibrate or update parameters to match the data. If the application conditions of interest are sufficiently similar to the conditions for which experimental data are available, it is claimed that adjusting the input parameters to compensate for the numerical error is a reasonable approach to predictive capability. Third, most simulation analysts recognize that even if the numerical error is estimated, there is generally no accepted procedure for mathematically including its effect on the SRQs of interest.

Our perspective is that the first of these three reasons for not addressing numerical solution error is unacceptable, whereas as the second and third reasons have some merit. The parameter calibration rationale is supported by the Bayesian estimation approach to predictive capability. As discussed in Section 13.6, there is a tenuous argument that this rationale is appropriate within the model validation domain. Outside the validation domain, the rationale is vacuous. The rationale that there is no generally accepted method for characterizing numerical error has been slowly eroding over the last two decades because of the development of PBA. In Section 13.6, a method is presented that explains how the estimated numerical uncertainties can be incorporated to increase the total uncertainty in the SRQs of interest.

13.2.3 Model Form Uncertainty

By *model form uncertainty*, we specifically mean uncertainty that is caused by the assumptions and approximations embedded in the formulation of the mathematical model, as opposed to uncertainty in inputs to the model. As discussed in Chapters 2 and 3, formulation of the model occurs in both the conceptual modeling and mathematical modeling phases. Model form uncertainty is the uncertainty that is due to the *entire aggregation of all components of the formulation of the system model and all submodels*, exclusive of model input uncertainty. For example, model formulation activities include: (a) assumptions concerning physical interactions or couplings that are included, as well as those that are ignored in the model of the system; (b) assumptions on how the boundary conditions and excitation conditions interact with the system; (c) approximations made in the mathematical formulation of the PDEs of the model; and (d) approximations made in the formulation of all of the submodels that complete the mathematical model. Therefore, model form uncertainty includes all assumptions, conceptualizations, abstractions, and mathematical formulations on which the complete model and all submodels rely. Some examples of submodels are material plasticity submodels, fracture mechanics submodels, submodels for stiffness and damping of structural joints, turbulence submodels, chemical-reaction submodels, chemical-diffusion submodels, and porous-media submodels.

For complex analyses with many submodels and lower-level supporting models, a tree-structure diagram can provide insight to show the make-up and interactions within the complete system model. Figure 13.4 gives an example of a complete system model that is composed of five submodels, and each model has anywhere from one to four auxiliary models. All five submodels interact in the complete system model, but only submodels 1 and 2, and submodels 3 and 4 interact directly with each other. That is, submodels 1 and 2, and submodels 3 and 4 are directly coupled, whereas submodel 5 is coupled through the complete system model. With this depiction, a better understanding can be obtained concerning the individual sources of model form uncertainty. Model form uncertainty is associated with the complete system model, each submodel, and each auxiliary model. Each model form uncertainty cascades upward through the tree structure, culminating in a convolved mixture in the complete system model. As a result, quantifying model form uncertainty using measurements from a validation experiment on the complete system only reveals a single facet of the disagreement between simulation and experiment.

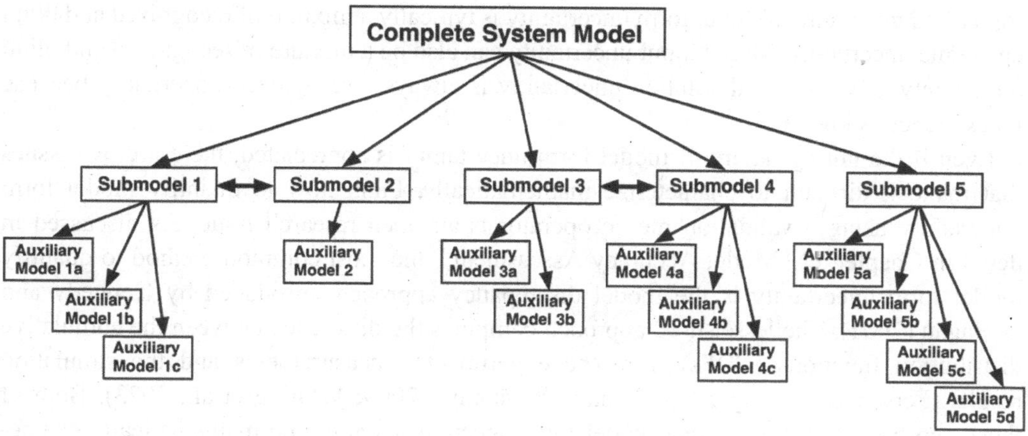

Figure 13.4 Tree structure for submodels and auxiliary models.

A tree-structure diagram of models and submodels can also be very useful in associating input uncertainties with specific submodels and auxiliary models. Furthermore, the diagram can then be used to create a table that specifically shows which uncertain and deterministic inputs are associated with each submodel and auxiliary model. This summary table is time consuming to compile but provides great value not only to the analysts conducting the simulation, but also to the analysis customers, stakeholders, and external reviewers. The table can help uncover inconsistencies, contradictions, and correlations in both uncertain and deterministic quantities. In a large-scale analysis, the list of uncertainty sources can number over a hundred, even over a thousand. If the project is exposed to an external review panel, it is highly beneficial to devise a method for clearly and quickly displaying considered uncertainties, the assumed dependencies between the uncertainties, and which inputs are considered deterministic.

Until the last two decades, model form uncertainty was commonly ignored in nondeterministic simulations. The primary reason for the neglect is that it is very troublesome to address mathematically. Model form uncertainty is *entirely and always* an epistemic uncertainty, i.e., it is completely due to lack of knowledge. This is in contrast to model input uncertainty, which can range from pure aleatory uncertainty, e.g., characterized by a random variable, to pure epistemic, which is characterized as an interval-valued quantity. Model form uncertainty is entirely due to inaccuracies in modeling of the physical system of interest. This can be due to our limited ability to formulate prescriptive rules for natural processes, perhaps due to our limited understanding of finer spatial scales and processes, or it can be due to our limited computer power to solve higher-fidelity models. As a result, model form uncertainty cannot be rigorously bounded by an interval because, by definition, every model of physics approximates reality. Recall from Chapter 2 that epistemic uncertainty is divided into two types: (1) *Recognized uncertainty*: an epistemic uncertainty for which a conscious decision has been made to either characterize or approximate it some way, or to ignore it for practical reasons; and (2) *Blind uncertainty*: an epistemic uncertainty for which it is *not* recognized that the knowledge is incomplete and/or it is *not* recognized that the knowledge is relevant to the issue at hand. Sometimes the phrase *unknown unknown* is used for a blind uncertainty. Model form uncertainty is typically a mixture of recognized and blind epistemic uncertainty. Model input uncertainty can also be a mixture of recognized and blind uncertainty, but numerical solution uncertainty is always a recognized uncertainty because its existence is known.

Even if the unruly nature of model form uncertainty is appreciated, there are two issues that make it difficult to characterize mathematically. First, how to quantify model form uncertainty using a validation metric operator is an open research issue. As discussed in detail in Chapter 12 "Model Accuracy Assessment," the most common method to quantify model form uncertainty is the model discrepancy approach introduced by Kennedy and O'Hagan (2001). The lesser-used approach computes the difference between the cumulative distribution functions (CDFs) from the experimental measurements and the simulation results (Ferson et al., 2008; Ferson and Oberkampf, 2009; Whiting et al., 2023). Both of these approaches recognize that model form uncertainty *cannot* be mathematically characterized solely as a random variable, but it must be treated similar to a systematic bias uncertainty. Second, even after the model form uncertainty is quantified for the conditions where experimental data are available, it is an open issue as to how to estimate and characterize model form uncertainty where there are no experimental data. The reason for this difficulty is that model form uncertainty does not need to be well behaved as a function of the identified dimensions of the input space of the model because it depends on physical processes. Furthermore, these physical processes and/or their coupling, may not even be included in the mathematical model of the system used in the simulation. These two issues of model form uncertainty are addressed in Section 13.5.

13.2.4 Example Problem: Step 2 for a Heat Transfer Analysis

All of the protective buildings for the electrical equipment are supposed to be fabricated according to the same specification for materials and construction. The electrical equipment and cinder-block walls are all attached to a square, concrete pad. The relevant physical characteristics of the buildings are:

- Concrete pad is 2.5 m × 2.5 m.
- Height of the cinder-block walls is 2.5 m.
- A thick wooden door provides access to the building.
- There is no ceiling in the building, only exposure to the underside of the roof.
- The roof of the building is metal and thermal insulation is attached to the underside of the roof.
- The cinder blocks are standard building material, sometimes referred to as a concrete masonry unit. They have two open cores where air is free to circulate inside the cores. Standard mortar is used to secure the cinder blocks in place.
- Thickness of the cinder blocks is 0.2 m.
- There is no insulation or covering on the inside or outside walls.
- An air vent was installed near the roof when the buildings were constructed, but these have all been sealed to prevent precipitation and vermin from entering the building.

As discussed in Step 1, the goal of the analysis is to estimate the highest temperature that could occur in an existing building for a normal operating environment, for example, a high atmospheric temperature expected on a hot summer day at any of the existing locations. Given the constraints on time and budget for the simulation, a number of major simplifications of the system of interest are required in the formulation of the conceptual and mathematical models for heat transfer. To avoid the 3D, coupled heat conduction and natural convection heat transfer simulation inside

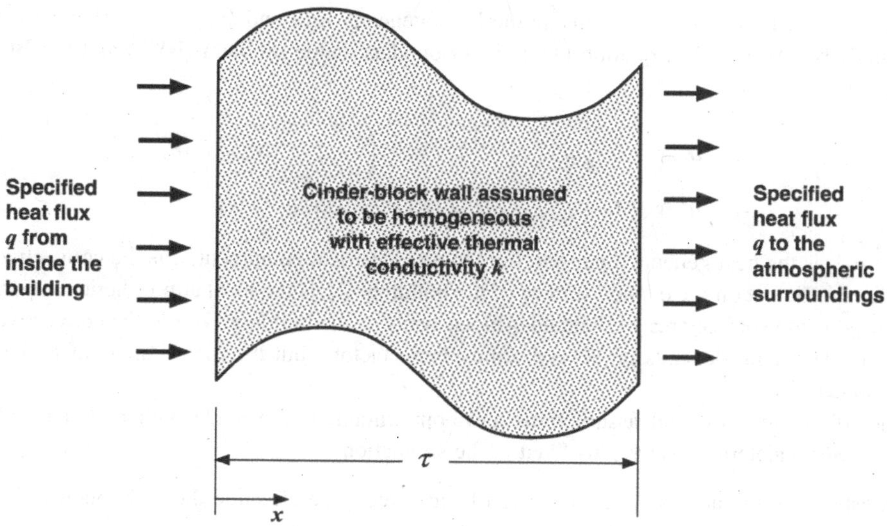

Figure 13.5 Simplified geometry for heat transfer through the cinder-block wall.

the wall, a homogeneous wall is assumed. Figure 13.5 shows the heat flux q through a cross-section of the simplified cinder-block wall. The following key assumptions and approximations are made for modeling the heat transfer through the walls:

- Assume a simplified homogeneous wall with an effective thermal conductivity to approximate the thermal transport through the solid material and movement of the air inside the cinder-block cores.
- Assume 1D heat conduction.
- The effective thermal conductivity will be obtained from available data in the literature for similar cinder-block walls.
- Ignore daily temperature variations and assume steady-state conditions at the hottest time of the day.
- The roof and the concrete pad are thermally insulated, and it is assumed that no heat escapes through the roof or the concrete pad.
- Heat transfer rate through the walls of the building is equal to the maximum heat generated per unit time by the electrical equipment.
- Assume that the air inside the building is well mixed due to electrical equipment cooling fans and natural convection such that the temperature inside is one temperature.
- Assume that the heat flux, i.e., heat transfer rate per unit area, is conducted uniformly through all of the walls.
- The outside wall is exposed to the surroundings allowing convective heat transfer to the atmosphere.

These assumptions and approximations are consistent with the goal of the analysis to estimate the highest temperature that could occur inside the building. Therefore, the ordinary differential equation (ODE) for the steady-state temperature distribution through the walls is given by

$$\frac{d}{dx}\left[k(T)\frac{dT}{dx}\right] = 0, \tag{13.1}$$

where T is the local temperature inside the homogeneous wall and $k(T)$ is the effective thermal conductivity of the wall. Equation (13.1) is a boundary value problem (BVP) and the BCs are given by

$$\begin{aligned} at\ x = 0, \quad q_{gen} &= -\left[k(x=0)\left(\frac{dT}{dx}\right)_{x=0}\right] S_{walls} \quad W, \\ at\ x = \tau, \quad q_{gen} &= h[T(x=\tau) - T_{\infty}]S_{walls} \quad W, \end{aligned} \tag{13.2}$$

where q_{gen} is the heat generated per unit time by the electrical equipment, h is the convective heat transfer coefficient on the outside surface of the building, T_{∞} is the local atmospheric temperature, and S_{walls} is the surface area of the cinder-block walls of the building. The h, the convective heat transfer coefficient, depends on several atmospheric factors, but the most important is the local wind speed.

The following additional assumptions and approximations are made with regard to what is considered nondeterministic versus fixed in the simulation:

- All aspects of the analysis are considered fixed, except the effective thermal conductivity $k(T)$ and the convective heat transfer coefficient h.
- In a literature search, a detailed 3D heat transfer analysis of various designs of cinder blocks was conducted by Fogiatto et al. (2016). The analysis included conduction through the concrete material, natural convection in the air cores, and radiation heat transfer between the walls of the cores. Using their results, the mean value of the effective thermal conductivity is approximated by

$$k(T) = 0.4 + 0.003T + 0.0002T^2 \quad \text{W/m} - {}^{\circ}\text{C}. \tag{13.3}$$

- The dependence on temperature approximates the increasing effect of convection and radiation as the temperature increases inside the cores. Equation (13.3) is only appropriate for the range of temperatures of interest in the present analysis and T is given in °C. Characterizing the uncertainty in $k(T)$ due to variability in building construction is discussed in Step 3.
- The worst-case scenario, i.e., the highest temperature inside the building, occurs when the thermal resistance due to convection on the outside wall of the building is at maximum. This occurs when the convective heat transfer coefficient is the lowest reasonably expected value. As a result, it is assumed that there is no atmospheric wind near the building, so that the convective heat transfer coefficient is only influenced by natural convection on the outside walls. Characterizing the uncertainty in h is discussed in Step 3.
- The maximum energy generated per unit time by the electrical equipment, q_{gen}, is 2,000 W.
- The worst-case scenario for atmospheric air temperature occurs for the building at the hottest location for the summer and for the hottest expected temperature at that location. The building fitting these criteria was determined and the highest recorded temperature at a nearby town was found to be 40°C.
- For simplicity, the heat transfer through the access door to the building is assumed to be equal to that through the cinder-block portion of the building.

A number of major simplifications have been made in the formulation of the conceptual and mathematical models for the heat transfer analysis. All of these assumptions and approximations will have an impact on the model form uncertainty, to be discussed later.

13.3 Step 3: Characterize Model Input Uncertainty

Characterizing a source of uncertainty means (a) assigning a mathematical structure to the uncertainty and (b) determining the numerical values for all of the needed elements of the structure so that a result can be computed. Thus, characterizing the uncertainty requires that a mathematical structure is given to the uncertainty and all parameters of the structure are numerically specified, such that an outcome can be computed. The primary decision to be made concerning the mathematical structure for each source is: "Should it be represented as a purely aleatory uncertainty, a purely epistemic uncertainty, or a mixture of the two?"

As discussed in Chapter 2, a purely aleatory uncertainty is one that is fully characterized by inherent randomness, i.e., purely chance. The classic examples are the roll of a die and Brownian motion. Purely epistemic uncertainty is one that characterizes complete lack of knowledge of the structure of the uncertainty for a quantity. Smithson's taxonomy of incompleteness includes vagueness, probability, and ambiguity (Smithson, 1989). As stated by Scott Ferson, "With an interval-valued uncertainty, there is almost no information there." If some type of likelihood or belief knowledge is added to the structure of the purely epistemic uncertainty, then the analyst has a mixture of epistemic and aleatory uncertainty. It has been proven that the total information uncertainty decreases from a maximum, when only an interval is known, to the state of a pure aleatory uncertainty (Klir, 2006).

How an uncertainty is characterized is a crucial topic in PBA because aleatory and epistemic uncertainty are treated in fundamentally different ways. Recall from Chapter 2 that the mathematical structure used in PBA to separately characterize aleatory and epistemic uncertainties is a p-box. A p-box expresses both aleatory and epistemic uncertainty in a way that does not confound the two. A p-box is a special type of cumulative distribution function that represents the set of all possible CDFs that fall within the prescribed bounds of the p-box. As a result, probabilities can be interval-valued quantities as opposed to a single probability, which contradicts a fundamental axiom in traditional probability theory. PBA is a marriage of interval analysis and traditional probability theory. The formal mathematical generalization of combining interval analysis and traditional probability theory is evidence theory (Kohlas and Monney, 1995; Bernardini and Tonon, 2010; Augustin et al., 2014).

PBA stresses the following aspects in a nondeterministic simulation:

- keep aleatory and epistemic uncertainties segregated throughout each step of the analysis;
- mathematically characterize aleatory uncertainties as probability distributions;
- characterize epistemic uncertainty as interval-valued quantities, i.e., all values over the range of the interval are possible, but no likelihood is associated with any value within the interval;
- if independence between uncertain quantities cannot be justified, then dependence should be considered as an epistemic uncertainty, which commonly increases the uncertainty in the SRQ; and
- separately propagate aleatory and epistemic uncertainties through the model which results in p-boxes for the SRQs.

13.3.1 Understanding Aleatory, Epistemic, and Mixed Uncertainties

It may seem rather easy to segregate uncertainties into aleatory or epistemic, but in reality, it can be confusing and ambiguous. The difficulty commonly arises because of three very different, both historical and pragmatic reasons. First, the risk assessment community has had a long tradition of *not* separately characterizing aleatory and epistemic uncertainties. Most practitioners in risk assessment use the concepts of aleatory and epistemic uncertainty, but they use subjective belief expressed as a *probability* for epistemic uncertainty. As a result, frequency of occurrence uncertainty (aleatory uncertainty expressing likelihood of occurrence over many observations) and subjective uncertainty (expressing personal belief) are convolved in the analysis so that there is *no actual distinction* between the two in the probabilistic results. Only during the last three decades have a number of leading risk analysts begun to stress the importance of different mathematical representations for aleatory and epistemic uncertainty.

Second, essentially all of the commercial risk assessment, UQ, and sensitivity analysis (SA) software available are focused on purely aleatory uncertainty, as well as epistemic uncertainty (i.e., subjective belief) expressed as probability distributions. If a simulation project group wanted to truly separate aleatory and epistemic uncertainties, they would have to either (a) build their own software, (b) build on existing open source software, or (c) use the limited commercial software available for imprecise probabilities (RAMAS, 2023; Sandia, 2023; Intelligent-Light, 2023).

Third, a minor change in perspective or a slight change in the information requested can change the mathematical structure of the uncertainty from aleatory to epistemic, and vice versa. For example, if one question was asked, the uncertainty could be characterized as an aleatory uncertainty. If a slightly different question was asked, it could be characterized as an epistemic uncertainty or as a mixed uncertainty. As a result, analysts must be very careful and clear in explaining the question to experts providing input information for an analysis or to experimentalists providing empirical data for a simulation. Furthermore, careful planning must go into what questions should be asked by the analyst so that they are aligned with the goals of the simulation.

Various examples are given to demonstrate how information content impacts epistemic and aleatory uncertainties, as well as what question is being asked. These examples will aid in understanding aleatory and epistemic uncertainty when noncontradictory and nonambiguous information is presented. For the more general situation when these occur, see (Kohlas and Monney, 1995; Bernardini and Tonon, 2010; Augustin et al., 2014).

Example 1

Consider the case of guessing the number of marbles inside of a jar. Suppose that the jar is transparent so that a person could see all of the marbles near the side of the jar. The number of marbles inside the jar is a pure epistemic uncertainty. It is *not* a random number, but a unique number that is simply unknown to the viewer. Depending on the motivation for guessing the right number of marbles, such as some type of wager, we may guess a single number. We could also guess an interval in which we believe the actual number may lie, or we may guess an interval with some type of personal belief structure for the likelihood of the correct number within the interval. For example, we may give a triangular belief structure over the range of the interval.

If we study the jar, we may start estimating the number of marbles and possibly make some measurements of the marbles and the jar. With time and effort, we are improving our knowledge. As a result, we may revise our interval estimate of the number of marbles, or we may revise our belief structure over the interval. If we spend a significant amount of time, and maybe even mathematical modeling of the marbles in the jar, we may significantly reduce the size of our interval estimate. If the jar is emptied and the number of marbles is counted, sufficient knowledge may have been added so that the number is exactly known. However, in engineering we seldom know the exact value of an uncertain quantity. Engineering is about making decisions based on incomplete and sometimes contradictory information.

Example 2

Consider the roll of a fair die. Before the die is rolled, the uncertainty is purely aleatory and the probability of each face of the die is 1/6. After the die is rolled, but *before* the die is observed, the uncertainty in the result is purely epistemic. That is, after the die is rolled, there is a fixed outcome, whether we know it or not. In this example, note that whether we consider the uncertainty as aleatory or epistemic depends on what question is asked. Are we asking an uncertainty question *before* the die is rolled or *after* the die is rolled?

Example 3

Suppose the safety of a design of a fleet of nuclear power plants is being analyzed. The question could be: "Based on our knowledge today, what is the estimated safety of the set of plants of similar design?" A similar question could also be posed *after* an accident has occurred at a given plant. Then, the question could be: "What is our estimate of the safety of the given plant after we have investigated the accident and after we have studied the inspection, maintenance, and operational procedures at the given plant?" If our estimate of the safety of the given plant with the accident has changed, then the change is due to the increased knowledge concerning this specific plant. With this valuable knowledge gained, we could then reevaluate the safety of the entire fleet of similarly designed plants.

Example 4

Consider the case of pseudo-random number generation. Suppose a person observes a long sequence of numbers and asks the question: "Is this a random sequence of numbers?" Various statistical tests may then be conducted, and the conclusion could be that the sequence is indeed random and that the next number is not knowable. Suppose now that the person was provided the algorithm that generated the sequence *and* the seed that started the sequence. With this knowledge, the person can determine, with perfect confidence, what the next number in the sequence will be. Without this knowledge, the sequence is characterized as aleatory. With this knowledge, a future outcome is completely deterministic.

Example 5

Consider the situation of a stranger approaching you in a gambling casino and asking if you would like to place a wager on the roll of a die. You're feeling lucky, so you agree. He reaches in his pocket and pulls out a die. He says he will pick a number between 1 and 6, and then you will pick another number between 1 and 6. He will then throw the die and the person's number that comes up first, wins the wager. How much do you want to bet? Before you answer, you start considering various ways to estimate the probability of winning or losing. If you were a Bayesian, you would assume a noninformative prior distribution, i.e., assume a uniform probability distribution that any number between 1 and 6 is equally likely.

Being a cautious Bayesian, you note that you have essentially *no basis* for assuming a uniform distribution. You have not seen the die, and you have never seen this person before in your life. So, you ask if you can see the die. You look at the die, it indeed has six unique faces, and it looks and feels *normal*. With this step, you have added significant knowledge to your decision process. There is now *some* evidence that the uniform distribution is reasonable.

Being a cautious and skeptical Bayesian, and gambling with your own money, you seek to add more knowledge before you characterize the uncertainty. You ask if you can roll the die a number of times to see if it appears to be fair. He agrees, and you start rolling the die. Each roll of the die adds knowledge concerning the probability of each number of the die appearing. You continue to roll the die a large number of times and, finally, conclude that the die *is* fair. About this time, the stranger shakes his head in frustration, takes the die, and walks off mumbling, "Must be an engineer."

This example shows that when the stranger initially asks you to make a wager, you can *only* defend a characterization of *pure* epistemic uncertainty. You know nothing about the die, and every belief of likelihood of an outcome is *presumption*. As you gather information, the uncertainty becomes a mixture of aleatory and epistemic uncertainty. At the final stage of gathering information after many rolls of the die, the uncertainty becomes purely aleatory. This example contrasts truly epistemic uncertainty and personal belief expressed as a probability.

Example 6

Suppose you are interested in the variability of the mass of manufactured parts that have just arrived from a new supplier. The contract with the supplier specifies the composition of the metal from which the parts are to be machined, the dimensional tolerance requirements for the parts, as well as the material properties, but nothing specifically dealing with the mass requirements of the parts. You have very little knowledge of the supplier's manufacturing process, the quality control processes, or the supplier's reputation. Suppose that all of the parts from the new supplier are dimensionally inspected and they are all found to satisfy the dimensional tolerances in the contract. The variability of the mass could be estimated by assuming a normal probability distribution for the mass uncertainty. The mean of the normal distribution could be estimated using all of the measured dimensions of the batch and assuming a nominal value for the mass density of the parts. The standard deviation of the mass could be estimated using the standard deviation of the measured volume combined with the assumed value for the mass density.

If the mass of some of the parts from the batch are measured, then the user has a good deal more information concerning the variability of the mass. To decide what theoretical family of distributions might be used to represent the mass variability, the user could conduct various statistical tests to determine which distribution appears to best characterize the variability. (A wide variety of commercial software exists for analyses such as this, including: JMP and STAT from SAS Inc., BestFit from Palisade Inc., Risk Solver from Frontline Systems, Inc., STATISTICA from StatSoft, Inc., and the statistical toolbox in MATLAB from The MathWorks, Inc.). Suppose that a two-parameter, log-normal distribution appears best, so it is chosen to characterize the variability. Using the samples available, the user could estimate the two parameters in the distribution as scalar quantities using various methods. If this were done, the user would be characterizing the variability as a purely aleatory uncertainty. Although this is common practice, it actually *over-represents* the true state of knowledge. If the number of mass measurements made is rather small, then the weakness of the argument for the characterization of the mass variability as a purely aleatory uncertainty is exposed.

The most informative approach to characterize the variability of the mass is to measure the mass of all of the parts in the batch. If the number of measurements is very large, then an accurately representative distribution can be chosen, and scalar values for the parameters of the distribution can be computed. However, if the number of measurements is small, a more representative characterization of the variability is to estimate interval-valued parameters for the distribution. This characterization of the variability would be a p-box, which contains both aleatory and epistemic uncertainties.

13.3.2 Model Input Uncertainty

For a preliminary simulation, an analyst may be able to roughly estimate the uncertainties for all of the input quantities. However, for most simulations, characterizing the uncertainty in model inputs is commonly a major effort in a nondeterministic simulation. For large-scale, high-consequence simulations, uncertainty characterization of the model input quantities, including obtaining experimental data, can take more time and financial resources than the development of the mathematical model of the system of interest. The information needed and the information sources may come from within the organization conducting the analysis, but it commonly comes from outside organizations and technical experts who have no association with the organization conducting the analysis.

Information obtained for the characterization of input quantities can come from one or more of the following three sources:

- experimentally measured data for quantities taken from the actual system or similar systems that are relevant to the primary system of interest;
- data generated by separate simulations that are constructed and computed solely to provide input data for the primary system of interest; and
- opinions expressed by technical experts familiar with the system of interest, or similar systems.

In using each of these information sources, the analyst is attempting to characterize the uncertainty in an input quantity. However, when any of these sources is used, the uncertainty due to the *source itself* is convolved with the uncertainty in the input quantity. Different procedures should be used to minimize the effect of the information source on the quantity of interest. For example, experimental measurements always contain random measurement uncertainties and systematic, or bias, uncertainties; even though they may be small.

The optimum situation would be using the experimental measurements of the needed input data for the system of interest. However, on complex systems or systems that have never been built, this is usually not possible. If systems similar to the one of interest are operational, then significant information can be obtained from these systems and their operational, inspection, and repair records. This route could require searching through old and incomplete records, as well as finding individuals who are familiar with the data in order to fill in gaps in the information and provide the proper interpretation. If the data are not owned by the organization conducting the simulation, then the value of the information must be weighed by both the simulation organization and the organization owning the data. An example of this tricky issue of who obtains the data and who owns the data is the modern concept of a *digital twin*, which is becoming popular with the commercial aviation industry. A vast amount of operational and performance data for the aircraft and major subsystems can be recorded while on the ground and during flight. The issue of ownership of the data, or certain aspects of the data such as repair records, is presently being debated and negotiated between the airlines, the aircraft manufacturers, and the suppliers of the major subsystems.

Input data can be provided by a simulation using a separate mathematical model that is specifically constructed so that the SRQs from the model are the input quantities needed for the simulation of the system of interest. This situation typically occurs when a sequence of models is used to address a time-dependent failure of a complex system. An example is the sequence of models and simulations that is used in analyzing the failure progression of a complex system. For example, in the investigation of the loss of Space Shuttle Columbia, a sequence of coupled models was constructed for analyzing the ascent phase of the flight when the wing leading edge damage occurred (Gehman et al., 2003). Simulations were conducted in a time sequence such that the SRQs from one model could be used for input data for the next simulation: (a) the flow around the external fuel tank, orbiter, and bipod ramp and support strut; (b) fracture of the insulating foam and ice formation around the bipod ramp; (c) trajectory of the foam and ice that dislodged from the external tank; (d) impact and fracture of the reinforced carbon-carbon (RCC) wing leading edge; (e) aerodynamic flow and heating outside and inside the perforated wing leading edge during reentry; and (f) melting and failure of the internal structure of the wing. Coupling of a sequence of models can also occur through spatial coupling, for example, by way of the SRQs of one model producing the BCs and excitation conditions for the next model.

Eliciting, analyzing, and characterizing expert opinion received a great deal of attention during the last two decades (Meyer and Booker, 2001; Ayyub, 2001; Vose, 2008; Augustin et al., 2014). This is due to (a) how often it must be done because no empirical data are available for the needed input quantities, and (b) the increased recognition of the difficulties in obtaining reliable expert opinion. Elicited experts should have two kinds of expertise: substantive expertise on the issue at hand, i.e., in-depth technical knowledge of the issue, and normative expertise, i.e., understanding of the methods of characterizing the uncertainty in the elicited information. The references given discuss a number of pitfalls that can occur in expert elicitation, as well as methods for reducing their impact. Three of the primary pitfalls are: (1) misinterpretation of expert opinion, (2) misrepresentation of expert opinion, and

(3) overconfidence of the expert. Here, *misinterpretation* means that either the experts misinterpreted the question being asked by the elicitor or the elicitor misinterpreted the information provided by the experts. *Misrepresentation* in this context means that the elicitor misrepresents, unintentionally or intentionally, the information from the experts. This most commonly occurs when the elicitor converts the experts' information into a mathematical structure that can be used as input to the model. With the increasing practical application of imprecise probability theories, misrepresentation is becoming less of a pitfall because of the enormous flexibility of characterizing information with imprecise probabilities as compared to traditional probability theory (Augustin et al., 2014). Overconfidence by the experts in their opinion of the accuracy of the information provided continues to be a significant problem. A number of researchers have studied this problem and how to potentially deal with it (Soll and Klayman, 2004; Lin and Bier, 2008).

Practitioners of PBA, more generally imprecise probabilists, have found that the most common difficulty is the experts' lack of understanding of aleatory, epistemic, and mixed uncertainties. It is the *responsibility of the elicitor* to clearly explain and give a number of examples to the experts before the elicitation process is initiated. After it appears that the experts understand each type of uncertainty, then specific questions dealing with the model inputs can be queried. After the experts provide answers to the questions, it is highly advisable that the elicitors describe to the experts their interpretation of what the experts seemed to have said. There is often found to be a miscommunication, primarily because of the subtle differences between aleatory, epistemic, and mixed uncertainties that are not fully grasped by the experts.

Concerning aleatory uncertainty, Cullen and Frey (1999) stress the importance of both the experts and elicitors understanding what kind of aleatory uncertainty (which is referred to here as simply variability) is being captured in the elicitation. Cullen and Frey (1999) point out that there are three types of variability: (1) temporal variability, (2) spatial variability, and (3) inter-individual variability. Temporal variability deals with the question of how a quantity varies as a function of the time scale of interest in the analysis. For example, suppose the variability of wind speed at a given location is needed in the simulation. Suppose that, because of the assumptions in the model that the simulation requires, the probability distribution of wind speed is averaged over the period of a month, for each month of the year. All of these aspects should be made clear to the experts so that no confusion or miscommunication occurs. For example, the experts may have never dealt with the wind speed variability over such a long time period.

Spatial variability refers to how a quantity varies in space. If a quantity varies over space, ignoring time dependence for the present, then the analyst must clarify for the experts what type of spatial averaging is appropriate, given the assumptions of the model. For example, suppose the model is dealing with the dispersal of a contaminant in the atmosphere. Suppose the finest scale for the discretization of space in the computational model is 1 m^3 of air and this occurs near the surface of the earth. As a result, the finest spatial scale of concentration of the mass of the contaminant is 1 m^3 of air. If the experts are questioned about their opinions of spatial scales of fluid dynamic turbulence, then it must be clearly stated that the smallest spatial scale of turbulence that exists in the model is 1 m^3.

Interindividual variability refers to how an outcome can vary over the sample space of all possible outcomes, i.e., the population. Outcomes can be the result of physical measurements, a simulation model result, or a sequence of observations. When a model for the nondeterministic simulation is constructed, a specific definition of a population is defined. For example, an analyst may be interested in the population of all parts manufactured by a supplier during a particular month or a particular year. If this information is not experimentally available, it may be elicited from experts who are knowledgeable about similar manufactured parts, but not necessarily from the same supplier. The elicitation process must be very clear to the experts concerning exactly what population is of interest for the model in the simulation.

Depending on the perspectives of the elicitors, they may require that the experts provide information in terms of very rigid mathematical structures, e.g., "We will not let the experts out of the room until they give us a precise probability distribution for the input." The present authors, of course, do not subscribe to this type of rigid requirement. The experts should not be pressured into providing more information than they feel that they can support. The following are examples of the types of mathematical structures that can be provided by experts for an input quantity, assuming they are allowed flexibility in the structure. The list is ordered in terms of the least information provided to the most information concerning the quantity of interest:

- A single (deterministic) value for the quantity is presumed to exist. The experts only claim to know this value to within an interval.
- A single (deterministic) value for the quantity is presumed to exist. The experts claim to know this value over an interval, but the experts have a higher level of confidence for the value over certain regions of the interval than others. As a result, they can provide a belief or confidence structure over the interval.
- The uncertain quantity is characterized as a continuous random variable. The experts claim that the sample space cannot be less than a certain value and it cannot be greater than a certain value. That is, support for the random variable is finite over a given range.
- The uncertain quantity is characterized as a continuous random variable. The experts claim that they are confident (a) in the family of the probability distribution, (b) the sample space cannot be less than a certain value, and (c) the sample space cannot be greater than a certain value.
- The uncertain quantity is characterized as a continuous random variable. The experts claim that they are confident (a) in the family of the probability distribution and (b) in the interval-valued range of all the parameters of the distribution.
- The uncertain quantity is characterized as a continuous random variable. The experts claim that they are confident (a) in the family of the probability distribution and (b) in the scalar values of all the parameters of the distribution.

Example 7

To contrast the level of probabilistic imprecision between the last two items in this list, consider a standard deviation for a normal distribution that is given by an interval versus a scalar quantity. For simplicity, consider a normal distribution with $\mu = 0$ with the reduced, or scaled, random variable $z = x/\sigma$. If the experts were to estimate that $\sigma = 1$, then the cumulative probability of the left tail of the normal distribution for $z = 5$ would be 2.87×10^{-7}. If the experts were to estimate that $\sigma = [1, 2]$, then the cumulative probability for $z = 5$ would be bounded by the interval $[6.21 \times 10^{-3}, 2.87 \times 10^{-7}]$. That is, there is an increase in the likelihood of a rare event by a factor of 21,600 because of the imprecision in the standard deviation. This level of sensitivity of probability for the normal distribution, as well as other similar distributions, for rare events comes as a surprise to most people. The reason for the sensitivity is that for unbounded distributions, i.e., infinite support, the probability distribution functions (PDF)s must produce extremely thin tails. This feature must be satisfied in traditional probability theory so that the integral over the entire range of the PDF, i.e., the CDF, yields a value of unity. If the experts are allowed to express a lack of knowledge in their opinions of a random process, probability bounds analysis has the flexibility to characterize this lack of knowledge.

Besides PBA, there are two distinctly different approaches to characterization of an input quantity: Frequentist and Bayesian. For an introduction to these approaches, see Raol (2016), Lambert (2018), and McElreath (2020). The Frequentist approach is also commonly called the classical approach, and it involves a number of methods. In the Frequentist's approach, probability is *strictly* interpreted as the relative frequency of an event occurring within the limit of a large number of random trials or observations from an experiment. The set of all possible outcomes of a random trial is called the sample space of the experiment. An event is defined as a particular subset, or realization, of the sample space. Some of the more common methods of characterization are point-estimation of a quantity, interval-estimation of the parameters of the distribution, hypothesis testing, and regression and correlation analyses. In the estimation of a point value for a quantity, it is assumed that the quantity is constant but imprecisely known because only a sample of the population is available. Various types of best-estimate methods are available to estimate the point value. There are also methods available for interval-estimates for the parameters of the distribution. When this is done, the uncertain quantity of interest becomes a mixture of aleatory and epistemic uncertainty. The two most commonly used methods are the method of moments and the maximum-likelihood method. Because of the restriction on the type of information that can be used and the larger quantity of information that is required, Frequentist methods are generally considered less capable for estimation of input quantities. The primary advantage of Frequentist methods is that they are widely viewed as simpler and easier to use than Bayesian methods.

Bayesian methods consider the distribution characterizing the uncertainty of a quantity as a function that (a) is initially unknown and (b) should be updated as more information becomes available concerning the quantity. The Bayesian paradigm uses the notion of subjective probability, or personal degree of belief, defined in the theory of rational decision making or rational choice theory by individuals. The initial postulated distribution for the quantity, referred to as the *prior distribution*, can come from any source or mixture of sources, including experimental measurements, expert opinion, or computational analyses. When no information is available for the prior distribution, the analyst can simply choose a prior distribution as their personal characterization. Every individual is considered a free agent and is not constrained to choose the same prior as another individual. There are serious criticisms of Bayesian estimation if little or no justification for prior distributions is available. The distribution for the quantity can be updated when any type of new information becomes available. After updating using Bayes' equation, the distribution is referred to as the *posterior distribution*. The primary defense of Bayesian inference is based on the argument that continually adding new information from all available sources minimizes the impact of the subjectivity in the initial choice of a prior distribution. Section 13.7 discusses Bayesian estimation in more detail.

Whenever there is uncertainty (aleatory, epistemic, or mixed) in more than one input quantity, correlation or dependencies commonly exist between the quantities. There are two basic types of dependency: aleatory dependency and epistemic dependency. Dealing with aleatory dependence is fairly well understood. For example, see Cullen and Frey (1999), Devore (2007), and Vose (2008). However, dealing with epistemic or mixed dependence is still a research topic because information dependence, correlation, ambiguity, coherence, and contradiction have many more attributes than pure randomness. (See Bernardini and Tonon, 2010; Augustin et al., 2014, for a discussion of this topic.) Deducing dependency between model inputs can be aided by experimental data, theoretical modeling information, and expert opinion. The task and complexity of determining dependency increases rapidly as the number of uncertain inputs in the nondeterministic simulation increases. As a result, the most common approach is to assume independence between all inputs and proceed with the simulation. This assumption, although expedient, can greatly underestimate the uncertainty in the outcomes of the simulation. It is beyond the scope of this text to deal with issues of characterization of dependency. For a detailed discussion, see the references just cited.

13.3.3 Example Problem: Step 3 for a Heat Transfer Analysis

As discussed in Step 2 above, only two input quantities are considered as uncertain: (1) the effective thermal conductivity $k(T)$ of the walls and (2) the convective heat transfer coefficient h for the outside of the building. Given Equation (13.3) for the mean value of the effective thermal conductivity, the analysis will include the variability of $k(T)$ due to fabrication of the buildings. Because of the very short time for conducting the analysis, reference books and a search of the internet are the only options possible for characterizing the uncertainty of each quantity.

Concerning the uncertainty in $k(T)$ due to construction variability, nothing could be quickly found in reference books or on the internet. To make a rough estimate of the variability, the article by Fogiatto et al. (2016) was examined from the perspective of how $k(T)$ varied for different concrete blocks of a similar design. The standard U-value (i.e., thermal transmittance) for block #1, which is the same design as used for the buildings being analyzed, is given as 5.342 W/m^2-K. The U-values for blocks #2 and #3, which have one and three air cavities, respectively, were 5.597 W/m^2-K and 5.185 W/m^2-K, respectively. Based on this somewhat relevant information, the CoV, σ/μ, for $k(T)$ is set to 0.1. The additional assumption is made that the construction variability is characterized by a normal (Gaussian) distribution.

Figure 13.6a shows the effective thermal conductivity as a function of temperature for the mean and $\pm 2\sigma$ values, which characterize the construction variability. Figure 13.6b shows the corresponding CDF of the effective thermal conductivity for temperatures of 0°C, 30°C, and 60°C. As can be seen in both figures, the variability or dispersion of the thermal conductivity increases as the temperature increases.

Concerning the convective heat transfer coefficient h, the uncertainty is commonly large due to its sensitivity to several factors, such as natural or forced convection, fluid density and viscosity,

Table 13.1 *Summary of the characterization of the model input data*

Model input data	Characterization	Value/Type
Width of building	Fixed	2.5 m
Depth of building	Fixed	2.5 m
Height of building	Fixed	2.5 m
Thickness of wall	Fixed	0.2 m
Heat generated, q_{gen}	Fixed	2,000 W
Air temperature, T_∞	Fixed	40°C
Wind speed	Fixed	0 m/s
Thermal conductivity, $k(T)$	Aleatory uncertainty	Normal distribution, $\sigma/\mu=0.1$
Heat transfer coefficient, h	Epistemic uncertainty	Interval, [7, 10] W/m^2-°C

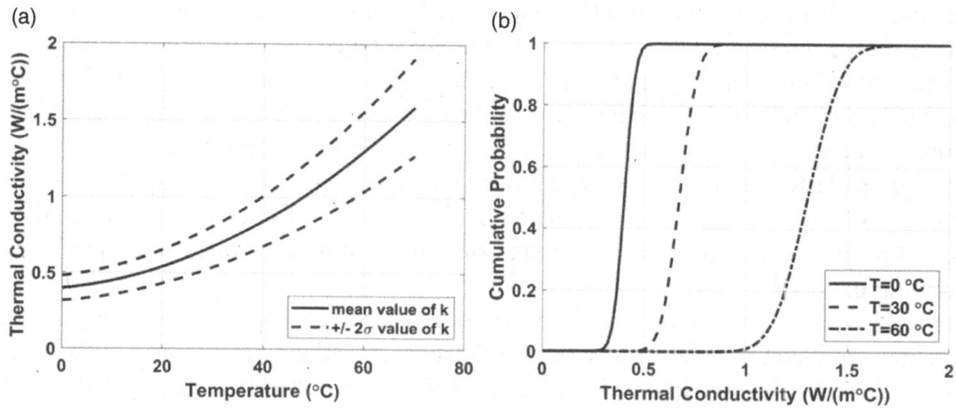

Figure 13.6 Temperature dependence and variability of effective thermal conductivity for the walls. (a) Thermal conductivity as a function of temperature. (b) Cumulative distribution functions for thermal conductivity.

laminar or turbulent flow, and surface roughness. For the worst-case analysis of interest here, we assume that there is no atmospheric wind so that only natural convection occurs on the outside of the building. Examining the reference text of Holman (2009), we find that for the conditions of this example, h could be as low as 7 W/m^2-°C. With slight variations in atmospheric and wall surface conditions, it could be as high as 10 W/m^2-°C. Because of this large range in uncertainty, and little knowledge of the atmospheric and wall surface conditions, h is assumed to be a purely epistemic uncertainty. The interval [7, 10] W/m^2-°C is assumed to bound h. Since h is considered as an interval-valued uncertainty, PBA will be used to properly capture this epistemic uncertainty.

Although there are only two uncertain input quantities explicitly addressed in this example, it is informative to list all of the input quantities for the simulation, their characterization, and their values. Table 13.1 shows the nine input quantities for the simulation. In any simulation of real-world systems, showing the customer of the analysis a table like this is very valuable because it helps reduce any miscommunication or misunderstanding with the customer concerning the analysis.

13.4 Step 4: Estimate Numerical Solution Uncertainty

As discussed in Chapter 2, numerical solution errors are characterized as uncertainties because they can only be estimated, as opposed to the error being known precisely. There are four error sources due to the approximate numerical solution of the mathematical model: (1) computer round-off, (2) iterative solution, (3) spatial and temporal discretization, and (4) statistical sampling. The error-estimation technique for each source most commonly relies on the concept of comparing one numerical solution result with a similarly obtained solution result. This general approach is referred to as *a posteriori* error estimation. As an example of this approach, consider the estimation of the impact of computer round-off error. First, a numerical solution is computed using single-precision arithmetic in all arithmetic operations that are used to compute an SRQ. Then the solution is recomputed using the computer complier option where all operations are computed using double-precision arithmetic. In this recalculation of the solution, *all* other model input data (including all numerical algorithm options and discretization procedures) are left *unchanged.* A comparison is then made between the two numerical results for the same SRQ. A relative error in the SRQ can be computed, but the usual procedure is simply to observe if there is any change in the significant digits of interest in the SRQ. If there is no significant change in any of the SRQs, particularly the SRQ that is most sensitive to round-off error, then it is concluded that the round-off error can be ignored. Similarly, if there is a significant change between the two levels of precision, it is concluded that round-off error is impacting the results, possibly even corrupting the results.

Iterative error, discretization error, and statistical sampling error are commonly the largest numerical error contributors in a nondeterministic simulation. Iterative and discretization errors have been discussed in detail in Chapters 8 and 9, respectively. As a result, this section only briefly discusses these two error sources and concentrates on the characterization of the errors as uncertainties. Statistical sampling error is the error due to the propagating of a finite number of input samples through the model to obtain the nondeterministic outputs, i.e., the SRQs. This section discusses statistical sampling error first, followed by iterative and discretization errors.

In computationally expensive simulations, it is common to use surrogate models, also referred to as response-surface models, meta-models, or model emulators. Surrogate models introduce *additional* numerical approximation error, beyond the four errors just mentioned. Note that the four errors just mentioned have a compounding effect on the construction of surrogate models. Hence, these four errors can individually and in combination distort and pollute the surrogate model. Although it is beyond the scope of this text to discuss surrogate model errors, they should be estimated in any simulation where they are used. For a detailed discussion of this topic, see Box and Draper (2007), Schilders et al. (2008), Myers et al. (2009), Smith (2014), Gramacy (2020), and Jiang et al. (2020) .

13.4.1 Propagation of Input to Output Uncertainty

Figure 13.7 illustrates the concept of propagating uncertain model inputs through the mathematical model to obtain uncertain model outputs. The figure begins with the specification of the environment of interest and the identification of the scenarios of interest discussed

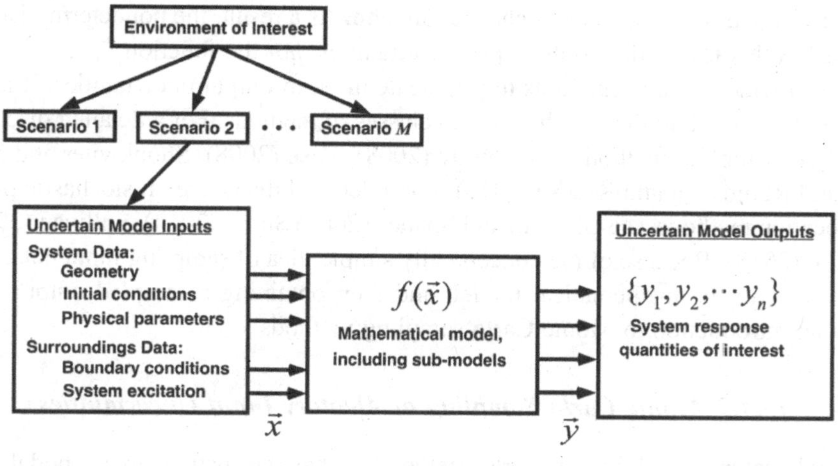

Figure 13.7 Propagation of uncertain model inputs through the mathematical model to obtain uncertain model outputs, for a given environment and scenario of interest.

in Step 1. The uncertain model inputs have been identified and characterized, as discussed in Steps 2 and 3, respectively. Let $\vec{x}$ be the vector of all uncertain input quantities. Let m be the number of uncertain elements in the input vector, so that

$$\vec{x} = \{x_1, x_2 \cdots x_m\}. \tag{13.4}$$

Let $\vec{y}$ be the vector of uncertain system response quantities. Let n be the number of elements in the output vector, each one being an SRQ, so that

$$\vec{y} = \{y_1, y_2 \cdots y_n\}. \tag{13.5}$$

In the simulation of most complex systems, it is common for $m >> n$. Some of the elements of the vector $\vec{y}$ are typically the dependent variables in the PDEs. However, some of the elements may be very different kinds of quantities, for example: (a) derivatives of dependent variables with respect to the independent variables, or derivatives with respect to input parameters to the model; (b) functionals such as spatial integrals, time-averages, or min and max operators on the dependent variables; and (c) PDFs or CDFs.

Let $f\left(\vec{x}\right)$ represent the mathematical model, including all of the submodels, indicating the dependence of the model on the uncertain input quantities $\vec{x}$. Therefore,

$$\vec{y} = f\left(\vec{x}\right). \tag{13.6}$$

The term $f\left(\vec{x}\right)$ can be thought of as a function that maps input uncertainties to output uncertainties. Note that this mapping of input to output is a mapping of *input functions to output functions*, whereas in deterministic simulation, it is a mapping of input scalars to output scalars. Using sampling techniques, the mapping is produced by using multiple samples of individual values of $\vec{x}$ to compute multiple samples of $\vec{y}$. Determining the effect of the input uncertainties, $\vec{x}$, on the response vector, $\vec{y}$, is termed *propagation of input uncertainties*. This text assumes that the mapping function $f(\cdot)$ is a *deterministic* function, as opposed to a nondeterministic or stochastic function. As a result, the nondeterministic nature considered in this text is due to the input uncertainties, *not* the function $f(\cdot)$.

There is a wide range of methods to propagate input to output uncertainties. It is beyond the scope of this book to discuss the various classes of methods. For a detailed discussion of Monte Carlo sampling methods, see Dimov (2008), Vose (2008), Shonkwiler and Mendivil (2009), and Rubinstein and Kroese (2016). For a detailed discourse on stochastic projection and collocation methods, see Ghanem and Spanos (2003) Smith (2014), Sullivan (2015), and McClarren (2018). Because of the conceptually simple idea of sampling input data, computing a solution to the mathematical model, and then obtaining a sampled output, this text provides an introduction to Monte Carlo sampling methods.

13.4.2 Monte Carlo Sampling of Aleatory Input Uncertainties

The central feature of all Monte Carlo methods is that the mathematical model $f\left(\vec{x}\right)$ is evaluated repeatedly using some type of random sampling of the input vector $\vec{x}$. Let $\vec{x}^k$ denote a random sample drawn from all of the components of the input vector $\vec{x}$. Let $\vec{y}^k$ be the response vector after evaluation of the model using the random sample $\vec{x}^k$. Then, Equation (13.6) can be written for the number of samples N as

$$\vec{y}^k = f\left(\vec{x}^k\right) \quad k = 1, 2, 3 \ldots\ N. \tag{13.7}$$

For Monte Carlo sampling (MCS), it can be proven that given a sufficiently large number of samples N drawn from all the input uncertainties, strong statistical statements can be made concerning the nondeterministic response $\vec{y}$. In general, there are *no assumptions* concerning: (a) the characteristics of the uncertainty structure of the input to the model, e.g., correlated versus uncorrelated and (b) the mathematical characteristics of $f\left(\vec{x}\right)$, e.g., smoothness or nonlinearity of the model. Individual Monte Carlo methods may have assumptions or approximations in order to improve the sampling convergence rate of the method. If $f\left(\vec{x}\right)$ is computationally expensive to evaluate, then an analyst may have to deal with less precise statistical conclusions concerning $\vec{y}$. If a parallel processing computer is used, the computational disadvantage of MCS is mitigated if each of the function evaluations, $f\left(\vec{x}^k\right)$, can be computed in parallel on multiple processors.

Figure 13.8 illustrates the basic concepts in MCS for a system of three aleatory uncertainties (x_1, x_2, x_3) resulting in one uncertain output, y. The first step is to draw uniformly distributed random deviates between zero and one. The uniformly distributed random deviates are obtained using a pseudo-random number (PRN) generator that is scaled to produce values between zero and unity. High-quality PRN generators are publicly available, as well as in commercial software packages, such as MATLAB or Mathematica. Figure 13.8 shows how these samples applied to the probability axis will generate the random deviates $(x_1, x_2, x_3)^k$, based on the particular CDF characterizing each input quantity. Each of these random deviates is used to evaluate the function f to compute one y. After many function evaluations using many random deviates, $k = 1, 2, 3 \ldots\ N$, the analyst can construct an empirical CDF of y as shown in the bottom panel of Figure 13.8.

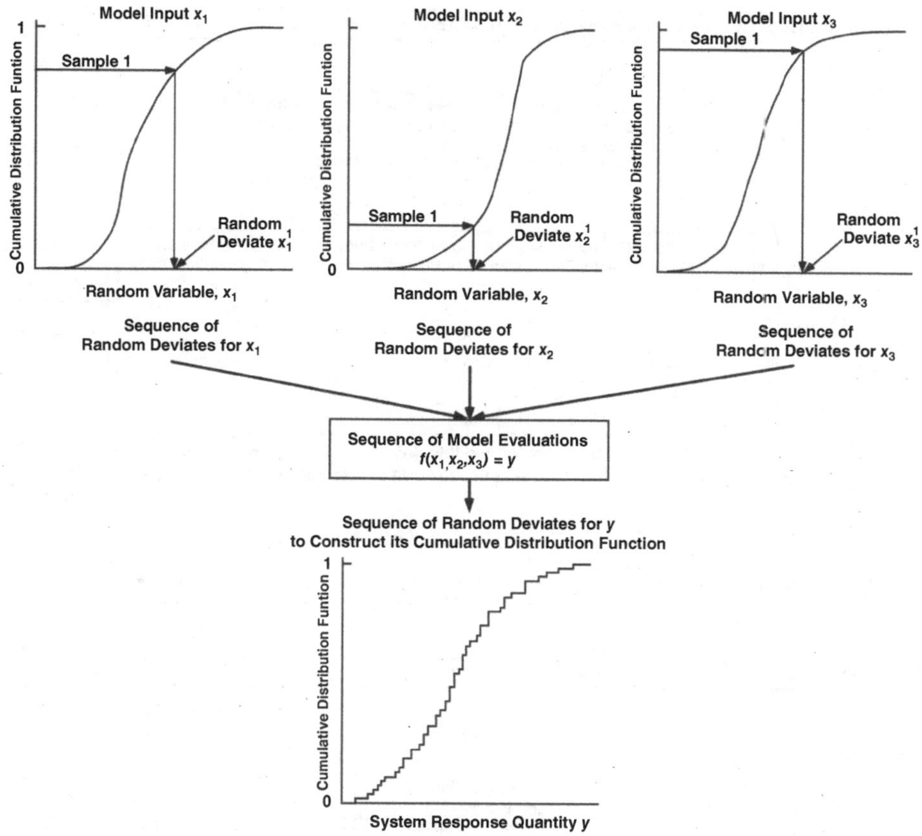

Figure 13.8 Schematic of MCS for three uncertain inputs and one output.

Figure 13.9 shows a flowchart for the various steps in simple, or standard, MCS for an arbitrary number of uncorrelated, uncertain inputs. Let the number of uncertain aleatory inputs be α. Each of the steps for simple MCS follows:

1. Choose the number of random samples N that will be used to sample each of the aleatory uncertainties.
2. Generate α sequences of N pseudo-random numbers, one for each of the uncertain aleatory inputs. Since each sequence must be independent of the other, each sequence must use a different seed for the PRN generator or be a continuation of a random sequence. As is common with many PRN generators, the numbers range between zero and unity.
3. Pick an individual number from each of the sequences of PRNs and create an array of numbers of length α. That is, take one number from the first sequence of PRNs, take one number from the second sequence, etc., until the array of length α is created. Once a number from the sequence has been used, it is not used again.
4. Each number in the array generated in Step 2 can be viewed as drawn from a uniform distribution over the range of (0,1). Given the probability integral transform theorem (Angus, 1994), a random deviate from the uniform distribution can be used to map the random deviate to any continuous distribution, as shown in the top panel of Figure 13.8.

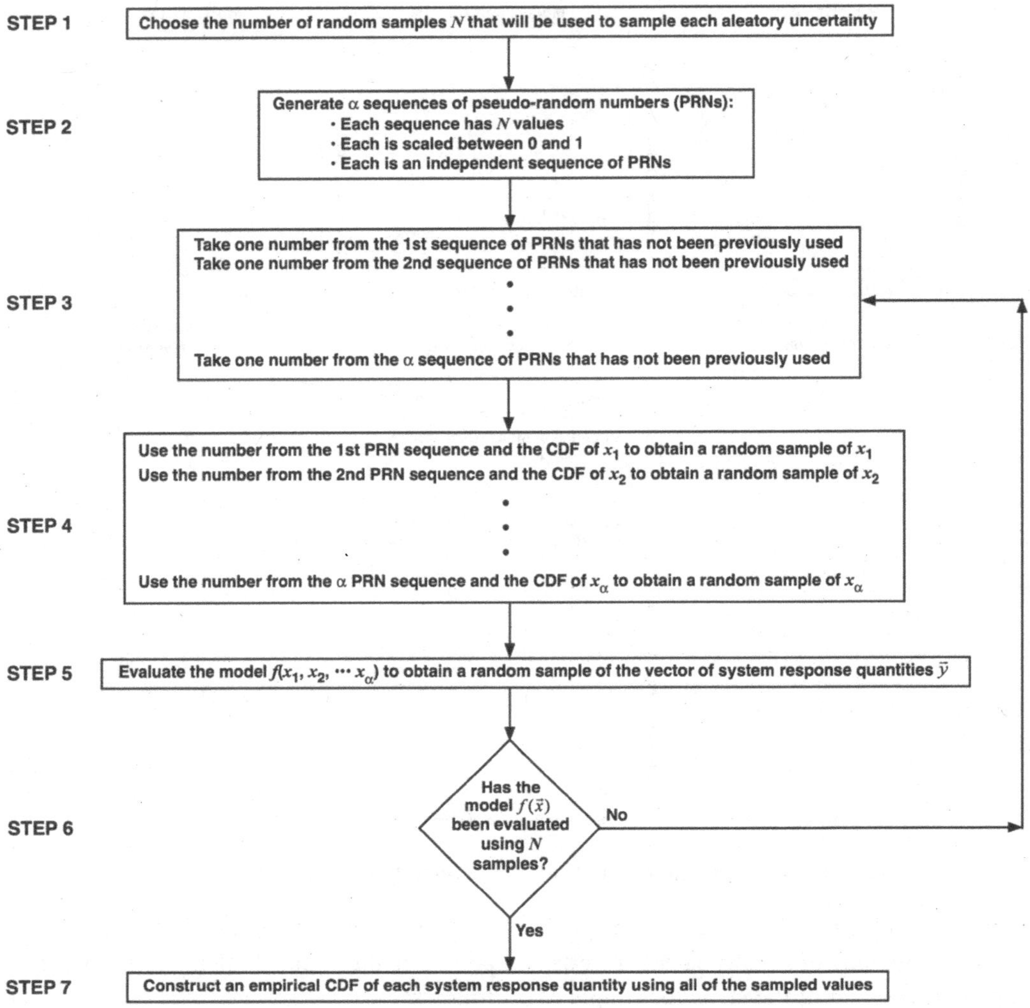

Figure 13.9 Flow chart for simple MCS with α aleatory input uncertainties and no correlation of input uncertainties.

5. Use the array $(x_1, x_2, \ \ldots \ x_\alpha)$ as input to evaluate the mathematical model. By solving the mathematical model, given the model input data, compute the array, $(y_1, y_2, \ \ldots \ y_n)$, of SRQs.
6. Test if all N PRNs have been used to evaluate the mathematical model. If *No*, return to Step 2. If *Yes*, go to Step 7.
7. Construct an empirical CDF for each element of the array, $(y_1, y_2 \ldots y_n)$. An empirical CDF of y_i is constructed using each of the samples from the Monte Carlo simulation. The y_i are ordered from the smallest to the largest. The abscissa of the empirical CDF is the observed (sampled) values of an SRQ, y_i, and the ordinate of the empirical CDF is the observed cumulative probability for all of the sampled values less than or equal to y_i. The empirical CDF is constructed as a nondecreasing step function with a constant

vertical step size of $1/N$. The locations of the steps correspond to the observed values of y_i. Such a distribution for the samples y^k for $k = 1, 2, 3 \ldots N$ is

$$S_N(y) = \frac{1}{N}\sum_{k=1}^{N} I(y^k, y), \tag{13.8}$$

where

$$I(y^k, y) = \begin{cases} 1 & y^k \le y, \\ 0 & y^k > y. \end{cases} \tag{13.9}$$

The term $S_n(y)$ is simply the fraction of sample values in the data set that are at or below each value of y^k. From Equation (13.8), it is clear that the total probability mass accumulated from the total of N empirical samples is unity. An example of an empirical CDF is shown on the bottom panel of Figure 13.8.There are three important topics that are not addressed in Figure 13.8 and Figure 13.9, but are briefly mentioned here. Readers should consult the following references for a detailed discussion of these topics: Dimov (2008), Vose (2008), Shonkwiler and Mendivil (2009) and Rubinstein and Kroese (2016). First, some type of correlation or dependence structure may exist between various uncertain inputs. If a correlation structure exists between two (or more) inputs, then it means that one input is statistically related to the other(s). For example, suppose an analyst considers variability in both thermal and electrical conductivity. For most materials, there is a strong correlation between thermal and electrical conductivity. If a correlation structure exists between two (or more) inputs, then a *dependency* relationship exists between the inputs. As a result, one input *cannot* be computed independently from the others. Determining a correlation or dependence structure is usually an involved task that is based primarily on large amounts of experimental data, as well as an understanding of the dependency relationship between the quantities. If correlation or dependence structures are quantified, then they are accounted for in Step 3 in Figure 13.9.

Second, the number of Monte Carlo samples needed for calculating various statistical quantities of the response depends on different factors. The most important of these is the probability value of interest for y_i. Ang and Tang (2007) give an estimate of the sampling error in simple MCS as

$$\% \text{ error in } P = 200\sqrt{\frac{1-P}{NP}}, \tag{13.10}$$

where P is the probability value of interest. It can be seen that the mean converges most rapidly, while low probability events converge very slowly. For example, if a probability of 0.01 is needed, then 100,000 samples are typically required to assure an error of less than 6%. It should be stressed, however, that one of the important advantages of MCS is that the convergence rate does *not* depend on the number of uncertain quantities or their variance.

Third, to improve the convergence rate of simple MCS, a wide range of variance reduction techniques are available. These techniques adjust the random samples of the inputs so that certain features of the output distribution converge more rapidly. The most well known is stratified Monte Carlo sampling, also referred to as Latin Hypercube Sampling (LHS). LHS stratifies the uniform distribution that is used with the inverse transform procedure into equal intervals of probability. The most common procedure is to let the number of intervals be equal to the number of samples that will be computed. Across all of these intervals, a PRN generator is used to obtain a separate random number for each interval so that a random sample is obtained from *each interval*. LHS almost always converges faster than simple MCS, so it is the most commonly used sampling technique. For a small number of uncertain input quantities, e.g., less than five, LHS converges much faster than simple MCS. For one uncertain input quantity, Dimov (2008) shows that

$$\% \text{ error in } P \sim N^{-3/2}. \tag{13.11}$$

All of the modern risk assessment software packages contain a number of sophisticated features, such as dealing with correlations, dependencies, and different methods for variance reduction in the sampling.

13.4.3 Sampling of Epistemic Input Uncertainties

Monte Carlo methods could also be used for sampling purely epistemic input uncertainties, but two issues critically alter the sampling strategy of the inputs. First, with interval-valued input uncertainties, we are *only* interested in the intervals of the SRQs. That is, we are only interested in the *minimum* and *maximum* of each SRQ as a function of the interval-valued inputs, nothing in between. Second, since there is *no likelihood structure* associated with the range of each of the input intervals, the entire range of each interval-valued input quantity must be sampled, as well as all combinations of each interval. Thus, *any possible combination* of input values over each interval-valued input quantity can produce the minimum and maximum of each SRQ. As a result, as the number of interval-valued inputs increases, the computational effort increases rapidly because of the increasing number of possible combinations of interval-valued inputs that must be sampled. This feature of rapidly increasing computational effort when increasing the number of interval inputs is radically different than sampling random variables using simple Monte Carlo sampling. As pointed out earlier, the convergence rate of simple Monte Carlo sampling does *not* depend on the number of input random variables.

The reason for this radical change in the sampling convergence rate as a function of the number of interval inputs is that random variables have a specified distribution of probability mass given by the PDF. However, intervals have no specified function of probability mass over the range of the interval. The probability mass of an interval is associated with the *entire interval, not a distribution function*. As a result, *any value within an interval* can have the entire probability mass of unity associated with it. Therefore, any combination of values within any of the input intervals is possible, so that *all combinations of possible values* must be sampled to ensure an estimate on the bounds on the response quantity.

Any type of sampling procedure over the range of each input interval could be used, as long as all possible combinations of values within and between each of the input intervals are sampled. Here, the simplest case of *no knowledge* of the dependence between each of the input intervals is addressed. Since all possible combinations of interval-valued inputs must be sampled, this is referred to as *all possible dependence* or *no knowledge of dependence*

Table 13.2 *Number of samples computed for the number of subintervals* $\mathcal{N}$ for each input interval and the number of input intervals β

$\mathcal{N}$	$\beta = 1$	$\beta = 2$	$\beta = 3$	$\beta = 4$	$\beta = 5$
1	2	4	8	16	32
2	3	9	27	81	243
3	4	16	64	256	1024

between the input intervals (Ferson and Kreinovich, 2006). If there is knowledge concerning the relationship between interval-valued inputs, this knowledge can be incorporated into the sampling strategy of the inputs. However, this topic is beyond the scope of this book. For a detailed discussion of the sampling strategy and for some knowledge of dependence between intervals, see Ferson and Kreinovich (2006), Ceberio et al. (2007) and Berleant et al. (2007).

Let the number of input intervals be β. Let the set of input intervals be given as

$$x_i \in [\underline{x}_i, \overline{x}_i] \quad i = 1, 2, \ldots, \beta, \tag{13.12}$$

where $\underline{x}_i$ is the lower value of the interval and $\overline{x}_i$ is the upper value of the interval. Equation (13.12) describes a β-dimensional hypercube in the input space x_i $i = 1, 2, \ldots \beta$. The minimum and maximum values of $f\left(\vec{x}\right)$ over this hypercube can be written, respectively, as

$$\begin{aligned} \underline{f} &= \min f(x_i) \quad \text{for} \quad x_i \in [\underline{x}_i, \overline{x}_i] \quad i = 1, 2, \ldots, \beta, \\ \overline{f} &= \max f(x_i) \quad \text{for} \quad x_i \in [\underline{x}_i, \overline{x}_i] \quad i = 1, 2, \ldots, \beta. \end{aligned} \tag{13.13}$$

It is seen that the estimation of $\underline{f}$ and $\overline{f}$ becomes an optimization problem, specifically convex optimization, over the sample space $x_i \in [\underline{x}_i, \overline{x}_i]$ $i = 1, 2, \ldots, \beta$. If $f(x_i)$ is linear in every dimension x_i , then $\underline{f}$ and $\overline{f}$ will occur on the boundaries of the hypercube. In general, however, $\underline{f}$ and $\overline{f}$ can occur at any combination of points within the set of intervals $[\underline{x}_i, \overline{x}_i]$ $i = 1, 2, \ldots, \beta$.

Over the last few years, a number of researchers have developed more efficient procedures for sampling intervals (Wang, Zhang, et al., 2018; Alvarez et al., 2018; Song et al., 2019). The present discussion will only address a simple sampling algorithm for intervals. Each interval is partitioned equally into the same number of subintervals. Let $\mathcal{N}$ be the number of partitions for each x_i, so that $\mathcal{N} = 1$ means that the input interval is not further divided into subintervals. Only the end points of each subinterval are sampled, so that the total number of samples propagated through the mathematical model is $(\mathcal{N} + 1)^{\beta}$. Table 13.2 shows how quickly the number of samples increases as a function of the number of partitions of each input interval and the number of input intervals. As discussed earlier, this is due to the requirement to sample every combination of end points from each partition.

Figure 13.10 shows a flowchart for the steps for sampling β uncertain input intervals by equally partitioning each interval. The following explains each of the steps in more detail:

1. Pick the number of partitions $\mathcal{N}$ that will be used to divide each of the input intervals. If the mathematical model $f\left(\vec{x}\right)$ is computationally expensive to evaluate, and β is large, then pick $\mathcal{N}$ =1 or 2.
2. Calculate an array of sample points from the combinations of end point for each partition. The total number of sample points to be evaluated is $(\mathcal{N} + 1)^{\beta}$. For example, if $\mathcal{N}$ =2 and

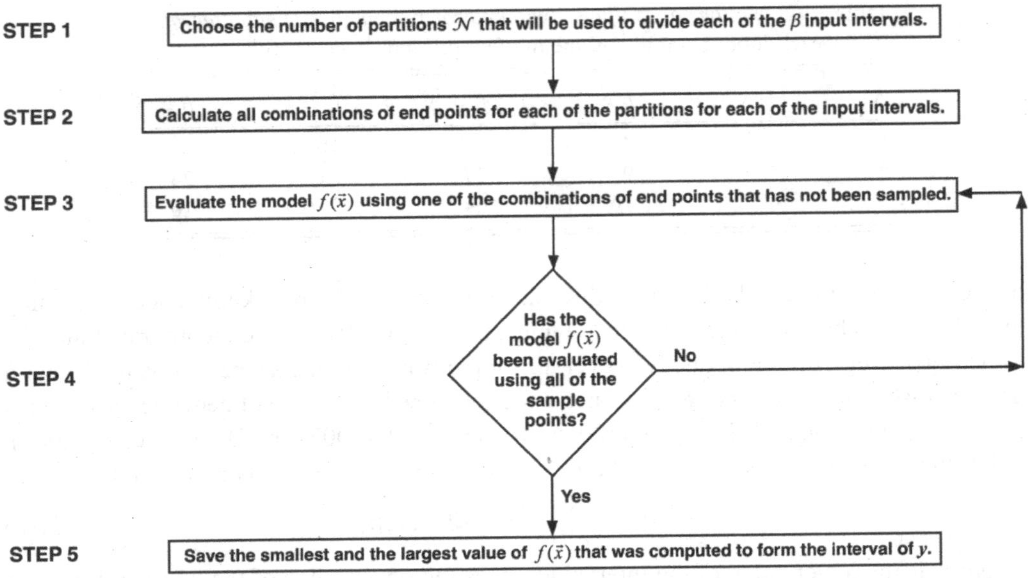

Figure 13.10 Flowchart for equal partition sampling of β uncertain input intervals.

$\beta = 2$, there would be nine elements in the array of sample points: $(\underline{x}_1, \underline{x}_2)$, $(\underline{x}_1, \hat{x}_2)$, $(\underline{x}_1, \bar{x}_2)$, $(\hat{x}_1, \underline{x}_2)$, $(\hat{x}_1, \hat{x}_2)$, $(\hat{x}_1, \bar{x}_2)$, $(\bar{x}_1, \underline{x}_2)$, $(\bar{x}_1, \hat{x}_2)$, $(\bar{x}_1, \bar{x}_2)$ where $\hat{x}_i = (\underline{x}_i + \bar{x}_i)/2$.

3. Evaluate the mathematical model $f(\vec{x})$ at one of the combinations of points that has not yet been sampled.
4. Test if all of the sample points in the array have been used in the evaluation of $f(\vec{x})$. If *No*, return to Step 3. If *Yes*, go to Step 5.
5. Search the array of function evaluations at each of the sample points and find the minimum and maximum values. These two points are the minimum and maximum of the interval of y, respectively, i.e., $(\underline{y}\bar{y})$.

If β is very large, for example on the order of hundreds, then the interval partitioning algorithm just described is impractical because of the number of function evaluations required to estimate $\underline{f}$ and $\bar{f}$. The most effective algorithm for this situation has been developed by Kreinovich et al. (2007). Kreinovich, et al. theoretically developed and demonstrated a sampling method based on Cauchy deviates. They show that if $f(\vec{x})$ is approximately linear within the set of intervals $[\underline{x}_i, \bar{x}_i]$ $\quad i = 1, 2, \ldots, \beta$, then the number of samples required, N, for a given accuracy is $N = 8/\Delta$, where Δ is the relative accuracy of the magnitude of the interval $(\underline{y}\bar{y})$, and N is large. Remarkably, the accuracy does *not* depend on the number of intervals β. Kleb and Johnston (2008) used the Cauchy deviates method on an aerothermodynamics analysis of a high-speed reentry vehicle into the Earth's atmosphere. For this analysis, there were 11,905 input interval uncertainties. These were due to poorly known chemical and radiation processes within the shock layer. They found that with 200 function evaluations, using the Cauchy deviates method, they were able to achieve a relative accuracy of ±20% on the SRQ intervals with a 90% confidence level.

13.4.4 Sampling of Combined Aleatory and Epistemic Input Uncertainties

In the previous two subsections, the propagation of aleatory uncertainties and epistemic input uncertainties through the mathematical model were discussed separately. Here, the discussion concerns the propagation of combined aleatory and epistemic input uncertainties through the model. Specifically, some of the input uncertainties are aleatory uncertainties, some are intervals, and some are characterized as a mixture of aleatory and epistemic uncertainty, i.e., a p-box. A p-box is a more general structure characterizing uncertainty because it can describe both aleatory and epistemic features in one structure. A p-box is a specific type of imprecise probability. It is one that is intuitively easy to understand and easy to use in computations because the aleatory and epistemic uncertainties are kept separate. The general structure of a p-box can be written as any nondecreasing function $F_X(x)$, such that $\underline{F}_X(x) \leq F_X(x) \leq \bar{F}_X(x)$, where $\underline{F}_X = \underline{P}(X \leq x)$ and $\bar{F}_X = \bar{P}(X \leq x)$ are the lower and upper bounds on cumulative probability, respectively. The difference between $\underline{F}_X(x)$ and $\bar{F}_X(x)$ for any value of x is the uncertainty due to lack of knowledge.

The propagation of a general input p-box through a mathematical model has rapidly advanced during the last decade (Eldred et al., 2011; Fetz and Oberguggenberger, 2016; Zhang et al., 2017). The most efficient sampling procedures are those that yield the minimum and maximum values of the output p-box, given the intervals in $\vec{x}$, with the minimum number of evaluations of the function $f(\vec{x})$. It has been shown that the propagation of a generalized p-box is more complex and computationally expensive than the propagation of a parameterized p-box (Bruns and Paredis, 2006; Ghosh and Olewnik, 2013; Xiao et al., 2016; Liu et al., 2018). A parameterized p-box is a special case of a general p-box, such that

$$F_X(x) = \left\{F_X\left(x; \vec{\theta}\right) \ : \ \vec{\theta} \in \left[\vec{\underline{\theta}}, \vec{\bar{\theta}}\right]\right\}. \tag{13.14}$$

The term $F_X(x; \theta)$ is a specified family of CDFs, $\vec{\theta}$ is a vector of the parameters of the specified family (sometimes referred to as hyperparameters), and $\left[\vec{\underline{\theta}}\ \vec{\bar{\theta}}\right]$ is the vector of intervals for the parameters of the specified family. There is a wide variety of two- and three-parameter family distributions that are commonly available in software packages to characterize an uncertain input quantity. The flexibility of a distribution increases dramatically as the number of parameters increases, which also makes them more delicate to compute with interval-valued parameters. As a result, an analyst should try to use a family with a lower number of parameters to characterize both aleatory and epistemic uncertainty. The interval-valued range of each parameter in the family can be estimated using a software package to approximate a set of experimental data or information provided by expert opinion.

Example 8

Consider a parameterized p-box formed using the family of normal distributions. The parameterized p-box is intended to characterize both the aleatory and epistemic uncertainty in an input quantity of interest. The epistemic uncertainty is due to the small amount of experimental data available for the quantity of interest. The software package used to fit the experimental data allows an interval to be computed for both the mean, μ, and the standard deviation, σ, of the normal distribution. The interval for μ and σ is computed based on the statistical confidence level chosen

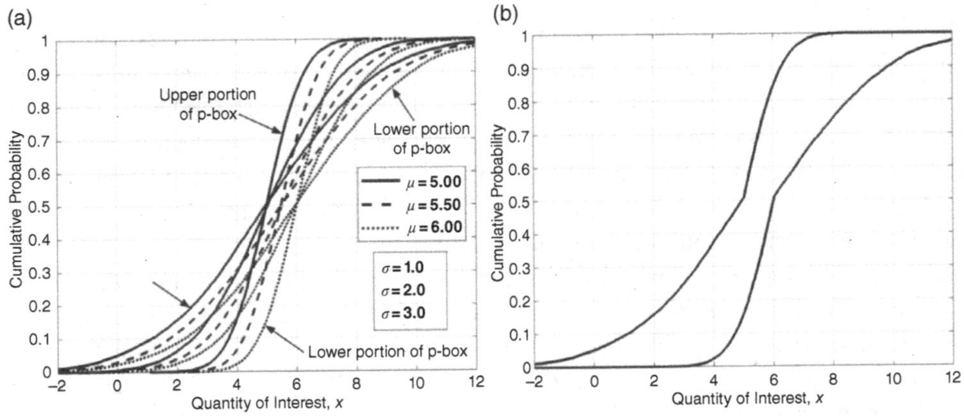

Figure 13.11 Parameterized p-box formed from a normal distribution with interval-valued parameters. (a) Family of normal distributions. (b) Envelope of p-box.

by the user. Assume that the fitting procedure and the confidence level desired resulted in the intervals: $\mu = [5, 6]$ and $\sigma = [1, 3]$. Figure 13.11a shows the set of normal distributions that is computed by sampling three values of μ and three values of σ for each value of μ. A total of nine distributions is shown in Figure 13.11a. The distributions shown indicate some of the possible distributions that exist, given the two interval-valued parameters of the normal distribution. Stated more generally, an infinite set of normal distributions exists, given the limited knowledge of the experimental data, which is expressed by interval-value parameters of the distribution.

The p-box of the quantity of interest is formed as the envelope of the set of all possible distributions, as shown in Figure 13.11b. Given that a p-box is commonly formed in this way, it can have rather unusual shapes. The interior region of the p-box characterizes the epistemic uncertainty in the quantity of interest. As a result, no interior structure is shown in a p-box to make it clear that no likelihood or belief structure exists in the interior. The epistemic uncertainty characterized in Figure 13.11b can be interpreted in two equivalent expressions. First, the likelihood of a given value of the quantity of interest can be stated as an interval-valued probability. For example, Figure 13.11b shows that the probability of the quantity of interest having a value of 4.0 or less can be as low as 0.025, but it can be as high as 0.37. Second, quantiles are interval-valued for a given probability level. For example, Figure 13.11b shows that the 0.9 probability quantile for the quantity of interest is [6.3, 9.9]. That is, the quantity of interest has at most a 0.9 probability of falling below 6.3 and at least a 0.9 probability of falling below 9.9. Since the second interpretation is alien to most analysts and decision makers, the first interpretation is the one most commonly used.

Two general approaches have been developed for sampling parameterized p-boxes and propagating them through a general mathematical model (Bruns and Paredis, 2006; Ghosh and Olewnik, 2013; Xiao et al., 2016; Liu et al., 2018): double-loop sampling and optimized parameter sampling. Only double-loop sampling is discussed here because of its simplicity. Double-loop sampling is a nested combination of the two previously described sampling methods for aleatory and epistemic uncertainty. The outer loop is typically used for sampling the epistemic uncertainties and the inner loop for sampling the aleatory uncertainties. Let the number of aleatory and mixed uncertainties be α, so that the vector of input random variables is $(x_1, x_2, \ldots, x_\alpha)$. The inner loop samples these random variables using some type of Monte Carlo sampling. Only simple MCS is considered here for sampling the inner loop. All of the epistemic, i.e., interval-valued, uncertainties are placed in the outer loop, whether they are intervals characterizing an uncertain model input x_i, or they are interval-valued hyper-parameters of a parameterized p-box. Let the total number of intervals be β, regardless of where they originate. Let the vector of input intervals be $\vec{\xi}$, so that the elements of this vector are $(\xi_1, \xi_2, \ldots, \xi_\beta)$. Since mixed input uncertainties have both aleatory and epistemic features, it is seen that $\alpha + \beta \geq m$, where m is the total number of uncertain model inputs.

Figure 13.12 shows a flow chart for the double-loop sampling strategy. The flow chart depicts the computational process for one SRQ, but the same procedure would be used for any number of SRQs. Use the following steps in the double-loop sampling strategy:

1. Choose the number of partitions, $\mathcal{N}$, that will be used to divide each of the input intervals. If the mathematical model $f\left(\vec{x}\right)$ is computationally expensive to evaluate and β is large, then pick $\mathcal{N} = 1$ or 2.
2. Calculate all the combinations of end points for each of the partitions for each of the input intervals. The total number of sample point combinations of end points to be evaluated is $(\mathcal{N} + 1)^\beta$. For example, if $\mathcal{N} = 1$ and $\beta = 2$, there would be 4 combinations of end points to sample: $\left(\underline{\xi}_1, \underline{\xi}_2\right), \left(\underline{\xi}_1, \overline{\xi}_2\right), \left(\overline{\xi}_1, \underline{\xi}_2\right), \left(\overline{\xi}_1, \overline{\xi}_2\right)$.
3. Choose the number of samples, N, to be used for simple MCS for each of the aleatory and mixed uncertainties.
4. Choose an end point combination from the interval partitions that has not been previously sampled.
5. Choose a random sample for each of the aleatory and mixed uncertainties that has not been previously used.
6. Evaluate the model $f\left(\vec{x}\right)$ using a random sample of the aleatory and mixed uncertainties that has not been previously used, along with the end-point combination from the interval partitions that was set in Step 4.
7. Test if all N samples of the aleatory and mixed uncertainties have been used to evaluate the mathematical model. If *No*, return to Step 5. If *Yes*, go to Step 8.
8. Construct an empirical CDF of the SRQ based on the N samples of the aleatory and mixed uncertainties.
9. Test if all $(\mathcal{N} + 1)^\beta$ samples of the end-point combinations of interval partitions have been used to evaluate the mathematical model. If *No*, return to Step 4. If *Yes*, go to Step 10.
10. Collect the $(\mathcal{N} + 1)^\beta$ empirical CDFs of the SRQ onto one plot to show the ensemble of all CDFs. Each CDF shows the distribution of variability of the SRQ for the given sample of the end-point combination of the partitions of the input intervals.
11. Compare each empirical CDF of the SRQ and store the smallest and largest value of probability for each sampled value of the SRQ. This procedure computes the envelope of all empirical CDFs computed.
12. Plot the smallest and largest value of probability for each sampled value of the SRQ over the range of sampled SRQs. This plot is a p-box because it shows the possible range in probabilities for all sampled values of the SRQ. Therefore, given the information known for the uncertain input quantities, the p-box shows smallest, i.e., the best possible, range of probabilities for the response of the system.

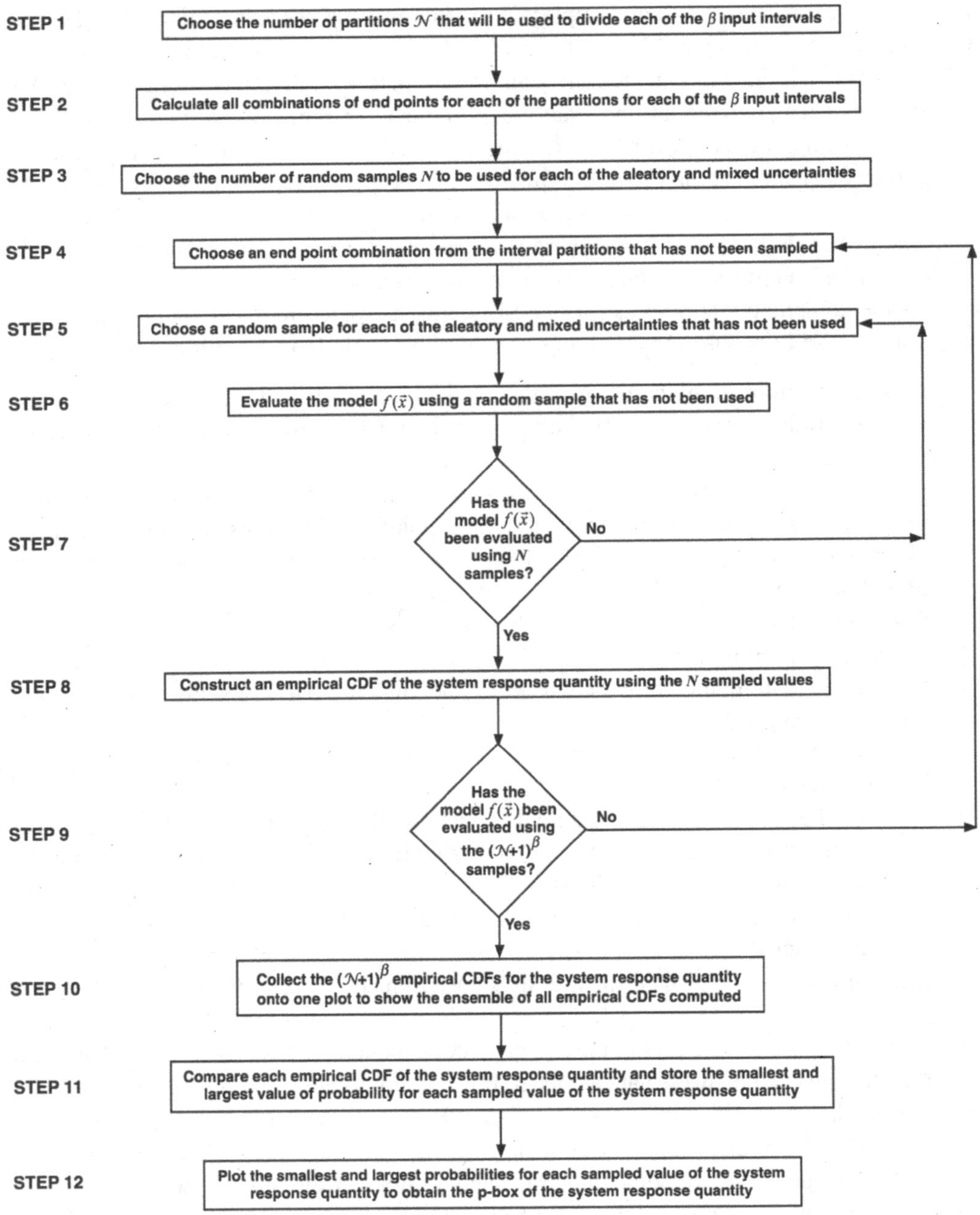

Figure 13.12 Flow chart for equal partition sampling of the epistemic input uncertainties and simple MCS for the independent aleatory input uncertainties.

13.4.5 Estimation of Iterative Error

13.4.5.1 Iterative Methods

As discussed in Chapter 8, iterative solution error arises due to incomplete convergence of the discrete equations approximating the mathematical model. Iterative methods are usually required for the solution of nonlinear discrete equations (except for explicit marching algorithms), as well as for large linear systems. Chapter 8 discusses various types of iterative error-estimation techniques using both stationary iterative methods and nonstationary (Krylov subspace) methods. Stationary methods tend to be more robust in applications compared to nonstationary methods. However, stationary methods tend to produce lower convergence rates compared to nonstationary methods.

Iterative methods are commonly applied in two different types of numerical solutions: initial value problems (IVPs) that use implicit methods and boundary-value problems (BVPs). For IVPs using implicit methods, fewer iterations are typically required for each marching step. This is because the initial guess for the iterative solution is based on the previously converged marching step of the solution, which is a good initial estimate for the present marching step. For BVPs, typically a larger number of iterations are required to obtain a converged solution. As a result, for BVPs, convergence characteristics should be closely monitored to determine if monotone, oscillatory, or mixed-convergence characteristics exist. The most convenient method of tracking the iterative convergence is to compute the L_2 norm of the local iterative residuals over the solution domain of each of the PDEs as a function of iteration number. However, the most reliable method is to monitor the L_2 norm of the local iterative residuals as a function of the number of iterations.

As discussed in Chapter 8, the iterative error should be driven at least as small as 1% of the discretization error when extrapolation-based discretization error estimators are used. However, it is advisable to reduce this error even smaller, e.g., 0.1%, because complex interactions can occur between iterative and discretization error. For example, if the iterative error is not driven small enough, it can result in a misleading or confusing trend in the observed order of accuracy as the mesh resolution is refined. An alternative to monitoring the L_2 norm of the local iterative error is to monitor the discrete residuals of each equation, i.e., the nonzero remainder that occurs when the current iterative solution is substituted into the discrete equations. However, relating residual convergence to iterative error must be done heuristically for a given class of problems. It is also emphasized that different SRQs typically converge at widely differing rates. The iterative convergence rate can be monitored on all SRQs, but this can become quite involved, and it is a computationally intensive process. A more efficient process is to identify the most slowly converging SRQs, and then monitor these for iterative convergence. It is stressed that monitoring the change in magnitude of iterates between successive iterations as an error indicator is *not* reliable and should always be avoided. That is, comparing the iterates in a sequence of iterations is *not* an iterative-error estimate; it is simply an indication of the iterative-convergence rate. As a result, the following discussion of implementation issues only deals with iterative methods that monitor the iterative convergence of SRQs.

13.4.5.2 Practical Implementation Issues

As defined in Equation (13.5), the vector of SRQs is given by $\vec{y}$. Some of the elements of the array are typically the dependent variables in the PDEs. However, some of the array elements may be very different kinds of quantities, for example: (a) derivatives of dependent variables with respect to the independent variables, or derivatives with respect to input parameters to the model; (b) functionals, such as integrals, time-averages, or min and max operators on the dependent variables; (c) cumulative distribution functions or specific probability values of a computed CDF; and (d) indicator functions, e.g., if the system is in a certain state, then an SRQ has the binary value of zero, but if the system is in another state, then an SRQ has a binary value of one. Given these diverse types of SRQs, the question is: "How can the iterative convergence error of each element of $\vec{y}$ be estimated?" If each of the elements of $\vec{y}$ can be computed using an individual numerical solution of the PDEs, then it is clear that the iterative convergence of the quantity can be monitored.

Two examples of this situation follow. First, suppose one element of $\vec{y}$ is the local value of a dependent variable of the PDE at some point in the solution domain, such as the local temperature or local strain energy. Then these quantities can be computed at each iteration and their convergence monitored. Second, suppose an element of $\vec{y}$ is an integral of some dependent variable over the solution domain of the PDE or over a portion of the solution domain, such as the total strain energy in a structure or the force on an object in a flow. These quantities can be computed at each iteration and the convergence monitored.

If the SRQ *cannot* be computed from simply knowing an individual numerical solution of the PDE, then it is more difficult to directly monitor the iterative convergence of the quantity. For example, suppose one of the elements of $\vec{y}$ is a statistic of the nondeterministic simulation or it is the CDF of a dependent variable of the PDE. An ensemble of hundreds or thousands of individual numerical solutions must be computed *before* the statistic or the CDF can be computed. As can be seen, this presents a more complex situation for monitoring the iterative convergence of the quantity. For this situation, a reasonable approach is to monitor the iterative convergence of each sample of the ensemble that contributes to the SRQ. For example, each nondeterministic solution sample that contributes to the statistic or the CDF should be monitored for iterative convergence. Monitoring the iterative convergence of individual samples is not a direct measure of iterative error of the statistic or the CDF, but it is a reasonable procedure in most situations. The procedure is more defensible if the most sensitive statistic of interest or the most sensitive portion of the CDF is monitored.

The iterative-convergence characteristics of the elements of $\vec{y}$, as well as the mesh-convergence characteristics discussed here, can change drastically over the sample space of the uncertain input quantities. For example, if a random sample is drawn from the tails of various probability distributions of uncertain input quantities, the physical interactions in the system of interest can change drastically compared to a sample drawn near the mean of each distribution. If the number of uncertain input quantities is quite large for a complex simulation, then this presents an additional computational expense for monitoring iterative convergence. A good understanding of the dominant physical effects on the various SRQs can greatly help in identifying the most troublesome, i.e., the slowest, iterative-convergence quantities. If an analyst is dealing with multi-physics simulations with large differences in temporal and spatial scales, extreme caution must be used. Two examples of these simulations are those that deal with chemically reacting flow or those that deal with phase changes of materials, such as solid to liquid, or liquid to gas phase.

Let the estimate of the magnitude of the iterative error for the ith SRQ be given as $\left[U_I\left(\vec{x}\right)\right]_{y_i}$ for $\vec{x} \in$ *input sample space* and $i = 1, 2, \ldots, n$. As discussed in Chapter 8, U_I is an uncertainty that has the character of a bias error (because of the lack of iterative convergence) with noise (because of the oscillation of the iterative error. However, U_I approaches zero as the number of iterations increases. The maximum of the nondeterministic sample set of iterative uncertainties for each SRQ can be written as

$$(U_I)_{y_i} = \sup\left\{\left[U_I\left(\vec{x}\right)\right]_{y_i} : \vec{x} \in input\, sample\ \ space\right\} \text{ for } i = 1, 2, \ldots, n, \tag{13.15}$$

where sup is the supremum bound, also referred to as the least upper bound. Since it is usually possible to achieve very small iterative uncertainties relative to discretization uncertainties, the common procedure is to impose a stringent iterative convergence criterion, such as 0.1%, so that the iterative uncertainty can be ignored. If this is *not done*, then Equation (13.15) should be used as one component of the estimate of total numerical uncertainty, which is discussed later in this section.

13.4.6 Estimation of Discretization Error

13.4.6.1 Temporal Discretization Error

There are two basic approaches for controlling temporal discretization errors in a simulation. First, the error is estimated at each time step for each SRQ, compared to some error criterion, and then adjustments are made either to the step size or the order of accuracy of the method. This approach is usually referred to as a variable time-step method or a variable-order method, respectively. It is strongly preferred over the second method because it directly controls the locally occurring, per-step numerical integration error in each SRQ. It should be noted that although this refers to integration in time, the integration could be in any independent variable of a hyperbolic system of PDEs. For example, in fluid dynamics, the boundary layer equations are integrated in a streamwise spatial direction (the downstream flow direction) to obtain a solution.

In the second method, a solution is computed with a fixed time step over the time of interest. Then a new solution is computed with a fixed time step that is either larger or smaller than the time step used in the previous computed solution. The two solutions are then compared over the time domain of interest for each SRQ. If significant or unacceptable differences between the two solutions are observed for any of the SRQs, then the entire solution should be recomputed with a smaller time step. It is clear that the fixed time-step method is indirectly responsive to local time-step integration error. As a result, the method is not a quantitative control of time-step integration error, but it is a qualitative control of integration error. The method is more akin to a sensitivity analysis of the impact of integration error. However, because of its simplicity, it is very commonly used when the IVP is given by a system of PDEs.

Practitioners of finite-element methods commonly use a fixed time-step Runge–Kutta integration method. Most Runge–Kutta methods are explicit methods of order 2, 3, or 4, and they are able to estimate the discretization error at each time step, if desired. The primary shortcoming of explicit Runge–Kutta methods, indeed for all explicit methods, is that they require relatively small time steps because they are conditionally stable schemes. Fixed time-step methods are more susceptible to round-off-error accumulation when a large number of steps are taken. When the mathematical model is given by a system of ordinary differential equations, Runge–Kutta methods with variable time step or variable order (or a combination of the two) are very popular because of their local control of numerical integration error and robustness.

Implicit time-integration methods provide a significant increase in numerical stability. This allows very large time steps to be taken, while damping both temporal and spatial modes of instability. Of course, this trade-off is at the expense of temporal discretization accuracy. Implicit methods provide significant advantages in the solution of stiff differential equations, i.e., differential equations with a wide range of time scales. Since most implicit methods are only second order, there is a rapid increase in discretization error as the time step is increased. Implicit Runge–Kutta methods have recently been developed and these have major advantages for stiff differential equations. The choice of implicit versus explicit methods should depend on both the stability restrictions and the accuracy needed to resolve the time scales in the simulation. (See Griffiths and Higham, 2010; Butcher, 2016, for a detailed discussion of a range of numerical methods for ordinary differential equations.)

13.4.6.2 Finite Element-Based Methods for Spatial Discretization Error

As discussed in detail in Chapter 9, methods for estimation of spatial discretization error use either *a priori* or *a posteriori* methods, that is, methods that do not require a numerical solution to the PDF and those that require one or more solutions, respectively. For certain types of linear PDEs, *a priori* methods have been developed for finite-element methods. However, for general linear and nonlinear PDEs, *a posteriori* error-estimation methods are the only practical methods. Only *a posteriori* methods are discussed in this chapter.

For finite-element methods, the most common *a posteriori* method is the Zienkiewicz-Zhu (ZZ) error estimator combined with the superconvergent patch recovery (SPR) method (Zienkiewicz and Zhu, 1992). The ZZ-SPR recovery method can obtain error estimates in terms of the global energy norm or in terms of local gradients, given two conditions: (1) finite-element types are used that have the super-convergent property, and (2) the mesh is adequately resolved. Finite-element recovery methods have focused on elliptic and parabolic PDEs for linear elasticity theory and very limited nonlinear PDEs (Ainsworth and Oden, 2000; Babuska et al., 2010; Yan, 2012). The method has recently been extended to finite-difference and finite-volume schemes in fluid dynamics (Verfurth, 2013). However, accurate error estimates can only be made in the global energy norm, which is a significant disadvantage for addressing many SRQs of interest, such as local strain energy.

Residual-based methods also provide error estimates in terms of the global energy norm for solid mechanics (Ainsworth and Oden, 2000; Babuska et al., 2010; Verfurth, 2013). Since the global energy norm is rarely an SRQ of practical interest, the adjoint system for PDEs must be solved along with the primal error PDE. As discussed in Chapter 9, an extension of the residual methods, referred to as adjoint methods, has recently been applied to various SRQs as well as to other discretization schemes (e.g., finite-volume and finite-difference methods). These approaches are very promising and are currently under investigation by a number of researchers.

13.4.6.3 Richardson Extrapolation for Spatial Discretization Error

Richardson extrapolation-based error estimators are quite popular, particularly in fluid dynamics, because they are extremely general in their applicability. These methods do not depend on the numerical algorithm used, whether it is an IVP or BVP, the nonlinearity of the PDEs, the submodels that are used, or whether the mesh is structured or unstructured. Standard Richardson extrapolation can be used if the numerical algorithm is second-order accurate and if the mesh is refined by a factor of two in each coordinate direction. Generalized Richardson extrapolation, however, can be used for any order accuracy algorithm and with arbitrary refinement factors. A key requirement of Richardson extrapolation is that the mesh must be uniformly refined or coarsened as the analyst changes from one mesh solution to another. Hence, the mesh refinement or coarsening ratio must be nearly constant over the entire mesh from one mesh to the other. For structured meshes, this does not cause any difficulty because a fine mesh can be easily coarsened by eliminating mesh points. However, for an unstructured mesh, refinement or coarsening requires significant capability of the mesh generator. This requirement should be kept in mind when considering the wide range of mesh generation software that is commercially available.

The crucial difficulty with using Richardson extrapolation is that the meshes must be sufficiently resolved so that all numerical solutions are within the asymptotic convergence region for the SRQ of interest. For many practical problems, achieving the asymptotic region for the SRQs of interest can require extraordinary computational resources. The spatial resolution required to attain the asymptotic region depends on the following factors: (a) the nonlinearity of the PDEs; (b) the range of spatial scales across the domain of the PDEs, i.e., the stiffness of the differential equations; (c) how rapidly the spatial mesh resolution changes from one mesh cell to the next; (d) the presence of singularities or discontinuities in the solution e.g., shock waves, flame fronts, or discontinuities in the first or second derivatives of the boundary geometry; and (e) lack of iterative convergence.

If only two mesh solutions are computed, the Grid Convergence Index (GCI) method (Roache, 1998) is commonly used with a safety factor of three. However, if there is no computational evidence that the meshes are in, or near, the asymptotic region, the GCI method, even with a safety factor of three, is not reliable for indicating discretization error. Computational analysts, even experienced analysts, commonly misjudge what spatial resolution is required to attain the asymptotic region. As discussed in detail in Chapter 9, a more reliable procedure is to obtain three mesh solutions and then compute the observed order of convergence based on the three solutions. If the observed order is relatively near the formal order of the numerical method, then the analyst has empirical evidence that the three meshes are in, or near, the asymptotic region for the SRQ of interest. A factor of safety of 1.25 or 1.5 would then be justified to yield a reliable uncertainty indication for the discretization error. An additional complication, as discussed in Chapter 9, is that the formal order of the numerical method *cannot* be achieved because of certain nonlinear features in the solution domain or on the boundaries. During the last few years, methods have been developed that do not require all of the solutions to be in the asymptotic convergence region in order to yield reasonable error estimates (Phillips and Roy, 2016; Rider et al., 2016; Eca et al., 2018).

13.4.6.4 Practical Implementation Issues

Three practical implementation issues were mentioned with regard to iterative convergence: (1) the wide range of types of SRQs in $\vec{y}$; (2) dealing with hundreds or thousands of samples (i.e., solutions) of $\vec{y}$; and (3) the sensitivity of certain elements of $\vec{y}$ to certain ranges of input quantities that are sampled in a nondeterministic simulation. Temporal and spatial discretization methods suffer from these same difficulties. In addition, there is a difference in the mathematical structure between iterative convergence as compared to temporal/spatial convergence. Iterative-convergence characteristics are typically monitored by examining the magnitude of the residuals during the iterations of each solution to the discrete equations. Obviously, this is monitored by tracking the norm of various quantities computed over the domain of the PDEs. With temporal/spatial discretization errors, however, the analyst must monitor the change in *local* quantities in the domain of the PDEs as the discretization error is reduced. Local quantities could include not only dependent variables of the differential equations, but also spatial or temporal derivatives of dependent variables. Derivatives, particularly higher-order derivatives, converge much more slowly with respect to discretization error than dependent variables of the differential equations.

If an analyst retreats to dealing with the L_2 norm of the discretization error as opposed to the local discretization error of the SRQs, then the risk is similar to those methods discussed earlier that only monitor error norms over the solution domain. Monitoring the L_∞ norm provides a much more sensitive indicator of temporal/spatial discretization error for each solution. However, the disadvantage is that a great deal more *noise* exists in the L_∞ norm compared to the L_2 norm. The noise in the L_∞ norm is due to the norm's feature that the maximum absolute value can occur at any point in the domain for one level of temporal/spatial resolution and then switch to another point in the domain for another level of temporal/spatial resolution. Thus, noise in the L_∞ norm should be expected because of the local nature of the mathematical operator. Even though more noise exists with the L_∞ norm, it is a recommended quantity to monitor because it directly indicates the magnitude of the largest local error in an SRQ over the domain of the PDEs. If a sufficiently refined mesh is not computationally affordable in order to achieve the asymptotic convergence region for the L_∞ norm of a local SRQ, then, as a minimum, the L_2 should be used.

A final recommendation is made when dealing with an important practical difficulty of estimating temporal/spatial discretization error, similar to one discussed earlier with regard to iterative convergence. Since there is commonly a very wide range of temporal/spatial discretization error over the solution domain, it is clumsy and inefficient to monitor the local error in each SRQ of interest. Identifying which SRQs are the most troublesome to converge in terms of temporal/spatial convergence is highly advisable. Local mesh clustering is intended to deal with this issue, but even when this is done, it still remains the most common difficulty and the largest solution error to control. The most troublesome SRQs are usually local quantities in the solution domain, particularly those that are derivatives of the dependent variables of the PDEs. Examples of these are local shear stress, local heat flux, local mass concentration gradient, and local energy gradient. If these types of quantities are important simulation results, then the temporal/spatial discretization error should be estimated and controlled in these quantities. By monitoring the discretization errors in these troublesome quantities, the discretization errors in other less sensitive SRQs are likely to be very small. Thus, these troublesome quantities are the least likely to be in the asymptotic convergence region for temporal/spatial discretization error compared to other SRQs of interest.

13.4.7 Characterization of the Total Numerical Solution Uncertainty

The approach for explicitly including the estimated numerical solution error is to consider each of the three contributors (iterative, temporal, and spatial) as an epistemic uncertainty for each of the SRQs of interest. Even though some numerical error estimators will provide a sign for the error, the sign will be ignored because it may not be reliable. For example, if

Richardson extrapolation is used to estimate the spatial discretization error, but the asymptotic convergence region has not been attained, the estimate could be significantly in error and even the sign could be incorrect. As a result, the absolute value of each estimated quantity is taken. Without assuming any dependence structure between the contributors, the total numerical solution uncertainty for each SRQ y_i is approximated as

$$(U_N)_{y_i} = |U_I|_{y_i} + |U_T|_{y_i} + |U_S|_{y_i} \quad \text{for } i = 1, 2, \ldots, n. \tag{13.16}$$

The temporal and spatial discretization error terms are defined similar to that for the iterative error, so that

$$\begin{aligned}(U_T)_{y_i} &= \sup\left\{\left[U_T\left(\vec{x}\right)\right]_{y_i} : \vec{x} \in \textit{input sample space}\right\} \quad \text{for } i = 1, 2, \ldots, n,\\ (U_S)_{y_i} &= \sup\left\{\left[U_S\left(\vec{x}\right)\right]_{y_i} : \vec{x} \in \textit{input sample space}\right\} \quad \text{for } i = 1, 2, \ldots, n.\end{aligned} \tag{13.17}$$

The term U_T is the estimate of the temporal discretization error over the solution time domain of interest, and U_S is the estimate of the spatial discretization error over the spatial solution domain of the PDEs. Note that the combination of nondeterministic input quantities that produces the least upper bound of each U_I, U_T, and U_S for one y_i is commonly different than the combination of input quantities that produces a least upper bound for a different y_i. This creates significant complexity in keeping track of what conditions produced the least upper bound of each U_I, U_T, and U_S. The term $(U_N)_{y_i}$, as given in Equation (13.16), is a set of n intervals, one for each SRQ. As a result, it is seen that even if there are no epistemic uncertainties in the inputs, there will be epistemic uncertainty in the outputs. The exception is the rare case where all of the numerical solution uncertainties are so small that they can be ignored.

The computational expense and complexity of computing U_I, U_T, and U_S for each y_i must be stressed. If any of the y_i are local quantities in the solution domain, all of the following solutions for these y_i must be monitored: all samples of the nondeterministic input uncertainties, all time steps of the temporal solution of these y_i for all mesh points over the spatial domain, and all mesh points over the spatial domain for each time step. Note that these three uncertainties are estimated at the simulation conditions of interest for the actual system being analyzed. That is, these uncertainties are *not extrapolated* to the actual operating conditions of the system of interest, as is the case with model form uncertainty. They are estimated at the actual system conditions of the input uncertainties that are sampled. As a final sobering comment on the present state of affairs in complex simulations, estimation of these three numerical uncertainties is commonly *not* done. If anything is documented regarding numerical uncertainties in a simulation, it is commonly a sensitivity analysis with regard to these uncertainties. However, this approach is *not* an uncertainty estimate, but an indication of the sensitivity of the results to the numerical errors.

It could be argued that Equation (13.16) is excessively conservative by adding the three largest numerical solution uncertainties. Hence, the argument can be made that since Equation (13.16) combines the largest of each of the three contributing uncertainties over all input samples, over the entire time domain, and over the entire spatial domain, it exaggerates the total numerical uncertainty in the SRQs. While Equation (13.16) is typically a conservative estimate of the three contributing numerical uncertainties in most situations, it is *not* a guaranteed bound on the combined numerical uncertainty. The reason for this is that local temporal and spatial discretization errors in the solution domain propagate to various portions of the solution domain and impact other SRQs in the solution, depending on the features of the PDEs. For example, for an IVP-BVP, local numerical errors propagate in time and space according to the local mathematical features of the PDEs, in addition to the local transport features of the physics in the PDEs. As a result, Equation (13.16) is *not* a guaranteed bound on the total numerical uncertainty in an SRQ.

As mentioned earlier, carefully estimating each of the three numerical uncertainties is rarely done in practice because of the computational expense and bookkeeping complexity. If a computational analyst is questioned by a customer of the simulation why these numerical uncertainties are not quantitatively estimated, there is usually one of two responses. First, the claim is made that their experience indicates that these uncertainties are small with regard to other uncertainties in the simulation. The numerical uncertainties may indeed be negligible, but there is no quantifiable evidence available that the claim is true for the simulation at hand. Second, the claim is made that any numerical uncertainties can be plausibly removed by calibration of various input parameters, or their probability distributions, so that agreement with experimental measurements is obtained. This perspective is quite common, particularly in the Bayesian estimation community. The technical weaknesses of this response, however, are that numerical uncertainties are inextricably entangled with physics-modeling uncertainties, i.e., model form uncertainties. Thus, after calibrating model input parameters to obtain agreement with experimental data, the simulation is getting the correct answer for the *wrong reasons*, which is dangerous.

For mathematical models that are given by ODEs, even large systems of ODEs, the situation just described does not occur, for two reasons. First, iterative methods for BVPs and numerical solution error estimators are extremely effective and reliable in modern software packages for the solution of ODEs. Second, the computational expense of solving ODE models is typically small compared to the cost of numerical solution of PDEs. For simulations described by complex PDE models, there are both significant technical and practical problems in attempting to control the numerical errors. However, in some situations it can be done by monitoring each of the numerical errors on the most sensitive SRQs and by controlling these errors. When this approach has been attempted on complex simulations, it is commonly found that the computational expense does not allow for reliable control of the numerical errors. Then the computational analyst is faced with the difficult choice of reducing the physical complexity of the mathematical model so that reliable error control is possible. Or, assume that the impact of numerical error can be mitigated by calibration of model input parameters using experimental data. This latter choice is almost always the path taken in computationally demanding simulations.

13.4.8 Example Problem: Step 4 for a Heat Transfer Analysis

As discussed earlier for this example of heat transfer through a cinder-block wall, the mathematical model is given by Equation (13.1) and boundary conditions are given by Equation (13.2). As is common for BVPs, they are numerically solved as an IVP, and then an iterative procedure is used to advance toward a solution that satisfies the given boundary values. There are two uncertain input quantities in the simulation: the thermal conductivity, $k(T)$, and the convective heat transfer coefficient, h, on the outside of the wall. The term $k(T)$ is an aleatory uncertainty given by a normal distribution (with $\sigma/\mu = 0.1$) and h is an epistemic uncertainty given by the interval [7, 10].

Figure 13.13 shows the flow chart for the numerical solution of the mathematical model, as well as the sampling procedure for the aleatory and epistemic uncertainties. The following gives a brief explanation of each step in the flow chart. Note that the flow chart does not include the estimation of model form uncertainty, which will be discussed in the next section.

1. Define all of the input data needed for the numerical solution of the mathematical model. This includes not only all input data needed for the physical system, but also all of the data needed for the numerical solution. For example, this may include the number of Monte Carlo samples, N, for the aleatory uncertainty, the number of partitions, $\mathcal{N}$, for sampling the epistemic uncertainty, the iterative convergence criterion, ε_{T_0}, for solving the BVP, and the per step, relative error criterion, ε_{ODE}, for the numerical integration of the ODE. The term T_0 is the wall temperature at $x = 0$, which is assumed to be the same as the temperature inside the building.
2. Pick a sample from one of the end points of the $\mathcal{N}$ equally spaced partitions of the interval $[h_1, h_2]$. The total number of samples over the interval is $\mathcal{N} + 1$. The sampling procedure of the end points of each partition was discussed in Section 13.4.3. The sampling procedure ensures that h_1 and h_2 are always sampled.
3. Compute an initial estimate of T_0 based on an approximate heat transfer analysis for the problem. An approximate analysis finds that the temperature drop through the wall is small compared to the temperature drop due to natural convection on the outside surface of the wall. The approximate analysis gives an initial estimate of T_0 as
$$T_0 = T_\infty + \frac{q}{S_{walls} h}.$$
4. Pick a random sample of k using simple Monte Carlo sampling, given the temperature T_0. As discussed in Sections 13.2.4 and 13.3.3, the mean value of k is a quadratic function of temperature, and there is variability in k due to building materials and construction. The variability of k is given by a normal distribution with a CoV, σ/μ, of 0.1. The simple Monte Carlo sampling procedure was discussed in Section 13.4.2 for sampling aleatory uncertainties.
5. Update the estimate of T_0 based on the newly computed numerical solution of the IVP. That is, the IVP must be solved iteratively until the computed heat transfer through the wall and from the outside wall is equal to the heat generated by the equipment inside the building, q. The equation used to update the value of T_0 is
$$(T_0)_{new} = (T_0)_{old} + R\left[\frac{(q - q_{old})\tau}{S_{walls} k_{old}}\right].$$
The term R is the iterative relaxation factor and the subscript *old* is the previously computed value of the quantity. Using numerical experimentation, a value of R should be found that produces rapid iterative convergence, but a value that does not produce excessive oscillation of iterative solutions of T_0.
6. Compute the difference between $(T_0)_{new}$ and $(T_0)_{old}$ to determine if the difference is less than the iterative error criterion, ε_{T_0}. If *No*, continue the iterative solution of the BVP by moving to Step 7. If *Yes*, jump to Step 12.
7. Update T_0 by setting $(T_0)_{old}$ equal to $(T_0)_{new}$.
8. Recompute the sampled value of k using the newly updated value of T_0. That is, keep the same random variability of k obtained from the Monte Carlo sampling of k, but shift the mean value of the sample distribution to the newly updated value of T_0.

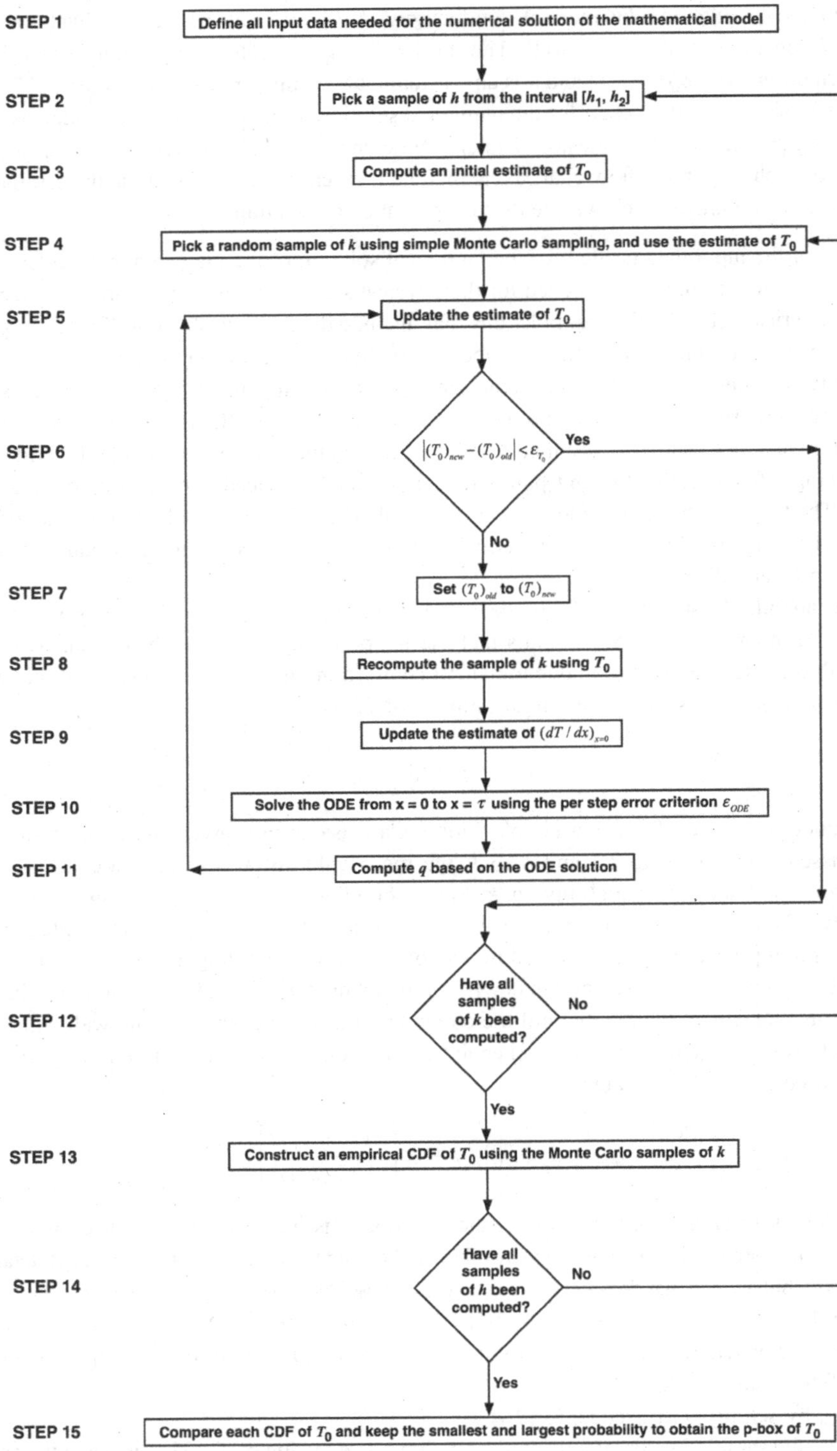

Figure 13.13 Flow chart for numerical solution of the heat transfer example problem.

9. Update the temperature gradient at $x = 0$ based on the new value of k. That is, the updated temperature gradient is

$$\left(\frac{dT}{dx}\right)_{x=0} = \frac{-q}{S_{walls}k(T_0)}.$$

10. Numerically integrate the ODE from $x = 0$ to $x = \tau$, using a numerical method that estimates the per-step numerical discretization error. Most software packages for solving ODEs require that the ODEs are first-order differential equations. As a result, the second-order ODE given by Equation (13.1) for the temperature distribution through the wall, T, is converted to two first-order ODEs. This is easily done by defining the temperature gradient, dT/dx, as the second dependent variable for the system of two first-order ODEs. The numerical method should also be able to adjust the integration step size or the order of the integration method so that the per-step, relative error, criterion, ε_{ODE}, is satisfied at each step. Note that this error criterion is imposed on *each* dependent variable, T and dT/dx, in the system of ODEs.
11. Compute q based on the numerical solution of the ODE. Return to Step 5 for continued iterations of the IVP. This value of q is referred to as q_{old} in Step 5.
12. Determine if all of the Monte Carlo samples of k have been computed. If *No*, return to Step 4. If *Yes*, go to Step 13.
13. Construct a CDF of T_0 using the Monte Carlo samples of k and the iteratively converged numerical solutions of the IVP. This CDF represents the variability of T_0 for the particular sample of h that was computed in Step 2.
14. Determine if all of the samples of h have been computed. If *No*, return to Step 2. If *Yes*, go to Step 15.
15. Compare each CDF of T_0 for all samples of k and h. Keep the smallest and largest value of T_0 for a given value of probability so as to yield the p-box of T_0, i.e., the envelope of all CDFs computed. The smallest value of T_0 is the upper bound of cumulative probability and the largest value of T_0 is the lower bound of cumulative probability.

The numerical solution procedure just described was implemented in MATLAB. Details of this numerical solution are given in the following:

- The number of Monte Carlo samples, N, for the aleatory uncertainty k was set to 100. The seed for the random samples was set to 0.
- The number of equally-spaced partitions, $\mathcal{N}$, of epistemic uncertainty, h, was set to 19, so that the number of samples of h was 20.
- The iterative convergence criterion for T_0, ε_{T_0}, was set to 1E-3.
- The per-step, relative discretization error criterion, ε_{ODE}, for the numerical integration of both T and dT/dx was set to 1E-6.
- The iterative relaxation factor R was set to 0.1.
- The numerical integration procedure used was MATLAB's Runge–Kutta 2(3) method, ode23. The procedure adjusts the step size so that the per-step, relative discretization error criterion, ε_T, is satisfied at each step for each dependent variable in the system of ODEs, i.e., both T_0 and dT/dx.

In the numerical solution procedure just described, the computer round-off error is small because a 64-bit computer was used, in addition to a small number of numerical integration steps. Furthermore, the iterative error was demonstrated to be small, and the numerical integration error at each step was demonstrated to be very small. As a result, these numerical solution errors are

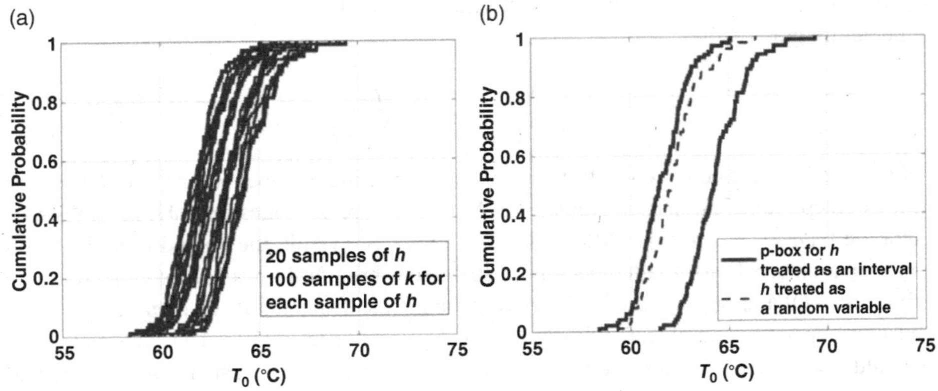

Figure 13.14 Nondeterministic simulation results for T_0 for the heat transfer example. (a) Cumulative distribution functions of T_0 for each sample of h. (b) P-box of T_0 and result for h treated as a random variable.

ignored when estimating the uncertainty in the simulation results. Since a probabilistic result with an accuracy of about 0.1 is considered satisfactory for this analysis, the number of aleatory and epistemic samples is considered adequate.

Using the model input data given in Table 13.1, nondeterministic simulation results are computed using the procedure diagrammed in Figure 13.13. Although several SRQs are computed for this simulation, the most important result is the temperature of the interior wall of the building, T_0. Figure 13.14a shows each CDF of T_0 that results from each of 20 samples of h. Each CDF is composed of 100 Monte Carlo samples of k. The CDFs show the variability of T_0 due to the variability of k, conditional on the sampled value of h over the interval of h. Figure 13.14b is the p-box of T_0 that is computed for the set of CDFs shown in Figure 13.14a. That is, the p-box shown is the envelope of all computed CDFs for T_0. Since only 20 samples of h were computed, the p-box shown is not guaranteed to be the largest p-box that could exist for a larger number of samples of k and h. Excluding this sampling limitation, the p-box can be described as the *best possible bounds on the uncertainty of* T_0, given the stated variability and the limited knowledge of the uncertain inputs.

The p-box in Figure 13.14b shows that the cumulative probability for any value of T_0 is an interval-valued quantity. For example, Figure 13.14b shows that for a value of $T_0 = 65^{\circ}$C *or less*, the probability is at least 0.72, but it can be as high as a probability of 0.98. Since the risk to electrical equipment and to maintenance workers inside the building is due to high temperatures, the more common method of presenting such results is to show the complementary CDF (CCDF), also referred to as the exceedance probability. The CCDF is simply computed as 1 – CDF. Although not shown, the CCDF of the p-box shows that for a temperature of $T_0 = 65^{\circ}$C *or greater*, the probability that this temperature could occur is no greater than 0.28, but it could be as low as 0.02.

To contrast the present results using PBA with the more common nondeterministic simulation procedure, Figure 13.14b shows the result if h were characterized as a uniform probability-density function over the interval [7, 10]. For this characterization, the simulation result is a *single* CDF near the left boundary of the p-box. Two important general observations are made contrasting PBA and the common nondeterministic procedure. First, when all uncertain input quantities are treated as random variables, there is *only one* resulting CDF. Second, the single CDF will always occur *within* the computed p-box using PBA. The single CDF shown in Figure 13.14b is a *very* different picture of simulation uncertainty for the decision maker compared to the p-box. The representation of the interval as a uniform distribution, as opposed to an interval, requires two unjustified assumptions in the statement of the uncertain character of h: (1) the possible value of h is entirely random over the range of the interval, and (2) all possible values of h are *equally likely* over the range of the interval. Therefore, when an epistemic uncertainty is characterized as a random variable, as opposed to an interval, the impact of the uncertainty on the SRQ is characterized as *increased dispersion/variance*. This mischaracterizes ignorance as randomness, implying to the decision maker that there is less total uncertainty than actually exists.

13.5 Step 5: Estimate Model Form Uncertainty

As discussed in Section 13.2.3, model form uncertainty is caused by the assumptions and approximations embedded in the formulation of the mathematical model and all submodels. Model form uncertainty is sometimes referred to as model discrepancy, structural uncertainty, or model bias uncertainty. Within the validation domain, the model form uncertainty can be estimated reasonably well using experimental measurements of the SRQs and a validation metric operator, as discussed in Chapter 12 "Model Accuracy Assessment." A validation metric operator is a mathematical operator that estimates the difference between experimental measurements of the SRQs and the simulation results for the SRQs for the conditions of the experiment. A well-constructed validation metric operator should account for both the estimated uncertainty in the experimental measurements of the SRQs, as well as the experimentally measured uncertainty in the model input data that is propagated through the mathematical model to obtain nondeterministic SRQs. Experimental measurements, particularly those that are independently repeated, are a much more reliable reflection of nature's response of the system, as opposed to what a model predicts. Therefore, the validation metric result at each set of measured conditions is a measure of the model form uncertainty, including the experimental measurement uncertainty and any uncertainty in the model input data that characterizes the actual experiment. This statement, of course, assumes that (a) code verification has been completed for the code features and options used in the simulation, and (b) solution verification has shown that all contributors to numerical solution errors are negligible.

Figure 13.15 captures the essence of the concepts of interpolation and extrapolation of the model form uncertainty for two input quantities. The terms α and β are parameters characterizing conditions of the system or the surroundings. Vs denote conditions where experimental data have been obtained and (α_i, β_i), $i = 1, 2, \ldots, 5$, denote the corners of the intended use domain for the system of interest, sometimes called the operating envelope of the system. The validation domain is defined as the convex hull of the set of model input data conditions where experiments are conducted. For an input dimensionality of three, the validation domain can be visualized as a volume. For higher input dimensionality, the validation domain is referred to as a hyper-volume. Computing the convex hull for high input dimensionality is computationally intensive. A validation metric result can be computed at each of the Vs in Figure 13.15. Within the validation domain, model form uncertainty can be estimated with relative confidence at conditions other than the Vs, using some type of

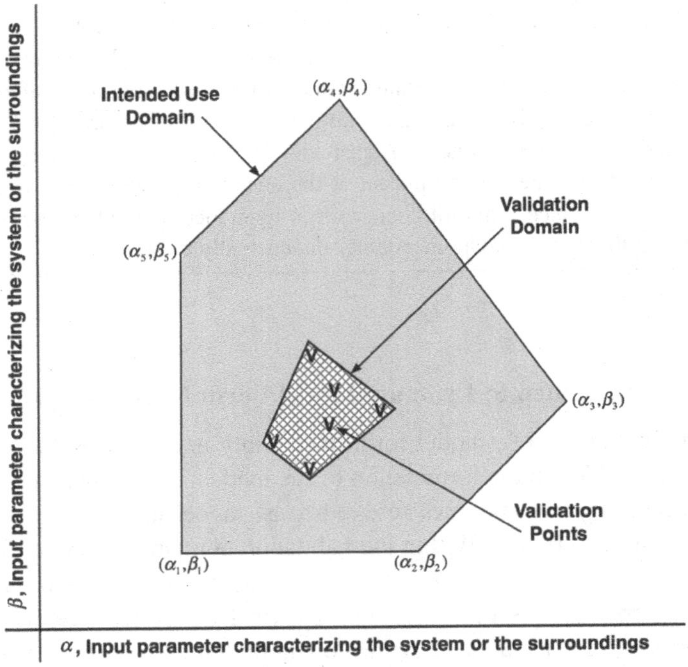

Figure 13.15 Validation domain and intended use domain in two dimensions. (adapted from Trucano et al., 2002)

interpolation function. Outside of the validation domain, model form is much more difficult to estimate with any confidence because it involves estimating a physics modeling error for conditions that have not been observed.

In Chapter 12, various approaches were discussed for estimating model form uncertainty using validation metrics. Two methods discussed in detail were those based on Bayesian inference using the Kennedy and O'Hagan (KOH) approach (Kennedy and O'Hagan, 2001) and those based on comparing CDFs from experiments and simulations. In this section, these two approaches are summarized and then a discussion is given concerning estimation of model form uncertainty outside the validation domain.

13.5.1 Estimation Using Bayesian Inference

Following the formulation originally developed by KOH for estimating model form uncertainty (also referred to as the model discrepancy), we write

$$\left(y_{\exp}\right)_i = f\left(\vec{x}_i; \vec{\theta}\right) + \delta\left(\vec{x}_i\right) + \varepsilon\left(\vec{x}_i\right) \quad i = 1,\ 2,\ \ldots,\ G. \tag{13.18}$$

This equation is written for multiple experimental measurements of one SRQ $\left(y_{\exp}\right)_i$ for $i = 1, 2, \ldots, G$. The term $f\left(\vec{x}_i; \vec{\theta}\right)$ is the mathematical model used to simulate the experiments, where $\vec{x}_i$ is the vector of model input quantities for the i conditions of the experiment, $\vec{\theta}$ is the vector of model input parameters to be calibrated, θ_j for $j = 1, 2, \ldots, H$. The term

$\delta(\vec{x}_i)$ is the model form uncertainty at condition i, and $\varepsilon(\vec{x}_i)$ is the experimental uncertainty in the measurements of $(y_{\text{exp}})_i$. The measurement uncertainty, $\varepsilon(\vec{x}_i)$, is typically assumed to be unbiased, i.e., no systematic uncertainty, and normally distributed for each measurement made. In general, $\vec{x}$ has m uncertain inputs. However, for a validation experiment, most of these elements are deterministic quantities because they are controlled and measured in a validation experiment. Using Bayesian estimation, $\delta(\vec{x}_i)$ can be inferred using the mathematical model and the experimental measurements $(y_{\text{exp}})_i$. Thus, the inverse problem is solved using the mathematical model in order to estimate $\delta(\vec{x}_i)$ such that the experimental measurements are reproduced in the updated simulation.

The most common approach to estimating $\delta(\vec{x}_i)$ is to assume that it can be characterized as a Gaussian process and written as

$$\delta(x) \sim GP(0, k). \tag{13.19}$$

GP is a Gaussian process model with a mean of zero over the range of experimental data, and k is the covariance or kernel function of the Gaussian process over the range of the data. The zero mean of the Gaussian process will result in the discrepancy term compensating for any mathematical modeling (mean) error at each of the experimental measurements in the updated simulation. That is, the discrepancy term is an inferred bias of the simulation result based on the mathematical model and the available experimental measurements. An additional benefit of using a Gaussian process model within the validation domain is that it also provides an estimate of the model form uncertainty between the conditions of the validation experiments. The Gaussian process provides a statistical interpolation procedure for estimating the model form uncertainty at any point within the validation domain. The form of the covariance function, $k(x, x')$, is usually assumed to be an exponential function based on the square of the Euclidean distance between points x and x'. As a result, two additional parameters of the exponential function need to be estimated as part of the Gaussian process model.

In addition to estimating the two parameters characterizing $\delta(x)$, the Bayesian procedure commonly estimates the standard deviation in $\varepsilon(\vec{x}_i)$, as well as the model calibration parameters, $\vec{\theta}$. All of these are estimated concurrently using a large number of samples of the mathematical model. The inverse procedure to estimate all of these parameters is depicted in Figure 13.16. If the standard deviation in $\varepsilon(\vec{x}_i)$ is independently estimated from the experimental procedure itself, then it does not need to be estimated using Bayesian inference. Prior distributions are assumed for each of the quantities to be estimated. Then the posterior distributions are calculated using Bayesian inference, given the experimental measurements and the mathematical model. The number of quantities to be estimated as part of $\delta(x)$ and ε_i is typically three, but the number of calibrated input parameters, $\vec{\theta}$, can be much larger. It is well known that the solution to inverse problems is more complex mathematically, more computationally intensive, and more error-prone than solving the forward mathematical problem, that is, propagating uncertain inputs through a model to obtain uncertain simulation responses. It is beyond the scope of this text to discuss the details of these issues. See the following texts for a detailed discussion: Gelman et al. (2013), Smith (2014), McClarren (2018), and Aster et al. (2019).

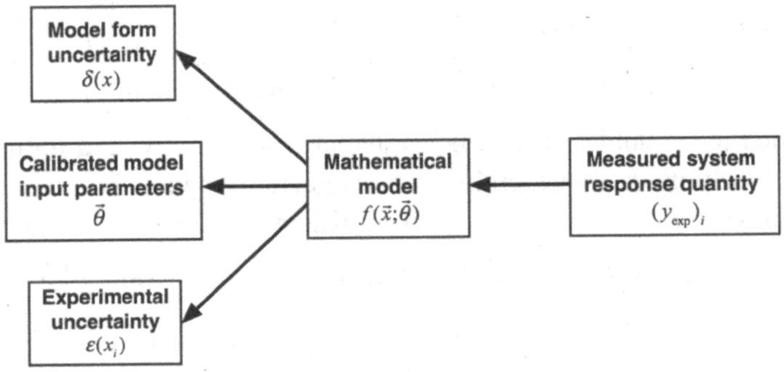

Figure 13.16 Solution of the inverse problem to infer model form uncertainty, model input parameters, and experimental measurement uncertainty.

During the last two decades, there have been many successful applications of the KOH approach published in diverse fields of study. However, there are several practical issues that can cause the approach to be unsuccessful or produce misleading results for the customer. Some examples are briefly discussed here.

- For each parameter estimated, a prior probability-density function must be assumed. There is substantial discussion in the literature concerning what type of prior PDF should be chosen when various types of independent information on the parameter is available. If very little information is available, a noninformative prior distribution is assumed, i.e., a uniform distribution over some assumed plausible range of the parameter. Although this may be a more defensible assumption than assuming a more informative prior, it does have the effect that more experimental data are required to produce an informative posterior distribution.
- The number of function evaluations, i.e., solutions of the mathematical model $f\left(\vec{x}_i; \vec{\theta}\right)$, required for the entire Bayesian procedure is extraordinary, e.g., tens of thousands of evaluations. As a result, a great deal of research has been devoted to the optimization of statistical sampling methods to reduce the number of function evaluations. Two general approaches have been used: (1) methods based on Markov chain Monte Carlo (MCMC) methods and (2) methods that construct surrogate models, also referred to as response surface models. The numerical accuracy of both approaches is dependent on the number of function evaluations of the mathematical model that are computed, the dimensionality of the model input space, the complexity of the response surface of the mathematical model, and the variability of the response surface. Unfortunately, the numerical accuracy of these approaches is rarely addressed in practice, relative to the exact stochastic response of the mathematical model itself.
- The number of parameters that can be informatively estimated depends heavily on the amount of experimental data available. If only a handful of physical experiments is available, very few parameters can be informatively updated. A great deal of experimental data are required if a large number of parameters must be inferred, particularly if noninformative priors are assumed. Otherwise, the posterior distributions for the estimated parameters will provide little likelihood of information for the quantification of the parameters. However, it should be noted that the amount of experimental data needed using Bayesian

estimation combined with a moderately accurate, physics-based, mathematical model is still *much less* than required for machine learning or data mining methods.

- The identifiability of the parameters to be estimated depends heavily on the number of parameters to be estimated, the amount of experimental data, the informativeness of the priors assumed, and the independence of each parameter relative to the other parameters to be identified. As is common with the solution of inverse problems, there can be multiple solutions for the identification of parameters. What this means in practice is that parameters can be misidentified in the sense that the updated model can mislead computational analysts and customers of the simulation results. For example, suppose an analyst has a given set of experimental data, a given mathematical model, and a given set of parameters to be updated. Different posterior distributions will be produced if different prior distributions are assumed for the various parameters. Furthermore, as the amount of experimental data decreases and the number of parameters increases, both the identifiability and the informativeness of the posterior distribution decreases.

13.5.2 Estimation Using CDFs

As discussed in detail in Chapter 12, model form uncertainty can be estimated by taking the difference between the CDF of the SRQ measured in the experiment and the CDF computed from the simulation. This type of difference measure is referred to as an *area validation metric* (AVM) because it computes the area between the CDF of the experimental measurements and the CDF of the simulation. Let the area between the two CDFs be defined as d. Even though d is always a positive quantity, it can indicate that the simulation result is either greater than the experimental result or less than the experimental result. The AVM was developed by Ferson et al. (Ferson et al., 2008; Ferson and Oberkampf, 2009) and it is interpreted as the *evidence for disagreement* between the simulation and the experiment. For example, if the CDFs from the simulation and the experiment are far apart, resulting in a large value of d, there is strong evidence that the simulation disagrees with the experiment. Similarly, if the CDFs from the simulation and the experiment are identical, then $d = 0$, indicating that there is *no* evidence for disagreement between the two.

It is clear that d can be used as a basis to estimate the observed *inaccuracy* of the model for future simulations where no experimental data are available. Including the estimated inaccuracy in the simulation results explicitly informs the computational analyst and the customer of the simulation results of the magnitude of the model form uncertainty. Including d in the simulation results is informative in the sense of a forthright recognition of structural uncertainty in the model, as opposed to an inferred bias correction in the model using the KOH approach. To account for this epistemic uncertainty, the area d is appended symmetrically to the left and right of the CDF of simulation results. If $Y(X)$ is the CDF from the simulation, then the model form uncertainty is characterized as a p-box, $[Y(X) - d, Y(X) + d]$. The AVM was used in various simulations so that model form uncertainty could be explicitly included as a contributor to the total prediction uncertainty of the simulation (for example, see Roy and Oberkampf, 2011; Wu et al., 2015; Choudhary et al., 2016).

In recent research (Whiting et al., 2023), it was recognized that the AVM had two important weaknesses in estimating model form uncertainty. First, the AVM did not indicate whether the simulation was less than or greater than the experimental measurements. That is, the AVM did not indicate the sign of the model form uncertainty because the absolute value of the difference between the CDFs was computed. Second, the AVM did not indicate the level of statistical confidence in the computed area between the two CDFs. For example, suppose for one case there was one experimental measurement result, but for another case there were a hundred measurements for the SRQ, yet the area between the two CDFs was identical in both cases. It is clear that there is a much higher statistical confidence in the area metric for the case of one hundred measurements, as opposed to one experimental measurement. To overcome these weaknesses in the AVM, Whiting et al. (2023) devised the *modified area validation metric* (MAVM). It should be emphasized that both the AVM and the MAVM are *not* inferences of model form uncertainty based on solution of the inverse problem, as discussed regarding Bayesian inference. Both the AVM and the MAVM solve the forward problem and then directly compare the simulation results with the experimentally measured results for the SRQ. Therefore, the AVM and the MAVM are measures of disagreement between simulation and experiment, and these measures have the dimensional units of the SRQ of interest.

The MAVM computes two area measures of the difference between the simulation and the experiment. One area measure is the difference between the simulation and the lower statistical confidence interval of the experimental measurements. The other area measure is the difference between the simulation and the upper statistical confidence interval of the experimental measurements. As a result, the MAVM is a dual-valued area that is expressed as an interval. The interval-valued characterization is appropriate because model form uncertainty is an epistemic uncertainty, specifically, a systematic error with only an estimated magnitude and no associated likelihood. The interval-valued area measure is written as

$$\left[d_{CL}^{-},\ d_{CL}^{+}\right]. \tag{13.20}$$

The term d_{CL}^{-} is the magnitude of the difference between the simulation and the CDF of the lower confidence interval of the experimental measurements. The term d_{CL}^{+} is the magnitude of the difference between the simulation and the CDF of the upper-confidence interval of the experimental measurements. There is a constraint on how each of these areas is calculated, which will be discussed shortly. The CDFs of the lower and upper-confidence intervals depend on the choice of the statistical confidence level desired, *CL*, the student *t*-distribution, the number of experimental measurements at a given set point, and the sample mean and variance of the experimental measurements.

If the simulation falls between the lower and upper experimental CDF confidence intervals, then d_{CL}^{-} indicates how much *larger* the simulation is than the CDF of the lower confidence interval. Furthermore, d_{CL}^{+} indicates how much *less* the simulation is than the CDF of the upper-confidence interval of the experiment. This situation is shown in Figure 13.17. The figure shows the CDF for five experimental samples and a smooth CDF for the simulation result. The smooth CDF for the simulation implies that a large number of samples of the aleatory input uncertainties characterizing the experiment were propagated through the mathematical model. The confidence intervals for the experimental measurements indicate the statistical confidence for capturing the *true (but unknown) mean* of the experimental measurements. The CDFs of the confidence intervals are constructed by

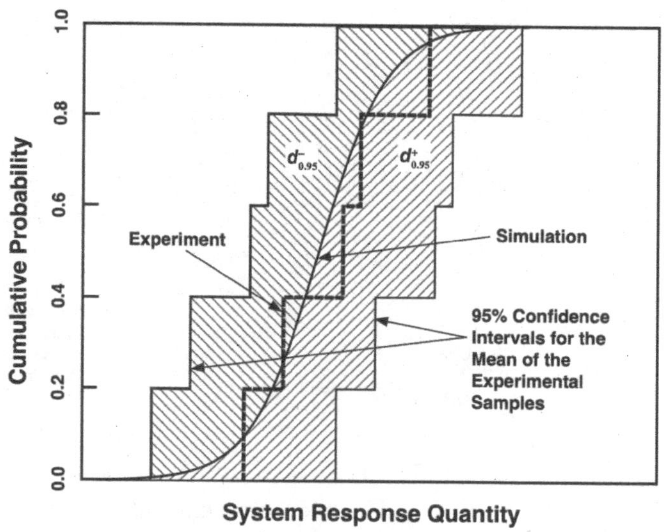

Figure 13.17 Example of the simulation falling within the ±95% confidence intervals for five experimental measurements.

shifting the experimental measurements to the left and to the right by equal amounts. The terms d^{-}_{CL} and d^{+}_{CL} approximate the bounds of the mismatch between the simulation and the *true (but unknown) distribution* of experimental measurements. For the case shown in Figure 13.17, d^{-}_{CL} and d^{+}_{CL} are composed of both model form uncertainty and statistical uncertainty in the experimental measurements. If a large number of experimental measurements is available, or the sample standard deviation of the measurements is small, then d^{-}_{CL} and d^{+}_{CL} are primarily composed of the model form uncertainty, as opposed to the sampling uncertainty in the experimental data. A final comment regarding Figure 13.17 is that it can be concluded that the model form uncertainty estimation is poorer in the tails of the experimental CDF, as opposed to the central portion of the distribution, because of the limited number of experimental samples.

If the simulation falls *outside* the confidence intervals of the experiment, then d^{-}_{CL} and d^{+}_{CL} are primarily composed of model form uncertainty, as opposed to experimental sampling uncertainty. This situation is shown in Figure 13.18. When the simulation falls outside the lower confidence interval, d^{-}_{CL} is defined to be zero. This is because there is *no experimental evidence* that the simulation is *larger* than the CDF of the lower confidence interval. As mentioned above, d^{+}_{CL} indicates how much *less* the simulation is than the CDF of the upper-confidence interval of the experiment. For this case, the magnitude of d^{+}_{CL} indicates that (a) the simulation is significantly biased such that it is *smaller* than the experimental measurements, and (b) it is primarily composed of model form uncertainty, as opposed to experimental sampling uncertainty. It can be shown that as the number of experimental measurements increases, and if the experimental and simulation CFDs do not cross, either d^{-}_{CL} or d^{+}_{CL} will tend to zero, and the other will tend to a constant nonzero value. This occurs because the width of the confidence interval tends to zero as the number of experimental measurements increases.

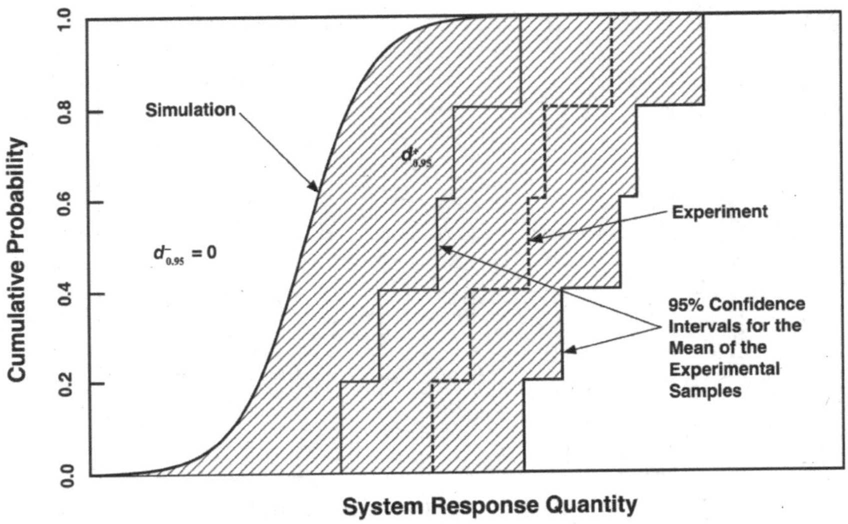

Figure 13.18 Example of the simulation falling outside the ±95% confidence intervals for five experimental measurements.

The discussion of the MAVM is concluded by summarizing the equations that are used to compute the confidence intervals of the experimental measurements. The sample mean and sample standard deviation of the experimental measurements are respectively given as

$$\bar{y}_{\text{exp}} = \frac{1}{N_{\text{exp}}} \sum_{i=1}^{N_{\text{exp}}} y^{i}_{\text{exp}}, \tag{13.21}$$

and $$s = \left[\frac{1}{N_{\text{exp}} - 1} \sum_{i=1}^{N_{\text{exp}}} \left(y^{i}_{\text{exp}} - \bar{y}_{\text{exp}} \right) \right]^{1/2}. \tag{13.22}$$

The term N_{exp} is the number of experimental measurements available at a given set of experimental conditions, and y^{i}_{exp} is the ith experimental measurement. From Equation (13.22), it is clear that N_{exp} must be 2 or greater for each set of experimental conditions used to compute a MAVM result. Note that, in general, $N_{\text{exp}} \neq G$, where G is the total number of experimental measurements for all set point conditions when using Bayesian inference to estimate model form uncertainty.

The confidence interval for the population mean of the experimental measurements, μ, is given by

$$\mu \sim \left[\bar{y}_{\text{exp}} - t_{\alpha/2,v} \cdot \frac{s}{\sqrt{N_{\text{exp}}}},\ \bar{y}_{\text{exp}} + t_{\alpha/2,v} \cdot \frac{s}{\sqrt{N_{\text{exp}}}} \right]. \tag{13.23}$$

The confidence level is given by $100(1 - \alpha)\%$, and $t_{\alpha/2,v}$ is the $1 - \alpha/2$ quantile of the student t-distribution for $v = N_{\text{exp}} - 1$ degrees of freedom. In the limit as $N_{\text{exp}} \to \infty$, the t-distribution approaches the standard normal distribution. For a detailed discussion of the MAVM, see Chapter 12.

13.5.3 Estimation over a Range of Conditions

First, consider the estimation of the model form uncertainty at any point *within* the validation domain, the central portion of Figure 13.15. From the above discussion of the KOH approach, it is seen that $\delta(\vec{x})$ can be used to estimate the model form uncertainty anywhere within the validation domain. However, the MAVM method does not provide a function for evaluation because it is a point estimate of model form uncertainty at the specific set of experimental conditions. To estimate the model form uncertainty over the entire validation domain using the MAVM, either an interpolation function or a regression fit of the point estimates must be constructed. The interpolation or regression function would yield an estimate of the model form uncertainty in between the Vs shown in Figure 13.15. If an estimate of model form uncertainty is *only* needed within the validation domain, then an interpolation function (analogous to the KOH procedure) would be appropriate because the interpolation function would reproduce the point estimates of model form uncertainty within the validation domain.

If an estimate of model form uncertainty is needed *both* inside and outside of the validation domain, which is the dominant situation in practice, then a separate issue must be considered in the choice between using an interpolation function or a regression fit of the point estimates of model form uncertainty. Outside of the validation domain, the issue of reasonableness of the trend of the extrapolation function is most important. As a result, use of an interpolation function, which is constrained within the validation domain but free outside of the validation domain, is *not* recommended. Since use of the same approximating function is desired both inside and outside the validation domain, a properly chosen regression function is recommended. Using a well-behaved regression function, the regression fit is computed based on the estimated model form uncertainty at points over the validation domain. Note that the arguments presented here for the choice of an approximating function for model form uncertainty are based on convenience, not physics, which is the source of the uncertainty.

Researchers using the AVM and the MAVM have approximated the model form uncertainty using a regression fit with low-order polynomials (Ferson et al., 2008; Roy and Oberkampf, 2011; Whiting et al., 2023). These types of functions have some flexibility to approximate the model form uncertainty inside the validation domain, and they are well behaved outside of the validation domain. More importantly, using the associated bounds of the statistical prediction intervals to estimate the model form uncertainty, as opposed to just the regression fit, will result in a continuously increasing estimate of the model form uncertainty as the application condition of interest moves further from the validation domain. Furthermore, the widths of the prediction intervals directly respond to the number and location of validation set points inside the validation domain, the statistical confidence level chosen, and the lack of fit by the regression function.

Prediction intervals should be used instead of confidence intervals because we are interested in estimating the uncertainty in a *single future value of the model form uncertainty*. A confidence interval is an estimate of the uncertainty in the *mean* of an ensemble of data. Prediction intervals are always larger than confidence intervals, and they do not approach zero as the number of samples becomes large (Devore, 2007). In MATLAB, prediction intervals are referred to as *nonsimultaneous observation bounds*. Although this extrapolation procedure for model form uncertainty has *no physics basis* for the approximation, it provides a simple method to reliably *increase* the uncertainty of a prediction due to lack of empirical evidence of the accuracy of the mathematical model. That is, the predictive capability of a simulation should, in general, *decrease* the further removed the application condition is from empirical evidence of the validation domain. Note, however, that the magnitude of the prediction intervals does *not* increase for model input dimensions that are not varied in the model validation experiments. For example, suppose that in addition to the two model input dimensions (α,) shown in Figure 13.15, there were additional model input dimensions that were not varied in the model validation experiment. The prediction intervals would, of course, have no knowledge of these additional dimensions. As a result, the prediction intervals would *not* increase in magnitude when these dimensions are varied in a simulation of the system.

Only first- or second-degree polynomials should be considered for the regression function because they are well behaved when used for extrapolation. For example, with two input dimensions, either a bilinear or biquadratic function should be used for the regression function. Note that a bilinear function is *not* a linear function in two-dimensional space because of the cross-product term, $x_1 \bullet x_2$. A bilinear function is a product of two linear functions in each dimension of the input space. For a 1, 2, 3, and 4-dimensional input space, the number of regression coefficients that are associated with a bilinear function is 2, 4, 7, and 11, respectively. The number of experimental set point conditions that is required to determine the regression coefficients must be larger than the number of regression coefficients, otherwise the procedure would devolve into a bilinear interpolation procedure. A natural benefit of a regression fit is that when the number of experimental conditions is only slightly larger than the number of regression coefficients, the statistical prediction interval is much larger than when more experimental conditions are available. Recent work by Whiting et al. (2023) on an example problem found that using bilinear and biquadratic regression functions with varying amounts of data in the validation domain produces highly reliable estimates of model form uncertainty both within and outside the validation domain. However, more research is needed to better understand the reliability of the procedure for diverse types of applications.

Example 9

Consider the estimation of model form uncertainty both within and outside the model validation domain using the MAVM. Figure 13.19 shows one input dimension and linear regression functions for d_{CL}^- and d_{CL}^+ extending beyond the validation domain. The figure shows four experimental measurement conditions for $x = 1, 2, 3,$ and 4. The associated values of the MAVM are $d_{CL}^- = [0.8, 0.7, 0.6, 0.0]$ and $d_{CL}^+ = [1.9, 2.3, 2.3, 2.7]$. Also shown in the figure are the associated prediction intervals for a 95% confidence level for each linear regression. Note that the 95% confidence level for the prediction interval for the regression fit is completely separate, i.e., independent, from the confidence level that is chosen for d_{CL}^- and d_{CL}^+. For example, suppose the confidence level used to compute d_{CL}^- and d_{CL}^+ in Figure 13.19 was 90%. Then it was decided to increase the confidence level of d_{CL}^- and d_{CL}^+ to 95%. This would significantly increase in magnitudes of both d_{CL}^- and d_{CL}^+; *independent* from the confidence level of the prediction interval for the regression fit. Even within the validation domain, the model form uncertainty associated

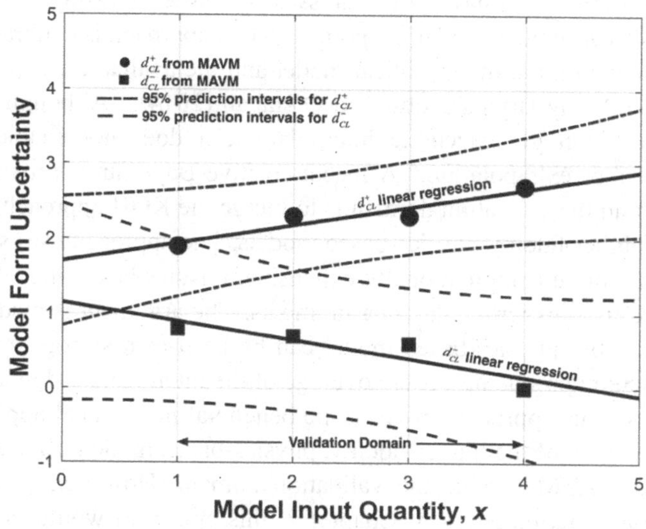

Figure 13.19 Example of model form uncertainty estimation within and outside the validation domain using the MAVM and linear regression of d^-_{CL} and d^+_{CL}.

with the prediction intervals of the regression fit should be used instead of the point values of d^-_{CL} and d^+_{CL}. For example, within the validation domain, the upper-prediction bound on each regression fit would be used for the model form uncertainty, as opposed to the values of d^-_{CL} and d^+_{CL} themselves. Outside of the validation domain it is clear that the upper-prediction bound for each regression fit should be used to estimate the model form uncertainty.

Let $(d^-_{CL})^+_{CL\oplus}$ be the upper bound of the prediction interval associated with the regression fit of d^-_{CL}, and $(d^+_{CL})^+_{CL\oplus}$ be the upper bound of the prediction interval associated with the regression fit of d^+_{CL}. The model form uncertainty would then be written as

$$\left[(d^-_{CL})^+_{CL\oplus},\ (d^+_{CL})^+_{CL\oplus}\right]. \tag{13.24}$$

As can be seen in Figure 13.19, the convex nature of $(d^-_{CL})^+_{CL\oplus}$ and $(d^+_{CL})^+_{CL\oplus}$ will result in the rapid growth of the estimated model form uncertainty as the distance from the validation domain increases. A physics-based argument could be made that this estimate may be much larger than the actual model form uncertainty. This may or may not be the case, but the argument could not be made based on the experimental measurements available and the observed performance of the model. Extrapolation of model form uncertainty far outside the validation domain should have the character such that the estimate of model form uncertainty should rapidly expand the estimated uncertainty of the simulation result. The mathematical model may actually have better predictive capability, i.e., smaller model form uncertainty, far outside the validation domain, but there is *no* direct experimental evidence that it does. For example, the physical processes and their coupling that occur within the validation domain can be significantly different far outside the validation domain. The mathematical model may, or may not, accurately characterize these processes. Unless there is some very convincing physics-based evidence of model accuracy far outside the validation domain, it is appropriate that the estimated model form uncertainty should be large.

Most Bayesian statisticians have not addressed the issue of extrapolation of the model discrepancy term because the focus of the popular KOH approach is calibrating all uncertain quantities, conditional on the mathematical model and the available experimental measurements. As commented by Brynjarsdottir and O'Hagan (2014), "Although it [estimation of model discrepancy] can give accurate interpolation, it does not enable learning about physical parameters or extrapolation." A recent creative Bayesian approach to the issue of extrapolation beyond the validation domain is to merge the KOH approach with the simulation results from higher-fidelity physics-based models. The approach uses simulation results from more reliable, more computationally expensive, physics-based models as if the simulation results were from experimental measurements. The argument is made that the results from the higher-fidelity physics-based model can be used as a surrogate for experimental measurements, thereby producing better coverage of the intended use domain for the system of interest. Whether this approach proves to be beneficial in practical applications depends entirely on the accuracy of the higher-fidelity, physics-based model. This approach can also be used with the MAVM outside the validation domain. However, as explained in the detailed discussion of terminology in Chapter 2, this approach would be referred to as a calibration procedure for model form uncertainty, *not* a validation procedure because there is no actual empirical evidence. For a more detailed discussion of multi-fidelity modeling from a Bayesian calibration perspective, see Park et al. (2017), Wang, Roy, et al. (2018), and Jiang et al. (2020).

A final comment should be made on the general topic of extrapolation in modeling and simulation. In the foregoing, the discussion has focused on the extrapolation of an uncertainty, specifically, model form uncertainty. However, the much more common example of extrapolation is the use of the mathematical model to make predictions for conditions *not previously observed, perhaps never observed.* For physics-based models, this type of extrapolation can be viewed as constrained by the principles of conservation of mass, momentum, and energy, and any other physics-based principles embedded in the model and submodels. For non-physics-based models, i.e., statistically based models, extrapolation would entail a much higher level of uncertainty because there are *no physical constraints* on the processes occurring during extrapolation. Two examples of statistical models are: (1) regression fits of experimental data over the validation domain and (2) models based on statistical inference from observed experimental data, such as machine learning and data mining. An interesting mixture of physics-based and statistical models are (a) nondimensional input scaling techniques that have been used for decades in nuclear reactor engineering (Bestion et al., 2017) and (b) physics-constrained machine learning models (Zhu et al., 2019; Geneva and Zabaras, 2020; Sun et al., 2020).

13.5.4 Example Problem: Step 5 for a Heat Transfer Analysis

The assumptions and approximations in the formulation of the mathematical model for the heat transfer example, as discussed in Section 13.2.4, result in model form uncertainty. The model form uncertainty is estimated using available experimental data and the modified area validation metric (MAVM) discussed earlier. Experimental measurements were made of the temperature inside four different buildings in different locations. Since the analysis and the model form uncertainty are only of interest for one external air temperature, 40°C, all four of the model

Table 13.3 *Temperature measurements in the model validation experiments*

Validation experiment	Temperature inside the building, T_0(°C)
1	58
2	54
3	59
4	56

validation measurements were conducted for this condition. As discussed in Section 13.5.3, if the validation experiments could not have been made for a temperature of 40°C, then the regression fit procedure of model form uncertainty would be needed. All of the model input data needed to compute a simulation result for these four experiments were characterized by the data given previously in Table 13.1, except for the convective heat transfer coefficient on the outside surface of the building, h. As discussed earlier regarding normal atmospheric conditions around a building, h is an epistemic uncertainty given by the interval [7, 10] W/m^2-°C. In order to have a precise value (as opposed to an interval-valued quantity) of model form uncertainty, h must be controlled and estimated (or measured) in the validation experiment. To achieve this, low-speed fans were situated around each building to generate forced air convection cooling on the outside of the building. A simple handheld anemometer was used to measure the wind speed produced by the fans near each wall of the building. Using the heat transfer correlations for turbulent flow over a flat plate (Holman, 2009) and the conditions measured during each validation experiment, the estimated value of h was 40 W/m^2-°C.

Each model validation experiment was conducted such that the air temperature inside the building attained a steady-state value. Furthermore, it was ensured that the measured temperature of each wall surface inside the building, T_0, was also at a steady-state temperature. Table 13.3 gives the temperature measurements of the inside wall in each of the four buildings for the four experiments. The experimental measurement uncertainty of T_0 is assumed to be negligible. The variability observed in the experimental measurements shown in Table 13.3 is due to the variability in all of the model input data, as well as those factors that were ignored in the mathematical modeling.

Using the model input data given in Table 13.1, *except* using $h = 40$ W/m^2-°C for the model validation experiments, nondeterministic simulation results were computed using the procedure discussed in Section 13.4.8. Although several SRQs are computed for this simulation, the important result for the model validation experiment is the temperature of the interior wall of the building, T_0. Figure 13.20 shows the CDF of the simulation result using 100 Monte Carlo samples of k and the fixed value of $h = 40$ W/m^2-°C. Also shown are the experimental measurements of T_0 from the four experiments, as well as the ±90% confidence bounds on the measurement. It is noted that the simulation stays within the confidence bounds of the experiment, except for the high temperature tail of the simulation. Using the procedure for the MAVM described in Section 13.5.2, the interval-valued model form uncertainty is calculated to be $\left[d_{0.90}^{-} = 3.139, d_{0.90}^{+} = 2.096\right]$ °C. Since $d_{0.90}^{-}$ is larger than $d_{0.90}^{+}$, the indication is that the simulation is predicting interior temperatures that are slightly higher than what are measured. However, Figure 13.20 shows that the largest contributor to the magnitude of the MAVM is that there are only four experimental samples.

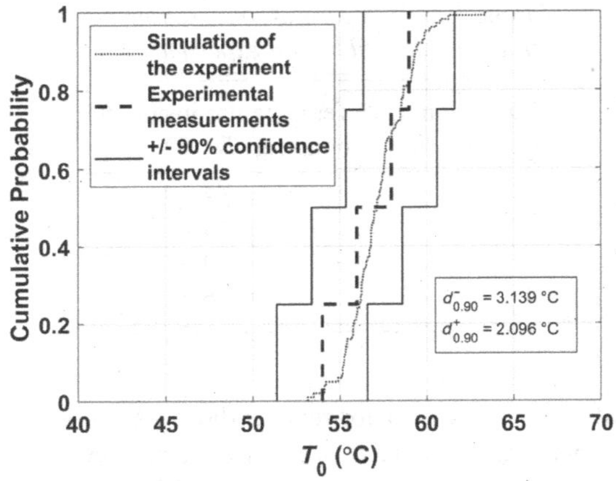

Figure 13.20 Nondeterministic simulation result, four model validation experimental measurements for T_0, and the ±90% confidence bounds on the measurements.

13.6 Step 6: Estimate Total Uncertainty in the Predicted System Responses

Previous sections have discussed how model input uncertainties are propagated through the model, how various numerical solution errors are estimated, and how model form uncertainty is estimated. The topic of this section is the question: "How should these sources of uncertainty be combined to estimate the total simulation uncertainty?" Since the groundbreaking work of Kennedy and O'Hagan (2001), significant progress has been made toward estimating model form uncertainty using an inverse solution procedure. However, combining numerical solution uncertainty with model input uncertainty and model form uncertainty is rarely addressed in the research literature or even in high-consequence simulations. The reason is that numerical solution uncertainty does not produce a purely random effect on the SRQs; instead, it primarily produces a systematic uncertainty with some level of randomness. As a result, numerical solution uncertainty is quite difficult to include, in the sense of explicitly estimating its impact on total simulation uncertainty, if traditional probability approaches are used. The impact of numerical solution uncertainty on the simulation results is very similar to that occurring in experimental measurements where both systematic error and random measurement variability exist together.

If the KOH approach is used, the numerical solution error (and numerical error from surrogate models if they are used) are convolved with the updating of the model input parameters and the estimation of the model discrepancy, $\delta(\vec{x})$. Although numerical solution uncertainty is *not* conceptually part of model form uncertainty, it tends to have more impact on the estimation of model form uncertainty because its primary feature is that of a systematic uncertainty in simulation results.

Because of the difficulty of explicitly including the impact of numerical solution uncertainty on the total simulation uncertainty, two practices are common. First, numerical solution uncertainty is ignored in the sense that its individual components are not even estimated. Second, each component of numerical solution uncertainty is quantitatively estimated, and it is shown that it has a negligible impact on all of the SRQs of interest. The latter practice is clearly preferable, but unfortunately, the first practice is more common.

During the last two decades, imprecise probability theory has been used to explicitly include both numerical solution uncertainty and model form uncertainty in the total simulation uncertainty. The reason this is possible in imprecise probability theory is that an explicit distinction is made between aleatory and epistemic uncertainty. The particular imprecise probability theory used in this text is probability bounds analysis. This section discussed how numerical solution uncertainty and model form uncertainty are included in total simulation uncertainty using p-boxes. For a comprehensive review of several approaches to the estimation of total predictive uncertainty, see Riedmaier et al. (2020).

13.6.1 Inclusion of Numerical Solution Uncertainty

As discussed earlier in this chapter, the estimated numerical error in the solution of the PDEs is an epistemic uncertainty, $(U_N)_{y_i}$, for each SRQ y_i for $i = 1, 2, \ldots n$. It is composed of three contributors and written as given in Equation (13.16)

$$(U_N)_{y_i} = |U_I|_{y_i} + |U_T|_{y_i} + |U_S|_{y_i} \quad \text{for } i = 1, 2, \ldots, n, \tag{13.25}$$

where U_I is the iterative solution uncertainty, U_T is the temporal discretization uncertainty, and U_S is the spatial discretization uncertainty. For completeness, equations given earlier for these three contributors are repeated here.

$$\begin{aligned}
(U_I)_{y_i} &= \sup\left\{\left[U_I\left(\vec{x}\right)\right]_{y_i} : \vec{x} \in \textit{input sample space}\right\} \quad \text{for } i = 1, 2, \ldots, n, \\
(U_T)_{y_i} &= \sup\left\{\left[U_T\left(\vec{x}\right)\right]_{y_i} : \vec{x} \in \textit{input sample space}\right\} \quad \text{for } i = 1, 2, \ldots, n, \\
(U_S)_{y_i} &= \sup\left\{\left[U_S\left(\vec{x}\right)\right]_{y_i} : \vec{x} \in \textit{input sample space}\right\} \quad \text{for } i = 1, 2, \ldots, n.
\end{aligned} \tag{13.26}$$

As can be seen from these equations, the supremum bound is computed over the entire sample space of the uncertain model inputs, $\vec{x}$. These three numerical uncertainties, $(U_I)_{y_i}$, $(U_T)_{y_i}$, and $(U_S)_{y_i}$, are coupled with the statistical sampling uncertainty due to propagating the uncertain model inputs through the PDEs. This coupling creates significant complexity in monitoring and quantitatively estimating each of these four contributing uncertainties. Practical methods to deal with these complexities were discussed in Section 13.4.

Although statistical sampling of input uncertainties to obtain output uncertainties generates a numerical error, it is not explicitly included in Equation (13.25) for two reasons. First, sampling input uncertainties is not part of the numerical solution of the mathematical model given by PDEs. As discussed in Chapter 3 "Modeling and Computational Simulation," sampling uncertainty is due to the approximate numerical solution of the nondeterministic simulation. Second, sampling input uncertainty is commonly controlled and estimated independently from the iterative, temporal, and spatial discretization errors. The most common procedure for propagating input uncertainties through the mathematical model is

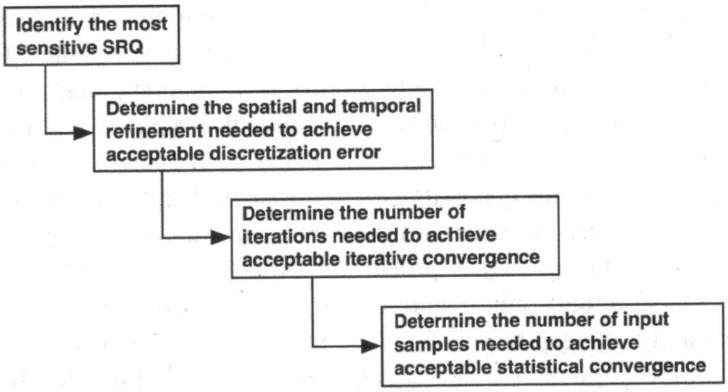

Figure 13.21 Recommended procedure for decoupling the estimation of the four sources of numerical uncertainty.

to place the sampling procedure as a wrapper or external numerical procedure *outside* of the numerical solution procedure for the PDEs.

Before discussing how the total numerical solution uncertainty, $(U_N)_{y_i}$, is incorporated into the SRQs, a procedure is recommended to partially decouple the estimation of the three numerical uncertainties shown in Equation (13.25), as well as the sampling uncertainty discussed in Section 13.4. Figure 13.21 shows a sequence of four steps that can be used to decouple the estimation of the four sources of numerical solution error. Note that the three lower elements in the figure mention *achieving acceptable error* for each of the four sources of numerical error. As discussed in Section 13.4, each of the error estimates is treated as a contributing uncertainty to the total simulation uncertainty. There is a wide range of sensitivity of SRQs to any source of numerical error or input uncertainty. The most sensitive SRQ should be identified, which is usually known from the nature of the simulation being performed. For example, it is well known that local solution quantities of interest, such as local temperature or local shear stress, are much more sensitive than global quantities of interest, such as total force of a fluid on an object or total strain energy in a structure.

Spatial and temporal discretization error is the first source of numerical error that should be addressed because it is commonly the most computationally demanding error to control and estimate. For spatial and temporal discretization error and iterative error, a quantitative error estimate of each source should be computed for the most sensitive SRQ. Even if a quantitative estimate of these errors proves computationally expensive, or even inaccurate, some type of quantitative estimate should be computed. Resorting to sensitivity analyses for these error sources has been used in some simulations, but this approach can be very misleading to the computational analyst and the customer of the simulation. Finally, the number of input samples should be set so that it is believed that acceptable statistical convergence of the most sensitive SRQ is achieved. Statistical sampling is placed last in the sequence because its convergence history is typically the most well behaved of the four sources. For statistical convergence, the most important question is: "What statistical feature of the SRQ is of interest?" For example, an extremely large number of Monte Carlo samples is required to converge the predicted likelihood of a rare event, as opposed to the number of samples needed for predicting the median response.

There is no general recommendation regarding how to decide on an acceptable magnitude for each of the four error sources in Figure 13.21. The two most common considerations that influence this decision are the following. First, the computational resources available and the consequences of inaccurate simulation results, because of large numerical errors, are always important factors. Second, estimators for discretization error are commonly inaccurate unless the asymptotic region of convergence has been achieved. As a result, achieving the asymptotic region for both spatial and temporal discretization error usually determines the requirement for accuracy. Using estimates of $(U_I)_{y_i}$, $(U_T)_{y_i}$, and $(U_S)_{y_i}$, Equation (13.25) can be used to compute $(U_N)_{y_i}$. Similar to the method for increasing the uncertainty of the simulation result due to model form uncertainty, as discussed in Sections 13.5.2 and 13.5.3, $(U_N)_{y_i}$ is used to generate a p-box for each y_i. This is written as

$$\left[Y_i(X) - (U_N)_{Y_i},\ Y_i(X) + (U_N)_{Y_i}\right] \quad \text{for } i = 1, 2, \ldots, n. \tag{13.27}$$

The term $(U_N)_{Y_i}$ equally expands the left and right boundaries of the CDF of $Y_i(X)$ to obtain a p-box. Note that the dimensional units of $(U_N)_{Y_i}$ are always consistent with the units of Y_i.

This method of approximating the impact of numerical solution uncertainty on the SRQs is *not* rigorous because of the coupling of the four error sources in the numerical solution. It is a simple, heuristic method that asserts that the true SRQ is likely to be contained within the p-box computed in Equation (13.27). The computed p-box does not claim to be a bound on the impact of the numerical solution error on the nondeterministic SRQs. A mathematical proof of *any* method attempting to compute a bound on the envelope of all possible CDFs for an arbitrary system of nonlinear PDEs and for arbitrary model input uncertainties is unlikely in the foreseeable future. The present method of taking the absolute value of the supremum bound over the model input sample space of each of the three numerical solution uncertainty contributors, as given in Equation (13.26), and generating the p-box as given in Equation (13.27), is a reasonable estimate of the impact of numerical error on each of the SRQs of interest.

Example 10

Consider the impact of large numerical uncertainty in computing a simulation result and how this could influence simulation-informed decision making. The center CDF in Figure 13.22 shows the SRQ obtained when numerical solution uncertainty is ignored, referred to as SRQ_C. Let U_N be the estimated total numerical solution uncertainty in SRQ_C using Equation (13.25). As can be seen from Figure 13.22, $U_N = 1$ in this example. Using the procedure just described and given in Equation (13.27), the CDF of SRQ_C is displaced equally to the left and right by the amount of unity. Since numerical uncertainty is an epistemic uncertainty, and only the magnitude of the total uncertainty is estimated, equal displacement of SRQ_C, both left and right, is warranted. This produces the p-box shown in Figure 13.22, which is bounded by both solid curves. That is, the p-box asserts that all possible CDFs due to combinations of the various sources of numerical solution uncertainty are enclosed within the p-box.

The interpretation of the impact of the numerical uncertainty on the simulation results shown in Figure 13.22 can be understood in two ways. First, the simulation result, including the numerical uncertainty, is given as the interval-valued p-box $[SRQ_C - 1, SRQ_C + 1]$. That is, at any specified value of probability, the SRQ will be less than, or equal to, $SRQ_C + 1$, given the present state of

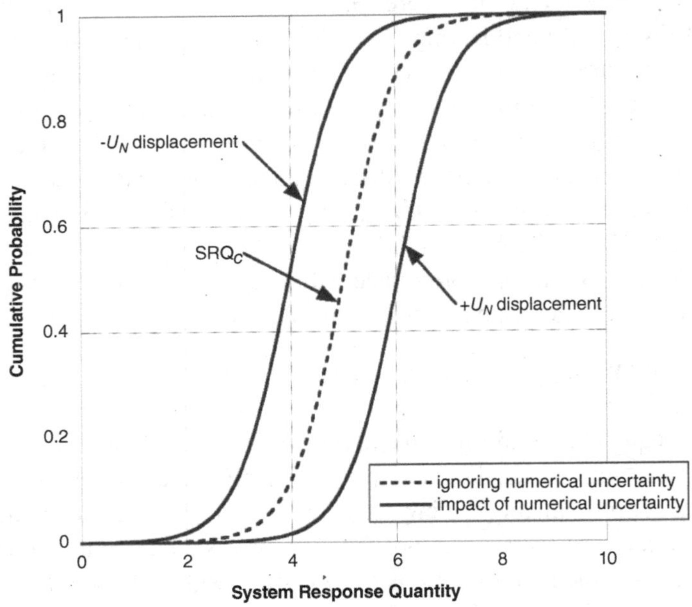

Figure 13.22 Example of the impact of numerical uncertainty in a nondeterministic simulation result using p-boxes.

knowledge of the numerical solution uncertainty. Second, for a given value of the SRQ, the probability of that value (or less) occurring is an interval-valued probability. For example, the likelihood of an SRQ ≤ 6 is the interval-valued probability [0.48, 0.98], given our present state of knowledge of the numerical uncertainty.

Consider the results shown in Figure 13.22 in the context of simulation-informed decision making. Suppose the SRQ shown is the predicted maximum internal pressure generated by a new product that is being designed. The safety specification on the product is that the internal pressure will not exceed a value of 8. Given the simulation results shown in Figure 13.22, the probability that the product will produce an internal pressure of 8 or less, can be as low as 0.9806, but as high as 0.9997. The probability of 0.9806 not achieving the safety requirement is deemed unacceptable to the project manager. Given the constraints of time and resources available, the manager has two options to remedy this situation. First, the design can be changed such that the maximum internal pressure is reduced, but the redesign time and cost is significant. Second, increased computing resources can be obtained such that the numerical solution uncertainty is reduced using a finer spatial discretization and iterative error convergence criterion, while keeping the system design and mathematical model the same. The project manager could be further informed which of the two sources of numerical uncertainty is most damaging to the predictive capability of the simulation. With the reduction in numerical solution uncertainty, a more precise probability interval can be obtained for the maximum predicted pressure of 8. The reduced numerical uncertainty will reduce the lack of knowledge due to numerical uncertainty, but it will not necessarily resolve the safety decision for the design.

13.6.2 Inclusion of Model Form Uncertainty

As discussed in Sections 13.5.2 and 13.5.3, the MAVM method is used to include the model form uncertainty. To estimate the model form uncertainty both within and outside the validation domain, a first- or second-degree polynomial regression fit is computed using the experimental set points within the validation domain. Using the concept of a p-box to include the model form uncertainty, we have

$$\left[Y_i(X) - \left(d_{CL,i}^{-}\right)_{CL\oplus}^{+},\ Y_i(X) + \left(d_{CL,i}^{+}\right)_{CL\oplus}^{+}\right] \quad \text{for } i = 1, 2, \ldots, n. \tag{13.28}$$

The term $Y_i(x)$ is the CDF from propagating model input uncertainties through the mathematical model to obtain the uncertainty in the ith SRQ. The term $d_{CL,\ i}^{-}$ is the magnitude of the difference between the simulation and the CDF of the lower confidence interval of the experimental measurements for the ith SRQ. The term $d_{CL,\ i}^{+}$ is the magnitude of the difference between the simulation and the CDF of the upper-confidence interval of the experimental measurements for the the ith SRQ. The magnitude of displacement of the lower and upper CDFs is determined by the choice of the statistical confidence level desired, CL, the student t-distribution, the number of experimental measurements at a given set point, and the mean and variance of the experimental measurements at a given set point. The term $\left(d_{CL,i}^{-}\right)_{CL\oplus}^{+}$ is the upper bound of the prediction interval associated with the regression fit of $d_{CL,i}^{-}$. The term $\left(d_{CL,i}^{+}\right)_{CL\oplus}^{+}$ is the upper bound of the prediction interval associated with the regression fit of $d_{CL,i}^{+}$. Note that CL and $CL^{\oplus}$ can each represent different levels of statistical confidence.

For two different situations, $Y_i(X)$ can itself be a p-box. First, if any model input quantity is an interval, i.e., an epistemic uncertainty, then $Y_i(X)$ would be a p-box. As discussed in Section 13.4.4, interval-valued inputs can occur in either model input data x_i , $i = 1, 2, 3, \ldots,$ m, associated with the system or surroundings or as an interval-valued hyperparameter that is associated with a probability distribution. This latter case is referred to as a mixed aleatory and epistemic input uncertainty and is commonly given as a parameterized p-box. Note that model input uncertainty and numerical solution uncertainty are *not extrapolated* to the simulation conditions of interest, as is model form uncertainty.

Second, if an estimate of the numerical solution uncertainty is included in the simulation result, then this will also generate a p-box. This procedure was discussed in the previous section and given by Equation (13.27). If both an interval-valued uncertainty exists in any model input, and numerical solution uncertainty is included in the simulation result, then there will be two displacements to construct the outermost p-box. The innermost p-box is always that which is due to the propagation of the uncertain input quantities through the mathematical model to obtain an envelope of all sampled distributions. This sampling and envelope procedure was discussed in Section 13.4.4. The next outer p-box shown is that which is due to numerical solution uncertainty, as discussed in the previous section. The outermost p-box shown is that which is due to model form uncertainty, which is appended to the numerical solution uncertainty p-box.

Example 11

Consider the impact of large model form uncertainty, in addition to that due to numerical solution uncertainty. Assume that only aleatory model input uncertainties exist and that these are propagated through the mathematical model. The innermost CDF in Figure 13.23 is the response of the system due to purely aleatory input uncertainties. The pair of CDFs surrounding the innermost CDF characterize the impact of the numerical solution uncertainty on the simulation. These two CDFs form the p-box due to explicitly accounting for numerical solution uncertainty. The next outer p-box is that which is due to the model form uncertainty using the MAVM. It is seen that the outermost envelope on the right-hand side of the p-box is displaced from the numerical solution uncertainty p-box by a larger amount than the displacement of the left p-box due to model form uncertainty. The asymmetric displacement of the p-box due to model form uncertainty is the most common situation because the comparison of the simulation with the experimental measurements is typically greater in one direction than the other.

As can be seen in this example, the combined impact of numerical solution uncertainty and model form uncertainty are crucial to correctly interpreting the results of the simulation. If the numerical solution uncertainty and the model form uncertainty are ignored, the customer's conclusion for the simulation is that it is more accurate than it actually is. For example, without explicitly accounting for numerical solution uncertainty and model form uncertainty, the probability of the SRQ = 4 or less, is 0.93. The impact of the numerical solution uncertainty increases the uncertainty such that the probability of the SRQ = 4 or less, is [0.74, 0.99]. The combined impact of numerical solution uncertainty and model form uncertainty for SRQ = 4 or less increases the uncertainty to [0.075, 0.995].

When such a large range of possible probabilities are presented to a decision maker, sometimes the response is, "The extremely large range of probability says that almost anything can happen!" Or, "With the stated range of probability, the simulation is uninformative." Our response to these comments is, "Given our present state of knowledge, the simulation can provide nothing more

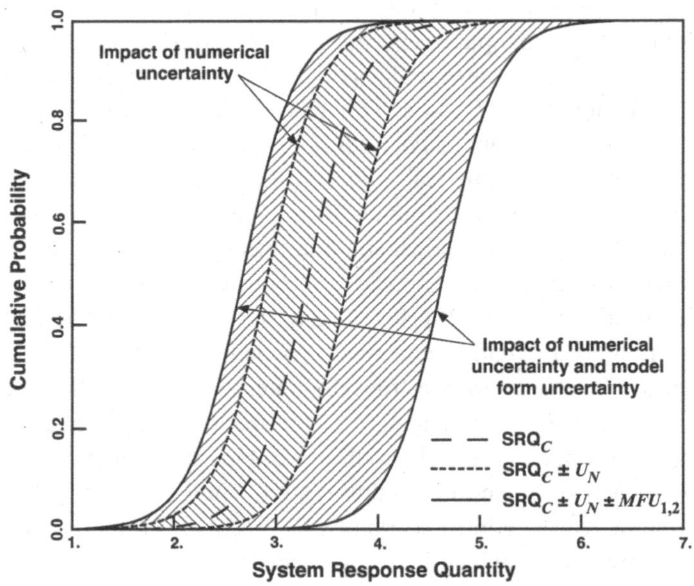

Figure 13.23 Example of the impact of both numerical uncertainty and model form uncertainty using p-boxes and characterized as a random variable.

precise than what is shown." Or, "The simulation clearly informs us that there is a very large uncertainty in the predicted system response." What *is clear* from the results shown in Figure 13.23 is that the largest contributor to predictive uncertainty is model form uncertainty. As a result, the most effective method of reducing uncertainty in the system response is to reduce this contributor. This could be done in one of two ways. First, if the specific condition of interest in the simulation is within the validation domain, then the accuracy of the model needs to be improved. Second, if the condition of interest in the simulation is outside the validation domain, then additional validation experiments should be conducted nearer to the conditions of interest for the simulation. If these experiments are not possible, alternative approaches are discussed in the next section.

13.6.3 Use of Alternative Mathematical Models

As pointed out in previous sections, the concept of extrapolating model form uncertainty based on observations within the validation domain produces large estimates of uncertainty when the simulation conditions of interest are far from the validation domain. Specifically, using any type of regression function combined with a prediction interval will dramatically increase the estimated model form uncertainty far from the validation domain. This same point can be made from a different perspective. Suppose the validation domain is a sphere in three-dimensional model input space. Assume that the intended use domain is the region outside the sphere, but *inside* a cube that circumscribes the sphere. As the dimensionality of the input space increases, suppose the validation domain remains spherical, typically referred to as a hypersphere. The Curse of Dimensionality forces the unpleasant realization that the ratio of the volume of the hypersphere to the volume of the hypercube that circumscribes it approaches *zero* as the dimensionality of the space increases. Therefore, as the dimensionality of the input space increases, the size of the validation domain becomes only a speck in the high-dimensional input space.

An approach to obviating large extrapolations of model form uncertainty was mentioned in Section 13.5.3 as part of Bayesian estimation. Multi-fidelity modeling, also referred to as hierarchical modeling, has proven effective when the simulation conditions of interest are very different from the available experimental data. Higher-fidelity models are used at the simulation conditions of interest to calibrate both model form uncertainty and model input parameters associated with the lower-fidelity model. For example, certain physics coupling that was not revealed in the validation domain, but is dominant at the simulation conditions of interest, *and is captured in the high-fidelity model*, can then be used to improve the calibration of model input parameters and model form uncertainty associated with the low-fidelity model. Multi-fidelity modeling can also be used with the MAVM; not in the mode of calibration of a low-fidelity model, but as a method to better estimate model form uncertainty far from the experimental data. This would remove the need for a large extrapolation of the model form uncertainty using a regression function and the associated prediction intervals. It is stressed that using multi-fidelity modeling with the MAVM approach requires a less demanding assumption than the Bayesian estimation approach because the MAVM approach does *not* assume that the multi-fidelity model is physically correct; only that it provides an improved estimate of the model form uncertainty.

A closely related approach is called multiscale-physics modeling (Berendsen, 2007; Bucalem and Bathe, 2011; Steinhauser, 2017). Here, the emphasis is directly on the use of distinctly different spatial and temporal scales of physics, for example, using microscale, mesoscale, and macroscale models. This approach should be effective in discovering how physical processes that are mathematically modeled at differing spatial and temporal scales can be used to extend the range of predictive capability. However, this approach is proving difficult to implement in practice because of the inconsistencies in assumptions between the different physical scales of each model. For example, there is a fundamental inconsistency between attempting to relate a mathematical model that assumes homogeneous and isotropic material modeling at the mesoscale to the assumptions of a mathematical model that addresses crystal and grain structure at the microscale.

When the simulation conditions of interest are strikingly different than the validation conditions, then a completely different modeling approach to estimating total simulation uncertainty is required. This situation most commonly occurs when different system environments and scenarios are considered. For example, as discussed in Chapter 3, experimental data may be available for the system under normal operating environments, but the simulation conditions of interest are for abnormal environments, such as mechanical damage to the system, loss of cooling fluid in the system, or the system being exposed to a lightning strike. For these types of modeling situations, very different conceptual and mathematical models are required. These types of models are referred to as *alternative plausible models* or the *method of competing models*. The approach is conceptually simple, but simulation analysts and project managers rarely use it because it is expensive and time consuming to develop radically different models of the system of interest. Only in simulations of very high-consequence systems, or where a system failure can last generations, have alternative models been developed and used. Typically, this magnitude of investment is only committed for simulations dealing with public safety or public policy at the national and international level. Three examples are nuclear power reactors (NRC, 2009), long-term underground storage of nuclear waste (Helton et al., 1995), and global warming (Pennell and Reichler, 2011; Frigg et al., 2014; Annan and Hargreaves, 2017; Knutti et al., 2019).

To use the approach of alternative plausible models, an analyst should have at least two independently developed models of the system and then a comparison made of the predictions from each model. In this approach, *none* of the models are presumed to be correct; each one is simply a postulated representation of reality. The only presumption is that each of the models is scientifically and logically defensible for the task at hand. This approach does *not* provide an estimate of total simulation uncertainty. It only provides an indication of the similar or dissimilar nature of each model prediction. Comparing results from alternative models should be thought of as a sensitivity analysis, with respect to conceptual modeling uncertainty, as opposed to an estimation of simulation uncertainty. The most important issue in using alternative models is the *independence of the teams* and the formulations of the conceptual models, including assumptions dealing with the scenarios, event trees, fault trees, surroundings, and possible human interaction with the system. In comparing the simulation results from each team, it is very often found that each team will have thought of important aspects affecting the conceptual modeling that the other team did not consider. Then discussions, as well as vigorous debates, can begin and analyses can be reformulated, improved, and possibly corrected for further comparisons.

In the prediction of hurricane or typhoon paths, it is a welcome sight for the news media to show all of the paths predicted by the various hurricane models. Sometimes these are called *spaghetti plots* of hurricane paths. The results of each of the models should be presented to the decision makers and stakeholders involved, *not averaged or combined in any way.* Some argue that the results from each of the models should be considered as equally likely and treated as a random variable. This would be treating total simulation uncertainty as an aleatory uncertainty, as if computational simulation is a random process. Model averaging destroys the individual creativity of each team and removes information from the decision makers and stakeholders. For example, if residents of a coastal region see a wide range of predicted paths of a hurricane, they intuitively recognize that there is great uncertainty in the prediction at this time. If the paths of all the models are averaged and only one path is shown with a title such as, "Most likely path," the residents have a very different perception of whether to begin preparations for the storm.

13.6.4 Example Problem: Step 6 for a Heat Transfer Analysis

The final step of the heat transfer example is the calculation of the total prediction uncertainty for the SRQs of interest. This calculation combines all of the uncertainties that impact the SRQs of interest, i.e., model input uncertainty, model form uncertainty, and total numerical solution uncertainty. As discussed in Section 13.4.8, the iterative convergence error in the solution to the boundary value problem was controlled so that the absolute error in T_0 was less than 1E-3. As a result, this error can be ignored. Furthermore, the numerical integration error per step was controlled so that the relative error in both T_0 and dT/dx was less than 1E-6. As a result, the discretization error can be ignored.

Figure 13.24 shows the total uncertainty in the prediction of T_0 using PBA. The model input uncertainty is due to purely aleatory uncertainty in k and purely epistemic uncertainty in h. As discussed in Section 13.5.4, the model form uncertainty is estimated using the MAVM, which yielded $\left[d^-_{0.90} = 3.139, d^+_{0.90} = 2.096\right]$ °C. As explained in Section 13.6.2, the left boundary of the p-box due to model input uncertainty is displaced 3.139°C to the left, and right boundary of the p-box due to model input uncertainty is displaced 2.096°C to the right. It can be seen in Figure 13.24 that the model form uncertainty is a larger contributor to total uncertainty than the epistemic uncertainty due to h. As discussed in Section 13.5.4, the largest contributor to model form uncertainty is the small number of model validation experiments. This situation commonly occurs in estimating total

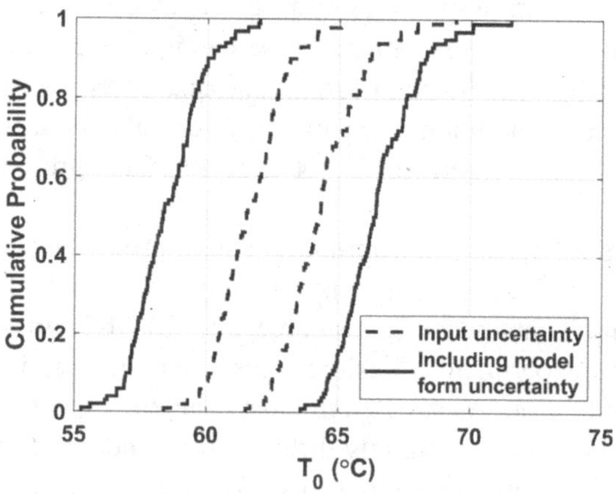

Figure 13.24 Total uncertainty in the prediction of T_0 for the heat transfer example.

predictive uncertainty using PBA because of the desire to minimize the number of model validation experiments.

When the computational analyst presents these results to the decision maker, the following interpretation of Figure 13.24 is given. The probability of the temperature inside the building being 70°C or less, could be as low as 0.97, but it could be as high as unity. Therefore, there is a high probability, given all of the uncertainties considered in the analysis, that the temperature inside the building will be 70°C or less on the hottest summer day. An additional method of presenting the results is to show the complementary CDF, or CCDF, also referred to as the exceedance probability. The CCDF is simply computed as 1 – CDF. Although not shown, the CCDF of the p-box shows that the probability of a temperature of 70°C or higher is *no greater* than 0.03. It can be concluded that this low probability of 70°C or higher is acceptable regarding the risk to electrical equipment and to maintenance workers inside the building.

If the decision maker asks the analyst what steps could be taken to reduce the uncertainty in the prediction, an appropriate response is to request that more model validation experiments be conducted. More experimental data would reduce the sampling uncertainty included in the MAVM. Attempting to reduce the epistemic uncertainty in h would not be productive because of the sensitivity of h to local atmospheric conditions and the surface roughness of the outside of the buildings, as well as the difficulty of measuring h in the field.

13.7 Step 7: Update Model Input Parameters

Model updating can take many different forms. For example, model updating could involve significant reformulation in scenarios, event trees, failure modes, changes in how the system interacts with the surroundings, changes in conceptual model assumptions, replacement of certain submodels that are performing poorly, and updating of model parameters. Although it is common practice to conduct different types of model updating, only updating of parameters that occurs *within* the mathematical model and submodels is discussed. That is, the model form or mathematical model structure does not change, but parameters that occur in the model are adjusted based on various types of new information that becomes available. These parameters could be part of the mathematical model for the system, such as geometry, initial conditions, and material properties, or part of the model for the surroundings, such as boundary conditions and system excitation. These parameters can be either a deterministic value (i.e., a scalar), or they can be nondeterministic quantities, such as a quantity given by a probability distribution. In addition, the parameters can also be a scalar field, a vector, or a tensor field, such as in the permeability tensor for underground transport of liquids and gases.

Updating parameters in models and submodels can improve the predictive capability of a simulation, and it is often a necessary activity in many simulations. Parameters are updated (a) when new model input data becomes available, and (b) when improvement in agreement between simulation and experiment is desired. This latter situation is commonly referred to as model parameter calibration. As stressed in most chapters, especially Chapter 12, the goal of model parameter calibration is distinctly different from model validation, which is model accuracy assessment. This section continues to stress the importance of keeping the concepts of parameter updating and model validation separate. However, there are shades of gray, or overlap, between some types of parameter updating versus model validation. The following section includes a discussion of the various types of model input parameters and how some are appropriate for updating, but not for calibration.

13.7.1 Types of Model Input Parameters

The various classes of parameters that occur in simulation activities have not been well developed in the literature. However, these classes need to be clarified in order to better understand which parameters should be updated, and which ones should not. Higdon et al. (2004) mention three important classes of parameters: (1) independently measurable parameters, (2) physically meaningful parameters that are not independently measurable, and (3) parameters that have little or no physical meaning. Morgan and Henrion (1990) discuss several classes of parameters that are quite informative for modeling, simulation, and risk assessment. Both of their basic classifications are used here, but certain important classes are divided into additional classes.

For simulation activities, model parameters are divided into the following six classes:

1. Measurable properties of the system or the surroundings,
2. Physical modeling parameters,
3. Ad hoc model parameters,
4. Numerical algorithm parameters,
5. Decision parameters, and
6. Parameters associated with a family of probability distributions.

Each of these classes is briefly discussed in this section, along with the type of information that is commonly available for updating each class of parameters.

(1) *Measurable properties* of the system or surroundings are physical quantities or characteristics that are independently measurable, at least in principle. Specifically, these are quantities that are measured *separately* from the model that is used to simulate the system of interest. Metrologists point out that measurements of empirical quantities always have some type of conceptual or mathematical model associated with the quantity to be measured (Rabinovich, 2005). Measurable properties of the system or surroundings generally rely on either (a) a simple, well-accepted model of the physical nature of the quantity or (b) some type of well-understood performance or reliability characteristic of the system or the surroundings. Several examples of system properties in this class are Young's modulus, tensile strength, mass density, electrical conductivity, dielectric constant, thermal conductivity, specific heat, phase change temperature, surface energy, chemical reaction rate, thrust of an engine, thermal efficiency of a power plant, and the failure rate of a valve in a system. Examples of properties of surroundings that are in this class are wind loading on a structure; external heating on a flight vehicle; atmospheric characteristics at a nuclear power plant that could exist at a given time; electrical, magnetic, and plasma characteristics of a lightning strike on a system; and characteristics of a physical or electronic attack on a system.

(2) *Physical modeling parameters* are those that are *not measurable* outside of the context of the mathematical model used to simulate the system under consideration. These

quantities are physically meaningful in the sense that they are associated with some type of well-accepted physical process occurring in the system. However, because of the complexity of the process or the interaction with other aspects of the system, the parameter *cannot be independently measured* separate from the system of interest. Hence, some type of physical process is known to occur in the system, but because of the complexity of the process, the physical effect of the process is represented as a simple mathematical model that *only exists* within the framework of the model of the system of interest. Quantifying these parameters relies on *mathematical inference* within the context of the assumptions and formulation of the model of the system. Examples of these types of parameters are (a) internal dynamic damping in a material, (b) aerodynamic damping of a complex structure, (c) stiffness and damping of assembled joints in a multielement structure, (d) effective chemical reaction rate in turbulent reacting flow, (e) effective surface roughness in turbulent flow along a surface, and (f) thermal contact and electrical resistance between material interfaces.

(3) *Ad hoc model parameters* are those that are introduced in models, or submodels, simply to provide a method for either (a) adjusting the results of a model to obtain better agreement with experimental data or (b) accommodating disparate features of submodels. These parameters have little or no physical justification in the sense that they do not characterize some specific feature of a physical process. They exist solely within the context of the model or submodel in which they are used. Some examples are (a) most parameters in fluid dynamic turbulence submodels, (b) most parameters in models for strain hardening in plastic deformation of materials, (c) most parameters in models for material fatigue, and (d) concocted parameters inserted into a complex model that are used solely to adjust model predictions to obtain agreement with experimental data. Sometimes ad hoc parameters are referred to as *kludging parameters* because there is no physical justification for the parameter (Lenhard and Winsberg, 2010; Lenhard, 2018).

(4) *Numerical algorithm parameters* are those that exist within the context of a numerical solution method that can be adjusted to meet the needs of a particular solution. These parameters typically have a recommended range of values, but they can be changed to give better performance, stability, or reliability of the numerical method. This *does not mean* discretization levels or iterative convergence levels, but quantities that typically appear in the formulation or execution of a numerical algorithm to make it work as intended. Some examples are (a) the relaxation factor in the successive over-relaxation method for the iterative solution of algebraic equations, (b) the number of iterations used on each level of a multigrid method, (c) artificial damping or dissipation parameters in numerical methods, and (d) parameters to control the hour-glassing phenomena in finite-element models for solids.

(5) *Decision parameters* are those that exist within an analysis and whose value is determined or controlled by the analyst, designer, or decision maker. Sometime these are referred to as system-operation variables, control variables, or policy-decision parameters. The question of whether a quantity is considered as a decision parameter or a measurable property of the system depends on the context and intent of the analysis. Decision parameters are commonly used in system design to optimize the performance, safety, or reliability of a system. Decision parameters are considered as changeable or adjustable, but they are not usually considered as uncertain in the sense of having

aleatory or epistemic uncertainty. In the analysis of abnormal or hostile environments, these parameters can be varied to determine the most vulnerable conditions that could affect the safety or security of the system. These parameters are also used, for example, in models constructed for aiding in public policy decisions, public health regulations, or environmental impact analyses. Some examples are (a) the value chosen for the maximum allowed operating speed of a vehicle with a feedback control system, (b) the minimum temperature allowed by an automatic control system before a manufacturing process can begin, and (c) the maximum concentration of specified pollutants allowed in public drinking water.

(6) *Parameters associated with a family of probability distributions* are those that are defined only in the context of being parameters of the family of probability distributions under consideration. These parameters are sometimes referred to as *hyperparameters*, and that term is used here. An example of hyperparameters would be the four parameters defined as part of a Beta distribution. A hyperparameter can be a scalar value, an interval, or a random variable. As a result, a hyperparameter can be either an adjustable value, an epistemic uncertainty, or an aleatory uncertainty. Hyperparameters can be used to characterize the uncertainty of any of the foregoing classes of parameters.

Of the six classes of parameters just discussed, numerical algorithm and decision parameters are not commonly viewed as being updated. They can certainly be changed and optimized during an analysis so as to better achieve the goals of the simulation. For example, numerical algorithm parameters are usually adjusted to obtain better numerical performance or stability for the various types of physics occurring in a given simulation. However, this type of adjustment of the parameters is not considered as updating because it does not fundamentally deal with the mathematical model itself. The remaining four classes of parameters are fundamental to the model, and they should be updated as new information becomes available.

13.7.2 Sources of New Information

New information to update a parameter can come in various forms. Four common sources of new information are:

- Experimental measurements of parameters,
- Expert opinion concerning parameters,
- Computational simulations or theoretical studies concerning parameters, and
- Inference from experimental measurements of the system of interest.

New experimental measurements on parameters can be combined with existing data, or the measurements can provide information on correlations for dependence on other quantities. Subject-matter experts can add new information based on experience with similar parameters or recommend certain families of distributions to characterize variability of the parameter. Independent computational simulations or theoretical analyses that are entirely focused on the parameter of interest can provide new information, e.g., the minimum and maximum range of the parameter. The first three sources of information provide information for updating model parameters that is *independent* from the system of interest. The last source of information, inference from the experimental responses of the system, provides

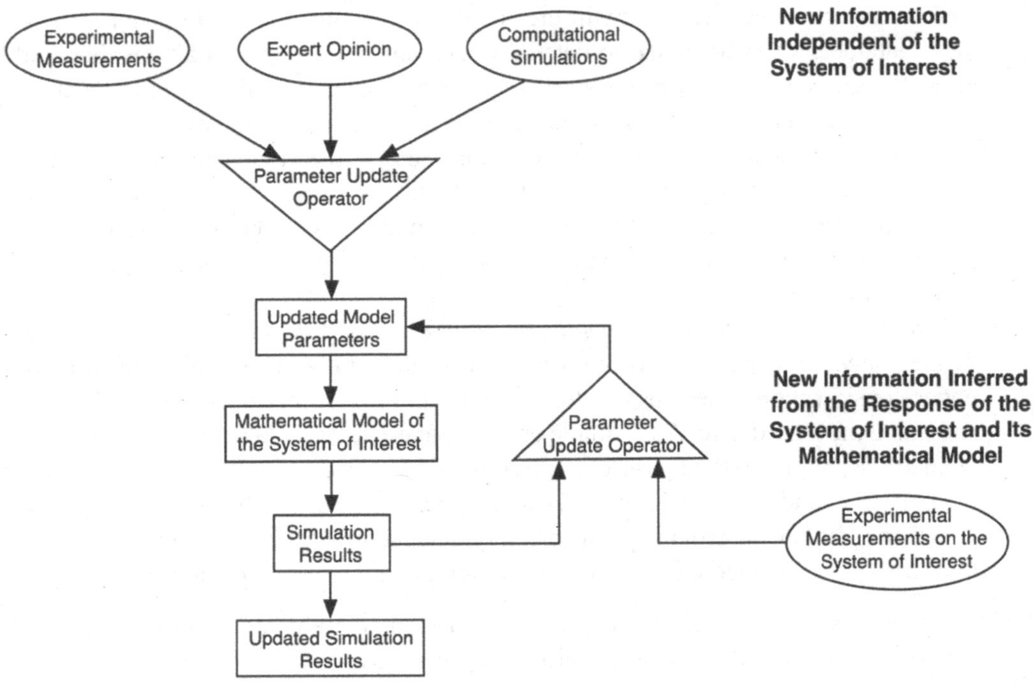

Figure 13.25 Independent and dependent sources of new information for parameter updating.

information that is *dependent* on the combination of the experimental responses of the system and the specific mathematical model of the system. Two examples follow. First, suppose two slightly different mathematical models of the system of interest are used, but the structure of each model, and even some of the parameters in each model, are the same. If the same experimental data for the system responses were used to update parameters in each model, the resulting inference of the same physical parameters in the two different models would be different. Second, suppose the same mathematical model of the system was used to update the model parameters. If there are different numerical solution errors in each simulation, the inferred parameters would be different.

The first three sources of information just mentioned are shown at the top of Figure 13.25. The parameter update operator used should depend on the characterization of each source of information. The characterization of the parameter can also be thought of as the mathematical model of the parameter. If the parameter is characterized as a scalar quantity, then various statistical Frequentist procedures can be used. These are types of optimization procedures such that new characterization minimizes the difference between the combined old and new information. If the parameter is characterized as a random variable, then the most common method of updating is Bayesian updating. If the new independent information comes from diverse sources, more general methods are needed for aggregating and merging the old and new information. For example, there may be limited experimental data on the parameter, but then subject-matter experts provide new information that is in conflict with the old experimental information. For a detailed discussion of update operators for fusing and assimilating diverse sources of information, see Mitchell (2012), Raol (2016), and Kedem et al. (2017).

After the parameters are updated based on new independent information, a new simulation for the system of interest can be computed. If experimental measurements for the response of the system become available, then an additional update of parameters can be computed. This new information is *entirely dependent* on the mathematical model of the system of interest, as well as the measured experimental responses of the system. There are various approaches to updating a parameter based on observations of the system of interest and the mathematical model of the system. If the parameter is a scalar, some type of optimization procedure could be used such that the response of the mathematical model with the updated parameters matches as closely as possible to the experimentally measured responses. If the parameter is a probability distribution, then the most common approach is Bayesian inference. No matter which approach is used, both require the solution of the inverse problem to infer the updated model parameters. For a detailed discussion of solution procedures for inverse problems, see Mueller and Siltanen (2012), Martin and Liu (2016), and Aster et al. (2019).

If Bayesian inference is used to update parameters, there are two key conceptual issues that arise. First is the question: "How broadly should the updating of different input parameters extend?" As discussed in Section 13.5.1, the quantities that are updated are the set of input parameters, θ_j for $j = 1, 2, \ldots, H$, where H is the number of input parameters to be updated. Within θ_j, there would be at least the physical model parameters and the ad hoc model parameters discussed above. Most Bayesian practitioners would also include the model form uncertainty, $\delta\left(\vec{x}_i\right)$, and the experimental measurement uncertainties, $\varepsilon\left(\vec{x}_i\right)$ for $i = 1, 2, \ldots, G$, where G is the total number of experimental measurements of the system of interest. Furthermore, some Bayesian practitioners would also choose to update the independently measurable properties of the system or the surroundings.

The perspective presented here is that the measurable input properties of the system should *not* be updated unless there is a strong physical argument that these parameters are somehow *different* than those that are independently measurable from the system of interest. The reason for this perspective is that Bayesian updating can improperly distort the posterior distribution of these parameters because the inverse problem must be solved. That is, Bayesian updating makes *no distinction* between independently measurable parameters and all other quantities that are updated. Bayesian inference *effectively convolves* all of the quantities that are updated, even though they appear in many different capacities within the complete mathematical model. For example, if any of the quantities to be updated is removed from the Bayesian updating process, and the updating is computed again, the posterior distributions for each of the remaining quantities will be different between the two updating procedures. The magnitude of the difference between the two updates will be contingent upon the prior distribution of each, the independence or identifiability of each of the quantities, and the amount of experimental data.

The second conceptual issue with Bayesian updating deals with the impact of numerical solution errors on the posterior distributions of the updated quantities. As discussed in Section 13.6, the iterative solution error, the temporal discretization error, and the spatial discretization error exist in every numerical solution. These errors are convolved with any quantities that are updated. Furthermore, because of the number of function evaluations that are required in Bayesian inference, the procedure is not practical unless surrogate or response-surface models are used. These mathematical models have their own numerical approximation errors. Each of these numerical errors fluidly moves among the others, as well as between each of the quantities that are updated. If each of these numerical errors is not explicitly estimated in a simulation, then the best that can be claimed is that they are insignificant. If a tough-minded decision maker is informed about this for a high-consequence, simulation-informed, decision, they may respond, "Your claim that the numerical errors are inconsequential is unsubstantiated, since no quantitative evidence has been presented. Either show me a sensitivity analysis regarding the impact of all of the numerical solution uncertainties or some procedure that explicitly indicates the impact of numerical solution uncertainties on the updated quantities and the simulation results." For a detailed discussion of statistical inference techniques, particularly Bayesian inference, see Gelman et al. (2013), Smith (2014), Martin and Liu (2016), McClarren (2018), and Santner et al. (2018). An alternative to Bayesian inference is *generalized fiducial inference*, which is showing significant advantages in certain applications (Hannig et al., 2016).

In this discussion, the focus has been on updating independently measurable parameters, physical modeling parameters, and ad hoc parameters that are characterized as random variables. A related question is: "If the parameters are epistemic uncertainties, e.g., intervals, or mixtures of aleatory and epistemic uncertainties, how can they be updated?" For example, suppose that an expert characterizes a system parameter as an interval. How should the parameter be updated or modified if new information becomes available for the parameter? Neither Frequentist nor traditional Bayesian approaches can update an interval-valued quantity because it characterizes the infinite set of probability distributions over the range of the interval. Recent work has derived a Generalized Bayes Rule for updating imprecise probabilities (Bernardini and Tonon, 2010; Augustin et al., 2014). The more general topic, usually referred to as information aggregation or data fusion, has also received increased attention during the last decade. The emphasis in this field has concentrated on dealing with contradictions and duplication of information from diverse sources. For an in-depth discussion of aggregation methods for diverse types of information using imprecise probabilities, see Beliakov et al. (2007), Torra and Narukawa (2007), Beliakov et al. (2015), and Raol (2016).

13.7.3 Parameter Updating, Validation, and Predictive Uncertainty

After parameters in a model have been updated based on new information, two related questions can be asked: "What is an authentic method for conducting model validation?" and "Has predictive uncertainty been reduced or predictive capability been improved?" The first question is directly related to the concept of assessment of model accuracy, whereas the second question is related to credibility and trustworthiness of a simulation used to foretell an outcome. The answer to both of these questions is critically important to understand because an understanding of model accuracy and model credibility is critically important to:

- Decision makers in assessing the safety, reliability, and performance of high-consequence systems or public projects;
- Government regulators in determining the public safety and potential environmental impacts of failure of high-consequence systems; and
- Policy makers using simulations to set governmental policies, environmental impact standards, and carbon offsets.

Most statisticians think of model validation in terms of the concept of *cross-validation*. Cross-validation, sometimes referred to as *out-of-sample testing*, is assessment of model accuracy relative to data that has *not been seen* by the model as part of the training set for model calibration. This is a reasonable approach to the question of model accuracy assessment, but then the question can be refined as: "How is new data defined?" From the perspective of cross-validation, the crucial assumption is that all the data that is used, whether for calibration or for model validation, must be drawn from the same *statistical population*. If only empirical or descriptive models are considered, such as regression models, machine learning, or data-mining models, this is a necessary assumption. Since this text is focused on validation of physics-based or explanatory models, the concept of a statistical population is inappropriate and unsuitable. For physics-based models, the concept of similarity or relatedness should be interpreted relative to whether the data is within, or outside, the validation domain of the system of interest. Two examples are given to describe the concept of similarity and dissimilarity of new data relative to the calibration data. The first example more directly addresses the issue of model validation, whereas the second example addresses the questions of predictive uncertainty and predictive capability. Both examples deal with the calibration and prediction of the vibrational modes of a multi-component built-up structure.

In the first example, the structure is assembled from a number of structural components that are bolted together at a large number of connection points. The structure is assembled with a given set of specifications for the preload torsion of all of the bolts used in the assembly. Fifty experiments on the fully assembled structure were conducted for diverse excitations of the structure. However, because of observations of the vibration of the structure, it was decided that the preload on all of the bolts used in the assembly of the structure should be doubled. Fifty additional experiments were conducted on the same structure using the higher torsional preload on each bolt. The mathematical model for the vibration of the structure includes physics-based submodels for the elastic deflection of the structural components, internal damping of the components, aerodynamic damping, and the stiffness and damping of the bolted joints of the structure. Parameters in the submodel for the bolted joints include the type of bolt and washers used and the torsional preload on the bolt. There are also a number of parameters in the bolted-joint submodel, as well as the internal and aerodynamic damping models, that need to be determined based on experimental data for vibration of the entire structure.

Given the 100 experimental measurements for the two somewhat different structures, the question is: "How should this set of data be partitioned between parameter-updating data and model validation data?" This question is set in the context of the upper and lower portions of Figure 13.26. Two diverse choices are: (1) choose 50 experiments that are randomly sampled from the set of 100 experiments and use the remaining data for the model validation data, and (2) use the 50 samples for the low bolt-preload condition for the parameter updating and use the 50 samples for the high bolt-preload for the validation data. If option (1) is used, then it is clear that model validation in the lower portion of Figure 13.26 would be a *very weak test* of model accuracy assessment. This option would follow the cross-validation strategy that the validation set of data is drawn from the same population as the training set that was used for parameter calibration. Thus, the model validation data is only unknown to the model in the sense of random variability in the various experiments used for calibration versus model validation.

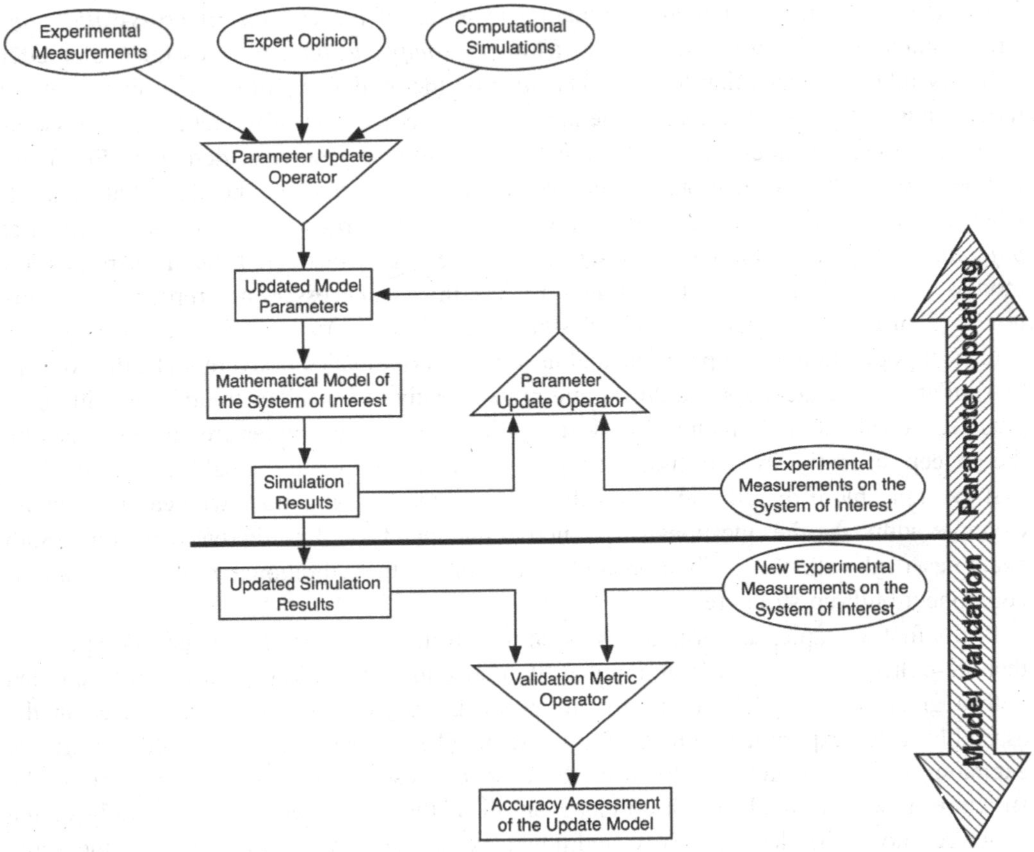

Figure 13.26 Parameter updating, model validation, and predictive uncertainty.

If option (2) is used, model validation would be a *very strong test* of model accuracy assessment. Specifically, this option would test the accuracy of the predictive capability of the submodel for stiffness and damping of the bolted joints, as well as the strong coupling of this submodel with the vibration of the entire structure. Therefore, this option would partition the validation domain into two disparate parts, one for the low-preload torque and one for the high-preload torque. Relative to Figure 13.26, the distinction between the data for the upper and lower portions of the figure is not meant to imply physics-based causality in the lower portion. The reason is that the updated model in the lower portion of the figure is heavily dependent on the upper portion of the figure because the upper portion relies on the inverse solution of the mathematical model.

Whether option (1) or option (2) is chosen, either the KOH validation approach or the MAVM validation approach can be used to compute a validation metric result. Whichever validation metric approach is used, it is obvious that *much better model accuracy* would be demonstrated for option (1) than option (2). Since this type of result is well known to experts in model calibration, a follow-up question can be asked: “Which approach would be used in various simulations?” Although it is impossible to answer in general, we recommend two perspectives that should be considered:

- What are the goals of the manager of the team who produced the simulation?
- What are the goals and expertise of the customer who will use the simulation results in a decision, presuming that the customer and manager of the simulation are different entities?

The second example concerning the vibration of the structure deals with the more complex question of predictive uncertainty and predictive capability of the model. Suppose that new requirements for the assembly of the structure require changing the bolted joints on the small components to riveted joints for assembly. Among the input options for submodels to fasten components for modeling the vibration the structure, there is a submodel available for riveted joints. Since no experimental data are available that are relevant to the present structure, it is decided to conduct several new vibration experiments on one small component of the complete structure using riveted joints. Predictions of the vibrational modes of the small component to be tested are made using the riveted joint option in the simulation package.

When these new experimental data on the small component become available, the data can be used in two very different ways relative to the prediction of the vibrational modes of the complete structure. The first method would be to update the parameters in the submodel for the stiffness and damping of the riveted joints, which would correspond to the top portion of Figure 13.26. Then use this updated submodel in the prediction of the vibrational modes of the entire structure. This approach would be referred to as a claim of *predictive capability* of the model for the entire structure. The second method would estimate the model form uncertainty computed, either using the KOH approach or the MAVM approach, in combination with the new experimental data on the small component. This model form uncertainty, in combination with the updated parameters in the submodel for the stiffness and damping of the riveted joints, would then be used in the prediction of the vibrational modes of the entire structure. Using this method, the total uncertainty would be greater, potentially much greater, than ignoring the model form uncertainty based on the vibrational data of the riveted component. This latter approach would be referred to as an estimate of the *predictive uncertainty* of the model for the entire structure.

As can be seen from these two examples, there is a very wide range of how parameters can be updated, how model accuracy can be assessed when experimental data are available, and how model form uncertainty can be used when a prediction extrapolates beyond the validation domain. As briefly discussed in Chapter 3, there is a conceptual distinction between empirical or descriptive models and physical law or explanatory models. Even if the conceptual model of a system is built directly on well-established physical laws of nature, the submodels or parameters in the submodels are built on experimental data, expert opinion, and other more detailed submodels. As a result, physics-based mathematical models are actually a combination of explanatory models *and* descriptive models. This means that physics-based models are not fully causal models, but a mixture of causal and correlation-based models. It is beyond the scope of this book to discuss this issue further, but important texts on the issue of causality include Pearl (2009), Morgan and Winship (2015), and Halpern (2019).

13.7.4 Bayesian Estimation vis-à-vis Probability Bounds Analysis

The Bayesian perspective of prediction is markedly different from the PBA perspective. Three central Bayesian principles are (1) all uncertainties are fundamentally dependent on individual subjective belief and they should be interpreted as subjective belief; (2) all uncertainties, whether due to randomness or lack of knowledge, are represented by subjective probability distributions; and (3) probability distributions are viewed as statements of subjective belief that should be updated as new information becomes available. Bayesians reason that the uncertainty results obtained using PBA will commonly yield such large uncertainties that the simulation results are of little value to decision makers. Therefore, the Bayesians reason that the resources spent on a PBA are wasted because the analysis commonly provides such a large range of uncertainty that the decision maker cannot make an informed decision based on the approach. They further emphasize that the time, resources, and effort spent on a PBA would be better spent on providing a *best estimated prediction* for the system of interest. Since the strategy of PBA is to represent lack of knowledge uncertainty as an interval, as opposed to representing the uncertainty as a probability distribution, it is evident that the uncertainty of the outcomes with PBA will be greater, commonly much greater, than a Bayesian approach. We argue that the Bayesian criticisms of PBA are without merit in most situations, and we give the following counterarguments.

First, PBA lucidly shows the decision makers how poorly the SRQs are predicted because of lack of knowledge. Thus, PBA does not bury or disguise any assumptions. If something is not known or is poorly known, it is simply included in the analysis as bounds on possible outcomes. Large uncertainty bounds in PBA usually cause a great deal of angst and sometimes frustration in the decision maker. It may even provoke them to the point of saying, "Why did we spend all of this money on modeling and simulation when you can't predict something any better than that?" One might give two responses: "The indicated uncertainty is the *smallest* that can be defended, given our present state of knowledge." Or, "Would you prefer that we make a number of questionable assumptions, mix frequency of occurrence with personal belief of likelihood, and then show much smaller uncertainties in the results?" Sometimes these responses don't seem to help the situation.

Second, in order to deal with the commonly large uncertainty predicted using PBA, a sensitivity analysis should also be provided with the PBA results. A sensitivity analysis tells the decision makers what the major contributors to the uncertainty in the SRQs of interest are. For example, if the p-box of the SRQ is very broad, then the SA can identify which epistemic uncertainties are the culprits. With this list of uncertainty contributors, the decision makers can determine which contributors they have control over and which ones they do not. For example, they may have significant control over the design of the system, but very little control over the environment or surroundings in which the system must operate. For those contributors that are under some control, a well-founded cost-benefit analysis can be conducted to determine the time, money, and resources needed to reduce certain contributors as compared to other large contributors. If the large contributors can then be reduced, a follow-up analysis can then be conducted so that the decision makers can see the impact on reducing the total uncertainty in the SRQs of interest.

As a final comment, we are not arguing that PBA is always the best approach for nondeterministic analyses. The goal of PBA is to present to the decision makers the total uncertainty of a simulation, given the present state of knowledge. This goal is aligned with the meaning of the term *predictive uncertainty* and the philosophy of "an abundance of caution." The goal of Bayesian estimation is to present to the decision maker the best estimate of uncertainty, given reasonable assumptions on the characterization of each uncertainty. In this context, lack of knowledge uncertainty is characterized as a uniform PDF over some reasonable range of the quantity. As a result, all uncertainties in the total uncertainty of an SRQ are *converted into variability or dispersion of the response*, as opposed to lack of knowledge. For the situation in which the decision maker is focused on the median of the outcomes (where half of the outcomes are less and half of the outcomes are greater) or on the mode of the outcomes (the most likely outcome), then a Bayesian approach can be the most informative or the most valuable to the decision maker. A recent, high-visibility example where the Bayesian approach was the most appropriate is the search for the wreckage of Malaysia Airlines Flight 370 (Davey et al., 2016).

13.8 Step 8: Conduct a Sensitivity Analysis

An SA is broadly defined as the determination of how a change in *any aspect* of the model changes the predicted responses of the system of interest (Morgan and Henrion, 1990). An SA is conducted for two fundamental reasons. First, an SA is conducted to improve the understanding of what factors most influence the simulation results and how the usability and credibility of the results might be assessed. Second, an SA is conducted to assist in planning future modeling, experimental, and computational activities, in addition to altering system design and to reduce uncertainty in system responses. The most common result of an SA is a prioritization of the factors that are changed according to magnitude of their impact on the response of the system. Figure 13.27 shows the common factors that are considered in an SA, as well as the computational sequence needed for an SA. The following are examples of common factors that could be changed in an SA, thereby producing a change in the response of the system: (a) assumptions in the conceptual model concerning which physical

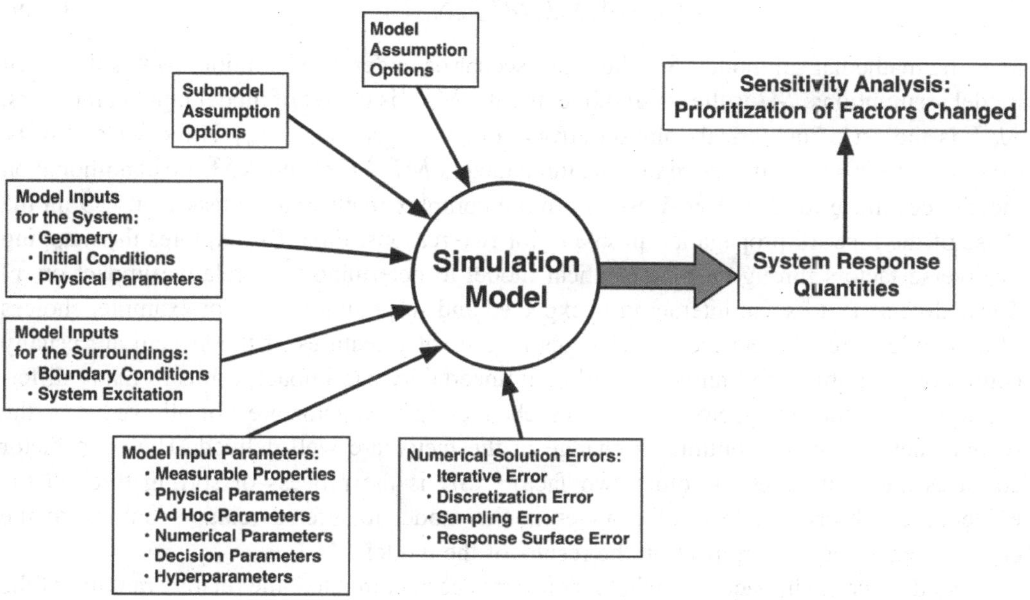

Figure 13.27 Common factors to consider in a sensitivity analysis.

interactions are included and which are excluded in the simulation; (b) choices of submodels for plastic deformation of a solid or choices for allowed chemical species in reacting flow; (c) geometrical features in the design of the system; (c) changes in boundary conditions on the system; (d) changes in the correlation structure for model input parameters; and (e) changes in spatial mesh resolution.

A UQ analysis differs from an SA in that the UQ analysis is the quantitative determination of the uncertainty in any SRQ as a function of any uncertainty in the simulation. Although UQ and SA are closely related, and they are commonly conducted together, they provide different information for the customer of the simulation. An SA is focused on the relative importance of the factors that are changed in the simulation result, as a function of the magnitude of change in the responses of interest. An SA can be viewed as a *what-if analysis*, i.e., an analysis to determine how the outputs change as a function of changes in any factors influencing the simulation. It is important to recognize that an SA can be conducted for *both* a deterministic simulation and a nondeterministic simulation. However, most SAs are conducted for nondeterministic simulations.

Excluded from Figure 13.27 are changes in the system environment (normal, abnormal, and hostile) because these simulations are usually very different in nature. Also excluded from the figure are changes in the scenario being analyzed, such as examining different branches of a fault tree or examining different actions taken by a human controller or computer controller of the system. Changes in scenarios in a simulation are commonly of interest, but they are not shown in the figure for simplicity. Note that in SA, the term *factors* includes *all of the possible changes in the simulation* indicated in Figure 13.27, in addition to changes in scenarios.

For a given system environment, the mapping of factors that can impact the SRQs can be written as

$$\mathbb{M}(S, \mathcal{MA}; \mathcal{MI}, \mathcal{MIP}, \mathcal{NSE}) \to \vec{y}. \tag{13.29}$$

$\mathbb{M}$ is the mathematical model, S is the set of scenarios under consideration, $\mathcal{MA}$ is the set of model assumptions, $\mathcal{MI}$ is the set of model inputs, $\mathcal{MIP}$ is the set of model input parameters, $\mathcal{NSE}$ is the set of numerical solution errors, and $\vec{y} = \{y_1, y_2, \ldots, y_n\}$ is the set of SRQs. Equation (13.29) is written such that the three factors $\mathcal{MI}$, $\mathcal{MIP}$, and $\mathcal{NSE}$ are conditional on the choices made for S and $\mathcal{MA}$. SA is a more complex mathematical task than UQ, in the sense of the forward propagation problem, for two reasons. First, SA explores the mapping of diverse factors through a mathematical model to determine their relative impact on $\vec{y}$. These diverse factors can interact in unexpected and surprising ways. For example, choices of submodels are discrete factors, changes in geometry features of the system are usually continuous variables, and numerical solution uncertainty can impact results in many different ways. For the first factor, continuous changes in the factor are not allowed. For the second factor, rates of continuous change in the factor are well defined. The third factor confuses and convolves the other two factors. SA is the process of sorting through the evidence of observed effects of changes in the model to infer a relative ranking of the changes regarding their impact on the results of the model.

Second, after the impact of the factors on $\vec{y}$ has been computed, the relative ranking of the factors changed can be problematic in certain situations. For example, suppose the two factors of interest that are changed are one submodel is used versus another, and the spatial discretization is changed in the numerical solution of the mathematical model. The impact of a change in these two factors on the elements of $\vec{y}$ can be computed and the magnitude of each change on $\vec{y}$ can be ordered. However, it is not clear how the magnitude of the change in one of these factors should be measured with regard to the other because they are completely dissimilar factors. That is, there is no measurement scale that can relate the magnitude of change in one factor relative to the other. The magnitude of impact on the results can be ordered, but the magnitude of change in one factor relative to the other cannot be measured. This is in sharp contrast to the definition of a derivative where the rate of change of the quantity of interest is measured with respect to the rate of change of the controlling quantity.

Results from an SA can be used in several ways, but the two most common are mentioned here. First, the results of an SA can be used to determine which input uncertainties cause the largest uncertainty in the elements of $\vec{y}$. Some examples of how this information can be used are (a) reducing the manufacturing variability in certain geometrical input quantities to improve the reliability or performance of the system; (b) altering the operating envelope of the system in order to improve the system performance, such as fuel consumption or range; and (c) altering the system design to improve system safety when it is discovered that certain input uncertainties due to the system excitation cannot be reduced and they must be tolerated.

Second, the results of an SA can be used as a planning tool for improved allocation of resources on a simulation project. For example, suppose that several submodels are used in the simulation and the computational cost of each submodel is quantified. An SA could be conducted to determine the ranking of which submodels have the most impact on each of the elements of $\vec{y}$. For simplicity, suppose that each submodel resulted in one output quantity that is then used in the complete model of the system. The output quantity from each submodel could be artificially changed, say by $\pm 10\%$, before the quantity was used as input to the complete model. This artificial change would be made to each submodel, one at a time. After the individual changes are made, the magnitude of the changes on the elements of $\vec{y}$ are rank ordered from largest to smallest. Note that the rank order for each element of $\vec{y}$ would typically be different than another element. If it is found that any of the computationally expensive submodels had very little effect on any of the elements of $\vec{y}$, then these submodels could be removed from the simulation. Furthermore, if model development and/or experimental resources are being spent on any of these expensive submodels, these activities could also be terminated.

There are two types of SAs: local and global. Each provides different types of information that can be used for various purposes, such as design optimization of the system performance or improvements in safety for abnormal or hostile environments. Both types of analyses are briefly discussed in the following section. For a detailed discussion of SA, see Saltelli et al. (2008), Smith (2014), McClarren (2018), Santner et al. (2018), and Da Veiga et al. (2021).

13.8.1 Local Sensitivity Analysis

In a local SA, the interest is in determining how outputs change locally as a function of a change in inputs. This is equivalent to computing the partial derivatives of the SRQs with respect to each of the input quantities. The concept of partial derivative applied to SA is only defined for input functions that are differentiable, either deterministic variables or random variables. For input quantities that are random variables, various methods can be applied, such as the adjoint sensitivity analysis procedure, automatic differentiation, and the use of Gateaux differentials. These topics are beyond the scope of this text (see Cacuci et al., 2005; Smith, 2014; McClarren, 2018).

For the purpose of illustration, here the discussion of local SA only deals with input quantities that are continuous, deterministic variables. Let the model input variables be $x_i = \{x_1, x_2, \ldots, x_m\}$ and the SRQs be $y_j = \{y_1, y_2, \ldots, y_n\}$. A local SA computes a matrix of $m \times n$ partial derivatives given by

$$\left(\frac{\partial y_j}{\partial x_i}\right)_{\vec{x}=\vec{c}} \quad i = 1, 2, 3, \ldots, m \quad \text{and} \quad j = 1, 2, 3, \ldots, n, \tag{13.30}$$

where $\vec{c}$ is a vector of constants $\{x_1 = c_1, x_2 = c_2, \ldots, x_m = c_m\}$. These constants are the set of values for the evaluation of the partial derivatives. One method of evaluating the partial derivatives is a finite difference approximation. For example, using central differencing for Equation (13.30), we have

$$\left(\frac{y_j(x_i + \Delta x_i) - y_j(x_i - \Delta x_i)}{2\Delta x_i}\right)_{\vec{x}=\vec{c}} \quad i = 1, 2, 3, \ldots, m \quad \text{and} \quad j = 1, 2, 3, \ldots, n. \tag{13.31}$$

It is clear from Equation (13.31) that the values of each element of the matrix of partial derivatives can vary significantly depending on the nonlinearity of $y_j(x_i)$.

Example 12

Consider the case where there is only one response quantity y and three deterministic input quantities $\vec{x} = \{x_1, x_2, x_3\}$. Figure 13.28 shows y as a function of x_1 and x_2, given specified values $x_2 = c_2$ and $x_3 = c_3$ for Figure 13.28a, and $x_1 = c_1$ and $x_3 = c_3$ for Figure 13.28b. Since $\vec{x}$ defines a three-dimensional input space, and y exists in a four dimensional space, Figure 13.28a can be viewed as y as a function of x_1, in the plane of $x_2 = c_2$ and the plane $x_3 = c_3$. Figure 13.28b can be

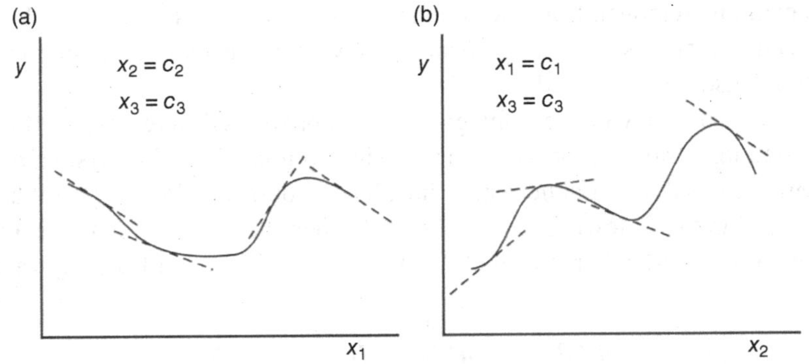

Figure 13.28 Example of a system response as a function of three input quantities, showing various derivatives in a local sensitivity analysis. (a) y vs. x_1. (b) y vs. x_2.

viewed as y as a function of x_2, in the plane of $x_1 = c_1$ and the plane $x_3 = c_3$. Each figure also shows partial derivatives $\left(\frac{\partial y}{\partial x_1}\right)_{x_2=c_2,\ x_3=c_3}$ and $\left(\frac{\partial y}{\partial x_2}\right)_{x_1=c_1,\ x_3=c_3}$ as dashed lines at various points for each input variable. It is emphasized that even though the input variable $\overline{x}$ is *not a random variable*, it still must be mapped point-by-point through the PDE to obtain $\vec{y}\left(\vec{x}\right)$. As a result, this mapping is susceptible to iterative error and discretization error. This creates additional difficulty in computing the partial derivatives, if they are evaluated using finite difference methods. This occurs because there is extraneous roughness in $\vec{y}$ due to the numerical errors, which is in addtion to the nonlinearity of $\vec{y}\left(\vec{x}\right)$. Therefore, if iterative and discretization errors are large, these will pollute any attempt to compute the local sensitivities.

Results from local SAs are most commonly used in system design and optimization studies. For example, consider how all the input parameters, from both the system and the surroundings, affect the performance, reliability, and safety of the system being simulated. The system designer might have some control of certain input parameters, whereas some are fixed by design constraints on the system, such as the operating envelope or the load-carrying requirement of the system. For those that can be controlled, a local SA can greatly aid in optimizing the design. For example, the performance of the system could be improved by altering the cruise speed of a vehicle or optimizing the center of gravity of a flight vehicle to reduce drag and fuel consumption. The two biggest limitations of deterministic local SA are (1) local rates of change of the response quantity can vary greatly with the model input values chosen, and (2) the computational cost increases rapidly as m, n, and the number of constants chosen for evaluation is increased.

13.8.2 Global Sensitivity Analysis

In a global SA, the interest is in determining how the outputs change globally as a function of changes in the inputs. A global SA always involves a nondeterministic simulation of at least some of the model inputs shown in Figure 13.27, and it may also involve investigating changes in model assumptions and numerical errors. Hence, a global SA is a prioritization of the impact of the input factor structure on the structural changes of the outputs. Two examples follow on how structural changes in the simulation change the structure of the outputs. First, consider a system with both human control and computer control in an abnormal environment, specifically, various failure scenarios are examined as part of a global SA. The ranges of scenarios considered are those involving various types of computer control system failures and how a human might respond to dealing with these failures. A simulation is computed for each of the scenarios envisioned. Each of these scenarios would also include all of the additional uncertainties shown in Figure 13.27, for example, variability of the initial conditions of the system at computer failure, variability of the physical parameters of the system, and variability of how the human controller might respond after computer failure. The global SA would attempt to determine if any of these scenarios would put the human or system at risk, as well as the risk of the system failure on the surroundings, including other humans.

For the second example, suppose a project manager or decision maker need information on what are the most important factors dominating the uncertainty of specific outputs. This type of information is especially needed when these outputs fall outside system performance or safety requirements, or when they exceed governmental regulatory requirements. With the information from a global SA, a project manager can determine the most important factors that cause the outputs of interest to deviate into unwanted or dangerous ranges of response. The information concerning the ordering of these factors based on their effect on the individual outputs would also include the effects of simulation-induced factors, such as numerical solution errors.

A common first step in a global SA is to construct some type of graphical display of the system responses as a function of the factors that are changed. The most common of these is a scatter plot for each of the $y_j = \{y_1,\ y_2, \ldots, y_n\}$ as a function of each $x_i = \{x_1,\ x_2, \ldots, x_m\}$. Each plot is a pairwise comparison relationship, resulting in a total of $m \times n$ plots. Let y_p be defined as the particular SRQ of interest that is being plotted and x_q be the factor of interest that is being plotted in the graph. Each plot of y_p vs. x_q can be thought of as the projection of *all of the varibilities* in x_i onto the $(y_p x_q)$ plane. The data points in the scatter plots are commonly generated from simple MCS or stratified MCS of the input quantities that are random variables. These samples are typically available as part of the UQ analysis, which precedes the SA. Note that the number of MCS needed for the scatter plots does *not* depend on the number of uncertain input factors, m. As a result, it is relatively inexpensive in terms of model evaluations relative to other, more quantitative SA methods. The purpose of the scatter plots is to determine if any clear trends exist in the output quantity, y_p as a function of each x_q. The trend of $y_p(x_q)$ in each of these plots can also be quantified with a regression fit of the data. If there is a trend in the data, there is commonly a great deal of scatter in the regression fit because the variability from *all* of the x_i show themselves in the plot.

Example 13

Consider the case where there is only one response quantity, y, and four input quantities, $\{x_1, x_2, x_3, x_4\}$, where each input is a random variable given by some distribution. Figure 13.29 shows the four scatter plots that are generated from a total of 100 randomly drawn samples of $\{x_1, x_2, x_3, x_4\}$. Stratified MCS, i.e., Latin Hypercube Sampling, is used to ensure that samples are drawn from a wide range of each distribution. That is, LHS ensures that samples are drawn from both tails of each distribution, whereas MCS does not ensure this with a relatively small number of samples. The scatter plots in Figure 13.29 are a projection of a five-dimensional uncertainty space, $\{y,\ x_1,\ x_2,\ x_3,\ x_4\}$, onto the $(y,\ x_1)$ plane, the $(y,\ x_2)$ plane, the $(y,\ x_3)$ plane, and the $(y,\ x_4)$ plane, respectively. The shape of the uncertainty clouds indicates that (a) there is no discernable trend of y with x_1, (b) there is a slightly decreasing trend of y with x_2, (c) there is a clear trend of y increasing with x_3, and (d) there is a strong trend of y increasing with x_4. Note that these observations of trends of y with the individual x_i should be viewed as correlations, not necessarily as causal factors because the variability of *all of the factors* is shown in each scatter plot.To determine which input quantity has the strongest global influence on y, a linear regression fit of each scatter plot is computed. The magnitude of the linear regression fit of each scatter plot is written as

$$\left|\frac{\partial y}{\partial x_1}\right|, \quad \left|\frac{\partial y}{\partial x_2}\right|, \quad \left|\frac{\partial y}{\partial x_3}\right|, \quad \text{and} \quad \left|\frac{\partial y}{\partial x_4}\right|. \tag{13.32}$$

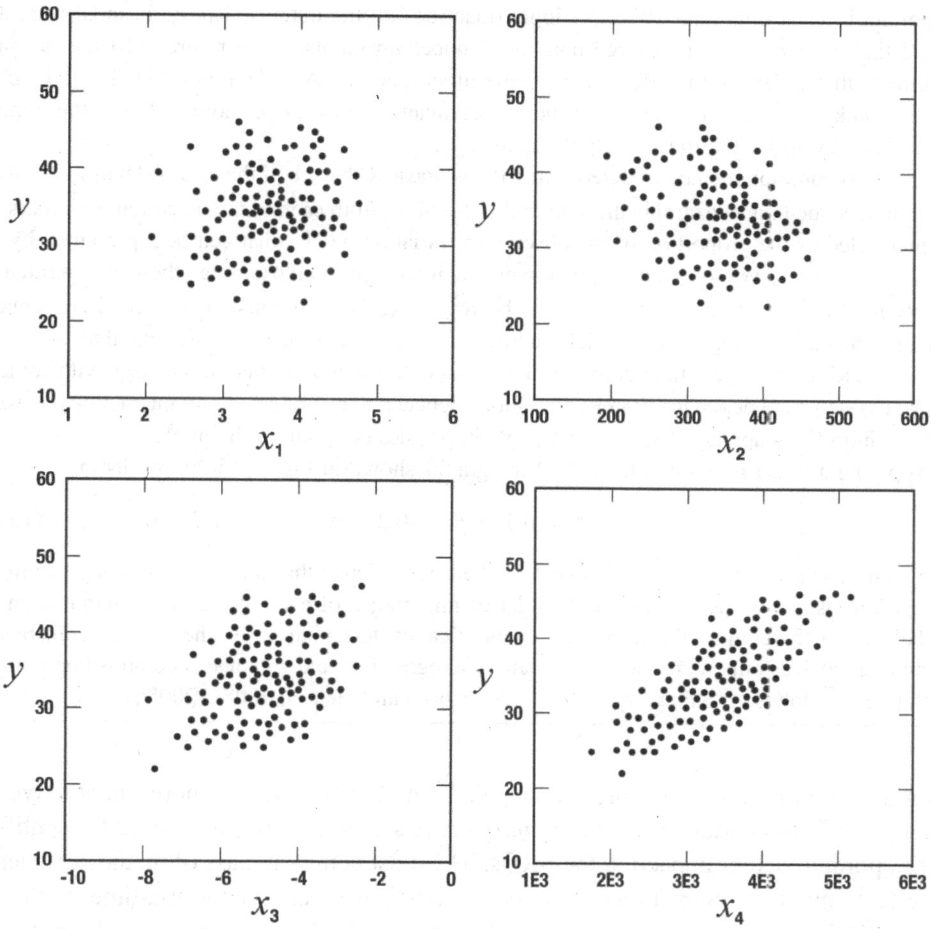

Figure 13.29 Scatter plots for one output quantity as a function of four input quantities. (a) Projection of y onto the x_1 plane. (b) Projection of y onto the x_2 plane. (c) Projection of y onto the x_3 plane. (d) Projection of y onto the x_4 plane.

However, comparing these quantities would be of limited value in a global SA because the magnitude of each derivative would depend on the dimensional units of each x_i. In addition, each derivative has no information concerning the magnitude of the uncertainty in each x_i.

It is standard practice to reformulate the scatter plots so that all the x_i are rescaled by their respective estimated means, $\bar{x}_i$, and y is rescaled by its estimated mean, $\bar{y}$. In this way, the magnitude of the regression slopes given in Equation (13.32) can be properly compared. Rescaling both x_i and y, yields

$$\left|\frac{\partial(y/\bar{y})}{\partial(x_1/\bar{x}_1)}\right|, \quad \left|\frac{\partial(y/\bar{y})}{\partial(x_2/\bar{x}_2)}\right|, \quad \left|\frac{\partial(y/\bar{y})}{\partial(x_3/\bar{x}_3)}\right|, \quad \text{and} \quad \left|\frac{\partial(y/\bar{y})}{\partial(x_4/\bar{x}_4)}\right|. \tag{13.33}$$

The terms in Equation (13.33) are either referred to as the sigma-normalized derivatives or Standardized Regression Coefficients (SRCs) (Saltelli et al., 2008; Saltelli et al., 2009). If the response y is nearly linear in all of the x_i, and the x_i are independent random variables, then the SRCs listed in Equation (13.33) can be rank ordered from largest to smallest to express the most important to least important effects of input quantities on the output quantity. In most SAs, it is found that even if there are a large number of uncertain inputs, x_i, there are only several input quantities that dominate the effect on a given output quantity. As a final point, it should be clear that the rank ordering of the SRCs for one output quantity, say y_1, does not need to be the same as the rank ordering for a different output quantity, y_1.

The most common method of determining if y is linear with each of the x_i, and determining if the x_i are independent, is to examine the sum of the R^2 values from each of the linear regression fits. R^2 is interpreted as the proportion of the observed y variation with x_i that can be represented by the regression model. For the four input quantities in the present example, these are written as $R_1^2, R_2^2, R_3^2, R_4^2$, respectively. If the sum of the R^2 values is near unity, then there is reasonable evidence that a linear regression model can be used to rank order the SRCs. Finding that the sum of the R^2 values is much less than unity, means there could be (a) nonlinear trends of y with some of the x_i; (b) statistical dependencies or correlations between the x_i; or (c) strong interactions between the x_i within the mapping of $x_i \rightarrow y$, e.g., certain physics coupling in the model.

Applying a linear regression to each of the graphs shown in Figure 13.29, results in

$$R_1^2 = 0.01, \quad R_2^2 = 0.11, \quad R_1^2 = 0.27, \quad \text{and} \quad R_1^2 = 0.55. \tag{13.34}$$

The sum of the R^2 values is 0.94, indicating there is evidence that the SRCs computed from the regression fits can properly estimate the relative importance of each of the input quantities in the global SA. If the sum of the R^2 values is significantly less than unity, then more sophisticated techniques such as rank regression, nonparametric regression, and variance decomposition must be used (Helton, Johnson, Sallaberry, et al., 2006; Storlie and Helton, 2008a, 2008b).

Another preliminary procedure for a global SA, but one that is more quantitative than scatter plots, is to conduct a *screening analysis*. A screening analysis is used to identify the most important factors impacting the SRQs. The most common class of methods is elementary effects methods. Sometimes these are referred to as one-factor-at-a-time methods or factorial design methods, and the most well known is the Morris method. If there is a very large number of factors considered in the simulation, the elementary effects method allows the computational analyst and the customer of the simulation to focus on the most important factors and ignore the less important factors. The number of model evaluations needed for an elementary effects method depends on the number of factors considered. The typical number of model evaluations ranges from two to four times the number of factors considered. The number of evaluations required also depends on the nonlinearity of the model responses as a function of the factors considered. For a detailed discussion of screening methods, see Saltelli et al. (2008), Smith (2014), Montgomery (2017), and Da Veiga et al. (2021).

The most quantitative global SA methods are *variance decomposition methods*. These methods can range from linear and rank regression methods to Sobol's method. These methods decompose the variance of the SRQs into quantitative portions that are attributed to each factor considered. This level of quantitative information about the factors influencing the response is only practical for considering roughly 10 to 20 factors. As a result, some type of screening analysis should be done first. The number of model evaluations required for Sobol's method is a hundred to a thousand times the number of factors considered. For a detailed discussion of variance decomposition methods, see Saltelli et al. (2008), Saltelli et al. (2009), Smith (2014), and Da Veiga et al. (2021).

The discussion of global SA up to this point has only applied to uncertainties that are either purely aleatory or factors that can be changed discreetly, such as changes in submodels and mesh resolution. When a factor includes both aleatory and epistemic uncertainties, global SA becomes more complex for two reasons. First, epistemic uncertainties given by intervals are more general informational structures, in the sense of greater ambiguity, than precise probability distributions. Intervals characterize the infinite set of *all probability distributions* that can exist over the range of the interval. For example, it makes no sense to try to show the impact of an interval-valued quantity on a scatter plot because all values over the range of the interval are possible. More advanced theories of uncertainty, such as evidence theory or the theory of upper and lower previsions, are required for SA (Walley, 2000; Bernardini and Tonon, 2010; Augustin et al., 2014).

Second, imprecise probability theory, of which traditional probability theory is a subset, explicitly incorporates lack of knowledge in the characterization of uncertainty. Therefore, ordering of the magnitude of impact of input uncertainties on an SRQ is, in general, no longer possible. The critical difference between imprecise probability theory and precise, or traditional, probability theory is that ordering of uncertainties is not always possible because ambiguity, duplication, and contradiction of information are allowed in imprecise probability. For example, consider an SRQ that is a mixture of aleatory and epistemic uncertainties because some of the model input uncertainties are aleatory and some are epistemic. The impact of the epistemic input uncertainties on the SRQ cannot always be ordered, relative to the aleatory input uncertainties, because the epistemic input uncertainties *only* express a lower and upper limit of their impact on the uncertainty in the SRQ. If there is any overlap in any of the intervals of the SRQ, then the magnitude of the impact of the input quantities on the SRQ cannot be ordered. When there is no overlap of the SRQ intervals, or even when there is some overlap, the most important input uncertainties can be identified because imprecise probabilities will identify the worst-case combinations of input uncertainties.

An approximate method of dealing with the impact of interval-valued inputs in practical applications has been suggested by Ferson and Tucker (2006). This method suggests "pinching" the interval-valued input uncertainty into a structure that can be dealt with in the framework of traditional global SA. One method of pinching, or reducing, the interval-valued uncertainty is to characterize the interval as a uniformly distributed random variable over the same range as the interval. With this change, existing global SA methods can be applied. However, an important caveat must be stressed regarding the interpretation of the results from a traditional global SA with this alteration. The impact of these factors in the global SA must be understood as, "The factor will be *at least* as important as indicated in the results, but it *may be much more important than indicated.*" As a final comment, recent work by Schobi and Sudret (2019) has extended the well-developed concept of Sobol indices for precise probabilities to Sobol indices for parametrized probability boxes (p-boxes). This approach does not require the alteration just mentioned.

13.9 Further Reading

A great deal of research and many new applications have been published on various aspects of predictive capability during the last 10 years. This section concentrates on citing the more recent work in both texts and journal publications.

For recent texts on uncertainty quantification and predictive capability using primarily traditional probabilistic methods, see Vose (2008), NASA (2010), Eisfeld et al. (2013), Weisberg (2014), Smith (2014), NaRC (2014), Haimes (2015), Sullivan (2015), Murray-Smith (2015), Engquist (2015), NaRC (2016), Modarres et al. (2016), Ghanem et al. (2017), McClarren (2018), Todinov (2018), and Beisbart and Saam (2019).

For recent texts on uncertainty quantification and predictive capability using primarily Bayesian methods, see NaRC (2009), Mitchell (2012), Gelman et al. (2013), Smith (2014), Kacprzyk and Pedrycz (2015), Raol (2016), Kedem et al. (2017), Lambert (2018), Halpern (2019), Aster et al. (2019), McElreath (2020), and Szabo and Babuska (2021a).

For recent texts on predictive capability and uncertainty quantification with an emphasis on imprecise probabilities, see Molchanov (2005), Ben-Haim (2006), Moller and Reuter (2007), Bernardini and Tonon (2010), Torra et al. (2014), Augustin et al. (2014), Torra et al. (2017), Salicone and Prioli (2018), and Raizer and Elishakoff (2022).

For recent texts on parameter estimation and inverse problems, see Hill and Tiedeman (2007), Biegler et al. (2011), Morgan and Winship (2015), Martin and Liu (2016), Pearl and Mackenzie (2018), Santner et al. (2018), and Aster et al. (2019).

For recent texts on sensitivity analysis, see Saltelli et al. (2008) Saltelli et al. (2009), Smith (2014), Montgomery (2017), McClarren (2018), Santner et al. (2018), and Da Veiga et al. (2021).

For recent texts on surrogate and response surface models, see Box and Draper (2007), Schilders et al. (2008), Smith (2014), Montgomery (2017), Jiang et al. (2020), and Gramacy (2020).

For recent research in model validation and predictive capability, see Hawkins and Kraker (2010), Stripling et al. (2011), Pace (2012), Grim et al. (2013); Hills (2013); Frigg et al. (2014), Morrison (2014), Adu et al. (2015), Wu et al. (2015), Kwasniewski and Bojanowski (2015), Xi et al. (2015), Pappenberger et al. (2015), Roy and Oberkampf (2016), Jatale et al. (2016), Rui et al. (2016), Xi and Yang (2016), Micic (2016), Allemang et al. (2016), Bi et al. (2017), Patterson and Whelan (2017), Xi et al. (2017), Zhao et al. (2017), Segalman et al. (2017), Jatale et al. (2017), Xiao et al. (2017), Toppila and Salo (2017), Stern et al. (2017), Xie et al. (2018), Maupin et al. (2018), Li and Lu (2018), Flage et al. (2018), Rocchetta et al. (2018), Peherstorfer et al. (2018), Crespo et al. (2018), Lenhard (2018), Schroeder et al. (2018), Gaymann et al. (2018), Oberkampf (2019), Knutti et al. (2019), Seibert et al. (2019), Martin (2019), Gardner et al. (2019), Frisch (2019), Szabo et al. (2019), Brune et al. (2020), Riedmaier et al. (2020), Glimm et al. (2020), Vanslette et al. (2020), Porter et al. (2021), Szabo et al. (2021), Szabo and Babuska (2021b), and Whiting et al. (2023).

For recent research on Bayesian approaches to uncertainty quantification, see Carter et al. (2006), Rutherford (2008), Broemeling (2009), Hemez and Atamturktur (2011), Oden et al. (2013), Brynjarsdottir and O'Hagan (2014), Ling et al. (2014), Li et al. (2014), Sankararaman and Mahadevan (2015), Atamturktur et al. (2015), Choudhary et al. (2016), Li and Mahadevan (2016), Li et al. (2016), Absi and Mahadevan (2016), Mullins et al. (2016), Phillips and Roy (2016), Yin et al. (2016), Nannapaneni and Mahadevan (2016), Zhu et al. (2016), Ao et al. (2017), Bi et al. (2017), Oden et al. (2018), Wang, Roy, et al. (2018), Matthies (2018), Beisbart (2019), Zhu et al. (2019), Jiang et al. (2019), Andrews and Fraser (2019), Bi et al. (2019), Geneva and Zabaras (2020), Sun et al. (2020), and Ghattas and Willcox (2021).

For recent research on sensitivity analysis, see Weirs et al. (2012), Zhao and Mousseau (2012), Garcia-Cabrejo and Valocchi (2014), Baroni and Tarantola (2014), Jiang et al. (2016), Ferretti et al. (2016), Borgonovo and Plischke (2016), Hale et al. (2017), Iooss and Saltelli (2017), Wang et al. (2017), Kaminski et al. (2018), Ribes et al. (2019), and Schobi and Sudret (2019).

For recent research on imprecise probability approaches, see Hall (2006), Helton, Johnson, Oberkampf, et al. (2006), Aughenbaugh and Paredis (2006), Hall et al. (2007), Guo and Du (2007), Ferson et al. (2008), Ghosh and Mujumdar (2009), Oberguggenberger et al. (2009), Helton et al. (2010), Augustin and Hable (2010), Coolen et al. (2011), Roy and Oberkampf (2011), Xiao et al. (2011), Helton and Johnson (2011), Roy and Balch (2012), Chen et al. (2013), Beer et al. (2013), Liu et al. (2014), Abdallah et al. (2014), Shah et al. (2015), Oberguggenberger (2015), Fetz and Oberguggenberger (2016), Shafer (2016), Rajabi and Ataie-Ashtiani (2016), Schobi and Sudret (2017a), Liu et al. (2017), Eek et al. (2017), Denoeux and Li (2018), Petturiti and Vantaggi (2018), Bradley (2019), Ewing et al. (2018), Balch et al. (2019), Crespo et al. (2019), Song et al. (2020), McKeand et al. (2021), Ben-Haim (2022), and Danquah et al. (2022).

For recent research on more efficient techniques for propagating imprecise probabilities through a mathematical model, see Regan et al. (2004), Boyd and Vandenberghe (2004), Kreinovich et al. (2007), Zhang et al. (2010), Chojnacki et al. (2010), Bertsekas (2015), Chen et al. (2015); Xiao et al. (2016); Chen et al. (2016), Alvarez et al. (2017), Schobi and Sudret (2017b), Wang, Zhang, et al. (2018), Wang, Matthies, et al. (2018), Liu et al. (2018), Alvarez et al. (2018), Nesterov (2018), Wei et al. (2019a, 2019b), Song et al. (2019), and Liu et al. (2020).

13.10 Problems

1. Consider the example problem of heat transfer through the cinder-block walls of the building discussed in this chapter. Reproduce all of the steps discussed for the heat transfer problem, in descriptive text, computer programs, graphical results, and include your interpretation of these results.
2. Complete Problem 1 and then extend this analysis by including model form uncertainty, as discussed in Section 13.5.2. This would involve introducing synthetic experimental data that would represent experimental measurements of heat transfer through a cinder-block wall in a laboratory environment. Use the Modified Area Validation Metric, MAVM, to estimate the model form uncertainty both within and outside the validation domain of the experimental data.
3. Complete Problems 1 and 2, and then extend this analysis by including an estimate of the total simulation uncertainty, as discussed in Section 13.6.
4. Write a computer program to solve Problem 1 in Oberkampf et al. (2004). Explain and interpret the results obtained.
5. Write a computer program to solve Problem 2 in Oberkampf et al. (2004). Explain and interpret the results obtained.
6. Write a computer program to solve Problem 3 in Oberkampf et al. (2004). Explain and interpret the results obtained.

7. Write a computer program to solve Problem 4 in Oberkampf et al. (2004). Explain and interpret the results obtained.
8. Write a computer program to solve Problem 5 in Oberkampf et al. (2004). Explain and interpret the results obtained.
9. Write a computer program to solve Problem 6 in Oberkampf et al. (2004). Explain and interpret the results obtained.
10. Write a computer program to solve the oscillation of the mass-spring-damper problem in Oberkampf et al. (2004). Explain and interpret the results obtained.
11. (Semester project) Write a computer program to solve the Sandia Thermal Challenge Problem in Dowding et al. (2008). Explain and interpret the results obtained.
12. (Semester project) Write a computer program to solve the Sandia Static Frame Challenge Problem in Babuska et al. (2008). Explain and interpret the results obtained.
13. (Semester project) Write a computer program to solve the Sandia Structural Dynamics Challenge Problem in Red-Horse and Paez (2008). Explain and interpret the results obtained.

13.11 References

Abdallah, N. B., N. Mouhous-Voyneau, and T. Denoeux (2014). Combining Statistical and Expert Evidence Using Belief Functions: Application to Centennial Sea Level Estimation Taking into Account Climate Change. *International Journal of Approximate Reasoning*, **55**(1), 341–354.

Absi, G. N., and S. Mahadevan (2016). Multi-fidelity Approach to Dynamics Model Calibration. *Mechanical Systems and Signal Processing*, **68–69**, 189–206.

Adu, S., I. Horvatovic, E. O. Darko, G. Emi-Reynolds, and F. D'Auria (2015). Application of Best Estimate Plus Uncertainty in Review of Research Reactor Safety Analyses. *Nuclear Technology & Radiation Protection*, **30**(1), 75–82.

Ainsworth, M., and J. T. Oden (2000). *A Posteriori Error Estimation in Finite Element Analysis*. New York, NY: John Wiley.

Allemang, R., M. M. Kolluri, M. Spottswood, and T. Eason (2016). Decomposition-Based Calibration/Validation Metrics for Use with Full-Field Measurement Situations. *Journal of Strain Analysis*, **51**(1), 14–31.

Alvarez, D. A., J. E. Hurtado, and J. Ramirez (2017). Tighter Bounds on the Probability of Failure Than Those Provided by Random Set Theory. *Computers and Structures*, **189**, 101–113.

Alvarez, D. A., F. Uribe, and J. E. Hurtado (2018). Estimation of the Lower and Upper Bounds on the Probability of Failure Using Subset Simulation and Random Set Theory. *Mechanical Systems and Signal Processing*, **100**, 782–801.

Andrews, S. A., and A. M. Fraser (2019). Estimating Physics Models and Quantifying Their Uncertainty Using Optimization with a Bayesian Objective Function. *Journal of Verification, Validation and Uncertainty Quantification*, **4**(1), 011002-1–011002-15.

Ang, A. H.-S., and W. H. Tang (2007). *Probability Concepts in Engineering: Emphasis on Applications to Civil and Environmental Engineering*, 2nd ed. New York, NY: John Wiley.

Angus, J. E. (1994). The Probability Integral Transform and Related Results. *SIAM Review*, **36**(4), 652–654.

Annan, J. D., and J. C. Hargreaves (2017). On the Meaning of Independence in Climate Science. *Earth System Dynamics*, **8**(1), 211–224.

Ao, D., Z. Hu, and S. Mahadevan (2017). Design of Validation Experiments for Life Prediction Models. *Reliability Engineering and System Safety*, **165**, 22–33.

Aster, R., B. Borchers, and C. Thurber (2019). *Parameter Estimation and Inverse Problems*, 3rd ed. Burlington, MA: Elsevier.

Atamturktur, S., Z. Liu, S. Cogan, and H. Juang (2015). Calibration of Imprecise and Inaccurate Numerical Models Considering Fidelity and Robustness: A Multi-Objective Optimization-Based Approach. *Structural and Multidisciplinary Optimization*, **51**, 659–671.

Aughenbaugh, J. M., and C. J. J. Paredis (2006). The Value of Using Imprecise Probabilities in Engineering Design. *Journal of Mechanical Design*, **128**(4), 969–979.

Augustin, T., F. P. A. Coolen, G. de Cooman, and M. C. M. Troffaes (eds.) (2014). *Introduction to Imprecise Probabilities*. Chichester: Wiley & Sons.

Augustin, T., and R. Hable (2010). On the Impact of Robust Statistics on Imprecise Probability Models: A Review. *Structural Safety*, **32**, 358–365.

Ayyub, B. M. (2001). *Elicitation of Expert Opinions for Uncertainty and Risks*. Boca Raton, FL: CRC Press.

Babuska, I., F. Nobile, and R. Tempone (2008). Formulation of the Static Frame Problem. *Computer Methods in Applied Mechanics and Engineering*, **197**(29–32), 2496–2499.

Babuska, I., J. R. Whiteman, and T. Strouboulis (2010). *Finite Elements: An Introduction to the Method and Error Estimation*. Oxford: Oxford University Press.

Balch, M. S., R. Martin, and S. Ferson (2019). Satellite Conjunction Analysis and the False Confidence Theorem. *Proceedings of the Royal Society A: Mathematical, Physical and Engineering Sciences*, **475** (2227), 20180565.

Baroni, G., and S. Tarantola (2014). A General Probabilistic Framework for Uncertainty and Global Sensitivity Analysis of Deterministic Models: A Hydrological Case Study. *Environmental Modelling & Software*, **51**, 26–34.

Beer, M., S. Ferson, and V. Kreinovich (2013). Imprecise Probabilities in Engineering Analyses. *Mechanical Systems and Signal Processing*, **37**(1–2), 4–29.

Beisbart, C. (2019). Simulation Validation from a Bayesian Perspective. In C. Beisbart and N. J. Saam (eds.), *Computer Simulation Validation – Fundamental Concepts, Methodological Frameworks, and Philosophical Perspectives*. New York, NY: Springer.

Beisbart, C., and N. J. Saam (eds.) (2019). *Computer Simulation Validation – Fundamental Concepts, Methodological Frameworks, and Philosophical Perspectives*. Berlin, Germany: Springer.

Beliakov, G., A. Pradera, and T. Calvo (2007). *Aggregation Functions: A Guide for Practitioners*. New York, NY: Springer.

Beliakov, G., H. B. Sola, and T. C. Sanchez (2015). *A Practical Guide to Averaging Functions*. New York, NY: Springer.

Ben-Haim, Y. (2022). Inferring Extreme Values from Measured Averages under Deep Uncertainty. *Journal of Verification, Validation and Uncertainty Quantification*, **7**(2), 021002-1–021002-12.

Ben-Haim, Y. (2006). *Information-Gap Decision Theory: Decision under Severe Uncertainty*, 2nd ed. New York, NY: Elsevier.

Berendsen, H. J. C. (2007). *Simulating the Physical World; Hierarchical Modeling from Quantum Mechanics to Fluid Dynamics*. Cambridge: Cambridge University Press.

Berleant, D., M. Ceberio, G. Xiang, and V. Kreinovich (2007). Towards Adding Probabilities and Correlations to Interval Computations. *International Journal of Approximate Reasoning*, **46**(3), 499–510.

Bernardini, A., and F. Tonon (2010). *Bounding Uncertainty in Civil Engineering*. New York, NY: Springer.

Bertsekas, D. P. (2015). *Convex Optimization Algorithms*. Nashua, NH: Athena Scientific Publishers.

Bestion, D., F. D'Auria, P. Lien, and H. Nakamura (2017). Scaling in System Thermal-Hydraulics Applications to Nuclear Reactor Safety and Design: A State-of-the-Art Report. Nuclear Energy Agency, NEA/CSNI/R(2016)14.

Bi, S., M. Broggi, and M. Beer (2019). The Role of the Bhattacharyya Distance in Stochastic Model Updating. *Mechanical Systems and Signal Processing*, **117**, 437–452.

Bi, S., S. Prabhu, S. Cogan, and S. Atamturktur (2017). Uncertainty Quantification Metrics with Varying Statistical Information in Model Calibration and Validation. *AIAA Journal*, **55**(10), 3570–3583.

Biegler, L., G. Biros, O. Ghattas, M. Heinkenschloss, D. Keyes, B. Mallick, et al. (eds.) (2011). *Large-Scale Inverse Problems and Quantification of Uncertainty*, New York, NY: Wiley.

Borgonovo, E., and E. Plischke (2016). Sensitivity Analysis: A Review of Recent Advances. *European Journal of Operational Research*, **248**(3), 869–887.

Box, G. E. P., and N. R. Draper (2007). *Response Surfaces, Mixtures, and Ridge Analyses*. New York, NY: John Wiley.

Boyd, S., and L. Vandenberghe (2004). *Convex Optimization*. Cambridge: Cambridge University Press.

Bradley, S. (2019). Imprecise Probabilities. In C. Beisbart and N. J. Saam (eds.), *Computer Simulation Validation – Fundamental Concepts, Methodological Frameworks, and Philosophical Perspectives*. New York, NY: Springer.

Broemeling, L. D. (2009). *Bayesian Methods for Measures of Agreement*. Boca Raton, FL: CRC Press.

Brune, A. J., T. K. I. West, and L. M. White (2020). Calibration of Probe Uncertainty and Validation for the Hypersonic Material Environmental Test System. *Journal of Thermophysics and Heat Transfer*, **34**(2), 404–420.

Bruns, M., and C. J. J. Paredis (2006). Numerical Methods for Propagating Imprecise Uncertainty. *Design Engineering and Technical Conferences and Computers and Information in Engineering Conference*, DETC2006–99237, Philadelphia, PA.

Brynjarsdottir, J., and A. O'Hagan (2014). Learning about Physical Parameters: The Importance of Model Discrepancy. *Inverse Problems*, **30**(11), 114007.

Bucalem, M. L., and K.-J. Bathe (2011). *The Mechanics of Solids and Structures – Hierarchical Modeling and the Finite Element Solution*. New York, NY: Springer.

Butcher, J. C. (2016). *Numerical Methods for Ordinary Differential Equations*, 3rd ed. New York, NY: John Wiley.

Cacuci, D. G., M. Ionescu-Bujor, and I. M. Navon (2005). *Sensitivity and Uncertainty Analysis, Vol. II: Applications to Large-Scale Systems*, Vol. II. Boca Raton, FL: CRC Press.

Carter, J. N., P. J. Ballester, Z. Tavassoli, and P. R. King (2006). Our Calibrated Model Has Poor Predictive Value: An Example from the Petroleum Industry. *Reliability Engineering and System Safety*, **91**, 1373–1381.

Ceberio, M., S. Fesron, V. Kreinovich, S. Chopra, G. Xiang, A. Murguia, et al. (2007). How to Take into Account Dependence between the Inputs: From Interval Computations to Constraint-Related Set Computations, with Potential Applications to Nuclear Safety, Bio- and Geosciences. *Journal of Uncertain Systems*, **1**(1),14–37.

Chen, N., D. Yu, and B. Xia (2015). Evidence-Theory-Based Analysis for the Prediction of Exterior Acoustic Field with Epistemic Uncertainties. *Engineering Analysis with Boundary Elements*, **50**, 402–411.

Chen, N., D. Yu, B. Xia, and M. Beer (2016). Uncertainty Analysis of a Structural–Acoustic Problem Using Imprecise Probabilities Based on P-Box Representations. *Mechanical Systems and Signal Processing*, **80**, 45–57.

Chen, Z., P. Wu, and B. Li (2013). A Strong Law of Large Numbers for Non-Additive Probabilities. *International Journal of Approximate Reasoning*, **54**(3), 365–377.

Chojnacki, E., J. Baccou, and S. Destercke (2010). Numerical Accuracy and Efficiency in the Propagation of Epistemic and Aleatory Uncertainties. *International Journal of General Systems*, **39**(7), 683–704.

Choudhary, A., I. T. Voyles, C. J. Roy, W. L. Oberkampf, and M. Patil (2016). Probability Bounds Analysis Applied to the Sandia Verification and Validation Challenge Problem. *Journal of Verification, Validation and Uncertainty Quantification*, **1**(1), 011003-1–011003-13.

Coolen, F. P. A., M. C. M. Troffaes, and T. Augustin (2011). Imprecise Probability. In M. Lovric (ed.), *International Encyclopedia of Statistical Sciences*. New York, NY: Springer.

Crespo, L. G., B. K. Colbert, S. P. Kenny, and D. P. Giesy (2019). On the Quantification of Aleatory and Epistemic Uncertainty Using Sliced-Normal Distributions. *Systems & Control Letters*, **134**, 104560-1–104560-10.

Crespo, L. G., S. P. Kenny, and D. P. Giesy (2018). Staircase Predictor Models for Reliability and Risk Analysis. *Structural Safety*, **75**, 35–44.

Cullen, A. C., and H. C. Frey (1999). *Probabilistic Techniques in Exposure Assessment: A Handbook for Dealing with Variability and Uncertainty in Models and Inputs*. New York, NY: Plenum Press.

Da Veiga, S., F. Gamboa, B. Iooss, and C. Prieur (2021). *Basics and Trends in Sensitivity Analysis: Theory and Practice in R*. Society for Industrial and Applied Mathematics.

Danquah, B., S. Riedmaier, and M. Lienkamp (2022). Potential of Statistical Model Verification, Validation and Uncertainty Quantification in Automotive Vehicle Dynamics: A Review. *Vehicle System Dynamics*, **60**(4), 1292–1321.

Davey, S., N. Gordon, I. Holland, M. Rutten, and J. Williams (2016). *Bayesian Methods in the Search for MH370*. Open Access, Springer Nature.

Denoeux, T., and S. Li (2018). Frequency-Calibrated Belief Functions: Review and New Insights. *International Journal of Approximate Reasoning*, **92**, 232–254.

Devore, J. L. (2007). *Probability and Statistics for Engineers and the Sciences*, 7th ed. Pacific Grove, CA: Duxbury.

Dimov, I. T. (2008). *Monte Carlo Methods for Applied Scientists*, 2nd ed. Singapore: World Scientific Publishing.

Dowding, K. J., M. Pilch, and R. G. Hills (2008). Formulation of the Thermal Problem. *Computer Methods in Applied Mechanics and Engineering*, **197**(29–32), 2385–2389.

Eca, L., G. Vaz, and M. Hoekstra (2018). A Contribution for the Assessment of Discretization Error Estimators Based on Grid Refinement Studies. *Journal of Verification, Validation and Uncertainty Quantification*, **3**(2), 021001-1–021001-10.

Eek, M., H. Gavel, and J. Olvander (2017). Definition and Implementation of a Method for Uncertainty Aggregation in Component-Based System Simulation. *Journal of Verification, Validation and Uncertainty Quantification*, **2**(1), 011006-1–011006-12.

Eisfeld, B., H. Barnewitz, W. Fritz, and F. Thiele (eds.) (2013). *Management and Minimisation of Uncertainties and Errors in Numerical Aerodynamics*. Berlin, Germany: Springer.

Eldred, M. S., L. P. Swiler, and G. Tang (2011). Mixed Aleatory-Epistemic Uncertainty Quantification with Stochastic Expansions and Optimization-Based Interval Estimation. *Reliability Engineering and System Safety*, **96**, 1092–1113.

Engquist, B. (ed.) (2015). *Encyclopedia of Applied and Computational Mathematics*. Heidelberg, Germany: Springer.

Ewing, M. E., B. C. Liechty, and D. L. Black (2018). A General Methodology for Uncertainty Quantification in Engineering Analyses Using a Credible Probability Box. *Journal of Verification, Validation and Uncertainty Quantification*, **3**(2), 021003-1–021003-12.

Ferretti, F., A. Saltelli, and S. Tarantola (2016). Trends in Sensitivity Analysis Practice in the Last Decade. *Science of the Total Environment*, **568**, 666–670.

Ferson, S., and V. Kreinovich. (2006). Modeling Correlation and Dependence Among Intervals. *Workshop on Reliable Engineering Computing: Modeling Errors and Uncertainty in Engineering Computations*, pp. 115–126, Savannah, Georgia.

Ferson, S., and W. L. Oberkampf (2009). Validation of Imprecise Probability Models. *International Journal of Reliability and Safety*, **3**(1–3), 3–22.

Ferson, S., W. L. Oberkampf, and L. Ginzburg (2008). Model Validation and Predictive Capability for the Thermal Challenge Problem. *Computer Methods in Applied Mechanics and Engineering*, **197**(29–32), 2408–2430.

Ferson, S., and W. T. Tucker (2006). Sensitivity Analysis Using Probability Bounding. *Reliability Engineering and System Safety*, **91**(10–11), 1435–1442.

Fetz, T., and M. Oberguggenberger (2016). Imprecise Random Variables, Random Sets, and Monte Carlo Simulation. *International Journal of Approximate Reasoning*, **78**, 252–264.

Flage, R., T. Aven, and C. L. Berner (2018). A Comparison Between a Probability Bounds Analysis and a Subjective Probability Approach to Express Epistemic Uncertainties in a Risk Assessment Context – A Simple Illustrative Example. *Reliability Engineering and System Safety*, **169**, 1–10.

Fogiatto, M. A., G. H. Santos, and N. Mendes (2016). Thermal Transmittance Evaluation of Concrete Hollow Blocks. *12th International Conference on Heat Transfer, Fluid Mechanics and Thermodynamics*, 1291–1296, Costa de Sol, Spain.

Frigg, R., S. Bradley, H. Du, and L. A. Smith (2014). Laplace's Demon and the Adventures of His Apprentices. *Philosophy of Science*, **81**(1), 31–59.

Frisch, M. (2019). Calibration, Validation, and Confirmation. In C. Beisbart and N. J. Saam (eds.), *Computer Simulation Validation – Fundamental Concepts, Methodological Frameworks, and Philosophical Perspectives*. New York, NY: Springer.

Garcia-Cabrejo, O., and A. Valocchi (2014). Global Sensitivity Analysis for Multivariate Output Using Polynomial Chaos Expansion. *Reliability Engineering and System Safety*, **126**, 25–36.

Gardner, P., C. Lord, and R. J. Barthorpe (2019). A Unifying Framework for Probabilistic Validation Metrics. *Journal of Verification, Validation and Uncertainty Quantification*, **4**(3), 031005-1–031005-11.

Gaymann, A., M. Pietropaoli, L. G. Crespo, S. P. Kenny, and F. Montomoli (2018). Random Variable Estimation and Model Calibration in the Presence of Epistemic and Aleatory Uncertainties. *SAE International Journal of Materials and Manufacturing*, **11**(4), 453–466.

Gehman, H. W., J. L. Barry, D. W. Deal, J. N. Hallock, K. W. Hess, G. S. Hubbard, et al. (2003). Columbia Accident Investigation Board Report Vol. I. *National Aeronautics and Space Administration*, Washington, DC.

Gelman, A. B., J. B. Carlin, H. S. Stern, D. B. Dunson, A. Vehtari, and D. B. Rubin (2013). *Bayesian Data Analysis*, 3rd ed. Boca Raton, FL: CRC Press.

Geneva, N., and N. Zabaras (2020). Modeling the Dynamics of PDE Systems with Physics-Constrained Deep Auto-Regressive Networks. *Journal of Computational Physics*, **403**, 109056-1–109956-32.

Ghanem, R., D. Higdon, and H. Owhadi (eds.) (2017). *Handbook of Uncertainty Quantification*. Berlin, Germany: Springer.

Ghanem, R. G., and P. D. Spanos (2003). *Stochastic Finite Elements: A Spectral Approach*, rev. ed. Mineola, NY: Dover Publications.

Ghattas, O., and K. Willcox (2021). Learning Physics-Based Models from Data: Perspectives from Inverse Problems and Model Reduction. *Acta Numerica*, **30**, 445–554.

Ghosh, S., and P. P. Mujumdar (2009). Climate Change Impact Assessment: Uncertainty Modeling with Imprecise Probability. *Journal of Geophysical Research*, **114**(D18113), 1–17.

Ghosh, D. D., and A. Olewnik (2013). Computationally Efficient Imprecise Uncertainty Propagation. *Journal of Mechanical Design*, **135**(5), 051002-1–051002-12.

Glimm, J., B. Cheng, D. H. Sharp, and T. Kaman (2020). A Crisis for the Verification and Validation of Turbulence Simulations. *Physica D*, **404**, 132346-1–132346-10.

Gramacy, R. B. (2020). *Surrogates: Gaussian Process Modeling, Design, and Optimization for the Applied Sciences*. Boca Raton, FL: Chapman & Hall.

Griffiths, D. F., and D. J. Higham (2010). *Numerical Methods for Ordinary Differential Equations*. New York, NY: Springer.

Grim, P., R. Rosenberger, A. Rosenfeld, B. Anderson, and R. E. Eason (2013). How Simulations Fail. *Synthese*, **190**(12), 2367–2390.

Guo, J., and X. Du (2007). Sensitivity Analysis with Mixture of Epistemic and Aleatory Uncertainties. *AIAA Journal*, **45**(9), 2337–2349.

Haimes, Y. Y. (2015). *Risk Modeling, Assessment, and Management*, 4th ed. New York, NY: John Wiley.

Hale, L., M. Patil, and C. J. Roy (2017). Sensitivity of Input Epistemic Uncertainty on Nondeterministic Performance Estimates Using Nondeterministic Simulations. *Journal of Verification, Validation and Uncertainty Quantification*, **2**(2), 021003-1–021003-7.

Hall, J. W. (2006). Uncertainty-Based Sensitivity Indices for Imprecise Probability Distributions. *Reliability Engineering and System Safety*, **91**, 1443–1451.

Hall, J., G. Fu, and J. Lawry (2007). Imprecise Probabilities of Climate Change: Aggregation of Fuzzy Scenarios and Model Uncertainties. *Climatic Change*, **81**(3–4), 265–281.

Halpern, J. Y. (2019). *Actual Causality*. Cambridge, MA: The MIT Press.

Hannig, J., H. Iyer, R. C. S. Lai, and T. C. M. Lee (2016). Generalized Fiducial Inference: A Review and New Results. *Journal of the American Statistical Association*, **111**(515), 1346–1361.

Hawkins, D. M., and J. Kraker (2010). Deterministic Fallacies and Model Validation. *Journal of Chemometrics*, **24**(3–4), 188–193.

Helton, J. C., D. R. Anderson, B. L. Baker, J. E. Bean, J. W. Berglund, W. Beyeler, et al. (1995). Effect of Alternative Conceptual Models in a Preliminary Performance Assessment for the Waste Isolation Pilot Plant. *Nuclear Engineering and Design*, **154**(3), 251–344.

Helton, J. C., and J. D. Johnson (2011). Quantification of Margins and Uncertainties: Alternative Representations of Epistemic Uncertainty. *Reliability Engineering and System Safety*, **96**(9), 1034–1052.

Helton, J. C., J. D. Johnson, W. L. Oberkampf, and C. J. Sallaberry (2006). Sensitivity Analysis in Conjunction with Evidence Theory Representations of Epistemic Uncertainty. *Reliability Engineering and System Safety*, **91**(10–11), 1414–1434.

Helton, J. C., J. D. Johnson, W. L. Oberkampf, and C. J. Sallaberry (2010). Representation of Analysis Results Involving Aleatory and Epistemic Uncertainty. *International Journal of General Systems*, **39**(6), 605–646.

Helton, J. C., J. D. Johnson, C. J. Sallaberry, and C. B. Storlie (2006). Survey of Sampling-Based Methods for Uncertainty and Sensitivity Analysis. *Reliability Engineering and System Safety*, **91**(10–11), 1175–1209.

Hemez, F. M., and S. Atamturktur (2011). The Dangers of Sparse Sampling for the Quantification of Margin and Uncertainty. *Reliability Engineering and System Safety*, **96**(9), 1220–1231.

Higdon, D., M. Kennedy, J. Cavendish, J. Cafeo, and R. D. Ryne (2004). Combining Field Observations and Simulations for Calibration and Prediction. *SIAM Journal of Scientific Computing*, **26**(2), 448–466.

Hill, M. C., and C. R. Tiedeman (2007). *Effective Groundwater Model Calibration*. New York, NY: John Wiley.

Hills, R. G. (2013). Roll-up of Validation Results to a Target Application. *Sandia National Laboratories*, SAND2013-7424, Albuquerque, NM.

Holman, J. P. (2009). *Heat Transfer*, 10th ed. Boston, MA: McGraw-Hill.

Huckle, T., and T. Neckel (2019). *Bits and Bugs: A Scientific and Historical Review of Software Failures in Computational Science*. Philadelphia, PA: Society of Industrial & Applied Mathematics.

Intelligent-Light (2023). SpectreUQ. www.ilight.com/open-source/spectreuq/.

Iooss, B., and A. Saltelli (2017). Introduction to Sensitivity Analysis. In R. Ghanem, D. Higdon, and H. Owhadi (eds.), *Handbook of Uncertainty Quantification*, 1103–1122. New York, NY: Springer.

Jatale, A., P. J. Smith, J. N. Thornock, S. T. Smith, and M. Hradisky (2016). A Validation of Flare Combustion Efficiency Predictions from Large Eddy Simulations. *Journal of Verification, Validation and Uncertainty Quantification*, **1**(2), 021001-1–021001-8.

Jatale, A., P. J. Smith, J. N. Thornock, S. T. Smith, J. P. Spinti, and M. Hradisky (2017). Multiscale Validation and Uncertainty Quantification for Problems with Sparse Data. *Journal of Verification, Validation and Uncertainty Quantification*, **2**(1), 011001-1–011001-10.

Jiang, P., Q. Zhou, and X. Shao (2020). *Surrogate Model-Based Engineering Design and Optimization*. New York, NY: Springer.

Jiang, Z., W. Chen, and B. J. German (2016). Multidisciplinary Statistical Sensitivity Analysis Considering Both Aleatory and Epistemic Uncertainties. *AIAA Journal*, **54**(4), 1326–1338.

Jiang, X., X. Cheng, and Y. Yuan (2019). Validation Using Bayesian Methods. In C. Beisbart and N. J. Saam (eds.), *Computer Simulation Validation – Fundamental Concepts, Methodological Frameworks, and Philosophical Perspectives*. New York, NY: Springer.

Kacprzyk, J., and W. Pedrycz (eds.) (2015). *Handbook of Computational Intelligence*. Heidelberg, Germany: Springer.

Kaminski, B., M. Jakubczyk, and P. Szufel (2018). A Framework for Sensitivity Analysis of Decision Trees. *Central European Journal of Operations Research*, **26**(1), 135–159.

Kedem, B., V. De Oliveira, and M. Sverchkov (2017). *Statistical Data Fusion*. Singapore: World Scientific Publishing.

Kennedy, M. C., and A. O'Hagan (2001). Bayesian Calibration of Computer Models. *Journal of the Royal Statistical Society Series B–Statistical Methodology*, **63**(3), 425–450.

Kimball, A. W. (1957). Errors of the Third Kind in Statistical Consulting. *Journal of the American Statistical Association*, **52**(278), 133–142.

Kleb, B., and C. O. Johnston (2008). Uncertainty Analysis of Air Radiation for Lunar Return Shock Layers. *AIAA Atmospheric Flight Mechanics Conference*, AIAA 2008-6388, Honolulu, HI.

Klir, G. J. (2006). *Uncertainty and Information: Foundations of Generalized Information Theory*. New York, NY: John Wiley.

Knutti, R., C. Baumberger, and G. H. Hadorn (2019). Uncertainty Quantification Using Multiple Models–Prospects and Challenges. In C. Beisbart and N. J. Saam (eds.), *Computer Simulation Validation – Fundamental Concepts, Methodological Frameworks, and Philosophical Perspectives*. New York, NY: Springer.

Kohlas, J., and P.-A. Monney (1995). *A Mathematical Theory of Hints – An Approach to the Dempster-Shafer Theory of Evidence*. New York, NY: Springer.

Kreinovich, V., J. Beck, C. Ferregut, A. Sanchez, G. R. Keller, M. Averill, et al. (2007). Monte-Carlo-Type Techniques for Processing Interval Uncertainty, and Their Potential Engineering Applications. *Reliable Computing*, **13**, 25–69.

Kwasniewski, L., and C. Bojanowski (2015). Principles of Verification and Validation. *Journal of Structural Fire Engineering*, **6**(1), 29–40.

Lambert, B. (2018). *A Student's Guide to Bayesian Statistics*. Los Angeles, CA: SAGE Publications.

Lenhard, J. (2018). Holism, or the Erosion of Modularity – A Methodological Challenge for Validation. *Philosophy of Science*, **85**(5), 832–844.

Lenhard, J., and E. Winsberg (2010). Holism, Entrenchment, and the Future of Climate Model Pluralism. *Studies in History and Philosophy of Modern Physics*, **41**(3), 253–262.

Li, C., and S. Mahadevan (2016). Role of Calibration, Validation, and Relevance in Multi-Level Uncertainty Integration. *Reliability Engineering and System Safety*, **148**, 32–43.

Li, L., and Z. Lu (2018). A New Method for Model Validation with Multivariate Output. *Reliability Engineering and System Safety*, **169**, 579–592.

Li, W., S. Chen, Z. Jiang, D. W. Apley, Z. Lu, and W. Chen (2016). Integrating Bayesian Calibration, Bias Correction, and Machine Learning for the 2014 Sandia Verification and Validation Challenge Problem. *Journal of Verification, Validation and Uncertainty Quantification*, **1**(1), 011004-1–011004-12.

Li, W., W. Chen, Z. Jiang, Z. Lu, and Y. Liu (2014). New Validation Metrics for Models with Multiple Correlated Responses. *Reliability Engineering and System Safety*, **127**, 1–11.

Lin, S.-W., and V. M. Bier (2008). A Study of Expert Overconfidence. *Reliability Engineering and System Safety*, **93**(5), 711–721.

Ling, Y., J. Mullins, and S. Mahadevan (2014). Selection of Model Discrepancy Priors in Bayesian Calibration. *Journal of Computational Physics*, **276**, 665–680.

Liu, H. B., C. Jiang, X. Y. Jia, X. Y. Long, Z. Zhang, and F. J. Guan (2018). A New Uncertainty Propagation Method for Problems with Parameterized Probability-Boxes. *Reliability Engineering and System Safety*, **172**, 64–73.

Liu, H. B., C. Jiang, and Z. Xiao (2020). Efficient Uncertainty Propagation for Parameterized P-Box Using Sparse-Decomposition-Based Polynomial Chaos Expansion. *Mechanical Systems and Signal Processing*, **138**, 106589-1–106589-16.

Liu, X., Z. Kuang, L. Yin, and L. Hu (2017). Structural Reliability Analysis Based on Probability and Probability Box Hybrid Model. *Structural Safety*, **68**, 73–84.

Liu, Z.-G., Q. Pan, and J. Dezert (2014). Classification of Uncertainty and Imprecise Data Based on Evidence Theory. *Neurocomputing*, **133**, 459–470.

Martin, R. (2019). False Confidence, Non-Additive Beliefs, and Valid Statistical Inference. *International Journal for Approximate Reasoning*, **113**, 39–73.

Martin, R., and C. Liu (2016). *Inferential Models*. Boca Raton, FL: CRC Press.

Matthies, H. G. (2018). Uncertainty Quantification and Bayesian Inversion. In E. Stein, R. de Borst, and T. J. R. Hughes (eds.), *Encyclopedia of Computational Mechanics*. New York, NY: John Wiley.

Maupin, K. A., L. P. Swiler, and N. W. Porter (2018). Validation Metrics for Deterministic and Probabilistic Data. *Journal of Verification, Validation and Uncertainty Quantification*, **3**(3), 031002-1–031002-10.

McClarren, R. G. (2018). *Uncertainty Quantification and Predictive Computational Science*. New York, NY: Springer.

McElreath, R. (2020). *Statistical Rethinking: A Bayesian Course with Examples in R and Stan*, 2nd ed. Boca Raton, FL: CRC Press.

McKeand, A. M., R. M. Gorguluarslan, and S.-K. Choi (2021). Stochastic Analysis and Validation Under Aleatory and Epistemic Uncertainties. *Reliability Engineering and System Safety*, **205**, 107258-1–107258-14.

Meyer, M. A., and J. M. Booker (2001). *Eliciting and Analyzing Expert Judgment: A Practical Guide*. New York, NY: Academic Press.

Micic, T. (2016). Risk Reality vs Risk Perception. *Journal of Risk Research*, **19**(10), 1261–1274.

Mitchell, H. B. (2012). *Data Fusion: Concepts and Ideas*. New York, NY: Springer.

Modarres, M., M. Kaminskiy, and V. Krivtsov (2016). *Reliability Engineering and Risk Analysis: A Practical Guide*, 3rd ed. Boca Raton, FL: CRC Press.

Molchanov, I. (2005). *Theory of Random Sets*. New York, NY: Springer.

Moller, B., and U. Reuter (2007). *Uncertainty Forecasting in Engineering*. New York, NY: Springer.

Montgomery, D. C. (2017). *Design and Analysis of Experiments*, 9th ed. New York, NY: John Wiley.

Morgan, M. G., and M. Henrion (1990). *Uncertainty: A Guide to Dealing with Uncertainty in Quantitative Risk and Policy Analysis*, 1st ed. Cambridge: Cambridge University Press.

Morgan, S. L., and C. Winship (2015). *Counterfactuals and Causal Inference*, 2nd ed. New York, NY: Cambridge University Press.

Morrison, M. (2014). Values and Uncertainty in Simulation Models. *Erkenntnis: An International Journal of Scientific Philosophy*, **79**(5), 939–959.

Mueller, J. L., and S. Siltanen (2012). *Linear and Nonlinear Inverse Problems with Practical Applications*. Philadelphia, PA: Society for Industrial and Applied Mathematics.

Mullins, J., Y. Ling, S. Mahadevan, L. Sun, and A. Strachan (2016). Separation of Aleatory and Epistemic Uncertainty in Probabilistic Model Validation. *Reliability Engineering and System Safety*, **147**, 49–59.

Murray-Smith, D. J. (2015). *Testing and Validation of Computer Simulation Models: Principles, Methods and Applications*. New York, NY: Springer.

Myers, R. H., D. C. Montgomery, and C. M. Anderson-Cook (2009). *Response Surface Methodology*, 3rd ed. New York, NY: John Wiley.

Nannapaneni, S., and S. Mahadevan (2016). Reliability Analysis under Epistemic Uncertainty. *Reliability Engineering and System Safety*, **155**, 9–20.

NaRC (2009). Evaluation of Quantification of Margins and Uncertainties: Methodology for Assessing and Certifying the Reliability of the Nuclear Stockpile. National Research Council of the National Academies, ISBN-13: 978-0-309-12853-7. Washington, DC.

NaRC (2014). *Lessons Learned from the Fukushima Nuclear Accident for Improving Safety and Security of U.S. Nuclear Plants*. Washington, DC: National Academies of Sciences, Engineering, and Medicine.

NaRC (2016). *Lessons Learned from the Fukushima Nuclear Accident for Improving Safety and Security of U.S. Nuclear Plants: Phase 2*. Washington, DC: National Academies of Sciences, Engineering, and Medicine.

NASA (2010). NASA Risk-Informed Decision Making Handbook. *National Aeronautics and Space Administration*, NASA/SP-2010-576. Washington, DC.

Nesterov, Y. (2018). *Lectures on Convex Optimization*, 2nd ed. New York, NY: Springer.

NRC (2009). Guidance on the Treatment of Uncertainties Associated with PRAs in Risk-Informed Decision Making. Nuclear Regulatory Commission, NUREG-1855. Washington, DC.

Oberguggenberger, M. (2015). Analysis and Computation with Hybrid Random Set Stochastic Models. *Structural Safety*, **52**, 233–243.

Oberguggenberger, M., J. King, and B. Schmelzer (2009). Classical and Imprecise Probability Methods for Sensitivity Analysis in Engineering: A Case Study. *International Journal for Approximate Reasoning*, **50**(4), 680–693.

Oberkampf, W. L. (2019). Simulation Accuracy, Uncertainty, and Predictive Capability: A Physical Sciences Perspective. In C. Beisbart and N. J. Saam (eds.), *Computer Simulation Validation – Fundamental Concepts, Methodological Frameworks, and Philosophical Perspectives*, 69–97. New York, NY: Springer.

Oberkampf, W. L., J. C. Helton, C. A. Joslyn, S. F. Wojtkiewicz, and S. Ferson (2004). Challenge Problems: Uncertainty in System Response Given Uncertain Parameters. *Reliability Engineering and System Safety*, **85**(1–3), 11–19.

Oden, J. T., I. Babuska, and D. Faghihi (2018). Predictive Computational Science: Computer Predictions in the Presence of Uncertainty. In E. Stein, R. de Borst, and T. J. R. Hughes (eds.), *Encyclopedia of Computational Mechanics*. New York, NY: John Wiley.

Oden, J. T., E. E. Prudencio, and P. T. Bauman (2013). Virtual Model Validation of Complex Multiscale Systems: Applications to Nonlinear Elastostatics. *Computer Methods in Applied Mechanics and Engineering*, **266**, 162–184.

Pace, D. K. (2012). Comprehensive Consideration of Uncertainty in Simulation Use. *Journal of Defense Modeling and Simulation: Applications, Methodology, Technology*, **10**(4), 367–380.

Pappenberger, F., M. H. Ramos, H. L. Cloke, F. Wetterhall, L. Alfieri, K. Bogner, et al. (2015). How Do I Know If My Forecasts Are Better? Using Benchmarks in Hydrological Ensemble Prediction. *Journal of Hydrology*, **522**, 697–713.

Park, C., R. T. Haftka, and N. H. Kim (2017). Remarks on Multi-Fidelity Surrogates. *Structural and Multidisciplinary Optimization*, **55**(3), 1029–1050.

Patterson, E. A., and M. P. Whelan (2017). A Framework to Establish Credibility of Computational Models in Biology. *Progress in Biophysics and Molecular Biology*, **129**, 13–19.

Pearl, J. (2009). *Causality: Models, Reasoning and Inference*, 2nd ed. Cambridge: University Press.

Pearl, J., and D. Mackenzie (2018). *The Book of Why: The New Science of Cause and Effect*. New York, NY: Basic Books.

Peherstorfer, B., K. Willcox, and M. Gunzburger (2018). Survey of Multifidelity Methods in Uncertainty Propagation, Inference, and Optimization. *SIAM Review*, **60**(3), 550–591.

Pennell, C., and T. Reichler (2011). On the Effective Number of Climate Models. *Journal of Climate Models*, **24**(9), 2358–2367.

Petroski, H. (2012). *To Forgive Design: Understanding Failure*. Cambridge, MA: Harvard University Press.

Petturiti, D., and B. Vantaggi (2018). Upper and Lower Conditional Probabilities Induced by a Multivalued Mapping. *Journal of Mathematical Analysis and Applications*, **458**(2), 1214–1235.

Phillips, T. S., and C. J. Roy (2016). A New Extrapolation-Based Uncertainty Estimator for Computational Fluid Dynamics. *Journal of Verification, Validation and Uncertainty Quantification*, **1**(4), 041006-1–041006-12.

Porter, N. W., K. A. Maupin, L. P. Swiler, and V. A. Mousseau (2021). Validation Metrics for Fixed Effects and Mixed-Effects Calibration. *Journal of Verification, Validation and Uncertainty Quantification*, **6**(1), 011005-1–011005-11.

Rabinovich, S. G. (2005). *Measurement Errors and Uncertainties: Theory and Practice*, 3rd ed. New York, NY: Springer.

Raizer, V., and I. Elishakoff (2022). *Philosophies of Structural Safety and Reliability*. Boca Raton, FL: CRC Press.

Rajabi, M. M., and B. Ataie-Ashtiani (2016). Efficient Fuzzy Bayesian Inference Algorithms for Incorporating Expert Knowledge in Parameter Estimation. *Journal of Hydrology*, **536**, 255–272.

RAMAS (2023). RAMAS Risk Calc. www.ramas.com/risk-calc-4-0.

Raol, J. R. (2016). *Data Fusion Mathematics: Theory and Practice*. Boca Raton, FL: CRC Press.

Red-Horse, J. R., and T. L. Paez (2008). Sandia National Laboratories Validation Workshop: Structural Dynamics Application. *Computer Methods in Applied Mechanics and Engineering*, **197**(29–32), 2578–2584.

Regan, H. M., S. Ferson, and D. Berleant (2004). Equivalence of Methods for Uncertainty Propagation of Real-Valued Random Variables. *International Journal of Approximate Reasoning*, **36**(1), 1–30.

Ribes, A., J. Pouderoux, and B. Iooss (2019). A Visual Sensitivity Analysis for Parameter-Augmented Ensembles of Curves. *Journal of Verification, Validation and Uncertainty Quantification*, **4**(4), 041007-1–041007-12.

Rider, W., W. Witkowski, J. R. Kamm, and T. Wildey (2016). Robust Verification Analysis. *Journal of Computational Physics*, **307**, 146–163.

Riedmaier, S., B. Danquah, B. Schick, and F. Diemeyer (2020). Unified Framework and Survey for Model Verification, Validation and Uncertainty Quantification. *Archives of Computational Methods in Engineering*, **28**. Doi: 10.1007/s11831-020-09473-7, 2655-2688.

Roache, P. J. (1998). *Verification and Validation in Computational Science and Engineering*. Albuquerque, NM: Hermosa Publishers.

Rocchetta, R., M. Broggi, and E. Patelli (2018). Do We Have Enough Data? Robust Reliability via Uncertainty Quantification. *Applied Mathematical Modeling*, **54**, 710–721.

Roy, C. J., and M. S. Balch (2012). A Holistic Approach to Uncertainty Quantification with Application to Supersonic Nozzle Thrust. *International Journal for Uncertainty Quantification*, **2**(4), 363–381.

Roy, C. J., and W. L. Oberkampf (2011). A Comprehensive Framework for Verification, Validation, and Uncertainty Quantification in Scientific Computing. *Computer Methods in Applied Mechanics and Engineering*, **200**(25–28), 2131–2144.

Roy, C., and W. L. Oberkampf (2016). Validation. In B. Engquist (ed.), *Encyclopedia of Applied and Computational Mathematics*. New York, NY: Springer.

Rubinstein, R. Y., and D. P. Kroese (2016). *Simulation and the Monte Carlo Method*, 3rd ed. New York, NY: John Wiley.

Rui, K., Z. Qingyuan, Z. Zhiguo, E. Zio, and L. Xiaoyang (2016). Measuring Reliability under Epistemic Uncertainty: Review on Non-Probabilistic Reliability Metrics. *Chinese Journal of Aeronautics*, **29**(3), 571–579.

Rutherford, B. M. (2008). Computational Modeling Issues and Methods for the "Regulatory Problem" in Engineering – Solution to the Thermal Problem. *Computer Methods in Applied Mechanics and Engineering*, **197**(29–32), 2480–2489.

Salicone, S., and M. Prioli (2018). *Measuring Uncertainty within the Theory of Evidence*. New York, NY: Springer.

Saltelli, A., K. Chan, and E. M. Scott (2009). *Sensitivity Analysis*. New York, NY: John Wiley.

Saltelli, A., M. Ratto, T. Andres, F. Campolongo, J. Cariboni, D. Gatelli, et al. (2008). *Global Sensitivity Analysis: The Primer*. New York, NY: John Wiley.

Sandia (2023). Sandia National Laboratories DAKOTA Software. https://dakota.sandia.gov/.

Sankararaman, S., and S. Mahadevan (2015). Integration of Model Verification, Validation, and Calibration for Uncertainty Quantification in Engineering Systems. *Reliability Engineering and System Safety*, **138**, 194–209.

Santner, T. J., B. J. Williams, and W. Notz (2018). *The Design and Analysis of Computer Experiments*, 2nd ed. New York, NY: Springer.

Schilders, W. H. A., H. A. van der Vorst, and J. Rommes (eds.) (2008). *Model Order Reduction: Theory, Research Aspects and Applications*. Berlin, Germany: Springer.

Schobi, R., and B. Sudret (2017a). Structural Reliability Analysis for P-Boxes Using Multi-Level Meta-Models. *Probabilistic Engineering Mechanics*, **48**, 27–38.

Schobi, R., and B. Sudret (2017b). Uncertainty Propagation of P-Boxes Using Sparse Polynomial Chaos Expansions. *Journal of Computational Physics*, **339**, 307–327.

Schobi, R., and B. Sudret (2019). Global Sensitivity Analysis in the Context of Imprecise Probabilities (P-Boxes) Using Sparse Polynomial Chaos Expansions. *Reliability Engineering and System Safety*, **187**, 129–141.

Schroeder, B. B., H. Silva, and K. D. Smith (2018). Separability of Mesh Bias and Parametric Uncertainty for a Full System Thermal Analysis. *Journal of Verification, Validation and Uncertainty Quantification*, **3**(3), 031006-1–031006-12.

Segalman, D. J., T. L. Paez, and L. E. Bauman (2017). A Robust Approach to Quantification of Margin and Uncertainty. *Journal of Verification, Validation and Uncertainty Quantification*, **2**(1), 011005-1–011005-10.

Seibert, J., M. Staudinger, and H. J. van Meerveld (2019). Validation and Over-Parameterization-Experiences from Hydrological Modeling. In C. Beisbart and N. J. Saam (eds.), *Computer Simulation Validation – Fundamental Concepts, Methodological Frameworks, and Philosophical Perspectives*. New York, NY: Springer.

Shafer, G. (2016). A Mathematical Theory of Evidence Turns 40. *International Journal for Approximate Reasoning*, **79**, 7–25.

Shah, H., S. Hosder, and T. Winter (2015). Quantification of Margins and Mixed Uncertainties Using Evidence Theory and Stochastic Expansions. *Reliability Engineering and System Safety*, **138**, 59–72.

Shonkwiler, R. W., and F. Mendivil (2009). *Explorations in Monte Carlo Methods*. New York, NY: Springer.

Smith, R. C. (2014). *Uncertainty Quantification: Theory, Implementation, and Applications*. Philadelphia, PA: Society for Industrial and Applied Mathematics.

Smithson, M. (1989). *Ignorance and Uncertainty: Emerging Paradigms*. New York, NY: Springer.

Soll, J., and J. Klayman (2004). Overconfidence in Interval Estimates. *Journal of Experimental Psychology: Learning, Memory, and Cognition*, **30**(2), 299–314.

Song, J., P. Wei, M. Valdebenito, S. Bi, M. Broggi, M. Beer, et al. (2019). Generalization of Non-Intrusive Imprecise Stochastic Simulation for Mixed Uncertain Variables. *Mechanical Systems and Signal Processing*, **134**, 106316-1–106316-17.

Song, Y., J. Mi, Y. Cheng, L. Bai, and K. Chen (2020). A Dependency Bounds Analysis Method for Reliability Assessment of Complex System with Hybrid Uncertainty. *Reliability Engineering and System Safety*, **204**, 107119-1–107119-13.

Steinhauser, M. O. (2017). *Computational Multiscale Modeling of Fluids and Solids: Theory and Applications*, 2nd ed. New York, NY: Springer.

Stern, F., M. Diez, H. Sadat-Hosseini, H. Yoon, and F. Quadvlieg (2017). Statistical Approach for Computational Fluid Dynamics State-of-the-Art Assessment: N-Version Verification and Validation. *Journal of Verification, Validation and Uncertainty Quantification*, **2**(3), 031004-1–031004-15.

Storlie, C. B., and J. C. Helton (2008a). Multiple Predictor Smoothing Methods for Sensitivity Analysis: Description of Techniques. *Reliability Engineering and System Safety*, **93**(1), 28–54.

Storlie, C. B., and J. C. Helton (2008b). Multiple Predictor Smoothing Methods for Sensitivity Analysis: Example Results. *Reliability Engineering and System Safety*, **93**(1), 55–77.

Stripling, H. F., M. L. Adams, R. G. McClarren, and B. K. Mallick (2011). The Method of Manufactured Universes for Validating Uncertainty Quantification Methods. *Reliability Engineering and System Safety*, **96**(9), 1242–1256.

Sullivan, T. J. (2015). *Introduction to Uncertainty Quantification*. New York, NY: Springer.

Sun, L., H. Gao, S. Pan, and J.-X. Wang (2020). Surrogate Modeling for Fluid Flows Based on Physics-Constrained Deep Learning Without Simulation Data. *Computational Methods in Applied Mechanics and Engineering*, **361**, 112732-1–112732-25.

Szabo, B., R. Actis, and D. Rusk (2019). Validation of Notch Sensitivity Factors. *Journal of Verification, Validation and Uncertainty Quantification*, **4**(3), 011004-1–011004-8.

Szabo, B., R. Actis, and D. Rusk (2021). Validation Test of a Predictor of Fatigue Failure under Biaxial Out-of-Phase Loading Conditions. *Computer Methods in Applied Mechanics and Engineering*, **378**, 113718-1–113718-13.

Szabo, B., and I. Babuska (2021a). *Finite Element Analysis: Method, Verification and Validation*, 2nd ed. New York, NY: John Wiley.

Szabo, B., and I. Babuska (2021b). Methodology of Model Development in the Applied Sciences. *Journal of Computational and Applied Mechanics*, **16**(2), 75–86.

Todinov, M. (2018). *Methods for Reliability Improvement and Risk Reduction*. New York, NY: John Wiley.

Toppila, A., and A. Salo (2017). Selection of Risk Reduction Portfolios under Interval-Valued Probabilities. *Reliability Engineering and System Safety*, **163**, 69–78.

Torra, V., A. Dahlbom, and Y. Narukawa (eds.) (2017). *Fuzzy Sets, Rough Sets, Multisets and Clustering*. Berlin, Germany: Springer.

Torra, V., and Y. Narukawa (2007). *Modeling Decision: Information Fusion and Aggregation Operators*. New York, NY: Springer.

Torra, V., Y. Narukawa, and M. Sugeno (eds.) (2014). *Non-Additive Measures*. Berlin, Germany: Springer.

Trucano, T. G., M. Pilch, and W. L. Oberkampf (2002). General Concepts for Experimental Validation of ASCI Code Applications. *Sandia National Laboratories*, SAND2002-0341, Albuquerque, NM.

Vanslette, K., T. Tohme, and K. Youcef-Toumi (2020). A General Model Validation and Testing Tool. *Reliability Engineering and System Safety*, **195**, 106684-1–106684-17.

Verfurth, R. (2013). *A Posteriori Error Estimation Techniques for Finite Element Methods*. Oxford: Oxford University Press.

Vose, D. (2008). *Risk Analysis: A Quantitative Guide*, 3rd ed. New York, NY: John Wiley.

Walley, P. (2000). Towards a Unified Theory of Imprecise Probability. *International Journal of Approximate Reasoning*, **24**(2–3), 125–148.

Wang, C., H. G. Matthies, M. Xu, and Y. Li (2018). Epistemic Uncertainty-Based Model Validation via Interval Propagation and Parameter Calibration. *Computer Methods in Applied Mechanics and Engineering*, **342**, 161–176.

Wang, C., H. Zhang, and M. Beer (2018). Computing Tight Bounds of Structural Reliability under Imprecise Probabilistic Information. *Computers and Structures*, **208**, 92–104.

Wang, J.-X., C. J. Roy, and H. Xiao (2018). Propagation of Input Uncertainty in Presence of Model-Form Uncertainty: A Multifidelity Approach for Computational Fluid Dynamics Applications. *ASCE-ASME*

Journal of Risk and Uncertainty in Engineering Systems, Part B; Mechanical Engineering, **4**(1), 011002-1–011002-8.

Wang, T., J. S. Dyer, and W. J. Hahn (2017). Sensitivity Analysis of Decision Making under Dependent Uncertainties Using Copulas. *EURO Journal of Decision Processes*, **5**(1–4), 117–139.

Wei, P., J. Song, S. Bi, M. Broggi, M. Beer, Z. Lu, et al. (2019a). Non-Intrusive Stochastic Analysis with Parameterized Imprecise Probability Models: I. Performance. *Mechanical Systems and Signal Processing*, **124**, 349–368.

Wei, P., J. Song, S. Bi, M. Broggi, M. Beer, Z. Lu, et al. (2019b). Non-Intrusive Stochastic Analysis with Parameterized Imprecise Probability Models: II. Reliability and Rare Events Analysis. *Mechanical Systems and Signal Processing*, **126**, 349–368.

Weirs, V. G., J. R. Kamm, L. P. Swiler, S. Tarantola, M. Ratto, B. M. Adams, et al. (2012). Sensitivity Analysis Techniques Applied to a System of Hyperbolic Conservation Laws. *Reliability Engineering and System Safety*, **107**, 157–170.

Weisberg, H. I. (2014). *Willful Ignorance: the Mismeasure of Uncertainty*. New York, NY: John Wiley.

Whiting, N. W., C. J. Roy, E. Duque, S. Lawrence, and W. L. Oberkampf (2023). Assessment of Model Validation, Calibration, and Prediction Approaches in the Presence of Uncertainty. *Journal of Verification, Validation and Uncertainty Quantification*, **8**(1), 011001-1–011001-21.

Wu, D., Z. Lu, Y. Wang, and L. Cheng (2015). Model Validation and Calibration Based on Component Functions of Model Output. *Reliability Engineering and System Safety*, **140**(1), 59–70.

Xi, Z., H. Pan, Y. Fu, and R.-J. Yang (2015). Validation Metric for Dynamic System Responses under Uncertainty. *SAE International Journal of Materials and Manufacturing*, **8**(2), 309–314.

Xi, Z., H. Pan, and R.-J. Yang (2017). Time Dependent Model Bias Correction for Model Based Reliability Analysis. *Structural Safety*, **66**, 74–83.

Xi, Z., and R.-J. Yang (2016). Reliability Analysis with Model Uncertainty Coupling with Parameter and Experiment Uncertainties: A Case Study of 2014 Verification and Validation Challenge Problem. *Journal of Verification, Validation and Uncertainty Quantification*, **1**(1), 011005-1–011005-11.

Xiao, H., J.-X. Wang, and R. G. Ghanem (2017). A Random Matrix Approach for Quantifying Model-Form Uncertainties in Turbulence Modeling. *Computer Methods in Applied Mechanics and Engineering*, **313**, 941–965.

Xiao, N.-C., H.-Z. Huang, Z. Wang, Y. Pang, and L. He (2011). Reliability Sensitivity Analysis for Structural Systems in Interval Probability Form. *Structural and Multidisciplinary Optimization*, **44**, 691–705.

Xiao, Z., X. Han, C. Jiang, and G. Yang (2016). An Efficient Uncertainty Propagation Method for Parameterized Probability Boxes. *Acta Mechanica*, **227**(3), 633–649.

Xie, C., G. Li, and F. Wei (2018). An Integrated QMU Approach to Structural Reliability Assessment Based on Evidence Theory and Kriging Model with Adaptive Sampling. *Reliability Engineering and System Safety*, **171**, 112–122.

Yan, N. (2012). *Finite Element Methods: Super-Convergent Analysis and a Posteriori Error Estimation*. Oxford: Alpha Science International.

Yin, S., D. Yu, H. Yin, and B. Xia (2016). Interval and Random Analysis for Structure–Acoustic Systems with Large Uncertain-but-Bounded Parameters. *Computer Methods in Applied Mechanics and Engineering*, **305**, 910–935.

Zhang, H., R. L. Mullen, and R. L. Muhanna (2010). Interval Monte Carlo Methods for Structural Reliability. *Structural Safety*, **32**(3), 183–190.

Zhang, Q., Z. Zeng, E. Zio, and R. Kang (2017). Probability Box as a Tool to Model and Control the Effect of Epistemic Uncertainty in Multiple Dependent Competing Failure Processes. *Applied Soft Computing*, **56**, 570–579.

Zhao, H., and V. A. Mousseau (2012). Use of Forward Sensitivity Analysis Method to Improve Code Scaling, Applicability, and Uncertainty (CSAU) Methodology. *Nuclear Engineering and Design*, **249**, 188–196.

Zhao, L., Z. Lu, W. Yun, and W. Wang (2017). Validation Metric Based on Mahalanobis Distance for Models with Multiple Correlated Responses. *Reliability Engineering and System Safety*, **159**, 80–89.

Zhu, S.-P., H.-Z. Huang, W. Peng, H.-K. Wang, and S. Mahadevan (2016). Probabilistic Physics of Failure-Based Framework for Fatigue Life Prediction of Aircraft Gas Turbine Discs under Uncertainty. *Reliability Engineering and System Safety*, **146**, 1–12.

Zhu, Y., N. Zabaras, P.-S. Koutsourelakis, and P. Perdikaris (2019). Physics-Constrained Deep Learning for High-Dimensional Surrogate Modeling and Uncertainty Quantification Without Labeled Data. *Journal of Computational Physics*, **394**, 56–81.

Zienkiewicz, O. C., and J. Z. Zhu (1992). The Superconvergent Patch Recovery and a Posteriori Error Estimates. *International Journal for Numerical Methods in Engineering*, **33**(7), 1365–1382.

Part V

Planning, Management, and Implementation Issues

This final section of the book primarily deals with the topic of how managers, including both line managers and project managers, plan, implement, and develop verification, validation, and uncertainty quantification (VVUQ) capabilities in their organizations. Chapter 14 deals with a process for allocating necessary resources for M&S to best achieve the goals of a project. Our discussion applies to projects within both industry and government. We emphasize engineering system projects, but the discussion also applies to the analysis of natural systems, such as environmental M&S.

Chapter 15 reviews several approaches for assessing the maturity and quality of an M&S effort. We then present a detailed discussion of the Predictive Capability Maturity Model (PCMM) developed at Sandia National Laboratories. The procedure has proven very beneficial in assessing progress in M&S activities, as well as helping identify activities where improvements are needed.

Chapter 16 discusses a range of staff and management responsibilities for VVUQ activities. The new topic of simulation governance for business organizations is explained with regard to responsibilities and recommended practices for executive management. The chapter closes with a discussion of the practical and technical issues for the development of V&V databases.

14

Planning and Prioritization in Modeling and Simulation

Without goals, and plans to reach them, you are like a ship that has set sail with no destination. —Fitzhugh Dodson

Contents of Chapter 14

This chapter makes a distinct change in topical direction from the earlier chapters in this text. Chapters 14, 15, and 16 discuss the organization, planning, and management of modeling and simulation (M&S) activities, particularly verification, validation, and uncertainty quantification (VVUQ). Although the importance and prevalence of simulation has grown astoundingly over the last few decades, relatively little work has been conducted in the planning and management of M&S activities. Most of the published work in planning and prioritization of M&S and experiments has been in the field of nuclear power plant safety.

Given the wide range of activities dealing with M&S and the credibility of simulation results, this chapter focuses on how resources should be allocated to achieve the goals of the simulation project in an efficient manner. The perspective here is that the simulation goals are primarily determined by the customer of the simulation project. For example, the customer could be a system designer, a project manager, or a regulatory authority. As discussed in Chapter 3 "Modeling and Computational Simulation," simulation provides two types of information: scientific knowledge and technological knowledge. The emphasis in this chapter is on planning and prioritization of M&S for technological knowledge, i.e., knowledge for the design of new systems or manufacturing processes, or modification of existing systems or processes. This chapter focuses on management responsibilities for large-scale, project-oriented activities, as opposed to scientific research. The discussion focuses on using the phenomena identification and ranking table (PIRT) and an M&S gap analysis process. Engineering system projects are emphasized, but the discussion also applies to the analysis of natural systems, e.g., underground storage of radioactive wastes, global climate change, and transport of contaminants or toxic materials due to atmospheric winds.

14.1 Methodology for Planning and Prioritization

For large-scale projects, planning and prioritization (P&P) are major activities that require a significant investment in time, money, and specialized personnel. The P&P discussed in this chapter includes both mathematical modeling, which is the engine for producing results for a customer, and VVUQ, which addresses the credibility of simulation results. Two related situations for P&P are discussed here. First, the M&S effort is intended to produce computational results, i.e., information, that will be used in simulation-informed decision making. For example, simulation results are used to design and optimize a new engineering system, or simulation results are used to improve the reliability of a manufacturing process. Second, the effort is intended to develop a simulation capability or purchase an M&S software package that will address planned applications in the future. For example, a corporation is interested in developing an M&S capability that can be used to reduce the time to market of new products or reduce the time and expense for physical testing in order to achieve product certification or regulatory approval.

Depending on the type of organization conducting the P&P, it can be challenging for management to assemble the personnel with the breadth and depth of talent needed for the task. For example, if a large-scale M&S project arrives in the midst of ongoing M&S activities for other projects, it can be difficult for management to enlist not only the M&S and VVUQ talent, but also people who are experienced with the new system or manufacturing process under consideration. Furthermore, the new project will probably require that the technical staff and managers change pre-existing viewpoints, traditions, and ingrained habits that may no longer apply or could be detrimental to the goals of the new project. If the organization is a government agency or a non-profit institute, the needed changes in perspective and priorities can be particularly difficult for management to affect.

There is a widely held view that the crucial importance of P&P is commonly ignored, almost always with dire consequences (Dorner, 1989; Vose, 2008; Stepanek, 2012; Huckle and Neckel, 2019). In an article in *IEEE Spectrum* in 2005, Charette (2005) states, "This year, organizations and governments will spend an estimated $1 trillion on IT hardware, software, and services worldwide. Of the IT projects that are initiated, from 5 to 15 percent

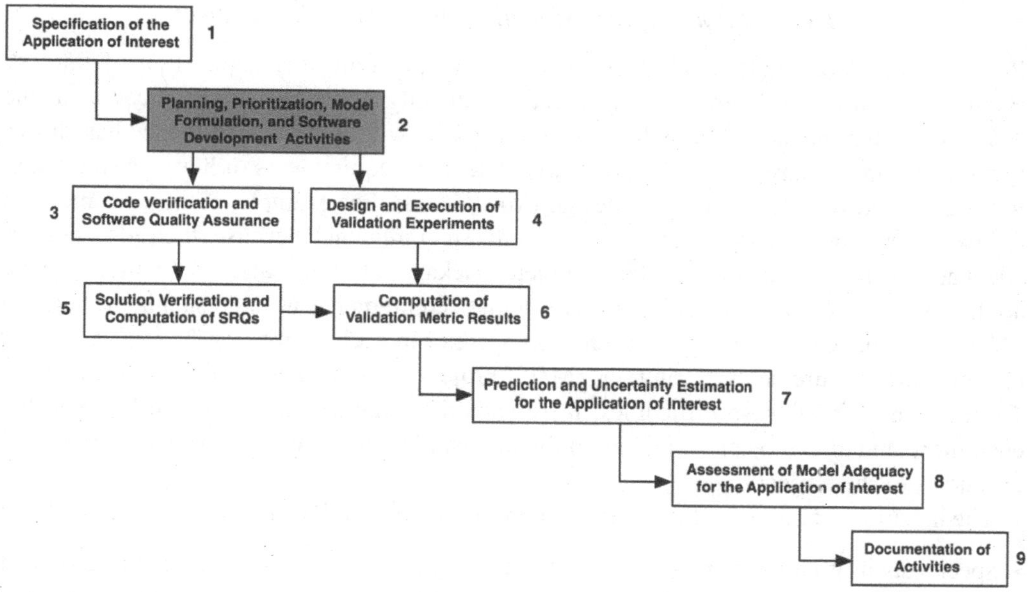

Figure 14.1 Integrated view of the elements of verification, validation, and uncertainty quantification.
(adapted from Trucano et al., 2002)

will be abandoned before or shortly after delivery as hopelessly inadequate. Many others will arrive late and over budget or require massive reworking. Few IT projects, in other words, truly succeed." Charette lists a dozen of the most common reasons for massive software failures, but several of these can be grouped as (a) unrealistic or unarticulated project goals, and/or (b) lack of P&P in the project. Most failures are poorly publicized and almost never document the root causes and lessons learned. The responsible organizations and government agencies just want the failures to disappear.

In Chapter 2 "Fundamental Concepts and Terminology" a summary discussion was given concerning an integrated view of VVUQ (Figure 14.1). Although this chapter addresses P&P depicted as Element 2 in Figure 14.1, P&P impacts all of the VVUQ activities shown in the other elements in the figure. P&P for M&S is not well-developed compared to P&P for general purpose software development projects. (See Sharp and Wood-Schultz, 2003, for early thoughts on how to build confidence in simulation for decision making as part of the Advanced Simulation and Computing [ASC] Program sponsored by the US National Nuclear Security Administration.) Large-scale software projects have been conducted for roughly three or four decades and various techniques are available for P&P for software projects (Galin, 2003; Wiegers and Beatty, 2013; Brechner, 2015; McConnell, 2019; Martin, 2019). However, large-scale M&S projects are quite new. The field of nuclear power plant safety has led the way in P&P of M&S, but much of this work has been focused on improved understanding and mitigating potentially serious nuclear accidents. As a result, most fields of scientific computing are not aware of this work. This chapter is primarily based on the work of the nuclear power safety field.

14.1.1 Planning for a Modeling and Simulation Project

Planning in a large-scale M&S project begins by specifying the application of interest, Element 1, shown in Figure 14.1. This activity directly relies on the specification of the customer requirements and how the simulation results will be used. Experience has shown that the customers may not be able to clearly state their requirements or how the simulation will be used in decision making or design (Vose, 2008). For example, if the customer is a regulatory authority, they cannot give specific requirements because their decision of adequacy is in the context of the complete package of information submitted by the applicant. The hesitancy of customers to define requirements is understandable because M&S results have traditionally *not* been a major factor in decision making. When a customer changes the requirements during the M&S project, this creates additional difficulties. To alleviate both of these difficulties, there should be significant interaction between the customers and the project team, not only in planning the project, but also during the execution of the project.

The most important topics that must be addressed *before* P&P activities can begin include:

- Specification of customer requirements. Most customers who use M&S results for decision making tend to think of requirements for the system or application of interest; *not* M&S requirements. M&S project management must work with the customer to map or transform the customer's information needs for the system or application of interest into M&S requirements.
- Specification of the application of interest. The key physical processes, engineering system, or event of interest should be well defined. The expected informational-contributions or value added from the simulation effort should also be specified.
- Specification of the environment for the system of interest. The application domain could be restricted to the normal environment, e.g., the range of operating conditions to which the system would normally be exposed. The environment of interest could be the system of interest exposed to certain abnormal or hostile environments. If abnormal or hostile environments are of interest, then it is particularly important for the customers to clarify their specific concerns. For example, there may be a concern that a system failure could place personnel at risk, or the simulation may be intended to determine if the system would fail in a safe sequence of events.
- Identification of important scenarios for the specified environment. The scenarios correspond to conditions that the system could be exposed to within a specified environment, various event sequences, or fault trees. Note that even if specific scenarios are identified, the degraded performance or failure modes and sequences can be quite difficult to identify before the P&P activity.
- Specification of the state of the system and its surroundings. Different specifications for the system and the surroundings could be given, depending on the environment and scenarios of interest. For example, in a normal environment, a scenario of interest may be that the system is near a scheduled service, or the scheduled service was not done as prescribed on the system.
- Specification of the system response quantities (SRQs) of interest. This should include all SRQs for the environment and scenarios of interest. For a normal environment, typical SRQs of interest are related to system performance and reliability. For abnormal and hostile environments, SRQs are usually related to some aspect of safety or security of the system.

- Specification of the accuracy requirements for each SRQ of interest. If the system design is at an early stage, or the environment and scenarios are not well defined, these requirements may be very poorly known. For example, accuracy requirements of some SRQs may only be known to within an order of magnitude, or only a qualitative feature of the SRQ may be of interest. Even though the customer may poorly specify accuracy requirements, it is of utmost importance that initial expectations be discussed with the M&S project management.

It should be noted that the term *customers* includes both those who will directly use the results of the simulation and those simulation engineers who will use the software developed or purchased by the project. Stakeholders or interested parties can also be important factors in the discussion of the tasks listed previously. The term *stakeholders* includes those who have any type of vested interest in the simulation results or development of a simulation capability, those who use the simulation results in follow-on activities, and those whose success is dependent on the success of the customer or the simulation capability. The funding source for the simulation results or the simulation capability is usually the customer, but in certain situations this is not the case. The most common example of this is funding of a private organization or institute by a governmental organization to conduct the simulation. When the funding source is *not* the customer, the likelihood of confusion, misunderstandings, delivering the wrong product, or misinterpretation of the results increases dramatically.

Completing the tasks listed previously requires significant effort by the M&S project managers and technical staff, the customers, and important stakeholders. It will require in-depth, forthright, and often delicate discussions and negotiation of trade-offs between all of the parties involved. Sometimes the customers only have a vague idea of their needed requirements because (a) they are inexperienced in using simulation results in decision making, (b) they have not carefully thought about their requirements, and (c) the application of interest has not yet been fully defined. Sometimes the customers will request unrealistic goals from a simulation effort, given the time and money available for the effort. The M&S project manager can also be unrealistic and inexperienced in what simulation results can be produced or developed/purchased, given the resources available.

The cause of many simulation project failures is that the M&S project manager either (a) is initially unrealistic in what could be accomplished, given the constraints of time and resources; or (b) allows significant changes or increases in requirements, features, or capabilities to be imposed during the simulation effort. During the negotiation of the specifications given in the previous bulleted list, there must be flexibility by the M&S project management team and the customers concerning costs for various capabilities desired. A significant aid in the negotiations is when the customers clearly express their priorities and value system concerning trade-offs of deliverables and time schedules. With this knowledge, M&S project management can suggest alternatives to the customers in order to satisfy the cost and schedule requirements.

14.1.2 Methods for Prioritization of Activities

This section addresses various methods that can be used for prioritizing work on activities such as: physics modeling, software development, software quality assurance (SQA), code verification, solution verification, physical tests associated with model validation experiments, construction of validation metrics, and UQ associated with predictive capability. The focus here is how to best allocate resources between various activities to achieve the highest possible level of success of a project-oriented M&S effort, given constraints on time and resources. It must be stressed that the success of the M&S effort is *not* synonymous with success of the engineering system project that is being assisted. The success of the engineering project typically depends on many other factors not of interest in this discussion. *Success of the M&S effort* means that the effort (a) achieves the M&S requirements and specifications discussed previously, and (b) contributes to the engineering system attaining its performance, safety, and reliability requirements.

Resources primarily refer to the time, money, personnel expertise, and facilities (both computational and experimental) needed to complete an activity. These are grouped together for convenience, but it must be stressed that there are crucial differences in the nature, development, and deployment of each of these. In addition, there are strong dependencies between each of these resources. For example, there is typically a strong connection between the personnel that conduct SQA activities and those that conduct code verification activities, i.e., they may be the same people. Likewise, there is usually overlapping expertise between physics model builders and software developers.

A specific computer code, or suite of codes, should be considered by the M&S project manager during the P&P. Alternatively, the M&S project manager may be comparing the (advertised) capabilities of competing commercial software packages before a license is purchased. There are various methods that can be used to prioritize project-oriented simulation activities. The fundamental difference between methods is the *value system* inherent in each method. The value system means the importance of the relationship between different features of (a) the expected simulation results, (b) the simulation capability at present in an organization or potentially available in a commercial software package, or (c) what is required to achieve the goals of the simulation project. In some simulation projects, the prioritization of activities has little to do with project-oriented goals, but more with the organizational power base of various groups within the organization conducting the M&S effort.

Since there are many different types of activities that are conducted in each of the elements shown in Figure 14.1, all of the activities need to be related to one common feature that links them. The most logical feature to use is the physics-modeling capability for the application of interest, i.e., mathematical modeling of the various physical processes occurring in the system of interest. Without this common feature, the remainder of the activities in the figure would have little meaning or effect on the goals of the M&S effort. Three of the more common value systems and when they may be appropriate in different situations are discussed below.

1. Rank the modeling of individual physical processes occurring in the system according to the expected or anticipated level of *inaccuracy*. In this approach, the highest rank would be given to those physical processes that are expected to be modeled the *least* accurately. This prioritization scheme is based on the present level of understanding of the physical processes and their interactions within the system. This approach would be appropriate for improving the understanding of complex, coupled systems and multi-physics processes. It is particularly effective for systems exposed to abnormal or hostile

environments, such as severe accident conditions, seriously damaged systems, an electromagnetic pulse attack on a computer-controlled system, and/or a terrorist attack on a public utility or transportation system. This approach would *not* be appropriate for engineering systems in normal environments, such as well-understood operating conditions.

2. Rank according to the expected impact of individual physical processes on SRQs of particular interest. In this approach, the highest rank would be given to those physical processes that have the *largest expected effect* on specific SRQs related to the goals of the simulation effort. This approach would be appropriate for improving the understanding of complex physics processes in abnormal and hostile environments. However, the focus here is on ranking the *impact of the physical processes* on specific system responses, *not* just on understanding the system and *not* on the expected inaccuracy of the simulation. The most well-developed method using this prioritization scheme is the Phenomena Identification and Ranking Table (PIRT). PIRT was developed by several organizations for the US Nuclear Regulatory Commission (NRC) to improve assessment of nuclear reactor safety in abnormal, i.e., accident, environments. The PIRT approach is discussed in Section 14.2.
3. Rank according to the *present ability* to predict the impact of individual physical processes on SRQs of interest. This is a two-step method in which the first step is conducting Approach 2 just mentioned, and then focusing on the important physical processes to determine if the present modeling capability can adequately address these phenomena. This method concentrates on the *perceived weaknesses* in the present modeling capability to predict the specified SRQs related to the goals of the M&S effort. As a result, this prioritization scheme is commonly referred to as a *gap analysis method*. Although this approach was initiated by several organizations for the NRC, it has recently been extended by Sandia National Laboratories as part of the Advanced Simulation and Computing (ASC) Program. The gap analysis method has been used for normal environments, but it is particularly effective for abnormal and hostile environments. This method is discussed in Section 14.4.

14.2 Phenomena Identification and Ranking Table (PIRT)

The nuclear reactor safety community, as it has done in many areas of risk and safety assessment, developed a new process for improving the understanding of nuclear plants in potential accident scenarios. In 1989, the NRC issued a revised emergency core cooling system rule for light water reactors that allows the use of best estimate plus uncertainty (BE +U) methods in safety analyses as an option (NRC, 1989a, 1989b). To support the licensing revision, the NRC and its contractors developed the code scaling, applicability, and uncertainty (CSAU) evaluation methodology to demonstrate the feasibility of the BE+U approach (Boyack et al., 1990). The phenomena identification and ranking table (PIRT) was developed as an important step in the CSAU methodology. The PIRT process was initially published by Shaw et al. in 1988 (Shaw et al., 1988) to help analyze the safety of pressurized water nuclear reactors during a loss of coolant accident. Since its initial development, the process has been developed further and applied many times to various nuclear reactor designs. For a detailed discussion of PIRTs, see Hanson et al. (1992), Rohatgi et al. (1997), Wilson and Boyack (1998), and Kroeger et al. (1998).

During the 1990s, the PIRT process was recognized as a much more powerful tool than originally envisioned (Wilson and Boyack, 1998). It was found that PIRT could also be used to aid in identifying critical future experiments and needed improvements in the understanding of important physical processes impacting the safety of a nuclear power plant in an accident environment. However, the additional objectives were still focused on improving the physical understanding of the plant behavior and possible interactions of subsystems during a specific accident scenario. The generalized PIRT process, as described by Wilson and Boyack (1998), is a 15-step process that requires significant time and personnel resources to complete.

As part of the ASC program, various researchers at Sandia National Laboratories shifted the emphasis of the PIRT process towards M&S of the important physical processes identified (Pilch et al., 2001; Tieszen et al., 2002; Trucano et al., 2002; Boughton et al., 2003). In addition, they simplified the PIRT process to five steps, so that it could be used by smaller-scale projects. The PIRT process has focused on abnormal and hostile environments because these environments typically involve large uncertainties in (a) knowing the condition, state, or geometry of the system; (b) the surroundings as they affect the system through the BCs or the excitation function; (c) strongly coupled physics that could occur; and (d) important human or computer-controlled intervention that could occur as part of the functioning and safety of the system. For the design of systems in abnormal or hostile environments, the primary goal is typically *not* to analyze the intended performance of the system, but rather to seek system designs that will behave in predictable ways. A common goal in these situations is that the system is fail-safe or fail-inoperable.

14.2.1 Steps in the PIRT Process for Modeling and Simulation

The five steps for a simplified PIRT process are:

1. Assembly of the team,
2. Definition of the objectives of the PIRT process,
3. Specification of environments and scenarios,
4. Identification of plausible physical phenomena, and
5. Construction of the PIRT.

The following description combines the work and recommendations of Wilson and Boyack (1998), the Sandia researchers referenced earlier, and the present authors.

14.2.1.1 Assembly of the Team

PIRT development is best accomplished using a team that has a broad base of knowledge and expertise in a number of areas. The most important areas for individual knowledge are (a) the range of relevant physics processes and phenomena that could occur, (b) the needs of the customer, (c) the operation of the system and subsystems of interest, (d) the environments and scenarios (E&S) of interest, and (e) the M&S of the system of interest or closely related systems for the given E&S. Additional expertise could be included, depending on the nature of the system and the E&S, such as possible actions of first responders and security personnel to an accident.

The number of people on the team should be roughly 5 to 10. Teams any larger than this become inefficient, unmanageable to keep on track, and meetings become difficult to schedule, as all team members should attend every meeting. The team members should have a demonstrated capability to work in a team environment, particularly the ability to control their individual agendas in the interest of the team effort. The efficiency and productivity of the team can also be significantly improved with the assistance of administrative support staff and possibly a team facilitator. Finally, and most importantly, the team should have a clearly defined leader who has the recognized authority and is ultimately responsible for the effort and its documentation. This person should also have participated in previous PIRT efforts and be familiar with all aspects of the PIRT process.

14.2.1.2 Definition of the Objectives of the PIRT Process

The primary objectives of the PIRT process are to creatively discover the possible physical processes that could occur in some event or sequence of events, and to rank the physical processes according to their importance on specified SRQs. The event or event sequence is typically postulated regarding what could possibly occur in the future, but sometimes a PIRT is constructed for a complex system failure that has already occurred. Typically, the objectives of the PIRT process are oriented toward the performance, safety, or reliability of a system in unusual conditions. However, in the Sandia approach to a PIRT, the orientation is toward activities that are required to simulate the important physical phenomena identified.

Prioritization is fundamentally about ordering activities; some things will get done and some will not. As part of defining the objectives, the team should specify what predictions are needed from the simulation capability in terms of only a few SRQs. If multiple physics codes are coupled together to produce the SRQs of ultimate interest, there may be several SRQs from intermediate codes that are needed to predict the ultimate SRQs of interest. Also, the SRQs of interest can be quantitative or qualitative in nature. For example, suppose the accident scenario of interest is concerned with the separation or fracturing of a fan blade attached to the rotor disk on a gas turbine engine. Two characteristics of interest would be: (1) What is the possibility of any fragments of the fan penetrating the engine cowling after the blade separates from the rotor? and (2) What is the possibility of fan or rotor fragments that are ingested in the compressor or turbine stages of the engine causing these stages to fracture, or possibly causing the entire engine to explode?

In defining the objectives of the PIRT process, there should be *no connection* to the existing M&S capabilities. The PIRT process is focused on discovering the most important physical phenomena that could affect a system and its response, *not* on whether an existing M&S capability can simulate the phenomena. Discussing existing or advertised M&S capability in the midst of a PIRT process completely derails the goal of constructing a PIRT. Excluding discussion of existing or advertised capabilities from the PIRT process requires exceptional discipline and leadership from the team.

During formulation of the objectives of the PIRT process, the team must estimate the resources required to complete the identified objectives, compared with the resources available for the effort. The team should be cautious and restrictive in setting objectives

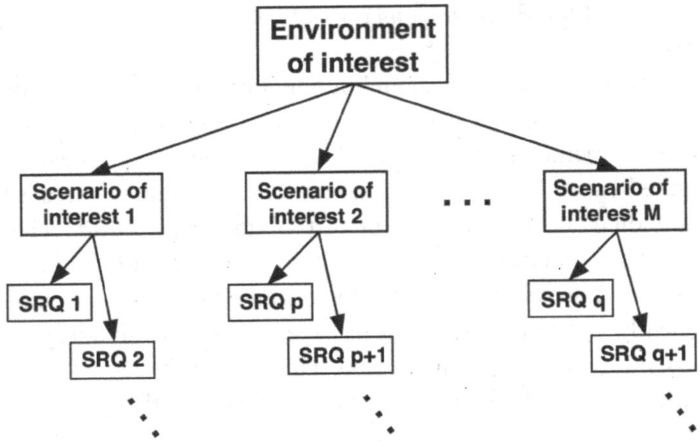

Figure 14.2 Environment-scenario SRQ tree.

because of the risk of expanding the time and resources required to conduct the effort and document the results.

14.2.1.3 Specification of Environments and Scenarios

Only one environment should be specified for analyzing in the PIRT process: normal, abnormal, or hostile. For the environment chosen, one or more scenarios are usually specified, but they should be closely related. The scenarios specified should match the technical expertise and background of the team. If the scenarios are unrelated, then separate PIRT process teams should be considered. In the specification of the E&S, key parameters characterizing the system and surroundings should be specified, along with their possible range. For example, these could be: (a) the quantity and physical properties of coolant available in an emergency cooling system, (b) the type and range of initial damage to a system or subsystem, (c) the quantity of fuel available in a fuel fire, and (d) the level of physical and electronic access to the computer system controlling a physical system that is under attack.

Figure 14.2 depicts the given environment chosen, the scenarios of interest, and the SRQs of interest for each scenario. Note that the SRQs for each scenario are not necessarily the same. It is seen from this tree structure that the total number of SRQs can become quite large, if the scope of the effort is not limited. In certain situations, the team may be given freedom to identify and pursue scenarios that were not identified at the beginning of the PIRT process. For example, as the PIRT process evolves, new failure modes or coupled subsystem interactions are commonly discovered that were not initially recognized. The option of pursuing new scenarios should be specified in setting the objectives in Step 1 because, as will be seen in Section 14.2.1.5, a PIRT will be constructed for *each* scenario. If PIRT analyses are not pursued for newly identified scenarios, they should be documented for possible future work.

Consider the following example of how one environment and scenario can lead to the identification of additional scenarios. Suppose the system of interest is a machine with considerable energy stored in high-speed rotating parts, such as a material-processing machine or a fluid pump. The specified environment of interest is the abnormal operating case where the machine is excessively vibrating due to either a rotationally asymmetric load, a damaged impeller, or a worn roller bearing. Suppose two scenarios of interest were initially identified: (1) a large vibration that causes a significant increase in the amplitude of vibrational modes in the structure, as well as the appearance of abnormal vibrational modes in the structure; and (2) concern that mechanical fasteners in the machine might possibly become loose, making matters worse. During the investigation of the scenarios, one of the team members, who is familiar with the operation and maintenance of the equipment, identifies a new concern. It is noted that large vibration will cause contaminants in the lubricating fluid to become dislodged and start circulating in the lubrication system. These contaminants could then clog the lubrication filters and starve the bearings of lubricant. This is not good.

14.2.1.4 Identification of Plausible Physical Phenomena

All plausible physical phenomena and interactions of phenomena should be considered for each scenario of interest, given that the phenomena could have an impact on the SRQs of interest. In many scenarios, it has proven beneficial to divide the scenario into time phases, since different physical phenomena tend to be important in different phases. For example, in loss of coolant accidents in nuclear reactors, the time phases are commonly divided into (a) initial rapid loss of coolant, (b) high heating of fuel rods before emergency coolant is circulated, (c) initial introduction of emergency coolant onto high temperature fuel rods, and (d) flooding of the entire system with emergency coolant. In some analyses, it is also useful to spatially divide the system or subsystem into regions where plausible phenomena can be better identified. For example, in a terrorist attack of a facility, it is useful to consider varying levels of physical access and computer control of a facility, as well as levels of armament and technical capability of the attackers.

Trying to discover all of the plausible phenomena that may occur in the system is a difficult and complex task. The team can never be sure that they have discovered all plausible phenomena; which is referred to as the problem of incompleteness or unknown unknowns. Having team members who are creative individuals and who think unconventionally can significantly aid in this step. Different approaches can be used to generate ideas about plausible phenomena. Depending on the makeup of the team, using a type of brainstorming approach has proven beneficial. In this approach, many different and apparently unrelated ideas are suggested. Sometimes these ideas are fruitful in themselves and sometimes the combination of ideas leads to other creative ideas that may not have been discovered.

Regardless of whether a team-brainstorming approach or a more separate individual approach is used, the key is that *no ideas* are criticized or ridiculed. The key element in phenomena identification is discovery, *not* evaluation, appraisal, or ranking of importance of phenomena or ideas. The team leader plays an important role here and must ensure that a proper and constructive atmosphere is maintained during the team's discussions. Simplified simulations of the scenarios, or order of magnitude analyses, can be used to aid in the discussions. Use of computational simulations can be used for a more quantitative analysis, but this approach is problematic for two reasons. First, the discussion begins to focus on details of the simulation, as opposed to discovery of new phenomena. Second, use of existing simulation capabilities tends to narrow the thinking of the team to what is observed given the present simulation capability, instead of creatively thinking about what unexpected and coupled physics are possible.

Table 14.1 *Example of an initial PIRT for a given environment and scenario*

SRQ / Phenomena	SRQ 1	SRQ 2	•••	SRQ m
Physical Phenomenon 1				
Physical Phenomenon 2				
Physical Phenomenon 3				
Physical Phenomenon 4				
⋮				
Physical Phenomenon n				

14.2.1.5 Construction of the PIRT

A PIRT is constructed for *each scenario* of interest by listing all of the plausible phenomena identified in the previous step, along with the SRQs of interest for the given scenario (see Table 14.1). The phenomena can be listed in any order in the table, but it is recommended that they be grouped into general classes of phenomena or processes. For example, in a heat transfer problem, they would be listed as conduction, convection, and radiation. All of the single-physics processes should be listed together, such as solid mechanics, fluid dynamics, and heat transfer. Physical processes that produce important effects because of coupling between single-physics processes should also be listed as separate entries. If a scenario is divided into multiple time phases or spatial regions in the system, then a separate PIRT would be given for *each* phase or region.

After the row and column headings of the PIRT are listed, the central issue of the PIRT process begins: the discussion and debate to rank-order the list of physical phenomena in terms of their importance for each SRQ. Various methods of ranking the phenomena can be used. The simplest and usually the most effective method is to define three relative ranks of importance on the SRQs: high, moderate, and low. A five-level scale can also be used: very high, high, moderate, low, and very low. No more than a five-level scale should be used because a higher level of precision is not appropriate for a qualitative effort such as this. If more than three levels of importance are used, it is substantially more difficult to obtain agreement by a group of individuals for ranking the order of physical phenomena.

Depending on the composition of the team and the resources planned for the effort, the discussion and debate on ranking can be quite lengthy and energetic. The team leader must show a great deal of discretion and patience in the debates. Unless the team leader has a very strong opinion on a certain ranking, he/she should avoid taking sides in a debate concerning a particular ranking. Taking sides in a debate can lead team members to conclude that the team leader is aligning with certain factions within the team. The team leader should serve

Table 14.2 *Example of a completed PIRT for a given environment and scenario*

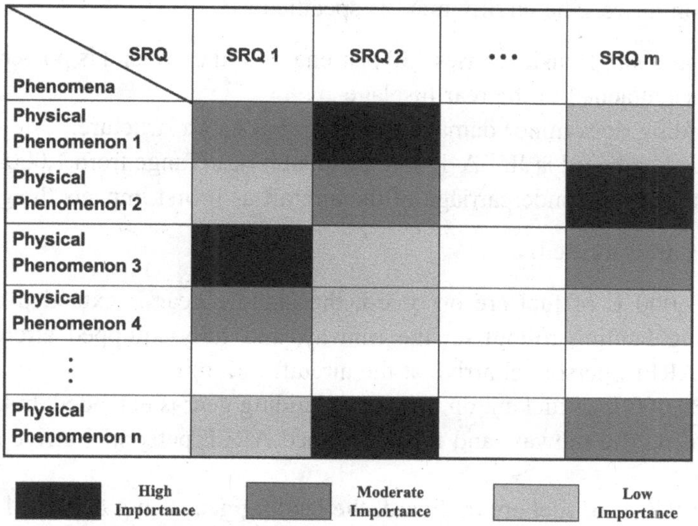

the role of an unbiased, rational observer and referee in the debate. If a consensus cannot be reached on the level of importance of each phenomenon for each SRQ, then the team leader should call for a formal vote to decide on the ranking. Furthermore, any team member should be allowed *not* to vote, if they so choose, including the team leader.

After agreement has been reached, or a vote has been taken concerning the agreed upon level for each phenomenon, the completed PIRT table would appear similar to that shown in Table 14.2. This table shows three levels of importance: (1) high is black, (2) moderate is medium gray, and (3) low is light gray. Note that if a particular physical phenomenon is a combination of two other single-physics phenomena, it could be ranked at a higher level than each of the single-physics phenomena. This could occur because of a strong nonlinear interaction of each of the contributing physical phenomena. One of the most common examples in fluid dynamics is the combination of chemical-diffusion reaction and fluid-dynamic turbulence, i.e., turbulent combustion. When multiple SRQs are listed in a PIRT table, it is stressed that the ranking must be consistent across *all of the SRQs*. Thus, the level of ranking is a function of the individual phenomena, *not* the importance of one SRQ relative to another. If ranking of the importance of the individual SRQs is needed based on the objectives of the PIRT process, this ranking should be done separately. However, this SRQ ranking should *not* be related to the ranking of importance of the physical phenomena.

14.3 PIRT Example Problem

Consider the crash landing of a commercial transport aircraft and the subsequent spread of a fire. The abnormal environment of interest is defined to be a survivable crash and the fuel fire that begins immediately after initial contact of the aircraft with the ground. The goal of the

PIRT analysis is to identify and rank the importance of the plausible physical phenomena that could occur regarding the survivability of the passengers to the fire and the generation of toxic combustion gases. The environment is specified as:

- The aircraft has a single aisle, carries 150 passengers and crew, and is powered by two gas-turbine engines attached to the rear fuselage.
- The crash landing does minor damage to the aircraft cabin structure.
- The fuel carried on board is JET A-1, and the quantity can range from 1,000 L to 20,000 L.
- The fire initiates in the undercarriage of the aircraft as it first impacts the ground.

Three scenarios are specified:

- Scenario 1: 1,000 L of fuel are on board, the landing gear is extended and has minor damage during landing rollout on the runway, and fully-equipped aircraft rescue and firefighting (ARFF) personnel arrive at the aircraft in 2 min.
- Scenario 2: 20,000 L of fuel are on board, the landing gear is extended, but it is destroyed during rollout on the runway, and fully-equipped ARFF personnel arrive at the crash in 5 min.
- Scenario 3: 1,000 L of fuel are on board, the landing gear is not deployed before ground impact, there is considerable damage to the underbelly of the aircraft during impact and deceleration on an open field, and no ARFF personnel are available.

A leader for the PIRT team is chosen and a team of experts is formed. The team is composed of experts from a diverse range of backgrounds and experience, including combustion modeling of a wide range of materials, toxicology, fluid dynamics, heat transfer, crash dynamics of aircraft structures, and tactics and materials used by ARFF personnel. The team is not given the SRQs of interest as part of the goals of the analysis, so these must be developed. A number of possible SRQs are considered by the team to try to determine the appropriate quantities for gauging the survivability and possible escape of the passengers and crew from the burning aircraft. The following quantities are chosen for all three scenarios:

- SRQ 1: gas temperature at mid-height of the cabin, along the center of the aisle, as a function of time.
- SRQ 2: soot temperature at mid-height of the cabin, along the center of the aisle, as a function of time.
- SRQ 3: carbon monoxide concentration at mid-height of the cabin, along the center of the aisle, as a function of time.
- SRQ 4: time of collapse of the emergency exit slide due to melting from the fire.
- SRQ 5: gas temperature along a vertical line from the top of each cabin exit to the ground, as a function of time.

Plausible physical phenomena are considered that could affect each of the scenarios and the prediction of the SRQs of interest. The primary difference between Scenario 1 and 2 is the amount on fuel on board at the time of the crash. Since both of these scenarios include ARFF personnel for controlling the fire and rescuing personnel, the PIRT analysis would need to address the physical phenomena related to various firefighting foams, firefighting techniques, and lifesaving tactics. The key difference between Scenario 3 and Scenarios 1 and 2 is that in

Table 14.3 *Completed PIRT for the aircraft fire environment and Scenario 3*

Physical Phenomena \ SRQ	SRQ 1	SRQ 2	SRQ 3	SRQ 4	SRQ 5
Convective: Buoyant turbulent mixing	Moderate	Moderate	Moderate	Moderate	High
Convective: Combusting turbulent flow	High	High	High	High	High
Convective: Wind effects	Moderate	Moderate	Moderate	High	High
Mass transport: Fuel evaporation	Low	Low	Low	Low	Low
Chemistry: Fuel combustion	Moderate	Moderate	Moderate	High	Low
Chemistry: Soot production	Low	High	Low	Low	Low
Chemistry: Cabin material combustion	High	High	High	Low	Low
Chemistry: Carbon monoxide production	Low	Moderate	High	Low	Low
Radiation: Emissive flux soot formation	Moderate	High	Moderate	Low	Low
Radiation: Emissive flux combustion chemistry	Low	Moderate	Low	Moderate	Moderate
Radiation: Mesoscale turbulent mixing	Low	High	Low	Moderate	Moderate
Radiation: Material properties in cabin	High	High	High	Low	Low
Radiation: Material properties of emergency slide	Low	Low	Low	High	Moderate
Conduction: Through aluminum structure	Low	Low	Low	Low	Low
Structure: Aluminum melting	Low	Low	Low	Low	Low
Structure: Emergency slide melting	Low	Low	Low	High	Moderate

High Importance — Moderate Importance — Low Importance

Scenario 3, there are no ARFF personnel to control the fire and to rescue personnel from inside and near the aircraft.

Because of time and resource constraints, only one branch of the scenario-SRQ tree can be pursued. The PIRT team leader consults with management, who commissioned the analysis, concerning which branch of the scenario-SRQ tree should be pursued. It was decided that Scenario 3 should be pursued because legal counsel advised that this scenario was the highest-liability risk to the organization. Table 14.3 lists all the physical phenomena considered by the PIRT team for Scenario 3. The table also shows the results of the team's analysis yielding three levels of importance-ranking of phenomena for each SRQ.

14.4 Gap Analysis Process

Sandia researchers familiar with the PIRT process for prioritizing physical processes recognized that the results of the PIRT could also be used in planning and prioritization of M&S activities. As part of the ASC program, they developed a procedure that could be used to identify weaknesses and needed improvements in M&S activities. This process is referred to as a *gap analysis* (Tieszen et al., 2002; Boughton et al., 2003). In the gap analysis, the emphasis shifts from developing physical understanding and prioritizing physical phenomena to an in-depth *assessment of the M&S capabilities that presently exist* for the E&S identified in the PIRT process. The key question that is answered in the gap analysis is: *Where do the existing M&S capabilities presently stand, relative to the phenomena and SRQs that have been identified as important?* This question is closely related to the issues discussed in more detail in Chapter 15 "Maturity Assessment of Modeling and Simulation."

To answer this question, a different mix of expertise is needed on the gap analysis team compared to the expertise needed for a PIRT team. In addition to experts familiar with the physical phenomena or system of interest, individuals are needed with knowledge of:

- the range of physics models available in the software,
- geometry and mesh generation options available in the software,
- the status of code verification for input options in the software,
- solution verification capabilities in the software,
- validation activities that could be relevant to the physical phenomena of interest, and
- uncertainty quantification capabilities that are applicable to the physical phenomena or system of interest.

This is a wide range of expertise needed, but most of these areas could be filled by developers of the available software package, computational analysts experienced with using the software, as well as individuals knowledgeable of the physical phenomena of interest. The new team members should replace some members of the PIRT team so that the team does not get larger than about 5 to 10 people. The PIRT team leader would normally also lead the gap analysis team.

By having software developers or any other proponents of the existing software on the team, a formidable dilemma is posed. The software developers or their proponents have a very real vested interest in espousing the capabilities of the software, instead of critically assessing its capabilities. There is no single method of dealing with this situation, because it depends on the individuals and organization involved. The managers responsible for the formation of the gap analysis team, as well as the team leader, must be cognizant of both the risk to the gap analysis, and the dilemma posed to the individuals themselves. For example, if the software developers or proponents of the software are unable to critically evaluate the existing capability, or are always defensive of criticisms of the software, it can greatly hinder the gap analysis, even render it useless. On the other hand, if software developers or proponents are critical of the existing capability, then they risk retribution from their peers on the software development team or their management. If the existing software being evaluated is from an external organization, for example, commercial software, different issues come into play. These issues are discussed in Section 14.5.

A different approach from the gap analysis to be described shortly has been recently published, but the approach has a similar goal. Unal et al. (2011) describe a proposed improvement to the BE+U approach for nuclear power reactors. The first step is conducting a PIRT analysis, and then various following steps are described that aid in identifying needed improvements in M&S activities and experimental data. Depending on the requirements for adequate safety margin in the power plant in an accident scenario, the approach guides needed improvements in verification, new experimental data for model validation, model parameter calibration, and UQ.

14.4.1 Construct the Gap Analysis Table

The gap analysis begins with the information generated in the completed PIRT, Table 14.2. For each SRQ of interest, the most important one or two levels of physical phenomena are singled out for a gap analysis. The most common activities for assessment of gaps are

Table 14.4 *Example of an initial gap analysis table for a given environment, scenario, and SRQ of interest*

Phenomena \ Gap Areas	Physics Modeling	Code Verification	Solution Verification	Model Validation	Uncertainty Quantification
Physical Phenomenon 1					
Physical Phenomenon 2					
Physical Phenomenon 3					
Physical Phenomenon 4					
Physical Phenomenon 5					
Physical Phenomenon 6					
Physical Phenomenon 7					

physics modeling, code verification, solution verification, model validation, and uncertainty quantification. Table 14.4 shows the initial table for a gap analysis for these groups for the top two levels of physical phenomena and physics coupling identified in the PIRT process. In this table, it is assumed that there are three phenomena in the top level (high importance, indicated by the dark stripe), and four in the second level (moderate importance, indicated by the gray stripe). The order of listing of the phenomena within a level is usually not important at this point.

The purpose of the gap assessment is to determine if each of the five M&S activities listed across the top of the gap analysis table is adequate for predicting the specified SRQ for each of the physical phenomena for the specified E&S. The detailed meaning of *adequate* depends on the physical phenomena listed, the specific M&S activity, and a general idea of what level of accuracy is needed for prediction of the SRQ. The general meaning of adequate for an activity indicates: *whether an activity is sufficiently mature for the software package to achieve a satisfactory prediction of the SRQ of interest.* The answer to this question depends on detailed knowledge of the present state of the specific activity, as well as some idea of the expected requirements for predictive capability for the particular application of interest. The former issue is more easily known than the latter issue.

In the following example, questions related to M&S-VVUQ adequacy are listed for each of the five activities listed in Table 14.4. For brevity, the term physical phenomenon or physics of interest will be used for any type of physics coupling:

1. *Physics modeling: Does the software have the capability to adequately address the phenomenon in question?* Examples of detailed questions are:
   - Does it have the ability to model the physical phenomenon over the entire physical parameter range of interest?
   - Does the software have all of the material models needed and are they operational over the needed range of parameters?
   - Does the software have alternative models that may be used to deal with the phenomenon of interest? For example, are there alternative models for turbulent flow, fracture mechanics, and large-material deformation and material recontact?
   - Are the models physics-based, or are they empirical, data-fit models? If they are entirely empirical models, they should not be used outside the range of data used to construct the model.
   - Do the models have the capability to address the needed spatial and temporal scales simulated in the physical phenomenon?
   - Do the models have the flexibility to deal with the geometries of interest, as well as geometry details that may be needed? For example, if material interfaces exist in the system of interest, can the physics occurring at the interfaces be addressed?
   - Is the SRQ of interest available as an output option?
2. *Code verification: Has the software undergone code verification testing for all of the software options that would be used for the application of interest?* Examples of detailed questions are:
   - Is there a regression-test suite that adequately tests the options that will be used and is it run routinely?
   - Is a SQA plan in place and followed?
   - Does the SQA plan address the various types of computer hardware, computer operating systems, and compilers that will be used?
   - Are there any code bugs reported that have not been resolved for the software options that will be used?
   - Has code verification testing been completed such that the observed order of convergence was computed for all of the software options, algorithms, and mesh structures that will be used?
   - In code verification testing, have only traditional analytical solutions been used, or have manufactured solutions also been used?
3. *Solution verification: Does the software have the capability to quantitatively estimate the various numerical solution errors in the SRQ of interest?* Examples of detailed questions are:
   - Are adequate methods available in the software to estimate or control round-off error?
   - Are adequate methods available in the software to estimate or control iterative solution error?
   - Are adequate methods available in the software to estimate or control spatial discretization error?
   - Are adequate methods available in the software to estimate or control temporal discretization error?
   - Are adequate methods available in the software to estimate or control statistical sample error in the SRQs of interest?

- Are adequate procedures available in the software for meshing the solution domain of interest, including spatial-mesh clustering and re-meshing?
- For estimating, either quantitatively or qualitatively, are methods available in the software for generating global mesh refinement with adequate user control, e.g., refinement of the mesh uniformly over the domain of interest?
- Does the software have the capability for adaptive re-meshing of the domain of interest, either for differing conditions or geometries of interest, or as a function of time?

4. *Model validation: Is the model validation domain adequate for the physical phenomenon and input conditions of interest, and have adequate quantitative accuracy assessments of the model form uncertainty been made for the SRQ of interest?* Examples of detailed questions are:
   - For the physical phenomenon of interest, have model validation experiments been conducted that are relevant to the application domain of interest?
   - At what tiers in the validation hierarchy for the system of interest has the model been compared to experimental measurements for the phenomena in question?
   - Has the model been compared to experimental measurements over the range of relevant parameters for the phenomenon and application of interest?
   - For the simulation and experimental comparisons that have been made, was there sufficient experimental data to adequately estimate model form uncertainty for the phenomena of interest?
   - Was adequate spatial and temporal discretization resolution attained for the comparisons between simulation and experiment that have been made? If inadequate spatial and temporal discretization was used in the comparisons, are methods available to address a mixture of model form uncertainty, numerical solution error, and model parameter calibration?
   - For the comparisons of simulation and experiment that have been made, were they made as a blind prediction or a postdiction?
   - Have the comparisons been made adequately related to the phenomena of interest in terms of system geometry and material properties?
5. *Uncertainty quantification: Is the accuracy of the predicted SRQ of interest, including its estimated uncertainty, adequate for the phenomena of interest and the needs of the application of interest?* Examples of detailed questions are:
   - Is the software capable of computing nondeterministic simulations of the phenomena of interest and the application of interest?
   - Can the software only compute traditional probabilistic solutions, or can it also compute imprecise probabilistic solutions, e.g., probability bounds analysis?
   - Can the prediction of the SRQ of interest include numerical solution uncertainties and model form uncertainties?
   - Can the software estimate model form uncertainty both inside and outside the validation domain for the phenomenon of interest and the application of interest?
   - Can the software extrapolate the predictive uncertainty to higher tiers of the validation hierarchy for the application of interest, if needed?
   - Are adequate computer resources available for nondeterministic simulations of the phenomenon of interest and the application of interest?
   - Is adequate UQ expertise among the technical staff available for the task at hand?

Table 14.5 *Example of a completed gap analysis table for a given environment, scenario, and SRQ of interest*

Phenomena \ Gap Areas	Physics Modeling	Code Verification	Solution Verification	Model Validation	Uncertainty Quantification
Physical Phenomenon 1	Adequate	Unknown (?)	Adequate	Inadequate	Inadequate
Physical Phenomenon 2	Adequate	Inadequate	Adequate	Adequate	Unknown (?)
Physical Phenomenon 3	Adequate	Adequate	Adequate	Inadequate	Inadequate
Physical Phenomenon 4	Inadequate	Unknown (?)	Adequate	Adequate	Unknown (?)
Physical Phenomenon 5	Adequate	Adequate	Adequate	Unknown (?)	Inadequate
Physical Phenomenon 6	Inadequate	Inadequate	Adequate	Inadequate	Inadequate
Physical Phenomenon 7	Unknown (?)	Inadequate	Adequate	Inadequate	Inadequate

Adequate | Inadequate | ? Unknown

- Is the software able to conduct a global sensitivity analysis for the phenomena of interest and the application of interest?

Answering these questions requires extraordinary knowledge of the details of the software package, forthright assessment of the past performance of software for the various phenomena of interest, and an understanding of the expected performance of the software for the physical phenomena and application of interest. The discussion of the five M&S activities among the team should be frank, constructive, and creative. The questions of software capability should be answered given the software status at the time of the gap analysis effort, not some anticipated or planned capability of the software. Many of the issues related to code verification, solution verification, and model validation are more factual than speculative. For example, they may include the following questions: (1) *Has the software been tested for this option?* Or (2) *Has this comparison of simulation and experiment been made for conditions relevant to the application of interest?* Most of the questions related to physics modeling and uncertainty quantification are more speculative. For example, they commonly include: (1) *Do we believe the physics models can properly compute the physics in this unusual combination of parameters and conditions?* Or (2) *How large do we believe the predictive uncertainty will be if large extrapolations are required outside the model validation domain?* On the more speculative issues, the team leader should expect a wide range of views and substantial debate among the team members.

After some level of consensus among the team has been achieved, the gap analysis table can be completed. Table 14.5 shows an example of a completed gap analysis table for a given E&S and for a given SRQ of interest. This table only shows three levels of assessment: adequate, inadequate, and unknown. When the gap analysis approach was developed at Sandia, only two levels were available: adequate and inadequate. However, the Sandia researchers found that there were a number of situations where there was little information or conflicting information concerning the adequacy or inadequacy of an activity. As a result, the unknown level was created to keep the gap analysis process moving instead of hosting a prolonged debate on adequacy or inadequacy. As is discussed in Section 14.4.3, new information can be added to a completed gap analysis table, resulting in an updated table.

Two observations are pointed out regarding the structure of assessments shown in Table 14.5. First, the most important activities that must be pursued in the future in order to close perceived gaps in needed capabilities are those that occur in the most important phenomena. The most important phenomena are denoted by the dark stripe on the left. For these three phenomena, there are five inadequate elements and two unknown status elements. Four of the inadequate elements are closely related, specifically, for Phenomena 1 and 2, inadequacy in UQ is directly dependent on inadequacy in model validation. The issue of which ones to pursue within this group of four is more than a technical issue, it is also a P&P issue, as well as an organizational responsibility issue. For example, an inadequacy in model validation can be attacked from two different approaches. The first approach is to develop a higher-fidelity physics model to possibly address the perceived weakness. The second approach is to obtain new experimental data that may expand the validation domain or better address the particular physical phenomenon of interest. If the new experiments can be accomplished within the institutional or corporate organization conducting the gap analysis, sufficient organizational pressure may be mustered to resolve the issue within the time scale needed for the application of interest. However, if an outside organization must be contracted or a new experimental facility must be built, the inadequacy may not be resolved within the time required.

The second observation concerning Table 14.5 is that solution verification and UQ are heavily dependent on capabilities within the software package. For example, if at least one method is available for estimating or controlling round-off error, iterative error, spatial discretization error, temporal discretization error, and statistical sampling error, then these capabilities can be applied to a wide range of physical phenomena and applications. Concerning UQ capability, if a method is available for extrapolating model form uncertainty beyond the validation domain, then this capability can be applied to a wide range of physical phenomena and applications.

14.4.2 Documenting the PIRT and Gap Analysis Processes

Although it is seldom a welcomed task, documentation of the team's work is an extremely important aspect of the project. The PIRT and gap analysis tables that are produced capture only the summary information from the discussions and processes that occurred within each team. A report should be completed that also captures the (a) objectives of both processes; (b) specification of the environment for the system; (c) description and possible discovery of new scenarios; (d) explanation as to why certain SRQs are important and why others are unimportant; (e) discovery and identification of the plausible physical phenomena and coupled physics; (f) explanation for ranking the importance of physical phenomena; and (g) explanation for categorizing the elements as indicated in the gap analysis table. If there are a number of elements ranked *inadequate* or *unknown*, the team may recommend steps to address these issues. The documentation should include the reasoning and arguments made for important decisions, as well as why certain aspects or issues were excluded from the processes.

As discussed earlier in this chapter, the purpose of the PIRT and gap analysis processes is to recommend how best to allocate resources between various activities to achieve success of a project-oriented simulation effort in an efficient manner. The PIRT and gap analysis processes should be viewed as a generation of information, which management can use for optimizing the allocation of resources. If the processes and explanations are not documented at some appropriate level of formality, then experience has shown that the value and impact of the effort will be minimal. Not only would the resources expended for the PIRT and gap analysis processes be wasted, but more importantly, there could be major inefficiencies and delays in the M&S effort directed toward the application of interest. This waste of information could possibly lead to the failure of the simulation effort, as well as failure of the engineering system to which the simulation is contributing.

14.4.3 Updating the PIRT and Gap Analyses

The results of the combined PIRT and gap analyses should be viewed as a state of knowledge of M&S activities for a given system in a specified E&S, as well as recommendations for an effective path forward. The recommended path forward contains two types of information: known deficiencies that need immediate attention and deficiencies that may prove to be important, but where information is missing. As management acts on the known deficiencies, the state of M&S activities changes. Some of these changes can be very slow to occur, such as building new experimental facilities to expand the model validation domain for the system of interest. However, other changes can happen much faster. For example, if there are a number of important gap analysis table elements that are *unknown*, some of these may be addressed relatively quickly. For example, it may be found that certain important code verification features that were considered as *unknown* are determined to *not* have been tested. This result would change this element in the gap analysis table from *unknown*, to *adequate* or *inadequate*.

When new information becomes available or improvements are made, the gap analysis table can be updated. For example, (a) coupled physics phenomena that were never exercised before may be found to be *adequate*; (b) new capabilities for solution verification are added to the software; (c) experimental data is obtained and model validation activities are conducted; or (d) new, more efficient, sampling procedures are added to the UQ capabilities. When appropriate for a large-scale M&S effort, a Gantt chart can be constructed and updated as new information becomes available and important gaps in capability are eliminated. This is particularly important when diverse portions of a large organization are involved in addressing gaps in M&S capabilities.

Some surprises found in the follow-up efforts related to each activity should be expected. Examples of some surprises, most of them bad, that have been observed in practice, are:

- A combination of software options that is needed for the physical phenomena of interest is known to be available, but it is discovered that the software would not run the needed combination.

- A manufactured solution for code verification is constructed to test the combination of options needed and it is found that the combination is zeroth-order accurate, i.e., the solution converged to the wrong answer.
- It is discovered that the numerical solution error estimator would not perform correctly on the physical phenomenon, i.e., it showed wild oscillations for the problems of interest.
- A new simulation or experiment is conducted, and it is found that a physical phenomenon that had been ranked as high importance, is changed to moderate importance.
- A previous comparison between simulation and experiment that showed good agreement is recomputed and it is found that when the mesh resolution is improved, the agreement becomes unacceptable.
- A new validation experiment is conducted, and it is found that a major recalibration of the model parameters is needed to obtain good agreement with all of the available data.
- When an alternative plausible model for a physical phenomenon is used to compute the SRQ of interest, it is found that there is a large disagreement compared to the traditional model used, which had been deemed *adequate*.

14.5 Planning and Prioritization with Commercial Software

This chapter on P&P has concentrated on large-scale M&S projects that are conducted primarily within an organization. Several comments were made concerning how P&P would be conducted differently by an organization that is using, or considering using, commercial software. The context of P&P when using commercial software must be understood as a business relationship between the organization doing the P&P and the software company, *not* a technical relationship. Technical issues are certainly important, especially to technical staff members, but the dominant issue is: *Can the organization build a strategic business relationship with a software company as a critical stakeholder?* If the organization only has relatively few software licenses from the software company, it is doubtful that a substantive relationship can be built. The software company may proclaim how important the customer is to them, but if few licenses are involved, there will be little authenticity to the claim. Only when the software company becomes an authentic stakeholder, which means the organization has a large number of software licenses, can the kind of honest discussions required for P&P be accomplished.

Management of the organization conducting the P&P needs to consider the amount of confidence they have that the software company can and will protect their proprietary information before a stakeholder relationship is attempted. The proprietary information would not only deal with existing products the organization sells, but also research information, proprietary experimental data, sensitive information concerning potential failure modes of existing products, and proposed designs for new products. The mindset toward this type of information within a software company is typically very different from a company that designs and sells hardware products. If the organization is dealing with a large software company, then another issue must be considered. The software company may also be working closely with a competitor of the organization. For example, the same software company may have proprietary information on, or be working on, two competing proposals from different clients. Similarly, the same technical staff of the software company may be working on competing proposals from different clients.

To conduct the gap analysis discussed here, the software company would need to be an intimate part of the discussions and be represented on the gap analysis team. The software company's team members would need to be very candid and constructively critical of their own company's software. When explicitly discussing shortcomings in their software, they would be concerned about potentially losing the organization's business, as well as detrimental information being passed on by the organization to other potential new customers for their software. It is clear this is a delicate business issue as well as technical issue. On the one hand, the organization needs to be certain it will obtain forthright and damaging information on the software company's product. On the other hand, the software company needs to be certain it will not lose its client, nor have its sensitive information or issues exposed for public consumption.

As a final topic, consider the issue of ownership of software capabilities. Suppose that after a gap analysis has been completed, it is found that a new capability is badly needed by the organization conducting the P&P. This feature could be developed by either the organization conducting the gap analysis or the software company. The organization may be willing to fund the development and testing of the new capability as an add-on, standalone feature, to the software. If the organization does not have sufficient expertise to develop and test the new capability, it may hire consultants to complete the task. If the organization decides to develop the new capability, it may require a higher level of support and information from the software company about their operations. It may even involve proprietary information from the software company, requiring negotiated contracts protecting the intellectual property of the software company.

Suppose the software company decides that it would be in its best long-term interest to develop and test the new capability needed by the organization conducting the P&P. The software company may recognize that it has other potential clients for this new capability. As a result, the software company may fund the development and testing of the new capability itself. Alternatively, the organization conducting the P&P may fund the software company to develop, test, and document the new capability. Then the question must be addressed: *Who will own the new capability?* The organization needing the capability would primarily view the issue as: *What is the anticipated return on investment for funding the new capability?* It may also want to protect its proprietary advantage by not allowing any other client of the software company to use the capability. The software company would view the issue as: *How can we increase our software license sales by advertising this new capability to existing or potential new customers?* Various solutions to these differing perspectives would need to be carefully negotiated between the organization needing the capability and the software company. One idea would be to allow the organization to have exclusive rights to use the new capability for a set period of time in the future. In addition, the software company would not be allowed to advertise or discuss the capability until after that time period.

14.6 Gap Analysis Example

After the previously described PIRT example dealing with an aircraft crash and fire was completed, documented, and presented to management, it is decided to conduct a gap

Table 14.6 *Completed gap analysis table for the aircraft fire environment, Scenario 3, and SRQ 1*

Phenomena \ Gap Areas	Physics Modeling	Code Verification	Solution Verification	Model Validation	Uncertainty Quantification
Convective: Combusting turbulent flow	Adequate	? Unknown	Inadequate	Adequate	Inadequate
Chemistry: Cabin material combustion	Inadequate	Inadequate	Inadequate	Inadequate	Inadequate
Radiation: Material properties in cabin	Inadequate	Inadequate	Inadequate	Inadequate	Inadequate
Convective: Buoyant turbulent mixing	Adequate	Adequate	Inadequate	? Unknown	Inadequate
Convective: Wind effects	Adequate	? Unknown	Inadequate	Adequate	Inadequate
Chemistry: Fuel combustion	Adequate	Inadequate	Inadequate	Adequate	Inadequate
Radiation: Emissive flux soot formation	Adequate	Inadequate	Inadequate	Inadequate	Inadequate

Adequate   Inadequate   ? Unknown

analysis. The goal of a gap analysis is to identify the important gaps in an existing M&S capability for predicting the specified SRQs. The team leader for the PIRT analysis is chosen to lead the gap analysis team. A new team of experts is formed that includes some of the PIRT team members, but now team members are added with detailed knowledge of the present software capabilities. The software being assessed is existing in-house software comprised of two coupled codes: a reacting flow, heat transfer, material-phase-change code, and a solid dynamics code. As a result, new expertise on the gap analysis team includes personnel knowledgeable of one or both codes. This expertise includes modeling options available in the software, code verification testing, solution verification capabilities, validation activities, UQ capabilities, and the coupling capability between the two codes.

The gap analysis begins with the physical phenomena ranked as high importance and moderate importance shown in Table 14.3. A gap analysis is performed on *each* of the SRQs shown in Table 14.3, but only the gap analysis for SRQ 1 is discussed here. The five areas of gap assessment of the existing software are (1) physics modeling, (2) code verification, (3) solution verification, (4) model validation, and (5) UQ. After considerable discussion and debate, a gap analysis table is constructed for SRQ 1 listed in Table 14.3. Table 14.6 shows the completed gap analysis table for SRQ 1 for the high and moderate importance levels. The three phenomena ranked as high importance are listed first (marked with a black vertical line), and the four phenomena ranked as moderate importance are listed next (marked with a gray vertical line).

As can be seen from all of the black boxes in Table 14.6, the existing software did not fare well when applied to the present problem of interest. Three features of the gap analysis shown in Table 14.6 are discussed. First, for the two phenomena *Chemistry: Cabin material combustion* and *Radiation: Material properties in cabin,* every element is assessed as

Inadequate. This occurred for the existing software because the software had been developed and applied to pooled-fuel fires in an open environment. The new application must address combustion of a very wide range of materials in an enclosure. This type of gap analysis result commonly occurs when existing software are assessed for new applications. If a commercial software package was assessed, based on broad claims and advertising of the software capabilities, project planners could be severely misled. Even though this experience is widely recognized, it can still be very difficult to convince project managers, especially for proposal efforts, to set aside time and resources to make informed planning decisions concerning needed capabilities and potential existing deficiencies.

Second, in Table 14.6, note that the entire column for *Solution Verification* is assessed as *inadequate*. This occurred for the existing software because there were no capabilities in place for estimating any of the numerical solution errors. This is a common situation in many internally developed software packages, as well as commercial software packages. Most mature software packages that solve systems of ordinary differential equations have this capability, but *not* packages that solve partial differential equations (PDEs). Present practice for solution verification in solving PDEs commonly relies on some type of sensitivity analysis. For example, two solutions are computed with different spatial meshes and the change in SRQs is qualitatively judged to be "sensitive" or "insensitive." This is not an adequate procedure, as discussed in detail in Chapter 9 "Discretization Error," because there is no indication whether the solution is in the asymptotic convergence region.

Third, in Table 14.6, it is seen that the entire column for *Uncertainty Quantification* is assessed as *inadequate*. This situation is common for both internally developed software and commercial software for most engineering and science fields of study. There are two distinct reasons for this present state of practice of computational simulation. First, the present tradition in computational simulation is deterministic simulation, as opposed to nondeterministic simulation. This situation is changing slowly as simulation-informed decision making becomes more common for engineering systems. Second, the dominant approach to nondeterministic simulation uses an external software package to analyze and steer computational samples needed from the package that solves the PDEs of interest. This structure is commonly referred to as a *wrapper code* for the PDE solver. There is an increasing number of both commercial and open-source wrapper codes to drive PDE solvers. These types of codes can potentially address the inadequacies of UQ shown in Table 14.6.

14.7 Further Reading

For a detailed discussion of BE+U and CSAU, see Boyack et al. (1990), Wilson et al. (1990), Wulff et al. (1990), Zhao and Mousseau (2012), and Wilson (2013).

For further study of the PIRT process, see Boyack et al. (2001), Ratnayake et al. (2002), Diamond (2006), Mousseau and Williams (2017), and Rider et al. (2017).

14.8 Problems

1. In your field of interest, identify an engineering system of interest in an abnormal environment. Identify three scenarios that could place the operators of the system or personnel near the system at risk.

2. In your field of interest, identify an engineering system of interest in a hostile environment. Identify three scenarios that could place the operators of the system or personnel near the system at risk.
3. (Semester project) Pick one scenario for the system of interest identified in Problem 1 and conduct a PIRT analysis. The PIRT analysis should follow a similar procedure described in the example problem discussed in this chapter.
4. (Semester project) For the system of interest, the scenario identified, and the PIRT analysis completed in Problem 3, conduct a gap analysis. For the gap analysis, pick a software package that you are familiar with. The gap analysis should follow a similar procedure described in the example problem discussed in this chapter.
5. (Semester project) Pick one scenario for the system of interest identified in Problem 2 and conduct a PIRT analysis. The PIRT analysis should follow a similar procedure described in the example problem discussed in this chapter.
6. (Semester project) For the system of interest, the scenario identified, and the PIRT analysis completed in Problem 5, conduct a gap analysis. For the gap analysis, pick a software package that you are familiar with. The gap analysis should follow a similar procedure described in the example problem discussed in this chapter.

14.9 References

Boughton, B., V. J. Romero, S. R. Tieszen, and K. B. Sobolik (2003). Integrated Modeling and Simulation Validation Plan for W80–3 Abnormal Thermal Environment Qualification – Version 1.0 (Official Use Only). *Sandia National Laboratories*, SAND2003-4152 (OUO), Albuquerque, NM.

Boyack, B. E., I. Catton, R. B. Duffey, P. Griffith, K. R. Katsma, G. S. Lellouche, et al. (1990). Quantifying Reactor Safety Margins Part 1: An Overview of the Code Scaling, Applicability, and Uncertainty Evaluation Methodology. *Nuclear Engineering and Design*, **119**(1), 1–15.

Boyack, B. E., A. T. Motta, K. L. Peddicord, C. A. Alexander, R. C. Deveney, B. M. Dunn, et al. (2001). Phenomenon Identification and Ranking Tables (PIRTs) for Rod Ejection Accidents in Pressurized Water Reactors Containing High Burnup Fuel. *US Nuclear Regulatory Commission*, NUREG/CR-6742, Washington, DC.

Brechner, E. (2015). *Agile Project Management with Kanban*. Redmond, WA: Microsoft Press.

Charette, R. N. (2005). Why Software Fails. *IEEE Spectrum,* September.

Diamond, D. J. (2006). Experience Using Phenomena Identification and Ranking Technique (PIRT) for Nuclear Analysis. PHYSOR-2006 Topical Meeting, BNL-76750-2006-CP, Vancouver, British Columbia, Canada.

Dorner, D. (1989). *The Logic of Failure; Recognizing and Avoiding Error in Complex Situations*. Cambridge, MA: Perseus Books.

Galin, D. (2003). *Software Quality Assurance: From Theory to Implementation*. London: Pearson.

Hanson, R. G., G. E. Wilson, M. G. Ortiz, and D. P. Grigges (1992). Development of a Phenomena Identification and Ranking Table (PIRT) for a Postulated Double-Ended Guillotine Break in a Production Reactor. *Nuclear Engineering and Design*, **136**(3), 335–346.

Huckle, T., and T. Neckel (2019). *Bits and Bugs: A Scientific and Historical Review of Software Failures in Computational Science*. Philadelphia, PA: Society of Industrial & Applied Mathematics.

Kroeger, P. G., U. S. Rohatgi, J. H. Jo, and G. C. Slovik (1998). Preliminary Phenomena Identification and Ranking Tables for Simplified Boiling Water Reactor Loss-of-Coolant Accident Scenarios. Brookhaven National Laboratory, NUREG/CR-6472, Upton, NY.

Martin, R. C. (2019). *Clean Agile: Back to Basics*. Upper Saddle River, NJ: Prentice Hall.

McConnell, S. (2019). *More Effective Agile: A Roadmap for Software Leaders*. Bellevue, WA: Construx Press.

Mousseau, V. A., and B. J. Williams (2017). Uncertainty Quantification in a Regulatory Environment. R. Ghanem, D. Higdon, and H. Owhadi (eds.), *Handbook of Uncertainty Quantification*, 1613–1648. New York, NY: Springer.

NRC (1989a). Best-Estimate Calculations of Emergence Core Cooling System Performance. Nuclear Regulatory Commission, RG 1.157. Washington, DC.

NRC (1989b). Quantifying Reactor Safety Margins. U.S. Nuclear Regulatory Commission, Office of Nuclear Regulatory Research, Division of Systems Research, NUREG/CR-5249. Washington, DC.

Pilch, M., T. G. Trucano, J. L. Moya, G. K. Froehlich, A. L. Hodges, and D. E. Peercy (2001). Guidelines for Sandia ASCI Verification and Validation Plans – Content and Format: Version 2. *Sandia National Laboratories*, SAND2000-3101, Albuquerque, NM.

Ratnayake, R. K., S. Ergun, L. E. Hochreiter, and A. J. Baratta. (2002). Identification and Ranking of Phenomena Leading to Peak Cladding Temperatures in Boiling Water Reactors during Large Break Loss of Coolant Accident Transients. *ICONE-10, 10th International Conference on Nuclear Engineering*, ICONE-22288, Virginia.

Rider, W. J., W. R. Witkowski, and V. Mousseau (2017). Uncertainty Quantification's Role in Modeling and Simulation Planning, and Credibility Assessment Through the Predictive Capability Maturity Model. In R. Ghanem, D. Higdon, and H. Owhadi (eds.), *Handbook of Uncertainty Quantification*, 1589–1612. New York, NY: Springer.

Rohatgi, U. S., H. S. Cheng, H. J. Khan, and W. Wulff (1997). Preliminary Phenomena Identification and Ranking Tables (PIRT) for SBWR Start-Up Stability. Brookhaven National Laboratory, NUREG/CR-6474. Upton, NY.

Sharp, D. H., and M. M. Wood-Schultz (2003). QMU and Nuclear Weapons Certification. What's under the hood? *Los Alamos Science*, **28**, 47–53.

Shaw, R. A., T. K. Larson, and R. K. Dimenna (1988). Development of a Phenomena Identification and Ranking Table (PIRT) for Thermal-Hydraulic Phenomena during a PWR LBLOCA. *EG&G, NUREG/CR-5074*. Idaho Falls, ID.

Stepanek, G. (2012). *Software Project Secrets: Why Software Projects Fail*. New York, NY: Apress.

Tieszen, S. R., T. Y. Chu, D. Dobranich, V. J. Romero, T. G. Trucano, J. T. Nakos, et al. (2002). Integrated Modeling and Simulation Validation Plan for W76–1 Abnormal Thermal Environment Qualification — Version 1.0 (Official Use Only). *Sandia National Laboratories*, SAND2002-1740 (OUO), Albuquerque, NM.

Trucano, T. G., M. Pilch, and W. L. Oberkampf (2002). General Concepts for Experimental Validation of ASCI Code Applications. *Sandia National Laboratories*, SAND2002-0341, Albuquerque, NM.

Unal, C., B. Williams, F. Hemez, S. H. Atamturktur, and P. McClure (2011). Improved Best Estimate Plus Uncertainty Methodology, Including Advanced Validation Concepts, to License Evolving Nuclear Reactors. *Nuclear Engineering and Design*, **241**(5), 1813–1833.

Vose, D. (2008). *Risk Analysis: A Quantitative Guide*, 3rd ed. New York, NY: John Wiley.

Wiegers, K., and J. Beatty (2013). *Software Requirements*, 3rd ed. Redmond, WA: Microsoft Press.

Wilson, G. E. (2013). Historical Insights in the Development of Best Estimate Plus Uncertainty Safety Analysis. *Annals of Nuclear Energy*, **52**, 2–9.

Wilson, G. E., and B. E. Boyack (1998). The Role of the PIRT in Experiments, Code Development and Code Applications Associated with Reactor Safety Assessment. *Nuclear Engineering and Design*, **186** (1–2), 23–37.

Wilson, G. E., B. E. Boyack, I. Catton, R. B. Duffey, P. Griffith, K. R. Katsma, et al. (1990). Quantifying Reactor Safety Margins, Part 2: Characterization of Important Contributors to Uncertainty. *Nuclear Engineering and Design*, **119**(1), 17–31.

Wulff, W., B. E. Boyack, I. Catton, R. B. Duffey, P. Griffith, K. R. Katsma, et al. (1990). Quantifying Reactor Safety Margins, Part 3: Assessment and Ranging of Parameters. *Nuclear Engineering and Design*, **119**(1), 33–65.

Zhao, H., and V. A. Mousseau (2012). Use of Forward Sensitivity Analysis Method to Improve Code Scaling, Applicability, and Uncertainty (CSAU) Methodology. *Nuclear Engineering and Design*, **249**, 188–196.

15

Maturity Assessment of Modeling and Simulation

The only thing more dangerous than ignorance is arrogance. —Albert Einstein

Contents of Chapter 15

This chapter discusses methods for assessing the maturity of the technical factors that contribute to the credibility of modeling and simulation (M&S) results. These factors are physics modeling fidelity, code verification, solution verification, model validation, and uncertainty quantification and sensitivity analysis. This chapter presents techniques for assessing the evidence available, indicating the maturity of these factors relative to the completeness of the details, and the rigor for each factor. Assessment of maturity is important to the staff conducting the M&S effort, but it is critically important for project managers and decision makers who use simulation results in support of decision making. Assessment of maturity is also important for simulation review committees, both internal and external to the organization conducting the simulation, who are tasked with assessing the credibility and reliability of simulation results. Regulatory authorities who are responsible for public and environmental safety of engineering systems have developed credibility assessment procedures over several decades. However, the past emphasis has been on experimental testing and completeness of experimental data for systems or subsystems; with M&S serving in a supporting role. With increased use of simulation results as part of the information package submitted to both regulatory authorities and decision makers, significant improvements in assessing simulation maturity have been developed over the last two decades. This chapter is based in part on Oberkampf et al. (2007), which introduced the Predictive Capability Maturity Model (PCMM) for assessing simulation maturity. This method is described in detail, along with improvements that have been published since then.

15.1 Survey of Maturity Assessment Procedures

Over the last four decades, many researchers have investigated how to measure the maturity and reliability of both software products and advanced technologies used in leading-edge engineering systems. As software systems and computational simulation have become prominent in essentially every aspect of modern life, a wide variety of maturity assessment procedures have been developed. *Confidence* that a software product or technology will meet its expectations is a concept that is in the mind of the individual. To reduce the subjectivism in this judgment, maturity assessment procedures have been developed in several application areas. The terminology used here is that *credibility* is a judgment based on evidence of maturity, reliability, correctness, and/or completeness. This section briefly reviews various developments in assessment procedures. For a listing of the most recent developments in maturity assessment of simulation-produced information, see Section 15.6 "Further Study."

15.1.1 Capability Maturity Model Integration

The best-known procedure for measuring the maturity of software product development and business processes is the *Capability Maturity Model Integration* (CMMI). CMMI is a successor to the Capability Maturity Model (CMM), whose development began in 1986 to improve software quality (Humphrey, 1989). Both CMM and CMMI were developed by the Software Engineering Institute operated by Carnegie Mellon University. CMMI, and other maturity assessment approaches discussed here, respond to the need for assessing the quality, reliability, maturity, and capability of a product or process regarding the following:

- Improve identification and understanding of the elements of the process.
- Determine the elements of the process that may need improvement so that the intended product of the process can be improved.
- Determine how time and resources can best be invested in elements of the process to obtain the maximum return on the investment expended.
- Better estimate the cost and schedule required to improve elements of the process.
- Improve the methods of aggregating maturity information from diverse elements of the process to better summarize the overall maturity of the process.
- Improve the methods of communicating to the decision maker the maturity of the process so that better decisions can be made.
- Measure the progress of improving the process so that managers of the process, stakeholders, and funding sources can determine the value added over time.
- Compare elements of the process across competitive organizations so that a collection of best practices can be developed and used.
- Measure the maturity of the process in relation to requirements imposed by the customer.

The most used and well-documented version of the CMMI is Version 1.3 (McMahon, 2010; Chrissis et al., 2011; Forrester et al., 2011). CMMI provides models for three application areas of interest: development (CMMI-DEV), acquisition (CMMI-ACQ), and services (CMMI-SVC). Five maturity levels are defined for each application area:

- *Level 1: Initial;* Processes are unpredictable, poorly controlled, and reactive.
- *Level 2: Managed;* Processes are characterized for projects and are often reactive.
- *Level 3: Defined;* Processes are characterized for the organization and are proactive.
- *Level 4: Quantitatively Managed;* Processes are measured and controlled.
- *Level 5: Optimizing;* Focus is on process improvement.

Note that the terms *verification* and *validation* are used in texts and reports on CMMI-DEV in a sense that is *different* than used in this text. In CMMI-DEV, these terms are used in the context of software quality assurance, as defined by the Institute of Electrical and Electronics Engineers (IEEE, 1989). For a detailed discussion of V&V terminology and its variations, see Chapter 2 "Fundamental Concepts and Terminology."

During years of CMM and CMMI development, several organizations, specifically the US Department of Defense and NASA, attempted to use these procedures for assessing the maturity of computational simulation models (Harmon and Youngblood, 2003, 2005). However, they concluded that CMM and CMMI are ineffective for this purpose. As a result, various organizations developed new maturity assessment procedures that are more effective for their needs.

15.1.2 Technology Readiness Levels

A maturity assessment procedure that has its origins in risk management for systems engineering is the *Technology Readiness Levels* (TRLs), pioneered by NASA in the late 1980s (Sadin et al., 1989; Mankins, 1995). The intent of TRLs is to lower acquisition risks of engineering systems by more precisely and uniformly assessing the maturity of advanced technologies that are planned for the system. The TRL procedure is used in NASA, the US

DoD, and private organizations around the world. TRLs consider nine levels of maturity in the evolution of technological systems. These levels are described by the DoD in (DoD, 2005) as follows:

- *TRL Level 1: Basic principles observed and reported.*
- *TRL Level 2: Technology concept and/or application formulated.*
- *TRL Level 3: Analytical and experimental critical function and/or characteristic proof of concept.*
- *TRL Level 4: Component and/or breadboard validation in a laboratory environment.*
- *TRL Level 5: Component and/or breadboard validation in a relevant environment.*
- *TRL Level 6: System/subsystem model or prototype demonstration in a relevant environment.*
- *TRL Level 7: System prototype demonstration in an operational environment.*
- *TRL Level 8: Actual system completed and operationally qualified through test and demonstration.*
- *TRL Level 9: Actual system proven through successful mission operations.*

The nominal specifications of TRLs as presented here are clearly aimed at assessing the maturity of technologies and hardware products, *not* software products or information produced by simulation software. A very recent development of the TRL approach has taken a broader view by examining the technology readiness assessment (TRA) of hardware, software, and process technologies (GAO, 2020).

Smith (2004) examined the difficulties in using TRLs for software, including commercial-off-the-shelf (COTS), government-off-the-shelf (GOTS), and open-source software. Clay et al. (2007) conducted a detailed study attempting to use TRLs to assess the maturity of computational simulation software and the reliability of information produced by the software. Both studies concluded that the TRL procedure is inappropriate for assessing the maturity of any type of software process or M&S software.

15.1.3 NASA Method for Simulation Maturity Assessment

After the loss of Space Shuttle Columbia and her crew in 2003, NASA began a fundamental review of decision making within the organization, specifically, how simulation results are used in decision making. One of the major initiatives was the formation of the Diaz team to investigate "A Renewed Commitment to Excellence" within NASA (NASA, 2004). As part of this work, a report was produced that called for NASA to: "Develop a standard for the development, documentation, and operation of models and simulations." Working through the NASA Office of Chief Engineer, a Technical Standards Working Group was formed under the leadership of Thomas Zang to develop a NASA standard for models and simulations used in decision making (Bertch et al., 2008). The work on this task formally began in April 2005 and the NASA Interim Standard was published in 2006 (NASA, 2006). (During this time period, Sandia National Laboratories was also developing the Predictive Capability Maturity Model, to be discussed in Section 15.2. Sandia and the NASA team exchanged ideas and perspectives concerning their approaches during this time.) After considerable discussion with all of the NASA Centers, as well as with NASA's private contractors, the NASA technical standard was completed in 2008 (NASA, 2008). The latest version of the standard is (NASA, 2016). An extensive handbook and implementation guide for the NASA standard has also been published (NASA, 2013). As can be seen from examining the procedures specified in *Standard for Models and Simulations* (NASA, 2016), formally applying the NASA standard to a large-scale engineering project is a major, time consuming, and costly endeavor.

The primary goal of the NASA standard is "to provide uniform engineering and technical requirements for processes, procedures, practices, and methods endorsed as standard for models and simulations (M&S) developed and used in NASA programs and projects" (NASA, 2016). The credibility assessment structure described in the standard is meant to aid decision makers in understanding the reliability of M&S results used in their decisions. The standard relates to all aspects of M&S activities, including development, maintenance, operation, supporting experimental activities, results analysis, training, credibility assessment, and reporting. The standard covers what types of activities could be undertaken and communicated related to M&S, but not specifically how this should be done within a project. Use of the standard in NASA projects is at the discretion of the NASA project manager, but it is highly recommended for M&S used in critical decisions on a project.

The NASA standard describes eight credibility factors for assessing M&S credibility:

1. *Data Pedigree:* Is the pedigree (and quality) of the data used to develop the model adequate or acceptable?
2. *Verification*: Were the models implemented correctly, per their requirements/specifications?
3. *Validation*: Did the M&S results compare favorably to the referent data, and how close is the referent to the real-world system?
4. *Input Pedigree*: Is the pedigree (and quality) of the data used to setup and run the model adequate or acceptable?
5. *Uncertainty Characterization*: Is the uncertainty in the current M&S result appropriately characterized? What are the sources of uncertainty in the results and how are they propagated through to the results of the analysis?
6. *Results Robustness*: How thoroughly are the sensitivities of the current M&S results known?
7. *M&S History*: How similar is the current version of the M&S to previous versions, and how similar is the current use of the M&S to previous successful uses?
8. *M&S Process/Product Management*: How well managed were the M&S processes and products?

These eight factors are grouped into three categories of M&S activities: (a) M&S Development (Data Pedigree, Verification and Validation); (b) M&S Operations (Input Pedigree, Uncertainty Characterization, and Results Robustness); and (c) Supporting Evidence (M&S History and M&S Process/Product Management).

Each of the eight factors listed has five levels of credibility or maturity that are numerically quantified as 0, 1, 2, 3, and 4. To achieve a credibility level in a factor, all of the descriptors given for that level of maturity, and all lower levels, *must be satisfied*. There is no fixed description or characterization of the requirements needed to attain a given level across each of the eight factors. That is, each factor has specific descriptors to characterize what is needed to achieve a specific level, except for Level 0. Level 0 means that no evidence exists or there is insufficient evidence for that factor to support any maturity. Increasing credibility levels must demonstrate increasing formality, traceability, completeness, or accuracy. The credibility level for each factor is essentially independent of other factors since each factor must have its own credibility evidence to achieve a given level. Furthermore, the credibility of the M&S activity is, as a whole, comprised of the assessment level for *all eight factors*, i.e., the individual levels cannot be combined or averaged in any way.

15.1.4 Additional Simulation Maturity Assessment Methods

A maturity assessment approach to M&S that is largely independent from CMMI, TRLs, NASA, and Sandia was developed by the operations research/systems engineering community. For a brief history of this perspective, see Sargent (2013), and Sargent and Balci (2017). A good description of this maturity assessment approach to M&S is given by Balci (2001, 2004), and Balci et al. (2002). They argue that M&S quality can be assessed based on indicators of product, process, and project. By *product,* he means either (a) the overall completed M&S application or (b) a work product created during the M&S development life cycle such as the conceptual model, M&S requirements specification, M&S design specification, or an executable M&S module. By *process*, he refers to the process used to create a work product during the M&S development life cycle, such as conceptual modeling, requirements, engineering, design, implementation, integration, experimentation, and presentation. *Project* refers to the quality indicators of the project plan, the capabilities and experience of the organization conducting the M&S, and the technical quality of the people tasked to develop the M&S application. Some of the attributes of the assessment of quality of product, process, and project are accuracy, verity, validity, clarity, completeness, acceptability, maintainability, timeliness, reliability, robustness, supportability, understandability, visibility, and maturity.

Another approach to maturity assessment of M&S that is reflective of the US DoD perspective has been expressed by (Harmon and Youngblood, 2003, 2005). They argue that M&S is a process that generates information for decision making and that V&V are the factors that provide a measure of the completeness, correctness, and credibility in the simulation results. They view the V&V process as using information from five contributing factors: (1) the conceptual model of the simulation, (2) verification results from intermediate development products, (3) the validation referent, (4) the validation criteria, and (5) the simulation results. The technique used by Harmon and Youngblood (2003, 2005) ranks each of these five factors into six levels of maturity. From lowest to highest, the six levels of maturity are:

1. We have no idea of the maturity.
2. It works, trust me.
3. It represents the right entities and attributes.
4. It does the right things; its representations are complete enough.
5. For what it does, its representations are accurate enough.
6. I'm confident this simulation is valid.

An approach that is based on the Sandia-developed Predictive Capability Maturity Model and the NASA standard has recently been developed by one of the standards subcommittees of the American Society of Mechanical Engineers (ASME, 2018). The Subcommittee on

Verification and Validation in Computational Modeling of Medical Devices (V&V 40) developed a risk-informed credibility assessment framework that could be used by the medical device industry and governmental regulators, e.g. the Food and Drug Administration. While the framework was developed for medical devices, the M&S maturity assessment procedure could be adapted to other types of engineering systems.

The V&V 40 procedure defines three factors that should be assessed for M&S maturity:

1. Verification factors, including:
   - Code verification, which encompasses software quality assurance and comparison with benchmark solutions; and
   - Calculation verification, which encompasses discretization and numerical solver error, as well as user errors associated with incorrect model inputs and outputs.
2. Validation factors, including:
   - Computational models, which encompass model form uncertainties and model input uncertainties;
   - Comparator, which encompasses number, range of characteristics, and uncertainty of measurements from laboratory or clinical trials; and
   - Assessment, which encompasses comparison of simulation results and comparator measurements.
3. Applicability factors, which encompass relevance of validation activities (both clinical and nonclinical) to the context of use.

Levels of maturity for each of the three factors are explicitly *not* included in the V&V 40 procedure. This is the primary difference from all of the procedures discussed previously. The level of maturity required for each of the factors should depend on the magnitude of influence of M&S on decision making and the potential adverse consequences of incorrect simulation-informed decision making.

15.2 Introduction to the Predictive Capability Maturity Model

15.2.1 Early Developments

In 2001, a team of researchers and computational analysts at Sandia began asking two questions related to the large-scale simulation program at the Department of Energy: Accelerated Strategic Computing Initiative (ASCI). First, "How do we assess maturity of M&S that is used in simulation-informed decision making?" Second, "How can we measure progress year to year in M&S activities?" Under the leadership of Martin Pilch, this team proposed a framework for assessing the maturity of M&S as it contributes to the nuclear weapons program (Pilch et al., 2004). They suggested that there are four key contributors to M&S impacting simulation-informed decision making: qualified computational analysts, qualified computer codes, qualified computational infrastructure, and appropriate levels of formality. As part of qualified codes, the team described nine factors that should be assessed:

1. request for service,
2. project plan development,
3. technical plan development,
4. technical plan review,

5. application-specific calculation assessment,
6. solution verification,
7. uncertainty quantification,
8. qualification and acceptance, and
9. documentation and archival.

For each of these factors, they described the key issues that should be addressed and the key evidence artifacts that should be produced to assess maturity. The artifacts and evidence must be relevant to an M&S application of interest, whether it is a system, subsystem, or manufacturing process. They also described four levels of formality that would apply over a wide range of M&S situations. The underlying feature of these four levels is the risk of adverse outcomes from simulation-informed decision making when simulation results proved to be unreliable, such that the system does not achieve performance requirements. The four levels of formality of these factors were related to the consequence of the system or decision being informed by M&S. These levels were defined as:

1. formality appropriate for research and development tasks, such as improving the scientific understanding of physical phenomena,
2. formality appropriate for nuclear weapon design support,
3. formality appropriate for nuclear weapon qualification support, i.e., confidence in component performance is supported by simulations, and
4. formality appropriate for qualification of nuclear weapon components, i.e., confidence in component performance is heavily based on simulations.

They then constructed a table with rows corresponding to the nine factors described and the columns corresponding to the four levels of formality just mentioned. In each entry in the table, the general characteristics that should be achieved for a given factor at a given level of maturity are listed. This table, or matrix, could then be used to assess the maturity of an M&S effort that is directed toward informing a specific decision.

15.2.2 Generalization of the Approach

As these fundamental concepts developed, it was recognized that the maturity assessment procedure could be generalized to serve the broader M&S community. During this time, close collaboration also began with the NASA Technical Standards Working Group developing the NASA interim standard for models and simulations (NASA, 2006). Following the work of Harmon and Youngblood (2003, 2005), Sandia made improvements based on the general concepts of information quality used in decision making. Wang and Strong (1996) questioned information consumers to determine what attributes of information quality were important to them when using information in decision making through an extensive survey. Wang et al. (2002) further developed the concepts of information quality from the perspective of Total Quality Management used in manufacturing physical products. From their survey and follow-on research, they concluded that the most constructive measure of information quality from a decision-making perspective should be *fitness for use by information consumers*. Wang et al. (2002) identified four primary categories of attributes for information quality:

- *Intrinsic information quality*: accuracy, objectivity, believability, and reputation;
- *Contextual information quality*: value added, relevance, timeliness, completeness, and amount of information;
- *Representational information quality*: interpretability, ease of understanding, consistent representation, and concise representation; and
- *Accessibility information quality*: accessibility and access security.

If the user of the information is not adequately satisfied with essentially all of these important attributes, then the user could (a) make minimal use of the information for the decision at hand; (b) completely ignore the information; or (c) misuse the information, either intentionally or unintentionally. These outcomes range from wasting information (and the time and resources expended to create it) to a potentially disastrous result caused by misuse of the information.

15.2.3 Factors Assessed in PCMM

The initial version of the *Predictive Capability Maturity Model* (PCMM) was documented in Oberkampf et al. (2007). This version of PCMM, with minor improvements, is discussed here. PCMM was developed to focus on physics-based computational models, as compared to the broader class of models considered in the NASA standard. There are similarities between PCMM and the NASA standard in terms of both the credibility factors assessed and the maturity levels for each factor. However, there are two differences between PCMM and the NASA standard that should be clarified. First, PCMM concentrates on technical factors related to the numerical solution of mathematical models that are given by partial differential equations (PDEs) and/or integral equations. Second, PCMM (as originally developed) is less formal, simpler, and easier to apply in practice than the NASA standard. The type of mathematical model considered and the level of assessment formality that is needed to support a decision determines which method to use for a given situation.

PCMM identifies six technical factors that fundamentally contribute to the credibility of a computational simulation. These factors, along with an example of an associated question, are:

- *Representation and Geometric Fidelity*: What is the geometric fidelity of the modeling relative to the physics being simulated?
- *Physics and Material Model Fidelity*: What are the relevant physics occurring within the system and what are the appropriate material models?
- *Code Verification*: What is the level of testing of the accuracy of the numerical solutions to the mathematical model?
- *Solution Verification*: Why do you believe that the numerical solution errors are not corrupting or biasing the solutions obtained for the application of interest?
- *Model Validation*: How well do simulation results compare with relevant experimental measurements for subsystems of the system of interest?
- *Uncertainty Quantification and Sensitivity Analysis*: Are all of the important uncertainties for the application conditions of interest represented in the mathematical model and what are the dominant contributors?

Each of these six factors is defined for minimal overlap or dependency between the other factors. Each factor contributes independent information concerning the M&S maturity for the application of interest. The six factors listed are the most important technical contributors to simulation credibility. Each of these factors is assessed relating to the four aspects of information quality identified by Wang and Strong (1996), and Wang et al. (2002). Two examples of *nontechnical contributors* are taken from the NASA standard (NASA, 2016). First, the factor identified as *M&S History* can be informative in evaluating communication to a decision maker. However, this factor does not answer the question, "How does past performance/reliability of the M&S project relate to the performance/reliability of the existing M&S project?" Second, the factor, *M&S Process/Product Management*, deals with management, control, and compliance of processes and simulation products. While these aspects are important, they relate more to process and compliance issues, as opposed to technical issues.

Of the six factors listed for inclusion in PCMM, only two, *Representation and Geometric Fidelity* and *Physics and Material Model Fidelity*, have not been discussed in detail earlier in this book. Thus, a brief discussion follows.

15.2.3.1 Factor: Representation and Geometric Fidelity

The factor, *Representation and Geometric Fidelity*, assesses the level of detail and precision included in the spatial and temporal definition of all constituent activities of the system being analyzed. Note that when we refer to a *system*, we mean *any* physical entity being simulated, e.g., a subsystem, component, manufacturing process, or naturally occurring entity such as an underground region. The representational and geometric definition of a system is commonly specified in a computer-aided design (CAD) software package. The traditional emphasis in CAD packages has been on manufacturing-related, dimensional fabrication, and assembly specifications. As M&S has matured, CAD vendors are now beginning to address issues that are specifically needed for the formulation and computation of a simulation for the system. Two examples of increased detail needed in M&S follow. First, there are important feature definition details of the system that are needed for various types of physics modeling, e.g., details in mechanical, thermodynamic, electrical, and optical properties and interfaces in a multicomponent structure. Second, there are important feature details that are needed for the surface definition of assembled structures to reduce the repair and cleanup needed for mesh generation. For example, there are many interface and geometric details between multiple components that are needed to produce a "water-tight" surface with properly proportioned discretization elements for the simulation.

Even though some progress has been made to ease the transition from traditional CAD files to those needed for detailed computational simulations, a great deal of work still needs to be done. For example, there is no general process for verifying that the CAD geometries and information loaded into simulations are appropriate or consistent with the physics modeling assumptions that are used in the simulation. A key issue that complicates the transfer of CAD information to a simulation package is that the level of detail of information needed *depends directly* on the particular type of physics to be modeled and the specific assumptions in the modeling. For example, a CAD package for joining two thin metal sheets that are spot welded may provide sufficient information for a computer-aided manufacturing (CAM) package, but not enough information for M&S of the spot weld joining process itself. The level of detailed information needed depends on the fidelity of the physics being modeled. The time-consuming task of addressing the level of detail needed for physical representation and geometry issues becomes the responsibility of technically trained staff with different backgrounds and experience, such as CAD/CAM package developers, computational modelers in specific disciplines, and mesh-generation experts. As commercial software developers move more vigorously into multi-physics simulations, these issues will become more complex.

15.2.3.2 Factor: Physics and Material Model Fidelity

It is well recognized that improvement in the fidelity of physics modeling has been the dominant theme pursued over many decades in computational simulations. The range of physics modeling fidelity can vary from empirical models that are based on the fitting of experimental data, which involve no physical principles, to what are typically called *physics-based models*. The three types of models in this range are referred to here as fully empirical models, semiempirical models, and physics-based models. Physical process models that are *completely* built on statistical fits of experimental data are referred to as *fully empirical models*. These fully empirical models typically have *no* relationship to physics-based principles. Consequently, the fully empirical models rest entirely on the calibration of responses to identified input parameters over a specified range. These calibrations should not be used beyond their calibration range (extrapolated), nor should they be used when other physical processes or parameters impact the response of interest.

A *semiempirical model* is one based on both calibrated models of processes and some simple physical principles. An example of a semiempirical model that has been heavily used in nuclear reactor safety is the control volume, or lumped parameter, model. Semiempirical models typically conserve mass, momentum, and energy, but at relatively large physical scales compared to the solution domain of the mathematical model. For example, an entire 3D thermal-hydraulic system may be divided into 10x10x10 control volumes in an analysis. Another common characteristic of semiempirical models is that they do not numerically converge as the discretization size is reduced. In addition, they rely heavily on fitting experimental data as a function of dimensional or nondimensional parameters, such as Reynolds or Nusselt numbers or void fraction, to calibrate the models.

With reference to *physics-based models*, we mean models that are characterized by PDEs or integrodifferential equations that conserve mass, momentum, and energy at infinitesimal length and time scales relative to the physical scales in the system. As discussed at length in this text, these types of models should demonstrate orderly numerical convergence for sufficiently refined discretization of the solution domain. However, it must be recognized that physics-based models (indeed every mathematical model of nature) include physics approximations or empiricism at some level. The empiricism should be based on spatial and temporal scales associated with the formulation of the mathematical model, *not* the discretization of the mathematical model. For example, the empirical approximations made in large eddy simulation of turbulent flow (usually referred to as a submodel) should be made based on fixed length and time scales, as opposed to discretization scales of space and time. If the empirical approximations rely on the discretization scale, then the mathematical model, i.e., the PDEs, changes every time the discretization scale is changed.

Another important aspect of physics-modeling fidelity that should be assessed in this factor is the degree to which various types of physics are coupled in the mathematical model of the

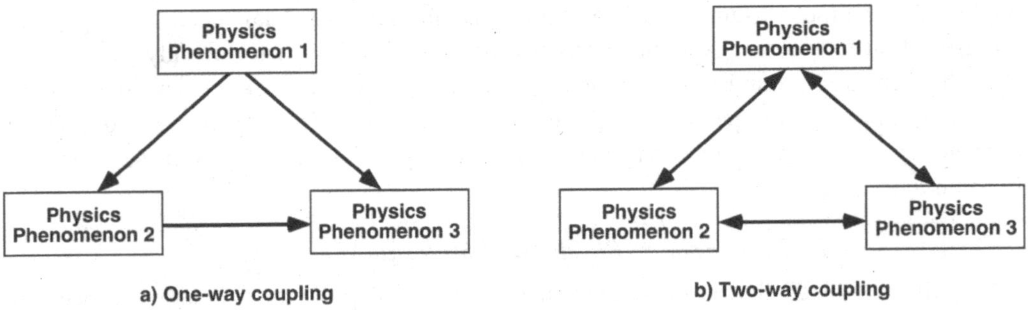

Figure 15.1 Example of two basic types of coupling of physical phenomena. (Oberkampf et al., 2007)

system. For fully empirical and semi-empirical models, strong assumptions are made to greatly simplify the physics considered, and little or no coupling of different types of physics is included. However, for physics-based models, the modeling assumptions must explicitly state the allowed types of physics coupling, all others being neglected. As shown in Figure 15.1, two basic approaches are used to couple different physics involved in the physical process:

- One-way causal effect, i.e., one physical phenomenon affects other phenomena, but the other phenomena *do not* affect the originating phenomenon (Figure 15.1a); and
- Two-way interactions, i.e., all physical phenomena affect all other physical phenomena (Figure 15.1b).

In physics-based modeling, each physical phenomenon is represented by a mathematical model with input data conditions, e.g., boundary conditions (BCs), excitation conditions, and initial conditions (ICs). In one-way coupling (Figure 15.1a), the physics of phenomenon 1 affect phenomena 2 and 3, but there is no direct feedback from phenomena 2 and 3 to phenomenon 1. Phenomenon 1 affects phenomena 2 and 3 either by way of BCs or through some type of system excitation, e.g., the addition of a source term on the right-hand side of the PDEs for phenomena 2 and 3. In addition, as show in Figure 15.1a, the BCs or source terms of phenomenon 3 are determined by both phenomena 1 and 2.

In two-way coupling (Figure 15.1b), all phenomena in the system affect *all* other phenomena. This two-way interaction is usually referred to as strong coupling, where two or more phenomena are modeled within the same mathematical model, or as weak coupling, where the interactions between phenomena occur through BCs or source terms between separate sets of mathematical models. For time-dependent mathematical models, this coupling can be either accounted for iteratively within each time step (in order to obtain a time-accurate solution) or accounted for in the next time step computed, i.e., time-lagged coupling. Time-lagged coupling typically reduces the order of accuracy of the numerical solution to first order or less.

15.2.4 Maturity Assessment of the Factors

15.2.4.1 Interpretation of Maturity Assessment Levels

All the maturity assessment procedures discussed previously have the common feature that higher maturity levels indicate some sense of increasing M&S credibility, reliability, and trustworthiness. The number of maturity levels in the various assessment methods ranges from four to nine. However, the number of gradations in the maturity scale is not a fundamental issue in how to interpret the meaning of a factor's maturity. A careful examination of the various methods shows that there is a wide range in the implication or interpretation of what a scale means in terms of operational significance or use as a decision-support tool. In terms of the critical concepts of information quality discussed by Wang and Strong (1996), and Wang et al. (2002), the particular concept that should be seriously examined for each of these methods is *representational information quality: interpretation*. For example, the TRL procedure explicitly relates a maturity level to a given status of achievement of the technology. Hence, the descriptor of the level itself *defines* the expected requirements that the technology must have achieved to attain that level. For instance, to achieve TRL Level 7, the system prototype must have been demonstrated in an operational environment.

The four-level scale of Pilch et al. (2004) is dominated by the concept of formality, magnitude of impact the M&S results have on the decision in question, importance of the decision to which the M&S results contribute, and higher reliability of M&S results with increasing levels of formality. Furthermore, in Pilch et al. (2006), they make it clear that the higher maturity levels in Pilch et al. (2004) are directly connected to increasing levels of risk *aversion*. The maturity scale is described as an ordinal ranking based on the decision maker's increasing *unwillingness* to accept risks by using increasing levels of M&S results as part of decision support. Therefore, increasing risk aversion is *inversely* related to the acceptable risk for the decision maker. This approach has some appealing features, but it also introduces additional complexities and ambiguities of interpreting the maturity scale. Two difficulties in using a risk-based scale are mentioned here, as well as how these difficulties have a critical impact on the interpretation of the assessment of maturity levels.

First, risk assessment is commonly defined as the product of two factors, likelihood of an adverse outcome and magnitude of the adverse effects of an outcome. It has proven difficult in practical risk assessment for major complex systems to: (a) estimate the likelihood of possible adverse outcomes because so little data is available for unusual events; (b) identify an extremely wide range of possible, unexpected, and unanticipated adverse outcomes; and (c) estimate the magnitude of the impact (e.g., loss of public confidence and loss of political support) of the identified adverse outcomes (Todinov, 2018). Combining these complex issues in the risk-based ranking of the maturity levels of an M&S effort can cause misinterpretation, confusion, and misuse of the maturity assessment results.

Second, the risk tolerance of a decision maker or group of decision makers is highly variable. It is also an impossible characteristic to quantify or measure in a decision maker or group. An indication of risk tolerance can be stated as: "What is the point at which a person considers the trade-off between potential loss and potential gain to be equal?" This point depends on both explicit factors, such as potential organizational and personal liability, and the implicit or personal factors of the decision maker, such as individual accountability for a decision as well as the short-term and long-term personal goals of the individual. For example, a given assessment of M&S maturity could be acceptable for one decision maker even though there was heavy reliance on M&S results. However, the same maturity assessment could be unacceptable for another decision maker who has a different level of experience or risk aversion. A classic example of contrasting risk tolerance is to compare individuals with little or no accountability for decision consequences, as opposed to governmental regulators who are responsible for public and environmental safety.

Contrary to the maturity assessment scales reviewed earlier, including the original concepts in Pilch et al. (2004), the NASA standard makes *no connection* of the levels of maturity assessment to a decision consequence or the risk tolerance of a decision maker (NASA, 2016). The levels of maturity in the NASA scale are simply an ordinal scale of increasing evidence, formality, attainment of recommended practice, and credibility of M&S activities. NASA clearly separated the idea of credibility assessment of M&S from the requirements for a given engineering application that uses M&S results in simulation-informed decision making. As stated in the NASA standard, "This NASA Technical Standard itself levies no requirements with respect to what levels to achieve (the sufficiency threshold levels), merely that the levels be determined and reported. (p. 29)" When the NASA standard discusses requirements, these are entirely focused on what must be achieved regarding the factors in their credibility assessment table to achieve a given level of maturity; *not what is required* for use of M&S in simulation-informed decision making. Inserting any concept of application requirements into the maturity assessment scale causes confusion and loss of representational information quality, even if the requirement is included in an indirect or inferential manner. As a result, the present version of PCMM only shows the level of maturity with no mention of the consequence level associated with a given maturity. With the increased popularity of simulation maturity assessment methods by industry, government, engineering societies, and regulators, the users and developers of these methods are cautioned *not* to include application requirements, such as perceived risk, into the maturity assessment scale.

The goal of characterizing maturity levels is to objectively track intellectual artifacts or relevant evidence obtained in an assessment of an M&S effort. Any piece of reproducible information or tangible evidence relating to the simulation effort can be considered an artifact. When moving to higher levels of maturity, the quality, quantity, and detail of the intrinsic and contextual information artifacts must increase. In the PCMM approach, four levels of general characteristics of maturity are defined. Sometimes these are referred to as *simulation readiness levels* (SRLs) (Clay et al., 2007; Jensen, 2021):

- *Level 0* – Little or no assessment of the accuracy and/or completeness has been made; little or no evidence of maturity exists; individual judgment, personal opinion, and personal anecdotes are dominant; convenience and expediency are the primary motivators.
- *Level 1* – Some informal assessment of the accuracy and/or completeness has been made; generalized characterization of the evidence is available; some evidence of maturity or reliability is available; an informal assessment of the evidence has been made by qualified colleagues.
- *Level 2* – Some formal assessment of the accuracy and/or completeness has been made; detailed characterization of the evidence is available; significant evidence of maturity or reliability exists; some formalized assessments of the evidence have been made by an internal peer-review group.
- *Level 3* – Formal assessment of the accuracy and/or completeness has been made; precise and accurate characterization of the evidence is available; very detailed and complete evidence of maturity or reliability exists; essentially all assessments have been made and formalized by an independent peer-review group.

It must be stressed that these levels represent an *ordinal scale* of maturity assessment. As discussed in detail in Allen and Yen (1979), an ordinal scale only represents a relative ordering of the attribute of interest, but it does *not* express a magnitude in an absolute or reference sense. For example, the four levels of maturity used in PCMM could have also been labeled with arbitrary symbols that did not imply any inherent magnitude, such as ☽, ♡, ♞, and ⚓. When arbitrary symbols are used instead of numerical values, it jarringly reminds the user that the maturity difference between each of the levels is *not* a constant.

15.2.4.2 What Constitutes Evidence?

In the original version of PCMM, as documented in Oberkampf et al. (2007), no description was made for what artifacts constituted acceptable evidence. Since then, Sandia National Laboratories has continued to develop and expand the concepts of PCMM, as well as apply PCMM to several high-consequence systems and decision making (Hills et al., 2013). Since the original interim version of the NASA standard was published in 2006, extensive development and broader use of the NASA standard has also occurred (NASA, 2016). These recent activities have demonstrated the important fact that evidence used in maturity assessment must be tangible, reproducible, and documented. At the lowest level of maturity, Level 0 in PCMM, little or no evidence is required. As the maturity level increases, there are increasing requirements for reproducibility, specificity, and documentation of the artifacts used as evidence. The present discussion of PCMM respects and accepts these recent developments in the concept of what is required for consideration as evidence in maturity assessment.

To be considered as evidence in the maturity assessment of the PCMM (for Levels 1 and higher), the artifact must be tangible, i.e., real in the sense that it must be documented and reproducible in some form. Verbal opinions or anecdotal recollections of individuals are not considered as evidence. An opinion or conclusion documented in a report, conference paper, or journal article could be considered as evidence, if the opinion or conclusion is quantitatively supported in the document. An audiovisual recording, e.g., of an experiment or test, is acceptable evidence if it is reproducible and has sufficient documentation to be correctly interpreted. When using audiovisual recordings at the higher levels of maturity, there are increasing requirements that quantitative data can be obtained from the recording. As the maturity levels increase, there is an increasing requirement for the detail, specificity, formality, control, and traceability of the evidence used to support an assessment of a factor. Since the amount of accumulated evidence increases significantly for increasing maturity levels, both Sandia and NASA have developed spreadsheet or database software packages for searching, retrieving, and managing the accumulated evidence for a maturity assessment. Sandia refers to the documented evidence for the maturity assessment of M&S as the *body of evidence* or the *evidence package* (Hills et al., 2013).

15.3 Conducting a PCMM Assessment

15.3.1 Structure of a PCMM Assessment

Six factors of M&S maturity are assessed in PCMM:

- *Representation and Geometric Fidelity,*
- *Physics and Material Model Fidelity,*

Table 15.1 *Table format for PCMM assessment*

MATURITY / FACTOR	Maturity level 0	Maturity level 1	Maturity level 2	Maturity level 3
Representation and geometric fidelity				
Physics and material model fidelity				
Code verification				
Solution verification				
Model validation				
Uncertainty quantification and sensitivity analysis				

- *Code Verification,*
- *Solution Verification,*
- *Model Validation, and*
- *Uncertainty Quantification and Sensitivity Analysis.*

Each of these factors is assessed regarding four maturity levels for M&S applied to: (a) a specific application of interest, which could be a system, subsystem, or manufacturing process and (b) if appropriate, a specific question or a set of closely related questions regarding the application of interest. That is, the PCMM maturity assessment must be focused on well-defined and restricted issues for the application of interest. These issues may include predicting the performance of a system in a normal operating environment, the safety of a system in a specific scenario in an abnormal environment, and the minimum functionality of the system in a specific scenario in a hostile environment. Requirements to achieve a maturity level are specified for each entry of a 6 × 4 matrix or table. Table 15.1 shows the structure of the PCMM maturity assessment table.

For each entry of the table, there is a set of predefined descriptors, which are discussed in the next section. For each factor, the descriptors of requirements become more demanding and specific in terms of evidence requirements as the maturity level increases for that factor. To achieve a given level of maturity, the evidence must satisfy not only the entry's descriptors, but *all of the descriptors in the lower levels of maturity* for that factor. If the evidence satisfies all of the descriptors for an entry, and all of the lower level descriptors, the factor is considered to be fully assessed for that level of maturity. A score is then assigned to that factor that is equal to the highest level of maturity descriptors that are satisfied for that factor. For example, if a factor is assessed so that it fully meets all of the predefined descriptors at both Levels 1 and 2, then the factor has a score of 2. The descriptors for Level 0 for all factors are satisfied automatically because essentially no evidence is required for this level. Furthermore, the assessed levels in the completed table should represent the actual status of the M&S enterprise at a given point in time, not some anticipated or expected future status. In other words, the table should measure the maturity of the actual status at a given point in time, not something that is nearly attained or a status that would look good in a program review.

Table 15.2 *Example of a PCMM table after M&S maturity assessment*

MATURITY / FACTOR	Maturity level 0	Maturity level 1	Maturity level 2	Maturity level 3	*Factor score*
Representation and geometric fidelity		Assessed			1
Physics and material model fidelity			Assessed		2
Code verification		Assessed			1
Solution verification	Assessed				0
Model validation			Assessed		2
Uncertainty quantification and sensitivity analysis		Assessed			1

Table 15.2 gives an example of a PCMM table after an M&S maturity assessment has been completed. For this example, the evidence satisfied all of the descriptors in the *Assessed* entry in the table, as well as all of the descriptors at the lower levels of the factor. The designator *Assessed* is then placed in the appropriate row and column in the table. For the two factors that are assessed at maturity Level 0, this could mean that either there was literally no evidence available, or the evidence did not satisfy all of the descriptors at Level 1. Once the assessment is completed, the set of scores for each factor is compiled. These scores can be shown to the right of the table, as shown in Table 15.2, or they can be shown as the set of scores [1, 2, 1, 0, 2, 1]. Upon completion of a PCMM assessment, there are six individual scores. *These scores cannot be combined or averaged in any way, i.e., all six scores are given in an assessment.*

In PCMM assessments, it is commonly found that some of the descriptors at a given maturity level have been satisfied, but not all. For example, only some of the descriptors for Level 2 have been satisfied, but others have not. For this situation, it has proven useful to give a fractional score for the maturity level instead of strictly assigning an integer score at the lower level of maturity. If this type of scoring is used, *all of the descriptors at the next lower maturity level must be satisfied.* For example, if a maturity score of 1.5 is assessed, roughly half of the descriptors at Level 2 are satisfied, but all of the descriptors at Level 1 are satisfied.

PCMM assessment has focused on the maturity of M&S for a particular engineering application of interest. Situations can exist where multiple PCMM assessments need to be aggregated into one score. For example, suppose a set of scores exists for multiple subsystems composing a complete system, and the M&S maturity of each subsystem is represented by six scores. Commonly, a decision maker would like an M&S maturity assessment for the entire system. The argument can be made that the M&S maturity of the complete system can be computed as the average of the maturity of each factor of each subsystem. This is a fallacious argument because it assumes that the M&S maturity of the complete system depends equally on all the subsystems and their corresponding factors to the complete system. The argument for using the average value would be analogous to someone claiming to compute the breaking strength of a chain by averaging the strength of each link in the chain. The following approach is defensible and recommended for the situation where M&S information is a significant factor for each subsystem of a complex system. The M&S maturity assessment of the complete system is computed by taking the *minimum* maturity score of each factor for each subsystem, resulting in a set of six scores for the complete system.

A final important aspect should be mentioned concerning the interpretation and use of scores from a PCMM maturity assessment. Users of a PCMM maturity assessment commonly interpret an increase in assessment of each of the factors over time as an improvement in the accuracy of the M&S results. Although increasing PCMM maturity tends to increase simulation accuracy, it is *not necessarily* true. While many people want to interpret the PCMM scores as a measure of M&S predictive accuracy, PCMM assesses the credibility, reliability, and trustworthiness of the simulation process factors, *not explicitly* the accuracy of the simulation results. The accuracy of the simulation results would commonly increase as the PCMM scores improve, but it should be understood that there is *not* a one-to-one correspondence.

To clarify why this is true, consider the example based on the assessment shown in Table 15.2. Suppose that the factor of *Uncertainty Quantification and Sensitivity Analysis* (UQ/SA) was improved from a value of 1 in the assessment to an assessment of 2 because significant improvements were made. This could occur during the following examples. First, suppose at Level 1, only one submodel of failure strength of welded joints in a fabricated structure was used. Furthermore, at Level 1, this submodel is only supported by individual judgment and personal opinion. However, at Level 2, multiple, higher-fidelity submodels of failure were used to estimate the magnitude of model form uncertainty, resulting in *increased* predictive uncertainty.

Second, suppose at Level 1, there was no quantitative estimate of numerical solution error included in the UQ analysis. Numerical solution error was either ignored or the computational analyst concluded it was small. However, at Level 2, a quantitative estimate of numerical solution error was made, and it was included in the UQ analysis. With these improvements, the estimated uncertainty in important system response quantities (SRQs) of interest increased noticeably. The argument could be made that the accuracy of the simulation results has improved, but the important reality to a designer or decision maker using the simulation results is that the estimated uncertainty *is larger*. The designer and the decision maker are now better informed regarding how to use simulation results because now they have an improved estimate of predictive uncertainty. As the maturity of certain factors increases, it is not uncommon that new physics is uncovered or new failure modes are recognized, such that predictive uncertainty *increases*.

15.3.2 Who Conducts the PCMM Assessment?

A crucial practical issue in completion of the PCMM assessment is: Who should conduct the assessment? The original Sandia procedure (Oberkampf et al., 2007) engaged a VVUQ expert who was also experienced with the PCMM table, and they conducted the assessment. The assessment was done in discussions with individuals who were familiar with the relevant evidence of the M&S project. Experience with this approach showed that disagreements and disputes often occurred between the assessor's results and the individuals familiar with the evidence artifacts. This most commonly occurred between mathematical model developers and software developers. These experts would often feel that the PCMM assessor was telling them how to do their job, or the PCMM descriptors were inappropriate or excessively vague in their specifications. As observed by Rider et al. (2017), this proved to be very counterproductive and confrontational in many situations. As stated by Rider et al. (2017), "If V&V allows itself to get into the role of judge and jury on quality, progress will be poor. V&V's job is to ask appropriate questions about quality as partners with M&S professionals interested in improving the quality of their work."

During the last decade of applying PCMM at Sandia, various refinements have been made to the assessment approach (Hills et al., 2013). A team approach to assessment is now used with a wide range of perspectives, which has proven to be much more effective and constructive. For example, Sandia recommends the following members of an assessment team: (a) a computational analyst, (b) an M&S software developer, (c) a customer or user of the M&S results, and (d) a subject matter expert on PCMM and VVUQ. For the *Model Validation* factor, and possibly other factors in the table, an experimentalist who is familiar with the relevant experimental activities should also be a member of the team. A team assessment approach has shown to promote better interpretation and application of the descriptors of the factors for each maturity level. A major benefit of this approach is to encourage discussion and learning among the team members as to the meaning and value of the six factors in the table. Furthermore, the value of a maturity assessment to the customer or user of the M&S results has proven most rewarding because they are usually not familiar with the six factors that build credibility.

15.4 Detailed Discussion of PCMM Factors

This section discusses each of the maturity assessment factors and the maturity levels in the PCMM table in more detail. This discussion is very similar to the original discussion in Oberkampf et al. (2007), but some improvements in clarity and specificity have been included. For each entry in Table 15.3, a brief description is given for the requirements to achieve the level of maturity for each factor in the PCMM table. This table can be used as summary statements of the basic descriptors for each factor. Note that the requirements of the descriptors at each maturity level accumulate when moving to higher maturity levels within an factor. For example, to attain a given maturity level for a given factor, the descriptors within the specific entry of the table must be satisfied, in addition to all descriptors at the lower levels in that factor. A detailed discussion of each factor of the table is given in the following sections.

Consistent with the recent work at Sandia, and the discussion given in Section 15.2.4, the maturity levels are now completely separate from application maturity requirements associated with each level. A discussion of how application maturity level requirements should be determined is discussed in Section 15.5.2.

15.4.1 Factor: Representation and Geometric Fidelity

This factor is directed primarily toward the level of physical and informational characterization of the system being analyzed and the specification of the geometrical features of the system. For fully empirical and semiempirical models, there may be little geometric fidelity,

Table 15.3 *General descriptions for factors of the PCMM table (Oberkampf et al., 2007)*

FACTOR \ MATURITY	Maturity level 0	Maturity level 1	Maturity level 2	Maturity level 3
Representation and geometric fidelity What features are neglected because of simplifications or stylizations?	• Judgment only • Little or no representational or geometric fidelity for the system and BCs	• Significant simplification or stylization of the system and BCs • Geometry or representation of major components is defined	• Limited simplification or stylization of major components and BCs • Geometry or representation is well defined for major components and some minor components • Some peer reviews conducted	• Essentially no simplification or stylization of components in the system and BCs • Geometry or representation of all components is at the detail of *as built*, e.g., gaps, material interfaces, fasteners • Independent peer review conducted
Physics and material model fidelity How fundamental are the physics and material models and what is the level of model calibration?	• Judgment only • Model forms are either unknown or fully empirical • Few, if any, physics-based models • No coupling of models	• Some models are physics based and are calibrated using data from related systems • Minimal or ad hoc coupling of models	• Physics-based models for all important processes • Significant calibration needed using separate effects tests (SETs) and integral effects tests (IETs) • One-way coupling of models • Some peer reviews conducted	• All models are physics based • Minimal need for calibration using SETs and some calibration using IETs • Sound physical basis for extrapolation and coupling of models • Full, two-way coupling of models • Independent peer review conducted
Code verification Are algorithm deficiencies, software errors, and poor SQE practices corrupting the simulation results?	• Judgment only • Minimal testing of any software components • Little or no SQE procedures specified or followed	• Code is managed by SQE procedures • Unit and regression testing conducted • Some comparisons made with benchmarks	• Some algorithms are tested to determine the observed order of numerical convergence • Some features & capabilities (F&C) are tested with benchmark solutions • Some peer reviews conducted	• All of the important algorithms are tested to determine the observed order of numerical convergence • All of the important F&Cs are tested with rigorous benchmark solutions • Independent peer review conducted

Solution verification Are numerical solution errors and human procedural errors corrupting the simulation results?	• Judgment only • Numerical errors have an unknown or large effect on simulation results	• Numerical effects on some SRQs are qualitatively estimated • Input/output (I/O) verified only by the analysts	• Numerical effects are quantitatively estimated to be small on some SRQs • I/O independently verified • Some peer reviews conducted	• Numerical effects are determined to be small on all important SRQs • Important simulations are independently reproduced • Independent peer review conducted
Model validation How carefully is the accuracy of the simulation and experimental results assessed at various tiers in a validation hierarchy?	• Judgment only • Few, if any, comparisons with measurements from similar systems or applications	• Quantitative assessment of accuracy of SRQs not directly relevant to the application of interest • Large or unknown exper-imental uncertainties	• Quantitative assessment of predictive accuracy for some key SRQs from IETs and SETs • Experimental uncertainties are well characterized for most SETs, but poorly known for IETs • Some peer reviews conducted	• Quantitative assessment of predictive accuracy for all important SRQs from IETs and SETs at conditions/geometries directly relevant to the application • Experimental uncertainties are well characterized for all IETs and SETs • Independent peer review conducted
Uncertainty quantification and sensitivity analysis How thoroughly are uncertainties and sensitivities characterized and propagated?	• Judgment only • Only deterministic analyses are conducted • Uncertainties and sensitivities are not addressed	• Aleatory and epistemic (A&E) uncertainties are propagated, but without distinction • Informal sensitivity studies conducted • Many strong UQ/SA assumptions made	• A&E uncertainties segregated, propagated, and identified in SRQs • Quantitative sensitivity analyses conducted for most parameters • Numerical propagation errors are estimated, and their effects known • Some strong assumptions made • Some peer reviews conducted	• All A&E uncertainties comprehensively treated and properly interpreted • Comprehensive sensitivity analyses conducted for parameters and models • Numerical propagation errors are demonstrated to be small • No significant UQ/SA assumptions made • Independent peer review conducted

e.g., lumped-mass representations or representations that simply deal with the functionality of system components. For physics-based models that solve PDEs or integral equations, significant geometric detail must be specified, which is then used to prescribe the ICs and BCs for such equations. For other mathematical models, such as for electrical circuits or agent-based models, other concepts of representation fidelity are needed. For example, in the case of electrical circuit models, physical characterization deals with the fidelity of the electrical circuit diagram and level of characterization of the electrical components in the system. For agent-based models modeling the movement and interaction of robots, for example, representation fidelity might be the geography and terrain over which agents move.

Geometric fidelity details increase proportionally as physical modeling fidelity increases because of the additional information that is required for the modeling. Thus, the lowest level of maturity related to geometric fidelity is based on convenience and expediency, consistent with the approximations made in the formulation of the mathematical model. The higher levels of geometric maturity provide increasingly detailed information that is more representative of the *as-built* geometry or *as-existing* geometry in an operational environment. These higher levels of detail must be consistent with the higher levels of spatial and temporal detail that are required by a mathematical model of the system and environment. For example, the geometry, material and surface characteristics, and mechanical assembly details of the system are typically specified in a CAD file. For systems that may be in a dangerous state of excessive wear, in a damaged condition, or have a crack existing in a material, the specification of the geometry and surface properties can become quite complex and detailed.

General descriptions of the maturity levels for *Representation and Geometric Fidelity* follow:

- *Level 0*: Simplicity, convenience, and functional operation of the system dominate the fidelity of the representation and the geometry for the system being analyzed. There is heavy reliance on personal opinion and convenience, with little or no expectation or quantification of the physical representation and geometry.
- *Level 1*: Quantitative specifications are applied to describe the geometry of the major components of the system being analyzed. Much of the real system remains stylized, approximated, or ignored. An informal assessment of the evidence and its appropriateness has been made by qualified colleagues.
- *Level 2*: Quantitative specifications are applied to assess the fidelity of the mathematical representation and geometry of most of the components of the real system. A small portion of the real system remains stylized or ignored. For example, important imperfections due to system assembly or defects due to wear or damage in the system are included. A formalized assessment of the evidence has been made by an internal peer-review group.
- *Level 3*: The mathematical and geometric representation in the model is *as built* or *as existing*, meaning that no important aspect of the geometry of the modeled real system is missing. The mathematical model of the system is of such detail that very fine scale features of the geometry are included. For example, details are included to characterize the geometric features of multiple phases of materials, i.e., solid, liquid, and gas phases within the system, as well as spatial and temporal details occurring as part of the boundary conditions. A formal assessment of the evidence has been made by an independent peer-review group.

15.4.2 Factor: Physics and Material Model Fidelity

This factor primarily addresses the following:

- The degree to which the mathematical model is physics based, i.e., whether the model is fully empirical, semiempirical, or primarily physics based.
- The degree to which the mathematical model is calibrated. (The reader is reminded that the term *model*, also includes *all submodels* that complete the mathematical model.)
- The physics fidelity basis with which the model extrapolates from the conditions in the validation domain, as well as the conditions for parameter calibration, to the conditions of the application of interest.
- The quality and degree of coupling of the multi-physics effects that exist in the mathematical model.

As the physical fidelity of the mathematical model increases, and the needed input data is provided to the model for a simulation, the model is increasingly able to provide physics-based explanatory power for the physical phenomena of interest. Within the broad class of physics-based models, there are important distinctions in the degree to which the model is calibrated. For example, does the model require recalibration (updating) even if there are relatively small changes in the system design or small changes in the BCs or ICs? Alternatively, does the model require calibration only at lower levels in the validation hierarchy, i.e., SETs, in order to yield accurate predictions? Or, does the model also require calibration and/or recalibration at higher levels of the validation hierarchy, i.e., IETs, to attain accurate predictions?

To clarify this point, consider two mathematical models yielding *the same* level of agreement with experimentally measured responses at the IET level. One of the models is calibrated only at the SET level and one model is calibrated at both the SET and IET levels. The model that only requires calibration at the SET level has more predictive capability than does the model that requires calibration at both the SET and IET levels. This statement is understandable by noting that the model calibrated with only SETs has demonstrated that it can yield the same prediction accuracy as the model calibrated with both SETs and IETs. In terms of explanatory power of the models, the SET-calibrated model must extrapolate further from its calibration domain than the SET-IET-calibrated model.

General descriptions of the various levels of *Physics and Material Model Fidelity* include:

- *Level 0*: The model is fully empirical, or the model form is not precisely known. There is little or no coupling of models representing multiple functional elements of the system, and the coupling that does exist is empirically based. Confidence in the model is strictly based on convenience and the personal opinion of the analyst.
- *Level 1*: The model is semiempirical in the sense that portions of the modeling are physics based; however, important features, capabilities, or parameters in the model are calibrated using data from very closely related physical systems. The coupling of functional elements or components is minimal or ad hoc, and not based on detailed physics. An informal assessment of the evidence and its appropriateness has been made by qualified colleagues.
- *Level 2*: All important physical process models and material models are physics based. However, calibration of important model parameters is necessary, using data from SETs and IETs. All model calibration procedures are implemented on the model input parameters, not

on adjustments to the SRQs. Important physical processes are coupled using physics-based models with couplings in at least one direction, and some two-way coupling may exist. A formalized assessment of the evidence has been made by an internal peer-review group.

- *Level 3*: All models are physics based with limited need for calibration using SETs. Calibration of some parameters using IETs is still required to predict certain SRQs. Where extrapolation of these models is required, the extrapolation is based on well-understood and well-accepted physical principles, which are reflected in the mathematical model. All physical processes are coupled in terms of physics-based models with two-way coupling and physical process effects on physical and material parameters, BCs, geometry, ICs, and forcing functions. A formal assessment of the evidence has been made by an independent peer-review group.

15.4.3 Factor: Code Verification

This factor focuses on the following:

- Correctness and fidelity of the numerical algorithms used to compute a solution in the code relative to the exact solution of the mathematical model, e.g., the PDEs.
- Correctness of source code to solve the discretized model to compute a solution relative to the exact solution of the mathematical model.
- Configuration management, version control, and regression testing of the software through SQE practices.

The correctness and fidelity of the numerical algorithms and the correctness of the source code are primarily determined by conducting various types of tests on the code. The most informative types of tests compare the numerical solution results from the code with highly accurate solutions of the mathematical model. The most rigorous and highly accurate solutions are analytical and manufactured solutions discussed in detail in Chapter 6 "Exact and Manufactured Solutions." Comparisons between the numerical solution produced by the code being tested and the exact/manufactured solution typically yield two types of numerical accuracy information. First, the error in the SRQs from the code being tested is compared with the exact/manufactured solution. Although this is useful, it provides no information concerning the numerical convergence characteristics of the code being tested. Second, using two or more solutions with uniformly refined discretizations for the code being tested, and using the exact/manufactured solution, the observed order of convergence of the code being tested can be computed. Observed order of convergence is a much more definitive statement of code verification than simply comparing an individual solution and the exact solution on a given mesh. The observed order of convergence procedure is discussed in detail in Chapter 5 "Code Order-of-Accuracy Verification." Comparing the solutions from the code being tested with benchmark numerical solutions can also be used, as discussed in Chapter 5 and Oberkampf et al. (2007), but assured accuracy of the benchmark numerical solution is difficult to obtain.

SQE practices should also be assessed as part of this factor. SQE practices measure the scope and rigor of configuration management, version control, and regression testing. Chapter 4 "Software Engineering for Scientific Computing," gave a detailed discussion of this topic.

General descriptions of the maturity levels of *Code Verification* are:

- *Level 0*: Code verification is based on the judgment, personal opinion, and experience of the software developers involved. There is little or no formal code verification testing of the software components. Little or no SQE practices are defined and practiced in the implementation, management, and use of the code.
- *Level 1*: Most associated software is implemented and managed with formal SQE practices. Unit and regression testing of the software is conducted regularly with a high percentage of line coverage attained. Code verification test suites using exact solutions are minimal and only error measures between the exact solution and the code being tested are obtained on a given mesh. An informal assessment of the evidence and its appropriateness has been made by qualified colleagues.
- *Level 2*: All associated software is implemented and managed with formal SQE practices. Code verification test suites are formally defined and systematically applied using exact/manufactured solutions to compute the observed order of convergence of some SRQs. Some features and capabilities (F&Cs) of the code (such as complex geometries, mesh generation, physics, and material models) have been tested with exact/manufactured solutions to determine the observed order of convergence for each of these options. A formalized assessment of the evidence has been made by an internal peer-review group.
- *Level 3*: All important algorithms have been tested using exact/manufactured solutions to compute the observed order of convergence. All of the important combinations of F&Cs of the code have been tested with exact/manufactured solutions to determine the observed order of convergence for each of these options. A formal assessment of the evidence has been made by an independent peer-review group.

15.4.4 Factor: Solution Verification

This factor deals with assessment of the following:

- Numerical solution errors in the computed results and
- Confidence in the computational results, as they may be affected by human error.

Estimation of numerical solution accuracy and procedural checking of possible human solution-related errors are the dominant components of the assessment of this factor. Numerical solution errors are any errors due to mapping the mathematical model to the discretized model and any errors due to solution of the discretized model on a computer. Numerical solution errors due to spatial and temporal discretization of the PDEs or integral equations, and the iterative solution error due to a linearized solution approach to a set of nonlinear discretized equations, are the primary concerns in this factor. Additional numerical solution errors that should be addressed are the potential detrimental effects of numerical parameters in solution algorithms; errors due to approximate techniques used to solve nondeterministic systems (e.g., error due to a small number of samples used in a Monte Carlo sampling method); and round-off error due to finite precision on a computer.

Human errors, which are blind epistemic uncertainties discussed in Chapter 2 "Fundamental Concepts and Terminology," are also a concern in the assessment of this factor. Some examples where human errors can be made in the numerical solution procedure are (a) preparing and assembling any of the aspects or procedures needed for constructing the discrete model; (b) executing and obtaining the numerical solution on a computer; and (c) post-processing, preparing, or interpreting the computational results obtained from the solution.

General descriptions of the maturity levels of *Solution Verification* are:

- *Level 0*: No formal or organized attempt is made to assess any of the possible sources of a numerical solution error. Any statement about the impact of numerical errors is based purely on the judgment and experience of the computational analyst. No formal or organized assessment of the correctness of software inputs or outputs has been conducted.
- *Level 1*: Some type of formal procedure is used to assess the influence of numerical solution errors on some SRQs. This could include *a posteriori* error estimation of global norms, iterative convergence studies, or sensitivity studies to determine how sensitive certain SRQs are to changes in mesh or temporal discretization. Some type of organized effort is made by the computational analyst to check the correctness of I/O data. An informal assessment of the evidence and its appropriateness has been made by qualified colleagues.
- *Level 2*: Quantitative error estimation methods are used to estimate numerical solution errors in some SRQs, and these estimates show that the errors are small for some conditions of the application of interest. Some of the assumptions required for the use of the quantitative error estimation methods have been investigated to determine if the assumptions are justified. The I/O quantities have been verified by knowledgeable computational analysts who have some level of independence from the simulation effort. A formalized assessment of the evidence has been made by an internal peer-review group.
- *Level 3*: Quantitative error estimation methods are used to estimate numerical errors on all important SRQs, and these estimates show that the errors are small over the entire range of conditions for the application of interest. All of the important assumptions required for the use of the quantitative error estimation methods have been verified to be true. Important computational simulations are reproduced using the same software and the same intended F&Cs for the simulation by an independent computational analyst. A formal assessment of the evidence has been made by an independent peer-review group.

Note that higher levels of maturity of *Solution Verification* do *not necessarily* imply higher levels of accuracy of the simulation results relative to physical reality. As previously discussed in this text, there is a clear difference between the goals of verification and the goals of model validation. Increased numerical solution accuracy is necessary to gain more confidence in the fidelity of the mathematical model mapping to the solution of the discrete model. However, we are *not* necessarily improving the comparison of the computational results with experimental data when maturity of *Solution Verification* is improved. High levels of maturity in *Code Verification* and *Solution Verification* are an indication of how well the numerical solution results represent the physics in the mathematical model, as opposed to a contaminated mixture of physics, numerical error, and possibly human error at low levels of maturity.

15.4.5 Factor: Model Validation

This factor focuses on the following:

- Completeness and precision of the characterization of the experimental conditions and measurements.

- Thoroughness and quantitative assessment of the computational results relative to the experimental measurements.
- Relevancy of the experimental conditions, physical hardware, and measurements in the validation experiments compared to the application of interest.

The focus of *Model Validation* in the PCMM is on the precision and completeness of the process of the model accuracy assessment, *not* on the accuracy of the mathematical model itself. The *precision* of validation means (a) how carefully and accurately the validation experiments are conducted, and how well the experimental uncertainties are estimated; and (b) how well understood and experimentally quantified are all the conditions of the experiment that are required as inputs for the mathematical model? By *completeness* of validation we mean: How well do the conditions (geometry, BCs, ICs, and system excitation) and actual physical hardware of the validation experiments conducted relate to the actual conditions and hardware of the application of interest?

For SETs, it is expected that there will be many dissimilarities between the SET experiments and the actual application of interest because they occur at lower levels in the validation hierarchy. However, for IETs, there should be a close relationship between the IET experiments and the application of interest, particularly with respect to the experimental hardware and the coupled physical phenomena occurring in each. For a more complete discussion of conducting a model validation experiment and the construction of a validation hierarchy, see Chapter 10 "Model Validation Fundamentals."

As discussed earlier in this chapter, the correctness and credibility of model validation fundamentally relies on assumptions that the numerical algorithms are reliable, that the computer program is correct, that no human procedural errors have been made in the simulation, and that the numerical solution error is small. These are major assumptions that are commonly discovered to be unfounded in assessing simulation maturity. Consequently, to properly inform the user of the information in the PCMM table about the veracity of these assumptions, it is recommended that the maturity level of the factors *Model Validation* and *Uncertainty Quantification and Sensitivity Analysis* can be *no higher than two levels above* the maturity levels assessed for both *Code Verification* and *Solution Verification*. This requirement places further restrictions on conducting the PCMM assessment and means that the maturity levels of *Code Verification and Solution Verification* must be assessed *before* the maturity levels of *Model Validation* and *Uncertainty Quantification and Sensitivity Analysis* are assessed. Assume that, as an example of the dependencies between factors, as shown in Table 15.2, *Code Verification* and *Solution Verification* were at Levels 1 and 0, respectively. Consequently, the *maximum maturity level* that the *Model Validation* factor and the *Uncertainty Quantification and Sensitivity Analysis* factor could be is Level 2. This maximum level is recommended even if the assessor (s) were to independently judge either or both of these factors at a level higher than 2.

General descriptions of the various maturity levels of *Model Validation* are:

- *Level 0*: Accuracy assessment of the model is based almost entirely on judgment, personal opinion, and experience. Few, if any, comparisons have been made between simulation results and experimental measurements of similar applications of interest.
- *Level 1*: Limited quantitative comparisons are made between simulation results and experimental results. Comparisons for SRQs between simulation and experiment that have been made are not directly relevant to the application of interest, or the experimental

conditions are not directly relevant to the application of interest. Experimental uncertainties, either in the SRQs and/or in the characterization of the conditions of the experiment, are largely undetermined, unmeasured, or based on opinion. An informal assessment of the evidence and its appropriateness has been made by qualified colleagues.

- *Level 2*: Quantitative comparisons between computational results and experimental results have been made for some key SRQs from SET experiments and limited IET experiments. Comparisons between simulation and experiment are directly relevant to the application of interest for some SETs, and of limited relevance for IET experiments. Experimental uncertainties are well characterized (a) for most SRQs of interest and (b) for experimental conditions for the SETs conducted. However, the experimental uncertainties are not well characterized or measured for the IETs. A formalized assessment of the evidence has been made by an internal peer-review group.
- *Level 3*: Quantitative comparisons between computational and experimental results have been made for all important SRQs from an extensive database of both SET and IET experiments. Most conditions of the SETs are directly relevant and representative of the application of interest. The conditions, hardware, and coupled physics of most of the IETs are similar to the application of interest. Most of the SET computational predictions and some of the IET predictions are *blind*. Experimental uncertainties and conditions are well characterized and measured for SRQs in both the SET and IET experiments. A formal assessment of the evidence has been made by an independent peer-review group.

15.4.6 Factor: Uncertainty Quantification and Sensitivity Analysis

This factor focuses on the following:

- Thoroughness and soundness of the uncertainty quantification (UQ) effort, including identification and characterization of all plausible sources of uncertainty for the application of interest.
- Accuracy and correctness of propagating uncertainties through the mathematical model and interpreting uncertainties in the SRQs of interest.
- Thoroughness and precision of a sensitivity analysis to determine the most important contributors to uncertainty in SRQs of interest.

Recognition of uncertainties refers to the factor of identifying and understanding all possible uncertainties within the application of interest (e.g., geometry, initial conditions, and physical modeling parameters) and in the surroundings (e.g., BCs and system excitation) for the application of interest. In order to limit the possible number of uncertainties, the application of interest should be restricted to a specific environment (normal, abnormal, or hostile) and possibly even specific scenarios for a given environment. Characterization of model predictive uncertainty primarily deals with the proper estimation and representation of all uncertainties that could exist as part of the prediction for the application of interest. A key aspect of characterization is the segregation of uncertainties into aleatory and epistemic components. This segregation of uncertainties has been discussed at length in this book.

A sensitivity analysis (SA) provides additional important information to the user of the computational simulation analysis beyond what is typically considered a part of an uncertainty quantification analysis. SA was briefly discussed in Chapter 13 "Predictive Capability," but for a detailed discussion, see Smith (2014), McClarren (2018), and Santner et al. (2018). An SA is typically directed toward one of two goals for understanding the response of the system to changes in model inputs. First, an analyst may be interested in determining how outputs locally change as a function of inputs. This is usually referred to as a local SA. The information obtained from a local SA is commonly used for system design and optimization, as well as for determination of the most advantageous operational conditions for maximizing system performance. Second, an analyst may be interested in determining how the uncertainty structure of all of the model inputs maps to the uncertainty structure of each of the outputs. This is usually referred to as a global SA. For example, the information from a global SA may be used to determine which manufacturing variabilities contribute most to variability in certain SRQs, or to determine what physical experiments should be conducted to maximize the reduction in the epistemic uncertainty that is due to poorly understood coupled-physics phenomena.

As previously discussed regarding the factor *Model Validation*, the maturity level of the *UQ/SA* factor can be *no higher than two levels above* the maturity levels assessed for both *Code Verification* and *Solution Verification*.

General descriptions of the various levels of *UQ/SA* are:

- *Level 0*: Only deterministic simulation analyses are conducted for the application of interest. Judgment, opinion, and experience are the dominant forms of UQ and SA. Informal *spot checks* or *what-if* studies for various conditions are conducted to determine their possible effect.
- *Level 1*: Model input uncertainties in the application of interest are identified, represented, and propagated through the mathematical model, but they are not segregated with respect to whether the uncertainties are aleatory or epistemic. Sensitivity of some system responses to some system and surrounding uncertainties are investigated, but the SA was primarily informal or exploratory rather than systematic. Many strong assumptions are made with respect to the UQ and SA; for example, most probability density functions are characterized as Gaussian, and uncertain model parameters are assumed to be independent of all other parameters. An informal assessment of the evidence and its appropriateness has been made by qualified colleagues.
- *Level 2*: Model input uncertainties in the application of interest are characterized as either aleatory or epistemic. The uncertainties are propagated through the computational model, while their character is kept segregated both in the input parameters and in the SRQs. Quantitative SAs are conducted for most input parameters, but epistemic uncertainties are characterized as a uniform distribution over an interval. Numerical approximation or sampling errors due to propagation of uncertainties through the model are estimated, and the effect of these errors on the UQ and SA results is understood and/or qualitatively estimated. Some strong UQ and SA assumptions are made, but qualitative results suggest that the effect of these assumptions is not significant. A formalized assessment of the evidence has been made by an internal peer-review group.
- *Level 3*: Aleatory and epistemic uncertainties are comprehensively treated (for both model input uncertainties and model form uncertainty) and their segregation is strictly maintained in the interpretation of the results. Detailed investigations are conducted to determine the

effect of uncertainty introduced due to model extrapolations (if required) to the conditions of the application of interest. A comprehensive SA is conducted for both model input uncertainty and model form uncertainty. Numerical approximation or sampling errors due to propagation of uncertainties through the model are carefully estimated, and their effect on the UQ and SA results is quantitatively demonstrated to be small. No significant UQ and SA assumptions are made, unless they are fully justified. A formal assessment of the evidence has been made by an independent peer-review group.

15.5 Use of PCMM in Simulation-Informed Decision Making

This section discusses various ways that a PCMM assessment can be used in M&S projects and in simulation-informed decision making. As discussed earlier in this chapter, conducting a PCMM assessment involves a significant investment in personnel time and the accumulation and documentation of evidence. On a small-scale M&S project, such as a preliminary design project, or a research activity in an academic environment, a PCMM assessment would probably not be conducted. For a large-scale M&S project and simulation-informed decision making, particularly if M&S information is a major contributor in the decision-making process, then the investment in a PCMM assessment would provide meaningful value added to the decision making.

15.5.1 Use of a PCMM Assessment in an M&S Project

A PCMM assessment is very useful in understanding the simulation maturity status, tracking M&S maturity progress, and planning future activities, such as reallocation of resources to factors that are lagging in maturity. If a large organization conducted the entire M&S project inhouse, the project would typically involve several technical staff members from different groups or departments. These staff would commonly report to their line or department managers, in addition to the M&S project manager. A more complex management structure occurs when some or all of the technical staff members on the M&S project are from an entirely different organization, e.g., they are subcontractors to the organization of the M&S project manager. An even more difficult situation for assessing and tracking progress occurs when an organization relies on multiple commercial software companies to develop or provide diverse M&S capabilities. For example, one software company would provide the physics-based simulation capability, and another software company would provide the uncertainty quantification and sensitivity analysis (UQ/SA) capability. This last situation is dominant at the present time because most software companies that license physics-based modeling capability remain focused on deterministic simulation, as opposed on nondeterministic simulation.

Some examples of how a PCMM assessment can be useful in various M&S project situations follow.

- When using a diverse team to conduct a PCMM assessment, most team members initially would not be familiar with many of the concepts and factors listed in the PCMM table. For example, computational analysts are commonly experienced in one or two factors in the table, but they are completely unfamiliar with other factors. If customers who intend to use the M&S results (such as decision makers) are members of the team, they are commonly not familiar with *any* of the factors in a PCMM assessment. However, by having them as members of the assessment team, they learn a great deal about how each factor contributes to credibility in the simulation results.

- Even if customers who intend to use the simulation results (such as decision makers, stakeholders, and line managers) are not team members conducting a PCMM assessment, they learn a great deal about how credibility is built in M&S by listening to the evidence presented in the PCMM assessment. It should be fully recognized that these diverse customers of simulation results typically have little or no experience in computational simulation and even less experience in the six factors listed in the PCMM table. A PCMM assessment engenders discussions that would *not* have occurred without a PCMM assessment. Therefore, the discussion of the evidence contributing to the factors becomes the new language of communication between all parties involved. This kind of communication is one of the most significant consequences of a simulation maturity assessment.
- PCMM assessments made over a period of time can be used to track the progress of a large-scale M&S development project. This is critical for an M&S project manager who must measure yearly progress for an M&S project. It is also very informative for customers who intend to use the simulation capability when it is available. This latter category is particularly important if the funding source for the simulation effort is *not* the customer who intends to use the M&S results. For example, this situation occurs when a government agency funds the development of a simulation capability, but the capability will be used by another government organization or a private government contractor.
- An increasing number of organizations that design and manufacture engineering systems or subsystems are obtaining M&S services through a contractor or consultancy. These organizations may have little experience in M&S; especially the six factors of the PCMM table. These organizations can use a PCMM assessment of the contractor-provided M&S effort to better inform them of the maturity of the simulation results they are provided in their business activities.

The biggest difficulty in conducting a PCMM maturity assessment of a large-scale M&S project is that the project typically has a broad range of application conditions to address, as opposed to a specific operating condition or a specific issue of an application of interest. As discussed in Section 15.3, a PCMM assessment should be focused on a well-defined issue regarding an application of interest. One approach to dealing with this difficulty on a large-scale M&S project is to restrict the range of application conditions that the M&S capability is expected to address. For example, conduct the PCMM assessment for the application of interest *only* in a normal operating environment. Although a normal operating environment for a system could have a wide range of operating conditions, it is still feasible to conduct a PCMM assessment if the range of conditions or scenarios is also restricted. This approach would be especially effective if the expectation of the assessment is to determine if the M&S capability is at a Level 0 or Level 1. However, for abnormal or hostile environments, conducting a PCMM assessment is more difficult because of the extremely broad range of conditions and scenarios that could exist for a system. For example, in an abnormal environment such as major mechanical damage to the system combined with an engulfing fire, it is doubtful that all of the most dangerous failure sequences and combinations of sequences could be identified.

15.5.2 Simulation Maturity Requirements for an Application

In the previous discussion of the PCMM, it has been stressed that the assessment of the maturity of each of the factors should *not* be associated in any way with the simulation maturity requirements needed for the application of interest. Although it should be clear conceptually that the maturity requirements as listed in the PCMM table are separate from the simulation maturity requirements needed for an application, these two issues are commonly confused or intertwined in practice. There are two reasons why this frequently occurs. First, some of the maturity assessment methods previously discussed connect, either explicitly or implicitly, the maturity assessment level with what is needed for the application requirement. The TRL procedure is a good example of explicitly relating the maturity level of the technology to the demonstrated status of achievement of the technology. Second, when a PCMM assessment is conducted, the application of interest and the issues addressed by the M&S activity set the context of the PCMM assessment. For example, consider a normal operating environment and the application of interest is a centrifugal turbopump for a liquid. Suppose the particular operating condition of interest is near the condition where serious vibrations might occur due to cavitation of the liquid. When conducting the PCMM assessment, it is natural for the team members to keep in mind how the maturity of the M&S capability compares to what is required to answer the question of interest. Since some of the team members conducting the assessment may have also developed the existing M&S capability, it is easy for them to become advocates of the M&S capability, as opposed to independent assessors of the capability.

If possible, it is best to begin discussion of the application maturity level requirements *after* an objective PCMM assessment has been made. The application maturity level requirements can be independently set for each of the six factors in the table. Experience at Sandia National Laboratories has shown that the application maturity level requirements are commonly *different* for each factor (Hills et al., 2013). Application maturity requirements may result from various sources; for example, (a) a customer who intends to use the M&S capability to design a system, (b) a customer who is purchasing the system of interest and some of the claims of performance, safety, and reliability are based on M&S results, and (c) a governmental regulatory authority that is evaluating the safety of a system of interest based heavily on M&S results provided by the applicant. These last two examples are closely related to the concept of Certification by Analysis (CbA).

As the PCMM procedure has developed over the years at Sandia (Hills et al., 2013), their experience has motivated them to add three high level factors to the PCMM table; *Customer Specification Completeness*, *Experimental Data for Constitutive Model Calibration*, and *Experimental Data for CompSim Model Validation*. Additionally, they have divided the high-level factors into finer-grained sub-factors. The finer-grained sub-factors have two advantages. First, more refined maturity assessment requirements can be defined within each of the sub-factors. For example, the *Code Verification* factor may have five sub-factors. Second, the sub-factors allow separate application maturity level requirements to be set for *each* sub-factor. For example, the Sandia PCMM table now has a total of 27 sub-factors spread across the nine primary factors, giving a total of 108 sub-entries in the table. This refinement approach, however, also has two disadvantages. First, the larger PCMM matrix requires a significantly increased investment in time and effort to assess maturity in each of the sub-factors. Second, the focus tends to shift from assessing M&S maturity to focusing on increments of "the grade." This last disadvantage is contrary to the goal of PCMM, as well as detrimental to M&S teamwork.

Setting an appropriate maturity level for the application of interest or issues of interest has proven to be a difficult task. The task is difficult because it becomes intertwined with many other management, business, and risk assessment issues that are *not* directly related to the M&S effort. This topic will be discussed in more detail in the next sub-section. Two general recommendations are made here on how an organization might go about determining appropriate application maturity level requirements. First, the customer who will use the M&S results and the M&S providers should both have input when setting the application maturity level requirements. This recommendation assumes that the customer is funding the development of the M&S capability. However, the practical difficulty is that the customer is typically not familiar with the technical aspects of the six factors in the PCMM table. Previous experience has shown that this group tends to set unrealistically high levels of maturity requirements in order to reduce *their* risk of potential misinformation from M&S results. Furthermore, this group rarely recognizes the rapid increase in time and resources needed to increase from one maturity level to another.

Second, if an independent external organization (such as a regulatory authority) will be setting the application maturity level requirements, it is recommended that it be done in close consultation with the organization developing the M&S capability. Even though M&S results are increasingly used in simulation-informed decision making, setting application maturity level requirements is a very new endeavor and it must be introduced slowly and carefully. Besides the extensive work at Sandia on this topic, the ASME VVUQ 40 Subcommittee is an excellent example of how to develop the concepts and factors of M&S maturity (ASME, 2018). Developing the ASME VVUQ 40 standard required roughly a decade of extensive discussions and negotiations between representatives of the US Food and Drug Administration and many medical device industry representatives. This was a slow, interactive, and constructive discussion/debate whose context was the development of an engineering standard, *not* the commonly adversarial setting of an industrial applicant submitting a request to a government regulator. More importantly, note that the ASME standard does *not* set maturity requirements on the credibility factors of M&S. It simply introduces what factors contribute to simulation credibility, what the context-of-use of the simulation is in decision making, and what the potential risks of a misinformed decision influenced by simulation are.

PCMM maturity assessment results and application maturity level requirements can be shown in various ways. Table 15.4 depicts an example of a PCMM maturity assessment, along with the application maturity level requirements for each factor. This table shows the assessed scores for the six factors are [1, 2, 1, 0, 2, 1], as indicted with the designator Assessed. The application maturity level requirements are [2, 2, 1, 1, 2, 2], as indicted with the designator Required. The grayed text in the table could be color-coded green to indicate that the assessed level has satisfied the application requirement, and red to indicate that the assessed level does not satisfy the requirement. Table 15.4 can be used as a project management tool to adjust funding and personnel resources to those factors that do not meet application maturity requirements. Similarly, the customer responsible for setting the requirements can also be informed about the estimated time and costs to attain the required level of application maturity.

Table 15.4 *Example of PCMM table assessment and project maturity requirements*

MATURITY / FACTOR	Maturity level 0	Maturity level 1	Maturity level 2	Maturity level 3
Representation and geometric fidelity		Assessed	Required	
Physics and material model fidelity			Assessed Required	
Code verification		Assessed Required		
Solution verification	Assessed	Required		
Model validation			Assessed Required	
Uncertainty quantification and sensitivity analysis		Assessed	Required	

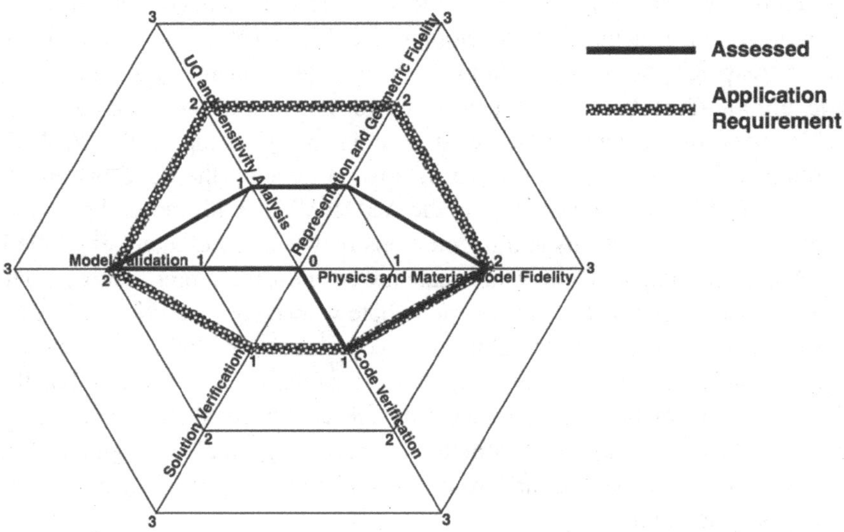

Figure 15.2 Example of a PCMM maturity assessment and application requirement results using a Kiviat diagram.

Both Sandia and NASA have found that a more effective and compact way of showing assessment and application requirement results is a Kiviat diagram, also called a radar plot or a spider web plot (Hills et al., 2013; NASA, 2016). Kiviat diagrams are a useful graphical display of results that depend on multiple quantities, particularly quantities that are discrete and independent of one another. Figure 15.2 shows the same assessment and application requirement results as shown in Table 15.4, but using a Kiviat diagram. Any shortcoming in the assessment relative to the application requirements is more apparent using the Kiviat diagram. Color could be added to the diagram for the factors that do not meet the requirement.

There are two strengths of using the Kiviat diagram for showing PCMM assessments and application requirements. First, each of the six factors assessed is independent of the other, but each one represents a critical aspect of M&S maturity. Second, the diagram shows a markedly different character for outliers, regardless of whether the outlier is small or large. Because of this last feature, the shape of the diagram can change drastically for the same set of assessments, depending on the ordering of the six factors. As a result, it is recommended that a fixed ordering of the six factors is used in all maturity assessments within an organization.

15.5.3 PCMM and Other Factors in Decision Making

As discussed in this chapter, a PCMM assessment is most effective if it is focused on an application of interest or a specific question related to an application of interest. If simulation maturity requirements have been set for the application or question of interest, then the decision maker has the additional advantage of knowing whether the PCMM assessment satisfies these requirements. Even if the PCMM assessment does *not* satisfy the application requirements, the decision maker may still be required to make a decision regarding the application or the question of interest. For example, if one of the assessed PCMM factors is lacking in maturity, but the design of the system of interest must still move forward, the decision maker may take other actions. For example, certain features of the system design may be altered to increase the design margin to ensure that the system achieves its required performance, safety, and reliability.

In the larger context of decision making, it must be stressed that the PCMM is only one factor that contributes to the simulation-informed decision. Figure 15.3 depicts a number of other important factors that commonly affect decision making for the design of an engineering system. The figure divides the factors into two major groups: *Technical Factors* and *Programmatic and Business Factors*. The figure demonstrates that there are diverse and complex factors to consider in decision making. Academic researchers, computational analysts, or design engineers working on a large-scale engineering system commonly do not recognize or appreciate these diverse factors. This text concentrates on many of the *Technical Factors* shown in the figure. PCMM, or a similar simulation maturity assessment method, is extremely valuable because it summarizes the credibility of the simulation information that enters into the decision making process. Sandia National Laboratories, as well as other organizations around the world, use PCMM in simulation-informed decision making.

Some of the *Technical Factors* in Figure 15.3 can be characterized fairly well, but others are based on qualitative judgment and experience. For example, predicted system performance, safety, and reliability can be compared using traditional or imprecise probability theory. The new approach referred to as Quantification of Margins and Uncertainties (QMU) focuses on this task with special emphasis on total predictive uncertainty (Helton and Pilch, 2011; Pilch et al., 2011). However, most of the other *Technical Factors* are not well characterized or even well understood. For example, an organization may have a simulation data management software system in place, but its value is limited if it is not well integrated with (a) a wide range of experimental data from system, subsystem, and component testing, and (b) operational, maintenance, and repair data of existing systems of similar design.

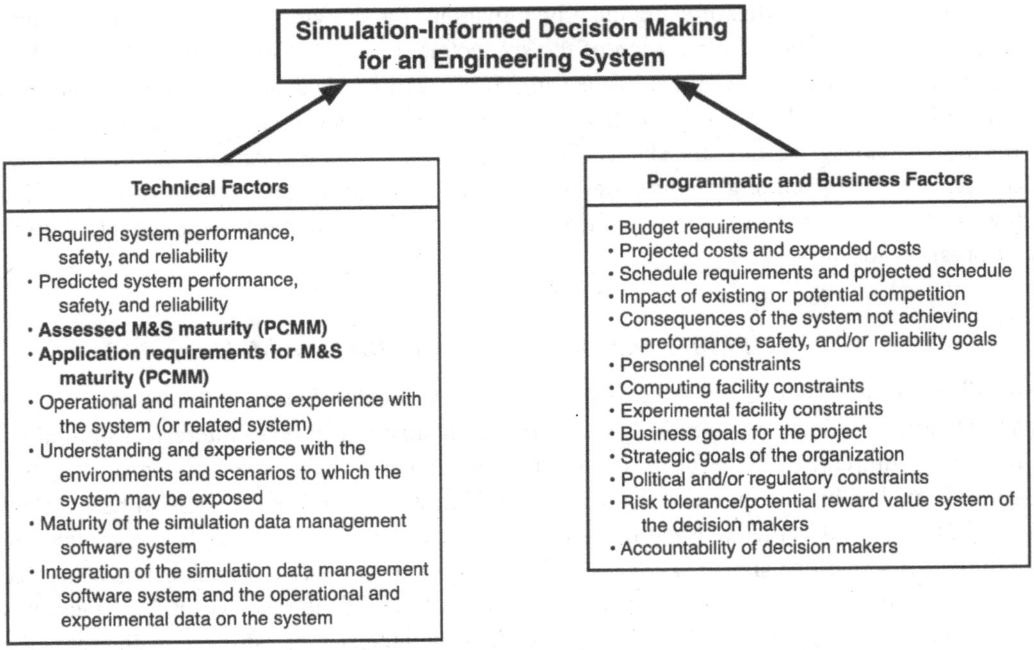

Figure 15.3 Factors influencing simulation-informed decision making for an engineering system.

Although it is beyond the scope of this text to discuss the *Programmatic and Business Factors* shown in Figure 15.3, they must be recognized by contributors to the engineering project. Some of these factors are well characterized, such as budget requirements, costs, and schedule. However, most of the remaining factors are poorly characterized or not characterized at all. For example, it is very difficult to estimate the consequences of poor system safety or reliability on financial liability and future business opportunities. Most of the *Programmatic and Business Factors* shown rely on the experience, judgment, and organizational and personal value system of the individuals who are a part of the decision-making process. Even though there may be one top manager making the final decision on an engineering project, there are many individuals with diverse perspectives, both project-related and personal, that contribute to the final decision.

15.6 Further Reading

For further study on information quality as it impacts decision making, see Wang et al. (2002), Batini and Scannapieco (2006), Ge and Helfert (2006), Sadiq (2013), DAMA-UK (2013), GAO (2020), and Hale and Hoheb (2020).

For further study on methodologies for assessing modeling and simulation maturity, see Roza (2004), Green et al. (2008), Blattnig et al. (2008), Steele (2008), Hemez et al. (2010), Pace (2012), Roza et al. (2012), Hills et al. (2013), Kamm et al. (2013), Kaizer et al. (2015), Beghini and Hough (2016), NASA (2016), Mousseau and Williams (2017), Rider et al. (2017), ASME (2018), Kaizer et al. (2019), Laili et al. (2019), Lin and Dinh (2020), and Walsh (2022).

15.7 Problems

1. (Semester project) In your field of interest, identify an engineering system of interest for a normal operating environment. Also, identify a few closely related issues dealing with SRQs of interest that need to be addressed by M&S activities in an engineering project. Pick a commercial software package or some other software package that you are familiar with to conduct a PCMM maturity assessment. The factors to assess are *Code Verification*, *Solution Verification*, and *Model Validation*. For each of these factors, conduct the assessments based on the documentation provided by the commercial software company or other software provider.
2. (Semester project) In your field of interest, identify an engineering system of interest in an abnormal operating environment and specify a scenario of interest. Also, identify a few closely related safety issues and the associated SRQs of interest that need to be addressed by M&S activities in an engineering project. Pick a commercial software package or some other software package that you are familiar with to conduct a PCMM maturity assessment. The factors to assess are *Code Verification*, *Solution Verification*, and *Model Validation*. For each of these factors, conduct the assessments based on the documentation provided by the commercial software company or other software provider.
3. (Semester project) In your field of interest, identify an engineering system of interest in a hostile operating environment and specify a scenario of interest. Also, identify a few closely related minimum system performance requirements and the associated SRQs of interest that need to be addressed by M&S activities in an engineering project. Pick a commercial software package or some other software package that you are familiar with to conduct a PCMM maturity assessment. The factors to assess are *Code Verification*, *Solution Verification*, and *Model Validation*. For each of these factors, conduct the assessments based on the documentation provided by the commercial software company or other software provider.
4. (Semester project) Pick a commercial software package or some other software package that you are familiar with where the focus is nondeterministic simulation. Conduct a maturity assessment on the PCMM factor *UQ/SA* for the chosen software package. Assume that the software package has the capability to request deterministic samples of any input quantity needed by the computational physics software package to compute a system response. Conduct the PCMM assessment based on the documentation provided by the chosen software package.

15.8 References

Allen, M. J., and W. M. Yen (1979). *Introduction to Measurement Theory*. Long Grove, IL: Waveland Press.

ASME (2018). Assessing Credibility of Computational Modeling Through Verification and Validation: Application to Medical Devices. *American Society of Mechanical Engineers*, ASME V&V 40-2018, New York, NY.

Balci, O. (2004). Quality Assessment, Verification, and Validation of Modeling and Simulation Applications. *2004 Winter Simulation Conference*, 122–129.

Balci, O. (2001). A Methodology for Certification of Modeling and Simulation Applications. *ACM Transactions on Modeling and Computer Simulation*, **11**(4), 352–377.

Balci, O., R. J. Adams, D. S. Myers, and R. E. Nance (2002). A Collaborative Evaluation Environment for Credibility Assessment of Modeling and Simulation Applications. *2002 Winter Simulation Conference*, 214–220.

Batini, C., and M. Scannapieco (2006). *Data Quality: Concepts, Methodologies and Techniques*. New York, NY: Springer.

Beghini, L. L., and P. D. Hough (2016). Sandia Verification and Validation Challenge Problem: A PCMM-Based Approach to Assessing Prediction Credibility. *Journal of Verification, Validation and Uncertainty Quantification*, **1**(1), 011002-1–011002-10.

Bertch, W. J., T. A. Zang, and M. J. Steele (2008). Development of NASA's Models and Simulations Standard. *2008 Spring Simulation Interoperability Workshop*, Paper No. 08S-SIW-037, Providence, RI.

Blattnig, S. R., L. L. Green, J. M. Luckring, J. H. Morrison, R. K. Tripathi, and T. A. Zang (2008). Towards a Credibility Assessment of Models and Simulations. *49th AIAA/ASME/ASCE/AHS/ASC Structures, Structural Dynamics, and Materials Conference*, AIAA 2008-2156, Schaumburg, IL.

Chrissis, M. B., M. Konrad, and S. Shrum (2011). *CMMI for Development: Guidelines for Process Integration and Product Improvement*, 3rd ed. Upper Saddle River, NJ: Addison-Wesley.

Clay, R. L., S. J. Marburger, M. S. Shneider, and T. G. Trucano (2007). Modeling and Simulation Technology Readiness Levels. *Sandia National Laboratories*, SAND2007-0570, Albuquerque, New Mexico.

DAMA-UK (2013). The Six Primary Dimensions for Data Quality Assessment. Data Management Association – United Kingdom Working Group, United Kingdom.

DoD (2005). Technology Readiness Assessment (TRA) Deskbook. Department of Defense, Washington, DC.

Forrester, E., B. Buteau, and S. Shrum (2011). *CMMI for Services: Guidelines for Superior Service*. Upper Saddle River, NJ: Addison Wesley.

GAO (2020). Technology Readiness Assessment Guide. *US Government Accountability Office*, GAO-20-48G, Washington, DC.

Ge, M., and M. Helfert (2006). A Review of Information Quality Research. *11th International Conference on Information Quality*, Cambridge, MA.

Green, L. L., S. R. Blattnig, J. M. Luckring, and R. K. Tripathi (2008). An Uncertainty Structure Matrix for Models and Simulations. *49th AIAA/ASME/ASCE/AHS/ASC Structures, Structural Dynamics, and Materials Conference*, AIAA 2008-2154. Schaumburg, IL.

Hale, J., and A. Hoheb (2020). INCOSE Model-Based Capabilities Matrix and User's Guide. *International Council on Systems Engineering (INCOSE), 2020-001*. San Diego, CA.

Harmon, S. Y., and S. M. Youngblood (2003). A Proposed Model for Simulation Validation Process Maturity. *Simulation Interoperability Workshop*, Paper No. 03S-SIW-127, Orlando, FL.

Harmon, S. Y., and S. M. Youngblood (2005). A Proposed Model for Simulation Validation Process Maturity. *The Journal of Defense Modeling and Simulation*, **2**(4), 179–190.

Helton, J. C., and M. Pilch (2011). Guest Editorial: Quantification of Margins and Uncertainties. *Reliability Engineering and System Safety*, **96**(9), 959–964.

Hemez, F., H. S. Atamturktur, and C. Unal (2010). Defining Predictive Maturity for Validation Numerical Simulations. *Computers and Structures*, **88**, 497–505.

Hills, R. G., W. R. Witkowski, A. Urbina, W. J. Rider, and T. G. Trucano (2013). Development of a Fourth Generation Predictive Capability Maturity Model. *Sandia National Laboratories*, SAND2013-8051, Albuquerque, NM.

Humphrey, W. S. (1989). *Managing the Software Process*. Reading, MA: Addison-Wesley.

IEEE (1989). IEEE Standard Glossary of Modeling and Simulation Terminology. *IEEE Std 610.3-1989*, New York.

Jensen, A. B. (2021). Industrial Implementation of a Simulation Maturity Scale. https://vimeo.com/568824295/4c3d27527f.

Kaizer, J. S., R. Anzalone, E. Brown, M. Panicker, S. Haider, J. Gilmer, et al. (2019). Credibility Assessment Framework for Critical Boiling Transition Models. US Nuclear Regulatory Commission. Washington, DC.

Kaizer, J. S., A. K. Heller, and W. L. Oberkampf (2015). Scientific Computer Simulation Review. *Reliability Engineering and System Safety*, **138**, 210–218.

Kamm, J. R., W. J. Rider, and T. G. Trucano (2013). The Application of the Predictive Capability Maturity Model for GTRF. *Sandia National Laboratories*, SAND2013-0651P.

Laili, Y., L. Zhang, and G. Yang (2019). A Comprehensive Method for Model Credibility Measurement. In L. Zhang, B. P. Zeigler, and Y. Iaili (eds.), *Model Engineering for Simulation*, 189–207. Cambridge, MA: Academic Press.

Lin, L., and N. Dinh (2020). Predictive Capability Maturity Quantification Using Bayesian Network. *Journal of Verification, Validation and Uncertainty Quantification*, **5**(3), 031001-1–031001-19.

Mankins, J. C. (1995). Technology Readiness Levels. National Aeronautics and Space Administration, Advanced Concepts Office, Office of Space Access and Technology, Washington, DC.

McClarren, R. G. (2018). *Uncertainty Quantification and Predictive Computational Science*. New York, NY: Springer.

McMahon, P. E. (2010). *Integrating CMMI and Agile Development*. Upper Saddle River, NJ: Addison Wesley.

Mousseau, V. A., and B. J. Williams (2017). Uncertainty Quantification in a Regulatory Environment. In R. Ghanem, D. Higdon, and H. Owhadi (eds.), *Handbook of Uncertainty Quantification*, 1613–1648. New York, NY: Springer.

NASA (2004). A Renewed Commitment to Excellence. *National Aeronautics and Space Administration*, Diaz Team Report. Washington, DC.

NASA (2006). Interim Technical Standard for Models and Simulations. *National Aeronautics and Space Administration*, NASA-STD-(I)-7009. Washington, DC.

NASA (2008). Standard for Models and Simulations. *National Aeronautics and Space Administration*, NASA-STD-7009. Washington, DC.

NASA (2013). NASA Handbook for Models and Simulations: An Implementation Guide for NASA-STD-7009. *National Aeronautics and Space Administration*, NASA-HDBK-7009. Washington, DC.

NASA (2016). Standard for Models and Simulations. *National Aeronautics and Space Administration*, NASA-STD-7009A. Washington, DC.

Oberkampf, W. L., M. Pilch, and T. G. Trucano (2007). Predictive Capability Maturity Model for Computational Modeling and Simulation. *Sandia National Laboratories*, SAND2007-5948, Albuquerque, NM.

Pace, D. K. (2012). Comprehensive Consideration of Uncertainty in Simulation Use. *Journal of Defense Modeling and Simulation: Applications, Methodology, Technology*, **10**(4), 367–380.

Pilch, M., T. G. Trucano, and J. C. Helton (2006). Ideas Underlying Quantification of Margins and Uncertainties (QMU): A White Paper. *Sandia National Laboratories*, SAND2006-5001, Albuquerque, NM.

Pilch, M., T. G. Trucano, and J. C. Helton (2011). Ideas Underlying the Quantification of Margins and Uncertainties. *Reliability Engineering and System Safety*, **96**(9), 965–975.

Pilch, M., T. G. Trucano, D. E. Peercy, A. L. Hodges, and G. K. Froehlich (2004). Concepts for Stockpile Computing (OUO). *Sandia National Laboratories*, SAND2004-2479 (Restricted Distribution, OUO), Albuquerque, NM.

Rider, W. J., W. R. Witkowski, and V. Mousseau (2017). Uncertainty Quantification's Role in Modeling and Simulation Planning, and Credibility Assessment Through the Predictive Capability Maturity Model. In R. Ghanem, D. Higdon, and H. Owhadi (eds.), *Handbook of Uncertainty Quantification*, 1589–1612. New York, NY: Springer.

Roza, M., J. Voogd, and D. Sebalj (2012). The Generic Methodology for Verification and Validation to Support Acceptance of Models, Simulations, and Data. *Journal of Defense Modeling and Simulation: Applications, Methodology, Technology*, **10**(4), 347–365.

Roza, Z. C. (2004). *Simulation Fidelity, Theory and Practice: A Unified Approach to Defining, Specifying and Measuring the Realism of Simulations*. Delft, The Netherlands: Delft University Press.

Sadin, S. R., F. P. Povinelli, and R. Rosen (1989). The NASA Technology Push Towards Future Space Mission Systems. *Acta Astronautica*, **20**, 73–77.

Sadiq, S. (ed.) (2013). *Handbook of Data Quality: Research and Practice*. Heidelberg, Germany: Springer.

Santner, T. J., B. J. Williams, and W. Notz (2018). *The Design and Analysis of Computer Experiments*, 2nd ed. New York, NY: Springer.

Sargent, R. G. (2013). Verification and Validation of Simulation Models. *Journal of Simulation*, **7**(1), 12–24.

Sargent, R. G., and O. Balci. (2017). History of Verification and Validation of Simulation Models *2017 Winter Simulation Conference*, 292–307.

Smith, J. (2004). An Alternative to Technology Readiness Levels for Non-Developmental Item (NDI) Software. Carnegie Mellon, Software Engineering Institute, CMU/SEI-2004-TR-013, ESC-TR-2004-013. Pittsburgh, PA.

Smith, R. C. (2014). *Uncertainty Quantification: Theory, Implementation, and Applications*. Philadelphia, PA: Society for Industrial and Applied Mathematics.

Steele, M. J. (2008). Dimensions of Credibility in Models and Simulations. *International Simulation Multi-Conference*, Paper No. 08E-SIW-076, Edinburgh, Scotland.

Todinov, M. (2018). *Methods for Reliability Improvement and Risk Reduction*. New York, NY: John Wiley.

Walsh, J. (2022). Understanding an Engineering Simulation Risk Model. *Analysis, Simulation & Systems Engineering Software Strategies Initiative, ASSESS Credibility Theme Strategic Insight 2022, version 3_0_0*.

Wang, R. Y., and D. M. Strong (1996). Beyond Accuracy: What Data Quality Means to Data Consumers. *Journal of Management Information Systems*, **12**(4), 5–34.

Wang, R. Y., M. Ziad, and Y. W. Lee (2002). *Data Quality*. New York, NY: Kluwer Academic Publishers.

16

Verification, Validation, and Uncertainty Quantification Responsibilities and Management

When there's no immediate threat to our understanding of the world, we change our beliefs. It's when that change contradicts something we've long held as important that problems occur. —Maria Konnikova

Contents of Chapter 16

This chapter discusses varying perspectives of responsibilities for the development, conduct, management, and delivery of verification, validation, and uncertainty quantification (VVUQ) work products. The topics considered here deal with technical, management, and cultural issues, but all are focused on the strategic goals of improved credibility and efficient use of modeling and simulation (M&S). Our experience, as well as the experience of others, has convinced us that while technical issues and computing resources are important, they are not the limiting factors in improving the credibility and value of scientific computing used in a decision-making environment. Business, organizational, and cultural traditions have constrained the value-added contributions of M&S and the improvement in credibility that VVUQ can provide. From a historical perspective, this is understandable because M&S are part of the *Information Revolution* that businesses and most societies around the world are struggling with. We are living in the early stages of this exciting, and sometimes frightening, revolution in all facets of society. Although it is too early to be certain, the indication is that the Information Revolution may have an impact on societies and world development that is greater than the Industrial Revolution beginning in the eighteenth century (Coggan, 2020).

This chapter discusses (a) staff and management responsibilities for conducting and assessing VVUQ activities, (b) mid-level management responsibilities for developing and requiring VVUQ practices, (c) executive-level management responsibilities for a business organization, and (d) development of verification and validation (V&V) databases. While these topics are far afield from the technical discussions in this book, they are possibly more important for M&S to fulfill its promised potential in business and society. As scientific computing continues to have an increasing impact on everyday life, economic competitiveness, public safety, global warming policies, and national security issues, the value-added by VVUQ practices must be better understood.

16.1 Staff and Management Responsibilities for VVUQ Activities

This text has explained at length the diverse technical tasks associated with VVUQ activities. For discussion in this section, these activities are defined as: (a) software quality assurance (SQA) and code verification, (b) solution verification, (c) model validation, and (d) uncertainty quantification. As discussed in Chapter 15 "Maturity Assessment of Modeling and Simulation," these activities are the primary technical contributors to the credibility and reliability assessment of simulation results. In this section, responsibilities for conducting these activities are discussed in terms of both "*Who conducts these activities?* "and "*Who has the role of assessing the adequacy of these activities*?" As pointed out by Walsh (2018), VVUQ activities build *confidence* from the perspective of who conducts these activities. VVUQ activities also build *credibility* from the perspective of the assessor or user of the results. The discussion of completeness and adequacy of these activities in this section is at a high level compared to the detailed discussion given in Chapter 15.

Conducting these activities is usually the responsibility of three entities: (1) the developer of the M&S software, (2) the computational analyst who produces the simulation, and (3) the experimentalist who provides the experimental/operational data used for model validation. As explained in Chapter 15, when viewing a simulation result as an information product, the developer, analyst, and experimentalist entities can be considered as *information producers* or suppliers. The recipient of the simulation information is an entity that is considered the *information customer* of the results. The information customer has the primary responsibilities of (a) properly interpreting the information, (b) assessing the completeness of the information for the task at hand, and (c) determining the adequacy of M&S and VVUQ activities for the intended use of the simulation. The information customer can be either internal to an organization, such as a project manager for the development of a new system; or external, such as a regulatory authority. Thus, the information customer has the primary responsibility of determining how and to what extent the simulation information is utilized and how it is used in a specific decision-making context. Common examples of the context of simulation-informed decision making are design or modification of a system to meet customer requirements; assessment or certification of system performance, reliability, or safety, sometimes referred to as Certification by Analysis (CbA); and governmental policy making.

For informal, small-scale, or research simulations, the roles of conducting and assessing the completeness of the VVUQ activities are commonly conducted by the same person. However, this section focuses on these two responsibilities for large-scale simulations where there are four responsible entities: software developer, computational analyst, experimentalist, and information customer for the results of the simulation. In this section, these four responsible entities are simply referred to as *developer*, *analyst*, *experimentalist*, and *customer*. The software developer may reside in the same organization as the computational analyst, but that is somewhat rare today. Presently, it is much more common for the computational analyst to use simulation tools that are developed and licensed from separate commercial software companies. The customer discussed here is primarily the direct user of the simulation results.

16.1.1 Software Quality Assurance and Code Verification

16.1.1.1 Who Conducts SQA and Code Verification?

SQA and code verification are primarily the responsibilities of the software developer. Software developers for major scientific computing capabilities commonly reside within commercial software companies or within large organizations that are developing specialized software. Most commercial software companies are subjected to demanding quality and reliability requirements from their customers. However, precisely what steps are taken and the level of effort that these companies devote to SQA and code verification is considered proprietary and not made available to customers.

From the perspective of software testing, SQA is typically divided into static testing and dynamic testing. For a more detailed discussion of SQA practices, see Chapter 4 "Software Engineering for Scientific Computing." Code verification is a special type of dynamic testing that is especially important in scientific computing. With every software release, software companies have SQA and code verification testing procedures that are set up under the supervision of a software quality group. These procedures include automatic reporting features for the results of the testing. In particular, the activities related to fixing, tracking, and managing software bugs are a critical part of the global SQA procedures of the software company.

Some large commercial software companies specializing in solid mechanics have developed an extensive set of code verification benchmarks that can be exercised by licensed users of their codes (ANSYS, 2020; SIMULIA, 2020). These documented benchmarks are intended to demonstrate the performance of the code on relatively simple benchmark problems, as well as serve as a training aid for users of the code. However, the primary weakness of essentially all of the documented code verification testing of commercial software is that it only demonstrates the "engineering accuracy" of the code. This type of verification testing is useful, but it does not quantify the observed order of convergence of the discretized numerical solution relative to the exact solution of the mathematical model. As discussed by many authors and at length in this book, the observed order of convergence of spatial discretization is the most demanding accuracy assessment of numerical performance of the code. Chapter 5 "Code Order-of-Accuracy Verification," describes this assessment as normally shown using a log-log plot of evaluated numerical error versus mesh discretization size. Note that the evaluated numerical error is the *true error* of the computational solution because for code verification testing, the exact solution must be known. Computational fluid dynamics (CFD) commercial software generally has not documented code verification benchmarks as well as commercial solid mechanics codes.

Large software development projects within an organization, separate from software companies, usually are one of two types: (1) corporations that develop either proprietary software or special purpose software or (2) research groups within universities, governmental laboratories, and institutes. There is considerable resistance to the implementation of extensive and formal SQA practices and code verification testing within many research groups. The degree of formality, software development constraints, ability of users to modify the code, and costs of SQA are key factors in their resistance. Furthermore, research groups operate in a well-established value system where new knowledge, particularly publishable results, is the valued product, *not* the reliability of the software, numerical algorithms, and simulation reproducibility. Attention is finally being drawn to the lack of reliability and reproducibility of published computational research results (Donoho et al., 2009; LeVeque et al., 2012; Stodden et al., 2016). Much more needs to be done to change the research culture so that the reproducibility and reliability of simulation results better match that of experimental results.

Secondary responsibility for SQA and code verification lies with the analyst conducting the simulation. When the software package is installed and operating properly, then the analyst is responsible for the SQA procedures and version control on the computer system being used. If there are a number of analysts across various groups using the software package, this responsibility may be designated to a single representative. Analysts expect, either explicitly or implicitly, that the software developer has achieved some level of code verification testing for a range of options and applications of the software. However, as discussed previously, observed order of convergence is rarely tested and documented by commercial software companies. The analyst should firmly request that this be done by the software developer. If the analysts' organization is paying for a number of commercial software licenses, then the analyst may have sufficient influence to expect that observed order of convergence testing will be completed and documented. If the analyst is not successful in motivating the software company to conduct the testing, then the analyst should consider conducting the observed order of convergence testing. However, convincing management of the analyst's organization to fund this type of software testing is typically a very difficult proposition because they presume that the license fee assures adequate software quality.

16.1.1.2 Who Assesses SQA and Code verification Adequacy?

Computational analysts are primarily responsible for assessing the adequacy of the SQA and code verification testing of the software package as applied to their computational applications of interest. The level of adequacy required should depend on (a) the magnitude of the impact of simulation results on decision making and (b) the consequence of the system of interest if it does *not* achieve its requirements for simulation-informed performance, safety, and reliability. For example, if the impact of the simulation results is the dominant factor in the design of a system, as opposed to a design based heavily on experimental measurements or operating experience, then there should be a higher requirement for software testing adequacy. If a formal assessment of adequacy is needed, then the reader should study Chapter 15 "Maturity Assessment of Modeling and Simulation," for a detailed discussion of this topic. For the vast majority of situations dealing with commercial software, the analyst is at the mercy of the software company for conducting the SQA and code verification testing. As discussed earlier, the amount of leverage that the analyst has in requesting additional software testing depends entirely on the number of software licenses that are owned by the analyst's organization.

The secondary responsibility for SQA and code verification testing lies with the simulation customer. The customer is the user of the information produced by the simulation and conducted by the analyst. As discussed in earlier chapters of this text, simulation results are considered an information product that is acted on by the customer. On small-scale simulations, the analyst and the customer are the same person. The advantage of this situation is that analysts are generally familiar with the concepts and value of SQA and code verification testing. For large-scale simulations, the difficulty is that the customer is commonly unfamiliar with how scientific computing is conducted, including the elements that are needed for credibility and reliability of the results. Some people have argued that as the use of simulation results in decision making becomes more prevalent, customers will become more knowledgeable of these credibility-building issues. The authors of this text do *not* believe this will be the case. In any hierarchical decision-making structure, whether in business or government, there are, and will always be, individuals without any concept of how simulation works or how credibility in simulation results is assessed. A more realistic and reliable approach for assessing simulation credibility and its adequacy is discussed in detail in Chapter 15.

16.1.2 Solution Verification

16.1.2.1 Who Conducts Solution Verification?

The primary responsible party for conducting solution verification is the computational analyst computing the simulation. As discussed in detail in Chapter 7 "Numerical Uncertainty Estimation," there are four primary contributors to numerical solution error: (1) discretization error, most often in space and time; (2) iterative solution error; (3) statistical sampling error; and (4) the numerical approximation error, due to using response surfaces to approximate the solution to the computational model. Solution verification, sometimes referred to as calculation verification, involves quantitatively estimating each of these errors. The most common procedure for estimating these errors is to compute multiple numerical solutions and then use these to estimate the numerical error. This approach is referred to as *a posteriori* error estimation.

Because of time and budget constraints on most analysts, they commonly rely on their experience to judge whether each of these error contributors is sufficiently small enough that they can be ignored. As discussed in this text, and by many other researchers, an analyst's experience has been found to be unreliable in many practical applications. An analyst's judgment is always influenced by the time and computing constraints under which they must operate. As a minimum responsibility, analysts should compute at least two numerical solutions that impact the relevant numerical error source to qualitatively determine the sensitivity of system response quantities (SRQ) of interest. If the SRQ is noticeably different for the two solutions, then the analyst knows that the error source has a significant impact on the computed results. As a result, additional investigation of the impact of the error source is warranted, such as by refinement of the spatial mesh or increasing the number of iterations to improve convergence.

The quantitative error estimation procedures that have been developed are a significant improvement over the qualitative procedure just described. These quantitative error estimation procedures can be relatively simple to use, such as Richardson extrapolation for estimating discretization error, described in Chapter 9. However, all of the error estimation procedures require some type of post-processing of the multiple numerical solutions obtained. If software tools incorporating the post-processing steps are not available to the analyst, then the analyst must build these tools. Experience has shown that engineering application analysts rarely have the time, interest, or background to develop such tools. A much better and more efficient approach to quantitatively estimating the various sources of numerical error is for the software developer to provide these software tools as part of the computational simulation software package. The availability of these tools within the software package should be a consideration in deciding what software package should be used by an organization.

The secondary responsibility for conducting solution verification lies with the software developer. This responsibility does not mean that the software developer conducts the solution verification, but the developer provides the needed software tools for solution verification. Some commercial software developers in solid mechanics have provided limited methods in their software packages to quantitatively estimate spatial discretization error. For example, the superconvergent patch recovery method (Zienkiewicz and Zhu, 1992) is very popular in solid mechanics software packages. It can provide an error estimate in the global energy norm, but the more relevant error estimates are for those SRQs representing *local quantities*, such as local stress. However, the authors are not aware of any software package that quantitatively estimates the discretization error on SRQs of interest in commercial CFD codes. As discussed earlier concerning SQA and code verification, the analyst should request that the commercial software developer include solution verification tools in their software packages.

16.1.2.2 Who Assesses Solution Verification Adequacy?

The analyst conducting the simulation has the primary responsibility for assessing solution verification adequacy. The analyst is the most familiar with the details of the computational solution and how to reduce the impact of the four numerical errors previously mentioned.

Even though the four numerical errors arise from different sources, there is still an interaction of these errors in a computational simulation. For example, if the analyst decides to refine the mesh to reduce the spatial discretization error, the time required to compute a solution on the finer mesh will increase. If the simulation is also nondeterministic, then multiple numerical solutions must be computed to characterize the stochastic nature of the simulation results. To address the increased computer time needed for the refined mesh, the analyst may need to reduce the number of solution samples needed for the stochastic simulation. As a result, the statistical sampling error will increase. Since the analyst most directly understands the trade-offs between the four sources of numerical error, the analyst is the most knowledgeable when deciding on an acceptable balance of errors that best meets the needs of the simulation's customer.

The customer of the simulation holds the secondary responsibility of assessing solution verification adequacy. Some people have argued that the complete responsibility for solution verification should rest with the analyst conducting the work. They argue that "The analysts are the experts in understanding the four error sources and the specific features of the solution, so they should be fully responsible." Or, "Customers of simulations are not trained in these matters, why should we have *any* responsibility for this technical matter?" As mentioned above, simulation results are an information product that is acted on by the customer. Even though customers, such as lawyers or policy makers, may have no background in computational simulation, they are responsible for inquiring about all of the issues that could impact the credibility of the information they receive.

16.1.3 Model Validation

16.1.3.1 Who Conducts Model Validation?

Model validation is primarily the responsibility of the computational analysts conducting the simulation. Model validation is conducted for the purpose of critically assessing the accuracy of the mathematical model for conditions where experimental data are available. Sometimes this is referred to as the *restricted view of model validation*. Whether the mathematical model compares well or poorly with the experimental measurements is *not* the primary issue in model validation; *accuracy assessment* is the primary goal (AIAA, 1998; Oberkampf and Trucano, 2008; ASME, 2022). As discussed in Chapter 10 "Model Validation Fundamentals," a strong distinction is made between model validation (i.e., model accuracy assessment) and model calibration (model improvement by way of optimizing model input parameters). Both activities require experimental data, but their goals are distinctly different. The most informative approach to model validation is having the analyst compute a blind prediction for the conditions of the validation experiment. All of the important model input data for the experiment should be provided by the experimentalist to the analyst for a blind prediction of the response of the system. If some model input quantities are not measured or estimated by the experimentalist, then the analyst should characterize these quantities as epistemic uncertainties. These epistemic uncertainties, as well as any aleatory uncertainties, are propagated through the mathematical model. The p-box results from the model are then compared with the experimentally measured SRQs. This procedure is discussed in detail in Chapter 12 "Model Accuracy Assessment."

Model validation experiments can be conducted at any tier of the model validation hierarchy, i.e., at any level of physical complexity of interest. For example, a model validation experiment can be conducted for a subsystem, a subassembly, or a component of the complete system. If important model input data cannot be measured by the experimentalist, then these input parameters must first be calibrated using appropriate experimental data and the inverse solution to the mathematical model. This situation most commonly occurs in systems where the physical processes of interest are poorly understood and/or mathematically modeled, e.g., stiffness and damping in joined parts in structural dynamics. As discussed in Chapter 13 "Predictive Capability," there are various strategies detailing how to segregate experimental data for use in input parameter calibration using Bayesian estimation, versus using data for model validation. Ultimately, the analyst is responsible for deciding on the appropriate model validation metric to measure the difference between simulation results and experimental measurements of an SRQ. Chapter 12 "Model Accuracy Assessment" discusses various approaches to validation metrics, specifically the area validation metric and Bayesian model validation.

Secondary responsibility for model validation lies with the experimentalist conducting the validation experiments. Model validation can be conducted without direct involvement from the experimentalist who obtained the experimental data for validation, but this approach commonly leads to calibration of model parameters as opposed to model validation. As discussed in Chapter 11 "Design and Execution of Model Validation Experiments," the primary reason for this is that important model input data needed to simulate the experiment are not measured or reported by the experimentalist conducting the experiment. For example, if model validation is conducted using published experimental data, it is almost certain that critical model input data are not measured or documented in the publication. The most valuable and informative model validation experiments are conducted when the experimentalist and the analyst jointly plan, design, and conduct the experiments. The experiments should be designed to (a) measure important model input data, (b) calibrate parameters that cannot be directly measured using specially designed experiments, and (c) decide what experimental conditions will challenge the mathematical model and have the computational analysts provide blind predictions for these experiments.

16.1.3.2 Who Assesses Model Validation Adequacy?

The computational analyst is primarily responsible for assessing the adequacy of model validation activities. This section discusses one aspect of the adequacy of model validation activities mentioned in the previous section. The model validation domain is the domain of model input parameter space over which experimental measurements are obtained. In a two-dimensional model input parameter space, the validation domain is the bounding area enclosing all conditions where experiments were conducted. For an input space of an arbitrary number of dimensions, which for real systems is a very high dimensional space, the validation domain is the convex hull of the entire space. A validation metric can be computed at each condition where experiments are conducted within this domain. It is very common that this domain does *not* include the application domain of the model input conditions where the system of interest operates. As a result, the model accuracy required at the available experimental conditions should be *more* demanding than at conditions where there are no experimental data for the system of interest. The reason for this increased accuracy requirement at the validation conditions is that the mathematical model must be used to interpolate or extrapolate to make predictions at the conditions where no experimental data are available. Since any mathematical model approximates physical reality, intuition and good judgment indicate that the model form uncertainty should be expected to increase as the extent of interpolation or extrapolation increases. The degree of stringency the model accuracy requirement should be at the validation conditions, relative to the accuracy required of the model at the system application conditions of interest, is an active area of research.

The customer who intends to use the simulation in some decision-making context has the secondary responsibility for assessing the adequacy of model validation activities. Customers of computational analyses are fully accustomed to this responsibility regarding model validation activities. This tradition exists because customers of computational analyses have consistently expected to see comparisons of model predictions and experimental measurements that are closely related to conditions experienced by their system of interest. The more closely related the validation conditions tested are to their system, and assuming there is adequate agreement of the simulations and the experiments, the more credibility they place in the simulation.

Customers, particularly customers who are skeptical of using M&S in decision making, as well as those who act in a regulatory capacity, are particularly focused on two perspectives of model validation. Their first concern is the concept of using the model to extrapolate to conditions outside of the validation domain, where no experimental data are available. Using the model to interpolate within the validation domain is less of a concern. The higher the consequence of system failure is for these customers, the more demanding the relevancy of the experimental conditions is to the actual operating or accident conditions of the system of interest. Second, these customers especially value seeing the comparison of a blind prediction of a model with measured results from a validation experiment. They want to see how well the model performs when there is *no opportunity* for parameter calibration or adjustment of the modeling assumptions to obtain good agreement with the measurements. Furthermore, these customers particularly value a blind prediction when the experiment is conducted at very demanding operating or accident conditions of the system of interest. Recent examples of sobering results from blind predictions are the World-Wide Failure Exercise for composite structures (Hinton et al., 2002; Kaddour and Hinton, 2013), the Dalmarnock Fire Test (Rein et al., 2009), the Sandia Fracture Challenge (Boyce et al., 2014; Boyce et al., 2016), and the Third International Benchmark Exercise for stratified flow in a pressure vessel (Andreani et al., 2016).

16.1.4 Uncertainty Quantification

16.1.4.1 Who Conducts Uncertainty Quantification?

The computational analyst has the primary responsibility for conducting uncertainty quantification, i.e., nondeterministic simulations. The nondeterministic predictions of interest are those that are focused on predictions of SRQs at the application conditions for the system of interest. The prediction of total uncertainty of the SRQs at the application conditions is the culmination of all of the efforts devoted to M&S and VVUQ. The uncertainty quantification

(UQ) activity is commonly the most demanding on time and resources because it brings together three diverse activities: (1) the modeling effort devoted to identifying and characterizing the environments and scenarios of the system of interest; (2) the V&V efforts directed toward understanding and quantifying all of the numerical and modeling sources of uncertainty that could impact the reliability of the simulation results; and (3) performing UQ for the system of interest operating at diverse conditions in normal, abnormal, or hostile environments. The level of effort devoted to UQ can vary widely, depending on the needs of the analysis relative to the customers' requirements. As a result, the burden of responsibility on the analyst can vary considerably. In order to stress the range of effort that may be required, there are two extremes of UQ activities described here: (1) very demanding and (2) minimally demanding.

For the first extreme, consider the emergency cooling of the core of a nuclear power reactor in an accident environment. This type of analysis involves a large team of people, commonly spread over several organizations, because it deals with a high-consequence system. For an accident environment, many possible scenarios are identified and several of the highest-risk scenarios are analyzed in detail. The UQ analysis includes many different types of mathematical models, experimental data from a wide variety of relevant experiments, and it may involve large model-based extrapolation from the existing experimental database to reactor conditions that have never been previously tested. As a result, the expectations are high for the analysts in order to support simulation-informed decision making related to the system's safety. This type of UQ analysis is exceptionally demanding in terms of time, resources, and the depth of technical expertise needed on the team, but it is appropriate for the public safety of high-consequence systems.

For the second extreme, consider the solid mechanics analysis of a non-safety-critical component of a gas turbine engine. Here, the UQ analysis is directed toward an incremental design change of a component, but the change is expected to have minimal impact on the performance, reliability, or safety of the engine. Furthermore, only a normal operating environment is considered for a well-understood set of engine-operating scenarios. As a result, one individual with experience in nondeterministic simulation conducts the UQ analysis. The analyst decides that because of schedule and resource constraints, only a meager UQ analysis is needed or is even possible. This situation is very common in industrial settings where competition between manufacturers is intense, time scales are very short, and a great deal of operating experience is available for the system of interest. Risks due to approximations and shortcuts in the UQ analysis are greatly mitigated because relatively little simulation information is required for this decision context. For this type of situation, the primary basis for decision making is the extensive operating experience and well-understood operating conditions of the gas turbine, as opposed to heavy reliance on simulation for conditions that have never been observed.

The software developer has the secondary responsibility for conducting the UQ analysis. Analogous to the earlier discussion of solution verification, this does not mean that the software developer conducts the UQ analysis, but that the software developer should provide the software tools needed by the analyst to conduct the UQ analysis. Developing a general purpose UQ software package is a massive endeavor. As a result, the analyst very rarely develops the software package. There has been a rapid growth in the availability of general purpose UQ software packages over the last two decades; some are open access, and some are proprietary. The most commonly available and widely applicable software packages are those that function as external drivers or software wrappers to traditional deterministic simulation software packages. Some examples of popular software packages available are: COSSAN (2020), OpenTURNS (2020), RAMAS (2023), Sandia (2023), and SMART-UQ (2020).

16.1.4.2 Who Assesses Uncertainty Quantification Adequacy?

The customer of the simulation results has the primary responsibility for assessing the adequacy of UQ. Some people argue that the analyst should have the primary responsibility for assessing the adequacy of the UQ analysis. In our view, this argument is weak because the user of the simulation results, i.e., the customer, makes the ultimate decision as to *how* and to *what extent* to use the results. For example, the customer, possibly in consultation with stakeholders, may decide on one of three options on how to use the UQ analysis. First, the customer decides that the UQ analysis fully satisfies all of the requirements needed to make a range of decisions based primarily on the total estimated uncertainty in the simulation results. For example, the customer, stakeholders, and computational analyst are satisfied that all conceivable scenarios in a given system environment have been adequately addressed by the UQ analysis. Second, the customer decides that the UQ analysis partially satisfies the requirements requested from the simulation. For example, the customer may decide to request that additional system scenarios be investigated in the UQ analysis or decide to reduce the expected performance or reliability of the delivered system. Third, the customer decides that the information does not satisfy what is needed for a simulation-informed decision and, as a result, chooses an alternative option. For example, the customer may decide to require additional major testing for the system of interest or decide that certain operating modes of the system will not be allowed because of critical safety or liability concerns. The customer, whether a project manager in a business organization, a regulatory authority, or a governmental policy maker, must weigh the positive and negative aspects of a wide variety of possible decision outcomes. The customer, serving as the simulation-informed decision maker, is ultimately responsible for how the simulations are used.

The computational analyst has the secondary responsibility for assessing the adequacy of the UQ analysis. As discussed briefly earlier, and at length in Chapter 13 "Predictive Capability," UQ analysts must not only be experienced with many technical aspects of nondeterministic simulation and V&V, but also must understand the simulation and UQ needs of the customer. If the customer is not familiar with UQ activities, many times the customer does not clearly communicate to the analyst how the UQ analysis might be used in decision making. Sometimes, the customer is not even clear how the UQ analysis will be used or how to interpret the UQ results. Furthermore, the customer may use the simulation results in a manner that was not initially communicated to the analyst, or even intended.

To best inform the customer of the information contained in the UQ analysis, and even diminish the likelihood that the simulation results might be used inappropriately, the analyst should *explicitly state the limitations and assumptions* made in the simulation and UQ analysis. Unfortunately, this is not a common occurrence in the presentation of simulation results, even for large scale simulation analyses. For example, UQ analyses consciously exclude some scenarios, and do not even imagine other scenarios, in the analysis. This is especially true in abnormal and hostile environments. Sometimes these scenarios are neglected because it is felt (by the UQ analyst) that they are so unlikely, or the adverse

Table 16.1 *Primary and secondary responsibilities for VVUQ activities*

Activity \ Entity	Developer	Analyst	Experimentalist	Customer
SQA and code verification	Conducts (Primary)	Conducts (Secondary) Assesses (Primary)	-	Assesses (Secondary)
Solution verification	Conducts (Secondary)	Conducts (Primary) Assesses (Primary)	-	Assesses (Secondary)
Model validation	-	Conducts (Primary) Assesses (Primary)	Conducts (Secondary)	Assesses (Secondary)
Uncertainty quantification	Conducts (Secondary)	Conducts (Primary) Assesses (Secondary)	-	Assesses (Primary)

consequences so insignificant, that they should be neglected. The scenarios, or sequences of scenarios, that are excluded in the UQ analysis should be pointed out to the customer. As discussed in Chapter 15 "Maturity Assessment of Modeling and Simulation," if this is *not* done, it is referred to as lack of contextual information quality, specifically, an unrecognized lack of completeness. Identification of possible scenarios in abnormal and hostile environments requires the most creativity and freedom in UQ analyses.

16.1.5 Summary of Responsibilities for VVUQ Activities

A summary of the responsibilities for conducting and assessing the four VVUQ activities is shown in Table 16.1. The four responsible entities are shown across the top of the table and the table entries show primary and secondary responsibilities for each entity. As discussed above, the distinction between who conducts the activity and who assesses the activity is distinctly separate in a large-scale simulation project. For small or informal simulation projects, one entity may be accountable for both primary and secondary responsibilities. For additional perspectives of the responsibilities for M&S and VVUQ activities, see Oberkampf and Pilch (2017) and Smith (2020b, 2020a).

Two final points should be noted regarding the responsibilities shown in Table 16.1. First, for large organizations that have comprehensively used M&S in their decision making over a period of time, these are the widely accepted responsibilities. For organizations that are just beginning to shift from test-based decision making to simulation-informed decision making, responsibilities between who conducts and who assesses activities have probably not been formally recognized or assigned. For example, consider an organization that does not have the resident technical staff to conduct the M&S and VVUQ activities and, as a result, they must contract with a consulting company for these services. Project management, or possibly line management within the organization who contracts with the consulting company, may not be familiar with VVUQ activities or responsibilities. This could put the organization at substantial risk in attempting to move toward simulation-informed decision making.

Second, if the customer is a regulatory authority, then the regulatory authority will commonly make itself the primary assessor on all four activities listed in the table. The effectiveness of discharging this responsibility depends heavily on the technical expertise of the regulatory staff regarding VVUQ activities. Since most regulatory authorities are traditionally focused on test-based certification of systems, the effectiveness of the assessor responsibilities shown in Table 16.1 is dubious at present. As regulatory authorities incorporate more M&S results in their evaluation and approval process, they recognize that they must improve the guidance and training of their regulatory staff in VVUQ activities (FDA, 2016).

16.2 Management of VVUQ Activities

This section discusses various actions that all levels of management can take to help the implementation, development, and effectiveness of good VVUQ practices in an organization. Most of the discussion is directed toward private businesses and governmental organizations, while some discussion is also appropriate for research institutes and universities. The topics discussed include the value added from VVUQ activities and what appropriate levels of VVUQ are for more reliable simulation-informed decision making. The section closes with recommendations for training technical staff in VVUQ and includes two approaches describing how organizations can be structured for implementing VVUQ activities.

The term *management* primarily refers to line management of an organization. Line managers are those appointed to the relatively permanent structure needed for the coordination and functioning of an organization. In certain instances, the discussion of actions and recommendations will also apply to project management. Project managers are those whose responsibilities are directly tied to the execution and completion of a project, typically an engineered system sold by the organization. Project managers in for-profit businesses are commonly under severe constraints related to the system of interest meeting specified requirements, schedule, and the cost of the project. We recognize and respect their goal-oriented perspectives and project responsibilities.

16.2.1 Value Added of VVUQ

From a business perspective, results from M&S are considered a *service*, because they involve the production of intangible goods, specifically, the generation of knowledge. As has been stressed throughout this book, VVUQ activities add credibility and reliability to the information provided by M&S. Businesses will assess the benefit of VVUQ activities in the same way any new technology or business process is introduced, asking: *"What is the value added compared to the resources and time expended for the activity?"* Some business sectors have devised innovative methods to measure the quality of the services they provide, including feedback surveys from customers concerning responsiveness, attention to personal needs, accessibility, and privacy. Since M&S and VVUQ activities are technical, their value can be better measured by their influence on both tangible and intangible assets, as opposed to personal preferences or the satisfaction of the receiver of the service. Tangible assets, such as reduced time to market, reduction in physical testing of new systems, and more comprehensive certification of system reliability, are easier to measure in terms of monetary value. However, intangible assets are very difficult to measure. These include intellectual property, business reputation, increased market share at present or in the future, ability of M&S to reduce liability risks when the system does not meet safety and reliability targets, and identification and elimination of potential system failure modes early in the design process.

For many M&S activities in a business setting, such as day-to-day simulations used in system design or optimization, formal VVUQ activities are unnecessary and inefficient. The time scales required for making the multitude of design decisions on a system are generally very short. For example, it is common for the design engineer to tell the computational analyst, "If you get the analysis results to me by tomorrow, I will use the results to make a design decision. If you don't, I will make the decision without your help." VVUQ activities take time to complete, commonly resulting in a degradation of the prompt response times required in a fast-paced, detailed design environment.

To deal with this business reality, there is a very long tradition in engineering design to use design rules or general guidelines. As stated by Szabo et al. (2017), design rules are of the form $SRQ_{max} < < SRQ_{allowed}$, where SRQ_{max} is the computed maximum of an SRQ (such as stress, fatigue life, or temperature), and $SRQ_{allowed}$ is the allowable value of an SRQ established by authorities charged with responsibility to formulate design rules. The formulation of design rules is a much more in-depth M&S process, which *would include* formal VVUQ activities. Furthermore, the design rules would incorporate adequate design margins to ensure reliability and safety for a wide range of general operating conditions of the system. These design margins should allow for various sources of uncertainty, including numerical solution errors in computing SRQ_{max}, model form uncertainties, variability in material properties, and diverse operating conditions. Even though use of design rules is very efficient and cost-effective in product design, technical experts who develop the design rules should continually monitor the adequacy of the design rules, particularly for unusual conditions of the system such as abnormal and hostile environments.

The level of formality and documentation of VVUQ activities should be commensurate with two practical issues: (1) the importance of M&S in decision making relative to other factors and (2) the near and long term adverse consequence of the system of interest *not* achieving its performance, reliability, or safety requirements. For example, in preliminary design or scoping studies, formal VVUQ activities should be minimal. In a situation where M&S is heavily relied upon for system performance, reliability, or safety, then the formality of VVUQ activities should be significant, explicit, and documented in the computational analyst's work product. Furthermore, when assessing the adequacy of VVUQ activities in producing the simulation results, the customer should assess if the activities were appropriate for the intended decision context. Thus, one size of VVUQ activities does *not* fit all applications of simulation or decision contexts.

The business question concerning VVUQ activities could be stated as: "What is the appropriate or optimum level of VVUQ activities for a given situation of simulation-informed decision making?" Recent work has attempted to quantify the net value of VVUQ activities in the context of business decision support (Paez et al., 2016). The net value of VVUQ is defined as the total benefit of a decision minus the total costs of the VVUQ activities. Paez et al. properly point out that VVUQ activities have no inherent value from a business perspective, i.e., *VVUQ activities have no tangible value*. VVUQ activities result in tangible value when they have a positive impact on a business decision, such as: reducing the time to market of a product; supporting the certification of the performance, reliability, or safety of a system, sometimes referred to as CbA; and reducing manufacturing costs, resulting in an increased profit margin on a delivered product. However, when inadequate VVUQ activities occur, i.e., M&S provides incorrect or misleading results for decision making, the *negative* impact of poor decision making can be extremely difficult to quantify. Hence, the total *negative* result, i.e., the total business loss, can be much greater than the cost of the VVUQ activities because the short-term or long-term negative impact may not be foreseen or even be imagined.

As discussed in Chapter 15 "Maturity Assessment of Modeling and Simulation" VVUQ activities are the primary technical contributors to the quality of the information produced by M&S. Their impact on information quality, particularly completeness and reliability, can be considered from two perspectives. First, M&S provides information that directly assists management in making good system design decisions. Some examples are system design optimization, reducing time and money spent on system testing to achieve system requirements, and identifying unusual failure modes that impact system performance and safety. Second, M&S provides information that aids in *avoiding* poor decisions. Estimating the monetary impact of avoiding poor decisions, poor outcomes, or poor designs is more difficult if penalties in deficient performance, safety, and reliability are not included in the product requirements. Recent studies have found that one of the major, if not the primary, advantage of model-based systems engineering (MBSE) is the early identification and resolution of system design defects (Carroll and Malins, 2016, 2020). Boehm et al. (2008) and Boehm and Sheard (2010) found that in software-intensive systems there can be a hundredfold cost reduction in identifying and fixing defects, including requirements defects, early in the design cycle.

Management, particularly project management, tends to focus on reducing short-term risks because they usually occur during the leadership of a single project manager. The impact of decisions that result in long-term negative outcomes can be much more difficult to estimate when the issue deals with inadequate system *safety* or *environmental impact* because these failures are usually rare and may even seem unimaginable. Examples of gross underestimation of adverse consequences of major system failures with loss of life, damage to the environment, or loss of customer or public confidence are: the Banqiao Dam Failure (1975), Three Mile Island Event (1979), Chernobyl Disaster (1986), loss of two US Space Shuttles (1986 & 2003), Fukushima Disaster (2011), and the Boeing 737 Max disasters (2018 & 2019).

Hu and Paez (2016) attempt to quantify the most cost-effective level of VVUQ activities in the use of M&S by considering various options in a decision tree. Their analysis is set in the context of the 2014 Sandia V&V Challenge Workshop, which considers the likelihood of fictional storage tank failures when varying levels of VVUQ are employed. They convincingly argue that when considering the value of VVUQ activities, "The question changes subtly from 'what is the value of doing V&V' to 'how does this V&V analysis impact my decision?'" They contend that "the value of V&V is equal to what the decision maker would pay to acquire V&V evidence – even when he or she knows that V&V may indicate that the model is not credible." This particular quantification of VVUQ value does not seem to be constructive because the *impact* of additional VVUQ evidence on the decision is difficult, if not impossible, to quantify. For example, what is the value of a disaster averted, especially when a decision maker is convinced that a disaster will not happen? Our perspective of the value of VVUQ could be expressed by noting the famous quote of George Box, "All models are wrong, but some are useful." and adding "*However, we don't know how useful or dangerous.*"

As discussed in this text, our perspective is to attempt to quantify the total uncertainty in the simulation using imprecise probability theory. Specifically, when using probability bounds analysis (PBA), an attempt is made to estimate the *bound* on the total uncertainty in a prediction for the relevant decision-support question. As a result, when additional VVUQ activities are conducted, the total uncertainty in a simulation result tends to *decrease*. For example, if additional model validation experiments are conducted that are more relevant to the application conditions of interest for the system, then the model form uncertainty would tend to decrease compared to the initially large estimate of model form uncertainty based on large extrapolation outside of the model validation domain. This means the model credibility and reliability tends to *increase*, resulting in improved simulation-informed decision making. Our perspective is described as an outside-in approach with PBA because as information is added, epistemic uncertainty tends to decrease, resulting in more reliable, i.e., less uncertain, M&S information for decision making.

Bayesian estimation, the widely accepted approach, can be thought of as an inside-out approach regarding uncertainty estimation. Bayesian estimation does not bound estimated total uncertainty, but it tends to best estimate uncertainty for conditions where substantial experimental data are available. The Bayesian calibration perspective can be summarized as: given the present model form and given the available experimental data, what are *a posteriori* probability distributions of model parameters such that the updated model best matches the available experimental observations? Bayesians refer to the approach as informing the model based on available data. However, at a fundamental level, it is a type of model optimization procedure, given the available experimental data. Using the example of large extrapolation outside of the model validation domain, Bayesian estimation could easily give *unreliable information*. Conversely, PBA could give *minimal information* because of very large estimates of uncertainty. As discussed in Chapter 13, when PBA is combined with a global sensitivity analysis, the largest contributors to predictive uncertainty can be identified.

16.2.2 How Much VVUQ Is Appropriate?

Given the PBA perspective, the total predictive uncertainty would tend to decrease as the time and resources devoted to VVUQ activities increase. This perspective of the value added by VVUQ activities to simulation-informed decision making is shown in Figure 16.1. Large total predictive uncertainty is characterized on the left side of the graph as *Low Simulation Credibility*. Similarly, small total predictive uncertainty is characterized as *High Simulation Credibility*. To achieve higher levels of simulation credibility and M&S reliability, time and resources must be expended, which are usually combined and measured in terms of financial cost, e.g., dollars. The right-hand side of Figure 16.1 relates the total value of M&S information to the concept of decision risk. Decision risk continually decreases as the time and resources devoted to VVUQ increase. Thus, increasing time and effort devoted to VVUQ

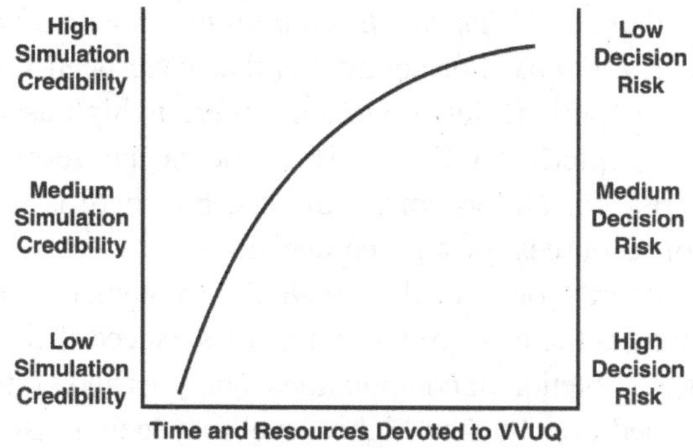

Figure 16.1 Trade-off between simulation credibility and risk as a function of time and resources devoted to VVUQ activities.

buys-down risk in the sense that VVUQ activities are a risk-mitigation strategy. Note that increasing the time and resources devoted to VVUQ includes an increase in experimental information through an increased effort in model validation and input parameter calibration.

It may appear that the three levels of simulation credibility shown in Figure 16.1 are analogous to the decision tree of Hu and Paez (2016) showing no V&V, moderate V&V, and higher V&V evidence. However, the present perspective is different from Hu and Paez (2016) in that increasing VVUQ activities tends to *decrease* the magnitude of the bound on total uncertainty. For example, when a new model validation experiment is conducted and compared with a blind prediction of the experiment, the simulation may have compared very poorly with the experimental measurements. In hindsight, it is recognized that the simulation underestimated the total predictive uncertainty. When moving forward, however, the cause of this uncertainty can be included in future predictive uncertainty. Hu and Paez (2016) takes the perspective that, "The main takeaway is that the outcome from performing V&V is uncertain." Whether the outcome of more V&V is uncertain, is *not* the issue. The critical issue is that increased VVUQ activities tend to *decrease* the total predictive uncertainty because of increasing knowledge of each element contributing to the credibility of the simulation and the responses of a complex system. As pointed out by Pilch (2019), M&S should also be thought of in terms of *knowledge discovery* about the system in terms of how it could respond to a wide variety of possible conditions and scenarios.

The business decision difficulty in using M&S is twofold: first, not knowing what mixture of M&S, experiment, and experience with an existing system is most cost effective; and second, assessing what level of VVUQ expenditures is associated with a given level of risk of making an incorrect or improper system decision. To aid in the business decision of increased use of M&S, it is very informative to remember that the traditional context of decision making is based primarily on experimental testing and experience with the system. For millennia, engineering tradition has been the paradigm: build-test-break-redesign (Petroski, 1985). This paradigm could be described as: *building high assurance that the design satisfies requirements for a few well-understood conditions*. The traditional engineering paradigm rarely asks the question: What is the expected margin of performance, safety, and reliability for *untested* conditions? Testing alone cannot answer this question because it is prohibitively time consuming and expensive. The context of simulation-informed decision making is quite different from engineering tradition. The M&S paradigm could be described as: *moderate assurance that the design satisfies requirements with an estimated margin for a wide range of untested conditions*. For conditions that have been tested, or where there is closely related operating experience of the system, there is high assurance that the system meets requirements. As depicted in Figure 16.1, increasing resources spent on VVUQ decreases the decision risk for all system conditions, but there is no magic answer to the question of adequacy of resources for a given decision.

Some managers will overtly or covertly thwart the implementation of VVUQ activities because they believe that: (a) the costs of these activities exceed their value added, and (b) it unnecessarily delays the completion of computational analyses and system design. If the resistance to VVUQ is expressed openly, then higher level managers must have frank discussions with these managers to clarify the value added by VVUQ. If resistance is hidden or surreptitious, then it is much more difficult for upper management to address. For example, hidden resistance by a project manager can be inferred by insisting that only the most experienced technical staff be assigned to their project, instead of "wasting time with VVUQ practices with less experienced staff." Project managers are probably the most difficult to convince of the value added by VVUQ because they are focused on the specific short-term goals of *their* project. Once a project is complete, the movement of management within a large organization, as well as organizational weaknesses in long-term accountability for poor leadership, make it almost impossible to implement continuous improvement in business operations. As wisely stated by Nassim Taleb, "Bureaucracy is a construction by which a person is conveniently separated from the consequences of his or her actions" (Taleb, 2018, 12).

Implementation of many of the VVUQ practices recommended in this book will be time consuming, expensive, and culturally difficult to implement. Furthermore, some of these approaches may even be technically or economically impractical in particular situations. In addition, some of the approaches and procedures have not been developed satisfactorily for implementation in a fast-paced, design-engineering environment. However, by increasing VVUQ activities, the reliability of M&S will be improved, resulting in increased credibility in the simulation results. Hence, incorporating VVUQ practices is a process requiring a change in culture; it is *not* a product. Attempting to *inspect in* VVUQ processes in a simulation culture has proven to be ineffective and commonly divisive. The technical complexities of M&S and VVUQ do not allow for a comprehensive set of rules to be laid down and then simply followed. Modeling is a distinctively creative endeavor that should *not* be prescriptive or constrained. Good VVUQ practices can be more prescriptive because they are akin to best practices in engineering and business. The development of good VVUQ practices requires a change in the culture of scientific computing. The culture must change from broad acceptance of simulation results to a culture of healthy skepticism and a desire for quantitative evidence of the details. Even though technological change can occur at a stunning rate, changes in organizational and educational cultures and traditions are exceedingly slow and problematic for leadership to transform.

16.2.3 Personnel Training

Presently, individuals trained and experienced in V&V are rare. V&V as a recognized field of research and application is actually quite new, relative to the rapid rise and broad deployment of computational simulation over the last four decades. There is much wider experience and training in the foundational fields of V&V: SQA, numerical analysis, and experimental uncertainty estimation. Some major universities in the US are now beginning to teach graduate courses specifically in the field of V&V. At the undergraduate level, some universities include certain concepts of V&V in elective courses in modeling and simulation. Consequently, it will be at least another decade or two before there will be a sizeable cadre of highly qualified and experienced individuals in the field of V&V.

Training and experience in the field of UQ is a different situation. There is significant depth of experience in UQ in two diverse communities. First, the field of probability and statistics has been well established for about a century and in-depth training at many universities has existed since the 1960s. In relation to the field of M&S, the most important field of probability and statistics is inferential statistics. Most major universities around the world have extensive courses of study and research in inferential statistics. The two dominant paradigms are classical (or frequentist) statistics and Bayesian statistics. University training in these approaches is usually taught in applied mathematics and statistics departments.

Second, a few specialty fields in science and engineering have extensive training and research in inferential statistics. Some examples are nuclear and particle physics, atomic and molecular physics, and certain fields of engineering, such as structural dynamics, statistical thermodynamics, and nuclear engineering. Apart from these specialty fields of science and engineering, the dominant tradition in university education and technical experience is *deterministic* simulation. Similar to university training and experience in V&V, it will require a decade or two for the university education system to change and meet the present need for nondeterministic simulation and uncertainty quantification.

During the near term, the best method for training staff and managers in V&V is through professional development short courses. These continuing education courses typically range from one to five days and are offered by professional societies, universities, and research institutes. They are also taught at professional society conferences, online, or upon request at an interested organization. At the present time, there are roughly five short courses taught in the field of VVUQ by different organizations.

16.2.4 Organizational Structures

There are two basic organizational approaches that can be used for deployment of VVUQ practices in a business, institute, or governmental agency. One involves the formation of a small organization within the parent organization that is composed of experienced VVUQ staff. This organization is responsible for developing and deploying VVUQ practices and training staff and management in other subgroups within the parent company. The second is a dispersed approach where the experienced staff are placed in various organizations that are conducting computational simulation and experimental validation activities. Since there are a relatively small number of individuals trained and experienced in VVUQ at the present time, and since VVUQ serves in a support role to simulation, it is an open question as to which approach is best suited for deployment of VVUQ practices. In fact, it is certainly reasonable that one approach may be optimum for one organization, while the other approach would be better in a different organization. There are many site-specific factors that managers should consider in deciding which approach would be best for their organization.

There are a number of advantages in forming a VVUQ group. With close interactions in the group, this can serve as a critical mass of ideas and expertise so that additional energy and learning is generated within the organization. This approach was used at Sandia National Laboratories beginning in 1999 with a large degree of success. The VVUQ group included staff members who, although most were not familiar with V&V practices, were experienced in UQ. This combination of both activities in the same group proved to be an excellent and effective approach. As V&V methodologies developed, the strong connection between model validation and UQ, particularly model form uncertainty, became better understood. Another benefit in the initial formation of the group was the inclusion of experienced individuals in nuclear reactor safety analyses. These individuals were experienced not only in each topic of VVUQ, but also in risk analysis techniques, e.g., fault tree analyses, the phenomena identification and ranking table, and the concept of a model validation hierarchy.

However, today the existence of a VVUQ group in an organization is an anomaly. If there are at least several experienced individuals in one or both fields of V&V and UQ within an organization, then management should consider forming a group of these individuals. This group can provide training and mentoring for other staff members in diverse parts of the organization, as well as lead VVUQ activities within a project. Although the principles of VVUQ are the same regardless of the application area, there are noticeable differences in some of the techniques depending on the application area. For example, in fluid dynamics, heat transfer, and solid mechanics, there are many applications that are steady state or involve slowly varying responses as a function of time. However, in structural dynamics, shock wave dynamics, neutron transport, and electrodynamics, the time-dependent nature of the responses is the dominant factor that significantly complicates VVUQ. Although the experience of the group would probably be focused on M&S, there should be technical staff skilled in experimental methods and diverse measurement techniques. Individuals who have experience in both computational and experimental methods are most valuable because they can serve in multiple roles within the group and in project applications. These individuals also quickly grasp the philosophy and goals of model validation experiments and the principles for the design and execution of validation experiments.

The responsibilities of the VVUQ group should include three areas. First, the group should develop VVUQ techniques and software tools that are useful not only for the needs within the group, but also for other groups within the parent organization. Development examples of needed software tools are: (a) manufactured solutions for code verification, (b) numerical solution error estimators, (c) validation metrics, and (d) propagation methods applicable to both aleatory and epistemic uncertainties. Second, the group should deploy VVUQ practices to other computational simulations and experimental groups. Deployment of practices can take several forms, such as: (a) explain and promote the concepts, procedures, and benefits of VVUQ; (b) apply VVUQ techniques to various simulation projects within the parent organization, i.e., serve as practitioner and problem solver instead of just a theorist; and (c) write and document software packages that are useful for various VVUQ tasks. Regarding this last suggestion, there is criticism that is sometimes posited by project organizations: "We don't have the expertise or the time to write the needed software packages to do what you are promoting." Third, the group should train staff and managers in other organizations concerning VVUQ practices. Relevant types of training include: (a) offering short courses or formal training on specific topics, (b) serving as consultants for computational and experimental projects in other organizations, and (c) serving as mentors to train and advise staff in VVUQ practices.

The second approach to the deployment of VVUQ practices is to disperse experienced VVUQ staff into various computational and experimental project groups. For smaller organizations, this is a more appropriate method. While the individuals could accomplish suggestions given in the previous paragraph, their focus would be more on tactical or near-term issues of the project group. The success of this approach would depend in part on how much time the project manager would allow the staff members to devote to longer-term VVUQ activities, such as development of software tools for VVUQ. If the manager allowed a significant portion of time to be devoted to longer-term VVUQ needs of the project, then this arrangement could be effective. However, if project tasks, such as those directed toward very specific project needs, were to always take priority over more strategic VVUQ needs, then little would be accomplished in moving the project group, or the parent organization, toward better VVUQ practices.

Whichever deployment method is used, one key factor is crucial to the success of VVUQ deployment in an organization. *Middle and upper management must provide unambiguous, genuine, and enduring leadership for VVUQ practices throughout the organization.* If organizational leadership does not provide authentic support for VVUQ, then the use of M&S in decision making will be ineffective at best; or at worst, will be a failed effort. As has been discussed in this and several other chapters, the key to long-term success of VVUQ is upper management's commitment to information quality and reliability provided by computational analyses. Through the funding and setting requirements for VVUQ activities, management demonstrates by word and action that they are dedicated to changing the culture regarding M&S in the organization.

16.3 Executive Management Responsibilities for VVUQ Activities

Executive management of an organization, especially a large organization, is responsible for an extraordinarily wide range of activities. Many of these responsibilities are focused on matters external to the organization itself, such as the relationship with the Board of Directors (BOD) and shareholders, raising new equity for the corporation, and dealing with politicians and news media. The discussion here is focused on only one of the many internal responsibilities for the strategic competitiveness and long-term profitability of the organization: *simulation governance.*

Simulation governance is a new management term, and the specific concepts behind it are now being developed. It can be broadly defined as top management's (a) recognition of the strategic importance of M&S in achieving the business goals of the organization and (b) developing and implementing processes to maximize the integrated value of M&S with traditional processes across the organization. Sections 16.1 and 16.2 discussed responsibilities and working level management of VVUQ, whereas this section focuses on the strategic recognition of the role of VVUQ and how to achieve the goals of the organization. The fundamental concepts of simulation governance are explained, including why executive management must take a leadership role in the planning and execution of simulation governance in the organization. Finally, this section discusses how executive management can promote the transformation from test-based decision making to simulation-informed decision making by promoting a culture of objectivity and completeness in the presentation of M&S results to management and decision makers.

16.3.1 Fundamental Concepts of Simulation Governance

The term *simulation governance* was first used by Szabo and Actis (2012). They described it as "procedures established for the purposes of ensuring and enhancing the reliability of predictions based on numerical simulation." In Szabo (2015), it was defined as an "exercise of command and control over all aspects of numerical simulation." The idea of simulation governance is built on the general concept of *corporate governance*, which is also a relatively new concept. After the failure of some major corporations in the US and Europe over the last few decades due to conflict of interest and accounting fraud, laws were passed to improve corporate operations. The emphasis has been on the enhancement and balance of responsibilities and accountability of the Board of Directors, executive management, shareholders, and stakeholders in the organization. Some view the most valuable long-term contribution of effective corporate governance policies as the increase in the authority and independence of the BOD. In a publicly held corporation, the BOD has the fiducial responsibility to represent the prolonged interest of the owners of the corporation. For a modern review of the wide-ranging concepts and impact of corporate governance, see Larcker and Tayan (2015), and Tricker (2019).

Corporate governance policies are aimed toward both external and internal responsibilities. Internal corporate governance policies focus on four areas:

1. Independent auditing of financial information;
2. Assurance of compliance with laws and government regulations;
3. Accountability, lines of responsibility, and independent reporting; and
4. Effectiveness and efficiency of business activities.

In the 1960s, business management began to recognize how computer systems and data processing were beginning to change business operations. However, it took decades to incorporate these new technologies into improvements in effectiveness and efficiency across the entire range of business activities. The telecommunications aspect of computer systems, specifically, internet-connected devices, is finally being recognized as a critical weakness in computer security. Until recently, many businesses and governmental organizations gave feigned support to the importance of computer security. To be more precise, a widely accepted risk analysis of the "optimum investment" in computer security concludes that only a small fraction of the expected loss of a security breach should be invested in computer security (Gordon and Loeb, 2002). As with many risk analyses, the difficulty lies with estimating the extent of negative impact of a security breach to the organization, especially the long-term impact. Organizations only took computer security seriously when massive breaches in proprietary information, classified information, personnel information, and customer information occurred.

Szabo and Actis (2012) recognized that M&S will become an ever increasing influence in corporate decision making and, as a result, will affect corporate competitiveness and profitability. A central part of executive management strategic responsibilities organizations must be to devise plans that integrate M&S into a wide range of business operations. Meintjes (2015) stated, "simulation governance is the communication and coordination of

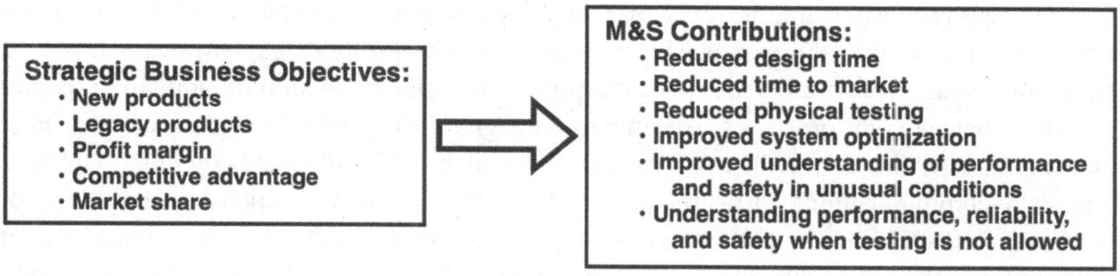

Figure 16.2 Examples of strategic business objectives that can be advanced by M&S.

overall strategy across the organization, supported by senior executive management." Used here, *executive management* refers to the chief executive officer, as well as other executive officers, such as chief financial officer, chief operating officer, and chief strategy officer. The integration of new technology is typically viewed by executive management from two perspectives: (1) return on investment (ROI) and (2) expected improvements in efficiency and speed of conducting business operations. Assessing the impact of M&S from these two perspectives is proving very difficult to measure, as was the monetary impact of computer systems and data processing in the mid twentieth century. The foundational reason for this difficulty is assessing the monetary value of information, not just M&S information, and its impact both within and outside an organization. For example:

- What is the ROI when converting marketing from paper and media advertising, and manual processing of sales, to an all-digital form, so that customers can search your website and purchase products immediately?
- What is the ROI of the ability to immediately search and retrieve the operational performance, inspection, and repair records of a company product, either purchased by a given customer or for all customers who purchased the product?

In many business sectors, numerous surveys over the last four decades have shown that the most important contributions of M&S to business operations are (a) reduced design time and time to market, (b) reduction of physical testing of the system, (c) improved system optimization, and (d) improved understanding of system performance and safety in abnormal environments (Brown et al., 2000). Figure 16.2 depicts how strategic business objectives can be advanced by M&S capabilities, most of which can only be achieved through data management systems. It is beyond the scope of this section to discuss all of the contributions listed on the right-hand side of Figure 16.2. The last two contributions are called out because they provide information on product design and performance that product testing does not, and cannot, provide. The two examples given are *only attainable* from M&S. First, M&S can be used to explore and mitigate risks to a patient when a medical device fails or a medical device is improperly implanted. Second, M&S can be used to understand and design for safety of high-consequence systems exposed to abnormal environments, such as a nuclear power plant responding to an earthquake, or hostile environments, such as a cyberattack of the electrical power grid.

Executive management should develop a simulation governance plan to achieve the high-level strategic business objectives of the organization. Vice presidents and lower-level management are responsible for execution and deployment of the plan across the organization. This latter responsibility is usually referred to as *simulation management*. Examples of simulation management tasks are (a) development and implementation of simulation quality and data management processes; (b) effective integration of M&S, including experimental and testing activities across the organization; (c) availability of adequate computer hardware and data communications infrastructure; and (d) training and competency assurance of technical staff (NAFEMS, 2019). As discussed in the next section, executive management must provide the leadership for envisioning and ensuring the execution of the simulation governance plan over the entire organization.

16.3.2 Necessity of Executive Management Leadership

Within a large organization, integration of all aspects of M&S is an arduous and long-term challenge for executive management for two reasons. First, it involves a very wide range of diverse activities, including research and technology development; product design and manufacturing; experimental testing and data assimilation during preliminary design, final design, and production; and integration of operating experience (including inspection, maintenance, and repair of the product) with M&S. Second, responsibility for these disparate activities is distributed across a number of constituent internal organizations. These constituent organizations are typically in different vice-presidencies in a large organization. By definition, vice presidents are strong leaders who have demonstrated their deep commitment to the organization, assertive leadership, and their independent opinions about how the entire organization should operate.

In order to clarify why simulation governance must be part of the responsibilities of executive management, consider the design and production of a new engineering product. Figure 16.3 shows the communication paths of M&S and experimental activities associated

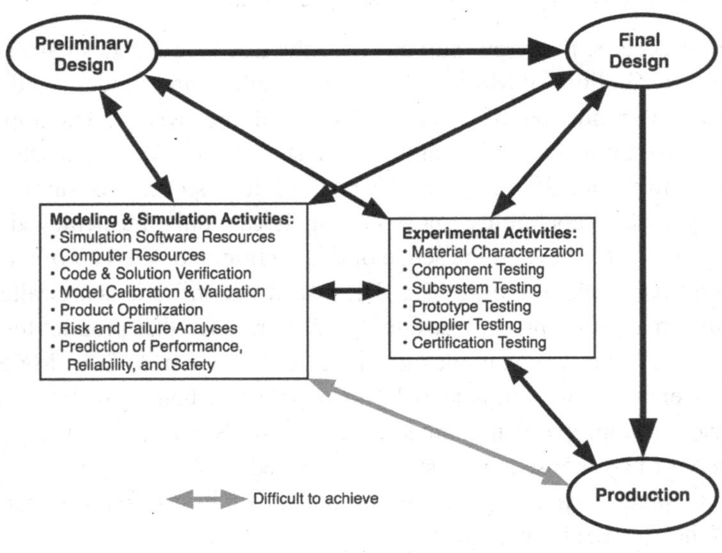

Figure 16.3 Communication of M&S and experimental activities in new product design and production.

with the design and production of a new system or subsystem. The M&S and experimental blocks show several important activities that occur in each block. The black arrows indicate the common exchanges of information and data that occur during preliminary design, final design, and production of the product. Perhaps the best example of the critical need for exchange of information between M&S and experimental activities is the field of structural dynamics. Structural dynamics is a field that is characterized by multi-physics phenomena that are poorly modeled, along with the strong coupling of the vibrational modes of every assembly, subassembly, and component of the system. The exchange of product information usually occurs using a product data management (PDM) software package. This software package keeps a searchable record of all aspects of the product and metadata associated with the product, such as geometric and material details, design changes, details of M&S and experimental activities, personnel involved, production and assembly details, and approval authorizations.

The gray arrow in Figure 16.3 shows the exchange of information that is commonly difficult to achieve. To be more precise, the information may be entered in the PDM software, but the information is commonly ignored or not acted upon by those involved in M&S. During the production phase of product development, the M&S management and personnel have usually changed their focus to other products that have been assigned to them. If a difficulty arises with producing or assembling the product, then M&S personnel are conscripted to address the problem. If production goes smoothly, then Figure 16.3 is representative of difficult or ineffective interactions between M&S activities and production operations.

Two examples of ineffective or wasted information follow. First, consider that staff and management in production operations find that the recommended assembly procedure provided by the design team does not work as anticipated. The production staff devise a modified procedure that does achieve the desired result for system assembly, which is documented in the PDM software. However, the reason that the original recommended procedure did not work was not sufficiently explained by the production staff. The result is that the M&S staff are not convinced that the recommended procedure would not work as intended with more effort. Consequently, the design and assembly practices devised by the production staff are not discussed with the M&S staff so that improvements can be made for the design of future products.

Second, consider the case where product certification staff in production operations discover that the variability of some vibrational modes of the completed structure are larger than expected. However, the completed structures meet all of the geometric, materials, joining specifications, and mass properties specified, as well as the functional requirements of the system. Production staff carefully document their findings in the PDM software. M&S management takes note of the findings but concludes that there is no adverse impact on product performance, reliability, or safety. As a result, no improvements in M&S practices are made for future products.

While understandable in a competitive and fast-paced production environment, both of these examples show how valuable information is lost that could have improved the design and production of future products. To make these two examples more relevant to a large original equipment manufacturer (OEM), consider these examples in the context where subcontractors supply assemblies and subassemblies. These two examples are much more

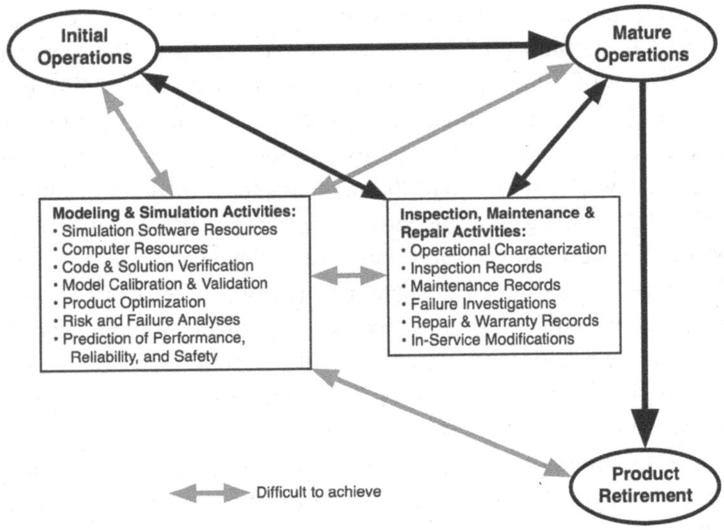

Figure 16.4 Communication of M&S and IM&R activities during system operation and product retirement.

likely to occur because of independent accountability and reward frameworks between contractors and OEMs. In fact, a subcontractor is highly motivated to keep its manufacturing and assembly procedures proprietary so that no competitor could replace them in their services to the OEM.

Figure 16.4 shows the communication paths of M&S and Inspection, Maintenance, & Repair (IM&R) activities associated with the life cycle of a delivered product. The constituent organization that is responsible for IM&R is typically part of the product support and customer service organization. This organization always reports to a different vice president than the M&S or product design organization. The value systems, background experience, and cultures of these two could not be more different. The organization responsible for product retirement and disposal at the end of service life may not be part of the customer service organization, and it may report to a different vice president. As on the previous figure, the black arrows indicate the exchanges of information and data that are commonly in place, whereas the gray arrows indicate difficult, even rare, exchanges of information and data. The experience of some readers from large OEMs may feel that some of the gray arrows should be black, or some may feel that most of the black arrows should be gray.

Consider the following example highlighting differences in perspectives, reward systems, and accountability between constituent organizations that typically results in the loss of valuable information to the organization. The product of interest is a standalone power generation system. The system generates electrical power that is derived from a gas turbine engine driving an electrical generator. This business sector is known for its extraordinarily high demands on performance, specifically fuel economy, and reliability of the system. The IM&R information gathered from in-service systems is highly proprietary to the OEM. This information is especially important if the purchase of the system included guarantees on fuel consumption, maximum time the system is out of service for inspection and maintenance, and failure rate of the system. All of this information can be extremely valuable to the M&S

activities of the OEM because it can be directly used to improve future products. For example, the data can be used to improve physics modeling assumptions of the system, calibrate physics-based models, and improve the prediction of failure modes of the system. Indeed, the improvement in failure mode prediction could be for modes that have *never* been seen in an operational system.

The vice president responsible for operations and service of the system may have little or no background or understanding of the value of the IM&R data to the M&S organization. For example, the vice president may have come from a test-based decision-making background or come from marketing and sales. Furthermore, gathering, compiling, and documenting the detailed information needed by M&S staff requires time and expense that would impact the profitability of operations and the service organization. This situation is made worse because there are always competing demands for shareholder stock dividends. It is also the common practice that vice-president bonuses are associated with their contribution to the corporation's profitability.

The purpose of Figures 16.3 and 16.4 and the discussion is to (a) exemplify the breadth of integration that is needed for M&S activities to be effective and impactful for present and future products and (b) to show the diverse perspectives and responsibilities of vice presidents that exist in a large corporation. The performance assessment measures of the vice presidents; corporate reward structure; value systems; and cultures and backgrounds of vice-presidents are remarkably different. Executive management has the responsibility to bind together and set expectations of each of these strong leaders in a way that is most beneficial to the entire organization for both the near term and the long term. When important subsystems or assemblies of the final product or portions of M&S activities are provided by contractors, the task of executive management becomes even more challenging. Developing and executing a simulation governance plan can aid in bringing together these heterogeneous contributors and help them achieve a common goal for the benefit of the organization and the stakeholders. Smaller organizations, such as startups, are more able to bridge these diverse perspectives and independent accountabilities.

16.3.3 Communications with Management

Every organization has a hierarchical structure for management authority, responsibilities, and decision making. The rigidity and level of communications within the hierarchy depends on the type of organization and the societal culture in which it operates. Some decision making is localized within a given level of management, such as operational or personnel decisions that impact staff at that particular level of the organization. However, for decisions on major products or services delivered by an organization, the decision-making process is distributed over multiple levels of management. To be more precise, lower levels of the organization typically provide higher levels of management with (a) M&S results and proposed system designs, (b) options for various paths forward, and (c) recommendations on which option to pursue. Consequently, a final decision is a result of a sequence of discussions, presentations, and decisions made as the issue in question moves upward through management. By necessity, the extent of detailed information presented to management decreases rapidly through the upward progression of communications. This section discusses what type of M&S and VVUQ information should be communicated through multiple levels of management for improved simulation-informed decision making.

Presentations to the first level of management are given by the technical staff producing simulation results on a new or modified system of interest. These presentations are the most detailed in terms of (a) identification of the normal, abnormal, and hostile environments that are analyzed for the system; (b) identification of which system scenarios are analyzed; (c) conceptual assumptions and approximations made to yield a mathematical model; (d) simulation results for design options considered and optimized; and (e) VVUQ activities that are conducted throughout the entire simulation process. Even at this first level, there are still a multitude of details that cannot be covered in presentations. They may be explained in detailed documentation of the analysis, but during the preliminary design and even final design stages of a system, this documentation has not yet been written. More importantly, by the time detailed documentation is prepared and published, usually during the system production phase of the project, most decisions are permanent. Most details should be captured in the PDM software, but detailed explanations do not usually exist for *why* or the *justification for* most analysis decisions that were made. When presentations on the system of interest move to higher levels of management, the extent of details presented declines precipitously, simply because of the extraordinary time demands on higher levels of management.

Every organization has traditions and culture concerning the character and content of these presentations. Most organizations use the approach that the project manager of the system of interest gives increasingly filtered versions of the presentation to higher levels of management. Sometimes the hierarchical structure of the organization is so rigid that only the immediately subordinate manager gives the presentation to the next level of management. The central question concerning objectivity and completeness in this sequence of presentations asks, "What type of message filtering occurs in the progression of presentations?" If the project manager gives the presentations, the tendency is that a positive or favorable filter is applied to the entire package of information. The project manager is understandably biased toward having the project approved because the project is his/her primary responsibility and emblem of achievement. If an immediately subordinate manager gives the presentation, this tendency is mitigated because of the decreasing association of the subordinate manager with the project.

In any summary presentation to management, the importance of ease of understanding and completeness of information cannot be overstated. Ease of understanding is in no way a criticism of management's technical abilities or backgrounds. As part of human nature, project management and technical staff who prepare the summary presentations tend to think that the various audiences have similar backgrounds and perspectives, technically, experientially, and culturally. In general, this is a serious misjudgment that can be devastating to the value and clarity of a presentation. As a result, upper management must stress to lower management and staff the crucial importance of interpretability and ease of understanding of the formulation, assumptions, approximations, and limitations of the analysis. An even larger condensation of information occurs when presentations are given to external review panels or to a governmental regulatory authority.

Two suggestions for the sequence of presentations made to management concerning completeness of information are recommended. First, the presentations should have a balance of not only the strengths, but also the *weaknesses* of the entire package of results. Furthermore, the pros *and* cons of the various decisions and options chosen should be presented regarding the system of interest. Presentations tend to overly stress the strengths of an analysis and minimize the discussion of weaknesses, if they are mentioned at all. A presentation of M&S results that does *not* discuss the effects of important assumptions and scenarios neglected in the analysis, should be highly suspect to managers at all levels. Furthermore, ignoring a discussion of the weaknesses and limitations of the simulation is, in our view, one of the single most damaging indicators of the incompleteness of information presented to management. Analyses that have not identified, or do not discuss their primary weaknesses, can expose decision makers to significant hidden risks. This situation is usually referred to as *risk ignorance*. If management chooses, either explicitly or implicitly, to *not* be informed about critical assumptions and weaknesses in the analysis or potential failure modes or liabilities, then this is referred to as *willful ignorance* or *intentional omission* (Weisberg, 2014; DeNicola, 2017; Hertwig and Engel, 2020). For an example of how complex mathematical modeling confused and misled financial investment managers in the worldwide market crash of 2008, see Salmon (2009) and Hubbard (2020). For a broad discussion of how complex computer simulations are impacting a wide range of social and political views, as well as decision making, see O'Neil (2016).

Second, the presentation should include a high-level summary of VVUQ activities that were conducted. If senior management asks detailed questions about any of these activities, then the presenter should be knowledgeable and prepared to answer the questions raised. If a formal assessment of M&S maturity was conducted, such as a Predictive Capability Maturity Model assessment, then the results of the assessment should be presented. As discussed in Chapter 15 "Maturity Assessment of Modeling and Simulation," the simulation maturity assessment should be shown for each activity, as well as how each of the component activity assessments compared with the requirements that were set for each activity. As part of the discussion on how the UQ analysis of the system was conducted, the discussion should include system failure scenarios analyzed, as well as those that were ignored. It may also be appropriate to include how these failure scenarios could impact corporate liability, potential loss of customer and regulatory confidence, and possible loss of political support, such that regulatory laws are changed for the entire business sector.

16.4 V&V Benchmark Databases

This section discusses the major V&V benchmark databases, how these databases can be technically improved using the concept of strong-sense benchmarks, and how new databases might be constructed to be more effective in assessing the accuracy and capability of M&S. Benchmark databases can be used by individuals and organizations in three distinct ways: (1) education and training, (2) code verification testing, and (3) model validation. Each of these goals is extraordinarily helpful in the improvement of the reliability and broader use of computational simulation. Only major databases that are either publicly available or available on a membership basis are discussed. As suspected, there are major proprietary databases built and maintained by for-profit and governmental organizations. Some examples of large-scale proprietary databases are those developed by (a) commercial software development companies; (b) manufacturers that use M&S in the design, optimization, and certification of their products; and (c) governmental organizations that produce engineering systems. The databases addressed here deal with the fields of solid mechanics, fluid dynamics, and heat transfer.

16.4.1 Existing Databases

During the last four decades, the International Association for the Engineering Modelling, Analysis and Simulation Community (NAFEMS) has developed some of the most widely known V&V benchmarks (NAFEMS, 2020). The majority of these benchmarks have targeted solid mechanics simulations and secondary fluid dynamics. Most of the code verification benchmarks use either an analytical solution or an accurate numerical solution of the mathematical model for the correct reference, or target, solution. The model validation benchmarks rely upon openly available experimental data from a wide range of sources. The V&V benchmarks show the numerical solution for a range of mesh sizes as an indication of sensitivity of the solution to the discretization error. However, none show a sufficient range of mesh convergence such that an observed order of accuracy could be computed. Furthermore, the validation benchmarks rarely indicate the experimental measurement uncertainty. To access the NAFEMS benchmark database, an organization must be a member of NAFEMS.

In the field of nuclear reactor engineering, the Nuclear Energy Agency, as part of the Organization of Economic Co-operation and Development (OECD), has devoted significant resources over many decades to developing model validation benchmarks. The Committee on the Safety of Nuclear Installations (CSNI) has been designing, conducting, and documenting International Standard Problems (ISPs) since 1977 (NEA/CSNI, 1977). The initial emphasis of ISPs was on loss-of-coolant accidents (LOCAs) for the reactor core. However, since that time, the breadth of the ISP validation benchmarks has been significantly expanded and the guidance for ISPs refined (NEA/CSNI, 2000, 2004). ISPs are performed as both blind predictions and postdictions, or retrodictions, meaning that all of the experimental results are known to the computational analysts at the beginning of the activity. Early on, the CSNI recognized the importance of issues such as (a) providing a detailed description of the actual operational conditions in the experimental facility, not simply those conditions that were requested or desired; (b) preparing careful estimates of the uncertainty in experimental measurements and informing the analyst of the estimates; (c) reporting the initial and boundary conditions that existed during the experiment; and (d) conducting a sensitivity analysis to determine the most important factors that affect the predicted system responses of interest. The primary goal of the ISPs remains the same for all types of benchmarks: "to contribute to a better understanding of postulated and actual events" that could affect the safety of nuclear power plants.

A number of efforts have been undertaken in the development of model validation databases for fluid dynamics. In the United States, the National Program for Applications-Oriented Research in CFD (NPARC) Alliance has developed a V&V database of more than 20 different flows (NPARC, 2020). NASA Langley Research Center has organized a Turbulence Modeling Resource that has a large number of benchmarks for both code verification and model validation (NASA, 2020). The activity is guided by the Turbulence Modeling Benchmarking Working Group of the American Institute of Aeronautics and

Astronautics. These benchmarks are well documented, freely available on their website, and are regularly expanded to address other flowfields.

Beginning in Europe in 1988, there had been a much more organized effort to develop V&V databases under the auspices of the European Research Community on Flow, Turbulence and Combustion (ERCOFTAC) (ERCOFTAC, 2020a). ERCOFTAC has over 40 special interest groups (SIGs) in a wide range of applications, e.g., internal and external aerodynamic flows; biomedical flows; and oil, gas, and petroleum applications. Most of the SIGs have V&V databases as part of the ERCOFTAC Knowledge Base (ERCOFTAC, 2020b). This knowledge base was originally developed by the European Union Network on Quality and Trust in the Industrial Application of Computational Fluid Dynamics (QNET-CFD). ERCOFTAC members come from industry, academia, and governmental organizations from roughly 15 countries. ERCOFTAC also produced a series of books starting in 1995, edited by Bernard Geurts and Maria Vittoria Salvetti, and published by Springer.

16.4.2 Strong-Sense Benchmarks

The existing validation databases, particularly those of the Committee on the Safety of Nuclear Installations and ERCOFTAC, address complex flows, sometimes referred to as *industrial applications*. Attempts to critically assess the accuracy of the mathematical model for a specific experimental condition for complex physical processes are often unsuccessful for two reasons. First, the experiment fails to adequately measure and document all of the important model input data needed to simulate the experiment. For example, detailed system characteristics, such as exhaustive geometric features and irregularities, boundary conditions, and initial conditions, are neither measured nor reported in the experimental database. Second, simulation results typically compare very poorly with the experimental measurements for some SRQs. For example, in a model validation experiment for flow over a bluff body, time-averaged quantities such as lift and drag are predicted fairly well, but time-dependent lift and drag are poorly predicted. As a result, the computational analysts using the validation database often do one of the following: (1) they engage in a model calibration activity, adjusting both physical modeling parameters and numerical parameters to obtain better agreement; (2) they reformulate the assumptions and approximations of their model, thereby changing the mathematical model to obtain better agreement; or (3) they start pointing accusatory fingers at the experimentalists about what they think is wrong with the experimental data. While these responses to using a validation database are understandable, they are actually focused on model and experiment improvement, as opposed to model accuracy assessment.

Oberkampf et al. (2004) introduced the concept of *strong-sense benchmarks* (SSBs) for inclusion in V&V databases. They argued that SSBs should be of sufficiently high quality that they could be viewed as *engineering reference standards*. They stated that SSBs are test problems that have the following four characteristics: (1) the purpose of the benchmark is clearly understood; (2) the definition and description of the benchmark is precisely stated; (3) specific requirements are stated for how comparisons are to be made with the results of the benchmark; and (4) acceptance criteria for comparison with the benchmark are defined. In addition, they required that information on each of these characteristics be promulgated, i.e., the information is well documented and publicly available. Although a few SSBs are available, a few of which were discussed above, they asserted that very few presently exist in science or engineering. They suggested that professional societies, academic institutions, and governmental or international organizations would be the most likely to construct and document SSBs.

Oberkampf and Trucano (2008) present an in-depth discussion of how SSBs should be constructed and describe the key features that are needed to qualify as an SSB for both verification and validation. Concerning verification benchmarks, the following elements are discussed that should be contained in the documentation of a verification benchmark: (a) conceptual description, (b) mathematical description, (c) accuracy assessment, and (d) additional user information. Examples are provided for applying these elements to the four types of code verification benchmarks: (1) manufactured solutions to the mathematical model, (2) analytical solutions to the mathematical model, (3) solutions to the mathematical model that are supplemented by using numerical solutions to ordinary differential equations (ODEs), and (4) numerical solutions to the mathematical model, sometimes referred to as code-to-code comparisons. Oberkampf and Trucano (2008) recommend that when a candidate code is compared with a verification benchmark, the results of the comparisons with benchmarks should *not* be included in the benchmark documentation. They also discuss how formal comparison results could be used and identify the types of information that should be included in the documentation of the comparisons.

Oberkampf and Trucano (2008) present four elements that should be contained in the documentation of a validation benchmark: (1) conceptual description, (2) experimental description, (c) uncertainty quantification of benchmark measurements, and (d) additional user information. They also discuss how candidate code results could be compared with the benchmark results, paying particular attention to issues related to (a) the computation of nondeterministic results to determine the uncertainty of SRQs due to uncertainties in input quantities, (b) the computation of validation metrics to quantitatively measure the difference between experimental and computational results, (c) the minimization of model calibration in comparisons with validation benchmarks, and (d) the constructive role of global sensitivity analyses in designing model validation experiments.

16.4.3 Implementation Issues of Databases

As the impact of computational simulation continues to rapidly grow in breadth and depth, V&V databases can also provide an excellent opportunity for education and training in computational simulation. The most effective V&V databases are built around individual technical fields, such as fluid dynamics and solid mechanics. The databases typically concentrate on specific industries, such as nuclear power plants, flight vehicles, gas turbines, and biomedical devices. For those industries in which the competitive pressure between manufacturers is intense, databases can still be constructed for simplified test cases, i.e., lower levels in the model validation hierarchy. For these simplified cases where competitors are willing to share data, it can be more cost effective and efficient to jointly build and share V&V databases rather than using proprietary databases. Some corporations feel that *all* of their V&V data is critical to their continued competitiveness. This perspective is based on the prejudiced view that any sharing or exchange of information with competitors is always a net loss of information. Unless the corporate V&V database is well documented, effectively catalogued, easily searched, and efficiently retrieved by many parts of the organization, the data are of little actual value. The business model of having key senior staff as the keepers of the data jewels of the organization is no longer realistic, and it can easily fail for a variety of reasons.

Rizzi and Vos (1998) and Vos et al. (2002) discuss how model validation databases could be built and used by a wide range of individuals and organizations. They stress the importance of close collaboration between corporations and universities in the construction, use, and refinement of a validation database. Recommendations for the construction, use, and functionality of a CFD model validation database are also discussed in Bai et al. (2006). The value of technical specialty workshops is stressed to improve the modeling efforts for model validation, but also to discuss how the workshops can initiate the development of a new database. A large number of workshops and initiatives have occurred in Europe sponsored by ERCOFTAC and QNET-CFD, primarily funded by the European Union. In the United States, two series of workshops have generated a significant V&V database in external aerodynamics. The first is the AIAA Drag Prediction Workshops, beginning in 2001 (Levy et al., 2003; Hemsch, 2004; Roy et al., 2018; Tinoco et al., 2018). The second is the AIAA High-Lift Prediction Workshops, beginning in 2010 (Rumsey et al., 2011; Rumsey et al., 2019). There is a wide range of successful workshops that have generated V&V databases in other fields, but these are not discussed here.

If the formation of a new V&V database is considered, a number of complex and difficult implementation and organizational issues should be recognized. The following are recommendations for critical issues that should be addressed:

- Agreement on the primary and secondary goals of the database;
- Minimum technical requirements of V&V data for inclusion in the database;
- Initial population of V&V data into the database;
- Review and approval procedures for entries into the database;
- Open versus restricted use of the database;
- Requirements for the software framework for searching and retrieving information from the database;
- Organizational, operational, and security control of the database;
- Relationship of the controlling organization of the database to private and governmental organizations and engineering societies; and
- Initial and long-term funding of the database.

These issues are of major importance to the broad community of individuals, corporations, commercial software companies, nonprofit organizations/institutes, engineering societies, universities, and governmental organizations with serious interest in improving scientific computing. However, each member of the community has diverse perspectives toward the value and use of V&V databases.

Initial construction of a database would be technically and organizationally complex, as well as costly. Instead of database participants funding an organization for initial construction of the database, a more workable approach would be for participants to volunteer technical staff for the initial work. Populating the database with relevant, high-quality benchmarks, such as SSBs, would require a wide-ranging effort that cuts across major communities of mathematical model building, experimentation, applied mathematics, and engineering applications. Putting this kind of collaborative effort together hinges on a careful plan that takes the long-term view for the perpetuation of the database. The construction of a database composed of SSBs is not feasible as a short-term task. Much of what is recommend here aims at a sustainable and long-term use of the database, with an implication that the quality and breadth of the database improves over time. The long-term success of the database requires a sound starting point, with broad consensus from all interested parties about goals, use, access, and funding over time.

Broad organizational issues must be addressed very early in the planning stage. For example, will a single organization (nonprofit, academic, or government) have responsibility for accepting entries in the database, configuration management, and day-to-day operation? Will the database have a role beyond its members, such as serve as an open-access database where data could be downloaded, but not uploaded? Potential financial supporters of the database would need to be convinced of the value returned to them for their investment. This returned value could be in many forms, such as improvements in their software products, the ability to attract new customers to their software products, and use of the database as a quality assessment tool for bidding on new projects.

If proprietary information is used in the database, it would, in our view, greatly diminish or possibly eliminate the ability to create the database. Some have argued that the database could be constructed so that proprietary information could be segregated from generally available information. Private corporations are understandably skeptical that such segregation would be accomplished and with high surety. Organizations of all types are finding it difficult to ensure computer and data security, even if they are entirely in control of the information and the computer system.

An internet-based system would provide the best vehicle for the development and use of a V&V database. Operators and users of the database must have the ability to build, modify, access, and download data from any location where they are working. For the ability to search the database for a particular application of interest, it must be possible for descriptive terms to be input to a search engine that could find all of the benchmarks that would contain those terms. The search engine could operate much like that found in Google or Wikipedia. Functionality could be expanded to include a relevancy-ranking feature that would further improve the search-and-retrieval capability. The overall system design would include configuration-management, document-management, and content-management elements. The benchmarks that were retrieved from a search could be sorted according to their relevance to the words input to the search. The user could then select the hyperlinks embedded within any of the benchmarks found. When a particular benchmark is displayed, it could have links from important words in the benchmark description to more detailed information in the benchmark.

The documentation of V&V benchmarks in the database should exist in an electronic format that is widely usable and robust across many computer operating systems. Adobe Portable Document Format (PDF) is the most commonly used and has many desirable features, which are continuing to expand. The database should also be supplemented with the use of additional file formats for specialized information. For example, tabular data could be stored in ASCI text files or in Microsoft Excel files; high-resolution digital photographs should be stored in easily usable formats such as TIFF, PNG, and JPEG; and digital video files should be stored in formats such as QuickTime, MPEG, or AVI. A number of commercial software packages are available with these capabilities.

An example of how a V&V database can be a natural extension of an existing industrial consortium is the Medical Device Innovation Consortium (MDIC) (MDIC, 2020). MDIC is a public-private partnership developed to work with government and industry stakeholders in an effort to advance solutions that promote patient access to innovative medical technologies. This consortium is unusual because it is formed from not only highly competitive companies, but also a governmental regulator, i.e., the Food and Drug Administration, and a governmental agency, i.e., the Centers for Disease Control and Prevention.

In the long term, new validation experiments should be funded either by the organization controlling the database or by a for-profit private, nonprofit, university, or governmental organization. The organization controlling the database could receive its funding from subscribing members to the organization and possibly from governmental funding. The funding could be directed to both operation and maintenance of the database and to constructing new V&V benchmarks. When new model validation results are entered into the database, there may be a unique opportunity for blind comparisons of simulations with the new data. As stressed throughout this text, blind comparisons are the real test of predictive-capability prowess. Identification and design of new validation experiments should be the responsibility of both the application community and the organization in charge of the database. The organizational role and facilitation of discussions regarding which experiments should be conducted is best served by the database organization. For example, the database organization could serve as an unbiased referee between for-profit corporations desiring more application-relevant experiments and mathematical model builders who are more interested in assessing and improving models for less complex systems.

16.5 Further Reading

For texts on organizational and personal decision making and multifarious risks, see Plous (1993), Tenner (1996), Reason (1997), Kahneman and Tversky (2000), Lawson (2005), Dekker (2007), Duffey and Saull (2008), Reason (2008), Mansfield (2010), Weisberg (2014), Gigerenzer (2014), Dekker (2014), Haimes (2015), Reason (2016), Taleb (2018), Dekker (2019), Hertwig and Engel (2020), and Hubbard (2020).

For diverse perspectives of the importance of simulation credibility and decision making, see Donoho et al. (2009), NASA (2010), Winsberg (2010), Stamatelatos and Dezfuli (2011), LeVeque et al. (2012), Cox (2012), Stodden et al. (2014), Boddy (2015), Paez et al. (2016), Hu and Paez (2016), Stodden et al. (2016), Barba and Thiruvathukal (2017), Oberkampf and Pilch (2017), Shortridge et al. (2017), Mesnard and Barba (2017), Winsberg (2018, 2019), and Walsh (2022).

For modern trends in digital engineering and model-based systems engineering related to simulation credibility, see Carroll and Malins (2016), Hale et al. (2017), Madni and Sievers (2018), DoD (2018), Carroll et al. (2019), Bone et al. (2019), Zimmerman et al. (2019), Baldwin (2019), Durak et al. (2019), Hale and Hoheb (2020), Carroll and Malins (2020), Sohier et al. (2021), Schindel (2022), McDermott et al. (2022), Smith et al. (2022), and AIAA et al. (2023).

For views on CbA, see Rushby (2002), Sharp and Wood-Schultz (2003), Goodwin and Juzaitis (2006), NaRC (2009), SISO (2012, 2013a, 2013b), Pellettiere and Moorcroft (2014), FDA (2016), Schlickenmaier et al. (2019), EASA (2020), AIAA (2021), Mauery et al. (2021), and FDA (2023).

16.6 References

AIAA (1998). Guide for the Verification and Validation of Computational Fluid Dynamics Simulations. *American Institute of Aeronautics and Astronautics*, AIAA-G-077-1998, Reston, VA.

AIAA (2021). Recommended Practice: When Flight Modelling Is Used to Reduce Flight Testing Supporting Aircraft Certification. *American Institute of Aeronautics and Astronautics*, AIAA R-154-2021, Reston, VA.

AIAA, AIA, and NAFEMS (2023). Digital Thread: Definition, Value, and Reference Model. *American Institute of Aeronautics and Astronautics, AIAA Public Policy Committee.*

Andreani, M., A. Badillo, and R. Kapulla (2016). Synthesis of the OECD/NEA-PSI CFD Benchmark Exercise. *Nuclear Engineering and Design*, **299**, 59–80.

ANSYS (2020). ANSYS Mechanical APDL Verification Manual. *ANSYS, Inc., 2020 R2*. Canonsburg, PA.

ASME (2022). Verification, Validation, and Uncertainty Quantification Terminology in Computational Modeling and Simulation. *American Society of Mechanical Engineers*, ASME VVUQ 1-2022.

Bai, W., L. Li, Z.-m. Li, and Y.-h. Liang (2006). CFD V&V and Open Benchmark Database. *Chinese Journal of Aeronautics*, **19**(2), 160–167.

Baldwin, K. (2019). Transforming the Engineering Enterprise – Applications of Digital Engineering and Modular Open Systems Approach. *Journal of Defense Modeling and Simulation: Applications, Methodology, and Technology*, **16**(4), 323–324.

Barba, L. A., and G. K. Thiruvathukal (2017). Reproducible Research for Computing in Science & Engineering. *Computing in Science & Engineering*, **19**(6), 85–87.

Boddy, C. R. (2015). Organisational Psychopaths: A Ten Year Update. *Management Decision*, **53**(10), 1–52.

Boehm, B., and S. Sheard (2010). What Is the Return on Investment for Using Systems Engineering. *Insight*, **13**(3), 44,48.

Boehm, B., R. Valerdi, and E. Honour (2008). The ROI of Systems Engineering: Some Quantitative Results for Software-Intensive Systems. *Systems Engineering*, **11**(3), 221,234.

Bone, M. A., M. R. Blackburn, D. H. Rhodes, D. N. Cohen, and J. A. Guerrero (2019). Transforming Systems Engineering Through Digital Engineering. *Journal of Defense Modeling and Simulation: Applications, Methodology, and Technology*, **16**(4), 339–355.

Boyce, B. L., S. L. B. Kramer, T. R. Bosiljevac, E. Corona, J. A. Moore, K. Elkhodary, et al. (2016). The Second Sandia Fracture Challenge: Predictions of Ductile Failure under Quasi-Static and Moderate-Rate Dynamic Loading. *International Journal of Fracture*, **198**(1–2), 5–100.

Boyce, B. L., S. L. B. Kramer, H. E. Fang, T. E. Cordova, et al. (2014). The Sandia Fracture Challenge: Blind Round Robin Predictions of Ductile Tearing. *International Journal of Fracture*, **186**(1–2), 5–68.

Brown, C. D., G. Grant, D. Kotchman, R. Reyenga, and T. Szanto (2000). Building a Business Case for Modeling and Simulation. *Acquisition Review Quarterly: Journal of the Defense Acquisition University*, **7**(4), 311–327.

Carroll, E. R., D. Grisham, N. Hayden, E. Rich, F. Salvatore, B. Schindel, et al. (2019). A Model-Based Engineering (MBE) Manifesto. *Sandia National Laboratories*, SAND2019-2911C, Albuquerque, NM.

Carroll, E. R., and R. J. Malins (2016). Systematic Literature Review: How Is Model-Based Systems Engineering Justified? *Sandia National Laboratories*, SAND2016-2607, Albuquerque, NM.

Carroll, E. R., and R. J. Malins (2020). What Questions Would a Systems Engineer Ask to Assess Systems Engineering Models as Credible? *Sandia National Laboratories*, SAND2020-9780, Albuquerque, NM.

Coggan, P. (2020). *More: A History of the World Economy from the Iron Age to the Information Age*. New York, NY: The Economist & Hachette Book Group.

COSSAN (2020). COSSAN Software. http://cossan.co.uk/.

Cox, L. A., Jr. (2012). Confronting Deep Uncertainties in Risk Analysis. *Risk Analysis*, **32**(10), 1607–1629.

Dekker, S. (2007). *Just Culture: Balancing Safety and Accountability*. Burlington, VT: Ashgate Publishing.

Dekker, S. (2014). *The Field Guide to Understanding 'Human Error'*, 3rd ed. Philadelphia, PA: Routledge.

Dekker, S. (2019). *Foundations of Safety Science: A Century of Understanding Accidents and Disasters*. Boca Raton, FL: CRC Press.

DeNicola, D. R. (2017). *Ignorance: The Surprising Impact of What We Don't Know*. Cambridge, MA: MIT Press.

DoD (2018). Digital Engineering Strategy. Office of the Deputy Assistant Secretary of Defense for Systems Engineering, Department of Defense, Washington, DC.

Donoho, D. L., A. Maleki, M. Shahram, I. U. Rahman, and V. Stodden (2009). Reproducible Research in Computational Harmonic Analysis. *Computing in Science & Engineering*, **11**(1), 8–18.

Duffey, R. B., and J. W. Saull (2008). *Managing Risk: The Human Element*. New York, NY: John Wiley.

Durak, U., I. Sturmer, T. Pawletta, and S. Mahmoodi (2019). Quality Assessment and Quality Improvement in Model Engineering. In L. Zhang, B. P. Zeigler, and Y. Iaili (eds.), *Model Engineering for Simulation*, 209–231. Cambridge, MA: Academic Press.

EASA (2020). Modelling & Simulation – CS-25 Structural Certification Specifications. *European Union Aviation Safety Agency*, EASA Proposed CM No.: CM-S-014 Issue 01.

ERCOFTAC (2020a). ERCOFTAC Association. www.ercoftac.org/.

ERCOFTAC (2020b). The ERCOFTAC Knowledge Base Wiki Home Page. www.kbwiki.ercoftac.org/w/index.php/Main_Page.

FDA (2016). Reporting of Computational Modeling Studies in Medical Device Submissions. *US Food and Drug Administration*, Guidance Document No. 1807.

FDA (2023). Assessing the Credibility of Computational Modeling and Simulation in Medical Device Submissions: Guidance for Industry and Food and Drug Administration Staff. *US Food and Drug Administration, Center for Devices and Radiological Health*, GUI01500056.

Gigerenzer, G. (2014). *Risk Savvy: How to Make Good Decisions*. New York, NY: Penguin Books.

Goodwin, B. T., and R. J. Juzaitis (2006). National Certification Methodology for the Nuclear Weapons Stockpile. *Lawerence Livermore National Laboratory*, UCRL-TR-223486. Livermore, CA.

Gordon, L. A., and M. P. Loeb (2002). The Economics of Information Security Investment. *ACM Transactions on Information and System Security*, **5**(4), 438–457.

Haimes, Y. Y. (2015). *Risk Modeling, Assessment, and Management*, 4th ed. New York, NY: John Wiley.

Hale, J., and A. Hoheb (2020). INCOSE Model-Based Capabilities Matrix and User's Guide. International Council on Systems Engineering (INCOSE), 2020-001. San Diego, CA.

Hale, J. P., P. Zimmerman, G. Kukkala, J. Guerrero, P. Kobryn, B. Puchek, et al. (2017). Digital Model-Based Engineering: Expectations, Prerequisites, and Challenges. *NASA Marshall Space Flight Center*, NASA/TM-2017-219633. Huntsville, AL.

Hemsch, M. (2004). Statistical Analysis of Computational Fluid Dynamic Solutions from the Drag Prediction Workshop. *Journal of Aircraft*, **41**(1), 95–103.

Hertwig, R., and C. Engel (eds.). (2020). *Deliberate Ignorance: Choosing Not to Know*. Cambridge, MA: The MIT Press.

Hinton, M. J., A. S. Kaddour, and P. D. Soden (2002). A Comparison of the Predictive Capabilities of Current Failure Theories for Composite Laminates, Judged Against Experimental Evidence. *Composites Science and Technology*, **62**(12–13), 1725–1797.

Hu, K. T., and T. L. Paez (2016). Why Do Verification and Validation. *Journal of Verification, Validation and Uncertainty Quantification*, **1**(1), 011008-1–011008-6.

Hubbard, D. W. (2020). *The Failure of Risk Management: Why It's Broken and How to Fix It*, 2nd ed. New York, NY: John Wiley.

Kaddour, A. S., and M. J. Hinton (2013). Maturity of 3D Failure Criteria for Fibre-Reinforced Composites: Comparisons between Theories and Experiments: Part B of WWFE-II. *Journal of Composite Materials*, **47**(6–7), 925–966.

Kahneman, D., and A. Tversky (eds.) (2000). *Choices, Values, and Frames*. Cambridge: Cambridge University Press.

Larcker, D., and B. Tayan (2015). *Corporate Governance Matters: A Closer Look and Organization Choices and Their Consequences*, 2nd ed. Old Tappan, NJ: Pearson Education.

Lawson, D. (2005). *Engineering Disasters – Lessons to be Learned*. New York, ASME Press.

LeVeque, R. J., I. M. Mitchell, and V. Stodden (2012). Reproducible Research for Scientific Computing: Tools and Strategies for Changing the Culture. *Computing in Science & Engineering*, **14**(4), 13–17.

Levy, D. W., T. Zickuhr, R. A. Wahls, S. Pirzadeh, and M. J. Hemsch (2003). Data Summary from the First AIAA Computational Fluid Dynamics Drag Prediction Workshop. *Journal of Aircraft*, **40**(5), 875–882.

Madni, A. M., and M. Sievers (2018). Model-Based Systems Engineering: Motivation, Current Status, and Research Opportunities. *Systems Engineering*, **21**(3), 172–190.

Mansfield, J. (2010). *The Nature of Change or the Law of Unintended Consequences*. London: Imperial College Press.

Mauery, T., J. Alonso, A. Cary, V. Lee, R. Malecki, D. Mavriplis, et al. (2021). A Guide for Aircraft Certification by Analysis. *NASA*, NASA/CR-2021-0015404.

McDermott, T., K. Henderson, A. Salado, and J. Bradley (2022). Digital Engineering Measures: Research and Guidance. *Insight*, **25**(1), 12–18.

MDIC (2020). Medical Device Innovation Consortium. https://mdic.org/.

Meintjes, K. (2015). Simulation Governance: Managing Simulation as a Strategic Capability. *Benchmark Magazine* (January), 10–14.

Mesnard, O., and L. A. Barba (2017). Reproducible and Replicable Computational Fluid Dynamics: Its Harder Than You Think. *Computing in Science & Engineering*, **19**(4), 44–55.

NAFEMS (2019). *What Is Simulation Governance and Management?* Vol. WT11. Hamilton: The International Association for the Engineering Modelling, Analysis and Simulation Community.

NAFEMS (2020). NAFEMS Website. www.NAFEMS.org.

NaRC (2009). Evaluation of Quantification of Margins and Uncertainties: Methodology for Assessing and Certifying the Reliability of the Nuclear Stockpile. National Research Council of the National Academies, ISBN-13: 978-0-309-12853-7. Washington, DC.

NASA (2010). NASA Risk-Informed Decision Making Handbook. *National Aeronautics and Space Administration*, NASA/SP-2010-576. Washington, DC.

NASA (2020). Langley Research Center – Turbulence Modeling Resource. https://turbmodels.larc.nasa .gov.

NEA/CSNI (1977). CSNI Loss of Coolant Accident Standard Problems. *OECD Nuclear Energy Agency, 17*. Paris, France.

NEA/CSNI (2000). CSNI International Standard Problems (ISP): Brief Descriptions (1975–1999). *Nuclear Energy Agency, Committee on the Safety of Nuclear Installations*, NEA/CSNI/R(2000)5. Paris, France.

NEA/CSNI (2004). CSNI International Standard Problem Procedures. *OECD Nuclear Energy Agency, NEA/CSNI/R(2004)5*, CSNI Report No. 17-Revision 4. Paris, France.

NPARC (2020). National Program for Applications-Oriented Research in CFD. www.grc.nasa.gov/www/wind.

O'Neil, C. (2016). *Weapons of Math Destruction: How Big Data Increases Inequality and Threatens Democracy*. New York, NY: Crown Publishers.

Oberkampf, W. L., and M. Pilch (2017). *Simulation Verification and Validation for Managers*, Vol. R0121. Hamilton: NAFEMS.

Oberkampf, W. L., and T. G. Trucano (2008). Verification and Validation Benchmarks. *Nuclear Engineering and Design*, **238**(3), 716–743.

Oberkampf, W. L., T. G. Trucano, and C. Hirsch (2004). Verification, Validation, and Predictive Capability in Computational Engineering and Physics. *Applied Mechanics Reviews*, **57**(5), 345–384.

OpenTURNS (2020). OpenTURNS Software. https://openturns.github.io/www/.

Paez, P. J., T. L. Paez, and T. K. Hasselman (2016). Economic Analysis of Model Validation for a Challenge Problem. *Journal of Verification, Validation and Uncertainty Quantification*, **1**(1), 011007-1–011007-13.

Pellettiere, J., and D. Moorcroft (2014). Certification by Analysis and Simulation Validation. *International Journal of Human Factors Modelling and Simulation*, **4**(3/4), 177–190.

Petroski, H. (1985). *To Engineer Is Human: The Role of Failure in Successful Design*. New York, NY: St. Martin's Press.

Pilch, M. (2019). Cautionary Tale When Using Computational Simulations to Support Regulatory Decisions. *ASME Verification and Validation Symposium*, Plenary Presentation, Las Vegas, NV.

Plous, S. (1993). *The Psychology of Judgment and Decision Making*. New York, NY: McGraw-Hill.

RAMAS (2023). RAMAS Risk Calc. www.ramas.com/risk-calc-4-0.

Reason, J. (1997). *Managing the Risks of Organizational Accidents*. Burlington, VT: Ashgate Publishing.

Reason, J. (2008). *The Human Contribution: Unsafe Acts, Accidents and Heroic Recoveries*. Burlington, VT: Ashgate Publishing.

Reason, J. (2016). *Organizational Accidents Revisited*. Boca Raton, FL: CRC Press.

Rein, G., J. L. Torero, W. Jahn, J. Stern-Gottfried, N. L. Ryder, S. Desanghere, et al. (2009). Round-Robin Study of a priori Modeling Predictions of the Dalmarnock Fire Test One. *Fire Safety Journal*, **44**(4), 590–602.

Rizzi, A., and J. Vos (1998). Toward Establishing Credibility in Computational Fluid Dynamics Simulations. *AIAA Journal*, **36**(5), 668–675.

Roy, C. J., C. L. Rumsey, and E. N. Tinoco (2018). Summary Data for the Sixth AIAA Computational Fluid Dynamics Prediction Workshop: Code Verification. *Journal of Aircraft*, **55**(4), 1338–1351.

Rumsey, C. L., J. P. Slotnick, M. Long, R. A. Stuever, and T. R. Wayman (2011). Summary of the First AIAA CFD High-Lift Prediction Workshop. *Journal of Aircraft*, **48**(6), 2068–2079.

Rumsey, C. L., J. P. Slotnick, and A. J. Sclafani (2019). Overview and Summary of the Third AIAA High Lift Prediction Workshop. *Journal of Aircraft*, **56**(2), 621–644.

Rushby, J. (2002). Modular Certification. *NASA*, NASA/CR-2002-212130. Washington, DC.

Salmon, F. (2009). Recipe for Disaster: The Formula That Killed Wall Street. *Wired.com*.

Sandia (2023). Sandia National Laboratories DAKOTA Software. https://dakota.sandia.gov/.

Schindel, W. D. (2023). Patterns in the Public Square: Reference Models for Regulatory Science. *Annals of Biomedical Engineering*, **51**(1), 225–240.

Schlickenmaier, H., M. G. Voss, and R. E. Wilkinson (2019). Certification Coordination Roadmap. *NASA*, NASA/CR-2019-220408. Washington, DC.

Sharp, D. H., and M. M. Wood-Schultz (2003). QMU and Nuclear Weapons Certification. What's under the Hood? *Los Alamos Science,* 47–53.

Shortridge, J., T. Aven, and S. Guikema (2017). Risk Assessment under Deep Uncertainty: A Methodological Comparison. *Reliability Engineering and System Safety*, **159**, 12–23.

SIMULIA (2020). Abaqus Benchmarks Guide. https://help.3ds.com/2020/English/DSSIMULIA_Established/SIMACAEBMKRefMap/simabmk-c-ov.htm?ContextScope=all&id=b9aebe13091547dba530d70dd0ad643a#Pg0.

SISO (2012). Guide for Generic Methodology for Verification and Validation (GM-VV) to Support Acceptance of Models, Simulation, and Data: Vol. 1, Introduction and Overview. Simulation Interoperability Standards Organization, SISO-Guide-001.1-2012. Orlando, FL.

SISO (2013a). Guide for Generic Methodology for Verification and Validation (GM-VV) to Support Acceptance of Models, Simulation, and Data: Vol. 2, Implementation Guide. Simulation Interoperability Standards Organization, SISO-Guide-001.2-2013. Orlando, FL.

SISO (2013b). Reference for Generic Methodology for Verification and Validation (GM-VV) to Support Acceptance of Models, Simulation, and Data: Vol. 3, Reference Manual. Simulation Interoperability Standards Organization, SISO-REF-039-2013. Orlando, FL.

SMART-UQ (2020). Predictive Analytics and Uncertainty Quantification Software Tools. www.smartuq.com/.

Smith, J. (2020a). *Engineering Simulation Quality Management Guidelines*, Vol. R0129. Hamilton: NAFEMS.

Smith, J. (2020b). *Engineering Simulation Quality Management Standard*, Vol. ESQMS:01. Hamilton: NAFEMS.

Smith, T., C. Payne, and J. Shackleton (2022). Constructing an Authoritative Source of Truth in a Changing Information Landscape. *Insight*, **25**(1), 29–33.

Sohier, H., P. Lamothe, S. Guermazi, M. Yagoubi, P. Menegazzi, and A. Maddaloni (2021). Improving Simulation Specification with MBSE for Better Simulation Validation and Reuse. *System Engineering*, **24**(6), 383–496.

Stamatelatos, M., and H. Dezfuli (2011). Probabilistic Risk Assessment Procedures Guide for NASA Managers and Practitioners. *NASA*, NASA/SP-2011-3421. Washington, DC.

Stodden, V., F. Leisch, and R. D. Peng (eds.) (2014). *Implementing Reproducible Research*. Boca Raton, FL: CRC Press.

Stodden, V., M. McNutt, D. H. Bailey, E. Deelman, Y. Gil, B. Hanson, et al. (2016). Enhancing Reproducibility for Computational Methods. *Science*, **354**(6317), 1240–1241.

Szabo, B. (2015). Simulation Governance: An Idea Whose Time Has Come. *IACM Expressions Magazine*, **37**, 6–8.

Szabo, B., and R. Actis (2012). Simulation Governance: Technical Requirements for Mechanical Design. *Computer Methods in Applied Mechanics and Engineering*, **249–252**, 158–168.

Szabo, B., R. Actis, and D. Rusk (2017). On the Formulation and Application of Design Rules. *Computers and Mathematics with Applications*, **74**(9), 2191–2202.

Taleb, N. N. (2018). *Skin in the Game: Hidden Asymmetries in Daily Life*. New York, NY: Random House.

Tenner, E. (1996). *Why Things Bite Back: Technology and the Revenge Effect*. New York, NY: Alfred A. Knopf.

Tinoco, E. N., O. P. Brodersen, S. Keye, K. R. Laflin, E. Feltrop, J. C. Vassberg, et al. (2018). Summary Data from the Sixth AIAA CFD Drag Prediction Workshop: CRM Cases. *Journal of Aircraft*, **55**(4), 1352–1379.

Tricker, B. (2019). *Corporate Governance: Principles, Policies, and Practices*, 4th ed. Oxford: Oxford University Press.

Vos, J. B., A. Rizzi, D. Darracq, and E. H. Hirschel (2002). Navier–Stokes Solvers in European Aircraft Design. *Progress in Aerospace Sciences*, **38**(8), 601–697.

Walsh, J. (2018). Engineering Simulation Credibility. *Analysis, Simulation & Systems Engineering Software Strategies Initiative, ASSESS Theme Positioning Paper, version 1_4*.

Walsh, J. (2022). Understanding an Engineering Simulation Risk Model. *Analysis, Simulation & Systems Engineering Software Strategies Initiative, ASSESS Credibility Theme Strategic Insight 2022, version 3_0_0*.

Winsberg, E. (2010). *Science in the Age of Computer Simulation*. Chicago, IL: University of Chicago Press.

Winsberg, E. (2018). *Philosophy and Climate Science*. Cambridge: Cambridge University Press.

Winsberg, E. (2019). Computer Simulations in Science. In E. N. Zalta (ed.), *Stanford Encyclopedia of Philosophy*. Stanford University: The Metaphysics Research Lab.

Weisberg, H. I. (2014). *Willful Ignorance: The Mismeasure of Uncertainty*. New York, NY: John Wiley.

Zienkiewicz, O. C., and J. Z. Zhu (1992). The Superconvergent Patch Recovery and a Posteriori Error Estimates. *International Journal for Numerical Methods in Engineering*, **33**(7), 1365–1382.

Zimmerman, P., T. Gilbert, and F. Salvatore (2019). Digital Engineering Transformation across the Department of Defense. *Journal of Defense Modeling and Simulation: Applications, Methodology, and Technology*, **16**(4), 325–338.

Index

Printed by Integrated Books International,
United States of America